Car.
Wen
**Highl ɹoorganic
Chemistry**

Further Reading from Wiley-VCH

Fuhrhop, J.-H., Li, G.

Organic Synthesis, 3rd Ed.

2003
3-527-30272-7 (Hardcover)
3-527-30273-4 (Softcover)

Waldmann, H., Janning, P.

Bioorganic Chemistry and Chemical Biology

2004
3-527-30778-8

Diederichsen, U., Lindhorst, T. K., Westermann, B., Wessjohann, L. A.

Bioorganic Chemistry
Highlights and New Aspects

1999
3-527-29665-4

Lindhorst, T. K.

Essentials of Carbohydrate Chemistry and Biochemistry, 2nd Ed.

2003
3-527-30664-1

Carsten Schmuck, Helma Wennemers (Eds.)

Highlights in Bioorganic Chemistry

Methods and Applications

WILEY-VCH Verlag GmbH & Co. KGaA

Editors:

Professor Dr. Carsten Schmuck
Department of Chemistry
University of Würzburg
Am Hubland
97074 Würzburg
Germany

Professor Dr. Helma Wennemers
Department of Chemistry
University of Basel
St. Johanns Ring 19
CH-4056 Basel
Switzerland

■ This book was carefully produced. Nevertheless, editors, authors and publisher do not warrant the information contained therein to be free of errors. Readers are advised to keep in mind that statements, data, illustrations, procedural details or other items may inadvertently be inaccurate.

Library of Congress Card No.: applied for
A catalogue record for this book is available from the British Library.
Bibliographic information published by Die Deutsche Bibliothek
Die Deutsche Bibliothek lists this publication in the Deutsche Nationalbibliografie; detailed bibliographic data is available in the Internet at http://dnb.ddb.de

© 2004 WILEY-VCH Verlag GmbH & Co. KGaA, Weinheim
All rights reserved (including those of translation in other languages). No part of this book may be reproduced in any form – by photoprinting, microfilm, or any other means – nor transmitted or translated into machine language without written permission from the publishers. Registered names, trademarks, etc. used in this book, even when not specifically marked as such, are not to be considered unprotected by law.

Printed in the Federal Republic of Germany.
Printed on acid-free paper.

Typesetting   Asco Typesetters, Hong Kong
Printing   Strauss Offsetdruck GmbH, Mörlenbach
Bookbinding   Litges & Dopf Buchbinderei GmbH, Heppenheim

ISBN   3-527-30656-0

Foreword

There are two different aspects to the large field of Bioorganic Chemistry. In one of them we use organic chemistry to learn about and influence biology. This area, part of which is sometimes called Chemical Biology, includes the use of organic chemistry to determine the chemical constitution of biological systems, and to determine the chemical structures of the components. There is intellectual flow from chemistry into biology, as we study the chemical properties of isolated and organized biological systems in order to explain their properties in terms of ordinary chemistry. This is in line with the general belief that we will "understand" biology when we can explain it in chemical terms. However, as we study the properties of biological molecules we also expand our understanding of chemistry itself, so in this sense there is also intellectual flow from biology to chemistry.

Another part of the effort to use chemistry to understand and influence biology is seen in the field of medicinal chemistry. Here too there is intellectual flow in both directions across the chemistry-biology interface, as chemists take what is known about the biology of disease and invent cures using chemicals, usually synthetic chemicals. Success in this field for chemists involves both "listening" to biology and "speaking" with chemistry.

There is another important aspect of bioorganic chemistry in which chemists observe the processes of natural biology and then invent new chemistry inspired by what they see, in a field I have named Biomimetic Chemistry. This activity has been going on for a long time, as we admire what goes on in biology and accept the challenge to imitate it. The imitation normally involves the principles that Nature uses, not the exact details of the process. It has been said that a jumbo jet is not just a scaled up pigeon. From birds we learned the principle of wings, but did not imitate the detail of flapping them. Thus in biomimetic chemistry we expand the scope of chemistry by taking some inspiration from biology. The intellectual flow is from biology to chemistry.

This book, Highlights in Bioorganic Chemistry, describes exciting recent advances in all the aspects of the field. Part 1 deals with Biomolecules and their Conformations. Chapters on the natural chemistry of RNA, of β-amino acids, on binding to DNA, on nucleic acid polymerases, on ribozymes and proteases, are concerned with using chemistry tools to help us understand biological chemistry. Part 2 deals with Non-Covalent Intermolecular Interactions. Here there is work on artificial receptors for natural molecules such as carbohydrates, amino acids, and peptides. Part 3 deals briefly with some aspects of Studies in Drug Development, addressing diseases such as cancer, Alzheimer's disease, and asthma. Part 4 of the book is labeled Studies in Diagnostic Developments, Part 5 is concerned with Catalysis by both natural and artificial enzymes, while part 6 covers Methodology, Bioengineering, and Bioinspired Assemblies, what we have called Biomimetic Chemistry. Its last chapter describes work on molecular motors, imitating the biological motors that drive the rotation of flagellae.

The editors Carsten Schmuck and Helma Wennemers are part of a new and exciting emphasis on bioorganic chemistry in Europe. The two met at Columbia University when Schmuck was in the Breslow lab and Wennemers was working with Clark Still, both in the area of biomimetic catalysis. The other authors are leaders in the field, largely from Germany but also including Switzerland and Austria and one each from Italy and Australia. They have produced a book that reflects the growing importance of bioorganic chemistry in Germany, Switzerland, and Austria. It should be required reading for students and others who want to see where chemistry is heading in the new century.

Ronald Breslow
Columbia University

Preface

Bioorganic Chemistry is a diverse research area that has attracted scientists from fields ranging from chemistry over biology, medicine to pharmaceutical science. The book presented here hence highlights research in nucleic acid chemistry, carbohydrates, peptides, molecular recognition, catalysis, biosynthesis and natural biosynthesis. It does not intend to give an exhaustive overview on the field of bioorganic chemistry but is a compilation of research interests by a new generation of scientists from Germany and the neighboring countries. All contributors are researchers from academia and industry who have attended the "Bioorganic Chemistry Symposium" during the years 1999–2002. This conference has been established 12 years ago as a platform to bring together young scientists working in the various fields of bioorganic chemistry and has ever since been an intriguing meeting with an open-minded exchange of ideas. Contributions from the earlier years of the conference have been published by Wiley-VCH in 1999 under the title "Bioorganic Chemistry" by U. Diederichsen, T. K. Lindhorst, B. Westermann and L. A. Wessjohann.

The main focus of bioorganic chemistry is the understanding of the underlying molecular basis of biological processes. There principles are the same for any class of biomolecules. We have therefore subdivided the present book in terms of structure, molecular interactions and function of biomolecules, rather then by the more traditional subdivision by different classes of biomolecules.

In addition to the research and review articles, more general topics and key words like "secondary structure of proteins", "split-and-mix synthesis", "molecular beacons", "SELEX", etc., that are usually regarded as "common knowledge" in bioorganic chemistry publications, are explained throughout the book in separate boxes. These definitions and background information are referenced with important reviews or seminal publications in the fields. They should allow students an easier access to the field of bioorganic chemistry and help all those teaching bioorganic chemistry. We hope that this book will be useful and inspiring to everybody interested in the ever growing and still developing field of bioorganic chemistry.

We thank all authors for their contributions to the book and our coworkers for their help in editing and merging the individual contributions into a single manuscript. Finally we would like to thank Dr. Peter Gölitz for encouraging us to undertake the challenge of editing this book and Drs. Karen Kriese and Elke Maase of Wiley-VCH for their continuous motivation and support in getting this book published.

Carsten Schmuck
Helma Wennemers

Contents

Foreword *v*

Preface *vii*

List of Contributors *xxiii*

Part 1 **Biomolecules and their Conformation** *1*

1.1 **Equilibria of RNA Secondary Structures** *3*
Ronald Micura and Claudia Höbartner
1.1.1 Introduction *3*
1.1.1.1 RNA Folding *3*
1.1.1.2 One Sequence – Two Ribozymes *4*
1.1.1.3 Nucleoside Methylation is Responsible for Correct Folding of a Human Mitochondrial tRNA *5*
1.1.2 Monomolecular RNA Two-state Conformational Equilibria *7*
1.1.3 The Influence of Nucleobase Methylations on Secondary Structure Equilibria, as Exemplified by the Ribosomal Helix 45 Motif *11*
1.1.4 Structural Probing of Small RNAs by Comparative Imino Proton NMR Spectroscopy *14*
Acknowledgments *15*
References *15*

1.2 **Synthesis and Application of Proline and Pipecolic Acid Derivatives: Tools for Stabilization of Peptide Secondary Structures** *18*
Wolfgang Maison
1.2.1 Introduction *18*
1.2.2 *syn-* and *anti-*Proline Mimics *20*
1.2.3 Templates for α-Helix Stabilization *25*
References *28*
B.1 Proline *syn–anti* Isomerization, Implications for Protein Folding *29*
Wolfgang Maison

1.3	**Stabilization of Peptide Microstructures by Coordination of Metal Ions** *31*
	Markus Albrecht
1.3.1	Introduction *31*
1.3.2	Dinuclear Coordination Compounds from Amino Acid-bridged Dicatechol Ligands: Induction of a Right- or a Left-handed Conformation at a Single Amino Acid Residue *34*
1.3.3	Peptide-bridged Dicatechol Ligands for Stabilization of Linear Compared with Loop-type Peptide Conformations *39*
1.3.4	Approaches Used to Stabilize Bioactive Conformations at Peptides by Metal Coordination *41*
1.3.5	Conclusions *43*
	References *43*
B.2	Conformational Analysis of Proteins: Ramachandran's Method *44*
	Markus Albrecht
B.3	Metals in Proteins – Tools for the Stabilization of Secondary Structures and as Parts of Reaction Centers *46*
	Markus Albrecht

1.4	**Conformational Restriction of Sphingolipids** *48*
	Thomas Kolter
	Summary *48*
1.4.1	Introduction *48*
1.4.1.1	Lipids *48*
1.4.1.2	Sphingolipids *49*
1.4.1.3	Signal Transduction *50*
1.4.2	Conformational Restriction *51*
1.4.2.1	Peptidomimetics *51*
1.4.2.2	Conformationally Restrained Lipids *52*
1.4.3	Conformational Restriction of Sphingolipids *54*
1.4.3.1	Rationale *54*
1.4.3.2	Present State of Knowledge *54*
1.4.4	Target Compounds *55*
1.4.4.1	Synthesis *55*
1.4.4.2	Analysis in Cultured Cells *55*
1.4.5	Discussion *57*
1.4.6	Outlook *58*
	References *59*
B.4	Lipids *60*
	Thomas Kolter

1.5	**β-Amino Acids in Nature** *63*
	Franz von Nussbaum and Peter Spiteller
1.5.1	Introduction *63*

1.5.2	β-Amino Acids and their Metabolites in Nature – Taxonomy of the Producer Organisms 64	
1.5.3	Common β-Amino Acids – Nomenclature 64	
1.5.3.1	β-Alanine 64	
1.5.3.2	Seebach's Nomenclature for β-Amino Acids 69	
1.5.3.3	(R)- and (S)-β-Aminoisobutyric Acid [(R)-β-AiB and (S)-β-AiB] 70	
1.5.4	β-Amino Acids *Related* to Proteinogenic α-Amino Acids 70	
1.5.4.1	Aliphatic β-Amino Acids – β-Lysine, β-Leucine, β-Arginine, and β-Glutamate 70	
1.5.4.2	Aromatic β-Amino Acids – β-Phenylalanine, β-Tyrosine, and β-3,4-Dihydroxyphenylalanine 72	
1.5.5	Miscellaneous β-Amino Acids 76	
1.5.5.1	β-Amino-L-alanine (L-Dap) 76	
1.5.5.2	β-Amino Acids *Related* to Cyanobacteria – Aboa, Adda, Admpa, Ahda, Ahmp, Ahoa, Amba, Amha, Amoa, Aoya, L-Apa, and Map 76	
1.5.5.3	Cispentacin as a Chemical Lead Structure – Interaction of β-Amino Acids with Natural α-Amino Acid-processing Systems 79	
1.5.6	Limiting the β-Amino Acid Concept 80	
1.5.7	Conclusion 80	
	Dedication 81	
	Acknowledgment 81	
	References 81	
1.6	**Biosynthesis of β-Amino Acids** 90	
	Peter Spiteller and Franz von Nussbaum	
1.6.1	Introduction 90	
1.6.2	Biosynthesis of β-Amino Acids by Catabolic Pathways 90	
1.6.2.1	β-Alanine 90	
1.6.2.2	Biosynthesis of β-Alanine from Uracil 91	
1.6.2.3	Biosynthesis of β-Alanine from L-Aspartic Acid 92	
1.6.2.4	Biosynthesis of β-Alanine from Spermidine and Spermine 92	
1.6.2.5	(R)- and (S)-β-Aminoisobutyrate 93	
1.6.3	Biosynthesis of β-Amino Acids by Aminomutases 93	
1.6.3.1	(S)-β-Lysine 93	
1.6.3.2	Properties of the Enzyme 94	
1.6.3.3	Stereochemical Aspects 94	
1.6.3.4	Reaction Mechanism 94	
1.6.3.5	(R)-β-Leucine 97	
1.6.3.6	(S)-β-Arginine 97	
1.6.3.7	(R)-β-Phenylalanine 98	
1.6.3.8	β-Tyrosine 99	
1.6.4	Other Aminomutases 100	
1.6.4.1	β-Lysine 5,6-Aminomutase (D-Lysine 5,6-Aminomutase) 101	
1.6.4.2	D-Ornithine 4,5-Aminomutase 102	

1.6.5	Discussion *102*
	Dedication *104*
	Acknowledgment *104*
	References *104*

Part 2   Non-Covalent Intermolecular Interactions *107*

2.1	Carbohydrate Recognition by Artificial Receptors *109*
	Arne Lützen
2.1.1	Introduction *109*
2.1.2	Design Principles and Binding Motifs of Existing Receptors *109*
2.1.3	Design, Synthesis, and Evaluation of Self-assembled Receptors *112*
2.1.4	Conclusions and Perspectives *117*
	References *118*
B.5	Molecular Basis of Protein–Carbohydrate Interactions *119*
	Arne Lützen, Valentin Wittmann

2.2	Cyclopeptides as Macrocyclic Host Molecules for Charged Guests *124*
	Stefan Kubik
2.2.1	Introduction *124*
2.2.2	Cation Recognition *124*
2.2.3	Anion Recognition *131*
	Acknowledgment *135*
	References *136*
B.6	Ion Transport Across Biological Membranes *137*
	Stefan Kubik

2.3	Bioorganic Receptors for Amino Acids and Peptides: Combining Rational Design with Combinatorial Chemistry *140*
	Carsten Schmuck, Wolfgang Wienand, and Lars Geiger
2.3.1	Concept *140*
2.3.2	Structural and Thermodynamic Characterization of the New Binding Motif *143*
2.3.3	Selective Binding of Amino Acids *145*
2.3.4	Binding of Small Oligopeptides *147*
2.3.5	Conclusion *151*
	References *152*
B.7	The Effect of Solvents on the Strength of Hydrogen Bonds *153*
	Carsten Schmuck

2.4	Artificial Receptors for the Stabilization of β-Sheet Structures *155*
	Thomas Schrader, Markus Wehner, and Petra Rzepecki
2.4.1	β-Sheet Recognition in Nature *155*
2.4.2	Artificial β-Sheets and Recognition Motifs *156*
2.4.3	Sequence-selective Recognition of Peptides by Aminopyrazoles *157*

2.4.4	Recognition of Larger Peptides with Oligomeric Aminopyrazoles	161
2.4.5	Recognition of Proteins with Aminopyrazoles 165	
	References 167	
B.8	Secondary Structures of Proteins 169	
	Thomas Schrader	

2.5 Evaluation of the DNA-binding Properties of Cationic Dyes by Absorption and Emission Spectroscopy   172

Heiko Ihmels, Katja Faulhaber, and Giampietro Viola

2.5.1	Introduction 172	
2.5.2	Binding Modes 173	
2.5.2.1	Groove Binding 174	
2.5.2.2	Intercalation 175	
2.5.3	Evaluation of the Binding 175	
2.5.3.1	UV–Visible Spectroscopy 176	
2.5.3.2	Emission Spectroscopy 179	
2.5.3.3	CD Spectroscopy 180	
2.5.3.4	LD Spectroscopy 183	
	Acknowledgment 186	
	References 186	
B.9	Binding of Small Molecules to DNA – Groove Binding and Intercalation 188	
	Heiko Ihmels, Carsten Schmuck	

2.6 Interaction of Nitrogen Monoxide and Peroxynitrite with Hemoglobin and Myoglobin   191

Susanna Herold

2.6.1	Biosynthesis, Reactivity, and Physiological Functions of Nitrogen Monoxide 191	
2.6.1.1	The Biological Chemistry of Peroxynitrite 192	
2.6.2	Interaction of Nitrogen Monoxide and Peroxynitrite with Hemoglobin and Myoglobin 192	
2.6.2.1	The NO·-mediated Oxidation of Oxymyoglobin and Oxyhemoglobin 193	
2.6.2.2	The Peroxynitrite-mediated Oxidation of OxyMb and OxyHb 195	
2.6.3	NO· as an Antioxidant 197	
2.6.3.1	The NO·-mediated Reduction of FerrylMb and FerrylHb 197	
2.6.4	Conclusion: A New Function of Myoglobin? 199	
	References 200	
B.10	Hemoglobin and Myoglobin 201	
	Susanna Herold	

2.7 Synthetic Approaches to Study Multivalent Carbohydrate–Lectin Interactions   203

Valentin Wittmann

2.7.1	Introduction 203	

2.7.2	Mechanistic Aspects of Multivalent Interactions	203
2.7.3	Low-valent Glycoclusters for "Directed Multivalence"	206
2.7.4	Spatial Screening of Lectin Ligands	208
2.7.4.1	Design and Synthesis of a Library of Cyclic Neoglycopeptides	209
2.7.4.2	On-bead Screening and Ligand Identification	209
2.7.5	Conclusion	212
	References	212

Part 3   Studies in Drug Developments   215

3.1   Building a Bridge Between Chemistry and Biology – Molecular Forceps that Inhibit the Farnesylation of RAS   217
Hans Peter Nestler

3.1.1	Prolog	217
3.1.2	RAS – The Good, The Bad and The Ugly	218
3.1.3	Bridging the Gap	220
3.1.4	Epilog	222
	References	224
B.11	Split-and-mix Libraries	225

Hans-Peter Nestler and Helma Wennemers

3.2   Inhibitors Against Human Mast Cell Tryptase: A Potential Approach to Attack Asthma?   227
Thomas J. Martin

3.2.1	Introduction	227
3.2.1.1	Asthma – Definition	227
3.2.2	Chemistry	229
3.2.3	Biological Results and Discussion	235
3.2.4	Conclusion	237
	Acknowledgment	238
	References	238
B.12	Serine Proteases	239

Thomas J. Martin

3.3   Preparation of Novel Steroids by Microbiological and Combinatorial Chemistry   242
Christoph Huwe, Hermann Künzer, and Ludwig Zorn

3.3.1	Introduction	242
3.3.2	Results	243
	References	246

3.4   Enantiomeric Nucleic Acids – Spiegelmers   248
Sven Klussmann
Abstract   248

3.4.1	Towards Nucleic Acid Shape Libraries	248
3.4.2	In-vitro Selection or SELEX Technology	249
3.4.3	Aspects of Chirality	250
3.4.4	Spiegelmer Technology	252
3.4.5	Examples and Properties of Mirror-image Oligonucleotides	252
3.4.5.1	Spiegelmers Binding to Small Molecules	252
3.4.5.2	Mirror-image DNA Inhibiting Vasopressin in Cell Culture	254
3.4.5.3	RNA and DNA Spiegelmers Binding to GnRH	256
3.4.5.4	In-vivo Data of GnRH Binding Spiegelmers	258
3.4.6	Conclusion	259
	Acknowledgments	259
	Appendix	261
	References	261

3.5 Aspartic Proteases Involved in Alzheimer's Disease 262
Boris Schmidt and Alexander Siegler

3.5.1	Introduction	262
3.5.2	β-Secretase Inhibitors	269
3.5.3	γ-Secretase Inhibitors	270
3.5.4	Outlook	273
	Acknowledgments	274
	References	274
B.13	Aspartic Proteases	276
	Boris Schmidt	

3.6 Novel Polymer and Linker Reagents for the Preparation of Protease-inhibitor Libraries 277
Jörg Rademann

3.6.1	A Concept for Advanced Polymer Reagents	277
3.6.2	Protease-inhibitor Synthesis – A Demanding Test Case for Polymer Reagents	278
3.6.3	The Development of Advanced Oxidizing Polymers	279
3.6.3.1	Polymer-supported Heavy-metal Oxides	279
3.6.3.2	Oxidation with Immobilized Oxoammonium Salts	279
3.6.3.3	Oxidations with Immobilized Periodinanes	282
3.6.3.4	Preparation of Peptide Aldehyde Collections	284
3.6.4	Polymer-supported Acylanion Equivalents [30]	285
3.6.5	Conclusions	288
	References	289
B.14	Polymer-supported Synthetic Methods – Solid-phase Synthesis (SPS) and Polymer-assisted Solution-phase (PASP) Synthesis	290
	Jörg Rademann	
B.15	Inhibition of Proteases	293
	Jörg Rademann	

| Part 4 | **Studies in Diagnostic Developments** 297 |

4.1 Selectivity of DNA Replication 299
Andreas Marx, Daniel Summerer, and Michael Strerath
4.1.1 Introduction 299
4.1.2 Biochemical and Structural Studies 300
4.1.3 Use of Tailored Nucleotide Analogs to Probe DNA Polymerases 303
4.1.3.1 Non-polar Nucleobase Surrogates 303
4.1.3.2 Analogs with Modified Sugar Moieties 305
4.1.4 Conclusions and Perspectives 307
References 308
B.16 Polynucleotide Polymerases 309
Susanne Brakmann

4.2 Homogeneous DNA Detection 311
Oliver Seitz
4.2.1 Introduction 311
4.2.2 Non-specific Detection Systems 311
4.2.3 Specific Detection Systems 312
4.2.3.1 Single Label Interactions 312
4.2.3.2 Dual Label Interactions 317
4.2.4 Conclusion 322
References 322
B.17 Melting Temperature T_M of Nucleic Acid Duplexes 323
Oliver Seitz
B.18 Molecular Beacons 325
Oliver Seitz
B.19 Peptide Nucleic Acids, PNA 327
Oliver Seitz

4.3 Exploring the Capabilities of Nucleic Acid Polymerases by Use of Directed Evolution 329
Susanne Brakmann and Marina Schlicke
4.3.1 Introduction 329
4.3.2 Directed Evolution of Nucleic Acid Polymerases 330
4.3.3 Practical Approaches to the Directed Evolution of Polymerase Function: Selection or Screening? 331
4.3.3.1 Selection of Polymerases with Altered Activity and Fidelity 331
4.3.3.2 Screening Polymerase Libraries for Altered Activity 331
4.3.4 Genetic Selection of an Error-prone Variant of Bacteriophage T7 RNA Polymerase 333
4.3.5 Screening for Polymerases with Altered Substrate Tolerance 335
4.3.6 Alternative Scenarios for Assaying Polymerase Activity 337
4.3.7 Concluding Remarks 338
References 339

B.20	Directed Molecular Evolution of Proteins 341

Petra Tafelmeyer, and Kai Johnsson

4.4	**Labeling of Fusion Proteins with Small Molecules in vivo** 344

Susanne Gendreizig, Antje Keppler, Alexandre Juillerat, Thomas Gronemeyer, and Kai Johnsson

4.4.1	Introduction 344
	Acknowledgment 350
	References 350

4.5	**Oxidative Splitting of Pyrimidine Cyclobutane Dimers** 352

Uta Wille

4.5.1	Introduction 352
4.5.2	Mechanism of the Oxidative Splitting of Pyr◇Pyr 354
4.5.3	Stereoselectivity of the Oxidative Splitting of Pyr◇Pyr 358
4.5.4	Conclusions 362
4.5.5	Experimental 363
4.5.5.1	Oxidative Cleavage of the 1,3-Dimethyluracil-derived Cyclobutane Dimers 1 by Nitrate Radicals ($NO_3^•$) 363
	References 363
B.21	DNA Damage 364

Uta Wille

4.6	**Charge Transfer in DNA** 369

Hans-Achim Wagenknecht

4.6.1	Introduction 369
4.6.2	Hole Transfer and Hole Hopping in DNA 369
4.6.2.1	Spectroscopic Studies 370
4.6.2.2	Biochemical Experiments 372
4.6.3	Protein-dependent Charge Transfer in DNA 373
4.6.4	Reductive Electron Transfer in DNA 379
	Acknowledgments 384
	References 384

Part 5	**Catalysis** 387

5.1	**Protease-catalyzed Formation of C–N Bonds** 389

Frank Bordusa

5.1.1	Optimization of Proteases for Synthesis: Selection of Current Techniques 389
5.1.2	Substrate Engineering 390
5.1.3	Classical Concept of Leaving-group Manipulation 390
5.1.4	Substrate Mimetics-mediated Syntheses 391
5.1.5	Enzyme Engineering 396
5.1.6	Chemical Enzyme Modifications 396

5.1.7	Genetic Enzyme Modifications	398
5.1.8	Conclusions	402
	References	402

5.2 Twin Ribozymes 404
Sabine Müller, Rüdiger Welz, Sergei A. Ivanov, and Katrin Bossmann

5.2.1	Introduction	404
5.2.2	Application of Ribozymes	404
5.2.3	Building Blocks for Twin Ribozymes	406
5.2.3.1	The Conventional Hairpin Ribozyme (HP-WT)	406
5.2.3.2	The Reverse-joined Hairpin Ribozyme (HP-RJ)	409
5.2.3.3	Three-way Junction Hairpin Ribozymes (HP-TJ)	411
5.2.3.4	Branched Reverse-joined Hairpin Ribozymes (HP-RJBR)	411
5.2.4	Design, Synthesis and Characterization of Twin Ribozymes	412
5.2.5	Application of Twin Ribozymes	416
5.2.6	Summary and Outlook	417
	References	419
B.22	Ribozymes	419
	Sabine Müller	

5.3 RNA as a Catalyst: the Diels–Alderase Ribozyme 422
Sonja Keiper, Dirk Bebenroth, Friedrich Stuhlmann, and Andres Jäschke

5.3.1	Introduction	422
5.3.2	Diels–Alder Reaction	423
5.3.3	In-vitro Selection	424
5.3.4	Sequence Analysis and Ribozyme Engineering	425
5.3.5	Mutation Analysis	427
5.3.6	True Catalysis	427
5.3.7	Kinetics	429
5.3.8	Stereoselectivity	430
5.3.9	Substrate Specificity and Inhibition	431
5.3.10	Conclusions	432
	References	433
B.23	SELEX: Systematic Evolution of Ligands by Exponential Enrichment	433
	Andres Jäschke and Sonja Keiper	

5.4 Combinatorial Methods for the Discovery of Catalysts 436
Helma Wennemers

5.4.1	Introduction	436
5.4.2	Testing of Parallel Libraries for Catalytic Activity	437
5.4.2.1	Colorimetric and Fluorescent Screening	437
5.4.2.2	IR–Thermography	439
5.4.3	Testing of Split-and-mix Libraries for Catalytic Activity	440
5.4.3.1	IR–thermography	440

5.4.3.2	Formation of Insoluble Reaction Products	441
5.4.3.3	Fluorescent pH Indicators	441
5.4.3.4	Gels as Reaction Media	443
5.4.3.5	Catalyst-Substrate Co-immobilization	443
5.4.4	Conclusions	444
	References	444

Part 6 Methodology, Bioengineering and Bioinspired Assemblies 447

6.1 Linkers for Solid-phase Synthesis 449
Kerstin Knepper, Carmen Gil, and Stefan Bräse

6.1.1	Introduction	449
6.1.2	General Linker Structures	451
6.1.2.1	Immobilization of Molecules	451
6.1.2.2	Spacers	452
6.1.3	Linker Families	452
6.1.3.1	Benzyl-type Linkers	453
6.1.3.2	Trityl Resins	455
6.1.3.3	Allyl-based Linkers	455
6.1.3.4	Ketal Linkers	456
6.1.3.5	Ester and Amide Linkers	457
6.1.3.6	Silicon- and Germanium-based Linkers	458
6.1.3.7	Boron Linkers	459
6.1.3.8	Sulfur Linkers	459
6.1.3.9	Stannane Linkers	460
6.1.3.10	Selenium Linkers	461
6.1.3.11	Triazene Linkers	461
6.1.4	Orthogonality Between Linkers	465
6.1.5	Cleavage of Linkers	465
6.1.5.1	Oxidative/Reductive Methods	466
6.1.5.2	Special Linkers	468
6.1.5.3	Metal-assisted Cleavage	468
6.1.6	Linker and Cleavage Strategies	472
6.1.6.1	Safety-catch Linkers	474
6.1.6.2	Cyclative Cleavage (Cyclorelease Strategy)	474
6.1.6.3	Fragmentation Strategies	476
6.1.6.4	Traceless Linkers	477
6.1.6.5	Multifunctional Cleavage	479
6.1.7	Conclusion, Summary, and Outlook	480
	References	481

6.2 Small Molecule Arrays 485
Rolf Breinbauer, Maja Köhn, and Carsten Peters

6.2.1	Introduction	485
6.2.2	Arrays	485

6.2.2.1 DNA Microarrays 485
6.2.2.2 Protein Microarrays 487
6.2.2.3 Cell Arrays 492
6.2.3 Small Molecule Arrays 493
6.2.3.1 Synthesis on Planar Supports 493
6.2.3.2 Spotting of Small Molecules 494
6.2.4 Outlook and Conclusions 497
References 497

6.3 Biotechnological Production of D-Pantothenic Acid and its Precursor D-Pantolactone 501
Maria Kesseler
6.3.1 Introduction 501
6.3.2 Fermentative Production of D-Pantothenic Acid 502
6.3.3 Biocatalytic Production of D-Pantolactone 504
6.3.3.1 Biocatalytic Asymmetric Synthesis 504
6.3.3.2 Resolution of *rac*-Pantolactone by Fungal Hydrolysis of D-Pantolactone 504
6.3.3.3 Resolution of *rac*-Pantolactone by Bacterial Hydrolysis of L-Pantolactone: The Development of a Novel Biocatalyst 505
6.3.4 Conclusions 508
References 509

6.4 Microbially Produced Functionalized Cyclohexadiene-*trans*-diols as a New Class of Chiral Building Block in Organic Synthesis: On the Way to Green and Combinatorial Chemistry 511
Volker Lorbach, Dirk Franke, Simon Eßer, Christian Dose, Georg A. Sprenger, and Michael Müller
6.4.1 Introduction 511
6.4.2 The Shikimate Pathway 511
6.4.3 Microbial Production of 2,3-*trans*-CHD 514
6.4.4 Application of 2,3-*trans*-CHD in Natural-product Syntheses 515
6.4.5 Regio- and Stereoselective Epoxidation 516
6.4.6 Nucleophilic Opening of the Epoxides Obtained 518
6.4.7 Regio- and Stereoselective Dihydroxylation 519
6.4.8 Microbial Production of 3,4-*trans*-CHD 520
6.4.9 Discussion 522
References 523
B.24 Metabolic Pathway Engineering 524
Volker Lorbach, Dirk Franke, Georg Sprenger, Michael Müller

6.5 Artificial Molecular Rotary Motors Based on Rotaxanes 526
Thorsten Felder and Christoph A. Schalley
Abstract 526
6.5.1 "Molecular Machines" – Reality or Just a Fashionable Term? 526

6.5.2	Tracing Back ATP Synthesis in Living Cells *527*	
6.5.3	Rotaxanes as Artificial Analogs to Molecular Motors? *529*	
6.5.4	Rotaxane Synthesis via Template Effects *530*	
6.5.5	How to Achieve Unidirectional Rotation in Artificial Molecular Motors? *531*	
6.5.6	The Fuel for Driving the Motor: Light, Electrons, and Chemical Energy *534*	
6.5.7	Conclusions *537*	
	References *538*	
6.6	**Chemical Approaches for the Preparation of Biologically-inspired Supramolecular Architectures and Advanced Polymeric Materials** *540*	
	Harm-Anton Klok	
6.6.1	Introduction *540*	
6.6.2	Ring-opening Polymerization of α-Amino Acid *N*-Carboxyanhydrides *541*	
6.6.3	Solid-phase Peptide Synthesis *544*	
6.6.4	Peptide Ligation *548*	
6.6.5	Summary and Conclusions *550*	
	References *553*	
B.25	Solid-phase Peptide Synthesis *554*	
	Harm-Anton Klok	
B.26	Peptide Ligation *557*	
	Harm-Anton Klok	

Index *561*

List of Contributors

Markus Albrecht
Institut für Organische Chemie RWTH-
  Aachen
Professor-Pirlet-Straße 1
52074 Aachen
Germany

Dirk Bebenroth
Ruprecht-Karls-Universität Heidelberg
Institut für Pharmazie und Molekulare
  Biotechnologie
Im Neuenheimer Feld 364
69120 Heidelberg
Germany

Frank Bordusa
Max-Planck Society
Research Unit "Enzymology of Protein
  Folding"
Weinbergweg 22
06120 Halle/Saale
Germany

Katrin Bossmann
Ruhr-Universität Bochum
Fakultät für Chemie
44780 Bochum
Germany

Stefan Bräse
Institut für Organische Chemie
Universität Karlsruhe (TH)
Fritz-Haber-Weg 6
76131 Karlsruhe
Germany

Susanne Brakmann
Institut für Zoologie
Universität Leipzig
Liebigstraße 18
04108 Leipzig
Germany

Rolf Breinbauer
Max Planck Institute of Molecular Physiology
Otto-Hahn-Straße 11
44227 Dortmund
Germany

Christian Dose
Max Planck Institute of Molecular Physiology
Otto-Hahn-Straße 11
44227 Dortmund
Germany

Simon Eßer
Institut für Biotechnologie
Forschungszentrum Jülich GmbH
52425 Jülich
Germany

Katja Faulhaber
Institute of Organic Chemistry University of
  Würzburg Am Hubland
97074 Würzburg
Germany

Thorsten Felder
Kekulé-Institut für Organische Chemie und
  Biochemie der Rheinischen Friedrich-
  Wilhelms-Universität Bonn
Gerhard-Domagk-Strasse 1
53121 Bonn
Germany

Dirk Franke
Institut für Biotechnologie
Forschungszentrum Jülich GmbH
52425 Jülich
Germany

List of Contributors

Lars Geiger
Institute of Organic Chemistry University of
  Würzburg Am Hubland
97074 Würzburg
Germany

Susanne Gendreizig
Swiss Federal Institute of Technology
  Lausanne (EPFL)
Institute of Molecular and Biological
  Chemistry
Institute of Biomolecular Sciences
1015 Lausanne
Switzerland

Carmen Gil
Kekulé-Institut für Organische Chemie und
  Biochemie der Rheinischen Friedrich-
  Wilhelms-Universität Bonn
Gerhard-Domagk-Strasse 1
53121 Bonn
Germany

Thomas Gronemeyer
Swiss Federal Institute of Technology
  Lausanne (EPFL)
Institute of Molecular and Biological
  Chemistry
Institute of Biomolecular Sciences
1015 Lausanne
Switzerland

Susanna Herold
Laboratorium für Anorganische Chemie
Eidgenössische Technische Hochschule ETH
  Hönggerberg
8093 Zürich
Switzerland

Claudia Höbartner
Institute of Organic Chemistry
University of Innsbruck
6020 Innsbruck
Austria

Christoph Huwe
Schering AG
Medicinal Chemistry Department
Research Center Europe
13342 Berlin
Germany

Heiko Ihmels
Institute of Organic Chemistry University of
  Würzburg Am Hubland
97074 Würzburg
Germany

Sergei A. Ivanov
Ruhr-Universität Bochum
Fakultät für Chemie
44780 Bochum
Germany

Andres Jäschke
Ruprecht-Karls-Universität Heidelberg
Institut für Pharmazie und Molekulare
  Biotechnologie
Im Neuenheimer Feld 364
69120 Heidelberg
Germany

Kai Johnsson
Swiss Federal Institute of Technology
  Lausanne (EPFL)
Institute of Molecular and Biological
  Chemistry
Institute of Biomolecular Sciences
1015 Lausanne
Switzerland

Alexandre Juillerat
Swiss Federal Institute of Technology
  Lausanne (EPFL)
Institute of Molecular and Biological
  Chemistry
Institute of Biomolecular Sciences
1015 Lausanne
Switzerland

Sonja Keiper
Ruprecht-Karls-Universität Heidelberg
Institut für Pharmazie und Molekulare
  Biotechnologie
Im Neuenheimer Feld 364
69120 Heidelberg
Germany

Antje Keppler
Swiss Federal Institute of Technology
  Lausanne (EPFL)
Institute of Molecular and Biological
  Chemistry
Institute of Biomolecular Sciences
1015 Lausanne
Switzerland

Maria Kesseler
BASF AG
Research Fine Chemicals
GVF/E -A30
67056 Ludwigshafen
Germany

Prof. Dr. Harm-Anton Klok
Ecole Polytechnique Fédérale de Lausanne
  (EPFL)
Institut des Matériaux, Laboratoire des
  Polymères
Bâtiment MX-D, 1015 Lausanne (Switzerland)

Sven Klussmann
NOXXON Pharma AG
Max-Dohrn-Straße 8-10
10589 Berlin
Germany

Kerstin Knepper
Kekulé-Institut für Organische Chemie und
  Biochemie der Rheinischen Friedrich-
  Wilhelms-Universität Bonn
Gerhard-Domagk-Strasse 1
53121 Bonn
Germany

Maja Köhn
Max Planck Institute of Molecular Physiology
Address
Otto-Hahn-Straße 11
44227 Dortmund
Germany

Thomas Kolter
Kekulé-Institut für Organische Chemie und
  Biochemie der Universität
Gerhard-Domagk Str. 1
53121 Bonn
Germany

Stefan Kubik
Institut für Organische Chemie und
  Makromolekulare Chemie
Heinrich-Heine-Universität
Universitätsstraße
40225 Düsseldorf
Germany

Hermann Künzer
Schering AG
Medicinal Chemistry Department
Research Center Europe
13342 Berlin
Germany

Volker Lorbach
Institut für Biotechnologie
Forschungszentrum Jülich GmbH
52425 Jülich
Germany

Arne Lützen
Universität Oldenburg
Institut für Reine und Angewandte Chemie
Postfach 2503
26111 Oldenburg
Germany

Wolfgang Maison
Institut für Organische Chemie der
  Universität Hamburg
Martin-Luther-King-Platz 6
20146 Hamburg
Germany

Thomas J. Martin
ALTANA Pharma AG
Department of Chemistry RDR/C2
Byk-Gulden-Str. 2
78467 Konstanz
Germany

Andreas Marx
Kekulé-Institut für Organische Chemie und
  Biochemie
Universität Bonn
Gerhard-Domagk-Str. 1
53121 Bonn
Germany

Ronald Micura
Institute of Organic Chemistry
University of Innsbruck
6020 Innsbruck
Austria

Michael Müller
Institut für Biotechnologie
Forschungszentrum Jülich GmbH
52425 Jülich
Germany

Sabine Müller
Ruhr-Universität Bochum
Fakultät für Chemie
44780 Bochum
Germany

H. Peter Nestler
Aventis Pharma Germany
Industrial Park Hoechst
Building G 879
65926 Frankfurt am Main
Germany

Franz von Nussbaum
Bayer AG
Pharma Research PH-R-EU-CR
Geb. 460/3

42096 Wuppertal
Germany

Carsten Peters
Max Planck Institute of Molecular Physiology
  Address
Otto-Hahn-Straße 11
44227 Dortmund
Germany

Jörg Rademann
Institute for Organic Chemistry
University of Tübingen
Auf der Morgenstelle 18
72076 Tübingen
Germany

Petra Rzepecki
Universität Marburg
Fachbereich Chemie
Hans-Meerwein-Strasse
35032 Marburg
Germany

Christoph A. Schalley
Kekulé-Institut für Organische Chemie und
  Biochemie der Rheinischen Friedrich-
  Wilhelms-Universität Bonn
Gerhard-Domagk-Strasse 1
53121 Bonn
Germany

Marina Schlicke
Universität Leipzig
Biotechnologisch-Biomedizinisches Zentrum
Ritterstraße 16
04109 Leipzig
Germany

Boris Schmidt
Clemens-Schöpf-Institute for Organic
  Chemistry and Biochemistry
TU Darmstadt
Petersenstr. 22
64287 Darmstadt
Germany

Carsten Schmuck
Institute of Organic Chemistry University of
  Würzburg
Am Hubland
97074 Würzburg
Germany

Thomas Schrader
Universität Marburg
Fachbereich Chemie
Hans-Meerwein-Strasse
35032 Marburg
Germany

Oliver Seitz
Fachinstitut für Organische und
  Bioorganische Chemie
Institut für Chemie der Humboldt-Universität
  zu Berlin
Brook-Taylor-Strasse 2
12489 Berlin
Germany

Alexander Siegler
Clemens-Schöpf-Institute for Organic
  Chemistry and Biochemistry
TU Darmstadt
Petersenstr. 22
64287 Darmstadt
Germany

Georg A. Sprenger
Institut für Mikrobiologie
Universität Stuttgart
Allmandring 31
70550 Stuttgart
Germany

Peter Spiteller
Institut für Organische Chemie und
  Biochemie
Technische Universität München
Lichtenbergstraße 4
85747 Garching
Germany

Michael Strerath
Kekulé-Institut für Organische Chemie und
  Biochemie
Universität Bonn
Gerhard-Domagk-Str. 1
53121 Bonn
Germany

Friedrich Stuhlmann
Chelona GmbH
Hermannswerder Haus 15
14473 Potsdam
Germany

Daniel Summerer
Kekulé-Institut für Organische Chemie und
  Biochemie
Universität Bonn
Gerhard-Domagk-Str. 1
53121 Bonn
Germany

Petra Tafelmeyer
Swiss Federal Institute of Technology
  Lausanne (EPFL)
Institute of Molecular and Biological
  Chemistry
Institute of Biomolecular Sciences
1015 Lausanne
Switzerland

Giampietro Viola
Department of Pharmaceutical Sciences
  University of Padova Via Marzolo 5
35131 Padova
Italy

Hans-Achim Wagenknecht
Institute for Organic Chemistry and
  Biochemistry
Technical University of Munich
Lichtenbergstr. 4
85747 Garching
Germany

Dr. Mark Wehner
Clariant GmbH
Werk Höchst/Geb. D569
65926 Frankfurt/Main
Germany

Rüdiger Welz
Ruhr-Universität Bochum
Fakultät für Chemie
44780 Bochum
Germany

Helma Wennemers
University of Basel
Department of Chemistry
St. Johanns-Ring 19
4056 Basel
Switzerland

Wolfgang Wienand
Institute of Organic Chemistry University of
  Würzburg
Am Hubland
97074 Würzburg
Germany

Uta Wille
School of Chemistry
The University of Melbourne
Victoria 3010
Australia

Valentin Wittmann
Universität Konstanz
Fachbereich Chemie
Universitätsstr. 10
78464 Konstanz
Germany

Ludwig Zorn
Schering AG
Medicinal Chemistry Department
Research Center Europe
13342 Berlin
Germany

Part 1
Biomolecules and their Conformation

1.1
Equilibria of RNA Secondary Structures

Ronald Micura and Claudia Höbartner

1.1.1
Introduction

1.1.1.1
RNA Folding

RNA plays a central role in the life of cells. It is, therefore, important to understand how RNA forms functional native structures endowed with properties such as catalysis, binding of small-molecular-weight ligands, or recognition of proteins. Termed the "RNA folding problem" [1], the question is how the primary structure of RNA, a linear polynucleotide, encodes its functional three-dimensional (tertiary) structure.

On the basis of early studies on tRNA and consistent in the first order with more recent studies on ribozymes, a hierarchical model of the folding of RNA has been established [2]. Implicit in this model is the assumption that there are two major structural changes on the way from an ensemble of unfolded RNA molecules to a native state. It is supposed that stable secondary structures, such as hairpins, form rapidly on a microsecond time scale. Subsequent assembly leading to tertiary folding occurs by bringing the secondary structural elements together. This procedure is slower – the time required for tertiary interactions is in the order of milliseconds up to seconds and minutes.

The hierarchical model, the commonly held picture of RNA folding, is not always correct. During recent years outlines of the mechanisms of folding of large RNA molecules are beginning to emerge thanks to several novel experiments in which more advanced techniques, e.g. single molecule-fluorescence spectroscopy [3], synchrotron hydroxyl radical footprinting [4], or temperature gradient gel electrophoresis [5] have been applied. There is growing evidence that the folding kinetics of RNA are complex, involving parallel pathways and kinetic traps that proceed via misfolded structures [6, 7].

In the following discussion, two outstanding discoveries concerning RNA folding are introduced briefly. These findings, which are of importance in the context of our own studies, can be described as:

Highlights in Bioorganic Chemistry: Methods and Applications. Edited by Carsten Schmuck, Helma Wennemers.
Copyright © 2004 WILEY-VCH Verlag GmbH & Co. KGaA, Weinheim
ISBN: 3-527-30656-0

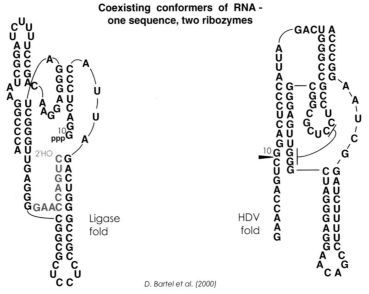

Fig. 1.1.1. Example of an RNA sequence that can adopt two different defined folds, each associated with a specific biological activity (HDV hepatitis delta virus).

- single RNA sequences that co-exist in different folds, and
- nucleoside modifications that cause the conversion of one RNA fold into another involving secondary structure rearrangements.

1.1.1.2
One Sequence – Two Ribozymes

In July 2000 Bartel and coworkers published the impressive example of a single RNA sequence that can adopt either of two ribozyme (Box 22) folds and catalyze two different reactions (Figure 1.1.1) [8]. One reaction is cleavage of RNA catalyzed by the hepatitis delta virus (HDV) ribozyme, which assists the replication of HDV viral RNA. The other is RNA ligation catalyzed by the class III ligase ribozyme, an activity obtained in the laboratory in in-vitro selection experiments. The two ribozyme folds are completely different and do not have a single base pair in common. Importantly, minor variants of this sequence are highly active for one or other catalytic activity and can be accessed from the prototype by a few nucleotide mutations only.

Experimental verification of the two different folds was obtained from gel shift assays by following the expected ligation versus cleavage products.

It is of outstanding significance that the intersection sequence of Schultes and

Bartel is the first example of a molecule that can adopt two different conformations, *each* associated with a distinct catalytic activity.

In principle, it is not unusual for large RNA to exist in more than one conformation [2] (compare misfolding of proteins that is often associated with misfunctions; see also Chapter 2.4). For example, it is well investigated that the three helixes of the hammerhead ribozyme change their three-dimensional arrangement upon addition of Mg^{2+} to reach the final active conformation which results in strand cleavage [9]. A vast source of alternative RNA conformations with secondary structure rearrangements is encountered in translational regulation mechanisms involving mRNA [10, 11]. Despite intensive research in this field, however, direct structural data or quantitative data, e.g. concerning the portions of the alternative mRNA folds involved, are usually not available. This is because most assays for translational regulation mechanisms are – for obvious reasons – activity or expression assays and do not focus on direct structural proof of the alternative folds [12, 13].

The rational design of very short RNA sequences (15 to 40 nucleotides) that are able to adopt alternative folds is one of our major efforts. Part of this program are RNAs in thermodynamic equilibrium with different defined secondary structures. This also includes the development of methods for direct experimental verification of RNA equilibria and, moreover, applications with respect to biologically relevant RNA switches.

We stress that RNA oligonucleotides 21 to 23 nucleotides in length are attributed a fundamental role in post-transcriptional regulation of gene expression in animals and plants [14]. Termed RNA interference (RNAi), double stranded RNA (dsRNA) effects the silencing of genes which are homologous to either of the RNA strands in the duplex [15–19]. This phenomenon results from degradation of the corresponding mRNA and can also be induced efficiently by very short duplex RNAs of 21 to 23 base pairs, which are called small interfering RNAs (siRNA). In a variety of organisms transfection with siRNA duplexes initiates a mechanism involving cellular components to silence the gene with sequence homology. siRNA is the upcoming gene-silencing methodology and, at this time, seems to be more promising than the alternatives including antisense and ribozyme-based strategies. The key components of this new technology are short RNA oligonucleotides.

1.1.1.3
Nucleoside Methylation is Responsible for Correct Folding of a Human Mitochondrial tRNA

The second aspect which is of concern in our own studies is the effect of naturally occurring modified nucleosides on RNA folding. With regard to RNA secondary structure and tertiary structure, modified nucleosides are generally understood to modulate the physicochemical properties of an existing RNA fold [20–22]. Predominantly, minor changes on the local conformation close to the incorporated modification are reported (e.g. ribose–phosphate backbone, glycosidic angle) [23–25].

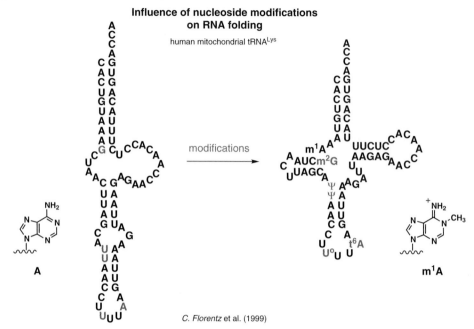

Fig. 1.1.2. Naturally occurring modified nucleoside 1-methyladenosine is necessary for adoption of the biologically active cloverleaf fold of mitochondrial tRNA (m^{1}A, 1-methyladenosine; m^{2}G, N^2-methylguanosine; Ψ, pseudouridine; t^{6}A, N^6-threonyl-carbamoyladenosine; U°, modified uridine structurally not characterized).

The first direct evidence for nucleoside modifications that cause conversion of one RNA fold into another involving secondary structure rearrangments was reported by Florentz's group in 1998 (Figure 1.1.2) [26]. Human mitochondrial tRNA^{Lys} has six modified nucleosides (m^{1}A9, m^{2}G10, Ψ27, Ψ28, and hypermodified nucleosides at positions U34 and A37). This tRNA folds into the expected cloverleaf. The solution structure of the corresponding in-vitro transcript unexpectedly did not fold into a cloverleaf but into an extended bulged hairpin.

This non-canonical fold, established according to chemical and enzymatic structure probing, includes an extended amino acid acceptor stem, an extra large loop instead of the T-stem and loop, and an anticodon-like domain. Hence, one or several of the six modified nucleosides are required and are responsible for its cloverleaf structure. In a further study a chimeric tRNA with the sole modification of 1-methyladenosine in position 9 was synthesized; it was demonstrated that this chimeric RNA folds correctly [27]. Thus, because of Watson–Crick base-pair disruption, a single methyl group is sufficient to induce the cloverleaf folding of this unusual tRNA sequence.

These findings encouraged us to search for further naturally occurring modification patterns which promise to be determining for the observed fold. We considered the helix 45 in rRNA of the small ribosomal subunit as an appropriate target.

1.1.2
Monomolecular RNA Two-state Conformational Equilibria

For the reasons outlined above, our efforts concentrated on the design and experimental verification of short RNA sequences which can adopt two different secondary structures. Implicit was the principal question "what is the minimum size of an RNA that is still able to appear in different defined shapes?" In this context, the closest study we found was a theoretical approach by C. Flamm et al. [28], who developed an algorithm to design multistable RNAs [29]. To the best of our knowledge, however, there is no experimental data verifying the multistable sequence examples suggested therein.

The original design of our bistable RNA molecules did not make use of the algorithm cited above. Our empirical approach was based on two common secondary structure motifs which we set in competition with each other [30]; for this purpose a GNRA- and UNCG-hairpin appeared suitable (Figure 1.1.3).

In particular, we designed a sequence with the following characteristics. rGAC**CGGAAGGUC** and r**CGGAAGGUC**CGCCUUCC provide an identical nucleotide partition r**CGGAAGGUC**. This nucleotide partition is the crucial feature, because it enables the two secondary structure motifs to be set in competition *within* a single sequence construct. For the sequence construct rGAC**CGGAAGGUC**CGCCUUCC we claimed an equilibrium situation, which had to be proven experimentally (Figure 1.1.3).

We first determined the thermodynamic stabilities of the individual submotifs. For the GGAA-hairpin rGAC**CGGAAGGUC** the melting temperature was 77 °C and ΔG^{298K} was -7.9 kcal mol^{-1}. This was derived from UV–melting profiles (150 mM NaCl, 10 mM Na$_2$HPO$_4$, pH 7.0) [30]. Although this hairpin consists of a four base pair stem, it is more stable than the UCCG-hairpin r**CGGAAGGUC**CGCCUUCC with a six base pair stem (Figure 1.1.3).

The latter hairpin melts at a temperature 5° lower. The difference $\Delta\Delta G^{UV-298K}$ is $+0.6$ kcal mol^{-1} and suggests an equilibrium ($\Delta\Delta G = -RT \ln K$) of 73%:27% in favor of the GGAA-hairpin. If the activation energy for the interconversion of one conformation to the other was high enough, it would not be unreasonable to expect a biphasic melting profile although the difference in melting is only 5 °C. Unfortunately, all the melting profiles – the melting profiles of the two reference sequences, rGAC**CGGAAGGUC** and r**CGGAAGGUC**CGCCUUCC, and the melting profile of the sequence construct, rGAC**CGGAAGGUC**CGCCUUCC – showed monophasic melting transitions and gel shift analysis under non-denaturing conditions did not favor two individual species.

We therefore considered NMR spectroscopy as a possible tool to investigate the

1.1 Equilibria of RNA Secondary Structures

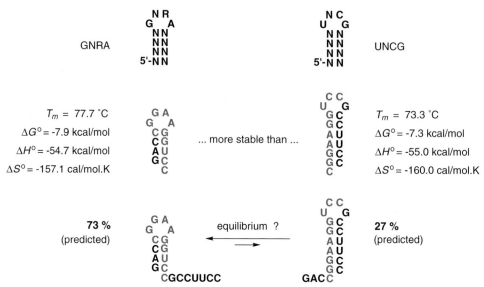

Fig. 1.1.3. The identical sequence partition CGGAAGGUCC (gray) of the GGAA- and the UCCG reference (center) allows the design of two competing secondary structures within a single sequence (below).

equilibria of short RNA sequences. The NH protons that are involved in Watson–Crick pairing ("imino protons") are easily detectable for RNA in submillimolar concentrations. In principle, each Watson–Crick base pair is assigned to a single signal. Because of base-pair fraying at the double helix end, the signal of the terminal base pair is broadened or not detectable at room temperature. Usually, the dispersion of the NH signals is good.

Figure 1.1.4 shows the NH spectra of our sequence ensemble. The spectrum of the sequence construct rGACCGGAAGGUCCGCCUUCC has the characteristics of slow exchange and reflects both sets of NH signals as measured for the individual reference hairpins rGACCGGAAGGUC and rCGGAAGGUCCGCCUUCC. This is clear evidence that rGACCGGAAGGUCCGCCUUCC indeed occurs as a two-state equilibrium of defined secondary structures [30]. It must also be stressed that the GGAA conformer is 3-fold less populated than the UCCG conformer, even though the GGAA-hairpin reference is a more stable secondary structure motif than the UCCG-hairpin reference. A likely explanation is that the GGAA conformer has an overhang which is fully complementary to the loop region and can therefore provide a higher forward rate constant.

Comparative imino proton NMR spectroscopy as described above proved a useful

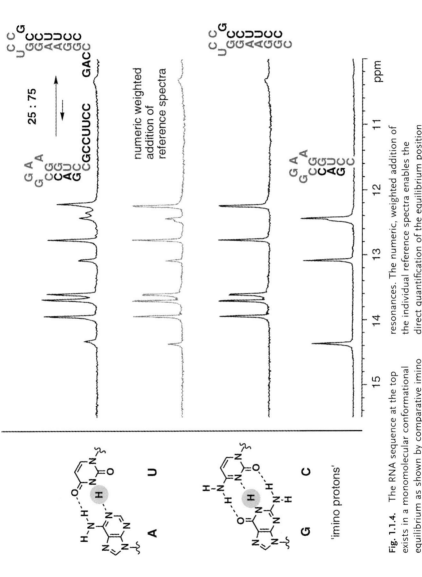

Fig. 1.1.4. The RNA sequence at the top exists in a monomolecular conformational equilibrium as shown by comparative imino proton NMR spectroscopy. The references are only able to adopt a single defined fold and allow the stem-wise assignment of the NH resonances. The numeric, weighted addition of the individual reference spectra enables the direct quantification of the equilibrium position (trace in gray). A 25:75 equilibrium is ascertained at 298 K, in 25 mM sodium arsenate buffer, pH 7.4, $H_2O/D_2O = 9:1$.

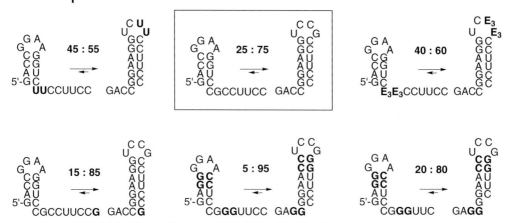

Fig. 1.1.5. The equilibrium position of short bistable RNAs is highly sensitive to nucleoside variations in the sequence (E_3 is triethylene glycol phosphate).

tool for probing RNA equilibrium sequences. We further investigated how single or double nucleobase mutations affected the position of the equilibrium (Figure 1.1.5). We found, e.g., that exchange of C13G14 within the loop by UU weakened the corresponding conformer as the extra stable UCCG loop was no longer available. Hence, the equilibrium shifted towards the GGAA conformer. An equilibrium shift of comparable extent was observed when C13G14 were replaced by non-nucleotide bis(triethylene glycol)phosphate linkers (E_3E_3). A shift towards the other direction was obtained for a guanosine additionally attached at the 3′ end of our original sequence construct; this strengthened the UCCG conformer by an additional base pair whereas it had no direct impact on the GGAA alternative. Moreover, if nucleosides were exchanged within the stem region, the RNA sequences obtained still existed in conformational equilibria.

Figure 1.1.6 illustrates a comparison of our results with the output of the program RNAfold (implemented in the ViennaRNAPackage 1.4) [28, 29]. The RNAfold algorithm generates a base pair probability matrix. The corresponding graphical output is termed a "dotplot". In the lower left triangle this reflects solely the base pairs of the minimum free energy structure of a given sequence. The upper triangle displays the frequency of each single base pair in the ensemble of all possible structures. For our type of sequence there is essentially *one* secondary structure alternative. In general, the algorithm recognizes well the bistable nature of the rationally designed sequences. The first examples of Figure 1.1.6 shows a perfect match of prediction and experimental result. The sequence example below demonstrates that prediction of the equilibrium position can deviate significantly from that found experimentally.

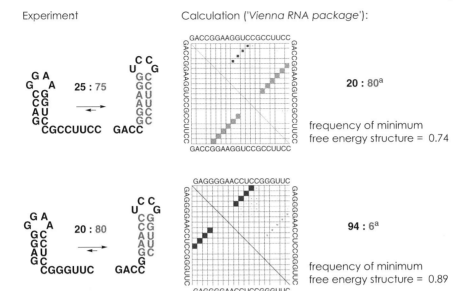

Fig. 1.1.6. Dotplots for two bistable RNA sequences generated by the RNAfold program (Vienna RNA package) [29]. The lower left triangle shows the base pairs of the minimum free energy structure. The upper right triangle represents the frequencies of each base pair within the ensemble of all possible structures. The area of the squares is proportional to the pairing probability. [a] Ratio predicted from calculated free energy differences between the two conformers #′ and #″ (RNAsubopt program [29], $\Delta\Delta G_{\#',\#''} = -RT \ln K$).

1.1.3
The Influence of Nucleobase Methylations on Secondary Structure Equilibria, as Exemplified by the Ribosomal Helix 45 Motif

As mentioned in the Introduction, we are especially interested in the effect of nucleoside modifications on RNA structure. The equilibrium sequences introduced above were selected for this reason. The position of the conformational equilibrium of the sequence constructs is expected to be highly sensitive to chemical alteration, e.g. methylation of nucleobases.

In particular, an additional lead for our design was the naturally occurring methylated sequence motif encountered in the terminal helix at the 3′ end of ribosomal RNA in the small subunit (Figure 1.1.7). This helix (helix 45) is capped by a hypermethylated GGAA tetranucleotide loop [31–35]. In most organisms the two successive adenosines are converted into N^6,N^6-dimethyladenosines and in some organisms one of the guanosines is also replaced by N^2-methylguanosine.

To make the required methylated RNA sequences available for our folding studies we synthesized the methylated nucleosides as 2′-O-(triisopropylsilyl)oxymethyl (TOM)-phosphoramidites and applied the building blocks to RNA synthesis and purification procedures [36, 46].

```
         E. coli                    B. stearothermophilus

        G   m⁶₂A                      G   m⁶₂A
    m²G       m⁶₂A               m²G      m⁶₂A
        G  C                          C  G
        G  C                          C  G
        A  U                          A  U
        U  G                          U  G
        G  C                          G  C
        C  G                          C  G
        C  G                          C  G
        A  U                          G  C
        A  U                          A  U
5'...U  GGAUCACCUCCUUA_OH     ...GU  GGAUCACCUCCUUUCUA_OH
```

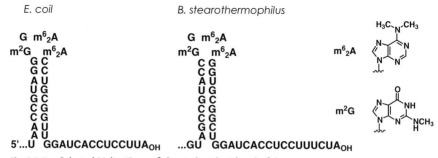

Fig. 1.1.7. Selected Helix 45 motifs located at the 3' end of the small subunit ribosomal RNA. The successive dimethylated adenosines within the tetraloop are conserved in most organisms. The reason for the methylation is unclear (m^6_2A is N^6,N^6-dimethyladenosine).

Wondering why these nucleotides are methylated we considered that, e.g., in *Bacillus stearothermophilus* the 3' overhang offers a sequence partition that is complementary to the loop region as long it is non-methylated (Figure 1.1.8). In principle, an interaction between the loop and the overhang is conceivable, and hypothetically, this interaction could result in a defined pseudoknot fold. So far, we have support for this hypothesis from the reduced sequence constructs depicted in Figure 1.1.8.

As demonstrated in the preceding section, the sequence rGAC**CGGAAGGUC**CGCCUUCC occurs in 75:25 equilibrium with the major conformer corresponding to the proposed pseudoknot fold. In general, this sequence family showed that extra stable GNRA loop structures can be broken up by complementary sequence sections. The question was, what happens to the conformational equilibrium when the sequence is methylated according to the naturally occurring methylation pattern. With a corresponding set of reference sequences the answer was again obtained by comparative imino proton NMR spectroscopy.

The methylated sequence rGAC**Cm²GGm⁶₂Am⁶₂AGG**UCCGCCUUCC reflected the NH resonances in accordance with the reference rGAC**Cm²GGm⁶₂Am⁶₂AGG**UC with the four base pair stem [30]. We can therefore conclude that the methylated sequence construct predominantly occurs in a single conformation and, importantly, this conformation accords to the former unfavorable one. At room temperature the other reference sequence r**Cm²GGm⁶₂Am⁶₂AGG**UCCGCCUUCC has no sharp signal. Evidence that the second reference r**Cm²GGm⁶₂Am⁶₂AGG**UCCGCCUUCC adopts predominantly a stem-loop structure at room temperature is obtained from the UV-melting profiles. The latter sequence melts at 54 °C, significantly lower than rGAC**Cm²GGm⁶₂Am⁶₂AGG**UC ($T_m = 72.7$ °C) [30]. The methylation of adenosine at the Watson–Crick base-pairing site disrupts the base pair, but does not necessarily prevent formation of

Hypothesis (Helix 45 in B. stearothermophilus):

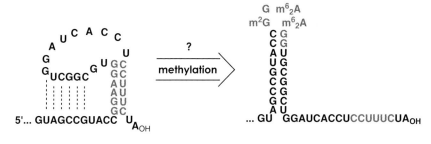

Experiment:

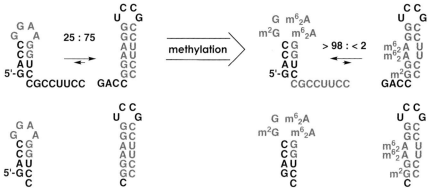

Fig. 1.1.8. Methylation of the helix 45 motif in the small ribosomal subunit might be responsible for the fold observed.

a double helical arrangement. The double interference of dimethyladenine with uracil seems to be well tolerated if there is no alternative to change the secondary structure to an energetically more favorable motif.

Similar to what we demonstrated for the non-methylated sequences we investigated the methylated sequences in the context of single or double nucleotide replacements. For all replacements tested we observed the occurrence of a single conformer comprising the $m^2GGm^6_2Am^6_2A$ loop.

In summary, we have evidence from the short model sequences that modification according to the naturally occurring methylation pattern of helix 45 selects the observed secondary structure. To test the behavior of the fortymer terminal rRNA sequence of *Bacillus stearothermophilus* is the target of ongoing research.

In the literature it is reported that in *E. coli* mutations in the rRNA dimethylase ksgAp block the $m^6_2Am^6_2A$ dimethylation of 16S rRNA; they do not, however, interfere with rRNA processing. Consequently, the mutants synthesize SSU rRNAs lacking the modification. These allow the growth of the mutant strains at a reduced

1.1.4
Structural Probing of Small RNAs by Comparative Imino Proton NMR Spectroscopy

Determining RNA secondary and tertiary structure is commonly based on enzymatic and chemical probing methods [37, 38]. These methods need only minor amounts of sample for analysis and provide a powerful tool for structure probing, especially that of large RNAs when NMR and crystallographic analysis become difficult. There are, however, also limitations to the specificities of the enzymes, and the reactivities of the probes are not always well understood. In particular, small RNA secondary structures that interconvert are not accessible by standard enzymatic and chemical probing methods. The conformers of a short bistable RNA sequence are expected to exist in a dynamic equilibrium with a significantly low energy barrier for their interconversion although this process involves breakage and formation of many different base pairs. The time required for conformer change is not long enough to enable investigation by gel-shift assays or UV–melt-

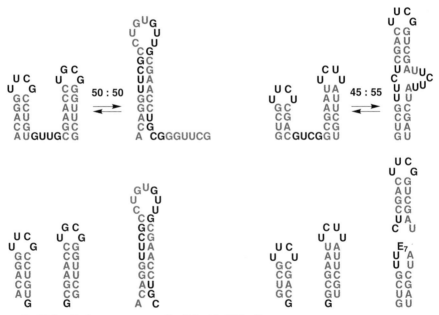

Fig. 1.1.9. Further sequence motifs of bistable RNAs. The equilibrium position was quantified by comparative imino proton NMR spectroscopy [45].

ing experiments; the time scale is, however, long enough for analysis of such equilibria by ^{1}H NMR spectroscopy (> milliseconds).

The proton resonances between 10 and 15 ppm directly reflect Watson–Crick base-paired double helical regions and are, in principle, sufficient to enable verification of a secondary structure model [13, 39–43]. The assignment of the resonances is a prerequisite and usually affords ^{15}N-labeled RNA samples and requires advanced NMR methods [13, 39]. We explore a different approach with a pronounced chemical viewpoint. The high thermodynamic stability of the secondary structure of an RNA compared with the tertiary structure enables segmentation into the individual Watson–Crick base paired double helixes – the separately synthesized reference oligonucleotides are easily defined as hairpins for terminal stem-loop segments; for internal double helical tracks of a larger RNA, reference oligonucleotides can also be defined by the double helical sequence homologs with non-nucleotide linker units such as hexa- or heptaethylene glycol phosphates [44, 45]. The spectra of the individual references provide a characteristic set of imino proton resonances (fingerprint) and enable a stem-wise assignment of resonances within the complex NH spectrum of the complete RNA. This approach which we term "comparative imino proton NMR spectroscopy", is very reliable for assignment and quantification of the equilibrium position of bistable RNAs (Figure 1.1.9) but should be expandable for verification of secondary structure models of larger RNAs also [45]. For small RNAs, labeling is not needed and one-dimensional NMR-spectroscopic methods are sufficient. In principle, the short reference sequences can be commercially ordered so that this fast and direct method should not be limited to a synthetic laboratory and represents a powerful alternative to chemical and enzymatic structural probing of small RNAs.

Acknowledgments

Financial support from the Austrian Science Fund (P15042) is gratefully acknowledged.

References

1 P. B. Moore, *The RNA World*, 2nd edn, R. F. Gesteland, T. R. Cech, J. F. Atkins, eds., CSHL Press, New York, **1999**, pp. 381–401.

2 P. Brion, E. Westhof, *Annu. Rev. Biophys. Biomol. Struct.* **1997**, *26*, 113–137.

3 X. Zhuang, L. E. Bartley, H. P. Babcock, R. Russell, T. Ha, D. Herschlag, S. Chu, *Science* **2000**, *288*, 2048–2051.

4 B. Scalvi, M. Sullivan, M. R. Chance, M. Brenowitz, S. A. Woodson, *Science* **1998**, *279*, 1940–1943.

5 A. A. Szewczak, E. R. Podell, P. C. Bevilacqua, T. R. Cech, *Biochemistry* **1998**, *37*, 11162–11170.

6 J. Pan, D. Thirumalai, S. A. Woodson, *J. Mol. Biol.* **1997**, *273*, 7–13.

7 J. Pan, T. R. Sosnick, *Nat. Struct. Biol.* **1997**, *4*, 553–558.

8 E. A. Schultes, D. P. Bartel, *Science* **2000**, *289*, 448–452.
9 D. B. McKay, J. E. Wedekind, *The RNA World*, 2nd edn, R. F. Gesteland, T. R. Cech, J. F. Atkins, eds., CSHL Press, New York, **1999**, pp. 265–286.
10 M. H. de Smit, *RNA Structure and Function*, R. W. Simons, M. Grunberg-Manago, eds., CSHL Press, New York, **1998**, pp. 495–540.
11 T. Platt, *RNA Structure and Function*, R. W. Simons, M. Grunberg-Manago, eds., CSHL Press, New York, **1998**, pp. 541–574.
12 K. A. LeCuyer, D. M. Crothers, *Proc. Natl. Acad. Sci. USA* **1994**, *91*, 3373–3377.
13 M. Wu, I. Tinoco Jr, *Proc. Natl. Acad. Sci. USA* **1998**, *95*, 11555–11560.
14 R. Micura, *Angew. Chem. Int. Ed. Chem.* **2002**, *41*, 2265–2269.
15 T. Tuschl, *ChemBioChem* **2001**, *2*, 239–245.
16 J. H. Hannon, *Nature* **2002**, *418*, 244–251.
17 P. D. Zamore, *Science* **2002**, *296*, 1265–1269.
18 G. Storz, *Science* **2002**, *296*, 1260–1263.
19 G. Plasterk, *Science* **2002**, *296*, 1263–1265.
20 B. G. Lane, *Modification and Editing of RNA*, H. Grosjean, R. Benne, eds., ASM Press, Washington, DC, **1998**, pp. 1–15.
21 H. Grosjean, G. Björk, B. E. H. Maden, *Biochimie* **1995**, *77*, 3–6.
22 P. F. Agris, *Prog. Nucleic Acid Res. Mol. Biol.* **1995**, *53*, 79–129.
23 M. Meroueh, P. J. Grohar, J. Qui, J. SantaLucia Jr, S. A. Scaringe, C. S. Chow, *Nucleic Acids Res.* **2000**, *28*, 2075–2083.
24 M. Sundaram, P. F. Crain, D. R. Davis, *J. Org. Chem.* **2000**, *65*, 5609–5614.
25 S. Derreumaux, M. Chaoui, G. Tevanian, S. Fermandjian, *Nucleic Acids Res.* **2001**, *29*, 2314–2326.
26 M. Helm, H. Brulé, F. Degoul, C. Cepanec, J.-P. Leroux, R. Giegé, C. Florentz, *Nucleic Acids Res.* **1998**, *26*, 1636–1643.
27 M. Helm, R. Giegé, C. Florentz, *Biochemistry* **1999**, *38*, 13338–13346.
28 C. Flamm, I. L. Hofacker, S. Maurer-Stroh, P. F. Stadler, M. Zehl, *RNA* **2001**, *7*, 254–265.
29 I. L. Hofacker, W. Fontana, P. F. Stadler, P. Schuster, http://www.tbi.univie.ac.at/~ivo/RNA/, **1994–2001** (Free Software).
30 C. Höbartner, M.-O. Ebert, B. Jaun, R. Micura, *Angew. Chem. Int. Ed. Engl.* **2002**, *41*, 605–609.
31 D. L. J. Lafontaine, D. Tollervey, *Modification and Editing of RNA*, H. Grosjean, R. Benne, eds., ASM Press, Washington, DC, **1998**, pp. 281–288.
32 R. van Charldorp, A. M. A. van Kimmenade, P. H. van Knippenberg, *Nucleic Acids Res.* **1981**, *19*, 4909–4917.
33 P. H. van Knippenberg, *Structure, Function, and Genetics of Ribosomes*, B. Hardesty, G. Kramer, eds., Springer, New York, NY, **1986**, pp. 412–424.
34 D. Nègre, C. Weizmann, J. Ofengand, *Proc. Natl. Acad. Sci. USA* **1989**, *86*, 4902–4906.
35 J. P. Rife, P. B. Moore, *Structure* **1998**, *6*, 747–756.
36 R. Micura, W. Pils, C. Höbartner, K. Grubmayr, M.-O. Ebert, B. Jaun, *Nucleic Acids Res.* **2001**, *29*, 3977–4005.
37 H. Moine, B. Ehresmann, C. Ehresmann, P. Romby, *RNA Structure and Function*, R. W. Simons, M. Grunberg-Manago, eds., Cold Spring Harbor Laboratory Press, NY, **1998**, pp. 77–115.
38 J. Hearst, *Nucleic Acids Structures, Properties, and Function*, V. A. Bloomfield, D. A. Crothers, I. Tinoco Jr, University Science Books, Sausalito, CA, **2000**, pp. 45–78.
39 Y. Tanaka, T. Hori, M. Tagaya, T. Sakamoto, Y. Kurihara, M. Katahira, S. Uesugi, *Nucleic Acids Res.* **2002**, *30*, 766–744.
40 J. Xu, J. Lapham, D. M. Crothers, *Proc. Natl. Acad. Sci. USA* **1996**, *93*, 44–48.

41 N. B. Leontis, P. B. Moore, *Biochemistry*, **1986**, *25*, 3916–3925.

42 H. A. Heus, O. C. Uhlenbeck, A. Pardi, *Nucleic Acids Res.* **1990**, *18*, 1103–1108.

43 O. Odai, H. Kodama, H. Hiroaki, T. Sakata, T. Tanaka, S. Uesugi, *Nucleic Acids Res.* **1990**, *18*, 5955–5960.

44 W. Pils, R. Micura, *Nucleic Acids Res.* **2000**, *28*, 1859–1863.

45 C. Höbartner, R. Micura, *J. Mol. Biol.* **2003**, *325*, 421–431.

46 C. Höbartner, C. Kreutz, E. Flecker, E. Ottenschlager, W. Pils, K. Grubmayr, R. Micura, *Monatsh.* **2003**, *134*, 851–873.

1.2
Synthesis and Application of Proline and Pipecolic Acid Derivatives: Tools for Stabilization of Peptide Secondary Structures

Wolfgang Maison

1.2.1
Introduction

De-novo design of peptides or peptidomimetics with defined conformational properties is a critical goal in bioorganic chemistry for elucidation of the mechanisms of protein folding and for development of new peptide or peptidomimetic drugs. Peptidomimetics can be used, for example, to initiate and stabilize certain elements of peptide secondary structure (Box 8) or as drugs derived from peptidic lead structures. In both cases, the initial event that is essential for stabilizing a secondary structure or eliciting a biological response is intra- or intermolecular interaction via H-bonding or electrostatic or hydrophobic interaction. The peptide backbone serves as a scaffold for the key functional groups involved in these interactions and can occasionally serve as a hydrogen-bond donor or acceptor (see also Chapter 2.3). As depicted in Figure 1.2.1, the backbone torsional angles ϕ, ψ, and ω, the side-chain angles χ, and the stereoelectronic properties of the side-chains determine the three dimensional shape of a peptide. If one wishes to predict the shape of a given peptide or peptidomimetic to be able to use it, for example, in rational drug design, one needs to understand the factors that determine its three-dimensional structure. It is thus essential to develop principles for predicting peptide backbone structures (Ramachandran space; ϕ, ψ and ω torsional angles; Box 2) and side-chain topographies (chi space; torsional angles).

A promising strategy for elucidating the mechanisms of peptide structure formation and its biological consequences is the design and synthesis of small peptide model systems in which a certain secondary structure element (e.g. a helix, turn, or sheet) can be investigated, in isolation from further complicating intra- or intermolecular interactions. The major advantage of these small molecule models is their accessibility by standard analytical methods such as IR, CD (circular dichroism), and NMR. Peptides of 30 amino acid residues or less, however, typically do not adopt well-defined conformations in solution. Biasing small peptides toward adopting a desired conformation therefore needs the help of a structure nucleating small molecule template [1] or a covalently fixed mimic of that structure (or the help of coordinating metal ions; see Chapter 1.3).

Highlights in Bioorganic Chemistry: Methods and Applications. Edited by Carsten Schmuck, Helma Wennemers.
Copyright © 2004 WILEY-VCH Verlag GmbH & Co. KGaA, Weinheim
ISBN: 3-527-30656-0

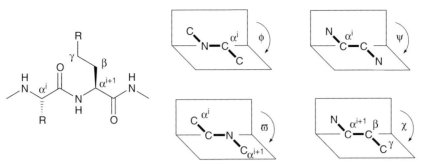

Fig. 1.2.1. Conventions for naming the dihedral angles ϕ, ψ, ω, and χ illustrated by a model dipeptide. Only certain backbone dihedral angles ϕ, ψ, and ω, which define the Ramachandran space, and χ (chi space) are populated by each amino acid.

A few selected examples for such templates or secondary structure mimics are shown in Figure 1.2.2.

Compounds of general formula **I** were designed to mimic turn structures, which are often key recognition sites in bioactive peptides. Lubell and Hanessian have developed efficient stereo-controlled synthetic routes to azabicycloalkanes **I** (X, Y = (CH$_2$)$_n$) starting from aspartate or glutamate. These routes provide access to a range of azabicycloalkanes with different side-chains, R, and ring sizes [2]. Some of these systems have been shown to adopt type II and type VI β-turn and γ-turn conformations. Similar azabicycloalkanes have been synthesized and analyzed by Scolastico and coworkers, starting from substituted proline derivatives [3]. Thia and oxa analogs of **I** (X = CH$_2$, Y = S or O) have been synthesized by several groups and some derivatives have been shown to adopt β-turn conformations [4]. Very recently, an interesting polyhydroxylated thia analog **II** was synthesized by

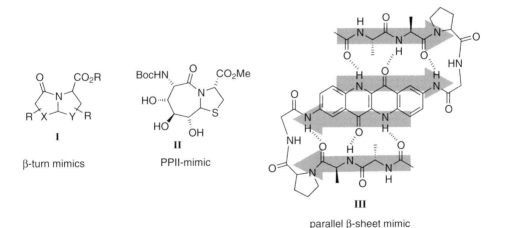

Fig. 1.2.2. Selection of known mimics for major secondary structure elements.

Tremmel and Geyer from glucuronolactone and cysteine methyl ester and demonstrated to adopt a PII helical conformation in oligomers [5].

Although successful attempts to develop templates for β-sheet structures are rare (see also Chapter 2.4), Kemp and coworkers introduced an epindolidione-based template that mimics the central strand of a β-sheet by appropriately orienting three hydrogen bonds to enforce an extended conformation on the attached peptide chains, as shown in Figure 1.2.2 with the model system **III** [6]. Direct attachment of peptide chains to the template leads to the formation of a parallel β-sheet mimic, whereas antiparallel β-sheet models can be obtained by incorporation of two urea groups for attachment of the peptide chains [7].

1.2.2
syn- and anti-Proline Mimics

Proline is unique among the proteinogenic amino acids in that its alkyl side-chain is covalently bonded to the α-amino function forming a five membered ring. Consequently, the peptide backbone has no amide hydrogen for use as a H-bond donor or in resonance stabilization of the amide bond of which it is part. The rigid cyclic backbone also imposes constraint on the dihedral angle ϕ and influences the geometry of the adjacent amide bond in peptides N-terminal to proline. Proline residues have thus significant effects on the conformation of polypeptide chains.

Although all proteinogenic amino acids form predominantly anti peptide bonds, a search in the Brookhaven Protein Database revealed that approximately 6–7% of all X-prolyl peptide bonds are found in the syn conformation in the native state of proteins [8]. The reason for this relatively frequent occurrence of *syn*-prolyl peptide bonds lies in steric repulsion of the proline δ protons and the adjacent N-terminal amino acid in the anti conformation, resulting in a low barrier of rotation and energetically similar syn and anti isomers (Figure 1.2.3).

Syn peptide bonds are primarily found in bends or turns and for syn imide bonds the high correlation suggests a specific role for these bonds in turn struc-

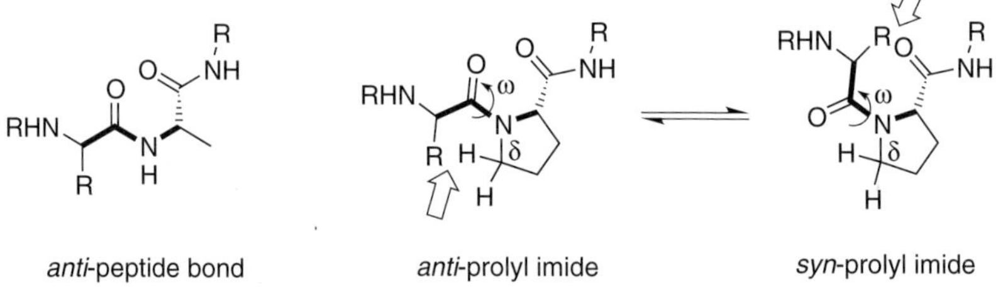

Fig. 1.2.3. Conventions for naming the syn and anti rotamers of peptide bonds. Rotation around the peptide bond is restricted by partial double-bond character caused by delocalization of the nitrogen electron lone pair and π-orbital electrons of the carbonyl group. Similar steric conflicts (arrows) for syn and anti isomers of the prolyl imide bond are shown.

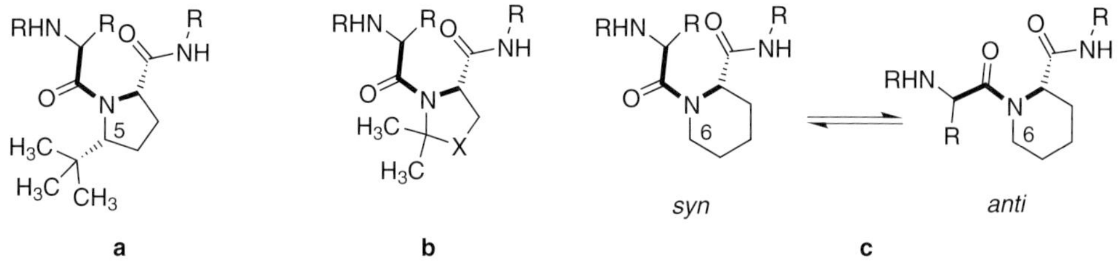

1.2.2 syn- and anti-Proline Mimics

Fig. 1.2.4. Bulky substituents attached to the heterocyclic scaffold make 5-*tert*-butylproline (a) and pseudoprolines (X = O or S) (b) efficient *syn*-proline mimics. (c) Imide isomerism N-terminal to pipecolyl residues.

tures (Box 1). Several bioactive peptides contain *syn*-proline, although it is not always clear whether the bioactive conformation is syn or anti. Because of the slow rotation around the proline amide bond it has been argued that syn–anti conversion of proline peptide bonds might be a rate-limiting step in protein folding [9]. Rotamase enzymes (FK506 binding proteins and cyclophilins) reduce the enthalpy barrier of rotation substantially in vivo and have been shown to act as protein foldases in vitro [10]. The ubiquitous occurrence of these enzymes suggests that labilization of the imide torsion might be important in vivo [11]. Moreover, prolyl isomerization might be involved in regulatory switches [12]. As a consequence, mimics of syn and anti proline peptide bonds are attractive target compounds for elucidation of the mechanisms of protein folding and structure–activity relationships of bioactive peptides.

Mimicking *syn*-prolines was achieved by different strategies, for example by substitution of the amide for a rigid alkene [13]. Maximizing steric repulsion between the proline ring and the adjacent N-terminal amino acid by attachment of bulky substituents to the heterocycle has also met with some success as exemplified by compounds **a** [14] and **b** [15] in Figure 1.2.4. Proline derivatives like **a**, for example, were incorporated into model peptides and shown to exist predominantly in syn form in water (44–84% syn isomer depending on the peptide sequence) [16].

Because of its larger six-membered ring scaffold, piperidine-2-carboxylic acid (pipecolic acid, also called homoproline), is known to have a stronger preference than proline for its syn isomer (Figure 1.2.4) [17]. It was therefore assumed that 6-substituted pipecolic acid derivatives should be even more effective mimics of syn prolyl peptide bonds then 5-substituted prolines in structure **a** in Figure 1.2.4. Synthetic routes to 6-substituted pipecolic acids are rare, however, and usually low-yielding multistep procedures. In addition, peptide coupling to these sterically hindered heterocycles can be extremely difficult, especially for the 2,6-trans diastereoisomers; this makes their incorporation into model peptides a problematic task. The multicomponent synthesis outlined in Scheme 1.2.1 was, therefore, the first efficient procedure for synthesis of model peptides (e.g. **1** and **2** in Scheme 1.2.1) incorporating 2,6-trans-substituted pipecolic acids [18]. Investigation of the

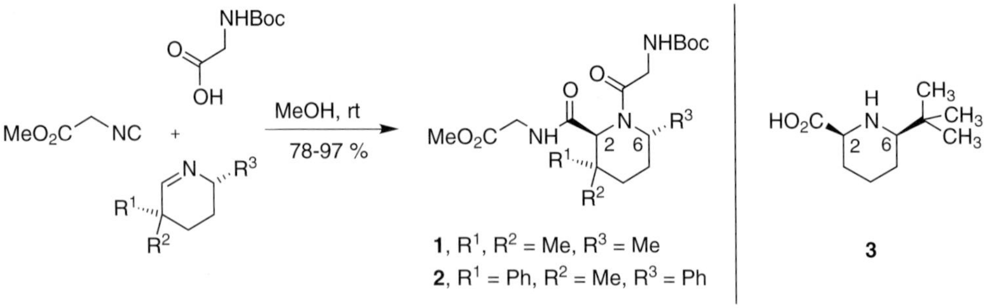

Scheme 1.2.1. Multicomponent synthesis of model tripeptides **1** and **2** and the structure of cis-6-tert-butylpipecolic acid **3**.

structural properties of these peptides by NMR spectroscopy and X-ray diffraction revealed that attachment of even small substituents in the 6 position of the pipecolic acid results in efficient *syn*-proline analogs like **1** in Scheme 1.2.1 [19].

The stereochemistry of these substituted cyclic amino acids is important, however, because 2,6-*trans*-substituted analogs **1** strongly favor the syn rotamer (> 99% syn isomer in methanol and DMSO), whereas 2,6-cis-substituted pipecolic acids like **3** do not favor the syn isomer (43% syn isomer in water) as demonstrated for a model bisamide incorporating 6-*tert*-butylpipecolic acid **3** [20]. The reason for these findings is the pseudo allylic 1,3-interaction of substituents attached to C2 and C6 of the piperidine ring. In 2,6-*trans*-substituted piperidines like **1** the smaller substituent (in this case the methyl group attached to C6) is in an equatorial position. Steric interactions between this substituent and the adjacent amino acid are therefore maximized. In 2,6-cis-substituted analogs like **3**, in contrast, both substituents at C2 and C6 prefer an axial position to avoid 1,3-allylic interactions with the neighboring amide bond. Consequently, steric interactions between the bulky *tert*-butyl group in **3** and the adjacent amino acid are minimized and induction of the syn rotamer is relatively weak. Evidence for this hypothesis can be drawn from NMR spectroscopic investigations and X-ray crystal structures for a range of different 2,6-substituted piperidines [19, 20].

Surprisingly, some 2,3,6-substituted pipecolic acid derivatives like **2** favor exclusively the anti amide isomer [19]. The anti preference of **2** has been demonstrated by NMR spectroscopy in solution (methanol and DMSO) and X-ray analysis in the solid state. The reason for this unexpected finding is the special arrangement of substituents on the pipecolic heterocycle. In contrast to compounds like **1**, the additional bulky substituent attached to C3 forces the carboxy group at C2 into an equatorial position and the substituent at C6 into an axial position.

The preferred geometry of the N-terminal peptide bond in pipecolic acids can therefore be triggered by choice of substituents attached to the heterocycle and their relative configuration.

2,3,6-Substituted piperidines like **2** were initially also synthesized in the racemic form by use of the multicomponent reaction mentioned above. A recently estab-

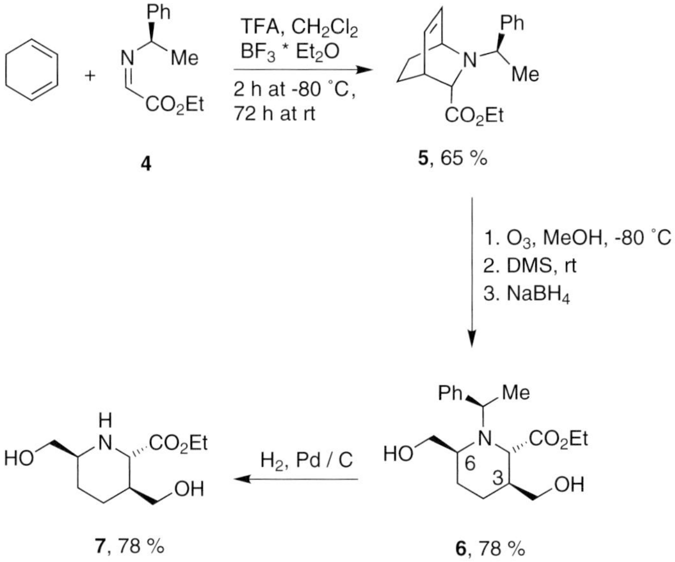

Scheme 1.2.2. Stereoselective synthesis of 3,6-disubstituted pipecolic acids.

lished synthesis, however, provides convenient access to enantiomerically pure analogs [21].

As shown in Scheme 1.2.2, the synthesis starts from a chiral imine **4**, which is prepared from ethylglyoxylate and (R)-phenylethylamine, by an *aza*-Diels–Alder reaction, leading to azabicyclooctene **5** in fair yield. This unsaturated precursor can be converted to pipecolic acid derivative **7** by ozonolysis and subsequent functional group transformations as depicted in Scheme 1.2.2.

Even more advantageous for use in peptide synthesis, azabicycloalkenes like **5** can be converted into aminodiols **8**, thus avoiding the difficult coupling to substituted pipecolic acids (Scheme 1.2.3). Bicyclics **8** are easy to couple N-terminally with additional amino acids by standard coupling procedures to give dipeptides **9**, which in turn can be cleaved oxidatively with sodium periodate. The resulting intermediate bisaldehydes like **10** cyclize intramolecularly, immediately, to give diazabicycloalkanes **11**, rigid analogs of *anti*-pipecolyl peptides, in quantitative yield. These can be transformed into covalently fixed *anti*-pipecolyl mimics like **13** (Scheme 1.2.4). Alternatively, the aminal in **11** or **12** might be cleaved, after conversion of the aldehyde attached to C3 to a side-chain of choice, to give 3,6-disubstituted pipecolic acid derivatives.

As illustrated by the synthesis of **13**, which can be viewed as a Gly–Hse (Hse = homoserine) mimic (Scheme 1.2.4), aminals of the general structure **11** serve as versatile educts for the preparation of pipecolyl-dipeptide analogs with fixed anti-peptide geometry. The stereochemistry of compound **13** was determined unambiguously by 2D NOESY NMR spectroscopy and is in accordance with a re-

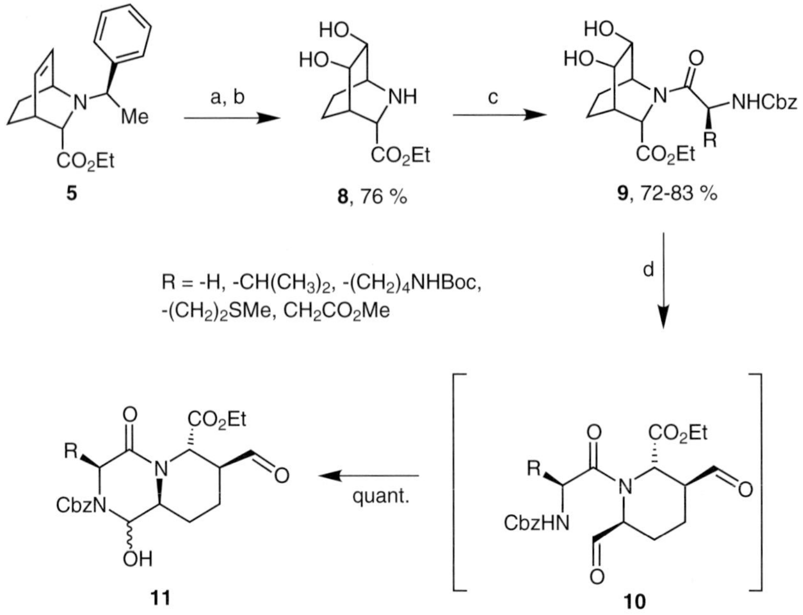

Scheme 1.2.3. Bicyclic diazabicycloalkanes **11**. Reagents and conditions: (a) cat. $K_2OsO_2(OH)_4$, $K_3Fe(CN)_6$, K_2CO_3, tert-BuOH/H_2O, 12 h, rt; (b) H_2 (1 atm), 5% Pd/C, EtOH, rt, 24 h; (c) Cbz-protected amino acid, DCC, HOBT, DMF, 12 h, rt; (d) $NaIO_4$, acetone/H_2O, 0 °C, 45 min.

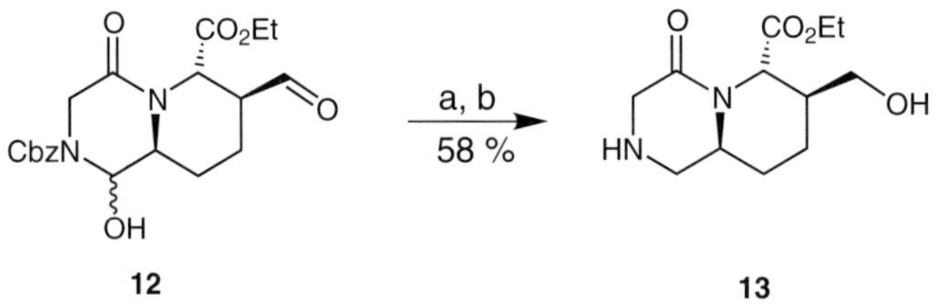

Scheme 1.2.4. Synthesis of dipeptide mimic **13**. Reagents and conditions: (a) $NaBH_4$, MeOH, 0 °C, 1 h; (b) H_2 (1 atm), 5% Pd/C, EtOH, rt, 24 h.

cently published X-ray structure of a pipecolic acid derivative, which was also synthesized from azabicyclooctene **5** [21].

Because the aldehyde in **11** or **12** can be converted into different side-chain functionalities, a range of different dipeptides is accessible. It should, furthermore, be noted that ring sizes of the cyclic amino acid scaffold, and stereochemistry in bicyclics like **13**, can be readily changed by varying the *aza*-Diels–Alder input. This in turn, enables one to systematically screen a certain set of backbone dihedral angles in dipeptide mimics.

1.2.3
Templates for α-Helix Stabilization

The above mentioned hypothesis that peptide primary sequences are translated into well defined three-dimensional conformations has stimulated many attempts to derive specific rules for secondary structure formation by short lengths of polypeptides [22]. The α-helix, one of the two most widely occurring secondary structures in proteins, has been the focus of several investigations aiming at analysis of effects of sequence on stability [23]. Helices are classified as repetitive secondary structure because their backbone dihedral angles, ϕ and ψ, have repeating values near the canonical value of ($-60°$, $-40°$). When the dihedral angles of a chain segment assume helical values, the carbonyl group of amino acid residue i is positioned to form a hydrogen bond with an NH function of an amino acid in position $i + 4$.

According to the Zimm–Bragg theory and related models of the helix–coil transition in polypeptides short helices are very unstable [24]. This assumption is verified by most experimental findings. Helix stabilization thus holds the promise of providing synthetically accessible model systems for studies of protein folding and can also enhance the potencies and/or specificities of bioactive peptides [25]. As a consequence, many efforts have been focused on this direction and several approaches have been reported for stabilizing α-helical peptides, including incorporation of salt bridges [26], metal chelates [27] or amide bonds [28] that bridge the i and $i + 4$ positions, incorporation of amino acids with high helix propensity [29], and the formation of amphiphilic helix bundles [30] and disulfide bridged peptides [31]. In this context, Kemp and coworkers described the conformationally constrained tripeptide mimic AcHel shown in Figure 1.2.5, that overcomes the nucleation penalty for helix formation by providing hydrogen-bond acceptors for the otherwise unsatisfied NH groups in the first helix turn at the N-terminus of a

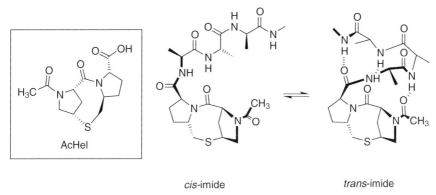

cis-imide   trans-imide

Fig. 1.2.5. The helix nucleating properties of AcHel depend on the two conformational states of the N-terminal imide bond.

helical peptide (Figure 1.2.5) [32]. This template makes use of the fixed ϕ angle in proline that assumes the helical value of approximately 60°.

Significant helix formation is observed for short peptides attached to this template, that is they nucleate preferentially from the template. In addition to this nucleation function, the template enables the direct measurement of two conformational states, which arise from the tertiary amide isomerism of the acetamide function, via ^{1}H NMR spectroscopy. These two states are correlated to the degree of helicity in the attached peptide (Figure 1.2.5), because only the trans imide is able to stabilize a helix. The nucleating template thus acts as a two-state switch, permitting direct observation and quantitation of the properties of various short helices [33]. AcHel has therefore proven useful for evaluation of the Zimm–Bragg parameters which characterize helix propagation. The applicability of AcHel in peptide engineering has, however, been somewhat hampered in the past, because the *N*-acetyl group blocks the N-terminus of the template and thus does not permit N-terminal elongation of the peptide chain. For convenient use of this helix template as a building block in peptide engineering, it should be compatible with standard automated peptide synthesis. Fmoc-protected analog **21** was therefore thought to be a valuable target compound. The last steps of its synthesis are outlined in Scheme 1.2.5 starting with the condensation of hydroxyproline derivative **14** and proline analog **15** to give thioester **16**. After Boc deprotection the thioester was intramolecularly aminolyzed to give diproline **17**. Cyclization to diproline analog **18** with a central eight membered ring system was then performed using potassium methylate in diluted acetonitrile solution. Standard protecting group manipulation finally gave Fmoc protected template **21** ready for use in automated peptide synthesis [34].

The versatility of this template was demonstrated with the synthesis of very short unusually helical polyalanine sequences stabilized by chaotrophic anions [35] and host systems for evaluation of the C-terminal helix capping propensities for nonpolar natural amino acids [36]. As model sequences WK$_4$Inp$_2$^{t}LG-Hel-A$_8$-**NH$_2$** for the primary C-terminal amide and WK$_4$Inp$_2$^{t}LG-Hel-A$_8$-**X**-Inp-NH$_2$ for candidate amino acids **X** were selected. In these sequences, Hel is the previously mentioned N-terminal helix template, Inp is 4-carboxypiperidine and ^{t}L is *tert*-leucine. In the N-terminal region tryptophan (W) provides a UV reporter, four lysines (K$_4$) are solubilizers, and Inp$_2$^{t}L a spacer element. The C-capping test region of these peptides is G-Hel-A$_8$-**X**-Inp, and its helicity is taken as proportional to $[\theta]_{222}$, the per-residue ellipticity derived from CD analysis [37].

Figure 1.2.6 reports experimental values for $[\theta]_{222}$ at different temperatures for the C-terminal primary amide and for six C-capping amino acids **X** (Gly, Ala, Leu, Val, Ile, and ^{t}Leu). The 2 °C ellipticites for Ala, Leu, Val, and Ile are essentially identical. This implies that relative to Ala the C-capping propensities of Leu, Val, and Ile are all 1.0; in other words, the natural amino acids with alkyl side-chains are equally efficient as helix C-caps. Relative to Ala the value for Gly is about 10% higher and that for ^{t}Leu is about 10% lower. These two represent the conformational extremes among α-amino acids. The ellipticity for the primary amide is 30% larger than that for alanine, implying a strong C-capping propensity for this function. In particular this last fact has important practical consequences as tempera-

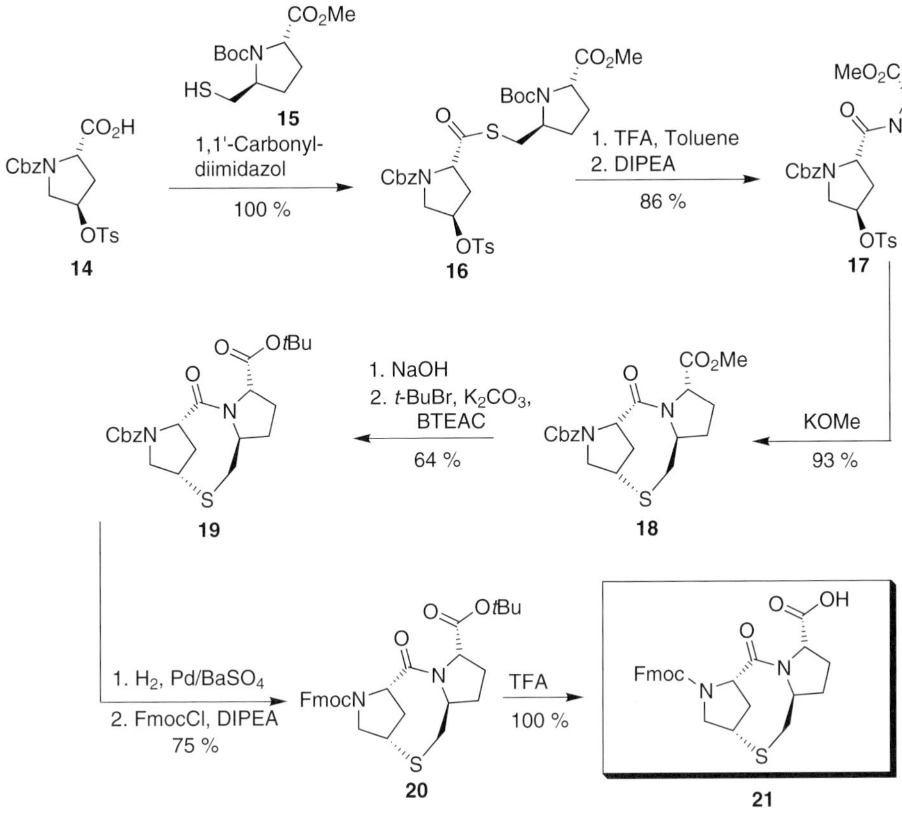

1.2.3 Templates for α-Helix Stabilization

Scheme 1.2.5. Synthesis of the Fmoc-protected helix template **21**.

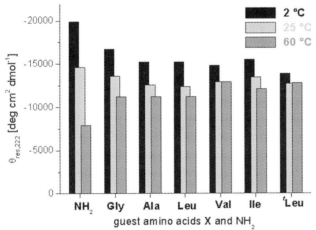

Fig. 1.2.6. Per-residue ellipticities at 2 °C, 25 °C, and 60 °C for model peptides WK$_4$Inp$_2$^{t}LG-Hel-A$_8$-**NH$_2$** and WK$_4$Inp$_2$^{t}LG-Hel-A$_8$-**X**-Inp-NH$_2$ in water at approximately 13 μmol concentration. The guest amino acids **X** and the primary amide are displayed on the x-axis in the diagram.

ture dependent helicity changes derived from model peptides with a C-terminal primary amide might be highly atypical and difficult to compare with conformational changes expected for larger peptides in a biological context.

Design and synthesis of peptide mimics with defined conformational properties is an increasing area of research in which proline and pipecolic acid play an important role. Because of their unique conformational properties these cyclic amino acids or derivatives thereof are versatile educts for a range of different constraint peptide mimics as exemplified within the last chapter. Because of the often unknown structural requirements for ligand–receptor interactions and incomplete understanding of peptide folding, however, many new analogs for special purposes such as mimics of peptide lead structures in medicinal chemistry or templates for certain peptide secondary structure elements remain to be developed. The challenge for bioorganic chemists is not only to design these mimics but also to provide the necessary methodology to synthesize them.

References

1 J. P. SCHNEIDER, J. W. KELLY, Chem. Rev. **1995**, 95, 2169–2187.
2 (a) F. POLYAK, W. D. LUBELL, J. Org. Chem. **2001**, 66, 1171–1180; (b) L. HALAB, F. GOSSELIN, W. D. LUBELL, Biopolymers **2000**, 55, 101–122.
3 (a) L. BELVISI, L. COLOMBO, M. COLOMBO, M. DI GIACOMO, L. MANZONI, B. VODOPIVEC, C. SCOLASTICO, Tetrahedron **2001**, 57, 6463–6473; (b) M. ANGIOLINI, S. ARANEO, L. BELVISI, E. CESAROTTI, A. CHECCHIA, L. CRIPPA, L. MANZONI, C. SCOLASTICO, Eur. J. Org. Chem. **2000**, 2571–2581.
4 S. HANESSIAN, G. MCNAUGHTON-SMITH, H.-G. LOMBART, W. D. LUBELL, Tetrahedron **1997**, 53, 12789–12854.
5 P. TREMMEL, A. GEYER, J. Am. Chem. Soc. **2002**, 124, 8548–8549.
6 (a) D. S. KEMP, B. R. BOWEN, Tetrahedron Lett. **1988**, 29, 5081–5082; (b) D. S. KEMP, B. R. BOWEN, Tetrahedron Lett. **1988**, 29, 5077–5080.
7 D. S. KEMP, B. R. BOWEN, C. C. MUENDEL, J. Org. Chem. **1990**, 55, 4650–4657.
8 D. E. STEWART, A. SARKAR, J. E. WAMPLER, J. Mol. Biol. **1990**, 214, 253–260.
9 (a) C. GRATHWOHL, K. WÜTHRICH, Biopolymers **1981**, 20, 2623–2633;
(b) J. F. BRANDTS, H. R. HALVORSON, M. BRENNAN, Biochemistry **1975**, 14, 4953–4963.
10 G. FISCHER, B. WITTMANN-LIEBOLD, K. LANG, T. KIEHABER, F. SCHMID, Nature **1989**, 337, 476–478.
11 (a) M. J. GETHING, J. SAMBROOK, Nature **1992**, 355, 33–45; (b) G. FISCHER, F. X. SCHMID, Biochemistry **1990**, 29, 2205–2212.
12 T. JAYARAMAN, A.-M. BRILLANTES, A. R. TIMERMAN, S. FLEISCHER, H. ERDJUMENT-BROMAGE, P. TEMPST, A. R. MARKS, J. Biol. Chem. **1992**, 267, 9474–9477.
13 S. A. HART, M. SABAT, F. A. ETZKORN, J. Org. Chem. **1998**, 63, 7580–7581.
14 E. BEAUSOLEIL, B. L'ARCHEVEQUE, L. BELEC, M. ATFANI, W. D. LUBELL, J. Org. Chem. **1996**, 61, 9447–9454.
15 P. DUMY, M. KELLER, D. E. RYAN, B. ROHWEDDER, T. WÖHR, M. MUTTER, J. Am. Chem. Soc. **1997**, 119, 918–925.
16 L. HALAB, W. D. LUBELL, J. Am. Chem. Soc. **2002**, 124, 2474–2484.
17 W.-J. WU, D. P. RALEIGH, J. Org. Chem. **1998**, 63, 6689–6698.
18 W. MAISON, A. LÜTZEN, M. KOSTEN, I. SCHLEMMINGER, O. WESTERHOFF, J. MARTENS, J. Chem. Soc. Perkin Trans. 1, **1999**, 3515–3525.
19 W. MAISON, A. LÜTZEN, M. KOSTEN, I.

Schlemminger, O. Westerhoff, W. Saak, J. Martens, *J. Chem. Soc. Perkin Trans. 1*, **2000**, 1867–1871.
20 M. E. Swarbrick, F. Gosselin, W. D. Lubell, *J. Org. Chem.* **1999**, *64*, 1993–2002.
21 W. Maison, G. Adiwidjaja, *Tetrahedron Lett.* **2002**, *43*, 5957–5960.
22 (a) P. Y. Chou, G. D. Fasman, *Ann. Rev. Biochem.* **1978**, *47*, 251–276; (b) J.-F. Gibrat, J. Garnier, B. Robson, *J. Mol. Biol.* **1987**, *198*, 425–443; (c) S. Kametkar, J. M. Schiffer, H. Xiong, J. M. Babik, M. H. Hecht, *Science* **1993**, *262*, 1680–1685; (d) B. Rost, C. Sander, *Curr. Opin. Biotechnol.* **1994**, *5*, 372–380; (e) R. Aurora, R. Srinivasan, G. D. Rose, *J. Biol. Chem.* **1997**, *272*, 1413–1416.
23 (a) H. X. Zhou, P. C. Lyu, D. E. Wemmer, N. R. Kallenbach, *Proteins: Struct. Funct. Genet.* **1994**, *18*, 1–7; (b) A. Horovitz, J. M. Matthews, A. R. Fersht, *J. Mol. Biol.* **1992**, *227*, 560–568; (c) A. Chakrabartty, R. L. Baldwin, *Advan. Protein Chem.* **1995**, *46*, 141–176; (d) M. Z. Blaber, X. Zhang, B. W. Matthews, *Science* **1993**, *260*, 1637–1640.
24 (a) B. H. Zimm, J. Bragg, *J. Chem. Phys.* **1959**, *31*, 526; (b) S. Lifson, A. Roig, *J. Chem. Phys.* **1961**, *34*, 1963–1974.
25 (a) A. M. Felix, E. P. Heimer, C.-T. Wang, T. J. Lambros, A. Fournier, T. F. Mowles, S. Maines, R. M. Campbell, B. B. Wegrzynski, V. Toome, D. Fry, V. S. Madison, *Int. J. Peptide Protein Res.* **1988**, *32*, 441–454; (b) G. Osapay, J. W. Taylor, *J. Am. Chem. Soc.* **1992**, *114*, 6966–6973; (c) M. R. Ghadiri, C. Choi, *J. Am. Chem. Soc.* **1992**, *114*, 4000–4002.
26 S. Marquesee, R. L. Baldwin, *Proc. Natl. Acad. Sci. USA* **1987**, *84*, 8898–8902.
27 (a) M. R. Ghadiri, A. K. Fernholz, *J. Am. Chem. Soc.* **1990**, *112*, 9633–9635; (b) F. Ruan, Y. Chen, P. B. Hopkins, *J. Am. Chem. Soc.* **1990**, *112*, 9403–9404.
28 M. Chorev, E. Roubini, R. L. McKee, S. W. Gibbons, M. E. Goldman, M. P. Caulfield, M. Rosenblatt, *Biochemistry* **1991**, *30*, 5968–5974.
29 S. Marquesee, V. H. Robbins, R. L. Baldwin, *Proc. Natl. Acad. Sci. USA* **1989**, *86*, 5286–5290.
30 W. F. Degrado, *Adv. Protein Chem.* **1988**, *39*, 51–124.
31 D. Y. Jackson, D. S. King, J. Chmielewski, S. Singh, P. G. Schultz, *J. Am. Chem. Soc.* **1991**, *113*, 9391–9392.
32 D. S. Kemp, J. G. Boyd, C. C. Muendel, *Nature* **1991**, *352*, 451–454.
33 D. S. Kemp, T. J. Allen, S. L. Oslick, *J. Am. Chem. Soc.* **1995**, *117*, 6641–6657.
34 W. Maison, E. Acre, P. Renold, R. Kennedy, D. S. Kemp, *J. Am. Chem. Soc.* **2001**, *123*, 10245–10254.
35 W. Maison, B. Kennedy, D. S. Kemp, *Angew. Chem.* **2001**, *113*, 3936–3938; *Angew. Chem. Int. Ed. Engl.* **2001**, *40*, 3819–3821.
36 W. Maison, R. Kennedy, D. S. Kemp, *Tetrahedron Lett.* **2001**, *42*, 4975–4977.
37 J. S. Miller, R. J. Kennedy, D. S. Kemp, *Biochemistry*, **2001**, *40*, 305–309.

B.1
Proline *syn–anti* Isomerization, Implications for Protein Folding

Wolfgang Maison

Syn-anti isomerization N-terminal to prolyl peptide fragments is one of the rate limiting steps in protein folding [1]. Prolyl fragments exist essentially either in the syn or anti form in native proteins, but as an equilibrium mixture of these two rotamers in unfolded proteins and small peptides. These two conformational states are almost isoenergetic (see Figure 1.2.3 for comparison), with the anti form being slightly more favorable. Both rotamers are separated by a significant energy barrier to rotation (~ 20 kcal mol^{-1}). The properties of X-Proline peptide fragments are not only essential for folding of globular proteins but also for homopolymers of proline that occur in two extreme forms. Polyproline I has a right-handed helical conformation with all amides in syn geometry and polyproline II is all-anti in a left-handed helical conformation.

Interconversion of syn and anti prolyl peptide fragments can be accelerated by disruption of the partial double-bond character of the imide bond. Protonation of the imide carbonyl with strong acids, for example, increases the interconversion significantly. Steric conflicts in non-natural proline analogs can also promote increased syn–anti interconversion by destabilizing the planar imide geometry. In addition, several enzymes such as peptidylprolyl isomerases (PPIases) catalyze prolyl syn–anti isomerization by mechanisms that involve disruption of the partial double-bond character of the peptide bond. PPIases were first discovered by Fischer and coworkers and are an ubiquitous class of enzyme [2]. Three families can be distinguished: (a) cyclophilines, (b) FK506-binding proteins, and (c) parvulines. The first two families are characterized by binding to natural products with an immunosuppressive activity and are therefore extremely interesting targets for pharmaceutical purposes. Cyclophilines bind to the cyclic undecapeptide cyclosporin A (CsA) which is an important pharmaceutical for the treatment of autoimmune diseases, whereas FK506-binding proteins bind to the macrolides FK506 and rapamycin. Parvulines do not bind to any of these natural products and are inhibited by juglon (5-hydroxy-1,4-naphthoquinone). CsA is an important pharmaceutical for immunosuppression, required, for example, during transplantations. CsA activity, however, is not a result of inhibition of PPIase activity but of an interaction of the CsA–Cyclophiline complex with the phosphatase calcineurin.

References

1 W. J. WEDEMEYER, E. WELKER, H. A. SCHERAGAV, *Biochemistry* **2002**, *41*, 14637–14644.

2 G. FISCHER, *Angew. Chem.* **1994**, *106*, 1479–1501; *Angew. Chem. Int. Ed. Engl.* **1994**, *33*, 1415–1437.

1.3
Stabilization of Peptide Microstructures by Coordination of Metal Ions

Markus Albrecht

1.3.1
Introduction

Molecular recognition processes play an essential role in the formation and function of biological systems (this is discussed in more detail in Chapter 2). In these processes non-covalent bonding enables reversible interaction between molecular species and the build up of large (supra)molecular structures with well-defined functions. Although we usually tend to visualize molecules as rather rigid structures, many compounds are highly flexible and have dynamic behavior. Thus, to some extent, molecules are able to adjust their structures to external stimuli ("induced fit"). The biological activity of a given molecule is very dependent on its ability to interact specifically with other molecules and, therefore, on the geometric and electronic complementarity between a substrate or a ligand and a receptor ("lock-and-key principle"; see also Chapter 1.4, Fig. 1.4.2). In nature, biologically important intermolecular interactions are maximized by appropriate conformational fixation of the three-dimensional structures of molecules.

For example, classes of natural products in which conformational fixation is very important are the peptides and proteins. They can adopt different secondary structures by intra- and/or interstrand hydrogen bonding, ranging from helical or sheet-type structures (with the most prominent being the α-helix or the β-sheet – Box 8) to different turn (e.g. γ-turn) or random coil structures (Figure 1.3.1) [1].

In medicinal chemistry we can use isolated, well-defined artificial peptides made from a few amino acid residues only, and which have a specific microstructure, to investigate the biological function of a selected peptide segment of a large protein (see also Chapter 1.2). It is, therefore, an important challenge in bioorganic chemistry to stabilize specific conformations of amino acids and small peptides to fix defined three-dimensional structures [2, 3].

Metal coordination is a very strong non-covalent interaction which is frequently used to selectively obtain large supramolecular architectures in self-assembly processes (see, for example, Chapter 2.1). The metal already induces three dimensional orientation of the ligands, and thus of their substituents, in space, because of the preferred coordination geometry at the metal (Figure 1.3.2) [4]. It would

Highlights in Bioorganic Chemistry: Methods and Applications. Edited by Carsten Schmuck, Helma Wennemers.
Copyright © 2004 WILEY-VCH Verlag GmbH & Co. KGaA, Weinheim
ISBN: 3-527-30656-0

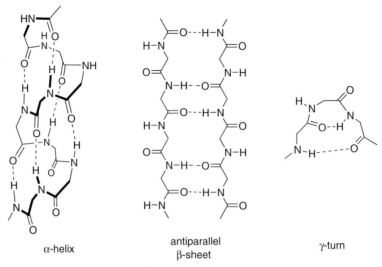

Fig. 1.3.1. Examples of three different types of peptide structure: α-helix, antiparallel β-sheet, and γ-turn.

therefore be of interest to use the geometrical conformation at a metal ion to induce three dimensional structure (= conformation) in amino acids (Nature also uses metal ions in this context – Box 3).

Examples have been reported of modified peptides in which metal bonding units are attached to different positions of the peptide and induce different structures. Starting from random coil peptides, examples have been reported in which the metal can induce either α-helical- (Figures 1.3.3A or B) or β-sheet-type structures (Figure 1.3.3C).

A helical structure can be stabilized by introducing ligand units to the amino acid side-chains. On addition of appropriate metal ions the α-helix (Figure 1.3.3A) is formed by use of the metal as a "cross-linking" unit [5]. Attachment of metal binding sites to the end of well-chosen decapentapeptides and coordination of the "random coil" peptide to appropriate metal ions leads to induction of an α-helix

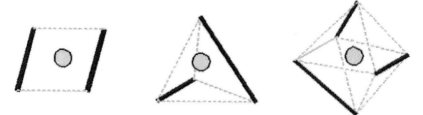

Fig. 1.3.2. Schematic presentation of metal complexes with bidentate ligands bonding to square-planar, tetrahedrally, or octahedrally coordinated metal centers.

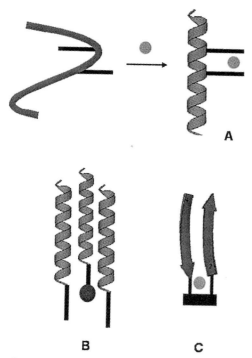

Fig. 1.3.3.

in the peptide so that – depending on the coordination properties of the metal – tris or quadruple helix bundles result (a schematic representation of a tris-helix bundle is given in Figure 1.3.3B) [6–9]. In such circumstances favorable interaction of non-polar side-chains supports helix formation and the helical twist is induced by the chiral metal centers. Incorporation of a metal binding site in an appropriate linear peptide, and binding of a metal, can lead to a conformational fixing of the peptide which induces an antiparallel β-sheet-type structure (Figure 1.3.3C) [10].

In structures like A–C rather large peptide domains are conformationally fixed by coordination to metals. Smaller domains can also be obtained, however (Figure 1.3.4) [11]. Extensive NMR studies of the metallomacrocycle D show that an α-helical structure is adopted by the Ac-His-Ala-Ala-Ala-His-NH$_2$ pentapeptide on binding to the (en)Pd-fragment (en = ethylene diamine). The peptidic part of D represents one single turn of the helix [12].

A smaller turn structure is induced in an N,N'-bis(2,3-dihydroxybenzoate)-substituted lysine derivative on coordination to the Mo(VI)O$_2$ fragment E. In E the conformation of the lysine residue is highly constrained, because of intramolecular hydrogen bonding in combination with coordination of the catechol units to the dioxomolybdenum(VI) moiety [13, 14].

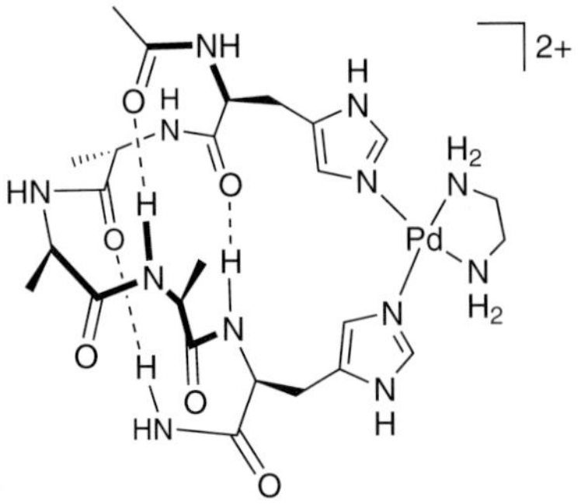

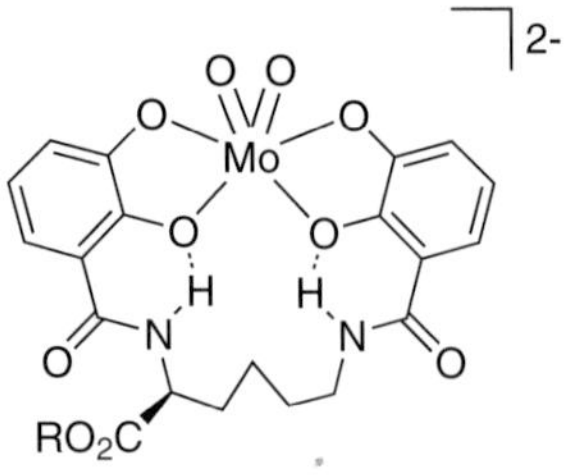

Fig. 1.3.4.

In this chapter selected examples from our group are discussed to show how metal coordination to ligand-modified amino acids or peptides can be used for induction or fixing of defined conformations in amino acid residues or di- and tripeptides. In this context Ramachandran's method for conformational analysis of peptide or protein structures will be introduced.

1.3.2
Dinuclear Coordination Compounds from Amino Acid-bridged Dicatechol Ligands: Induction of a Right- or a Left-handed Conformation at a Single Amino Acid Residue

The amide bond is the dominating structural feature in peptide chemistry. Because of resonance between the CO π-system and the lone pair in the π-orbital at the nitrogen the amide moieties in peptides are planar (or close to planar) with s-trans orientation of the NH and the CO units. Thus, the conformation of the amino acid residue of a protein mainly is defined by the dihedral angles Φ (C(O)-N-Cα-C(O)) and Ψ (N-Cα-C(O)-N) which were introduced by Ramachandran (Figure 1.3.5) [15]. Typical dihedral angles for an amino acid located in the α-helical part of a protein are in the region $\Psi = -47°$ and $\Phi = -57°$, angles for an antiparallel β-sheet are close to $\Psi = 135°$ and $\Phi = -139°$. In general, amino acids prefer a dihedral angle, Φ, with a negative sign, because here a right-handed helical conformation is adopted by the amino acid residue. Such a right-handed conformation at L-amino

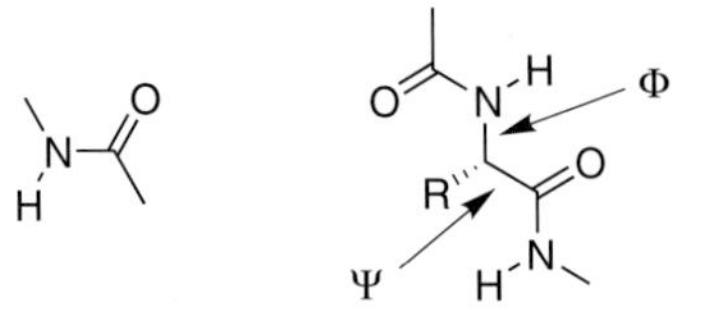

Fig. 1.3.5.

1.3.2 Dinuclear Coordination Compounds from Amino Acid-bridged Dicatechol Ligands

acids (with the exception of glycine) is favored over the left-handed, because of minimization of the steric interaction between the substituent at the α-carbon atom and the amino acid (peptide) backbone [1].

A starting point for investigation of the effect of metal coordination on the conformation of a single amino acid is the synthesis of amino acid-bridged ligands with two metal binding sites and study of their coordination chemistry [13, 14]. The synthetic sequence used to obtain amino acid-bridged dicatechols **1a–1e**-H_4 is shown in Scheme 1.3.1 [16].

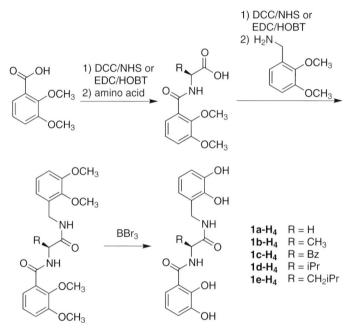

Scheme 1.3.1. Preparation of the amino acid-bridged dicatechol ligands **1a–1e**-H_4.

Standard peptide coupling chemistry is performed (Box 25). The acid component is transformed to an activated ester derivative and then treated with the amine [17–19]. This procedure is repeated twice. In the final step the ligands **1a–1e**-H_4 are deprotected by aryl ether cleavage with BBr_3. By use of this simple reaction sequence derivatives with glycine, alanine, phenylalanine, valine, leucine, and other amino acids as spacers were obtained [16].

The thus prepared ligands **1a–1e**-H_4 are introduced into coordination studies with titanium(IV) ions in the presence of base (M_2CO_3: M = Li, Na, K) with methanol as solvent. Negative ESI–MS (electron spray ionization mass spectrometry) of the obtained orange–red solids in methanol reveals that coordination compounds of general formula $M_2[\mathbf{1}_2(OCH_3)_2Ti_2]$ are formed (Scheme 1.3.2) [20, 21].

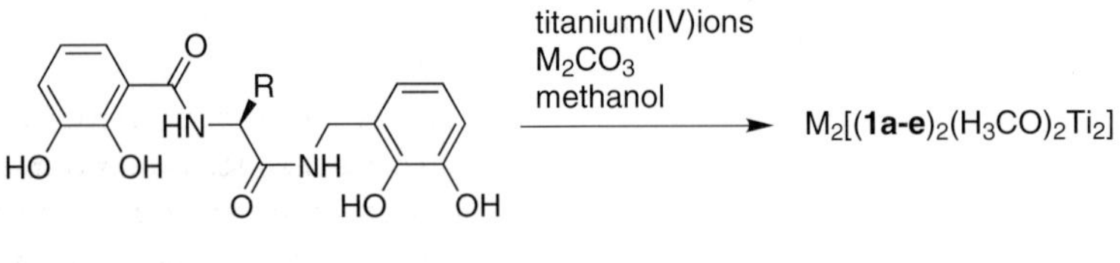

Scheme 1.3.2. Preparation of dinuclear titanium(IV) complexes $M_2[1_2(OCH_3)_2Ti_2]$.

Different orientations of the directional amino acid-bridged ligands (parallel or anti-parallel; helical or non-helical, Figure 6) can, however, lead to up to seven isomers **I–VII** of complexes with the composition $[1_2(OCH_3)_2Ti_2]^{2-}$ (Figure 1.3.6) [20, 21].

NMR spectroscopy of methanol-d_4 solutions of the complexes $[1_2(OCH_3)_2Ti_2]^{2-}$ shows that initially all seven isomers **I–VII** are formed. On standing of the solution, however, some of the signals disappear and after approximately 14 days at room temperature one dominating species and minor side products can be detected. Thus the reaction first leads to a mixture of products which is formed under kinetic reaction control and at room temperature this is slowly transformed into the one thermodynamically favored species. A first indication on the nature of this species is obtained by following the specific rotation of a freshly prepared solution of the complexes for 18 days. Relatively high optical rotation is measured initially; this slowly drops on standing of the solution. As a working hypothesis (which is supported by modeling studies on a PM3 level) it is assumed that the thermodynamically favored species have a non-helical structure. In the non-helical complexes four stabilizing intramolecular hydrogen bonds between amide protons and

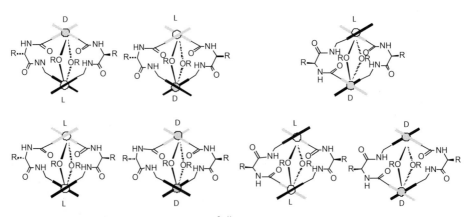

Fig. 1.3.6. Schematic representations of all seven isomers of the complexes $[1_2(OCH_3)_2Ti_2]^{2-}$. Catecholate units at the N-termini are indicated as gray bars, those at the C-termini as black bars.

internal catecholate oxygen atoms are formed whereas in a helical arrangement two of those bonds have to be broken. Thus stabilization by intramolecular hydrogen bonding is significantly greater for non-helical arrangement of the ligands (meso relation between the complex units). The chiral helical complexes formed at the beginning should lead to high optical rotation $[\alpha]_D$ values, whereas in the non-helical compounds the strong metal complex chromophores have a meso-type arrangement which reduces the "chirality introduced by the metal centers" (the expression "meso-type" corresponds to the configuration at the metal centers while chirality is still introduced by the spacers).

Finally, the structure of the thermodynamically favored isomer of $[\mathbf{1}_2(OCH_3)_2Ti_2]^{2-}$ was deduced from conformational analysis of X-ray structural data of some of the complexes using Ramachandran's method. The dihedral angles Φ and Ψ of the amino acid residues observed in the X-ray structures were determined and were correlated in a Ramachandran diagram.

Four X-ray structures were obtained for the complexes $[\mathbf{1}_2(OR)_2Ti_2]^{2-}$ (R = Me, H) with alanine, phenylalanine, leucine, or valine residues in the spacer, representing a total of three different isomers. (For valine the structure of the meso complex $[(\mathbf{1d})(\mathbf{1d'})(OCH_3)_2Ti_2]^{2-}$ ($\mathbf{1d'}$ = R-$\mathbf{1d}$) was obtained, because of epimerization of the ligand during its synthesis.) The Ramachandran plot in Figure 1.3.7 shows that the derivatives with alanine and leucine as spacer adopt conformations which are typical for amino acid residues of right handed α-helical peptides. One of the valine spacers and one of the phenylalanine spacers adopt structures typical of a right handed twisted sheet.

Surprisingly, two of the data points obtained (Phe-1, R-Val) are found in the region of left-handed helical conformations. For the R-valine bridged ligand this is not a surprise. It is, however, very unusual for the S-phenylalanine derivative. As already mentioned, amino acids prefer to adopt a right-handed helical conformation. In the left-handed arrangement there is repulsive steric interaction between the substituent on the α-C and the amino acid backbone. A closer look at the X-ray structures reveals that all ligands with the favored right-handed helical twist are bound with their more rigidly connected N-terminal ligand moiety to the Λ-configured metal center of the meso-type complex. The N-termini of the left-handed twisted ligands, on the other hand, bind to a Δ-configured metal. Thus, Λ-configuration at the N-terminus induces the more favored right-handed helicity at the ligand whereas Δ-configuration at the N-terminus induces the unfavored left-handed twist.

In the thermodynamically favored species, therefore, both ligand strands must bind with their N-termini to the Λ-configured metal center of the meso-type dinuclear complex to adopt the sterically more favored right-handed helical conformation. Thus, under thermodynamically controlled conditions the dominating species in solution is isomer **I** (Figure 1.3.8). The results and considerations discussed show that stereochemical communication between a metal center and an amino acid residue can control the microstructure at the amino acid. In the example presented this leads, in a thermodynamically controlled system, after initial formation of a complex mixture, to only one final dominating species [20, 21].

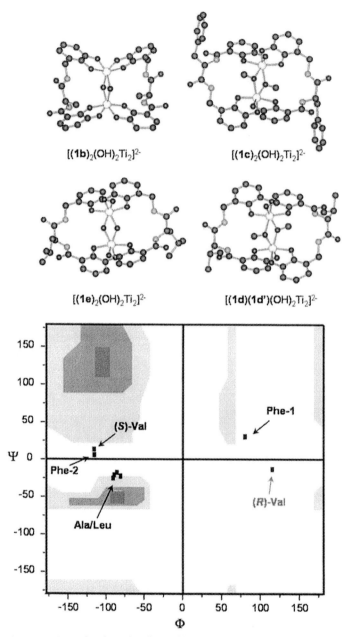

Fig. 1.3.7. Ramachandran plot obtained by correlation of Φ and Ψ of the amino acid residues observed in the solid-state structures of $[(1b)_2(OH)_2Ti_2]^{2-}$, $[(1c)_2(OCH_3)_2Ti_2]^{2-}$, $[(1e)_2(OH)_2Ti_2]^{2-}$, and $[(1d)(1d')(OCH_3)_2Ti_2]^{2-}$.

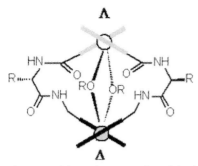

Fig. 1.3.8. Schematic presentation of the thermodynamically favored isomer **I** with the N-terminal catecholates of both ligands binding to a Λ-configured metal and the C-termini binding to a Δ-configured metal.

1.3.3
Peptide-bridged Dicatechol Ligands for Stabilization of Linear Compared with Loop-type Peptide Conformations

Although it is important to control the configuration at selected amino acids, the real challenge is structural fixing of small peptidic domains. Thus, to investigate the effect of metal coordination on the microstructure of small peptides, the dipeptide- and tripeptide-bridged ligands 2-H_4 and 3-H_4 (Figure 1.3.9) were prepared by use of a standard peptide-coupling procedure [22].

Valine is an amino acid with a bulky isopropyl substituent. This leads in oligovalins to repulsion of the isopropyl groups and the preference of a β-sheet-type structure in which the substituents at the peptide are separated as far as possible [1].

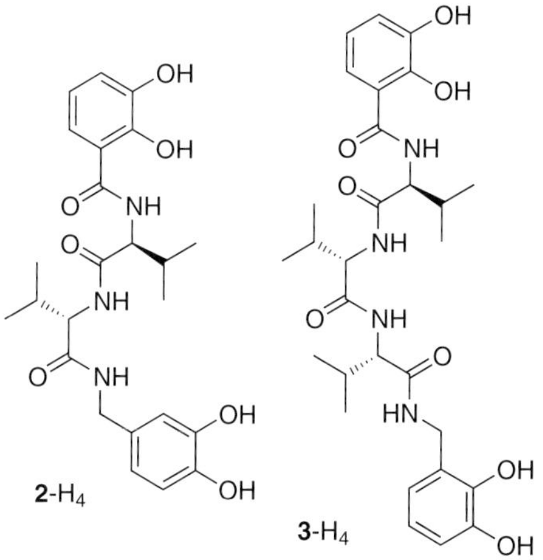

Fig. 1.3.9. Val–Val- and Val–Val–Val-bridged dicatechol ligands 2-H_4 and 3-H_4.

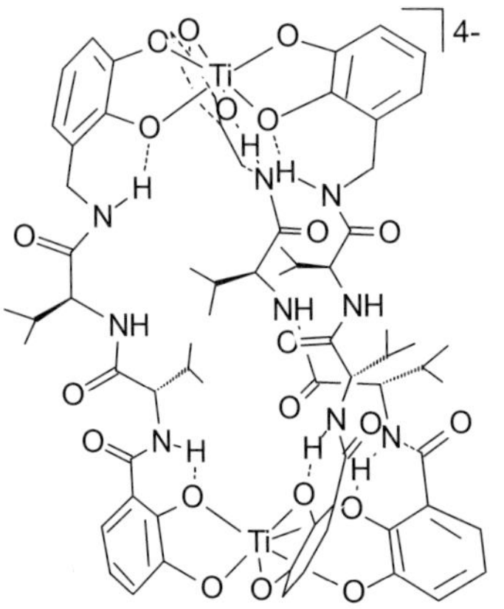

[(2)₃Ti₂]⁴⁻

Fig. 1.3.10.

The Val–Val-bridged ligand 2-H$_4$ has a spacer which, because of steric constraints, prefers the stretched "sheet type" structure. This forces the metal binding sites to be located far away from each other. On coordination to titanium(IV) ions this leads to the formation of a triple-stranded helicate-type complex [2$_3$Ti$_2$]$^{4-}$ (initially as a mixture of up to eight different regio- and stereoisomers). One dominating thermodynamically favored isomer is formed when a solution of the complex is left for several days at room temperature. NMR spectroscopy shows that the main isomer has C$_3$-symmetry and, therefore, the three ligand strands must all be orientated in the same direction with the three N-termini binding to one metal and the C-termini to the other (Figure 1.3.10). The ligand strands are thus forced to adopt a stretched conformation, probably leading to a β-sheet type microstructure at each ligand [23].

Ligand **3** has more flexibility than ligand **2**, because of the third amino acid and the additional flexibility at the further σ-bonds. In a coordination study with titanium(IV) ions and base, **3** did not lead to one specific triple-stranded dinuclear complex [3$_3$Ti$_2$]$^{4-}$ but to a mixture which also contains a single-bridged species. In the latter two metal centers are connected by one ligand **3** whereas the two remaining ligands **3** bind with both binding sites to the same metal. In contrast to the studies with ligand **2**, an unspecific mixture of structurally diverse coordination compounds is formed. It would, however, be of some interest to obtain selectively loop-type structures as found in the single bridged isomer of [3$_3$Ti$_2$]$^{4-}$ [23].

This can be achieved by using the cis molybdenum(VI)dioxo ion instead of titanium(IV). In this ion two corners of the octahedron at the molybdenum are already blocked by oxygen atoms and only two catechol units can be bound to the metal. Thus, reaction of ligand 3-H$_4$ with MoO$_2$(acac)$_2$ in the presence of potassium carbonate leads to a mononuclear macrocyclic complex [3MoO$_2$]$^{2-}$ in which a loop-type conformation is stabilized at the peptide (Scheme 1.3.3) [23]. (Similar

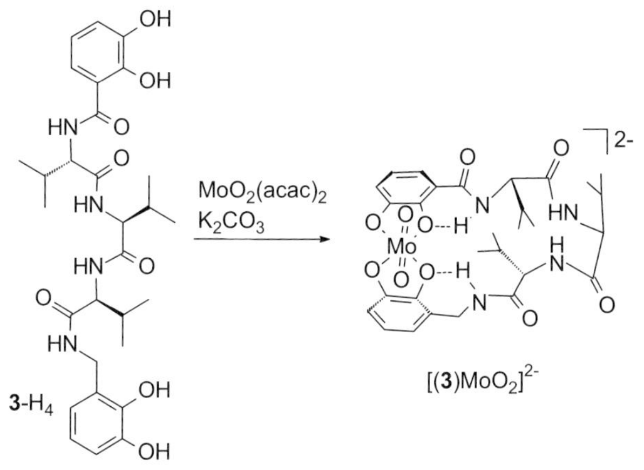

Scheme 1.3.3. Stabilization of a loop-type structure at a tripeptide by coordination of ligand **3** to the MoO_2^{2+} fragment.

experiments with ligand **2**-H_4 and molybdenum(VI)dioxo do not lead to defined complexes. In this case the spacer seems to be sterically too hindered to enable formation of a short-turn structure.)

These coordination studies with titanium(IV) or molybdenum(VI)dioxo and oligo-valine bridged dicatechol ligands show that the preferred conformation of geometrically constrained ligands with bulky amino acids as building blocks can control the specificity of complex formation. With titanium(IV) either the selective formation of one triple-stranded complex (ligand **2**) or the unspecific formation of mixtures of valence isomers (ligand **3**) can be observed. On the other hand, cis-MoO_2, which can bind both catechol units of one ligand strand **3** can be used to obtain macrocyclic complexes with a loop-type peptide front. Here the metal complex fragment acts as a clip and stabilizes the peptidic loop-type microstructure [23].

1.3.4
Approaches Used to Stabilize Bioactive Conformations at Peptides by Metal Coordination

Protein–protein interactions usually occur on the surface of proteins where mainly loops or turns are present [24]. It is, therefore, of interest to stabilize specific peptide turns and test them for their biological activity [25–28]. As discussed, metal coordination seems to be a powerful tool for obtaining macrocycles with a "bent" peptidic domain simply by mixing metal ions and peptide-bridged ligands.

Examples of naturally occurring bioactive cyclopeptides are Segetalins A and B (Figure 1.3.11), which were isolated from *Vaccaria segetalis* (*Caryophyllacea*) and were shown to have estrogen-like activity. The seeds of *Vaccaria segetalis* are used

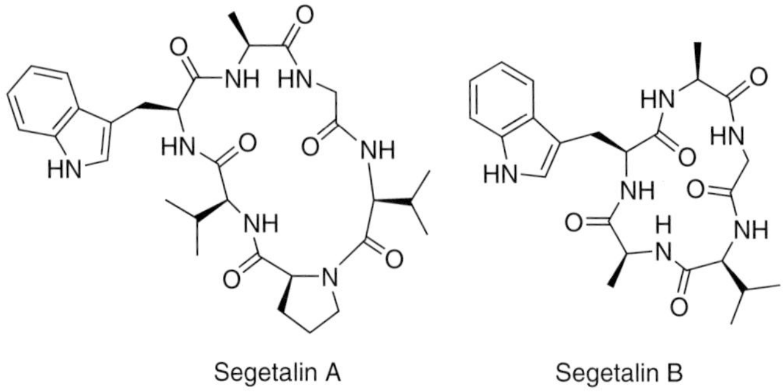

Segetalin A   Segetalin B

Fig. 1.3.11. Segetalin A and Segetalin B isolated from *Vaccaria segetalis* (*Caryophyllacea*).

in Chinese folk medicine to activate blood flow, to promote milk secretion, and to treat amenorrhea and breast infections [29–31].

That Segetalin A and Segetalin B have the Try-Ala-Gly-Val (WAGV) sequence in common indicates this is the biologically active part of the molecules [32]. NMR studies indicate, however, that the orientation of the Val residue is different in the molecules. The effect of this amino acid on the activity of the compounds might not be very important [29–31].

The WAG sequence can easily be introduced as a spacer between two catechol moieties by following the FMOC strategy for the preparation of peptides. An N-terminal FMOC-protected amino acid is attached to an amine using HBTU as coupling reagent. The protecting group is removed and another FMOC-protected amino acid can be attached, to elongate the peptide chain. By use of this procedure, then cleavage of the methyl ethers by BBr$_3$, the ligand **4-H$_4$** is prepared in eight steps (Scheme 1.3.4) [33].

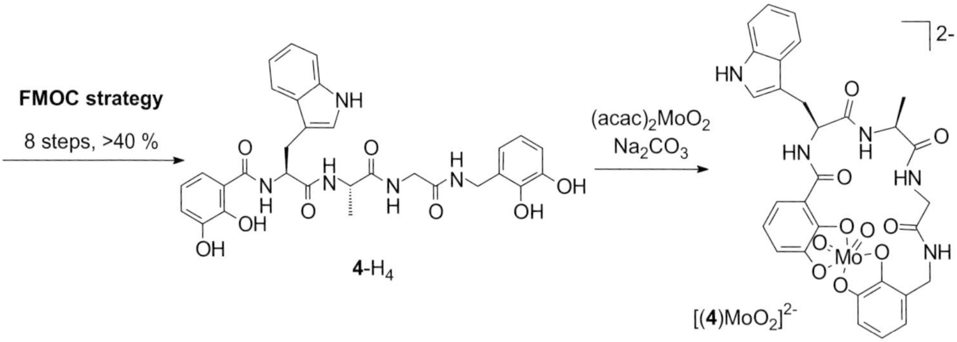

FMOC strategy

8 steps, >40 %

(acac)$_2$MoO$_2$
Na$_2$CO$_3$

4-H$_4$

[(**4**)MoO$_2$]$^{2-}$

Scheme 1.3.4. Preparation of the WAG-bridged ligand **4-H$_4$** and formation of the macrocyclic complex [**4**MoO$_2$]$^{2-}$.

Reaction of the ligand **4**-H$_4$ with (acac)$_2$MoO$_2$ should lead to the macrocyclic complex in which the biologically active WAG-sequence of the cyclopeptidic Segetalins A/B is stabilized in a loop-type fashion.

Clipping of the linear derivative **4**-H$_4$ to fix the WAG-sequence in a bent conformation proceeds smoothly by addition of O$_2$Mo(acac)$_2$ and K$_2$CO$_3$ in methanol. The *cis*-dioxomolybdenum(VI) complex K$_2$[**4**MoO$_2$] is characterized by spectroscopic methods like NMR or ESI–MS. The results obtained indicate that only one of the two possible stereoisomers is formed. This shows that the amino acids of the WAG sequence are able to induce chirality perfectly at the metal complex unit. The good resolution and signal dispersion in the ^{1}H NMR spectrum indicate the peptide turn adopts a well defined conformation. From the present results, however, it cannot be deduced if this conformation is the biologically active one [34].

1.3.5
Conclusions

The stabilization of well-defined peptide microstructures is an important challenge in bioorganic chemistry. As the presented results show, metal coordination can be a simple but very effective tool for reaching this goal and fixing three dimensional molecular structures. The conformational (and stereochemical) information which is embedded in configurationally stable metal complex units can be transferred to amino acid residues or even to relatively large peptides and can induce helical-, sheet- or as discussed turn-type structures.

The next challenge is to investigate the interaction of the metal-complex-fixed peptides with natural receptors to gain deeper understanding of receptor–ligand interactions and to learn something about the stereochemical and conformational requirements at small peptides to enable development of new and superior inhibitors or ligands for naturally occurring binding sites or reactive centers.

References

1 A. FERSHT, *Structure and Mechanism in Protein Science*, W. H. FREEMAN, New York, **1998**.
2 R. HAUBNER, D. FINSINGER, H. KESSLER, *Angew. Chem.* **1997**, *109*, 1440; *Angew. Chem. Int. Ed.* **1997**, *36*, 1374.
3 N. VOYER, *Top. Curr. Chem.* **1996**, *184*, 1.
4 M. ALBRECHT, *Angew. Chem.* **1999**, *111*, 3671; *Angew. Chem. Int. Ed.* **1999**, *38*, 3463.
5 F. RUAN, Y. CHEN, P. B. HOPKINS, *J. Am. Chem. Soc.* **1990**, *112*, 9403.
6 M. LIEBERMAN, M. TABET, T. SASAKI, *J. Am. Chem. Soc.* **1994**, *116*, 5035.
7 M. LIEBERMAN, T. SASAKI, *J. Am. Chem. Soc.* **1991**, *113*, 1470.
8 M. R. GHADIRI, M. A. CASE, *Angew. Chem.* **1993**, *105*, 1663; *Angew. Chem. Int. Ed.* **1993**, *32*, 1594.
9 M. R. GHADIRI, C. SOARES, C. CHOI, *J. Am. Chem. Soc.* **1992**, *114*, 4000.

10 J. P. Schneider, J. W. Kelly, *J. Am. Chem. Soc.* **1995**, *117*, 2533.
11 A. Torrado, B. Imperiali, *J. Org. Chem.* **1996**, *61*, 8940.
12 M. J. Kelso, H. N. Hoang, T. G. Appleton, D. P. Fairlie, *J. Am. Chem. Soc.* **2000**, *122*, 10488.
13 A.-K. Duhme, Z. Dauter, R. C. Hider, S. Pohl, *Inorg. Chem.* **1996**, *35*, 3059.
14 A.-K. Duhme, *J. Chem. Soc., Dalton Trans.* **1997**, 773.
15 G. N. Ramachandran, V. Sasiskharan, *Adv. Protein Chem.* **1968**, *23*, 283.
16 M. Albrecht, M. Napp, M. Schneider, *Synthesis* **2001**, 468.
17 P. Li, J.-C. Xu, *Chin. J. Chem.* **2000**, *18*, 456.
18 A. Speicher, T. Klaus, T. Eicher, *J. Prakt. Chem.* **1998**, *340*, 581.
19 M. Bodanszky, A. Bodanszky, *The Practice of Peptide Synthesis*, Springer, Berlin, **1984**.
20 M. Albrecht, M. Napp, M. Schneider, P. Weis, R. Fröhlich, *Chem. Commun.* **2001**, 409.
21 M. Albrecht, M. Napp, M. Schneider, P. Weis, R. Fröhlich, *Chem. Eur. J.* **2001**, *7*, 3966.
22 M. Albrecht, O. Spieß, M. Schneider, *Synthesis* **2002**, 126.
23 M. Albrecht, O. Spieß, M. Schneider, P. Weis, *Chem. Commun.* **2002**, 787.
24 D. P. Fairlie, M. L. West, A. K. Wong, *Curr. Med. Chem.* **1998**, *5*, 29.
25 K. Burgess, *Acc. Chem. Res.* **2001**, *34*, 826.
26 A. G. Cochran, R. T. Tong, M. A. Starovasnik, E. J. Park, R. S. McDowell, J. E. Theaker, L. J. Skelton, *J. Am. Chem. Soc.* **2001**, *123*, 625.
27 M. MacDonald, J. Aubé, *Curr. Org. Chem.* **2001**, *5*, 417.
28 J. A. Robinson, *Synlett* **1999**, 429.
29 H. Morita, Y. S. Yun, K. Takeya, H. Itokawa, *Tetrahedron Lett.* **1994**, *35*, 9593.
30 H. Morita, Y. S. Yun, K. Takeya, H. Itokawa, K. Yamada, *Tetrahedron* **1995**, *51*, 6003.
31 H. Morita, Y. S. Yun, K. Takeya, H. Itokawa, M. Shiro, *Tetrahedron* **1995**, *51*, 5987.
32 P. Sonnet, L. Petit, D. Marty, J. Guillon, J. Rochette, J.-D. Brion, *Tetrahedron Lett.* **2001**, *42*, 1681.
33 J. Jones, *Synthese von Aminosäuren und Peptiden*, VCH, Weinheim, **1995**.
34 M. Albrecht, P. Stortz, P. Weis, *Supramol. Chem.*, in press.

B.2
Conformational Analysis of Proteins: Ramachandran's Method

Markus Albrecht

Because the three dimensional structure of proteins is important for their specific function, it is, crucial to find ways of understanding, analyzing, and describing the folding of protein structures. In the 1960s Ramachandran introduced a method for analyzing the structure of proteins, starting from the results obtained by X-ray analysis in the solid state [1]. Today this method is also used for description of protein structures obtained by NMR in solution or by molecular modeling.

Ramachandran pointed out that the folding of a protein depends on the conformation adopted by the two central σ-bonds of each amino acid and that this conformation can be described by using the dihedral angles Φ and

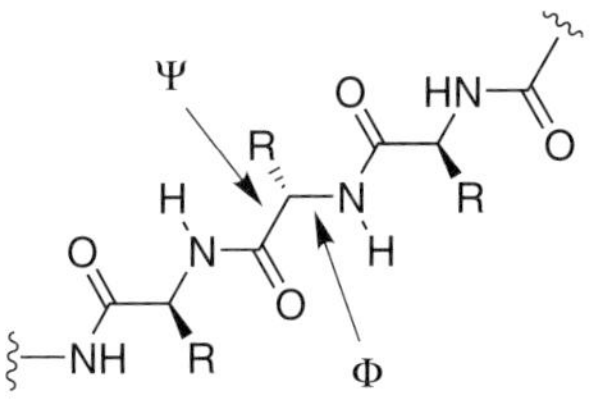

Fig. B.2.1.

Ψ (Figure B.2.1). It is assumed the amide unit "always" prefers a planar arrangement with the NH and CO orientated in an s-trans fashion.

Plotting the dihedral angle Ψ against Φ produces a diagram in which more or less preferred conformations can be found. Regions which represent favored conformations of the amino acids in the diagram are shown in dark whereas less favored regions are shown in gray and unfavored regions in white (Figure B.2.2). Two major regions of favored conformation are apparent; these represent the areas which are observed in a β-sheet (Ψ = 125°, Φ = −120°, approx.) or a right handed α-helical structure (Ψ = −50°, Φ = −90°, approx.) [2].

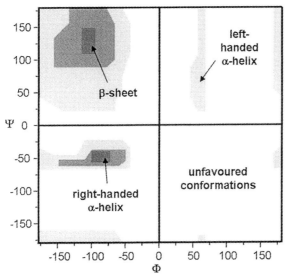

Fig. B.2.2.

In contrast with the right-handed α-helical structure of proteins a similar left-handed folding is theoretically possible in which the dihedral angles would be approximately Ψ = 50° and Φ = 90°. Because of the substituent and the S configuration of "naturally dominating" amino acids, however, the left-handed form is less favored. Steric interaction occurs between the substituent and the peptide backbone if a left-handed helical structure is adopted.

Only the achiral glycine can readily adopt conformations different from those in the favored regions; proline, because of, the ring system, is highly constrained in its conformation.

References

1 G. N. Ramachandran, V. Sasiskharan, *Adv. Protein Chem.* **1968**, *23*, 283.

2 A. L. Morris, M. W. MacArthur, E. G. Hutchinson, J. M. Thornton, *Proteins,* **1992**, *12,* 345.

B.3
Metals in Proteins – Tools for the Stabilization of Secondary Structures and as Parts of Reaction Centers

Markus Albrecht

Very often metal ions are found encapsulated in the interior of a protein (or located on its surface). These metal ions can undergo two different functions:

- The three-dimensional structure (secondary structure) of a protein is controlled by non-covalent interactions between different parts of the peptide strand. Metal ions form very strong "non-covalent" coordination bonds and can therefore fix peptide structures very effectively. A prominent example is the zinc finger motif (Figure B.3.1). A random coil peptide binds to zinc(II) ions through two cysteine and two histidine residues inducing formation of a β-sheet structure by the cysteines and an α-helix by the histidines (Figure B.3.1). The biological activity of the resulting conformationally strained peptide is "switched on" and the peptide can now interact with DNA [1].
- In addition to structure control, metal ions can act as reactive centers of proteins or enzymes. The metals can not only bind reaction partners, their special reactivity can induce chemical reaction of the substrate. Very often different redox states of the metal ions play a crucial role in the specific chemistry of the metal. Non-redox-active enzymes, e.g. some hydrolytic enzymes, often react as a result of their Lewis-acid activity [2]. Binding of substrates is, however, important not only for their chemical modification but also for their transport. Oxygen transport by hemoglobin is an important example of this [3].

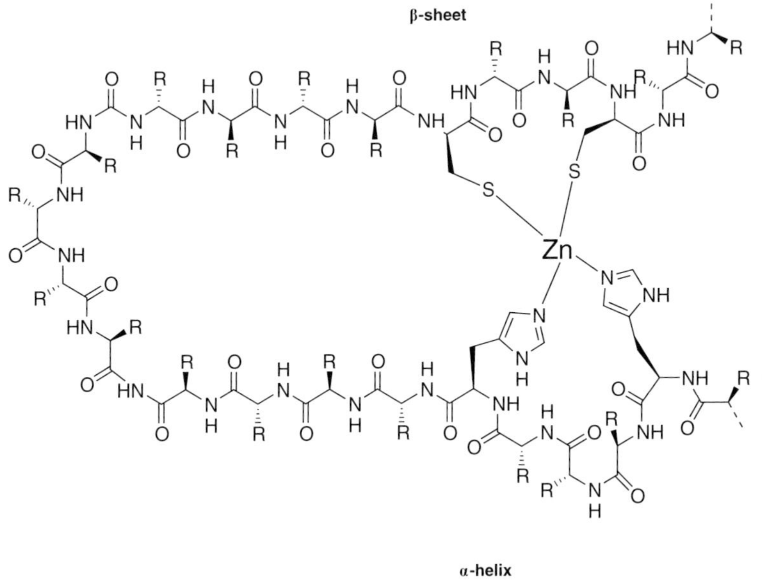

Fig. B.3.1.

References

1. J. M. BERG, *J. Biol. Chem.* **1990**, *265*, 6513.
2. S. J. LIPPARD, J. M. BERG, *Principles of Bioinorganic Chemistry*, University Science Books, Mill Valley, USA, **1994**.
3. M. F. PERUTZ, G. FERMI, B. LUISI, B. SHAANON, R. C. LIDDINGTON, *Acc. Chem. Res.* **1987**, *20*, 309.

1.4
Conformational Restriction of Sphingolipids

Thomas Kolter

Summary

Conformational restriction of biologically flexible molecules is a successful strategy in drug development. Application of this concept to ceramide, the membrane anchor of sphingolipids, led to the unexpected inhibition of glycosyltransferases involved in the combinatorial biosynthesis of gangliosides. This discovery offers a new approach to cell surface engineering.

1.4.1
Introduction

Bioorganic chemistry can be understood as a discipline that addresses relevant biochemical problems with synthetic organic chemistry as the key technique [1]. A substance class that has been accompanied by enigmas since its discovery by L. W. Thudichum in 1884 is that of the sphingolipids [2], which he named after the Egyptian sphinx. This chapter gives an overview on the application of conformational restriction as a bioorganic technique to ceramide, the structural component of most sphingolipids.

1.4.1.1
Lipids

Biological membranes are composed of integral membrane proteins and of amphiphilic lipids (Box 4), which can be classified according to the structure of their lipophilic backbone as glycerolipids, sphingolipids, and sterols [3]. In addition to their structural role, a variety of additional physiological functions has been attributed to these lipids. The abundance of various molecular lipid species within the membranes of different cells and organelles can be different and is tightly regulated. Even both leaflets of a lipid bilayer differ in their lipid composition. This is also true for the apical and the basolateral membrane of polarized cells and also for the lateral lipid composition within a given membrane. Microdomains within cel-

Highlights in Bioorganic Chemistry: Methods and Applications. Edited by Carsten Schmuck, Helma Wennemers.
Copyright © 2004 WILEY-VCH Verlag GmbH & Co. KGaA, Weinheim
ISBN: 3-527-30656-0

lular membranes, so called rafts, are characterized by a high content of sphingolipids, cholesterol, and glycosylphosphatidylinositol- (GPI-) anchored proteins [4]. Single membrane lipids can be essential for the function of membrane-bound enzymes and receptors. For example, the activity of β-hydroxybutyrate dehydrogenase of the internal mitochondrial membrane is completely dependent on the presence of the glycerolipid phosphatidylcholine [5]. Furthermore, mitochondrial cytochrome c reductase requires cardiolipin [6], and the functional trimer of bacteriorhodopsin dissociates without a central glycolipid [7].

1.4.1.2
Sphingolipids

Sphingolipids are composed of sphingosine, a $(2S,3R)$-2-aminooctadec-4-en-1,3-diol, which is acylated with a fatty acid to a ceramide (Figure 1.4.1). In most mammalian glycolipids, the carbohydrate part is linked to ceramide as the membrane anchor. With the exception of sphingosine, sphingosine-1-phosphate, and sphingosine-1-phosphorylcholine, most sphingolipids are characterized by the presence of two fatty acid residues within the hydrophobic part of the molecule. Figure 1.4.1 shows the structures of ceramide **1**, sphingomyelin **2**, and of ganglioside GM1 **3** as representative sphingolipid structures. Ceramides of different alkyl chain length, desaturation, and hydroxylation status, occur in living cells at low concentrations as metabolic intermediates and, at higher concentrations, in the human skin [8]. Because of variations in the type, number, and linkage of carbohydrate

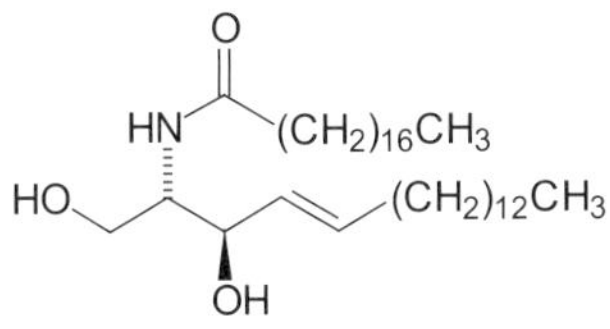

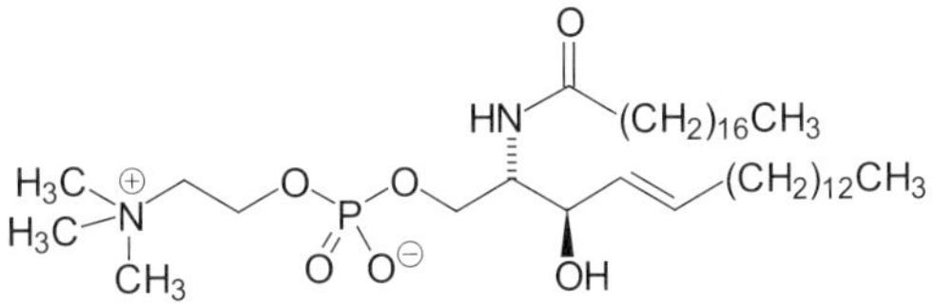

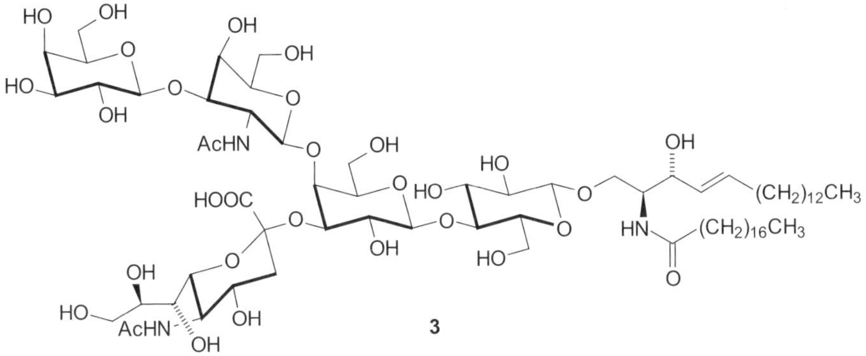

Fig. 1.4.1. The structures of ceramide **1**, sphingomyelin **2**, and ganglioside GM1 **3**.

residues within the oligosaccharide part of glycosphingolipids, a variety of these structures is found on the surface of living cells. There they form complex patterns, that change during biological processes like cell growth, differentiation, and viral and oncogenic transformation. The function of these complex glycolipid patterns on cellular surfaces is largely unclear. Sialic acid containing glycosphingolipids of the ganglio series, e.g. GM1, are biosynthetically formed in a combinatorial manner [9]. Several physiological functions have been attributed to glycosphingolipids. They can serve as binding sites for bacteria, viruses, and toxins and mediate recognition events, e.g. as blood group substances. GM1, for example, is the receptor for cholera toxin in intestinal cells and occurs at high concentrations on neuronal cells of the brain. In mammals sphingolipids are essential for embryogenesis and for the permeability barrier of the skin. Because of their various biological functions, bioorganic strategies have been proposed for the development of conceptionally new therapeutic agents [8, 10].

Sphingolipid Metabolism The formation of ceramide is catalyzed by membrane-bound enzymes on the cytosolic leaflet of the endoplasmic reticulum (ER) [11]. Starting from the amino acid L-serine and two molecules of coenzyme A (CoA)-activated palmitic acid, dihydroceramide is formed in three steps. This *N*-Acyl-2-aminoalkyl-1,3-diol (*N*-acylsphinganine) is dehydrogenated to ceramide with a 4,5-trans double bond by a dihydroceramide desaturase. In the Golgi apparatus hydrophilic head groups are attached to ceramide leading to sphingomyelin, galactosylceramide, and glucosylceramide. Higher glycosphingolipids are synthesized mainly from glucosylceramide by stepwise addition of monosaccharides. Glycosphingolipid biosynthesis is coupled to exocytotic vesicle flow through the Golgi apparatus to the plasma membrane.

Constitutive degradation of sphingolipids occurs in the endosomes and lysosomes [8, 12]. Parts of the plasma membrane are endocytosed and transported via the endosomal compartment to the lysosomal compartment. The carbohydrate residues are cleaved by hydrolytic enzymes in a stepwise manner. Glycosphingolipids with short oligosaccharide chains, but also ceramide, require the additional presence of activator proteins and negatively charged lysosomal lipids for degradation [13]. In humans inherited defects of glycosphingolipid and sphingolipid catabolism give rise to lysosomal storage diseases, the sphingolipidoses [14, 15].

1.4.1.3
Signal Transduction

Ceramide has been recognized as a signaling substance that can be released from sphingomyelin in response to various stimuli, for example tumor necrosis factor α, or platelet-derived growth factor [16]. It causes antimitogenic effects in most cell types, for example cell-cycle arrest, cell differentiation, and apoptosis. The molecular details of these processes are not unequivocally clear [17]. They might be mediated by binding proteins or by formation of ceramide-rich domains or channels within the plasma membrane [18]. The metabolic coupling of ceramide to other signaling substances, for example sphingosine-1-phosphate or diacylglycerol (**6**, Figure 1.4.4), prevents clear conclusions from cell culture studies on the downstream effects of this lipid. For example, ceramide can be hydrolyzed to sphingosine, which can be phosphorylated to sphingosine-1-phosphate. This behaves as a mitogenic regulator in most cell types and is recognized by extracellular receptors of the EDG family. The ratio of the concentrations of ceramide and sphingosine-1-phosphate can be viewed as a rheostat that directs the cell cycle towards differentiation or cell division [19].

1.4.2
Conformational Restriction

The restriction of conformationally flexible ligands is a successful strategy in drug development [20] (for possible approaches see Chapters 1.2 and 1.3). This strategy can lead to analogs with enhanced affinity (1), selectivity (2), and metabolic stability (3) compared with the parent compound. (1) Conformational restriction reduces entropy loss on receptor binding and contributes in this way to receptor affinity. The contributions of the different types of interaction to the change of free enthalpy on ligand binding have been estimated. Typical values are: loss of translational and rotational entropy, $+5.4$ kJ mol^{-1}; ideal neutral hydrogen bond, -4.7 kJ mol^{-1} (however the actual value strongly depends on the microenvironment; Box 7); ideal ionic interaction, -8.3 kJ mol^{-1}; lipophilic contact, -0.17 kJ mol^{-1} × Å^2; arrest of a rotatable bond, $+1.4$ kJ mol^{-1} [21]. For other values compare Ref. [22]. The first and the last of the positions mentioned above are favorably influenced by conformational restriction. (2) Also binding selectivity can be enhanced when alternative receptors recognize alternative conformations, which are not accessible for the restricted system (Figure 1.4.2). Numerous examples of this bioorganic strategy can be found in pharmacology [23], e.g. the early differentiation between acetylcholine receptors based on their affinity for agonists such as nicotine or muscarine, or the application of kainic acid as agonist of glutamate receptors on neuronal cells. (3) Also, drug-metabolizing enzymes recognize distinct conformations of their substrates that can differ from those bound by other receptors. Therefore, conformational restriction can also be used to enhance the biological half life of a ligand and improve its pharmacokinetic properties.

1.4.2.1
Peptidomimetics

Naturally occurring peptides are substrates of enzymes and ligands of receptors. Many peptides serve as neurotransmitters and hormones. Because they control a variety of physiological processes, they are also implicated in different diseases. For several reasons peptides are usually unsuitable as drugs, so many efforts in medicinal chemistry aim at the development of peptidomimetics with improved properties. In contrast with proteins most peptides of small to medium size – 30 to

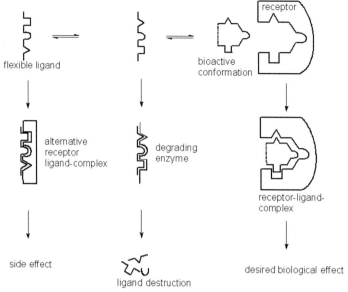

Fig. 1.4.2. Conformational equilibrium of a flexible ligand. Modified from Ref. [20].

50 amino acids – exist in a multitude of conformations in dilute aqueous solution. A generally applicable and successful method for the development of peptidomimetics is the synthesis of suitable conformationally restricted analogs that imitate the receptor-bound conformation of the endogenous peptide ligands. In the case of enzyme inhibitors peptidomimetics often resemble the transition state of the peptide substrate in enzyme-catalyzed reactions (a more detailed discussion and examples are given in reference [20]). Both can be achieved either by modification of the structure of the parent peptide, or by searching for compounds of non-peptide structure that can imitate or block the function of a peptide on the molecular level [24]. The opioid alkaloids are classic examples of non-peptide ligands that have later been discovered to be mimetics of endogenous peptides. For example, morphine **5** (Figure 1.4.3) imitates the effect of flexible peptides, for example Met-enkephalin **4**, at the common recognition site of the receptor for both compounds [25].

1.4.2.2
Conformationally Restrained Lipids

With the exception of sterols most membrane lipids are conformationally flexible molecules. The question arises whether the concept of conformational restriction can be extended from peptides to the field of sphingolipids. The observation that conformationally restricted derivatives of myristic acid have selective toxicity towards trypanosomes demonstrates that this strategy is also applicable

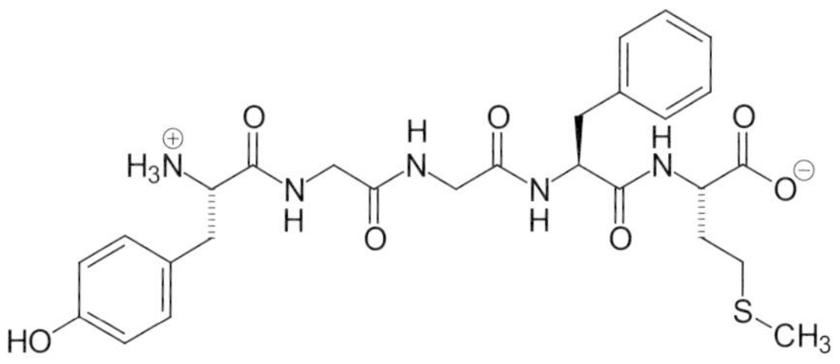

Fig. 1.4.3. The structures of Met-enkephalin **4** and morphine **5**.

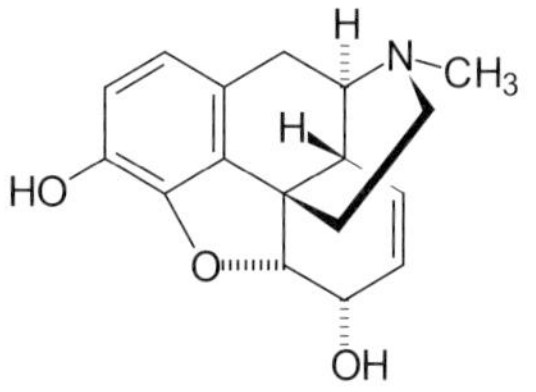

to lipophilic compounds [26]. In contrast to mammals, trypanosomes use glycosylphosphatidylinositol anchors with myristic acid as fatty acid component. Conformationally restricted derivatives inhibit the remodeling reaction that is required for formation of appropriate membrane anchors of GPI-bound proteins. Other examples are conformationally rigid natural compounds, that are able to competitively replace another lipid, diacylglycerol (**6**, Figure 1.4.4), from its binding

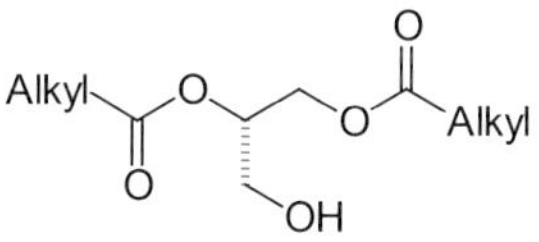

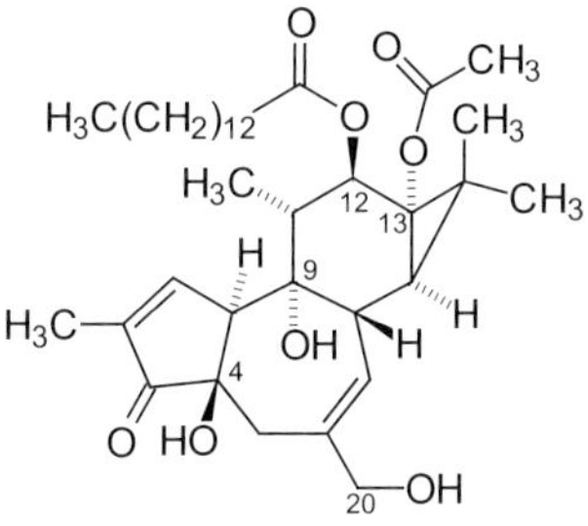

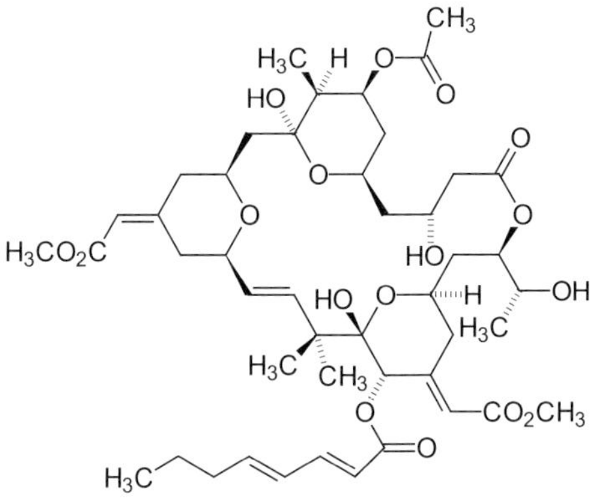

Fig. 1.4.4. The structures of DAG **6**, phorbol esters **7**, and bryostatin **8**.

site on protein kinase C. Diacylglycerol plays a major role in one of the most prominent signal-transducing systems, the phosphoinositide cycle: in response to extracellular signals G-proteins are activated that lead to the rapid hydrolysis of the membrane lipid phosphatidylinositol-4,5-bisphosphate by a phospholipase C (PLC). This leads to liberation of the hydrophilic head group, inositol-1,4,5-trisphosphate from the lipid backbone, 1,2-diacylglycerol (DAG). Inositol-1,4,5-trisphosphate interacts with an intracellular membrane receptor and leads to a mobilization of Ca^{2+} ions, whereas DAG binds directly to, and activates, protein kinase C (PKC), and thus initiates a separate signal-transduction cascade [27]. PKC phosphorylates serine and threonine residues in target proteins and requires Ca^{2+} ions and phosphatidylserine for full activity. The affinity for Ca^{2+} ions is enhanced by DAG so that PKC becomes active at physiological Ca^{2+} concentrations. PKC consists of a catalytic and a regulatory domain which resembles PKC substrates. Binding of DAG removes this intramolecular blockade so that PKC substrates have access to the active site. To the natural products that are known to replace DAG **6** from its protein receptor belong the esters of phorbol esters like **7**, which have attracted attention as tumor promoters. Other natural products are the teleocidins, and bryostatin **8** [28, 29]. These natural products can be regarded as conformationally rigid analogs of DAG and have been used as lead compounds in the development of analogs of more simple structures [30–32].

1.4.3
Conformational Restriction of Sphingolipids

1.4.3.1
Rationale

We have applied the concept of conformational restriction to the synthesis of sphingolipid analogs. We report results obtained with ceramide analogs. One aim of this strategy was to reduce the variety of ceramide-mediated effects. Ceramide analogs are potential inhibitors of pharmacologically important enzymes, for example ceramide glucosyltransferase, inositolphosphorylceramide synthase, and others [8]. In addition, they might serve as agonists or antagonists of binding proteins which are thought to be involved in ceramide-mediated signal transduction.

1.4.3.2
Present State of Knowledge

Few compounds have yet been regarded as conformationally rigid ceramide analogs. Isoquinoline derivatives **9** (Figure 1.4.5) have been investigated as ligands of protein phosphatase 2A, a ceramide-binding protein that has been implicated in the transmission of the ceramide-signal into the cell. Slight modulation of the enzyme has been reported [33]. Uracil and thiouracil derivatives **10** have been synthesized as conformationally restricted ceramide analogs. They show moderate effects in terms of toxicity and anti-tumor activity in vitro and in vivo [34].

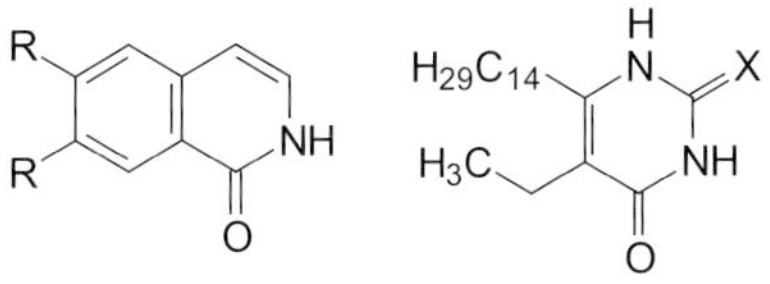

9 (R = Alkyl)   **10** (X = O, S)

Fig. 1.4.5. The structures of isoquinoline derivatives **9** and uracil and thiouracil derivatives **10**.

1.4.4
Target Compounds

Because ceramide-metabolizing enzymes have to recognize and to modify the head group of this lipid, we were especially interested in conformational restriction of the head group. We therefore designed a series [35] of ceramide analogs that represent remote points within the conformational space available to the head group of this lipid. Several derivatives have been synthesized according to this strategy by our group [36]. Two members of the series that turned out to be active in our assay systems are the heterocyclic derivatives **11** and **12** (Figure 1.4.6).

1.4.4.1
Synthesis

Compound **11** has been prepared in a sequence of ten reaction steps in 0.9% total yield starting from D-galactose [37] or, alternatively, in nine steps starting from Garners serine aldehyde [38] with a total yield of 5.5% (not shown [39]). In both instances, two C–C bond-formation reactions were required; ring formation was achieved by a Heck reaction on both routes. Compound **12** was also prepared starting from Garners aldehyde, by modification of a reported route [40] with a Wittig reaction and an iodolactonization as key steps (not shown [37]). The diastereomer with opposite stereochemistry at C-6 was also obtained and investigated.

1.4.4.2
Analysis in Cultured Cells

An important question in bioorganic chemistry is that of the appropriate assay system [1]. We investigated the effects of the designed compounds on sphingolipid

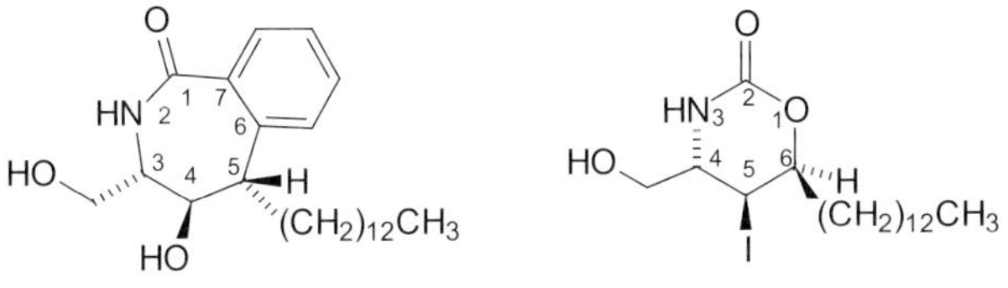

11   **12**

Fig. 1.4.6. Structures of the active heterocyclic derivatives **11** and **12**.

metabolism in cultured cells, especially in primary cultured neurons which are rich in complex gangliosides. In contrast with the current paradigm in medicinal chemistry, we applied few substances of defined structure to complex cell culture systems. Primary cell cultures are artificial compared with multicellular organisms [41], but are sufficiently complex compared with cell lines or enzyme assays. As will become clear below, the results observed would have not been obtained in enzyme assays.

Because the target compounds can be expected to interfere with the metabolism of endogenous ceramide, we investigated the effect of the ceramide analogs **11** and **12** on incorporation of biosynthetic precursors L-serine and D-galactose into ceramide-containing lipids of primarily cultured murine cerebellar granule cells. Serine-incorporation reflects de-novo biosynthesis of sphingolipids. On the other hand, galactose is, in part after intracellular metabolism to other carbohydrates, enzymatically added to ceramide derived from the de-novo pool and from the salvage pool, which can contribute up to 80% of glycosphingolipid biosynthesis. Exogenously added ceramide analogs can also be metabolically labeled by galactose [42]. In brief [43, 44], primary neurons were prepared from the cerebellum of 6-day-old mice and the cells were treated with different concentrations of the target compounds for 24 h. Radiolabeled 3-[^{14}C]serine or [^{14}C]galactose was added to the culture medium and incorporation into newly synthesized sphingolipids was analyzed after labeling for 24 h. Lipids were extracted, separated by thin-layer chromatography, and visualized with a phosphoimager. Radioactivity found in the selected lipids is expressed in relation to untreated cells.

L-[3-^{14}C]serine labeling of sphingolipids in the presence of **11** led to a concentration-dependent decrease of sphingomyelin and glucosylceramide levels whereas the amount of ceramide increased. Unexpectedly, [^{14}C]galactose labeling indicated inhibition of sialyltransferase II (SAT II, GD3-synthase) at much lower concentrations of **11** of 10 μM in the culture medium and, at higher concentrations, also of GalNAc-transferase (GM2-synthase, Figure 1.4.7). Even more exciting results were obtained with **12**. As determined by [^{14}C]galactose labeling, concentrations of 10 μM in the culture medium caused dramatic – up to 40-fold – elevations in the levels of LacCer, GM3, and GD3, which indicated inhibition of GM2-synthase. These findings were confirmed by metabolic labeling with L-[3-^{14}C]serine and [4,5-^{3}H]sphinganine. The compound with opposite stereochemistry at the carbon bearing the alkyl chain had nearly no effect, indicating specificity.

To determine whether **12** inhibits GalNAc-transferase directly, we applied 100 μM **12** in an enzyme assay with the homogenate of insect Sf21-cells over expressing murine GalNAc-transferase [45]. No inhibition was detected, so we had to conclude that **12** operates indirectly, either after metabolism, or by interfering with intracellular membrane transport in the same way as, for example, the macrolide brefeldin A [46]. Brefeldin A specifically binds to a usually transient intermediate of the reaction between an ARF-protein (ARF = ADP-ribosylating factor), guanosine diphosphate, and the Sec-7 domain of a guanine-nucleotide-exchange factor (GEF). This disturbs the activation cycle of ARF and exocytotic membrane flow is interrupted [47]. Because other ceramide analogs not mentioned here interfere with

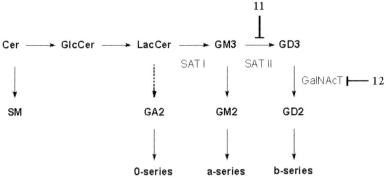

Fig. 1.4.7. Early steps in combinatorial ganglioside biosynthesis [9]. The sites of (indirect) inhibition by **11** and **12** are indicated. Abbreviations: Cer, ceramide, N-acylsphingosine; Glc, glucose; Gal, galactose; GalNAc, N-acetylgalactosamine; NeuAc, N-acetylneuraminic acid; SAT, sialyltransferase; GalNAcT, N-acetylgalactosaminyltransferase; GlcCer, Glcβ1Cer; LacCer, Galβ1,4Glcβ1Cer; GM3, NeuAcα2,3Galβ1,4Glcβ1Cer; GD3, NeuAcα2,8NeuAcα2,3Galβ1,4Glcβ1Cer; GA2, GalNAcβ1,4Galβ1,4Glcβ1Cer; GM2, GalNAcβ1,4(NeuAcα2,3)Galβ1,4Glcβ1Cer; GD2, GalNAcβ1,4(NeuAcα2,8NeuAcα2,3)·Galβ1,4Glcβ1Cer. The next a-series ganglioside would be ganglioside GM1 (Figure 1.4.1).

other glycosyltransferases, we do not consider **12** as a brefeldin mimic. The results with galactose labeling were qualitatively confirmed by incorporation of L-[3-^{14}C]serine and [4,5-^{3}H]sphinganine in the sphingolipids in the presence of **12** and its diastereomer. Analysis of other ceramide analogs indicates that several glycosyltransferases within the ganglioside biosynthesis pathway can be inhibited by this approach [10].

1.4.5
Discussion

We reported on the extension of conformational restriction, which was particularly successful in the field of peptidomimetics, to sphingolipids as another class of conformationally flexible molecules. The major effect of this approach is the unexpected inhibition of glycosyltransferases that do not accept ceramide itself but instead accept glycolipids of more complex structure as substrates. Inhibitors are available for glucosylceramide synthase, a glycosyltransferase that accepts ceramide as substrate [8, 10]. This prodrug mechanism offers the possibility of modifying the lipid composition of cellular surfaces. A somehow related effect has been achieved by interference with carbohydrate metabolism. Analogs of N-acetyl-D-mannosamine are biosynthetically incorporated into sialic acids, which in turn are transformed into glycocolipids and glycoproteins of the cell surface and modify their properties like cell adhesion [48, 49]. This process is also known as cell-surface engineering [50].

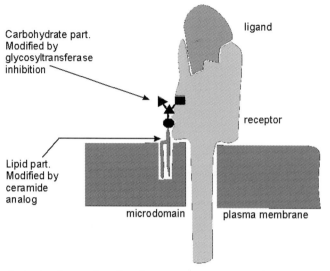

Fig. 1.4.8. Concept for cell surface engineering: exogenously added ceramide analogs can lead to modification of the hydrophobic and the hydrophilic part of membrane sphingolipids and alter the properties of membrane proteins.

Because different glycosyltransferases bound to the membrane of the Golgi apparatus can be indirectly inhibited by ceramide-analogs at non-toxic concentrations, this should lead to modified glycolipid pattern (Figure 1.4.8) on cell surfaces and is the pharmacological equivalent to mice with the corresponding genetically engineered defects [9]. It has been shown that chemically modified ceramides are metabolized to glycolipids and to sphingomyelin [10]. These artificial metabolites should be transported to the plasma membrane where they can modify the activity of membrane proteins, similar to inhibition of glycosyltransferases in the Golgi apparatus. Neoglycolipids of this type might alter the colligative properties of microdomains or influence receptor activity as single molecules (Figure 1.4.8). In addition to the lipids mentioned in the introduction, glycosphingolipids can also drastically modify the activity of pharmacologically relevant receptors. For example, the receptors for insulin [51] and epidermal growth factor [52] are down-regulated by ganglioside GM3, and the activity of the nerve growth receptor is increased by ganglioside GM1 [53]. Therapeutic applications of this approach are reviewed elsewhere [10].

1.4.6
Outlook

Conformational restriction of ceramide led to the development of cell-permeable, indirectly acting inhibitors of glycosyltransferases. Together with other ceramide analogs [10, 36] this concept demonstrates that, in principle, it is pharmacologically possible to achieve effects comparable with those obtained in genetically engineered mice with glycosyltransferase deficiencies [9]. Further efforts are required to develop heterocyclic lipidomimetics by this approach.

References

1 H. Waldmann, M. Famulok, ChemBioChem. **2001**, *2*, 3–6.
2 A. H. Merrill Jr, K. Sandhoff, *Biochemistry of Lipids, Lipoproteins and Membranes*, 4th edn, D. E. Vance and J. E. Vance, eds., Elsevier Science, Chapter 14, in press.
3 D. E. Vance and J. E. Vance, eds., *Biochemistry of Lipids, Lipoproteins and Membranes*, 4th edn, Elsevier Science, in press.
4 G. van Meer, Q. Lisman, *J. Biol. Chem.* **2002**, *277*, 25855–25858.
5 M. D. Houslay, K. K. Stanley, *Dynamics of Biological Membranes*, J. Wiley and Sons, **1982**, p. 97.
6 C. Hunte, J. Koepke, C. Lange, T. Rossmanith, H. Michel, *Struct. Fold. Des.* **2000**, *8*, 669–684.
7 M. Weik, H. Patzelt, G. Zaccai, D. Oesterhelt, *Mol. Cell* **1998**, *1*, 411–419.
8 T. Kolter, K. Sandhoff, *Angew. Chem.* **1999**, *111*, 1633–1670, *Angew. Chem. Int. Ed.* **1999**, *38*, 1532–1568.
9 T. Kolter, R. L. Proia, K. Sandhoff, *J. Biol. Chem.* **2002**, *277*, 25859–25862.
10 S. Brodesser, P. Sawatzki, T. Kolter, *Eur. J. Org. Chem.*, **2003**, 2021–2034.
11 A. H. Merrill Jr, *J. Biol. Chem.* **2002**, *277*, 25843–25846.
12 K. Sandhoff, T. Kolter, *Trends Cell. Biol.* **1996**, *6*, 98–103.
13 O. Macheleidt, T. Kolter, K. Sandhoff, *Sphingolipid activator deficiencies and Niemann-Pick Type C*, F. Platt, S. U. Walkley, eds., *Lysosomal disorders of the brain*, Chapter 8.
14 T. Kolter, K. Sandhoff, *Brain Pathol.* **1998**, *8*, 79–100.
15 T. Kolter, K. Sandhoff, *Sphingolipidosen*, D. Ganten, K. Ruckpaul, eds., *Handbuch der Molekularen Medizin*, Band 6, Monogen bedingte Erbkrankheiten 1, Kap. 2.3, Springer, **2000**, pp. 195–234.
16 Y. A. Hannun, L. M. Obeid, *J. Biol. Chem.* **2002**, *277*, 25847–25850.
17 A. Huwiler, T. Kolter, J. Pfeilschifter, K. Sandhoff, *Biochim. Biophys. Acta* **2000**, *1485*, 63–99.
18 L. J. Siskind, M. Colombini, *J. Biol. Chem.* **2000**, *275*, 38640–38644.
19 S. Spiegel, S. Milstien, *J. Biol. Chem.* **2002**, *277*, 25851–25854.
20 A. Giannis, T. Kolter, *Angew. Chem.*, **1993**, *105*, 1303–1326; *Angew. Chem. Int. Ed. Engl.* **1993**, *32*, 1244–1267.
21 H. J. Böhm, *Ligand Design*, H. Kubinyi, ed., *3D QSAR in Drug Design*, Escom, Leiden, **1993**, pp. 386–405.
22 D. F. V. Lewis, M. Dickins, *Drug Discovery Today* **2002**, *7*, 918–925.
23 W. C. Bowman, M. J. Rand, *Textbook of Pharmacology*, 2nd edn, Blackwell Scientific Publications, Oxford, **1980**.
24 R. M. Freidinger, *Curr. Opinion Chem. Biol.* **1999**, *3*, 395–406.
25 J. Frackenpohl, *Chemie uns. Zeit* **2000**, *34*, 99–112.
26 T. L. Doering, T. Lu, K. A. Werbovez, G. W. Gokel, G. W. Hart, J. I. Gordon, P. T. Englund, *Proc. Natl. Acad. Sci. USA* **1994**, *91*, 9735–9739.
27 M. J. O. Wakelam, *Biochim. Biophys. Acta* **1998**, *1436*, 117–126.
28 P. A. Wender, J. DeBrabander, P. G. Harran, J.-M. Jimenez, M. F. T. Koehler, B. Lippa, C.-M. Park, C. Sidenbiedel, G. R. Pettit, *Proc. Natl. Acad. Sci. USA* **1998**, *95*, 6624–6629.
29 P. A. Wender, J. De Brabander, P. G. Harran, J.-M. Jimenez, M. F. T. Koehler, B. Lippa, C.-M. Park, M. Shiozaki, *J. Am. Chem. Soc.* **1998**, *120*, 4534–4535.
30 Y. Endo, M. Hirano, P. E. Driedger,

S. Stabel, K. Shudo, *Bioorg. Med. Chem. Lett.* **1997**, *7*, 2997–3000.
31 Y. Endo, A. Yokoyama, *Bioorg. Med. Chem. Lett.* **2000**, *10*, 61–66.
32 K. Nacro, B. Bienfait, N. E. Lewin, P. M. Blumberg, V. E. Marquez, *Bioorg. Med. Chem. Lett.* **2000**, *10*, 653–655.
33 L. M. Leoni, H. C. Shih, L. Deng, C. Tuey, G. Walter, D. A. Carson, H. B. Cottam, *Biochem. Pharmacol.* **1998**, *55*, 1105–1111.
34 M. Macchia, S. Barontini, S. Bertini, V. Di Bussolo, S. Fogli, E. Giovannetti, E. Grossi, F. Minutolo, R. Danesi, *J. Med. Chem.* **2001**, *44*, 3994–4000.
35 V. Austel, *Chemometric Methods in Molecular Design*, H. van der Waterbeemd, ed., VCH, Weinheim, **1995**, p. 49–62.
36 T. Kolter, *Habilitationsschrift*, University of Bonn, **2002**.
37 P. Sawatzki, *Doctoral Thesis*, University of Bonn, **2002**.
38 P. Garner, J. M. Park, *J. Org. Chem.* **1987**, *52*, 2361–2364.
39 P. Sawatzki, M. Skowron, G. van Echten-Deckert, T. Kolter, submitted for publication.
40 J. M. Jorda-Gregori, M. E. Gonzalez-Rosende, J. Sepulveda-Arques, R. Galeazzi, M. Orena, *Tetrahedron Asymmetry* **1999**, *10*, 1135–1143.
41 T. Kolter, T. Magin, K. Sandhoff, *Traffic* **2000**, *1*, 803–804.
42 L. Elsen, R. Betz, G. Schwarzmann, K. Sandhoff, G. van Echten-Deckert, *Neurochem. Res.* **2002**, *27*, 717–727.
43 G. van Echten-Deckert, A. Zschoche, T. Bär, R. R. Schmidt, A. Raths, T. Heinemann, K. Sandhoff, *J. Biol. Chem.* **1997**, *272*, 15825–15833.
44 G. van Echten, K. Sandhoff, *J. Neurochem.* **1989**, *52*, 207–214.
45 M. Wendeler, H. Reilaender, J. Hoernschemeyer, G. Schwarzmann, T. Kolter, K. Sandhoff, *Recombinant Ganglioside GM2-Synthase – Expression in Insect Cells and Enzyme Assay*, Y. C. Lee, R. T. Lee, eds., *Methods in Enzymology, Recognition of Carbohydrates in Biological Systems*, submitted for publication.
46 G. van Echten, H. Iber, H. Stotz, A. Takatsuki, K. Sandhoff, *Eur. J. Cell Biol.* **1990**, *51*, 135–139.
47 C. L. Jackson, J. E. Casanova, *Trends Cell Biol.* **2000**, *10*, 60–67.
48 H. Kayser, R. Zeitler, C. Kannicht, D. Grunow, R. Nuck, W. Reutter, *J. Biol. Chem.* **1992**, *267*, 16934–16938.
49 O. T. Keppler, R. Horstkorte, M. Pawlita, C. Schmidt, W. Reutter, *Glycobiology* **2001**, *11*, 11R–18R.
50 C. R. Bertozzi, *Science* **2001**, *291*, 2357–2364.
51 S. Tagami, J. Inokuchi, K. Kabayama, H. Yoshimura, F. Kitamura, S. Uemura, C. Ogawa, A. Ishii, M. Saito, Y. Ohtsuka, S. Sakaue, Y. Igarashi, *J. Biol. Chem.* **2002**, *277*, 3085–3092.
52 G. Zhou, S. Hakomori, K. Kitamura, Y. Igarashi, *J. Biol. Chem.* **1994**, *269*, 1959–1965.
53 T. Mutoh, A. Toluda, T. Miyadai, M. Hamaguchi, N. Fujiki, *Proc. Natl. Acad. Sci. USA* **1995**, *92*, 5087–5091.

B.4
Lipids

Thomas Kolter

Lipids are hydrophobic substances or their derivatives that are poorly soluble in water but much more soluble in organic solvents. Triacylglycerols have long been known as a means of storage of metabolic energy and amphiphilic lipids, for example phospholipids, glycolipids, and cholesterol, have been recognized as constituents of biological membranes. More recently, lipid-based signaling systems and the function of lipid moieties covalently bound to proteins have been discovered. The regulation of membrane protein function by their lipid surroundings will be an important area of research in the future.

Selected Lipid Structures

- *Fatty acids* are long-chain monocarboxylic acids, for example palmitic acid ($H_3C(CH_2)_{14}COOH$) and stearic acid ($H_3C(CH_2)_{16}COOH$). They are most often found as building blocks of complex lipids and can also be branched or hydroxylated. Many fatty acids found in lipids, for example oleic acid, (Z)-$H_3C(CH_2)_7HC=CH(CH_2)_7COOH$, contain Z-configured double bonds. Because of their detergent-like properties free fatty acids are not usually found in living cells, only bound to binding proteins.
- *Fatty alcohols* are found as components of waxes and, e.g., as pheromones in insects (cf. other lipids).
- *Eicosanoids*, e.g. prostaglandins, prostacyclins, and the leukotrienes are signaling substances formed from arachidonic acid which, in turn, is released from phospholipids in response to different stimuli.
- *Terpenes* and *steroids* are biosynthetically derived from mevalonate. They can be hydrocarbons, e.g. carotene, alcohols, e.g. dolichol, esters, e.g. dolicholphosphate, or cholesterol esters, and others. Cholesterol is essential for membrane function in higher animals.
- *Esters and amides.* *Waxes* are esters of fatty acids with long chain alcohols. Together with other hydrophobic substances, they occur on exterior surfaces of plants. *Fats* and *oils* are triesters of glycerol and serve as fuels. Other esters and amides are signaling substances in insects or higher animals; e.g. N-Arachidonoylethanolamine is an endogenous ligand of the cannabinoid-receptor in the human brain.
- *Glycerolipids* are glycerol derivatives with one, two, or three hydrocarbon chains. The lipid bilayers found in most biological membranes are essentially composed of derivatives of phosphatidic acid, a diacylglycerol phosphate. Attachment of hydrophilic "head groups" by ester linkages to phosphatidic acid give rise to many phospholipids (Figures B.4.1 and B.4.2). Phosphatidylcholine (lecithin) is the main glycerolipid found in eukaryotic membranes; phosphatidylethanolamine is abundant in bacterial membranes. More than 1500 different molecular phospholipid species have been detected in biological membranes, but little is known about their precise molecular function.
- *Sphingolipids* are derivatives of ceramide (N-acylsphingosine, compare Chapter 1.4). Most eukaryotic glycolipids and the phospholipid sphingomyelin contain ceramide as the hydrophobic backbone.

1.4 Conformational Restriction of Sphingolipids

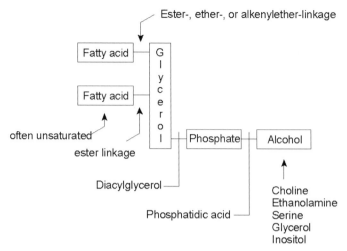

Fig. B.4.1. Schematic representation of membrane phospholipids such as phosphatidylcholine, phosphatidylethanolamine, phosphatidylserine, phosphatidylglycerol, and phosphatidylinositol.

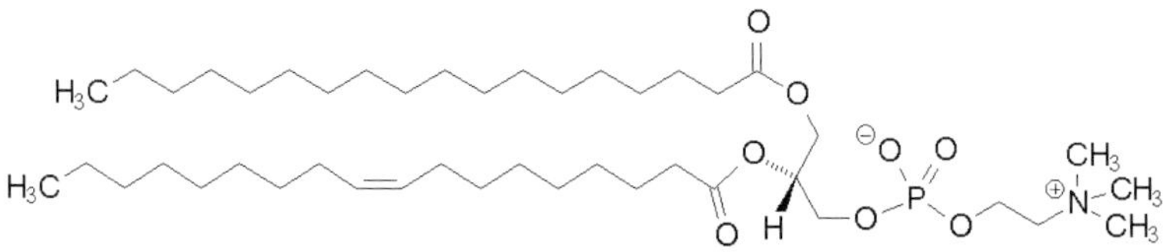

Fig. B.4.2. Structure of a phosphatidylcholine with stearic acid and oleic acid side chain.

General References

D. E. METZLER, *Biochemistry – The Chemical reactions of living cells*, 2nd edn, Academic Press, **2001**.

D. E. VANCE, J. E. VANCE, eds., *Biochemistry of Lipids, Lipoproteins and Membranes*, 4th edn, Elsevier Science, in press.

1.5
β-Amino Acids in Nature

Franz von Nussbaum and Peter Spiteller

1.5.1
Introduction

β-Amino acids occur in nature in form of the free amino acids and as substructures of peptides and alkaloids. They have been encountered in terrestrial and marine species of prokaryotic (Eubacteria, Archaeabacteria) and eukaryotic (Plantae, Fungi, Animalia, and various groups of protists) organisms [1]. Many natural β-amino acids are formed by 2,3-aminomutases [2], enzymes that seem to be present in bacteria [3], plants [4], and fungi [5]. Although some research has been published on the biological *activity* [6] of natural free β-amino acids, little is still known about their "original" biological *function*. The regulation of osmosis seems to be one task within plants [7] and (marine) bacteria [8, 9]. Furthermore, it is not surprising that β-amino acids interact with the natural α-amino acid-processing systems, such as the active (trans-membrane) transport of α-amino acids or protein biosynthesis, especially translation [10].

More β-amino acids are, apparently, *special building blocks* within natural (cyclo)peptides and depsipeptides [11]. It is not clear whether Nature aimed at an increase of proteolytic stability or at conformational modification. Chemists, however, followed this example of natural β-amino acid chemistry decades ago and started to *modify* known α-peptides by substitution of *solitary* α-amino acids with β-amino acids [12]. One major issue of this *synthetic* analoging was to increase the proteolytic stability of peptides [13, 14] with regard to peptidases [15]. A second goal was the *controlled* conformational tuning of cyclo-α-peptides with β-amino acid modifications.

To the best of our knowledge, life has never entered the chemistry of "pure" β-amino acid oligomers (β-peptides). However, recent work [16] by Gellman [17], Seebach [18] and others [19] has tried to catch up by detailed investigation of secondary structures and biological properties of β-peptides. These investigations have triggered a profound current interest in β-amino acids in organic synthetic chemistry [20], bioorganic chemistry [21], medicinal chemistry [22], and natural product chemistry [23]. As Seebach [18] and Gellman [17] showed, hexamers of β-amino acids form discrete secondary structures, for example helices, sheets, and turns

Highlights in Bioorganic Chemistry: Methods and Applications. Edited by Carsten Schmuck, Helma Wennemers
Copyright © 2004 WILEY-VCH Verlag GmbH & Co. KGaA, Weinheim
ISBN: 3-527-30656-0

("foldamers") [16]. Though β-peptides have a high proteolytic stability towards peptidases [24], which could be a major advantage for their use in medicinal chemistry, their biodegradation under the action of a *consortium of microorganisms* was observed; this might be important in their acceptance by members of the public with concern for the environment [25]. Bioorganic investigations have demonstrated that β-peptides can *in principle* interact specifically with the natural world of α-peptides:

1. a cyclic β-peptide that mimics the cyclopeptide octreotide (a somatostatin analog) and binds to human somatostatin receptors, was disclosed as a first "β-peptidic" (α-)peptidomimetic [26]; and
2. oligomers of β-amino acids have antimicrobial activity against Gram-positive and Gram-negative bacteria [27].

1.5.2
β-Amino Acids and their Metabolites in Nature – Taxonomy of the Producer Organisms

This article will focus on β-amino acids [28] as natural products. A survey will be made of their role as precursors or substructures of secondary metabolites, for example alkaloids and peptide related compounds, including some information on the gross taxonomy of the producer organisms [29]. In Tables 1.5.1 and 1.5.2 the organisms are grouped into prokaryotic (AR = Archaea; B = Eubacteria) and eukaryotic (F = Fungi; A = Animalia; P = Plantae; PR = other groups of monocellular organisms formerly summarized in Protista) sources. The biological activity of selected natural products linked to β-amino acids, and synthetic derivatives thereof, will be discussed briefly.

1.5.3
Common β-Amino Acids – Nomenclature

1.5.3.1
β-Alanine

β-Alanine **1** is a widespread β-amino acid found in plants, fungi, animals, and bacteria (Schemes 1.5.1 and 1.5.2). This is not surprising, as β-Ala **1** is a precursor of pantothenic acid, an intermediate in coenzyme A biosynthesis [30], that is essential for *primary metabolism* in all kingdoms. Furthermore β-Ala moieties are found in the widespread non-proteinogenic amino acids carnosine (β-Ala-L-His) and anserine (3-methylcarnosine) [31]. Several peptides, cyclopeptides, and cyclodepsipeptides of *secondary metabolism* contain β-Ala **1** as a non-proteinogenic building block, often combined with hydroxylated amino acids and D-amino acids. These peptides and peptolides are *secondary metabolites* from non-ribosomal protein biosynthesis that is typical for microorganisms. Accordingly, the Ca^{2+} channel

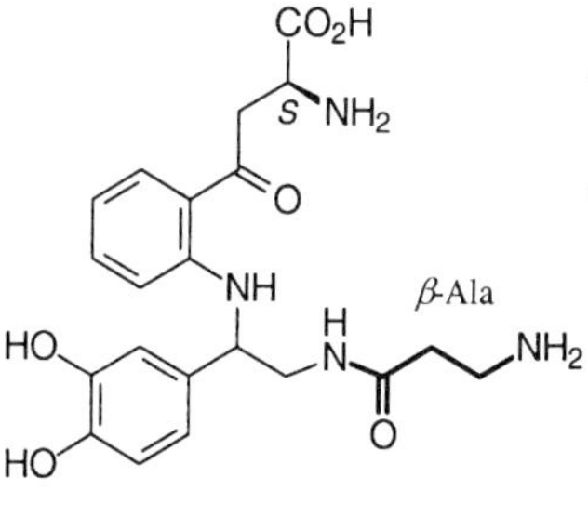

1.5.3 Common β-Amino Acids – Nomenclature

papiliochrome II (**11**)

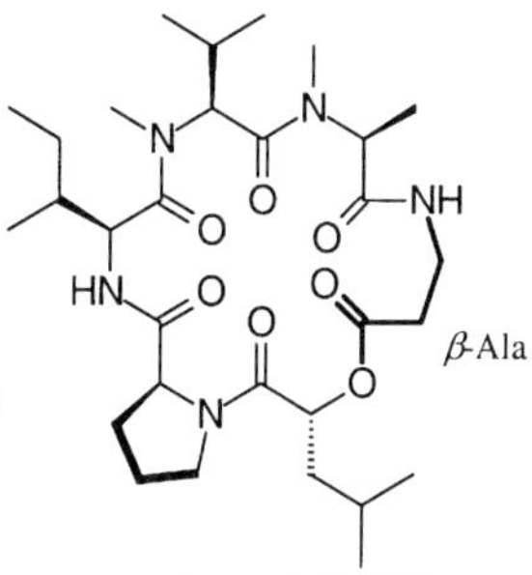

destruxin B (**12**)

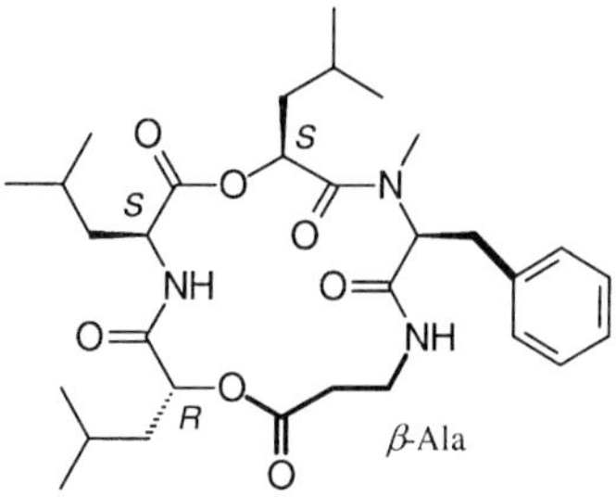

leualacin (**13**)

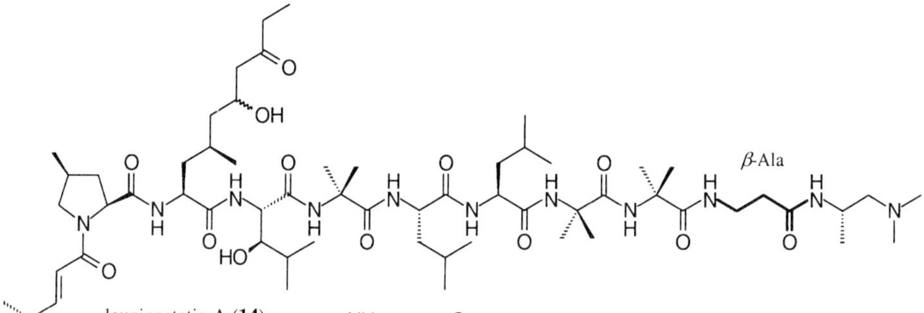

leucinostatin A (**14**)

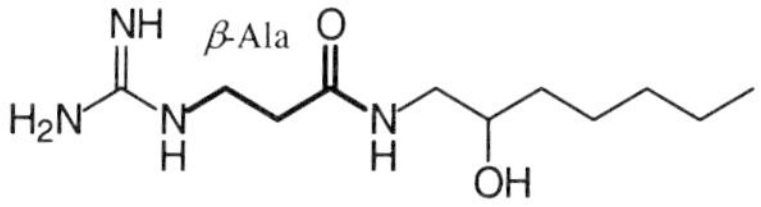

phascoline (**15**)

Scheme 1.5.1. Natural peptides, cyclopeptides, and cyclodepsipeptides related to β-Ala **1**.

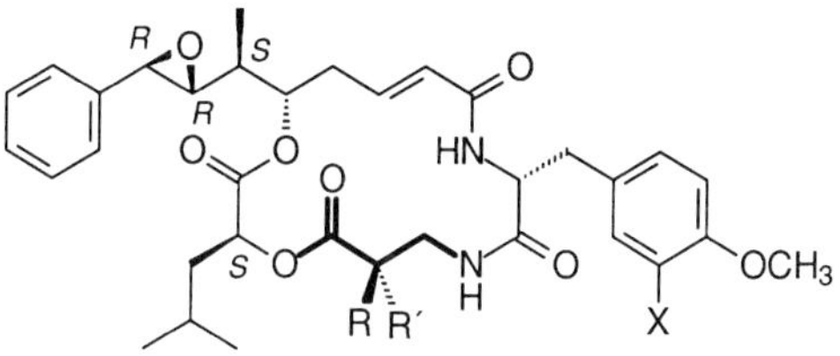

cryptophycin 1 (R = CH$_3$, R´= H, X = Cl) (**16**)
arenastatin (cryptophycin 24, R, R´ = H, X = H) (**17**)
{*synthetic* cryptophycin 52 [R, R´ = (CH$_3$)$_2$, X = Cl] (**18**)}

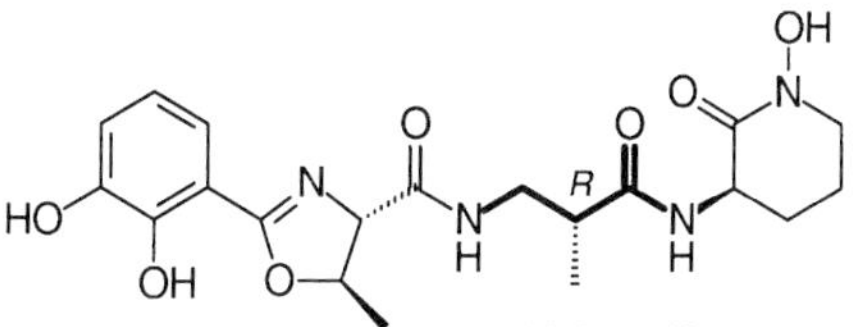

L654040 (**19**)

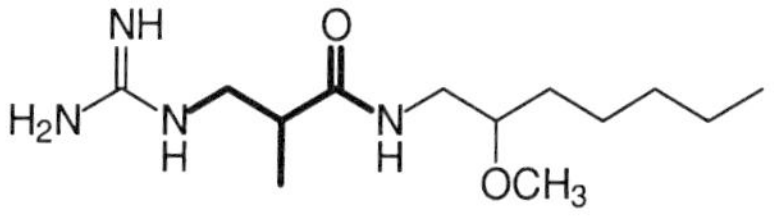

phascolosomine (**20**)

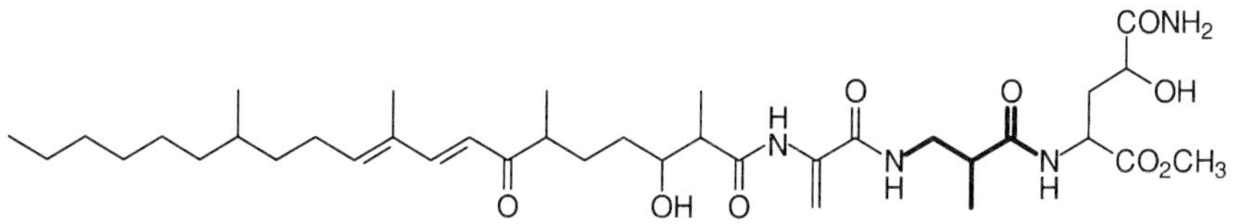

YM-170320 (**21**)

Scheme 1.5.2. Natural products related to β-AiB **2**.

Tab. 1.5.1. Occurrence of natural β-amino acids related to proteinogenic α-amino acids.

β-Amino Acid [28]	Type	Source(s) of the Free Amino Acid[a] and/or Related Natural Products[b,c]
H$_2$N–$\overset{3}{\beta}$–$\overset{2}{\alpha}$–CO$_2$H β-alanine **1** (β-Ala)	–	B: Cyanobacteria (yanucamides A, B) [37]; marine bacteria[a], marine sediments[a] [38]. PR: Dinoflagellata (dehydro-β-Ala, palytoxin) [36]. F[a]: Ascomycota *Alternaria brassicae*, *Metarhizium anisopliae* (destruxin B) [32]; *Hapsidospora irregularis* (leualacin) [39]; *Paecilomyces marquandii* (leucinostatins A–K) [40]. P[a]: Plumbaginaceae (*N,N,N*-trimethyl-β-Ala) [7]. A: Marine animals[a,b], Mollusca *Crassostrea virginica* [41]; Porifera *Dysidea arenaria*[c] (arenastatin a, cryptophycin 24) [42]; *Theonella swinhoei*[c] (barangamides A–D [43], theonellapeptolides [44], theonellamides C–F [75]); Hexapoda *Papilio demolus* (papiliochromes) [45]; Sipunculida *Phascolion strombi* (phascoline) [46]; Mammalia[a,b] [1].
H$_2$N–$\overset{|}{R}$–CO$_2$H (R)-β-aminoisobutyric acid[e] [(R)-**2**, D-β-AiB]; [(S)-**2**, L-β-AiB]	β^2	B: Cyanobacteria *Nostoc* sp. (D-β-AiB: cryptophycins 1–6, A–D) [47]; Actinobacteria *Streptomyces* sp. (D-β-AiB: L654040) [48]. PR: Chlorophyceae[a] [49]; Laminariales (DGTA) [50]; Prasinophyceae[a] [50]. F: Basidiomycota *Agaricus campestris*[a] (L-β-AiB) [51]; Ascomycota *Candida tropicalis* (YM-170320) [52]. P: Bryopsida[a] [53] {DGTA [50]}; Brassicaceae *Lunaria annua*[a], Iridaceae *Iris tingitana*[a] (D-β-AiB). A: Sipuncula *Phascolion strombi* (phascolomine) [46]; Mammalia[a] (D,L-β-AiB).
NH$_2$ $\overset{}{R}$–CO$_2$H (R)-β-leucine **3** (L-β-Leu)	β^3	B: Firmicutes *Clostridium sporogenes*[a] [54]. P[a,f] [54]. A: Mammalia[a] {human serum [54], rat liver}.
NH$_2$ H$_2$N–$\overset{}{S}$–CO$_2$H (S)-β-lysine **4** (L-β-Lys)	β^3	AR: *Methanogenium cariaci*, *Methanococcus thermolithotrophicus* (N^ε-acetyl-β-lysine) [8]. B: Firmicutes *Clostridium* sp.[a] [2]; *Bacillus subtilis*[a] [2]; Proteobacteria *Pseudomonas fluorescens* (sperabillin C) [55]; Actinobacteria (albothricin) [56]; *Streptomyces* ssp. {lavendothricin [57], negamycin [58], streptothricin [59]}; *Streptomyces nashvillensis* [(R)-bellenamine] [60]; *Streptomyces capreolus* (capreomycins) [95]; *Streptomyces puniceus* {tuberactinomycins [61], viomycin [62]}; *Streptomyces stramineus* (LL-BO1208β) [63]; *Nocardia* sp. {LL-BM782α1–2 [64], myomycin B [65]}; *Micromonospora pilospora* (lysinomycin) [66]; Bacteroidetes *Flexibacter* spp. (TAN-1057 A–D) [67].
NH NH$_2$ H$_2$N–$\overset{}{N}$–$\overset{}{S}$–CO$_2$H H (S)-β-arginine **5** (L-β-Arg)	β^3	B: Actinobacteria *Streptomyces griseochromogenes* (blasticidins H, S) [68]; *Nocardia* sp. (LL-BM 547β) [69].

1.5.3 Common β-Amino Acids – Nomenclature

Tab. 1.5.1 (continued)

β-Amino Acid [28]	Type	Source(s) of the Free Amino Acid[a] and/or Related Natural Products[b,c]
β-glutamate **6** (β-Glu) HO$_2$C–CH(NH$_2$)–CH$_2$–CO$_2$H	β^3	AR: Methanogenium cariaci[a], Methanococcus thermolithotrophicus[a] [8, 9]; B: Proteobacteria Pseudoalteromonas luteoviolacea[a] [70]; Cyanobacteria Prochloron didemni (N-methyl-β-Glu) [96]; marine sediments[a,b] (DFAA) [71]. PR: Rhodophyta Chondria armata[a] [72].
(S)-isoserine **7** [L-α-OH-β-Ala, L-iso-Ser] H$_2$N–CH$_2$–CH(OH)–CO$_2$H	"β^2"	B: Firmicutes Bacillus brevis [edeines A, B, D, F] [89]. A: Porifera Theonella sp.[c] {D-iso-Ser: keramamides F–H, J [73]; theonegramide [74]; theonellamides A, B [75]; theopalauamide [76]}.
(R)-β-phenylalanine **8** (L-β-Phe); (S)-β-phenylalanine **8** (D-β-Phe)	β^3	B: Firmicutes Bacillus brevis (D-β-Phe: edeine D–F) [89]; Proteobacteria Enterobacter sp. (D-β-Phe: andrimid) [77]; Pseudomonas fluorescens (D-β-Phe: moiramides a–c) [78]. F: Ascomycota Penicillium islandicum (cyclochlorotine) [79]. P: Azollaceae Azolla caroliniana (γ-L-Glu-D-β-Phe) [80]; Celastraceae Peripterygia marginata [(S)-periphylline] [100]; Fabaceae Phaseolus angularis (γ-L-Glu-L-β-Phe) [81]; Scrophulariaceae Verbascum sp. [(S)-verbascenine]; Asteraceae Aster tataricus (astins A–I) [82]; Gymnospermae Taxaceae {β-Phe: nicotaxines [83]; N-methyl-β-Phe: nicotaxines; N,N-dimethyl-β-Phe (Winterstein's acid) [84]: nicaustrine [83], taxine B [85]; (2R,3S)-3-phenylisoserine: paclitaxel [86]; N,N-dimethyl-3-phenylisoserine: taxine A [87]}; Taxus brevifolia[a] (cell free extracts) [88].
(R)-β-tyrosine **9** (L-β-Tyr); (S)-β-tyrosine **9** (D-β-Tyr)	β^3	B: Firmicutes Bacillus brevis [(S)-β-Tyr: edeines A, B] [89]; Proteobacteria, Myxobacteria Chondromyces crocatus (chondramides A–D) [90]. F: Basidiomycota Cortinarius violaceus[a] [5]. P: Acanthaceae Aphelandra spp. (aphelandrine, orantine [102]); Veronicaceae Chaenorhinum sp. [(S)-β-Tyr: chaenorhine, chaenorpine] [103]. A: Porifera Geodia sp. (geodiamo-lides H–I) [107]; Jaspis sp. (jaspamides) [106].
(R)-β-Dopa **10** (L-β-Dopa)[d]	β^3	B: Actinobacteria Streptomyces globisporus [(S)-β-Dopa: C-1027] [108]. F: Basidiomycota Cortinarius violaceus[a,b] (cortiferrins) [5, 111]. P: Veronicaceae Chaenorhinum sp. (chaenorhine) [103].

AR = Archae, B = Eubacteria; F = Fungi, A = Animalia, P = Plantae;
PR = other monocellular organisms (formerly Protista)
[a] Within this source the *free* β-amino acid has been found
[b] Within this source the β-amino acid has been found as a *substructure* of a natural product (e.g. alkaloid, peptide, depsipeptide); only indicated if the related natural product is not mentioned
[c] The source organism mentioned *might* not be the real producer
[d] Although L-Dopa is not a proteinogenic α-amino acids, L-β-Dopa is discussed here, because of its close structural and biosynthetic relationship with L-β-Tyr and L-β-Phe
[e] α-AiB is not a proteinogenic amino acid

1.5 β-Amino Acids in Nature

Tab. 1.5.2. Natural β-amino acids *related* to marine and freshwater cyanobacteria.

β-Amino acid [28]	Type	Source(s)[a] of related natural products
Amba **67**	$\beta^{2,3}$	A: Gastropoda *Dolabella auricularia*[a] (dolastatin D) [118].
Map **68**	$\beta^{2,3}$	B: Cyanobacteria *Lyngbya majuscula* (majusculamide C) [119]; *Lyngbya majuscula/Schizothrix calcicola* (dolastatin 12, 15-*epi*-dolastatin 12, lyngbystatin 1, 15-*epi*-lyngbystatin 1) [134]. A: Gastropoda *Dolabella auricularia*[a] (dolastatins 11–12) [120].
Amha **69**	$\beta^{2,3}$	B: Cyanobacteria *Symploca laete-viridis* (malevamide B) [121]; *Lyngbya* sp. [(2R,3R)-Amha: ulongamides A–C; (2S,3R)-Amha: ulongamides D–F] [122]. A: Gastropoda *Philinopsis speciosa*[a] [(2R,3R)-Amha: kulokekahilide-1] [123].
Admpa **70**	$\beta^{2,3}$	B: Cyanobacteria *Lyngbya majuscula* (dolastatin 16) [124]. A: Gastropoda *Dolabella auricularia*[a] (dolastatin 16) [125].
L-Apa **71**	β^{3}	B: Cyanobacteria *Lyngbya confervoides* (obyanamide) [126].
Aoya (Doy) **72**	β^{3}	A: Gastropoda *Dolabella auricularia*[a] (dolastatin 17) [127].
Amoa (Amoya) **73**	$\beta^{2,3}$	B: Cyanobacteria *Symploca laete-viridis* (malevamide C) [121]. A: Gastropoda *Onchidium* sp.[a] (onchidin) [128].
Adda **74** [114]	$\beta^{2,3}$	B: Cyanobacteria (microcystins) [129]; *Nodularia* sp. {nodularin [129], nodularin-Har [130]}. A: Porifera *Theonella swinhoei*[a] (motuporin) [131].
Ahoa **75**	"$\beta^{2,3}$"	B: Cyanobacteria *Nostoc* sp. (nostophycin).

1.5.3 Common β-Amino Acids – Nomenclature | 69

Tab. 1.5.2. (continued)

β-Amino acid [28]	Type	Source(s)[a] of related natural products
Ahda **76**	"$\beta^{2,3}$"	B: Cyanobacteria *Scytonema* sp. (scytonemin A) [132].
Ahmp (Apoa) **77**	β^3	A: Porifera *Theonella swinhoei*[a] {theonegramide [74], theopalauamide [76]}; *Theonella* sp.[a] (theonellamides A–B) [75].
Aboa **78**	β^3	A: Porifera *Theonella* sp.[a] (theonellamides C–F) [75].

B = Eubacteria; A = Animalia
[a] The source organism mentioned does not seem to be the real producer.

blocker leualacin **13**, which has been isolated from the ascomycete *Hapsidospora irregularis*, contains D- and L-leucic acid and a β-Ala moiety [39].

Destruxin B **12** [32] is a cyclodepsipeptide isolated from the phytopathogenic fungus *Alternaria brassicae*. This phytotoxin elicits the biosynthesis of secondary defense substances (phytoalexins and phytoanticipins [33]) within a variety of plants, e.g. the production of the isothiazoloindole sinalexin by white mustard (*Sinapsis alba*) [34]. Such complex interactions of plants and fungi are of great current interest. They range from symbiosis (mycorrhiza) to mutualism or parasitism and often seem to be a result of a parallel evolution within closely related biotops. Secondary metabolites are believed to act as signal compounds in these environments, even though their actual biological function is still widely unknown [35]. Research efforts that aim at an understanding *on a molecular level* involve many areas between chemistry and biology, such as chemical ecology, molecular biology, or genetics.

Dehydro-β-Ala has been found to be a substructure of the extremely poisonous palytoxin (LD_{50} 10 ng kg^{-1}, i.v. in mice) [36]. This cytolytic compound belongs to a group of natural marine toxins that are responsible for the malicious fish poisoning.

1.5.3.2
Seebach's Nomenclature for β-Amino Acids

Shifting from β-Ala **1** to substituted derivatives, we will use a nomenclature introduced by Seebach [91]. β-Alanines that are substituted at C2 will be termed β^2-

amino acids (2,2-disubstituted acids $\beta^{2,2}$), whereas the β congeners of natural α-amino acids, that have the side-chain at C3, will be called β^3-amino acids (Table 1.5.1). Thus the β-congener of L-Leu [(S)-2-amino-4-methylpentanoic acid] will be termed L-β-Leu [(R)-3-amino-4-methylpentanoic acid, (R)-**3**. An example of a natural β^2-amino acid is (R)-β-aminoisobutyrate [(R)-β-AiB, (R)-**2**] [1].

1.5.3.3
(R)- and (S)-β-Aminoisobutyric Acid [(R)-β-AiB and (S)-β-AiB]

(R)- and (S)-β-AiB **2** are considered to be common intermediates of primary metabolism in all kingdoms (see biosynthesis section). (R)-β-AiB [(R)-**2**] has been detected as the free β-amino acid in animals, bacteria, and plants whereas (S)-β-AiB [(S)-**2**] has so far been found in bacteria and fungi (Scheme 1.5.2).

The cryptophycins (e.g. **16**) form a group of depsipeptide antibiotics with a (R)-β-AiB [(R)-**2**] or β-Ala **1** substructure. They have already been isolated from cyanobacteria (*Nostoc* sp., see below) [47]; arenastatin A **17** has been isolated from a marine sponge [42]. The cryptophycins are new lead structures within cancer drug research for their potent tumor-selective cytotoxicity [92]. The *synthetic analog* cryptophycin-52 ($\equiv$ 6-methylcryptophycin-1, LY-355703, **18**) has reached clinical phase II trials [93].

1.5.4
β-Amino Acids *Related* to Proteinogenic α-Amino Acids

Naturally β-amino acids that are not intermediates of the *primary* metabolism, e.g. β-Ala **1**, (R)-, or (S)-β-AiB **2**, which have been discussed previously, are less widespread in nature. Whereas a variety of natural β-amino acids has been found as substructures of *secondary* metabolites, for example alkaloids, peptides, or depsipeptides, their free counterparts, the free β-amino acids, seem to be less common (see concluding remarks).

1.5.4.1
Aliphatic β-Amino Acids – β-Lysine, β-Leucine, β-Arginine, and β-Glutamate

Several bacteria (e.g. *Clostridium* sp.) catabolize L-Lys yielding acetate, butyrate, and ammonia; the first step of this process is the isomerization of L-α-Lys to L-β-Lys [(S)-**4**] [94]. (S)-β-Lys **4** is frequently found in microorganisms, e.g. as a solubilizing side-chain of aminoglycosides **31**–**37**, e.g. in myomycin B **44** [65] (Scheme 1.5.4), or within the periphery of cyclopeptides (Scheme 1.5.3) encountered in the capreomycins **22** and the tuberactinomycins **23**, **24** [95].

Researchers from Takeda have isolated the antibiotics TAN-1057 A–D **29**, **30** from *Flexibacter* sp. [67]. These (S)-β-Lys-derived dipeptides have promising activity against clinically problematic methicillin-resistant strains of *Staphylococcus aureus* (MRSA). The β-amino acid moiety is essential for the biological activity of TAN-1057 A **29**. The bisguanidine **29** seems to have a dual mode of action, involving

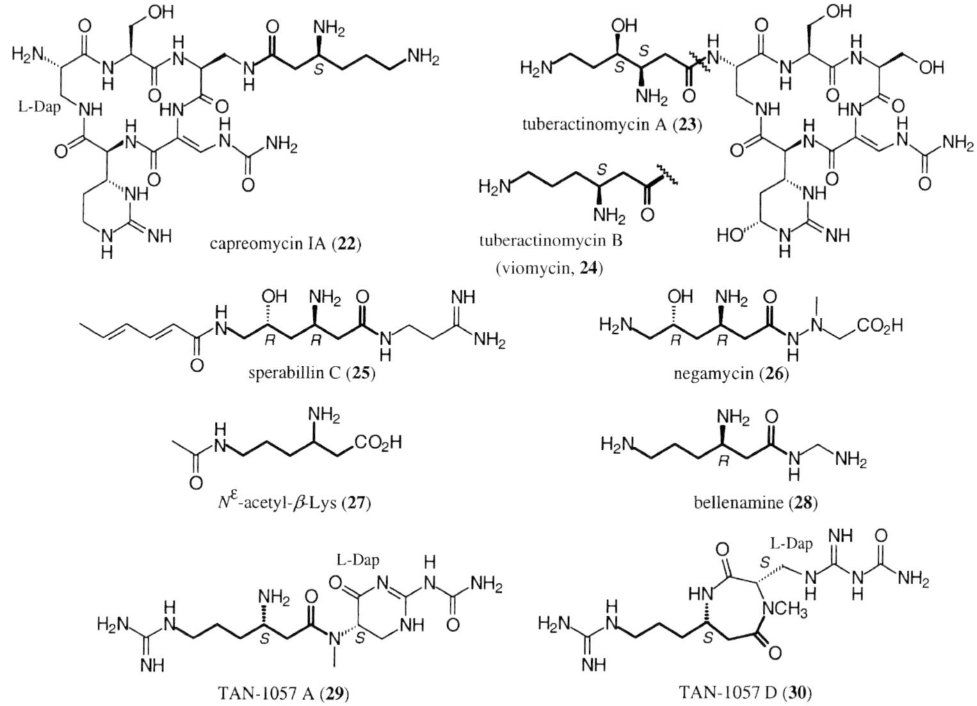

1.5.4 β-Amino Acids Related to Proteinogenic α-Amino Acids

Scheme 1.5.3. Peptidic natural products related to β-Lys **4**.

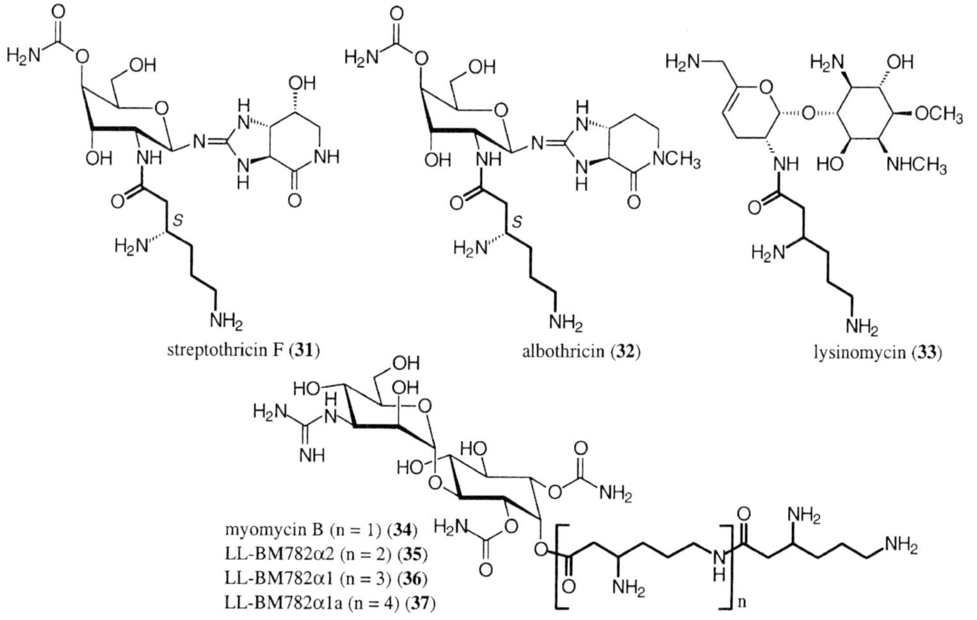

Scheme 1.5.4. Glycopeptide antibiotics related to β-Lys **4**.

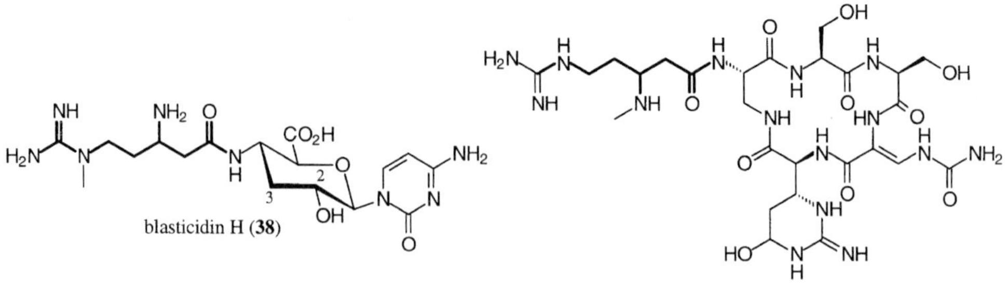

Scheme 1.5.5. Natural products related to β-Arg **5**.

interference with translation and an inhibitory effect on 50S ribosomal subunit formation [67].

Also (*R*)-β-Lys **4** is found to be a substructure of several antibiotics. Examples are the hydrazinodipeptide negamycin **26** from *Streptomyces purpeofuscus*, bellenamine **28** from *Streptomyces nashvillensis* (Actinobacteria), and sperabillins A–D **25** from *Pseudomonas fluorescens* (Proteobacteria).

Of all the β-amino acids, β-Lys **4** seems to be the one most frequently found in bacterial antibiotics. In fact, synthetic analogues of tuberactinomycin and capreomycin have been investigated as potential lead structures for new antibacterial drugs at Pfizer [95].

L-β-Arg **5** is a biosynthetic precursor of the antifungal antibiotic blasticidin H **38** that has been obtained, by fermentation, from the bacterium *Streptomyces griseochromogenes* [68] (Scheme 1.5.5).

β-Glu **6**, a major constituent of dissolved free amino acids (DFAA) in marine sediments [71], has been detected in red algae (Rhodophyta) [72] and as a potential osmolyte in several marine bacteria, for example obligatory aerobic heterotrophs [70] and methanogenic archaebacteria [8]. *N*-methyl-β-glutamate has been reported solely as a natural product from the cyanobacterium *Prochloron didemni* [96].

1.5.4.2
Aromatic β-Amino Acids – β-Phenylalanine, β-Tyrosine, and β-3,4-Dihydroxyphenylalanine

Several valued terpene alkaloids [86] from Taxaceae bear β-Phe **8** or α-hydroxy-β-Phe (phenylisoserine) side-chains at C13 and C5. Often these aromatic β-amino acids are *N*-methylated, as exemplified by α-hydroxylated Winterstein's acid, found to be a side-chain of taxine A **42** (Scheme 1.5.7). The *N*-benzoylphenylisoserine ester moiety of paclitaxel (taxol) has been found to be essential for its crucial microtubuli-stabilizing activity. Accordingly *synthetic* taxol derivatives [97] that have already reached the market (e.g. *non-natural* docetaxel (**40**, taxotere)) or are still undergoing clinical/preclinical trials, e.g. *non-natural* BAY 59-8862/IDN5109 **41**

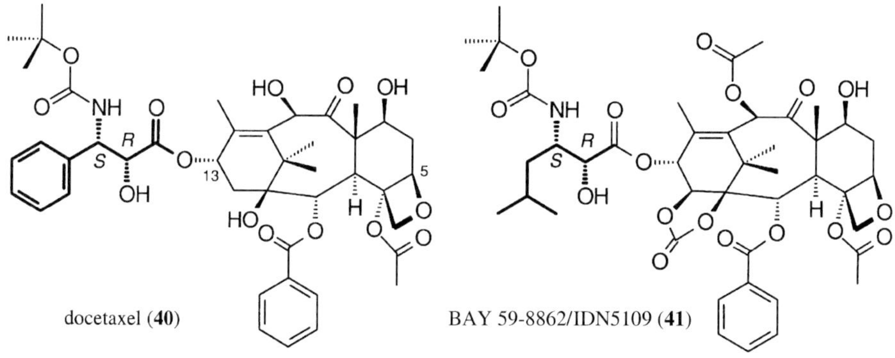

Scheme 1.5.6. *Synthetic* taxoides.

[98, 99], still bear an α-hydroxy-β-amino acid side-chain at C13 (Scheme 1.5.6). Clearly, chemistry programs aiming at modification of the taxol phenylisoserine side-chain have significantly stimulated β-amino acid chemistry in general [20]. Up to now free β-Phe **8** has not been isolated from natural sources [88].

The (S)-β-Phe or (S)-β-Tyr substructures of several spermidine alkaloids *from plants*, for example (S)-periphylline **43** [100, 101] or the spermine derivatives (S)-verbascenine **51** [101], aphelandrine **52** [102], chaenorhine **56** [103] and chaenorpine [104], hint at Michael addition of the biogenic amine to cinnamic acid **48** (or some oxidized equivalent, e.g. **49**) within the β-amino amide-forming step

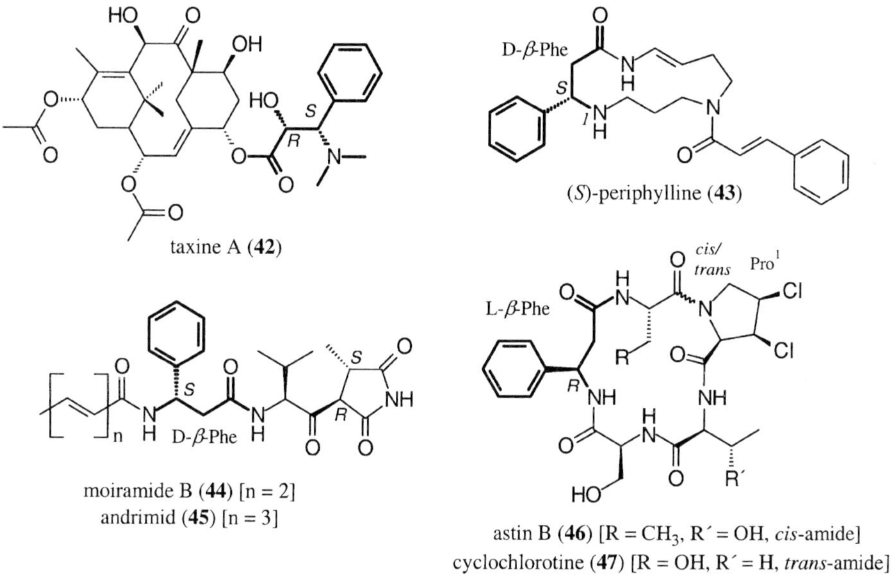

Scheme 1.5.7. Natural products related to β-Phe **8**.

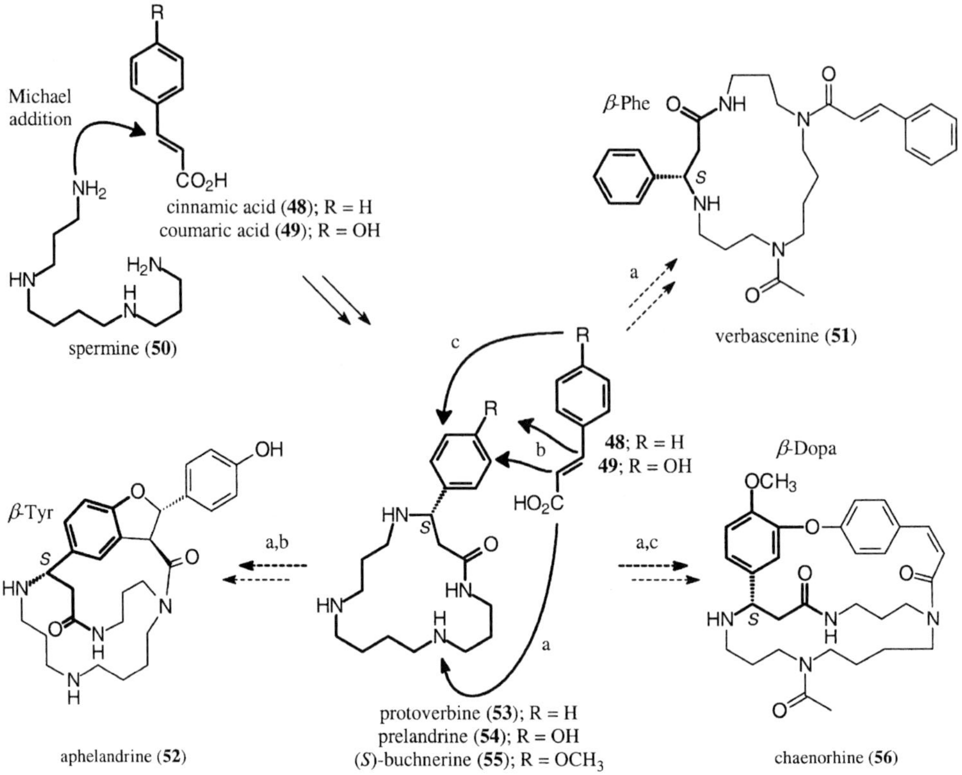

Scheme 1.5.8. Spermine alkaloids and their proposed [105] biosynthesis via Michael addition.

(Scheme 1.5.8). In fact, as recent investigations by Hesse [105] have shown, protoverbine **53** and its hydroxy analog prelandrine **54** seem to be exemplary key intermediates within a not yet fully characterized ensemble of closely related biogenetic pathways that are governed by alternating Michael additions and P-450 oxidation steps. The alkaloids **50**–**56** demonstrate Nature's impressive ability to set up a multitude of highly diverse core structures by selective C–C, C–N and C–O bond-forming steps, only using simple starting materials that are not at all diverse.

β-Tyr substructures are, moreover, found within several peptidic antibiotics, for example the edeines **57** from *Bacillus brevis*, or within cyclodepsipeptides that have been isolated from marine sponges, for example jaspamides A–C **59** [106] or geodiamolides H–I **58** [107] (Scheme 1.5.9). The chondramides **60** are closely related antibiotics that have been obtained from the myxobacterium *Chondromyces crocatus*. Within the edeine biosynthesis D-β-Tyr **9** seems to be a real precursor [89]. The biosynthesis of the jaspamides, the geodiamolides, or the chondramides has not yet been investigated.

Although L-Dopa is not a proteinogenic α-amino acid, L-β-Dopa **10** is discussed here, because of its close structural and biosynthetic relationship with L-β-Tyr **9** and L-β-Phe **8**.

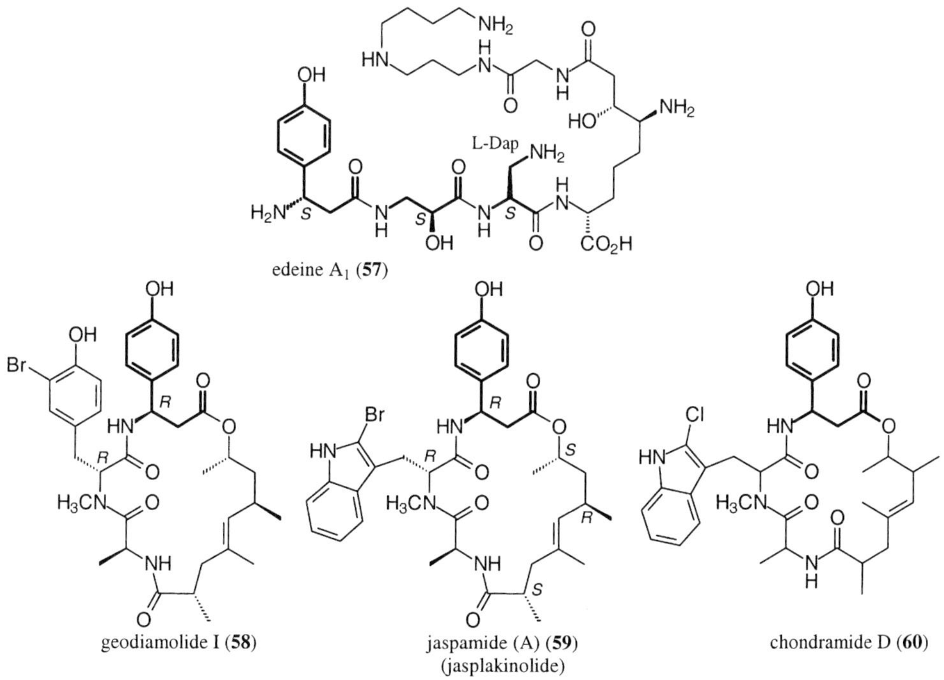

1.5.4 β-Amino Acids Related to Proteinogenic α-Amino Acids | **75**

edeine A₁ (**57**)

geodiamolide I (**58**)

jaspamide (A) (**59**)
(jasplakinolide)

chondramide D (**60**)

Scheme 1.5.9. Natural peptides and cyclodepsipeptides related to β-Tyr **9**.

The chromoproteine antibiotic C-1027 [108] from *Streptomyces globisporus* has antitumor activity, unfortunately correlated with strong cytotoxicity that hampers the use of this lead structure class as a drug; which is a general problem in cancer drug development (Scheme 1.5.10). The labile chromophor **61** (C-1027-Chr) con-

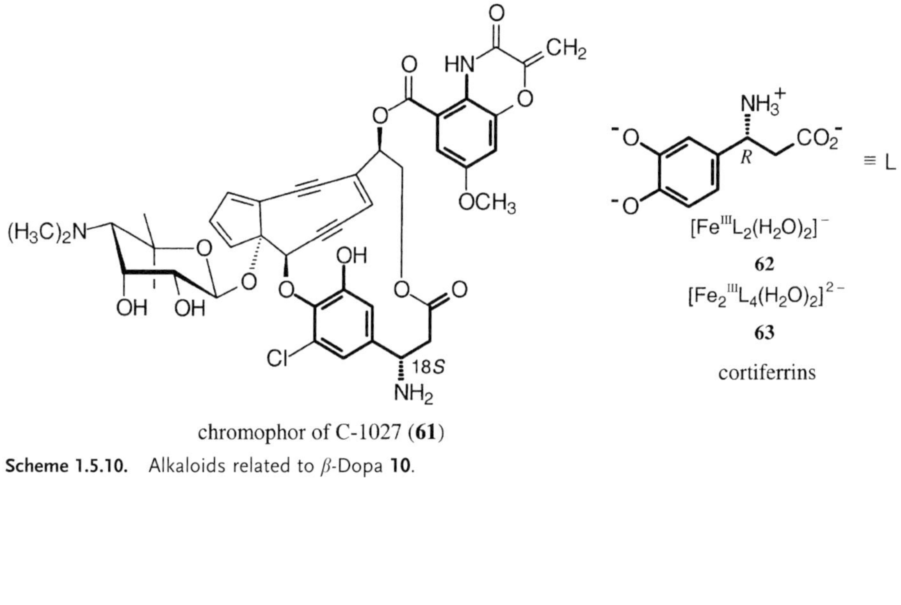

chromophor of C-1027 (**61**)

$[Fe^{III}L_2(H_2O)_2]^-$
62
$[Fe_2^{III}L_4(H_2O)_2]^{2-}$
63

cortiferrins

Scheme 1.5.10. Alkaloids related to β-Dopa **10**.

sists of a highly strained nine-membered enedyine ring, the DNA-cleaving toxophor [109], that is clamped by an (*S*)-3-chloro-β-Dopa moiety. Recent investigations of the C-1027 biosynthesis gene cluster [110] show that (*S*)-3-chloro-β-Dopa really is the biosynthetic precursor of C-1027-Chr **61**. The β-amino acid seems to be activated for attachment (with the enedyine alcohol) as a thio ester of a peptidyl-carrier-protein (3-chloro-β-Dopa-*S*-PCP).

Steglich [111] recently isolated significant amounts of *free* (*R*)-β-Dopa **10** from *Cortinarius violaceus*, a mushroom globally found in the temperate climate zones. Analogous to classical iron ink, the blue violet color of this fungus is created by iron(III) complexes of (*R*)-β-Dopa **62** and **63**, that seem to form a pH-dependent equilibrium. The biosynthesis of (*R*)-β-Dopa **9** will be discussed in chapter 1.6.3 [5].

1.5.5
Miscellaneous β-Amino Acids

1.5.5.1
β-Amino-L-alanine (L-Dap) [112]

β-Lactam antibiotics [113] are still the most prominent class of therapeutic antibacterial agents. They have been isolated from fungi (penicillins, cephalosporins) and from bacteria (cephalosporins) (Scheme 1.5.11). As penicillins **65** and cephalosporins **66** contain an L-Dap moiety **64**, they are clearly the most important group of drugs with a β-amino acid substructure. Though there have been *chemically demanding* research programs in abundance in the context of β-lactam antibiotics, these efforts, that began in the 50s and are still in progress, never seemed to trigger a *general chemical and biological investigation* of β-amino acids, apart from their use as valuable starting materials for β-lactam synthesis. β-Lactam antibiotics will not be discussed here any further [113].

1.5.5.2
β-Amino Acids *Related* to Cyanobacteria – Aboa, Adda, Admpa, Ahda, Ahmp, Ahoa, Amba, Amha, Amoa, Aoya, L-Apa, and Map [28]

Cyanobacteria are a rich source of β-amino acids (Table 1.5.2). A multitude of peptides and depsipeptides, with substructures ranging from "simple" aliphatic

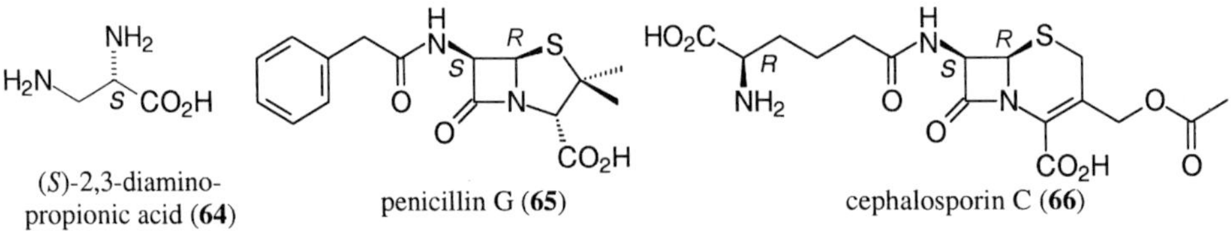

(*S*)-2,3-diamino-propionic acid (**64**) penicillin G (**65**) cephalosporin C (**66**)

Scheme 1.5.11. L-Dap **64** and related β-lactam antibiotics.

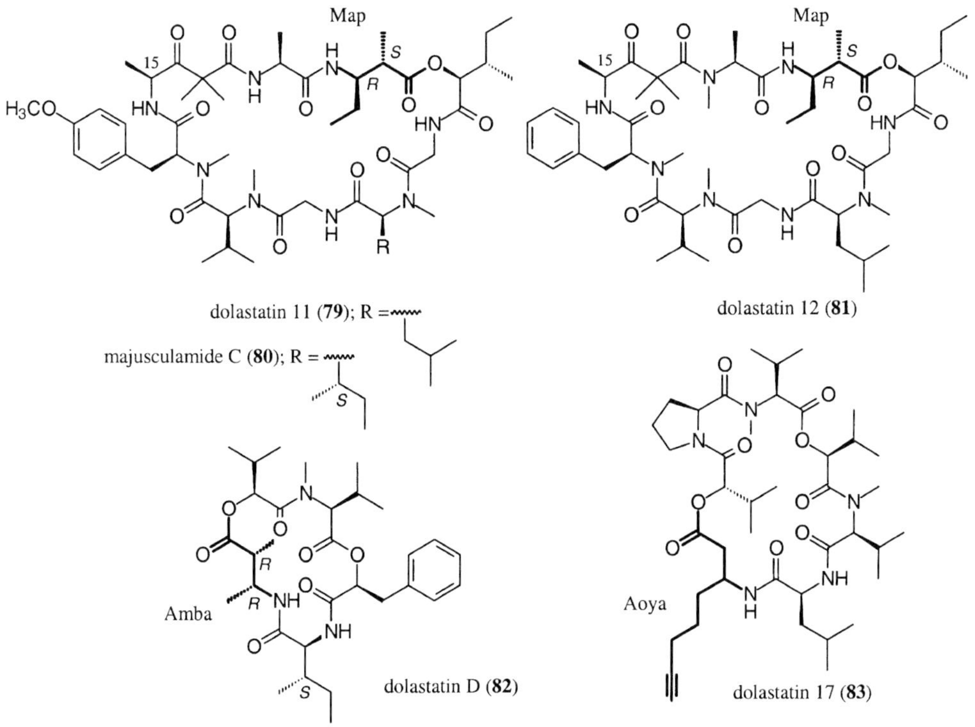

1.5.5 Miscellaneous β-Amino Acids

Scheme 1.5.12. Cyanobacterial cyclodepsipeptides with aliphatic β-amino acid substructures.

β-amino acids, for example Map **68**, to "complicated" aromatic compounds, such as Adda **74** [114], has been isolated from freshwater and marine cyanobacteria (Schemes 1.5.12 and 1.5.13) [115]. α-Methylation and α, γ or δ-hydroxylations are typical features of these cyanobacterial β-amino acids.

Often marine peptides or depsipeptides with β-amino acid substructures were originally derived from invertebrates, for example sponges, tunicates, or mollusks [116]. Thus, initially it was not clear whether the isolated organism, symbionts thereof, or prey species were the real producers of the isolated secondary metabolites [117]. Because of the complex networks within marine ecosystems the biosynthesis of natural products does not have to take place in a single organism. These interspecies networks might lead to multistep *interspecies metabolism* of bioactive natural products, as they move through the "marine food web" [133]. It must generally be stated that numerous unique secondary metabolites (or their biogenetic precursors) isolated from marine invertebrates may well be produced by microbial (fungal, bacterial) symbionts. For many β-amino acid-containing compounds, however, cyanobacterial biosynthesis has already been proven by microbiological investigations. A second indicator of such non-host biosynthesis is structural similarity of the natural product in question to further metabolites originating without doubt from cyanobacteria.

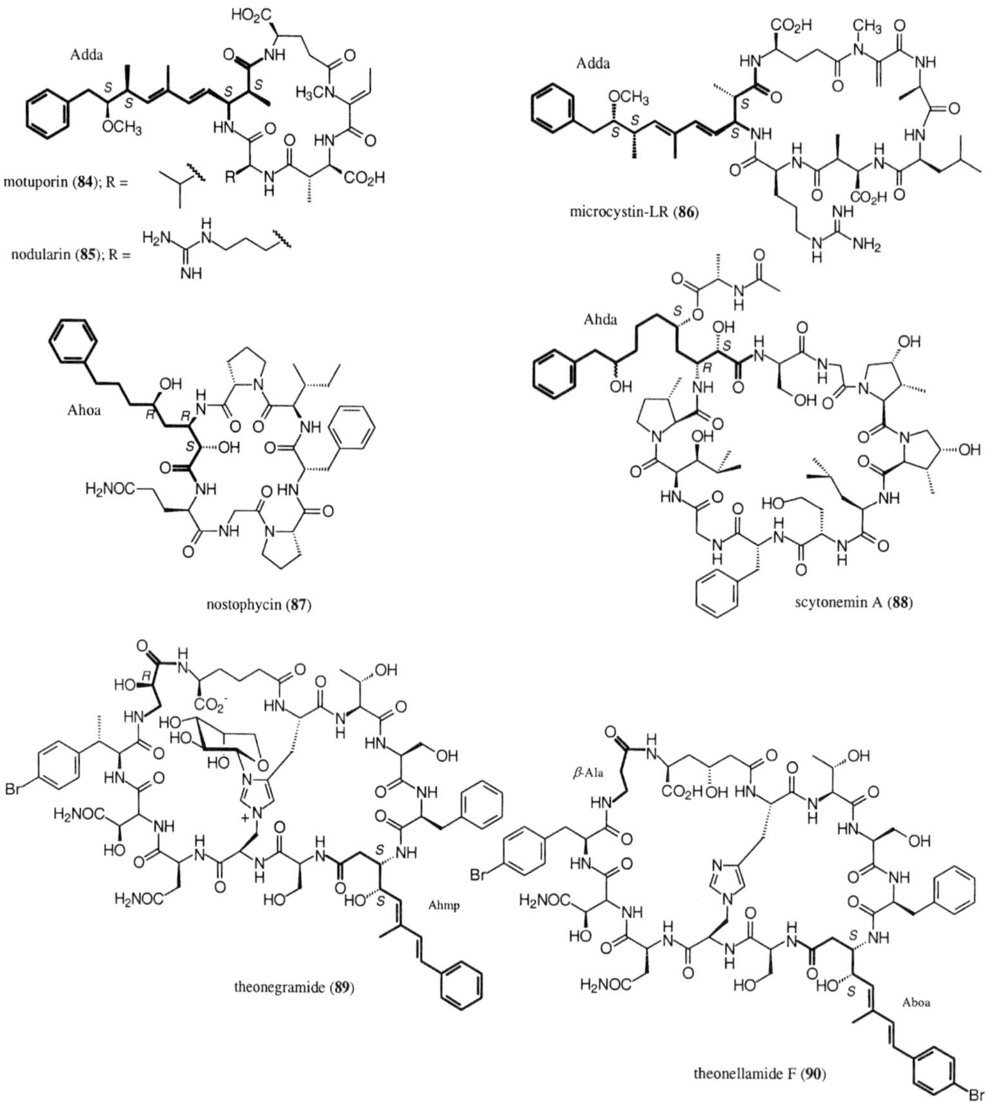

Scheme 1.5.13. Cyclopeptides *related* to cyanobacteria (Adda, Ahoa, Ahda, Ahmp and Aboa).

Dolastatins 11, 12 **79**, **81**, first isolated from the sea hare *Dolabella auricularia* and now clearly considered to be of dietary cyanobacterial origin, can be mentioned as examples [133]. The structurally closely related cyclodepsipeptide majusculamide C **80** has been isolated from defined cyanobacterial sources [119]. This was one and a half decades before dolastatin 12 **81** was furthermore detected in extracts from a marine cyanobacteria assemblage (*Lyngbya majuscula*/ *Schizothrix calcicola*)

[134]. Dolastatins 11, 12 **79**, **81** and majusculamide C **80** contain the aliphatic β-amino acid Map **68**.

Ahoa **75** and Ahda **76** are related β-amino-α,δ-dihydroxy acids that have been found within cyanobacteria as substructures of the cyclopeptides nostophycin **87** and scytonemin A **88**, respectively [11].

The aromatic β-amino acid Adda **74** [114] is a substructure of several cyclopeptides from cyanobacteria, i.e. the microcystins **86** and nodularin **85**. More than 60 microcystin and nodularin homologs have already been isolated from the genera *Microcystis*, *Anabaena*, and *Oscillatoria* (microcystins) as well as *Nodularia* (nodularins). A multitude of literature is available on the biosynthesis and biological activity of these hepatotoxins, which can constitute a serious health hazard in fresh water [135].

Further aliphatic α-methyl-β-amino acids, for example Amha **69**, Admpa **70**, and Amoa **73**, have been found in peptidic natural products from marine animals and from cyanobacteria. Although dolastatin D **82** and dolastatin 17 **83** have so far been obtained from sea hare isolates only, these peptides should now be considered to be of cyanobacterial origin, as indicated by their β-amino acid substructures, i.e. Amba **67** and Aoya **72**.

Adda **74** has also been encountered in an utterly different source as a substructure of the highly potent protein phosphatase 1 (PP1)-inhibitor motuporin **84**. Cyclopeptide **84** was obtained from the marine sponge *Theonella swinhoei*, a species of the sponge order Lithistida – "star performers or hosts to the stars" in natural product synthesis, as Faulkner [136] put it. Further remarkable bicyclic structures of this group are theonegramide **89** and theonellamide F **90**, which contain Ahmp **77** or its bromo derivative Aboa **78**. Recent microbiological investigations substantiate a correlation between symbiotic filamentous bacteria (not clearly cyanobacteria) within *Theonella* sp. and the occurrence of cyclopeptides, especially of those that contain β-amino acids [136].

1.5.5.3
Cispentacin as a Chemical Lead Structure – Interaction of β-Amino Acids with Natural α-Amino Acid-processing Systems

For cispentacin **91**, a metabolite of *Bacillus cereus*, antifungal activity has been described for the *free β-amino acid*; this is remarkable for such a simple molecule [137]. The anti-*Candida* activity of **91** [138] initiated a chemistry program aimed at finding derivatives with superior (oral) efficacy for the treatment of yeast infections (Scheme 1.5.14) [10]. This lead optimization yielded the synthetic methylene derivative BAY 10-8888/PLD-118 **92** that acts by a dual mode of action. First, **92** is accumulated in yeast cells by active transport via permeases, specific for branched-chain α-amino acids (e.g. L-Ile). Second, the isoleucyl-tRNA is specifically inhibited, leading to inhibition of protein synthesis and cell growth [10]. In contrast, active transport and inhibition of protein synthesis of natural cispentacin **91** seems to be correlated with the corresponding enzymes specific for L-Pro (and not L-Ile) [138].

This example illustrates how *β-amino acids might generally interfere with natural α-*

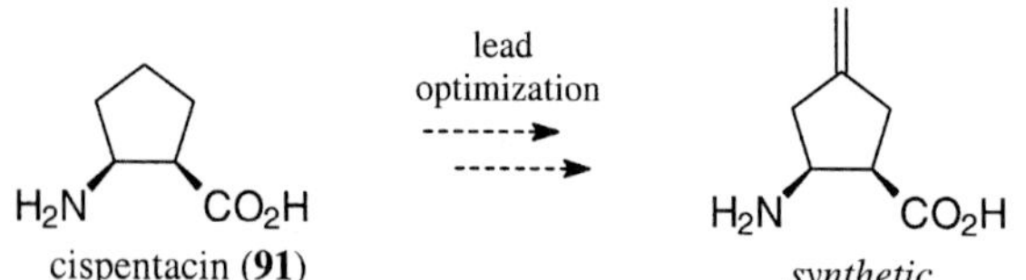

cispentacin (**91**)
from *Bacillus cereus*
(accumulation via
L-Pro permeases)

synthetic
BAY 10-8888/PLD-118 (**92**)
(accumulation via L-Ile transporters)

Scheme 1.5.14. Natural cispentacin **91** as the chemical lead for BAY 10-8888/PLD-118 **92**.

amino acid processing systems (e.g. active transport, protein biosynthesis). So far little work has been published on the general biological properties of β-amino acids, e.g. their physicochemistry and pharmacokinetics (e.g. data on absorption, distribution, metabolism, and excretion – ADME).

1.5.6 Limiting the β-Amino Acid Concept

Obviously a substructure search within the >15000 alkaloids [23] known will uncover many compounds with a 1,4-relationship between a nitrogen atom and a carboxylic acid functionality, a carboxamide, or a carboxylic acid ester (~ 1100 examples) [23]. Also anthranilic acid might be regarded as a 2,3-dehydro β-amino-acid. However, these "formal β-amino acids" [139] will not be discussed further, because we do not see a general biogenetic relationship with the "real β-amino acids" (vide infra).

1.5.7 Conclusion

β-Amino acids have been found within secondary and primary metabolism of all kingdoms. Although quantitative ranking is tricky, at this point bacteria seem to be the most active "β-amino acid chemists", followed by fungi, plants, and animals; we exclude protists from this grading list as a consequence of their lower examination status.

A *natural β* counterpart has not yet been obtained for every proteinogenic α-amino acid (Table 1.5.1). β-Ala **1** and (*R*)- or (*S*)-β-Aib **2** seem to be present in all kingdoms, with no obvious preferences. Selected β-amino acids, for example β-Lys **4**, have been found only within one kingdom (i.e. bacteria), although they might be quite common there. Accordingly, the various natural products that contain β-Lys substructures have been isolated exclusively from bacteria. On the other hand, alkaloids and peptides linked to β-Phe **8** have been obtained from bacteria, plants, and fungi, whereas the free β-amino acid β-Phe **8** has not yet been isolated from natural sources.

It is apparent that most known *natural* β-amino acids have not yet been isolated in form of the free β-amino acid, but obtained only as substructures of alkaloids, peptides, and depsipeptides. Certainly, this general trend does not automatically prove a general preference of β-amino acid *substructures* with regard to free β-amino acids within natural products.

Rather up-to-date isolation methodology might play a role here, because it results in *general discrimination of small molecular weight, polar metabolites*. Small, polar compounds, for example free β-amino acids, typically have non-characteristic UV–visible spectra and masses which are often too small for the LC–ESI–MS range commonly scanned; they also elute "with the solvent front" under standard reversed-phase chromatography conditions. Finally, they might have only a weak or no biological activity in a bioassay set up "randomly" with regard to their "original" biological function.

Not only within industrial natural-product research, LC–ESI–MS-guided, UV-guided, and activity-guided searching for substances is performed out *in the middle polarity range* under more or less standardized conditions. Certainly this established methodology leads to the desired higher output of new natural products. A cut within the structural [140] and sensitivity windows seems to be unavoidable, however, and we are blinded by these methods for polar compounds with low molecular weights.

Several natural alkaloids, peptides, and depsipeptides related to β-amino acids are active antibacterial, antifungal, or cytotoxic compounds (Tables 1.5.1 and 1.5.2). Accordingly, β-amino acids already have implications within medicinal chemistry as lead structures or as analogs of α-amino acids within peptides or peptidomimetics. Selected examples have been discussed.

Dedication

Dedicated to Professor Wolfgang Steglich on the occasion of his 70th birthday.

Acknowledgment

We are grateful to Dr Norbert Arnold and Dr Marc Stadler for many valuable hints and suggestions and to the Alexander von Humboldt-Stiftung for a Feodor-Lynen-Stipendium to F.v.N. and P.S.

References

1 For previous reviews dealing with natural β-amino acids see: (a) O. W. GRIFFITH, *Ann. Rev. Biochem.* **1986**, 55, 855–878; (b) C. N. C. DREY, *Beta and Higher Homologous Amino Acids* in *Chemistry and Biochemistry of the Amino Acids*, G. C. BARRETT, ed., Chapman and Hall, London, **1985**,

pp. 25–54; (c) C. N. C. Drey, *The Chemistry and Biochemisty of β-Amino Acids*, G. Weinstein, ed., *The Chemistry and Biochemistry of the Amino Acids*, Dekker, New York, **1976**, *4*, 241–299.

2 P. A. Frey, C. H. Chang, *Aminomutases* in *Chemistry and Biochemistry of B_{12}*, R. Banerjee, ed., Wiley, New York, **1999**, pp. 835–857. For further references on Frey's work see biosynthesis section.

3 J. J. Baker, T. C. Stadtman, *Aminomutases* in B_{12}: *Biochemistry and Medicine*, Vol. 2, D. Dolphin, ed., Wiley, New York, **1982**, pp. 203–232.

4 (a) P. E. Fleming, U. Mocek, H. G. Floss, *J. Am. Chem. Soc.* **1993**, *115*, 805–807; (b) K. D. Walker, H. G. Floss, *J. Am. Chem. Soc.* **1998**, *120*, 5333–5334.

5 P. Spiteller, M. Rüth, F. von Nussbaum, W. Steglich, *Angew. Chem. Int. Ed.* **2000**, *39*, 2754–2756.

6 J. Tamariz, *Biological Activity of β-Amino Acids and β-Lactams* in *Enantioselective Synthesis of β-Amino Acids*, E. Juaristi, ed., Wiley–VCH, New York, **1997**, 45–66.

7 A. D. Hanson, B. Rathinasabapathi, J. Rivola, M. Burnet, M. O. Dillon, D. A. Gage, *Proc. Natl. Acad. Sci. USA* **1994**, *91*, 306–310.

8 Marine methanogenic archaebacteria respond to osmotic stress by accumulating N^ε-acetyl-β-lysine and β-Glu as metabolism "complatible solutes": (a) K. R. Sowers, D. E. Robertson, D. Noll, R. P. Gunsalus, M. F. Roberts, *Proc. Natl. Acad. Sci. USA* **1990**, *87*, 9083–9087; (b) D. E. Robertson, D. Noll, M. F. Roberts, *J. Biol. Chem.* **1992**, *267*, 14893–14901.

9 β-Glu is a major soluble component of *Methanococcus thermolithotrophicus*. (a) E. E. Robertson, S. Lesage, M. F. Roberts, *Biochim. Biophys. Acta* **1989**, *992*, 320–326; (b) D. D. Martin, R. A. Ciulla, P. M. Robinson, M. F. Roberts, *Biochim. Biophys. Acta* **2000**, *1524*, 1–10.

10 (a) K. Ziegelbauer, P. Babczinski, W. Schönfeld, *Antimicrob. Agents Chemother.* **1998**, *42*, 2197–2205; (b) J. Mittendorf, D. Häbich, F. Kunisch, M. Matzke, H.-C. Militzer, K.-M. Mohrs, A. Schmidt, W. Schönfeld, Poster, ICAAC, San Diego, **2002**; (c) J. Mittendorf, J. Benet-Buchholz, P. Fey, K.-H. Mohrs, *Synthesis* **2003**, 136–140.

11 (a) G. Cardillo, C. Tomasini, *Chem. Soc. Rev.* **1996**, 117–128; (b) N. Fusetani, S. Matsunaga, *Chem. Rev.* **1993**, *93*, 1793–1806; (c) C. E. Ballard, B. Wang, *Curr. Med. Chem.* **2002**, *9*, 471–498.

12 (a) N. Sewald, *Bioorganic Chemistry*, U. Diedrichsen, T. K. Lindhorst, B. Westermann, L. A. Wessjohann, eds., Wiley–VCH, Weinheim, **1999**; (b) M. North, *J. Peptide Sci.* **2000**, *6*, 3001–3313.

13 E. Abderhalden, R. Fleischmann, *Fermentforschung* **1928**, *10*, 173.

14 A selection of recent activity: (a) S. Reinelt, M. Marti, S. Dédier, T. Reitinger, G. Folkers, J. A. López de Castros, D. Rognan, *J. Biol. Chem.* **2001**, *276*, 24525–24530; (b) R. A. Lew, E. Boulos, K. M. Steward, P. Perlmutter, M. F. Harte, S. Bond, M.-I. Aguilar, A. I. Smith, *J. Peptide Sci.* **2000**, *6*, 440–445.

15 Furthermore, the incorporation of D-amino acids or other unusual (often hydroxylated) amino acids, the *N*-methylation of amide functionalities or the usage of cyclic peptides and depsipeptides limits these problems to a certain extent.

16 (a) R. P. Cheng, S. H. Gellman, W. F. DeGrado, *Chem. Rev.* **2001**, *101*, 3219–3232; (b) K. Gademann, T. Hintermann, J. V. Schreiber, *Curr. Med. Chem.* **1999**, *6*, 905–925; (c) S. H. Gellman, *Acc. Chem. Res.* **1998**, *31*, 173–180; (d) D. Seebach, J. L. Matthews, *Chem. Commun.* **1997**, 2015–2022.

17 (a) D. H. Appella, J. J. Barchi Jr, S. R. Durell, S. H. Gellman, *J. Am. Chem. Soc.* **1999**, *121*, 2309–2310; (b) D. H. Appella, L. A. Christianson, D. A. Klein, D. R. Powell, X. Huang, J. J. Barchi Jr, S. H. Gellman, *Nature* **1997**, *387*, 381–384.

18 (a) D. Seebach, K. Gademann, J. V. Schreiber, J. L. Matthews, T.

Hintermann, B. Jaun, L. Oberer, U. Hommel, H. Widmer, *Helv. Chim. Acta* **1997**, *80*, 2033–2038; (b) M. Rueping, J. V. Schreiber, G. Lelais, B. Jaun, D. Seebach, *Helv. Chim. Acta* **2002**, *85*, 2577–2593.

19 (a) J. M. Fernańdez-Santín, J. Aymamí, A. Rodríguez-Galán, S. Muñoz-Guerra, J. A. Subirana, *Nature* (London) **1984**, *311*, 53; (b) A. M. De Ilarduya, C. Alemań, M. García-Alvarez, F. López-Carrasquero, S. Muñoz-Guerra, *Macromolecules* **1999**, *32*, 3257.

20 (a) M. Liu, M. P. Sibi, *Tetrahedron*, **2002**, *58*, 7991–8035; (b) S. Abele, D. Seebach, *Eur. J. Org. Chem.* **2000**, 1–15; (c) *Enantioselective Synthesis of β-Amino Acids*, E. Juaristi, ed., Wiley–VCH, New York, **1997**; (d) D. C. Cole, *Tetrahedron* **1994**, *50*, 9517–9582; (e) E. Juaristi, D. Quintana, J. Escalante, *Aldrichim. Acta* **1994**, *27*, 3–11.

21 Recent examples: (a) I. Karle, H. N. Gopi, P. Balaram, *Proc. Natl. Acad. Sci. USA* **2002**, *99*, 5160–5164; (b) R. Günther, H.-J. Hofmann, *J. Am. Chem. Soc.* **2001**, *123*, 247–255.

22 (a) E. Juaristi, H. López-Ruiz, *Curr. Med. Chem.* **1999**, *6*, 983–1004; (b) A. F. Abdel-Magid, J. H. Cohen, C. A. Marynoff, *Curr. Med. Chem.* **1999**, *6*, 955–970.

23 (a) *Dictionary of Natural Products*, Chapman and Hall/CRC, CD-ROM, 1982–2003; (b) *Römpp Encyclopedia – Natural Products*, W. Steglich, B. Fugmann, S. Lang-Fugmann, eds., Georg Thieme, Stuttgart, **2000**.

24 T. Hintermann, D. Seebach, *Chimica* **1997**, *51*, 244–247.

25 J. V. Schreiber, J. Frackenpohl, F. Moser, T. Fleischmann, H.-P. Kohler, D. Seebach, *ChemBioChem.* **2002**, *3*, 424–432.

26 (a) K. Gademann, M. Ernst, D. Hoyer, D. Seebach, *Angew. Chem. Int. Ed.* **1999**, *38*, 1223–1226; (b) D. Seebach, M. Rueping, P. I. Arvidsson, T. Kimmerlin, P. Micuch, C. Noti, *Helv. Chim. Acta* **2001**, *84*, 3503–3510.

27 E. A. Porter, B. Weisblum, S. H. Gellman, *J. Am. Chem. Soc.* **2002**, *124*, 7324–7330.

28 (3S,4S,5E,7E)-3-Amino-4-hydroxy-6-methyl-8-(4-bromophenyl)-octa-5,7-dienoic acid [Aboa ≡ Br-Ahmp, Br-Apoa]. (2S,3S,8S,9S)-3-Amino-9-methoxy-2,6,8-trimethyl-10-phenyl-deca-4,6-dienoic acid (Adda). 3-Amino-2,4-dimethylpentanoic acid (Admpa). (2S,3R,5S)-3-amino-2,5,9-trihydroxy-10-phenyldecanoic acid (Ahda). β-Aminoisobutyric acid (β-AiB). (3S,4S,5E,7E)-3-Amino-4-hydroxy-6-methyl-8-phenyllocta-5,7-dienoic acid [Ahmp (Faulkner) or Apoa (Fusetani)]. (2S,3R,5R)-3-Amino-2,5-dihydroxy-8-phenyloctanoic acid (Ahoa). 3-Aminopentanoic acid (Apa ≠ β-Apa). (2R,3R)-3-Amino-2-methylbutanoic acid (Amba). 3-Amino-2-methylhexanoic acid (Amha). 3-Amino-2-methyl-7-octynoic acid (Amoa, Amoya). 3-Amino-7-octynoic acid (Aoya). (2S,3R)-3-Amino-2-methylpentanoic acid (Map).

29 http://www.ncbi.nlm.nih.gov/Taxonomy/tax.html

30 T. P. Begley, C. Kinsland, E. Strauss, *Vitam. Horm.*, **2001**, *61*, 157–171.

31 A. A. Boldyrev, *Int. J. Biochem.* **1990**, *22*, 129–132.

32 (a) M. S. C. Pedras, L. Irina Zaharia, D. E. Ward, *Phytochemistry* **2002**, *59*, 579–596; (b) Y. Kodaira, *Agr. Biol. Chem.* **1961**, *25*, 261–262; (c) H. S. Kim, M. H. Jung, S. Ahn, C. W. Lee, S. N. Kim, J. H. Ok, *J. Antibiot.* **2002**, *55*, 598–601.

33 A. Farooq, S. Tahara, *Curr. Top. Phytochem.* **2000**, *3*, 95–111.

34 S. C. Pedras, K. C. Smith, *Phytochemistry* **1997**, *46*, 833–837.

35 J. C. Frisvad, U. Thrane, O. Filtenborg, *Role and Use of Secondary Metabolites in Fungal Taxonomy*, in *Chemical Fungal Taxonomy*, J. C. Frisvad, P. D. Bridge, D. K. Arora, eds., Marcel Dekker, New York, **1998**, pp. 289–313.

36 R. E. Moore, G. Bartolini, *J. Am. Chem. Soc.* **1981**, *103*, 2491–2494.

37 N. Sitachitta, R. T. Williamson, W. H. Gerwick, *J. Nat. Prod.* **2000**, *63*, 197–200.

38 (a) R. G. Keil, E. Tsamakis, J. C. Giddings, J. I. Hedges, *Geochim. Cosmochim. Acta* **1998**, *62*, 1347–1364; (b) H. Kawahata, T. Ishizuka, *Oceanol. Acta* **1993**, *16*, 373–379.

39 Isolation: (a) K. Hamano, M. Kinoshita, K. Furuya, M. Miyamoto, Y. Takamutsu, A. Hemmi, K. Tanzawa, *J. Antibiot.* **1992**, *45*, 899–905. Structure elucidation: (b) K. Hamano, M. Konoshita, K. Tanzawa, K. Yoda, Y. Ohki, T. Nakamura, T. Kinoshita, *J. Antibiot.* **1992**, *45*, 906–913.

40 (a) C. Rossi, L. Tuttobello, M. Ricci, C. G. Casinovi, L. Radics, *J. Antibiot.* **1987**, *40*, 130–133; (b) G. A. Strobel, W. M. Hess, *Chem. Biol.* **1997**, *4*, 529–536.

41 T. M. Soniat, M. L. Koenig, *J. Shellfish Res.* **1982**, *2*, 25–28.

42 Isolation of arenastatin A (cryptophycin-24) from a marine sponge (*Dysidea arenaria*): (a) M. Kobayashi, S. Aoki, N. Ohyabu, M. Kurosu, W. Wang, I. Kitagawa, *Tetrahedron Lett.* **1994**, *35*, 7969–7972. Synthesis: (b) M. Kobayashi, M. Kurosu, W. Wang, I. Kitagawa, *Chem. Pharm. Bull.* **1994**, *42*, 2394–2396; (c) M.-J. Eggen, S. K. Nair, G. I. Georg, *Org. Lett.* **2001**, *3*, 1813–1815.

43 M. C. Roy, I. I. Ohtani, T. Ichiba, J. Tanaka, R. Satari, T. Higa, *Tetrahedron* **2000**, *56*, 9079–9092.

44 Compare cited literature: M. Doi, T. Ishida, M. Kobayashi, Y. Katasuya, Y. Mezaki, M. Sasaki, A. Terashima, T. Taniguchi, C. Tanaka, *Biopolymers* **2000**, *54*, 27–34.

45 (a) H. Rembold, Y. Umebachi, *Progress in Tryptophan and Serotonin Research* **1984**, 743–746; *Chem. Abstr.* **1984**, *101*, 87654s; (b) P. B. Koch, B. Behnecke, M. Weigmann-Lenz, R. H. Ffrench-Constant, *Pigment Cell Research, Suppl.* **2000**, *8*, 54–58.

46 Y. Guillou, Y. Robin, *J. Biol. Chem.* **1973**, *248*, 5668–5672.

47 (a) R. E. Schwartz, C. F. Hirsch, D. F. Sesin, J. E. Flor, M. Chartrain, R. E. Fromtling, G. H. Harris, M. J. Salvatore, J. M. Liesch, K. J. Yudin, *J. Ind. Microbiol.* **1990**, *5*, 113–123; (b) G. V. Subbaraju, T. Galakoti, G. M. L. Patterson, R. E. Moore, *J. Nat. Prod.* **1997**, *60*, 302–305. Structure: (c) T. Golakoti, J. Ogino, C. E. Heltzel, T. L. Husebo, C. M. Jensen, L. K. Larsen, G. M. L. Patterson, R. E. Moore, S. L. Mooberry, T. H. Corbett, F. A. Valeriote, *J. Am. Chem. Soc.* **1995**, *117*, 12030–12049. Synthesis: (d) J. D. White, J. Hong, L. A. Robarge, *J. Org. Chem.* **1999**, *64*, 6206–6216.

48 S. A. Currie, T. W. Miller, J. P. Springer, M. E. Valiant, S. B. Zimmerman, S. M. Del Vai (Merck and Co. Inc.), EP 332248, **1989**.

49 N. Sato, M. Furuya, *Phytochemistry* **1984**, *23*, 1625–1627.

50 Isolation of DGTA [1,2-diacylglyceryl-O-$2'$-(hydroxymethyl)-N,N,N,N-trimethyl-β-Ala] from marine algae, such as green algae and brown algae: (a) V. E. Vaskovsky, S. V. Khotimchenko, B. Xia, L. Hefang, *Phytochemistry* **1996**, *42*, 1347–1356; (b) M. Kato, M. Sakai, K. Adachi, H. Ikemoto, H. Sano, *Phytochemistry* **1996**, *42*, 1341–1345.

51 M. R. Altamura, F. M. Robbins, R. W. Andreotti, L. Lond, T. Hasselstrom, *J. Agr. Food Chem.* **1967**, *15*, 1040–1043.

52 T. Sugawara, A. Tanaka, K. Tanaka, K. Nagai, K. Suzuki, T. Suzuki, *J. Antibiot.* **1998**, *51*, 435–438.

53 Distribution of DGTA in non-vascular green plants, such as mosses (Bryopsida) and green algae (Chlorophyceae): N. Sato, M. Furuya, *Plant Sci.* **1985**, *38*, 81–85.

54 J. M. Poston, *J. Biol. Chem.* **1980**, *255*, 10067–10072. For further references see biosynthesis section.

55 Sperabillin C has a δ-OH-β-D-Lys substructure: T. Hida, S. Tsubotani, Y. Funabashi, H. Ono, S. Harada, Setsuo, *Bull. Chem. Soc. Jpn.* **1993**, *66*, 863–869.

56 K. Ohba, H. Nakayama, K. Furihata, K. Furihata, A. Shimazu, H. Seto, N. Ōtake, Y. Zhao-Zhong, X. Li-Sha, X. Wen-Si, *J. Antibiot.* **1986**, *39*, 872–875.

57 Lavendothricin is related to the

strepthricins. Isolation: N. DOBREVA, B. VASILEVA, *Farmatsiya* (Sofia) **1977**, *27*, 25–29.
58 (+)-Negamycin contains δ-OH-β-D-Lys. Isolation: (a) M. HAMADA, T. TAKEUCHI, S. KONDO, Y. IKEDA, H. NAGANAWA, K. MAEDA, Y. OKAMI, H. UMEZAWA, *J. Antibiot.* **1970**, *23*, 170–171. Structure: (b) S. KONDO, S. SHIBAHARA, S. TAKAHASHI, K. MAEDA, H. UMEZAWA, M. OHNO, *J. Am. Chem. Soc.* **1971**, *93*, 6305–6306. For a recent synthesis see: (c) R. P. JAIN, R. M. WILLIAMS, *J. Org. Chem.* **2002**, *67*, 6361–6365.
59 Isolation: (a) R. L. PECK, A. WALTI, R. P. GRABER, E. FLYNN, C. E. HOFFHINE JR, V. ALLFREY, K. FOLKERS, *J. Am. Chem. Soc.* **1946**, *68*, 772–774. Biosynthesis: (b) K. J. MARTINKUS, C.-H. TANN, S. J. GOULD, *Tetrahedron* **1983**, *39*, 3493–3505. Review: (c) H. THRUM, *Drugs Pharm. Sci.* **1984**, *22*, 367–386.
60 (a) Isolation and structure elucidation: Y. IKEDA, S. KONDO, D. IKEDA, K. YOKOSE, H. FURUTANI, T. IDEKA, M. HAMADA, M. ISHIZUKA, T. TAKEUCHI, H. UMEZAWA, *J. Antibiot.* **1986**, *39*, 476–478. Biosynthesis via an L-Lys-2,3-aminomutase: (b) Y. IKEDA, H. NAGANAWA, S. KONDO, T. TAKEUCHI, *J. Antibiot.* **1992**, *45*, 1919–1924. Synthesis: (d) Y. IKEDA, D. IKEDA, S. KONDO, *J. Antibiot.* **1992**, *45*, 1677–1680.
61 T. WAKAMIYA, T. SHIBA, *J. Antibiot.* **1974**, *27*, 900–902.
62 T. KITAGAWA, Y. SAWADA, T. MIURA, T. OZASA, H. TANIYAMA, *Tetrahedron Lett.* **1968**, *1*, 109–113.
63 J. H. MARTIN, J. P. KIRBY, D. B. BORDERS, A. A. FANTINI, R. T. TESTA (American Cyanamide), EP 58838 **1983**.
64 W. J. MCGAHREN, B. A. HARDY, G. O. MORTON, F. M. LOVELL, N. A. PERKINSON, R. T. HARGREAVES, D. B. BORDERS, G. A. ELLESTAD, *J. Org. Chem.* **1981**, *46*, 792–799.
65 J. C. FRENCH, Q. R. BARTZ, H. W. DION, *J. Antibiot.* **1973**, *26*, 272–283.
66 M. W. JACKSON, R. J. THERIAULT, A. C. SINCLAIR, E. E. FAGER, J. P. KARWOWSKI (Abbot Laboratories), DE 3003497.
67 (a) N. KATAYAMA, S. FUKUSUMI, Y. FUNABASHI, T. IWAHI, H. ONO, *J. Antibiot.* **1993**, *46*, 606–613. Structure: (b) Y. FUNABASHI, S. TSUBOTANI, K. KOYAMA, N. KATAYAMA, S. HARADA, Setsuo *Tetrahedron* **1993**, *49*, 13–28; (c) R. M. WILLIAMS, C. YUAN, V. J. LEE, S. CHAMBERLAND, *J. Antibiot.* **1998**, *51*, 189–201; (d) V. V. SOKOLOV, S. I. KOZHUSHKOV, S. NIKOLSKAYA, V. N. BELOV, M. ES-SAYED, A. DE MEIJERE, *Eur. J. Org. Chem.* **1998**, 777–783. Mode of action: (e) W. S. CHAMPNEY, J. PELT, C. L. TOBER, *Curr. Microbiol.* **2001**, *43*, 340–345. Biology: (f) M. BRANDS, M. ES-SAYED, D. HÄBICH, S. RADDATZ, J. KRÜGER, R. ENDERMANN, R. GAHLMANN, H.-P. KROLL, F.-U. GESCHKE, A. DE MEIJERE, V. N. BELOV, V. SOKOLOV, S. KOZHUSHKOV, M. KORDES, (Bayer A.G.), WO-0012484-A1, **2000**.
68 Isolation: (a) S. TAKEUCHI, K. HIRAYAMA, K. UEDA, H. SAKAI, H. YONEHARA, *J. Antibiot.* **1958**, *11A*, 1–5. Biosynthesis: (b) H. SETO, H. YONEHARA, *J. Antibiot.* **1977**, *30*, 1019–1021; (c) H. SETO, H. YONEHARA, *ibid.* **1977**, *30*, 1022–1024.
69 Stereochemistry of LL-BM 547β according to the structurally related viomycins. W. J. MCGAHREN, G. O. MORTAM, M. P. KUNSTMANN, A. G. ELLSTAD, *J. Org. Chem.* **1977**, *42*, 1282–1286.
70 S. M. HENRICHS, R. CUHEL, *Appl. Environ. Microbiol.* **1985**, *50*, 543–545.
71 β-Glu is a major constituent of dissolved free amino acids (DFAA) in marine sediments. (a) S. M. HENRICHS, J. W. FARRINGTON, *Nature* **1979**, *179*, 319–322; (b) D. J. BURDIGE, C. S. MARTENS, *Biogeochemistry* **1989**, *8*, 1–23; (c) J. C. COLOMBO, N. SILVERBERG, J. N. GEARING, *Organic Geochemistry* **1998**, *29*, 933–945.
72 E. FATTORUSSO, M. PIATELLI, *Amino Acids from Marine Algae* in *Marine Natural Products*, P. J. SCHEUER, eds., Academic Press, New York, **1980**, pp. 95–140.

73 (a) F. Itagaki, H. Shigemori, M. Ishibashi, T. Nakamura, T. Sasaki, J.-i. Kobayashi, *J. Org. Chem.* **1992**, *57*, 5540–5542; (b) J.-i. Kobayashi, R. Itagaki, H. Shigemori, T. Takao, Y. Shimonishi, *Tetrahedron* **1995**, *51*, 2525–2532. Synthesis: (c) J. A. Sowinski, P. L. Toogood, *Chem. Commun.* **1999**, 981–982.

74 C. A. Bewley, D. J. Faulkner, *J. Org. Chem.* **1994**, *59*, 4849–4852; ibid. **1995**, *60*, 2644.

75 Theonellamides A–F, isolation: (a) S. Matsunaga, N. Fusetani, K. Hashimoto, M. Walchli, *J. Am. Chem. Soc.* **1989**, *111*, 2585–2588. Theonellamides B, C: (b) S. Matsunaga, N. Fusetani, *J. Org. Chem.* **1995**, *60*, 1177–1181. Biological activity: (c) S.-I. Wada, S. Matsunaga, N. Fusetani, S. Watabe, *Mar. Biotechnol.* **2000**, *2*, 285–292.

76 E. W. Schmidt, C. A. Bewley, D. J. Faulkner, *J. Org. Chem.* **1998**, *63*, 1254–1258.

77 Isolation from a bacterial symbiont (*Enterobacter* sp.) of a brown planthopper (*Nilaparvata lugens*): (a) A. Fredenhagen, S. Y. Tamura, P. T. M. Kenny, H. Komura, Y. Naya, K. Nakanishi, K. Nishiyama, M. Sugiura, H. Kita, *J. Am. Chem. Soc.* **1987**, *109*, 4409–4411; (b) M. P. Singh, M. J. Mroczenski-Wildey, D. A. Steinberg, R. J. Andersen, W. M. Maiese, M. Greenstein, *J. Antibiot.* **1997**, *50*, 270–273.

78 Isolation: (a) J. Needham, M. T. Kelly, M. Ishige, R. J. Andersen, *J. Org. Chem.* **1994**, *59*, 2058–2063. Synthesis: (b) S. G. Davies, D. J. Dixon, *J. Chem. Soc. Perkin Trans. 1* **1998**, *17*, 2635–2644.

79 (a) H. Yoshioka, K. Nakatsu, M. Sato, T. Tatsuno, *Chem. Lett.* **1973**, 1319–1322; (b) T. Tatsuno, M. Tsukioka, Y. Sakai, Y. Suzuki, Y. Asami, *Chem. Pharm. Bull.* **1955**, *3*, 476–477.

80 J. L. Corbin, B. H. Marsh, G. A. Peters, *Phytochemistry* **1986**, *25*, 527–528.

81 (a) M. Koyama, Y. Obata, *Agr. Biol. Chem.* **1966**, *30*, 472–477; (b) M. Koyama, Y. Obata, *Agr. Biol. Chem.* **1967**, *31*, 738–742.

82 (a) S. Kosemura, T. Ogawa, K. Totsuka, *Tetrahedron Lett.* **1993**, *34*, 1291–1294; (b) K. K. Schumacher, D. B. Hauze, J. Jiang, J. Szewczyk, R. E. Reddy, F. A. Davis, M. M. Joullié, *Tetrahedron Lett.* **1999**, *40*, 455–458.

83 L. Ettouati, A. Ahond, O. Convert, C. Poupat, P. Potier, Pierre, *Bull. Soc. Chim. Fr.* **1989**, 687–694.

84 (*R*)-*N*,*N*-dimethyl-β-phenylalanine: (a) E. Winterstein, A. Latrides, *Hoppe-Seyler's Z. Physiol. Chem.* **1921**, *117*, 240–283; (b) S. G. Davies, J. Dupont, R. J. E. Easton, *Tetrahedron Asymmetry* **1990**, *1*, 279–280.

85 G. Appendino, S. Tagliapietra, H. C. Ozen, P. Gariboldi, B. Gabetta, E. Bombardelli, *J. Nat. Prod.* **1993**, *56*, 514–520.

86 (a) G. M. Cragg, *Med. Chem. Rev.* **1998**, *18*, 315–331; (b) K. C. Nicolaou, W. M. Dai, R. K. Guy, *Angew. Chem. Int. Ed.* **1994**, *33*, 15–44.

87 E. Graf, A. Kirfel, G.-J. Wolff, E. Breitmaier, *Liebigs Ann. Chem.* **1982**, 376–381.

88 Free β-Phe has only been detected in cell-free extracts of *T. brevifolia* (Floss et al.).

89 Structure: (a) T. P. Hettinger, L. C. Craig, *J. Am. Chem. Soc.* **1968**, *7*, 4147–4153; (b) T. P. Hettinger, Z. Kurylo-Borowska, L. C. Craig, ibid. **1968**, *7*, 4153–4160. Biosynthesis: (c) Z. Kurylo-Borowska, T. Abramsky, *Biochim. Biophys. Acta* **1972**, *264*, 1–10. Edeine D, F: (d) H. Wojciechowska, W. Zgoda, E. Borowski, K. Dziegielewski, S. Ulikowski, *J. Antibiot.* **1983**, *36*, 793–798.

90 The absolute and relative stereochemistry of the α-methoxy-β-Tyr moiety was not described: R. Jansen, B. Kunze, H. Reichenbach, G. Höfle, *Liebigs Ann.* **1996**, 285–290.

91 (a) D. Seebach, M. Overhand, F. N. M. Kühnle, B. Martinoni, L. Oberer, U. Hommel, H. Widmer, *Helv. Chim. Acta* **1996**, *79*, 913–941; (b) D. Seebach, S. Abele, K.

Gademann, B. Jaun, *Angew. Chem. Int. Ed.* **1999**, *111*, 1595–1597; (c) T. Hintermann, D. Seebach, *Synlett*, **1997**, 437–438.
92 M. J. Eggen, G. I. Georg, *Med. Res. Rev.* **2002**, *22*, 85–101.
93 T. Li, C. Shih, *Front. Biotechnol. Pharm.* **2002**, *3*, 172–192.
94 T. C. Stadtman, *Adv. Enzymol. Relat. Areas Mol. Biol.* **1970**, *38*, 413–448.
95 Isolation: (a) E. B. Herr, *Antimicrob. Agents Chemother.* **1962**, 201–212. Structure elucidation: (b) S. Nomoto, T. Teshima, T. Wakamiya, T. Shiba, *J. Antibiot.* **1977**, *30*, 955–959. Synthetic analogues: (c) R. G. Linde II, N. C. Birsner, R. Y. Chandrasekaran, J. Clancy, R. J. Howe, J. P. Lyssikatos, C. P. MacLelland, T. V. Magee, J. W. Peptitpas, J. P. Rainville, W. G. Su, C. B. Vu, D. A. Whipple, *Bioorg. Med. Chem. Lett.* **1997**, *7*, 1149–1152 and literature cited therein.
96 E. Graf, A. Kirfel, G.-J. Wolff, E. Breitmaier, *Liebigs Ann. Chem.* **1982**, 376–381.
97 (a) I. Ojima, R. Geney, I. M. Ungureanu, D. Li, *IUBMB Life* **2002**, *53*, 269–274; (b) H. M. Deutsch, J. A. Glinski, M. Hernandez, R. D. Haugwitz, V. L. Narayanan, M. Suffness, L. H. Zalkow, *J. Med. Chem.* **1989**, *32*, 788–792; (c) I. Ojima, X. Geng, S. Lin, P. Pera, R. J. Bernacki, *Bioorg. Chem. Lett.* **2002**, *12*, 349–352.
98 G. Taraboletti, G. Micheletti, M. Rieppi, M. Poli, M. Turatto, C. Rossi, P. Borsotti, P. Roccabianca, E. Scanziani, M. I. Nicoletti, E. Bombardelli, P. Morazzoni, A. Riva, R. Giavazzi, *Clin. Cancer Res.* **2002**, *8*, 1182–1188.
99 M. A. Jordan, I. Ojima, F. Rosas, M. Distefano, L.Wilson, G. Scambia, C. Ferlini, *Chem. Biol.* **2002**, *9*, 93–101.
100 Structure: (a) R. Hocquemiller, A. Cavé, H.-P. Husson, *Tetrahedron*, **1977**, *33*, 645–651. Synthesis: (b) R. Hocquemiller, A. Cavé, H.-P. Husson, *Tetrahedron*, **1977**, *33*, 653–656.
101 (a) K. Seifert, S. Johne, M. Hesse, *Helv. Chim. Acta* **1982**, *65*, 2540–2547; (b) H. H. Wasserman, H. Matsuyama, R. P. Robinson, *Tetrahedron* **2002**, *58*, 7177–7190. Synthesis or (±)-verbascenine: (c) H. H. Wasserman, R. P. Robinson, *Tetrahedron Lett.* **1983**, *24*, 3669–3672.
102 ($17R,18R$)-aphelandrine ≡ orantine; (a) A. Guggisberg, R. Prewo, M. Hesse, *Helv. Chim. Acta* **1986**, *69*, 1012–1016; (b) L. Nezbodavá, M. Hesse, K. Drandarov, C. Werner, *Tetrahedron Lett.* **2001**, *42*, 4139–4141. Isolation: (c) P. Dätwyler, H. Bossardt, S. Johne, M. Hesse, *Helv. Chim. Acta* **1979**, *62*, 2712–2723.
103 (a) H. O. Bernhard, I. Kompiš, S. Johne, D. Gröger, M. Hesse, H. Schmid, *Helv. Chim. Acta* **1973**, *56*, 1266–1303; (b) H. H. Wasserman, R. P. Robinson, C. G. Carter, *J. Am. Chem. Soc.* **1983**, *105*, 1697–1698.
104 Isolation from *Chaenorhinum minus*: J.-p. Zhu, A. Guggisberg, M. Hesse, *Helv. Chim. Acta*, **1988**, *71*, 218–223.
105 (a) K. Drandarov, A. Guggisberg, M. Hesse, *Helv. Chim. Acta* **2002**, *85*, 979–989; (b) V. Dimitrov, H. Geneste, A. Guggisberg, M. Hesse, *Helv. Chim. Acta* **2001**, *84*, 2108–2118; (c) A. Guggisberg, K. Drandarov, M. Hesse, *Helv. Chim. Acta* **2000**, *83*, 3035–3042; (d) L. Nezbedová, K. Drandarov, C. Werner, M. Hesse, *Helv. Chim. Acta* **2000**, *83*, 2953–2960.
106 (a) T. M. Zabriskie, J. A. Klocke, C. M. Ireland, A. H. Marcus, T. F. Molinski, D. J. Faulkner, C. Xu, J. C. Clardy, *J. Am. Chem. Soc.* **1986**, *108*, 3123–3124; (b) P. Crews, L. V. Manes, M. Boehler, *Tetrahedron Lett.* **1986**, *27*, 2797–2800; (c) J. P. Konopelski, *Asymmetric Synthesis of β-Aryl- and β-Alkyl-β-Amino Acids via Enantiomerically Pure Dihydropyrimidinones, Enantioselektive Synthesis of β-Amino Acids*, E. Juaristi, ed., Wiley–VCH, New York, **1997**, pp. 249–259; (d) A. Zampella, C. Giannini, C. Debitus, C. Roussakis, M. V. D'Auria, *J. Nat. Prod.* **1999**, *62*, 332–334.
107 (a) J. E. Coleman, R. Van Soest, R. J. Andersen, R. G. Kelsey, *J. Nat. Prod.* **1999**, *62*, 1137–1141; (b) W. F. Tinto,

A. J. Lough, S. McLean, W. F. Reynolds, M. Yu, W. R. Chan, *Tetrahedron* **1998**, *54*, 4451–4458.

108 (a) Y.-s. Zhen, S.-y. Ming, Y. Bin, T. Otani, H. Saito, Y. Yamada, *J. Antibiot.* **1989**, *42*, 1294; (b) K.-i. Yoshida, Y. Minami, T. Otani, *Tetrahedron Lett.* **1994**, *35*, 5253–5256; (c) L. Yu, S. Mah, T. Otani, P. Dedon, *J. Am. Chem. Soc.* **1995**, *117*, 8877–8878; (d) T. Sasaki, M. Inoue, M. Hirata, *Tetrahedron Lett.* **2001**, *42*, 5299–5303.

109 The real reactive, DNA-cleaving, toxophore is a phenylene-1,4-diyl radical that is formed via a Bergman cyclization.

110 Biosynthesis: (a) W. Liu, S. D. Christenson, S. Standage, B. Shen, *Science*, **2002**, *297*, 1170–1173; (b) J. S. Thorson, B. Shen, R. E. Whitwam, W. Liu, Y. Li, J. Ahlert, *Bioorg. Chem.* **1999**, *27*, 172–188.

111 F. von Nussbaum, P. Spiteller, M. Rüth, W. Steglich, G. Wanner, B. Gamblin, L. Stievano, F. E. Wagner, *Angew. Chem. Int. Ed.* **1998**, *37*, 3292–3295.

112 For a review on Dap (≡ Dpr) see: R. Andruszkiewicz, *Pol. J. Chem.* **1995**, *69*, 1615–1629.

113 C. J. Schofield, M. W. Walter, *β-Lactam Chemistry*, in *Amino Acids, Peptides, and Proteins* **1999**, *30*, 335–397.

114 (a) K. L. Rinehart, K. Harada, M. Namikoshi, C. Chen, C. A. Harvis, M. H. G. Munro, J. W. Blunt, P. E. Mulligan, V. R. Beasley, A. M. Dahlem, W. W. Carmichael, *J. Am. Chem. Soc.* **1988**, *110*, 8557–8558. Synthesis of Adda: (b) C. Pearson, K. L. Rinehart, M. Sugano, J. R. Costerison, *Org. Lett.* **2000**, *2*, 2901–2903; (c) J. S. Panek, T. Hu, *J. Org. Chem.* **1997**, *62*, 4914–4915.

115 (a) W. H. Gerwick, L. Tong Tan, N. Sitachitta, *The Alkaloids*, **2001**, *57*, 75–184; (b) A. M. Burja, B. Banaigs, E. Abou-Mansour, J. G. Burgess, P. C. Wright, *Tetrahedron* **2001**, *57*, 9347–9377.

116 D. J. Faulkner, *Nat. Prod. Rep.* **2002**, *19*, 1–48.

117 D. J. Faulkner, H. Y. He, M. D. Unson, C. A. Bewley, M. J. Garson, *Gazz. Chim. Ital.* **1993**, *123*, 301–307.

118 H. Sone, T. Nemoto, H. Ishiwata, M. Ojika, K. Yamada, *Tetrahedron Lett.* **1993**, *34*, 8449–8452.

119 D. C. Carter, R. E. Moore, J. S. Mynderse, W. P. Niemczura, J. S. Todd, *J. Org. Chem.* **1984**, *49*, 236–241.

120 R. B. Bates, K. G. Brusoe, J. Burns, S. Caldera, W. Cui, S. Gangwar, M. R. Gramme, K. J. McClure, G. P. Rouen, H. Schadow, C. C. Stessman, S. R. Taylor, V. H. Vu, G. V. Yarick, J. Zhang, G. R. Pettit, R. Bontems, *J. Am. Chem. Soc.* **1997**, *119*, 2111–2113.

121 F. D. Horgen, W. Y. Yoshida, P. J. Scheuer, *J. Nat. Prod.* **2000**, *63*, 461–467.

122 H. Luesch, P. G. Williams, W. Y. Yoshida, R. E. Moore, V. J. Paul, *J. Nat. Prod.* **2002**, *65*, 996–1000.

123 Though kulokekahilide-1 has been isolated from the marine mollusk *Philinopsis speciosa*, this cytotoxic depsipeptide might originate from dietary cyanobacteria. J. Kimura, Y. Takada, T. Inayoshi, Y. Nakao, G. Goetz, W. Y. Yoshida, P. J. Scheuer, *J. Org. Chem.* **2002**, *67*, 1760–1767.

124 L. M. Nogle, W. H. Gerwick, *J. Nat. Prod.* **2002**, *65*, 21–24.

125 G. R. Pettit, J. Xu, F. Hogan, M. D. Williams, D. L. Doubek, J. M. Schmidt, R. L. Cerney, M. R. Boyd, *J. Nat. Prod.* **1997**, *60*, 752–754.

126 P. G. Williams, W. Y. Yoshida, R. E. Moore, V. J. Paul, *J. Nat. Prod.* **2002**, *65*, 29–31.

127 G. R. Pettit, J.-P. Xu, F. Hogan, R. L. Cerny, *Heterocycles* **1998**, *47*, 491–496.

128 J. Rodríguez, R. Fernandez, E. Quiñoa, R. Riguera, C. Debitus, P. Bouchet, *Tetrahedron Lett.* **1994**, *35*, 9239–9242.

129 (a) K. Sivonen, W. W. Carmichael, M. Namikoshi, K. L. Rinehart, A. M. Dahlem, S. I. Niemela, *Appl. Envrion. Microbiol.* **1990**, *56*, 2650–2657; (b) D. P. Botes, A. A. Tuinman, P. L. Wessels, C. C. Viljoen, H. Kruger,

D. H. Williams, S. Santikarn, R. J. Smith, S. J. Hammond, *J. Chem. Soc., Perkin Trans. 1* **1984**, 2311–2318; (c) G.-B. Trogen, A. Annila, J. Eriksson, M. Kontteli, J. Meriluoto, I. Sethson, J. Zudnek, U. Edlund, *Biochemistry* **1996**, *35*, 3197–3205; (d) J. M. Humphrey, J. B. Aggen, A. R. Chamberlin, *J. Am. Chem. Soc.* **1996**, *118*, 11759–11770.

130 M. Saito, A. Konno, H. Ishii, H. Saito, F. Nishida, T. Abe, C. Chen, *J. Nat. Prod.* **2001**, *64*, 139–141.

131 Isolation: (a) E. Dilip de Silva, D. E. Williams, R. J. Andersen, H. Klix, C. F. B. Holmes, T. M. Allen, *Tetrahedron Lett.* **1992**, *33*, 1561–1564. For a recent synthesis see: (b) T. Hu, J. S. Panek, *J. Am. Chem. Soc.* **2002**, *124*, 11368–11378; (c) S. M. Bauer, R. W. Armstrong, *J. Am. Chem. Soc.* **1999**, *121*, 6355–6366; (d) R. Samy, H. Y. Kim, M. Brady, P. L. Toogood, *J. Org. Chem.* **1999**, *64*, 2711–2728. First Synthesis: (e) R. L. Valentekovich, S. L. Schreiber, *J. Am. Chem. Soc.* **1995**, *117*, 9069–9070.

132 G. L. Helms, R. E. Moore, W. P. Niemczura, G. M. L. Patterson, K. B. Tomer, M. L. Gross, *J. Org. Chem.* **1988**, *53*, 1298–1307.

133 H. Luesch, G. G. Harrigan, G. Goetz, F. D. Horgen, *Curr. Med. Chem.* **2002**, *9*, 1791–1806.

134 G. G. Harrigan, W. Y. Yoshida, R. E. Moore, D. G. Nagle, P. U. Park, J. Biggs, V. J. Paul, S. L. Mooberry, T. H. Corbett, F. A. Valeriote, *J. Nat. Prod.* **1998**, *61*, 1221–1225.

135 Y. Unedo, S. Nagata, T. Tsutsumi, A. Hasegawa, M. F. Watanabe, H.-D. Park, G.-C. Chen, G. Chen, S.-Z. Yu, *Carcinogenesis* **1996**, *17*, 1317–1321.

136 C. A. Bewley, D. J. Faulkner, *Angew. Chem. Int. Ed.* **1998**, *37*, 2162–2178.

137 (a) M. Konishi, M. Nishino, K. Saitoh, T. Miyaki, T. Oki, K. Kawaguchi, *J. Antibiot.* **1989**, *42*, 1749–1755; (b) D. Jethwaney, M. Hofer, R. K. Khaware, R. Prasad, *Microbiology* **1997**, *143*, 397–404; (c) J. Capobianco, D. Z. Zakula, M. L. Coen, R. C. Goldman, *Biochem. Biophys. Res. Commun.* **1993**, *190*, 1037–1044.

138 (a) D. Jethwaney, M. Hofer, R. K. Khaware, R. Prasad, *Microbiology* **1997**, *143*, 397–404; (b) J. Capobianco, D. Z. Zakula, M. L. Coen, R. C. Goldman, *Biochem. Biophys. Res. Commun.* **1993**, *190*, 1037–1044.

139 In these cases the "formal" β-amino acid relationship often is a result of late stage condensation or cyclization reactions (e.g. Mannich-type, Pictet–Spengler) within the biosynthesis: Typical examples are cocaine and correlated tropane alkaloids, Catharanthus alkaloids or Iboga alkaloids like heyneanine.

140 T. Henkel, R. M. Brunne, H. Müller, F. Reichel, *Angew. Chem. Int. Ed.* **1999**, *38*, 643–647.

1.6
Biosynthesis of β-Amino Acids

Peter Spiteller and Franz von Nussbaum

1.6.1
Introduction

The biosynthetic pathways leading to genesis of β-amino acids that are structurally related to proteinogenic α-amino acids can be divided roughly into two groups. Most β-amino acids are derived from α-amino acids by a 2,3-shift of the α-amino group to the β position. In contrast, β-alanine and β-aminoisobutyrate are generated by degradation reactions. These fundamentally different biosynthetic pathways are discussed in the following sections.

1.6.2
Biosynthesis of β-Amino Acids by Catabolic Pathways

1.6.2.1
β-Alanine

Precursors of β-alanine **3** biosynthesis are uracil **2**, L-aspartate **4**, and polyamines which are subjected to degradation in distinct ways – whereas uracil is metabolized by hydrogenation followed by hydrolysis, polyamines like spermine **1** are oxidized and L-aspartate is decarboxylated (Scheme 1.6.1).

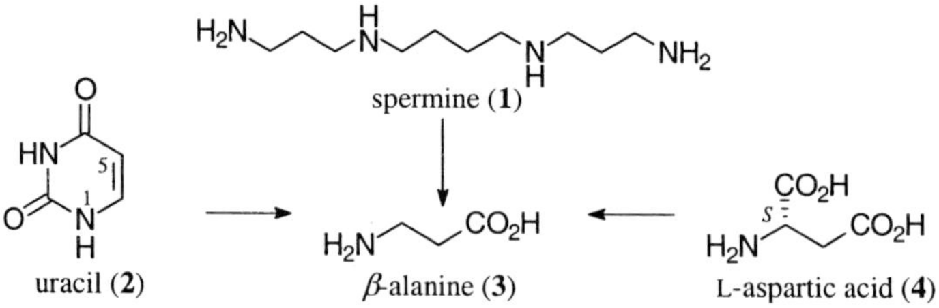

Scheme 1.6.1. Biosynthetic pathways to β-alanine **3**.

Highlights in Bioorganic Chemistry: Methods and Applications. Edited by Carsten Schmuck, Helma Wennemers.
Copyright © 2004 WILEY-VCH Verlag GmbH & Co. KGaA, Weinheim
ISBN: 3-527-30656-0

1.6.2.2
Biosynthesis of β-Alanine from Uracil

In mammals, uracil **2** and its precursor cytosine are transformed in a three step reaction to β-alanine **3** [1]. The enzymes involved are not very specific and also accept as substrates different 5-substituted pyrimidines, e.g. thymine which is similarly transformed to (R)-β-aminoisobutyric acid [(R)-**10**], as uracil is to β-alanine **3**. This was shown by feeding experiments with ^{14}C-labeled uracil and thymine, respectively [2, 3].

In the first step, the pyrimidine is hydrogenated at the double bond by dihydrouracil dehydrogenase (EC 1.3.1.2) to a dihydropyrimidine (Scheme 1.6.2). The enzymes obtained from rat [4] and human [5] liver have been purified and characterized. They were later subjected to molecular cloning [6]. The hydrogens are added to the double bond at the *Si* face of C5 and C6 in an anti-addition reaction. This was deduced from NMR spectra recorded from the isolated degradation products after administration of [5-^{2}H]- and [6-^{2}H]uracil and ^{2}H$_2$O to a mammalian enzyme system [7].

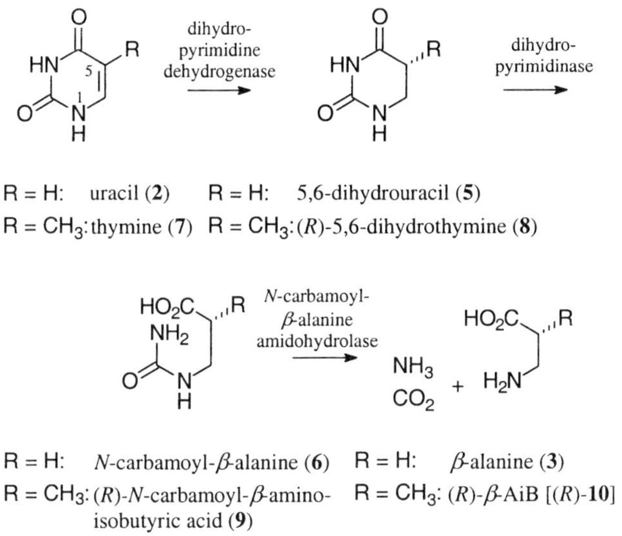

R = H: uracil (**2**)   R = H: 5,6-dihydrouracil (**5**)
R = CH$_3$: thymine (**7**)   R = CH$_3$: (R)-5,6-dihydrothymine (**8**)

R = H: N-carbamoyl-β-alanine (**6**)   R = H: β-alanine (**3**)
R = CH$_3$: (R)-N-carbamoyl-β-amino-   R = CH$_3$: (R)-β-AiB [(R)-**10**]
isobutyric acid (**9**)

Scheme 1.6.2. Biosynthesis of β-alanine **3** and (R)-β-aminoisobutyric acid [(R)-**10**], respectively by degradation of pyrimidines.

In the second step the dihydropyrimidines obtained are hydrolyzed by β-dihydropyrimidinase (EC 3.5.2.2) to N-carbamoyl-β-alanine **6** and N-carbamoyl-β-aminoisobutyrate **9**, respectively (Scheme 1.6.2). The enzyme isolated from rat liver was purified, characterized [8] and cloned [9].

1.6 Biosynthesis of β-Amino Acids

In the third step, the carbamoyl group is cleaved, with loss of CO_2 and NH_3, to generate β-alanine **3** and (R)-β-aminoisobutyric acid [(R)-**10**], respectively, by N-carbamoyl-β-alanine amidohydrolase (EC 3.5.1.6) (Scheme 1.6.2) [10]. The enzyme purified from rat liver is a hexamer [11].

1.6.2.3
Biosynthesis of β-Alanine from L-Aspartic Acid

In contrast to mammals, β-alanine **3** is generated in *Escherichia coli* [12] mainly by decarboxylation of L-aspartate **4** [13] (Scheme 1.6.3). The tetrameric enzyme, L-aspartate-α-decarboxylase (EC 4.1.1.15), was isolated from *E. coli* [13], cloned [14], and its crystal structure [15] was determined. In bacteria, pantothenate synthase converts β-alanine to pantothenate, a constituent of coenzyme A [16].

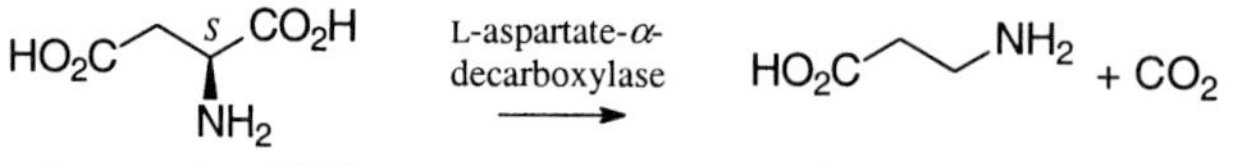

Scheme 1.6.3. Biosynthesis of β-alanine **3** by decarboxylation of (S)-aspartic acid **4**.

1.6.2.4
Biosynthesis of β-Alanine from Spermidine and Spermine

A third biosynthetic pathway to β-alanine **3** by oxidation of spermidine **13** and spermine **1** has been recently elucidated in yeast [17]. This reaction starts by transformation of L-methionine **15** to decarboxyadenosylmethionine (dcAdoMet, **16**). The latter alkylates putrescine **12** at one or both nitrogens generating spermidine **13** and spermine **1** which are oxidized to 3-aminopropanal **14** and further to β-alanine (Scheme 1.6.4).

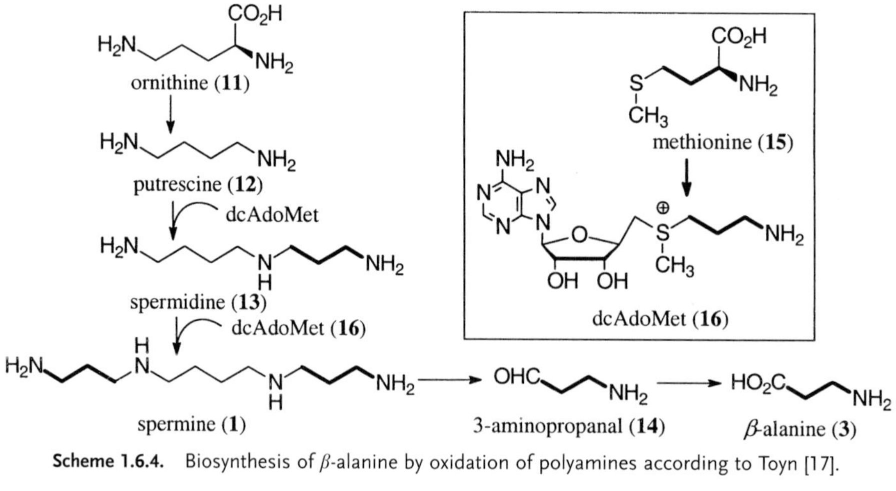

Scheme 1.6.4. Biosynthesis of β-alanine by oxidation of polyamines according to Toyn [17].

1.6.2.5
(R)- and (S)-β-Aminoisobutyrate

In mammals, thymine is transformed to (R)-β-aminoisobutyric acid [(R)-10] analogously to the conversion of uracil to β-alanine 3 (Scheme 1.6.2). (S)-β-aminoisobutyric acid [(S)-10] is a degradation product of L-valine 17 in mammals, generated in a multistep reaction via (S)-2-methylmalonic acid semialdehyde 23 [1] (Scheme 1.6.5).

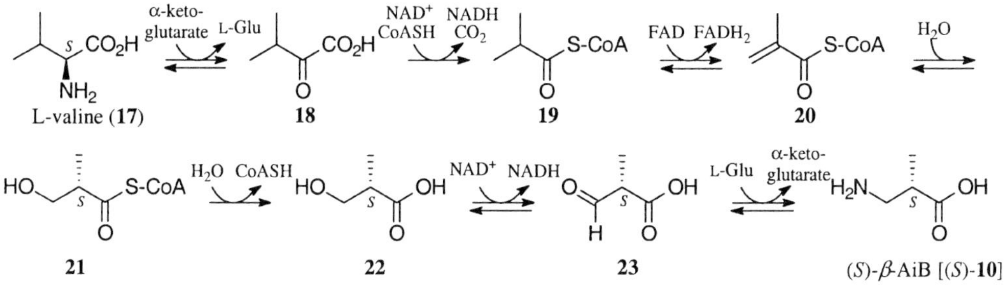

Scheme 1.6.5. Biosynthesis of (S)-β-aminoisobutyric acid [(S)-10].

1.6.3
Biosynthesis of β-Amino Acids by Aminomutases

In contrast to the catabolic nature of β-alanine and β-aminobutyric acid biosynthesis, the generation of higher β-amino acids usually starts from the corresponding α-amino acid. The shift of the amino group is catalyzed by different types of aminomutase.

The occurrence of aminomutases in nature was first recognized as a result of investigations by Stadtman [18] and Barker [19] with the aim of elucidating the metabolism of lysine 24 in *Clostridium subterminale* SB4 and *C. sticklandii* – lysine is transformed by rearrangement of the α-amino group to the β position, yielding L-β-lysine 25, which is then converted to 3,5-diaminohexanoic acid, as shown by use of [^{15}N]lysine [20, 21]. These findings lead both to the detection of a lysine 2,3-aminomutase and of a β-lysine 5,6-aminomutase in *C. sticklandii*. Interestingly, each enzyme represents a different type of aminomutase, because it was observed that lysine 2,3-aminomutase depends on SAM as cofactor whereas β-lysine 5,6-aminomutase depends on B_{12}. Several aminomutases are known; they will be discussed in detail in the following sections.

1.6.3.1
(S)-β-Lysine

In the bacteria *Clostridium subterminale* SB4 and *C. sticklandii* (S)-β-lysine 25 is generated from (S)-lysine by (S)-lysine 2,3-aminomutase [22–24].

1.6.3.2
Properties of the Enzyme

The action of a lysine 2,3-aminomutase (EC 5.4.3.2) was recognized as early as 1966 by Barker and coworkers in the bacterium *C. subterminale* SB4 [25, 26]. This enzyme was recently cloned, sequenced, and overexpressed in *E. coli* [27]. The enzyme has a molecular weight of 285 kDa and is a hexamer [28]. Activation of the enzyme requires S-adenosylmethionine (SAM), pyridoxal phosphate (PLP), and iron ions. It is rapidly inactivated by oxygen. The enzyme contains three [4Fe–4S] clusters [29] and six molecules of PLP per hexamer [28]. Comparison of the amino acid sequence with that of other SAM binding proteins revealed the presence of a glycine-rich highly conserved region that might represent a SAM binding domain. Three highly conserved cysteines are also present in the protein; these might constitute three of the four ligands of the iron–sulfur cluster.

Recently, a new SAM dependent lysine 2,3-aminomutase was detected and characterized in *Bacillus subtilis*. Unlike the enzyme from *C. subterminale* SB4, the enzyme in *B. subtilis* apparently consists of four identical subunits each with a molecular mass of 54 kDa [30]. A PLP binding motif was identified in this aminomutase that is also highly conserved in other lysine 2,3-aminomutases [31].

1.6.3.3
Stereochemical Aspects

The stereochemistry of the reaction catalyzed by the lysine 2,3-aminomutase in *C. subterminale* SB4 was elucidated in detail by Aberhardt and Gould [32]. Incubation experiments with cell-free extracts of *C. subterminale* SB4 and (2RS)-[3-^{13}C,2-^{15}N]lysine and NMR spectroscopy of the isolated β-lysine as the di-N-phthaloyl ethyl ester derivative revealed that the amino group migrates in an intramolecular reaction to position 3S in β-lysine **25**.

Incubation with (2RS,3R)-[3-^{2}H]lysine and with (2RS,3S)-[3-^{2}H]lysine, respectively, and subsequent ^{2}H NMR spectroscopy of the isolated β-lysine proved that the *pro-3R* hydrogen of α-lysine **24** is transferred to the *pro-2R* position in **25** whereas the *pro-3S* hydrogen is retained at C3 (Scheme 1.6.6).

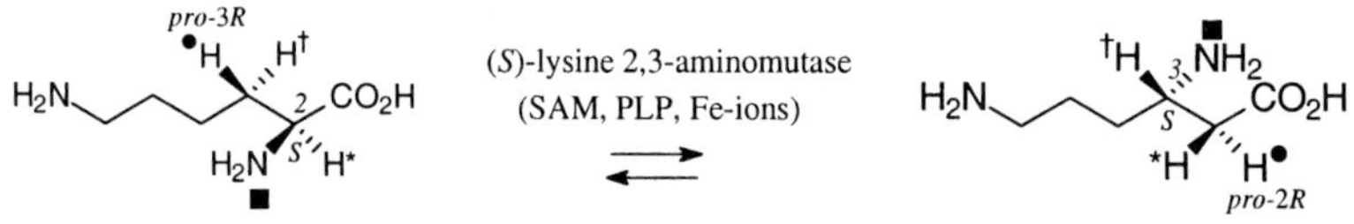

Scheme 1.6.6. Stereochemical aspects of the reaction catalyzed by (S)-lysine 2,3-aminomutase.

1.6.3.4
Reaction Mechanism

Further insight into the reaction mechanism was gathered by Frey, who found that SAM plays a similar role to adenosylcobalamin for generation of radicals, recognized by proving the existence of a 5′-deoxyadenosyl radical as intermediate. Because SAM is a much simpler molecule than cobalamin, Frey refers to SAM as "a poor man's adenosylcobalamin" [33, 34]. To confirm the role of SAM, non-stereospecifically labeled [5′-^{3}H]SAM was applied to the incubation mixture [35]. The tritium label was found to be completely transferred to the substrate and the product indicating that the 5′-deoxyadenosyl moiety **31** is cleaved from SAM. The resulting active species **31** should be involved in the transfer of the *pro*-3R hydrogen of α-lysine **24** to the *pro*-2R position in β-lysine **25**. Frey [36] suggests that the reaction proceeds after cleavage of SAM and generation of the 5′-deoxyadenosyl radical **31** by abstraction of the *pro*-3R hydrogen from α-lysine bound as aldimine **26** to PLP. In the next step the α-lysyl radical **27** is isomerized to the β-lysyl radical **29**. Abstraction of a hydrogen from the methyl group of 5′-deoxyadenosine generates a β-lysyl aldimine **30**, that is finally hydrolyzed to β-lysine (Scheme 1.6.7). This reaction sequence implies regeneration of the 5′-deoxyadenosyl radical.

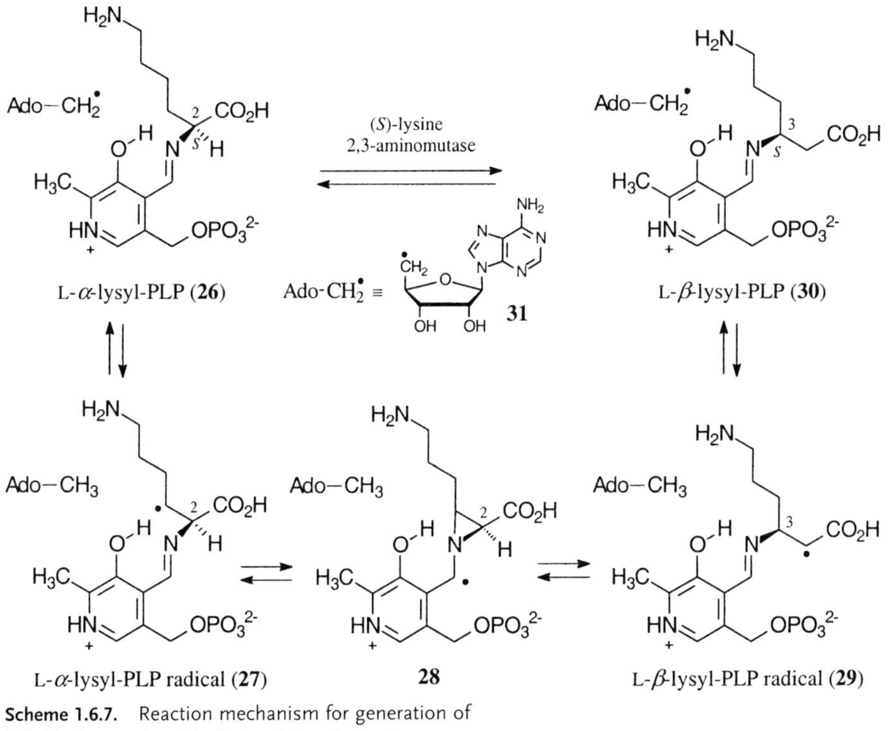

Scheme 1.6.7. Reaction mechanism for generation of (S)-β-lysine catalyzed by lysine 2,3-aminomutase.

The reaction mechanism outlined was confirmed by direct observation of the β-lysyl-PLP radical **29** by EPR methods [37]. A strong signal was detected in the EPR spectrum by incubation of lysine 2,3-aminomutase with α-lysine and SAM and subsequent freezing in the steady state with liquid N_2. The presence of a β-lysine-

PLP radical was proved by application of [2-^{2}H]lysine and [2-^{13}C]lysine instead of lysine **24** to the reaction mixture [38]. In the first experiment the EPR signal was narrowed; in the second it was broadened. The involvement of PLP in the reaction was proven by ESEEM spectroscopy. By incubation of the aminomutase with lysine, SAM, and [4′-^{2}H]PLP a prominent doublet centered at the Lamour frequency for ^{2}H was recognized [39], in accordance with the structure of an external aldimine. These findings establish a new role for PLP in enzyme reactions – PLP facilitates the radical isomerization.

Other radicals could not be detected directly by EPR, because of their instability. By use of analogous molecules which stabilize these radicals, however, their existence was rationalized. Stabilization of an α-lysyl radical analog was achieved by use of 4-thialysine **32** [40] (Scheme 1.6.8), with *trans*-4,5-dehydrolysine [41] as substrate.

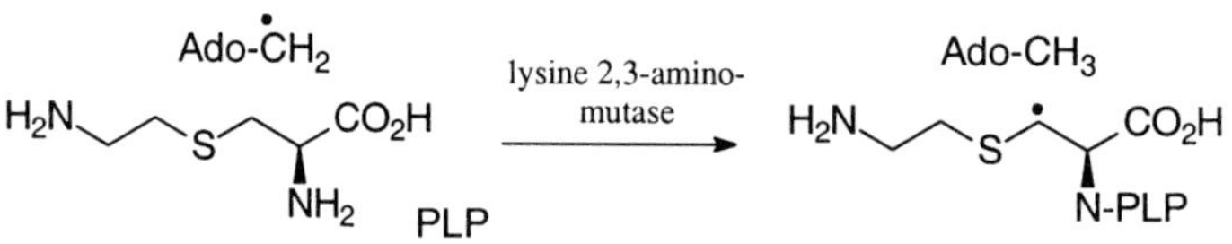

4-thialysine (**32**) 4-thialysyl-PLP radical (**33**)

Scheme 1.6.8. Incubation of 4-thialysine with lysine 2,3-aminomutase generates 4-thialysyl-PLP radicals.

It was also possible to confirm by EPR the existence of the 5′-deoxyadenosyl radical species by use of 3′,4′-anhydroadenosylmethionine **34** as allyl analog of SAM [42] (Scheme 1.6.9).

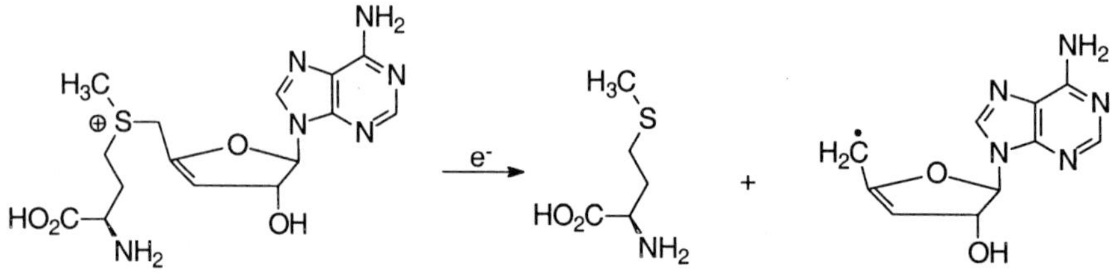

3′,4′-anhydroadenosylmethionine (**34**) methionine (**15**) 3′,4′-anhydro-5′-deoxyadenosyl radical (**35**)

Scheme 1.6.9. Generation of the allyl analog **35** of the 5′-deoxyadenosyl radical **31**.

This reaction mechanism seems not to be restricted to the lysine 2,3-aminomutase itself. The cobalamin-dependent lysine 5,6-aminomutase and the ornithine 4,5-aminomutase from *C. sticklandii* follow apparently the same reaction mechanism except that they need B_{12} instead of SAM as cofactor.

1.6.3.5
(R)-β-Leucine

Overton and coworkers discovered a leucine 2,3-aminomutase in plant tissue cultures of *Andrographis paniculata* that converts (S)-leucine in (R)-β-leucine [43] (Scheme 1.6.10). The enzyme activity was investigated in cell free extracts by incubation with (S)-[U-^{14}C]leucine and by measuring the radioactivity of the methyl ester camphanamide derivatives of the reaction mixtures by radio-GC. The stereochemistry of the β-amino acid was determined by radio-GC comparison of the enzyme reaction product as methyl ester camphanamide derivative with an authentic sample. The enzyme is not dependent on cobalamin, because addition of intrinsic factor does not induce its inhibition.

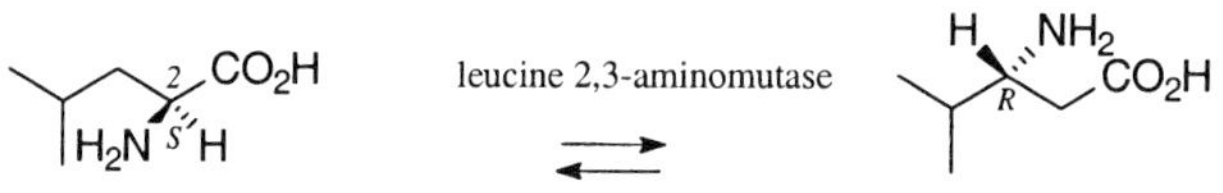

L-α-leucine (**36**) → leucine 2,3-aminomutase → L-β-leucine (**37**)

Scheme 1.6.10. Biosynthesis of (R)-β-Leucine **37** in *Andrographis paniculata*.

The presence of B_{12}-dependent leucine 2,3-aminomutases in bacteria, mammals, and plants has also been reported [44–46]. Their existence was deduced only indirectly. B_{12} itself has not yet been detected in plants, however. Later Stabler [47] and Aberhart [48] reported independently they were unable to detect β-leucine and leucine 2,3-aminomutase activity in human blood and rat liver. The presence in these of B_{12}-dependent leucine 2,3-aminomutases is therefore questionable.

1.6.3.6
(S)-β-Arginine

β-Arginine **39** itself has not yet been found in nature, although a β-arginine moiety is incorporated into the antibiotics blasticidin S and H [49] from *Streptomyces griseochromogenes*, and in LL-BM547β [50]. The conversion of arginine to β-arginine in *S. griseochromogenes* has been investigated in detail by chemical methods [51, 52]. Administration of (rac)-[3-^{13}C,2-^{15}N]arginine to bacteria cultures and NMR spectroscopic analysis of the isolated blasticidin S revealed the retention of the α-amino group and its intramolecular migration to the β position, suggesting the presence of an arginine 2,3-aminomutase. Further feeding experiments with (3R,2RS)-[3-^{2}H]arginine and (3S,2RS)-[3-^{2}H]arginine revealed that the *pro*-3R hydrogen from arginine migrates to position 2 in β-arginine whereas the *pro*-3S hydrogen remains at C3. (Scheme 1.6.11). These results are in perfect agreement with those derived from incubation experiments with lysine 2,3-aminomutase. The arginine 2,3-aminomutase might therefore belong to the same enzyme family as the lysine 2,3-aminomutase. The cofactor requirements of the enzyme have not been investigated.

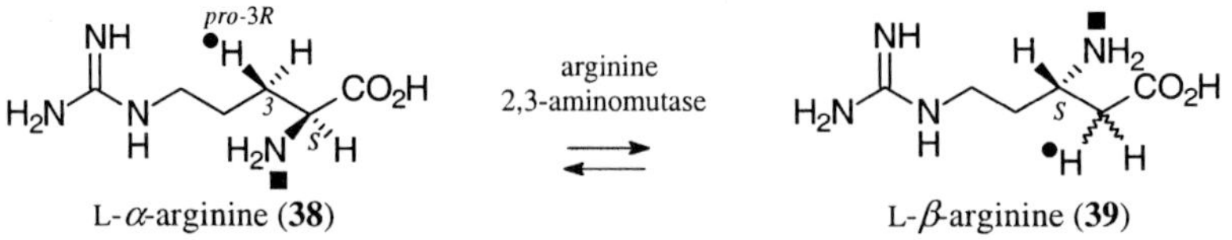

L-α-arginine (**38**)   L-β-arginine (**39**)

Scheme 1.6.11. Biosynthesis of (S)-β-arginine **38** in *Streptomyces griseochromogenes*.

1.6.3.7
(R)-β-Phenylalanine

Biosynthesis of (R)-β-Phenylalanine in *Taxus brevifolia* The phenylisoserine side-chain of taxol in *Taxus brevifolia* originates from (R)-β-phenylalanine **41** [53, 54]. Floss and coworkers elucidated the stereochemistry of the phenylalanine 2,3-aminomutase reaction in *Taxus brevifolia* in detail by incubating cell-free extracts of *T. brevifolia* with a series of isotopically labeled phenylalanines [55]. The configuration of β-phenylalanine **41** was established as *R* by HPLC comparison of the (1S)-camphanate methyl ester of the aminomutase reaction product with authentic samples. The configuration of **41** corresponds to that of Winterstein's acid (see below). Incubation with (S)-[2-^{15}N,ring-^{2}H$_5$]phenylalanine generated, according to results from GC–MS analysis of the *N*-benzoyl methyl ester derivative, exclusively (R)-[3-^{15}N,ring-^{2}H$_5$]-β-phenylalanine, indicating a strictly intramolecular shift of the amino group. Incubation experiments with (2S,3R)-[ring,3-^{2}H$_6$]phenylalanine and (2S,3S)-[ring,3-^{2}H$_6$]phenylalanine revealed that the *pro*-3*R* hydrogen atom remains at C3 whereas the *pro*-3*S* hydrogen atom of **40** migrates to position 2 in **41** (Scheme 1.6.12). In this regard, the phenylalanine 2,3-aminomutase differs from the lysine 2,3-aminomutase in *C. subterminale* SB4; the mutase in *T. brevifolia* might therefore constitute a new type of aminomutase.

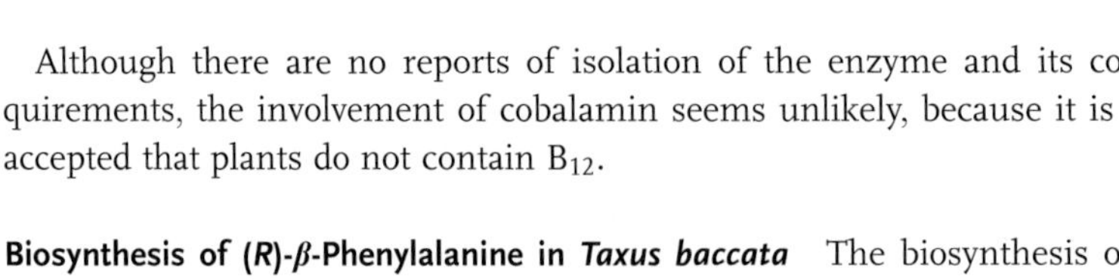

L-α-phenylalanine (**40**)   L-β-phenylalanine (**41**)

Scheme 1.6.12. Biosynthesis of (R)-β-phenylalanine **41** in *Taxus brevifolia*.

Although there are no reports of isolation of the enzyme and its cofactor requirements, the involvement of cobalamin seems unlikely, because it is generally accepted that plants do not contain B$_{12}$.

Biosynthesis of (R)-β-Phenylalanine in *Taxus baccata* The biosynthesis of the (R)-3-(dimethylamino)-3-phenylpropionic acid (Winterstein's acid) moiety of taxine A [56] and taxine B [57] in *Taxus baccata* has been reported by Haslam and coworkers [58]. They investigated whether β-phenylalanine is generated by an aminomutase or by addition of ammonia to cinnamic acid. Their findings provide evidence for the action of a phenylalanine 2,3-aminomutase *in T. baccata*, because (2*S*)-[2-^{14}C]phenylalanine is incorporated 10 to 100 times more effectively into the Winterstein's acid moiety than either [2-^{14}C]- or [3-^{14}C]cinnamic acid. The stereochemistry of Winterstein's acid was determined to be *R* [59]. According to Haslam [58], the aminomutase reaction in *T. baccata* is combined with loss of 91% of the radioactivity if the *pro*-3*R* was labeled with tritium and loss of 33% of the *pro*-3*S* hydrogen atom if that atom was labeled, giving a hint that the *pro*-3*S* hydrogen is retained on C-3 (Scheme 1.6.13). This conclusion would, however, be in contradiction to results from the aminomutase in *T. brevifolia* [55]. Reinvestigation is therefore desirable to clarify whether different types of aminomutase are present in the closely related species *T. baccata* and *T. brevifolia*.

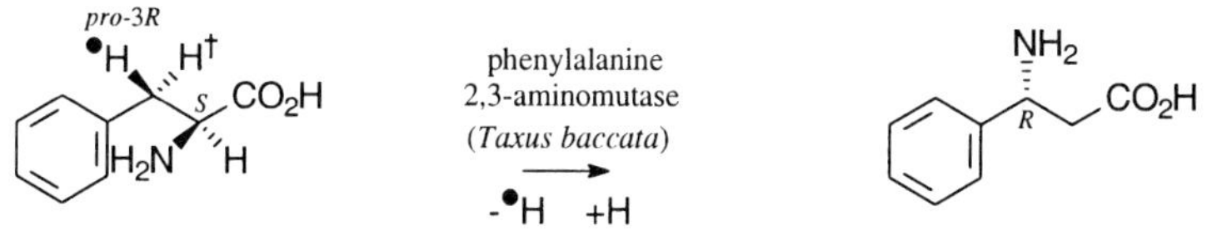

L-α-phenylalanine (**40**) → phenylalanine 2,3-aminomutase (*Taxus baccata*) → L-β-phenylalanine (**41**)

Scheme 1.6.13. Biosynthesis of (*R*)-β-phenylalanine **41** in *Taxus baccata* according to Haslam.

1.6.3.8
β-Tyrosine

Biosynthesis of (S)-β-Tyrosine in *Bacillus brevis* Vm4   β-Tyrosine **43** is a constituent of the peptide antibiotics edeine A and B [60] obtained from cultures of *Bacillus brevis* Vm4. β-Tyrosine is derived from α-tyrosine **42** by use of a tyrosine 2,3-aminomutase [61]. The purified enzyme has properties fundamentally different from those of all other aminomutases so far mentioned. It requires ATP and Mg^{2+} ions, but no other cofactors.

The stereochemistry of **43** was established to be *S* by comparing the optical rotation of an authentic sample derived from (*R*)-β-tyrosine hydrochloride with β-tyrosine hydrochloride isolated from edeine A and B [62]. Incubation experiments with [^{15}N]tyrosine revealed that the [^{15}N]amino group in β-tyrosine is lost, suggesting a reaction mechanism similar to that of an ammonia lyase [62]. This finding is supported by further incubation experiments with (2*RS*,3*R*)-[3-^{3}H]tyrosine and (2*RS*,3*S*)-[3-^{3}H]tyrosine in combination with (2*RS*)-[3-^{14}C]tyrosine. Determination of the ^{3}H/^{14}C ratios of the isolated β-tyrosine **43** leads to the conclusion that the *pro*-3*S* hydrogen is lost in the course of the reaction whereas the *pro*-3*R* hydrogen is retained (Scheme 1.6.14). A similar incubation experiment with (2*S*)-[2-^{3}H]tyrosine proceeds with a loss of most of the tritium label from C2.

Biosynthesis of (R)-β-Tyrosine and (R)-β-Dopa in *Cortinarius violaceus*   Fruit bodies of the higher fungus *Cortinarius violaceus* produce (*R*)-β-dopa **44** [63]. The biosyn-

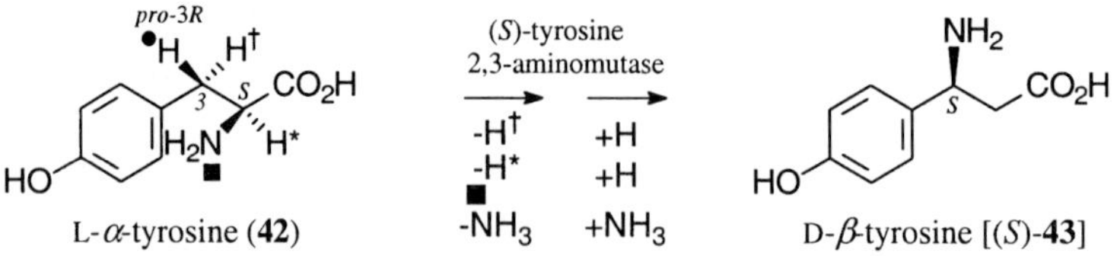

1.6 Biosynthesis of β-Amino Acids

Scheme 1.6.14. Biosynthesis of (S)-β-tyrosine in *Bacillus brevis* Vm4.

thesis of (R)-**44** was investigated by in-vivo feeding experiments mainly with fluorine-labeled precursors [64]. The incorporation was monitored by GC–MS analysis of the pertrimethylsilyl derivatives of β-dopa. The latter is synthesized in the mushroom by hydroxylation of β-tyrosine **43**, as demonstrated by incorporation of (S)-3-fluorotyrosine and (RS)-3-fluoro-β-tyrosine; incorporation of (RS)-6-fluoro-β-dopa was not observed, suggesting the presence of a tyrosine 2,3-aminomutase. The stereochemistry of β-tyrosine and β-dopa was determined to be R by GC–MS comparison of a Mosher amide derivative of natural **44** with a corresponding authentic sample. Administration of (rac)-3-fluoro[^{15}N]tyrosine showed that the shift of the amino group proceeds with retention of the nitrogen, indicating the involvement of a mutase acting similarly to the lysine 2,3-aminomutase. This deduction was confirmed by feeding of 3-fluoro-[3′,3′-^{2}H$_2$]tyrosine resulting in formation of 5-fluoro-[2′,3′-^{2}H$_2$]-β-dopa, and revealed an (at least partial) intramolecular transfer of a hydrogen from C3 to C2 (Scheme 1.6.15).

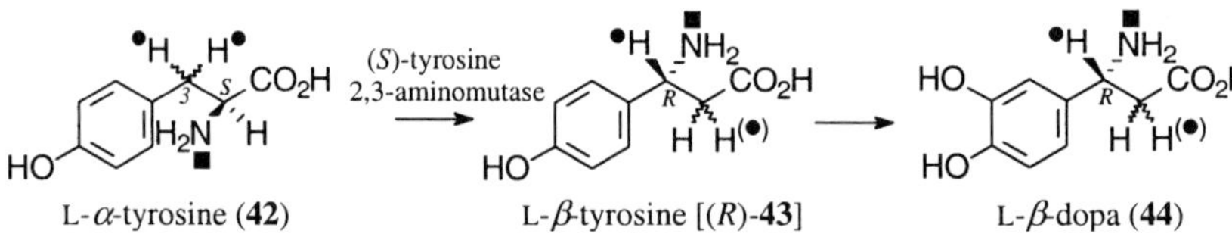

Scheme 1.6.15. Biosynthesis of (R)-β-tyrosine and (R)-β-dopa in *Cortinarius violaceus*.

These experiments do not, nevertheless, enable conclusions to be drawn about which of the hydrogen atoms at C3 is transferred to C2. It therefore remains an open question whether the mutase of the mushroom resembles more the phenylalanine 2,3-aminomutase from *Taxus brevifolia* or the lysine 2,3-aminomutase from *Clostridium subterminale* SB4. Because of the lack of any evidence of the occurrence of B$_{12}$ in higher fungi [65], involvement of B$_{12}$ in the aminomutase reaction is unlikely.

1.6.4
Other Aminomutases

Some other aminomutases, for example β-lysine 5,6-aminomutase (EC 5.4.3.3), D-lysine 5,6-aminomutase (EC 5.4.3.4), and D-ornithine 4,5-aminomutase (EC

5.4.3.5), are not involved in the shift of the amino group from C2 to C3 in amino acids. The should, nevertheless, be mentioned briefly, because β-amino acids are partially substrates for this class of enzyme. In addition, they were the first examples of the action of B_{12}-dependent aminomutases [66, 67].

1.6.4.1
β-Lysine 5,6-Aminomutase (D-Lysine 5,6-Aminomutase)

Enzyme Properties An enzyme complex containing a β-lysine 5,6-aminomutase (EC 5.4.3.3) and a D-lysine 5,6-aminomutase (EC 5.4.3.4) has been isolated from *Clostridium sticklandii*. The enzyme was purified as a single protein complex that accepts two different substrates, both L-β-lysine and D-lysine. The products of the mutase reaction are (3S,5S)-diaminohexanoic acid **45** and (2R,5S)-diaminohexanoic acid, respectively. Approximately 35 years ago the enzyme was discovered by Stadtman and coworkers [68]. Recently the two genes encoding the aminomutase have been cloned, sequenced, and heterologously expressed in *E. coli* [69]. The enzyme is a 170 kD complex consisting of two 55 kDa and 30 two kDa subunits. The aminomutase is stimulated by B_{12} and pyridoxal phosphate. A rapid turnover-associated inactivation occurs both for the enzyme purified from *C. sticklandii* and in the recombinant protein. Analysis of the 5,6-aminomutase amino acid sequence reveals a region in the small subunit that has similarities with other cobalt-dependent mutases probably responsible for the binding of B_{12} [69]. In the D-lysine 5,6-aminomutase from *Porphyromonas gingivalis* an essential lysine residue was detected that is part of a PLP binding motif [70].

Stereochemical Aspects The configuration of the product of the β-lysine 2,3-aminomutase reaction was established to be 3S,5S by NMR spectroscopic comparison of the δ-lactam of the 3,5-diaminohexanoic acid **45** from *C. sticklandii* with authentic samples of both pairs of diastereomers [71]. The stereochemical mechanism of the β-lysine 5,6-aminomutase reaction was investigated by means of tritium labeling. It turned out that the *pro*-5S hydrogen atom migrates to position 6 in the 3,5-diaminohexanoic acid [72] (Scheme 1.6.16).

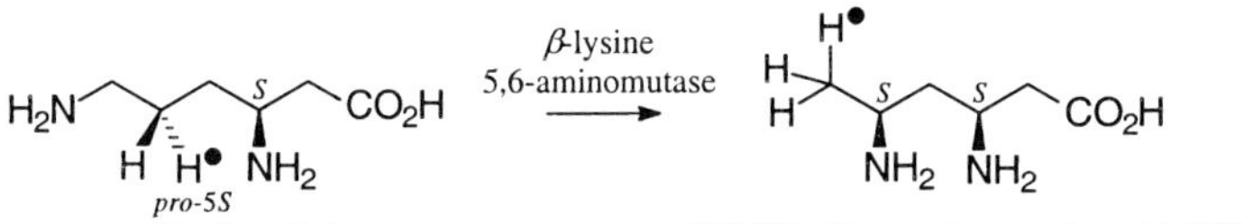

Scheme 1.6.16. Biosynthesis of (3S,5S)-diaminohexanoic acid in *Clostridium sticklandii*.

Reaction Mechanism Although the β-lysine 5,6-aminomutase requires cobalamin as cofactor instead of SAM its reaction mechanism seems to be similar to that of the lysine 2,3-aminomutase [73]. Experiments with tritium-labeled lysine and B_{12} showed that B_{12} is directly involved in the hydrogen shift from position 5 in D-lysine to position 6 in 2,5-diaminohexanoic acid [74]. Also PLP is required for enzyme activity [75]. The task of PLP in the mutase reaction is recognized to stabilize the radical intermediates by imine formation [76, 77].

1.6.4.2
D-Ornithine 4,5-Aminomutase

Besides a lysine 5,6-aminomutase, *Clostridium sticklandii* also has a D-ornithine 4,5-aminomutase (EC 5.4.3.5) [78, 79]. D-Ornithine is generated from L-ornithine by ornithine racemase [80]. The two genes encoding D-ornithine 4,5-aminomutase have been cloned, sequenced, and expressed in *E. coli* [81]. The enzyme is an $\alpha_2\beta_2$-heterotetramer, consisting of 12 800 Da and 82 900 Da subunits. The protein requires B_{12} and pyridoxal phosphate as cofactors. Similar to the lysine 5,6-aminomutase, a conserved base-off/histidine-on cobalamin binding motif is present in the 82 900 Da protein.

D-ornithine 4,5-aminomutase catalyzes the reversible interconversion of D-ornithine **11** to (2R,4S)-diaminopentanoic acid **46** [78] (Scheme 1.6.17).

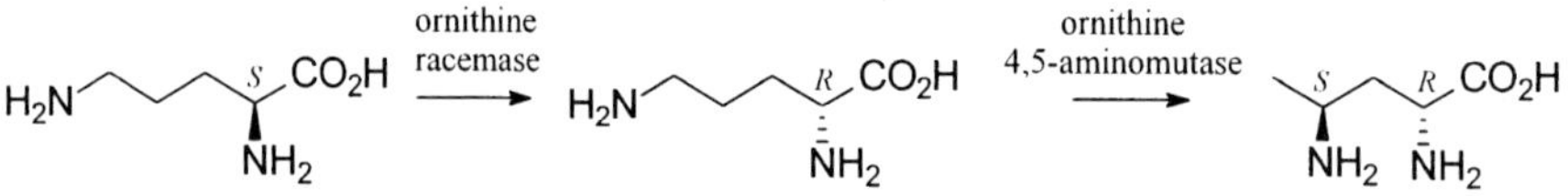

Scheme 1.6.17. Biosynthesis of (2R,4S)-diaminopentanoic acid in *Clostridium sticklandii*.

1.6.5
Discussion

Considering that few biosynthetic pathways to β-amino acids have yet been investigated, it seems nevertheless possible to conclude that most β-amino acids that are structurally related to proteinogenic α-amino acids are generated from α-amino acids by aminomutases (Table 1.6.1).

So far, two types of aminomutase have been investigated in detail. Lysine 2,3-aminomutase from *Clostridium subterminale* SB4 is the *example par excellence* for the SAM-dependent type of aminomutase. Several other enzymes belonging to the same family are known. Examples are biotin synthase [82], pyruvate formate lyase [83, 84], and anaerobic ribonucleotide reductase [85].

The second class of aminomutase is represented by D-ornithine 4,5-aminomutase and β-lysine 5,6-aminomutase. Although cobalamin is used as cofactor instead of SAM, the mechanism of the aminomutase reaction is similar.

Interestingly, all known cobalamin-dependent aminomutases seem to catalyze a shift of the ω amino group whereas the SAM dependent lysine 2,3-aminomutase catalyzes the shift of the α amino group. The latter is true for phenylalanine 2,3-aminomutase in *Taxus brevifolia* and tyrosine 2,3-aminomutase in *Cortinarius vio-*

Tab. 1.6.1. Characteristic properties of aminomutases.

Enzyme	Substrate	Product	Cofactor	NH$_2$ migrates to	H migrates From	H migrates To
Lysine 2,3-aminomutase	L-Lysine	L-β-Lysine	SAM	3S ≡ 3L	pro-3R	pro-2R
Leucine 2,3-aminomutase (A. p.)	L-Leucine	L-β-Leucine	?[a]	?	?	?
Arginine 2,3-aminomutase	L-Arginine	L-β-Arginine	?	3S ≡ 3L	pro-3R	2
Phenylalanine 2,3-aminomutase (T. br.)	L-Phenylalanine	L-β-Phenylalanine	?[a]	3R ≡ 3L	pro-3S	2
Phenylalanine 2,3-aminomutase (T. ba.)	L-Phenylalanine	L-β-Phenylalanine	?[a]	3R ≡ 3L?	pro-3R?	Solvent?
Tyrosine 2,3-aminomutase (B. b.)	L-Tyrosine	D-β-Tyrosine	ATP	Solvent[b]	pro-3S	Solvent
Tyrosine 2,3-aminomutase (C. v.)	L-Tyrosine	L-β-Tyrosine	?[a]	3R ≡ 3L	3	2
β-Lysine 5,6-aminomutase	L-β-Lysine	(3S,5S)-Diaminohexanoic acid	B$_{12}$	5S ≡ 5L	pro-5S	6
D-Lysine 5,6-aminomutase	D-Lysine	(2R,5S)-Diaminohexanoic acid	B$_{12}$	5S ≡ 5L	pro-5S	6
D-Ornithine 4,5-aminomutase	D-Ornithine	(2R,4S)-Diaminopentanoic acid	B$_{12}$	4S ≡ 4L	?	?

A. p. = Andrographis paniculata; T. br. = Taxus brevifolia; T. ba. = Taxus baccata; C. v. = Cortinarius violaceus; B. b. = Bacillus brevis Vm4
[a] Involvement of B$_{12}$ is unlikely
[b] An ^{15}N label in tyrosine is lost in the product β-tyrosine

laceus. These enzymes are unlikely to use B_{12}. These enzymes have, on the other hand, not yet been isolated nor investigated in detail, so their classification remains speculative. The tyrosine 2,3-aminomutase of *Bacillus brevis* Vm4 differs fundamentally from all the aminomutases mentioned and its properties are more like those of an ammonia lyase.

In recent years rapid progress was made in the study of aminomutase enzymes at the genetic level and in heterologous expression of aminomutase genes in *E. coli*. So far, lysine 2,3-aminomutase, D-ornithine 4,5-aminomutase and β-lysine 5,6-aminomutase have been cloned and overexpressed. By this method large amounts of enzymes become available. This will facilitate the detailed study of the three-dimensional structure and the mode of action of the enzymes. Bioinformatic methods are already used to identify new aminomutases and will be applied increasingly in the future [86]. The detection of biosynthetic gene clusters is also revealing the existence of new aminomutases – the gene cluster for the biosynthesis of the endiyne antibiotic C-1027 in *Streptomyces globisporus* was recently detected. According to sequence information the 3-chloro-β-dopa moiety in C-1027 is generated by the gene *sgcC5* that was assigned to function as an aminomutase [87].

Dedication

Dedicated to Professor Wolfgang Steglich on the occasion of his 70th birthday.

Acknowledgment

We are grateful to Dr Norbert Arnold, Dr Marc Stadler, Dr Thomas Koch, and Professor Dr Tin-Wein Yu for many valuable hints and suggestions, and to the Alexander von Humboldt-Stiftung for financial support by providing a Feodor-Lynen-Stipendium to P.S. and F.v.N.

References

1 Review: O. W. Griffith, *Annu. Rev. Biochem.* **1986**, *55*, 855–878.
2 K. Fink, R. E. Cline, R. B. Henderson, R. M. Fink, *J. Biol. Chem.* **1956**, *221*, 425–433.
3 P. Fritzson, A. Pihl, *J. Biol. Chem.* **1957**, *226*, 229–235.
4 T. Shiotani, G. Weber, *J. Biol. Chem.* **1981**, *256*, 219–224.
5 Z.-H. Lu, R. Zhang, R. B. Diasio, *J. Biol. Chem.* **1992**, *267*, 17102–17109.
6 X. Wei, G. Elizondo, A. Sapone, H. L. McLeod, H. Raunio, P. Fernandez-Saiguero, F. J. Gonzalez, *Genomics*, **1998**, *51*, 391–400.
7 D. Gani, D. W. Young, *J. Chem Soc. Perkin Trans I*, **1985**, 1355–1362.
8 M. Kikugawa, M. Kaneko, S. Fujimoto-Sakata, M. Maeda, K. Kawasaki, T. Takagi, N. Tamaki, *Eur. J. Biochem.* **1994**, *219*, 393–399.
9 K. Matsuda, S. Sakata, M. Kaneko, N. Hamajima, M. Nonaka, M. Sasaki, N. Tamaki, *Biochim. Biophys. Acta*, **1996**, *1307*, 140–144.
10 M. M. Matthews, T. W. Traut, *J. Biol. Chem.* **1987**, *262*, 7232–7237.

11 N. Tamaki, N. Mizutani, M. Kikugawa, S. Fujimoto, C. Mizota, *Eur. J. Biochem.* **1987**, *169*, 21–26.
12 J. E. Cronan Jr, *J. Bacteriol.* **1980**, *141*, 1291–1297.
13 J. M. Williamson, G. M. Brown, *J. Biol. Chem.* **1979**, *254*, 8074–8082.
14 M. K. Ramjee, U. Genschel, C. Abell, A. G. Smith, *Biochem. J.* **1997**, *323*, 661–669.
15 A. Albert, V. Dhanaraj, U. Genschel, G. Khan, M. K. Ramjee, R. Pulido, B. L. Sibanda, F. von Delft, M. Witty, T. L. Blundell, A. G. Smith, C. Abell, *Nat. Struct. Biol.* **1998**, *5*, 289–293.
16 R. Zheng, J. S. Blanchard, *Biochemistry* **2001**, *40*, 12904–12912.
17 W. H. White, P. L. Gunyuzlu, J. H. Toyn, *J. Biol. Chem.* **2001**, *276*, 10794–10800.
18 Review: T. C. Stadtman, *Adv. Enzymol. Relat. Areas Mol. Biol.* **1973**, *38*, 413–448.
19 V. Zappia, H. A. Barker, *Biochim. Biophys. Acta*, **1970**, *207*, 505–513.
20 R. C. Bray, T. C. Stadtman, *J. Biol. Chem.* **1968**, *243*, 381–385.
21 L. Tsai, T. C. Stadtman, *Arch. Biochem. Biophys.* **1968**, *125*, 210–225.
22 Review: P. A. Frey, *Curr. Opin. Chem. Biol.* **1997**, *1*, 347–356.
23 Review: P. A. Frey, S. J. Booker, *Adv. Protein Chem.* **2001**, *58*, 1–46.
24 Review: P. A. Frey, *Annu. Rev. Biochem.* **2001**, *70*, 121–148.
25 R. N. Costilow, O. M. Rochovansky, H. A. Barker, *J. Biol. Chem.* **1966**, *241*, 1573–1580.
26 T. P. Chirpich, V. Zappia, R. N. Costilow, H. A. Barker, *J. Biol. Chem.* **1970**, *245*, 1778–1789.
27 F. J. Ruzicka, K. W. Lieder, P. A. Frey, *J. Bacteriol.* **2000**, *182*, 469–476.
28 K. B. Song, P. A. Frey, *J. Biol. Chem.* **1991**, *266*, 7651–7655.
29 R. M. Petrovich, F. J. Ruzicka, G. H. Reed, P. A. Frey, *Biochemistry*, **1992**, *31*, 10774–10781.
30 D. Chen, F. J. Ruzicka, P. A. Frey, *Biochem. J.* **2000**, *348*, 539–549.
31 D. Chen, P. A. Frey, *Biochemistry* **2001**, *40*, 596–602.
32 D. J. Aberhart, S. J. Gould, H.-J. Lin, T. K. Thiruvengadam, B. H. Weiller, *J. Am. Chem. Soc.* **1983**, *105*, 5461–5470.
33 P. A. Frey, M. D. Ballinger, G. H. Reed, *Biochem. Soc. Trans.* **1988**, *26*, 304–310.
34 P. A. Frey, *FASEB J.* **1993**, *7*, 662–670.
35 J. Baraniak, M. L. Moss, P. A. Frey, *J. Biol. Chem.* **1989**, *264*, 1357–1360.
36 P. A. Frey, G. H. Reed, *Arch. Biochem. Biophys.* **2000**, *382*, 6–14.
37 M. D. Ballinger, G. H. Reed, P. A. Frey, *Biochemistry* **1992**, *31*, 949–953.
38 M. D. Ballinger, P. A. Frey, G. H. Reed, *Biochemistry* **1992**, *31*, 10782–10789.
39 M. D. Ballinger, P. A. Frey, G. H. Reed, R. LoBrutto, *Biochemistry* **1995**, *34*, 10086–10093.
40 W. Wu, K. W. Lieder, G. H. Reed, P. A. Frey, *Biochemistry* **1995**, *34*, 10532–10537.
41 W. Wu, S. Booker, K. W. Lieder, V. Bandarian, G. H. Reed, P. A. Frey, *Biochemistry* **2000**, *39*, 9561–9570.
42 O. T. Magnusson, G. H. Reed, P. A. Frey, *Biochemistry* **2001**, *40*, 7773–7782.
43 I. Freer, G. Pedrocchi-Fantoni, D. J. Picken, K. H. Overton, *J. Chem. Soc. Chem. Commun.* **1981**, 80–82.
44 J. M. Poston, *J. Biol. Chem.* **1976**, *251*, 1859–1863.
45 J. M. Poston, *Science* **1977**, *195*, 301–302.
46 J. M. Poston, *Phytochemistry* **1978**, *17*, 401–402.
47 S. P. Stabler, J. Lindenbaum, R. H. Allen, *J. Biol. Chem.* **1988**, *263*, 5581–5588.
48 D. J. Aberhart, *Anal. Biochem.* **1988**, *169*, 350–355.
49 H. Yonehara, N. Ōtake, *Tetrahedron Lett.* **1966**, *32*, 3785–3791.
50 W. J. McGahren, G. O. Morton, M. P. Kunstmann, G. A. Ellestad, *J. Org. Chem.* **1977**, *42*, 1282–1286.
51 P. C. Prabhakaran, N.-T. Woo, P. S. Yorgey, S. J. Gould, *J. Am. Chem. Soc.* **1988**, *110*, 5785–5791.
52 P. C. Prabhakaran, N.-T. Woo, P. S. Yorgey, S. J. Gould, *Tetrahedron Lett.* **1986**, *27*, 3815–3818.

53 P. E. Fleming, U. Mocek, H. G. Floss, *J. Am. Chem. Soc.* **1993**, *115*, 805–807.

54 P. E. Fleming, A. R. Knaggs, X.-G. He, U. Mocek, H. G. Floss, *J. Am. Chem. Soc.* **1994**, *116*, 4137–4138.

55 K. D. Walker, H. G. Floss, *J. Am. Chem. Soc.* **1998**, *120*, 5333–5334.

56 E. Graf, A. Kirfel, G.-J. Wolff, E. Breitmaier, *Liebigs Ann. Chem.* **1982**, 376–381.

57 L. Ettouati, A. Ahond, C. Poupat, P. Potier, *J. Nat. Prod.* **1991**, *54*, 1455–1458.

58 R. V. Platt, C. T. Opie, E. Haslam, *Phytochemistry* **1984**, *23*, 2211–2217.

59 S. G. Davies, J. Dupont, R. J. C. Easton, *Tetrahedron: Asymmetry* **1990**, *1*, 279–280.

60 T. P. Hettinger, L. C. Craig, *Biochemistry* **1970**, *9*, 1224–1232.

61 Z. Kurylo-Borowska, T. Abrahamsy, *Biochim. Biophys. Acta*, **1972**, *264*, 1–10.

62 R. J. Parry, Z. Kurylo-Borowska, *J. Am. Chem. Soc.* **1980**, *102*, 836–837.

63 F. von Nussbaum, P. Spiteller, M. Rüth, W. Steglich, G. Wanner, B. Gamblin, L. Stievano, F. E. Wagner, *Angew. Chem. Int. Ed.* **1998**, *37*, 3292–3295.

64 P. Spiteller, M. Rüth, F. von Nussbaum, W. Steglich, *Angew. Chem. Int. Ed.* **2000**, *39*, 2754–2756.

65 J. R. Roth, J. G. Lawrence, T. A. Bobik, *Annu. Rev. Microbiol.* **1996**, *50*, 137–181.

66 Review: P. A. Frey, C. H. Chang, Aminomutases, in *Chemistry and Biochemistry of B_{12}*, R. Banerjee, ed., Wiley–Interscience, New York, **1999**, pp. 835–857.

67 Review: J. J. Baker, T. C. Stadtman, Aminomutases, in *B_{12}: Biochemistry and Medicine*, Vol. 2, D. Dolphin, ed., Wiley–Interscience, New York, **1982**, pp. 203–232.

68 J. J. Baker, C. van der Drift, T. C. Stadtman, *Biochemistry* **1973**, *12*, 1054–1063.

69 C. H. Chang, P. A. Frey, *J. Biol. Chem.* **2000**, *275*, 106–114.

70 K.-H. Tang, A. Harms, P. A. Frey, *Biochemistry* **2002**, *41*, 8767–8776.

71 F. Kunz, J. Rétey, D. Arigoni, L. Tsai, T. C. Stadtman, *Helv. Chim. Acta* **1978**, *61*, 1139–1145.

72 J. Rétey, F. Kunz, D. Arigoni, T. C. Stadtman, *Helv. Chim. Acta* **1978**, *61*, 2989–2998.

73 K.-H. Tang, C. H. Chang, P. A. Frey, *Biochemistry* **2001**, *40*, 5190–5199.

74 C. G. D. Morley, T. C. Stadtman, *Biochemistry* **1971**, *10*, 2325–2329.

75 C. G. D. Morley, T. C. Stadtman, *Biochemistry* **1972**, *11*, 600–605.

76 S. D. Wetmore, D. M. Smith, L. Radom, *J. Am. Chem. Soc.* **2000**, *122*, 10208–10209.

77 S. D. Wetmore, D. M. Smith, L. Radom, *J. Am. Chem. Soc.* **2001**, *123*, 8678–8689.

78 R. Somack, R. N. Costilow, *Biochemistry* **1973**, *12*, 2597–2604.

79 Y. Tsuda, H. C. Friedmann, *J. Biol. Chem.* **1970**, *245*, 5914–5926.

80 H.-P. Chen, C.-F. Lin, Y.-J. Lee, S.-S. Tsay, S.-H. Wu, *J. Bacteriol.* **2000**, *182*, 2052–2054.

81 H.-P. Chen, S.-H. Wu, Y.-L. Lin, C.-M. Chen, S.-S. Tsay, *J. Biol. Chem.* **2001**, *276*, 44744–44750.

82 D. Guianvarc'h, D. Florentin, B. Tse Sum Bui, F. Nunzi, A. Marquet, *Biochem. Biophys. Res. Commun.* **1997**, *236*, 402–406.

83 J. Knappe, H. P. Blaschkowsky, P. Gröbner, T. Schmitt, *Eur. J. Biochem.* **1974**, *50*, 253–263.

84 J. Knappe, S. Elbert, M. Frey, A. F. V. Wagner, *Biochem. Soc. Trans.* **1993**, *21*, 731–734.

85 P. Reichard, *Science*, **1993**, *260*, 1773–1777.

86 H. J. Sofia, G. Chen, B. G. Hetzler, J. F. Reyes-Spindola, N. E. Miller, *Nucl. Acid Res.* **2001**, *29*, 1097–1106.

87 W. Liu, S. D. Christenson, S. Standage, B. Shen, *Science*, **2002**, *297*, 1170–1173.

Part 2
Non-Covalent Intermolecular Interactions

Highlights in Bioorganic Chemistry: Methods and Applications. Edited by Carsten Schmuck, Helma Wennemers.
Copyright © 2004 WILEY-VCH Verlag GmbH & Co. KGaA, Weinheim
ISBN: 3-527-30656-0

2.1
Carbohydrate Recognition by Artificial Receptors

Arne Lützen

2.1.1
Introduction

Given the importance of the different processes involving carbohydrate recognition, it is not surprising that over the last 15 years much effort in supramolecular chemistry has been devoted to developing efficient artificial receptors for these substrates [1]. The recognition of carbohydrates is, however, one of the biggest challenges for structural and preparative chemists, because they have a very complex three-dimensional array of functional groups that is an interesting problem for the design of suitable receptors (for a similar discussion on peptides see Chapter 2.3). Moreover, the thermodynamic forces driving the binding observed in natural systems, for example the force driving the desolvation of substrate and receptor binding sites, are still not completely understood. Thus, supramolecular "bottom up" approaches could help further elucidation of these processes although one has to admit that effective, truly biomimetic carbohydrate recognition is still far away at the moment (how Nature does it is explained in Box 5). Nonetheless, non-biomimetic systems also promise very interesting (future) applications in medicinal chemistry in that they could possibly be used to prevent bacterial and viral infections, to monitor cells' health, in the identification of malignant cells, for instance cancer cells, as transport vehicles for saccharides or related pharmaceuticals, or as sensors for saccharides in biological fluids.

2.1.2
Design Principles and Binding Motifs of Existing Receptors

Like lectins, which undergo few if any changes in their global conformation on binding to sugars, most of the receptors reported so far have a rather rigid structure that provides a more or less well preorganized binding site for the substrate. Because of the high density of hydroxy groups on carbohydrates it is not surprising that most receptors are designed to target these functions to achieve binding. Although there is a growing number of very effective receptors that use boronic acid functions to form boronate esters with two appropriately orientated hydroxy

Highlights in Bioorganic Chemistry: Methods and Applications. Edited by Carsten Schmuck, Helma Wennemers.
Copyright © 2004 WILEY-VCH Verlag GmbH & Co. KGaA, Weinheim
ISBN: 3-527-30656-0

1

(Aoyama, 1988)

2

(Anslyn, 1992)

3

(Mazik, 2001)

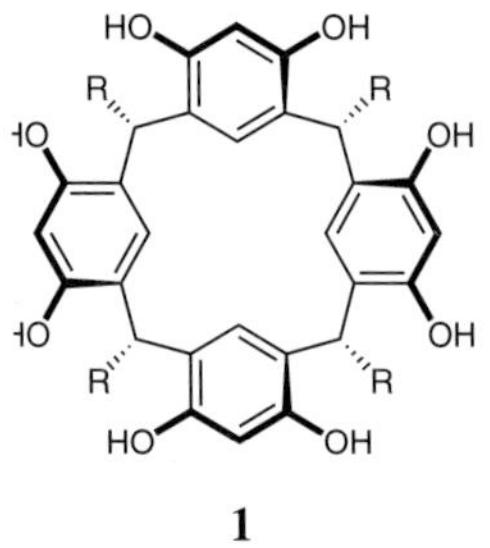

Fig. 2.1.1. Achiral artificial receptors for the recognition of carbohydrates via hydrogen bonds [1, 3h].

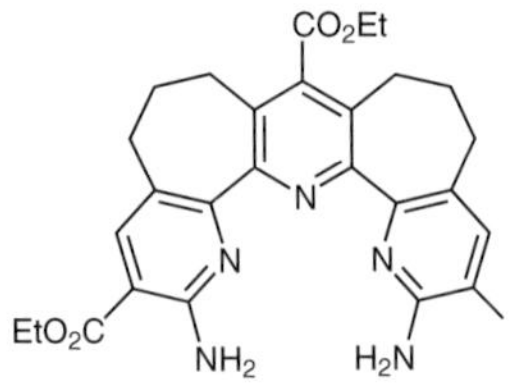

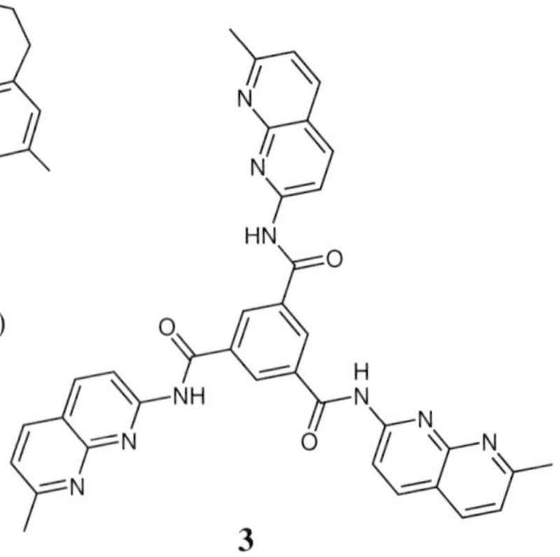

groups, this approach is neither biomimetic nor supramolecular since – although only moderately strong and reversible – covalent bonds are formed during binding [2]. Non-covalent, intermolecular binding can be achieved by hydrogen bonding through (an)ionic or neutral hydrogen bond-donor and -acceptor groups, as in natural examples [1, 3]. Other than natural models, however, most of the receptors reported so far have been designed to be effective in organic solutions to profit not only from the directionality of hydrogen bonds but also to obtain a maximum energetic contribution to overall binding (Box 7). Figures 2.1.1–2.1.3 show some

4

(Hamilton, 1997)

5

(Diederich, 1995)

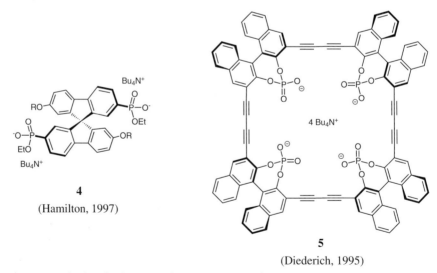

Fig. 2.1.2. Chiral artificial receptors for the recognition of carbohydrates via ionic hydrogen bonds involving anionic acceptors [1].

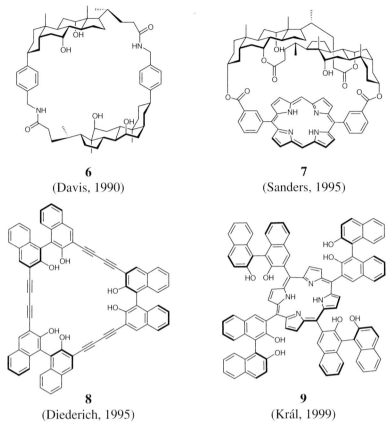

Fig. 2.1.3. Chiral artificial receptors for the recognition of carbohydrates via neutral hydrogen bonds [1, 3d, 3g].

synthetic examples that were designed to bind carbohydrates via hydrogen bonds [1, 3d,g,h].

Like the natural examples almost all artificial receptors offer an array of functional groups (pyridines, amide NH and CO groups, amino, or phenolic hydroxyl groups) to exploit co-operative hydrogen bonding in which the sugar hydroxyl groups can act simultaneously as a hydrogen-bond donor and acceptor. Only when phosphonates or phosphate groups (**4** and **5**) are used to bind carbohydrates via strong ionic hydrogen bonds can the sugar OH groups act only as hydrogen-bond donors.

In addition, most of the receptors bear hydrophobic patches in their binding site, to bind complementary areas of carbohydrates. Although the energetic contributions of this interaction to overall binding are obviously negligible in organic solutions these contacts might play a more prominent role in terms of specificity and in recognition in more polar solvents like DMSO, acetonitrile/methanol, or even mixtures containing water, as observed for instance for **5**, **7**, and **9**.

The results obtained with these receptors show that diastereoselective binding of single sugar epitopes is – to some extent – possible even with non-chiral structures although the order of preferred binding often reflects the monosaccharides' tendency to self-aggregate in the order α-galactosides ≤ α-glucosides < β-glucosides < α-mannosides. Improved diastereoselectivity and, of course, enantioselectivity can be observed with chiral receptors where **4** was found to show the highest enantioselectivity of approx. 5:1 for the binding of *n*-octyl β-glucopyranosides.

Association constants for the binding of mono- or disaccharides are usually in the range 1×10^2 M^{-1} (**2** and **8**) to 3×10^4 M^{-1} (**3** and **6**) in organic solvents like chloroform or dichloromethane, but can also reach 1×10^4 M^{-1} (**4**, **5**, **7** and **9**) when measured in more competitive solvents like acetonitrile, acetonitrile/methanol, DMSO, or dichloromethane/water.

2.1.3
Design, Synthesis, and Evaluation of Self-assembled Receptors

The synthesis and isolation of many of these receptors proved to be quite tedious, however, although occasionally convergent strategies could be used. We therefore looked for an alternative way to build up such structures and decided to make use of self-assembly processes of metal complexes [4] to build up self assembled oligo(BINOL) analogs [5]. (How similar metal complexes can be used to stabilize peptide microstructures is discussed in Chapter 1.3.)

We therefore designed and synthesized two ligands **10** and **11** following the basic idea to employ a convergent modular approach, where elaborated building blocks can be combined in a rather fast and flexible manner. These ligands **10** and **11** were thought to form di- and mononuclear coordination complexes in self-assembly processes with suitable metal ions thereby orienting the BINOL groups in a fashion potentially useful for the molecular recognition of monosaccharide derivatives (Figures 2.1.4 and 2.1.5).

Having accomplished the synthesis (Figure 2.1.6), the next task was to prove the formation of the self-assembled metal coordination complexes. This was achieved by NMR and ESI–MS experiments and elemental analysis and in one instance we were even able to perform X-ray structure analysis (Figures 2.1.7–2.1.9).

To learn more about the effect of the metal ion on the binding of these receptors we also synthesized a covalently linked analog of the [M**11**$_2$]$^+$ complex in which the central metal bis(bipyridine) complex was substituted by a spirobifluorene moiety [6]. Figure 2.1.10 shows the structures of both, which we obtained from molecular modeling studies.

Although we strictly followed our convergent building block approach again, the synthesis of **12** (Figure 2.1.11) proved to be considerably more demanding, because its isolation by chromatography turned out to be quite onerous; this again demonstrated the advantages of employing self-assembly processes of smaller molecules to obtain access to larger functional aggregates, rather than synthesizing covalently linked structures.

With the aggregates and the covalent analog available we started to evaluate the potential of these structures to act as receptors for monosaccharides in organic

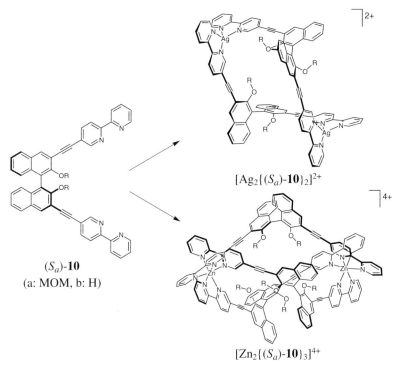

Fig. 2.1.4. Bis(2,2′-bipyridine)-substituted BINOL **10** and its dimeric and trimeric dinuclear metal complex.

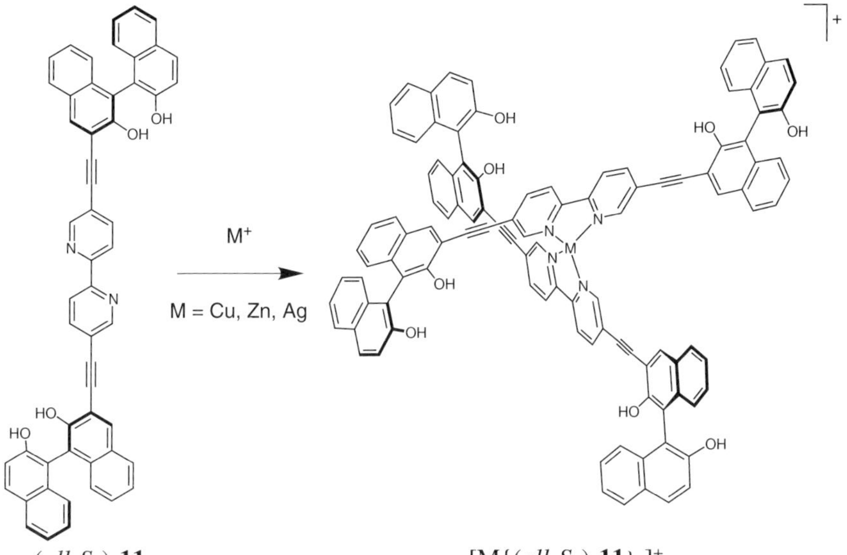

Fig. 2.1.5. Bis(BINOL)-substituted 2,2′-bipyridine **11** and its dimeric metal complex.

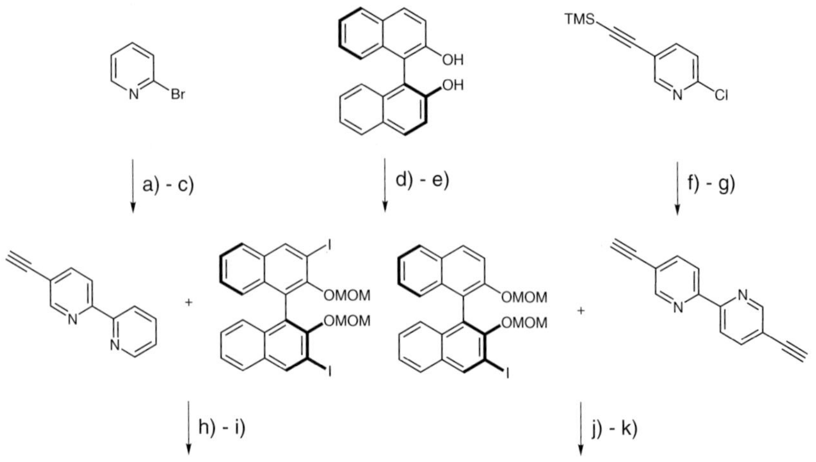

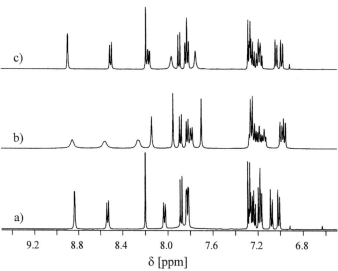

Fig. 2.1.6. Synthesis of (S_a)-**10** and (all-S_a)-**11**. (a) t-BuLi, THF, −78 °C, then ZnCl$_2$, −78 °C to rt, 2 h; (b) 2-chloro-5-trimethylsilylethynyl-pyridine, Pd$_2$dba$_3$·CHCl$_3$, t-Bu$_3$P, THF, 83%; (c) KF, MeOH, 97%; (d) NaH, DMF, MOMCl, 80%; (e) n-BuLi, Et$_2$O, rt, then I$_2$, Et$_2$O, −78 °C, 42% monoiodo and 40% diiodo compound; (f) Pd(OAc)$_2$, t-Bu$_2$(2-biph)P, Zn, DMF, rt, 62%; (g) KF, MeOH, rt, 79%; (h) Et$_3$N, CuI, Pd$_2$dba$_3$·CHCl$_3$, dppf, 50 °C, 77%; (i) conc. HCl, THF, MeOH, 95%; (j) Et$_3$N, CuI, Pd$_2$dba$_3$·CHCl$_3$, dppf, 50 °C, 96%; (k) conc. HCl, THF, MeOH, 97%.

Fig. 2.1.7. ^{1}H NMR spectra (500.1 MHz, [**11**]$_0$ = ca. 10 mmol L^{-1} in THF-d$_8$/CD$_3$CN at 300 K) (a) (all-S_a)-**11**, (b) (all-S_a)-**11** + 1/2 equiv. [Cu(CH$_3$CN)$_4$]BF$_4$, (c) (all-S_a)-**11** + 1/2 equiv. [Ag(CH$_3$CN)$_2$]BF$_4$.

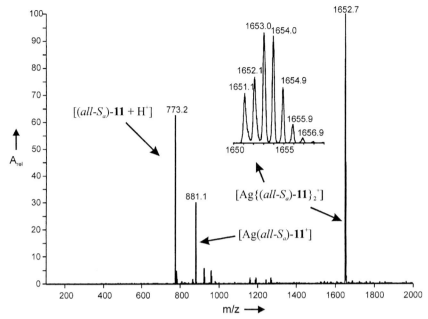

Fig. 2.1.8. Positive ESI MS of ca. 5×10^{-4} mol L^{-1} solution of [Ag{(all-S_a)-**11**}$_2$]BF$_4$ complex in CH$_2$Cl$_2$/CH$_3$CN.

solutions. We first performed qualitative NMR studies with alkyl glycosides, for example *n*-octyl- and methyl hexopyranosides. These preliminary experiments gave very promising results indicating our aggregates can indeed function as receptors for monosaccharide derivatives [7]. Unfortunately, the solubility of all the dinuclear

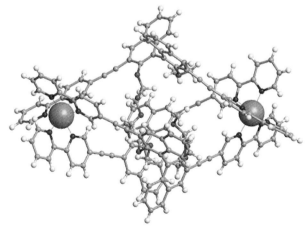

Fig. 2.1.9. Crystal structure of (Δ,Δ)-[Zn$_2${(S_a)-**10a**}$_3$](BF$_4$)$_4 \cdot 2.5$ THF·5 CH$_3$CN (counter-ions and solvent molecules omitted).

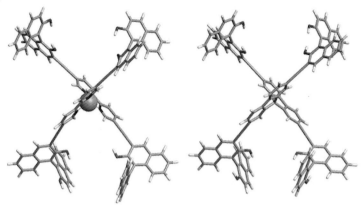

Fig. 2.1.10. Energy minimized structures of self-assembled [Cu{(all-S_a)-**11**}$_2$]$^+$-complex (*MMFF*-minimized, left) and of analogous covalently assembled tetra (BINOL) substituted spirobifluorene (all-S_a)-**12** (*MM2*-minimized, right).

metal complexes of **10b** so far investigated turned out to be too low to perform quantitative binding studies also. Thus, we are currently trying to improve solubility by using different counter-ions and introducing further groups that facilitate dissolution in different solvents.

The solubility of the coordination complexes of **11** and its covalent analog **12** proved to be much higher in organic solvents, however, so we could start to per-

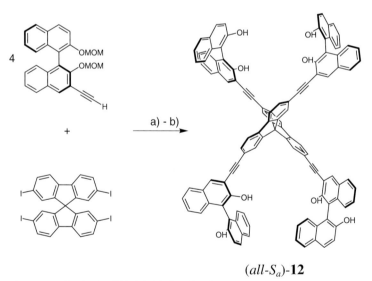

Fig. 2.1.11. Synthesis of **12**. (a) Pd$_2$dba$_3$·CHCl$_3$, Mes$_3$P, CuI, *n*-Bu$_4$NI, DMF, THF, *i*-Pr$_2$NEt, 62%; (b) conc. HCl, MeOH, THF, 86%.

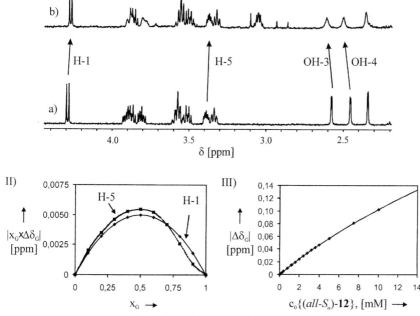

Fig. 2.1.12. NMR studies of the binding of **12** to n-octyl β-D-glucopyranoside. (I) ^{1}H NMR spectra (500.1 MHz in CDCl$_3$) (a) 0.4 mM sugar, (b) 0.4 mM sugar + 1.6 mM (all-S_a)-**12**; (II) Job-plot ($c_{Total} = c_{Guest} + c_{Host} = 2$ mM); (III) titration curve (c_{Sugar} was kept constant at 0.1 mM).

form quantitative studies with (all-S_a)-**12**. Figure 2.1.12 shows results from one of these NMR binding studies (a more detailed explanation of NMR binding studies is given in Chapter 2.3). These indicate that (all-S_a)-**12** and n-octyl β-D-glucopyranoside form a 1:1 complex. As expected from the results obtained for similar receptors, for example **8** and **9**, the binding is rather weak and we calculated an association constant of $K_A = 25 \pm 1$ M^{-1} by non-linear regression analysis of the titration curve. We are currently extending these studies to several n-octylhexopyranosides and methylhexopyranosides in solvents such as chloroform, benzene, and THF, to evaluate the hosts' affinity and diastereo-, and enantioselectivity when binding with these substrates.

2.1.4
Conclusions and Perspectives

Self-assembly processes have been demonstrated to be an excellent means of building up sophisticated oligo(BINOL) structures [M$_2$**10**$_2$]$^{2+}$, [M$_2$**10**$_3$]$^{4+}$, and [M**11**$_2$]$^+$. These provide an array of functional groups that could be used for the recognition of monosaccharide derivatives, as proven by qualitative NMR binding studies. Preliminary quantitative binding studies revealed that the association constants are of the same order of magnitude as for similar receptors, for example **8**, that have previously been reported. However, more studies with different sugar epitopes have to be performed to elucidate further the self-assembled receptors' affinity and their diastereo- and enantioselectivity in the recognition of carbohydrates. In this context it will also be interesting to examine whether these aggregates could also be used in more polar solvents, as has been shown very successfully with receptor **9** for example. It should also be tested if the binding affinity can be increased by converting the phenolic hydroxyl functions into phosphates, as demonstrated with similar host **5**.

Much work still needs to be done to develop artificial receptors that combine sufficient orientation through an array of hydrogen bond contacts with favorable hydrophobic effects to provide sufficient binding energy to reach the ultimate goal – effective and biomimetic recognition of carbohydrates in water.

References

1 A. P. DAVIS, R. S. WAREHAM, *Angew. Chem.* **1999**, *111*, 3160–3179; *Angew. Chem. Int. Ed.* **1999**, *38*, 2978–2996.

2 In fact, boronic acids have been extensively used in carbohydrate chemistry, for example in chromatographic separations since the early 1970s. However, Wulff and coworkers were the first to employ boronic esters in imprinted polymer carbohydrate receptors: (a) G. WULFF, *Pure Appl. Chem.* **1982**, *54*, 2093–2102; (b) G. WULFF, H.-G. POLL, *Makromol. Chem.* **1987**, *188*, 741–748. Moreover, many solution phase receptors have been reported recently: some reviews: (c) T. D. JAMES, P. LINNANE, S. SHINKAI, *Chem. Commun.* **1996**, 281–288; (d) T. D. JAMES, K. R. A. S. SANDANAYAKE, S. SHINKAI, *Angew. Chem.* **1996**, *108*, 2038–2050; *Angew. Chem. Int. Ed.* **1996**, *35*, 1910–1922; Some recent examples not covered by Refs. (c) and (d): (e) L. A. CABELL, M.-K. MONAHAN, E. V. ANSLYN, *Tetrahedron Lett.* **1999**, *40*, 7753–7756; (f) C. J. WARD, P. PATEL, T. D. JAMES, *J. Chem. Soc., Perkin Trans. 1* **2002**, 462–470; (g) S. ARIMORI, M. L. BELL, C. S. OH, K. A. FRIMAT, T. D. JAMES, *J. Chem. Soc., Perkin Trans. 1* **2002**, 803–808; (h) N. DICESARE, J. R. LAKOWICZ, *Org. Lett.* **2001**, *3*, 3891–3893; (i) N. DICESARE, J. R. LAKOWICZ, *Tetrahedron Lett.* **2001**, *42*, 9105–9108; (j) N. DICESARE, J. R. LAKOWICZ, *Tetrahedron Lett.* **2002**, *43*, 2615–2618; (k) J. C. NORRILD, I. SØTOFTE, *J. Chem. Soc., Perkin Trans. 2* **2001**, 303–311; (l) J. C. NORRILD, *J. Chem. Soc., Perkin Trans. 2* **2001**, 719–726; (m) H. KIJIMA, M. TAKEUCHI, S. SHINKAI, *Chem. Lett.* **1998**, 781–782; (n) M. TAKEUCHI, T. IMADA, S. SHINKAI, *Bull. Chem. Soc. Jpn.* **1998**, *71*, 1117–1123; (o) M. YAMAMOTO, M. TAKEUCHI, S. SHINKAI, *Tetrahedron* **1998**, *54*, 3125–3140; (p) T. MIZUNO, M. TAKEUCHI, S. SHINKAI, *Tetrahedron* **1999**, *55*, 9455–9468; (q) M. TAKEUCHI, T. MIZUNO, S. SHINKAI, S. SHIRAKAMI, T. ITOH, *Tetrahedron: Asymmetry* **2000**, *11*, 3311–3322; (r) M. IKEDA, S. SHINKAI, A. OSUKA, *Chem. Commun.* **2000**, 1047–1048; (s) A. SUGASAKI, K. SUGIYASU, M. IKEDA, M. TAKEUCHI, S. SHINKAI, *J. Am. Chem. Soc.* **2001**, *123*, 10239–10244; (t) C. J. DAVIS, P. T. LEWIS, M. E. MCCARROLL, M. W. READ, R. CUETO, R. M. STRONGIN, *Org. Lett.* **1999**, *1*, 331–334; (u) M. HE, R. J. JOHNSON, J. O.

Escobedo, P. A. Beck, K. K. Kim, N. N. St. Luce, C. J. Davis, P. T. Lewis, F. R. Fronczek, B. J. Melancon, A. A. Mrse, W. D. Treleaven, R. M. Strongin, *J. Am. Chem. Soc.* **2002**, *124*, 5000–5009.

3 Some recent examples not covered by Ref. [1]: (a) A. S. Droz, F. Diederich, *J. Chem. Soc., Perkin Trans. 1* **2000**, 4224–4226; (b) A. Bähr, B. Felber, K. Schneider, F. Diederich, *Helv. Chim. Acta* **2000**, *83*, 1346–1376; (c) A. S. Droz, U. Neidlein, S. Anderson, P. Seiler, F. Diederich, *Helv. Chim. Acta* **2001**, *84*, 2243–2289; (d) O. Rusin, V. Král, *Chem. Commun.* **1999**, 2367–2368; (e) V. Král, O. Rusin, J. Charvátová, P. Anzenbacher, Jr, J. Fogl, *Tetrahedron Lett.* **2000**, *41*, 10147–10151; (f) V. Král, O. Rusin, F. P. Schmidtchen, *Org. Lett.* **2001**, *3*, 873–876; (g) O. Rusin, K. Lang, V. Král, *Chem. Eur. J.* **2002**, *8*, 655–663; (h) M. Mazik, H. Bandmann, W. Sicking, *Angew. Chem.* **2000**, *112*, 562–565, *Angew. Chem. Int. Ed.* **2000**, *39*, 551–554; (h) M. Mazik, W. Sicking, *Chem. Eur. J.* **2001**, *7*, 664–670; (i) H.-J. Kim, Y.-H. Kim, J.-I. Hong, *Tetrahedron Lett.* **2001**, *42*, 5049–5052; (j) Y.-H. Kim, J.-I. Hong, *Angew. Chem.* **2002**, *114*, 3071–3074; *Angew. Chem. Int. Ed.* **2002**, *41*, 2947–2950; (k) J. Bitta, S. Kubik, *Org. Lett.* **2001**, *3*, 2637–2640; (l) D. W. P. M. Löwik, C. R. Lowe, *Eur. J. Org. Chem.* **2001**, 2825–2839; (m) S.-i. Tamura, M. Yamamoto, S. Shinkai, A. B. Khasanov, T. W. Bell, *Chem. Eur. J.* **2001**, *7*, 5270–5276.

4 (a) J.-M. Lehn, *Supramolecular Chemistry*, VCH, Weinheim, **1995**, pp 139–190; (b) A. von Zelewsky, *Stereochemistry of Coordination Compounds*, Wiley, Chichester, **1995**, pp. 177–201; (c) J. W. Steed, J. L. Atwood, *Supramolecular Chemistry*, Wiley, Chichester, **2000**, pp 463–571. Some recent reviews: (d) M. Albrecht, *Chem. Rev.* **2001**, *101*, 3457–3497; (e) S. Leininger, B. Olenyuk, P. J. Stang, *Chem. Rev.* **2000**, *100*, 853–908; (f) G. F. Swiegers, T. J. Malefetse, *Chem. Rev.* **2000**, *100*, 3483–3537; (g) M. Fujita, K. Umemoto, M. Yoshizawa, N. Fujita, T. Kusukawa, K. Biradha, *Chem. Commun.* **2001**, 509–518; (h) B. D. Holliday, C. A. Mirkin, *Angew. Chem.* **2001**, *113*, 2076–2098; *Angew. Chem. Int. Ed.* **2001**, *40*, 2022–2043.

5 (a) A. Lützen, M. Hapke, J. Griep-Raming, D. Haase, W. Saak, *Angew. Chem.* **2002**, *114*, 2190–2194; *Angew. Chem. Int. Ed.* **2002**, *41*, 2086–2089; (b) A. Lützen, M. Hapke, S. Meyer, *Synthesis*, **2002**, 2289–2295.

6 A. Lützen, F. Thiemann, S. Meyer, *Synthesis*, **2002**, 2771–2778.

7 A. Lützen, M. Hapke, F. Thiemann, *unpublished results*; preliminary results have been presented at the *35th Int. Conf. Coordination Chemistry (ICCC 35)*, Heidelberg, 21–26 July, **2002**, Abstract O 5.18, and at *ORCHEM 2002*, Bad Nauheim, 12–14 September, **2002**, Abstract B-005.

B.5
Molecular Basis of Protein–Carbohydrate Interactions

Arne Lützen, Valentin Wittmann

Nature has developed a huge diversity of independent protein architectures for recognition of carbohydrates but, despite this broad variety, in accordance with the numerous events involving carbohydrate binding processes, these structures do share some common key features. Irrespective of

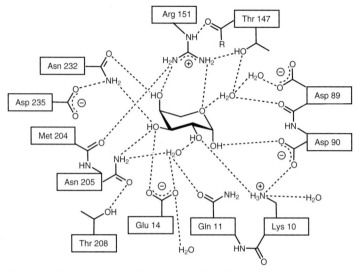

Fig. B.5.1. Crystal structure of α-L-arabinose bound to the binding site of the *periplasmic ABP* of *Escherichia coli* (PDB entry 1ABE [5]). Dashed lines represent direct and indirect hydrogen bonds.

whether the protein framework provides binding sites in deep clefts, as in most enzymes and bacterial periplasmic carbohydrate binding proteins, or in relatively shallow depressions in the protein surface, as is typical for lectins [1–4], the molecular basis of carbohydrate binding involves a complex array of non-covalent interactions such as hydrogen bonds, hydrophobic interactions, and, occasionally, an essential contribution of divalent cations to protein structure fixing and/or direct coordination to the substrate.

Hydrogen Bonding

As shown in Figure B.5.1, the protein usually exploits co-operative hydrogen bonding in which a single sugar hydroxyl group acts simultaneously as a hydrogen-bond donor and acceptor to distinguish between different carbohydrate epitopes.

One acidic amino acid side-chain is usually used as a hydrogen-bond acceptor for one or two sugar OH groups. The most common hydrogen bonding scheme is:

Protein $(NH)_n$ → sugar OH → protein C=O

Particularly effective in this context are pairs of vicinal sugar OH groups or one OH and the ring oxygen atom interacting with two functional groups in a single amino acid side-chain or with consecutive main-chain amide

groups, especially when the vicinal OH groups have an either equatorial/
equatorial or equatorial/axial configuration. Examples include the interaction of Asn 232 with OH-3 and OH-4 or Arg 151 with OH-4 and the ring oxygen atom in the complex of α-L-arabinose with the *periplasmic ABP* of *Escherichia coli* [5] (Figure B.5.1). Also water was observed to mediate (indirect) hydrogen bonds between amino acid residues and the saccharide OH groups. In these circumstances the water molecules act as fixed structural elements, equivalent to hydrogen-bonding groups of the protein, and can therefore be regarded as a part of the binding-site architecture.

Non-polar Interactions

Although carbohydrates have many polar functional groups, most biologically relevant saccharides still have significant non-polar patches on their surface that can interact with complementary hydrophobic amino acid residues. This is particularly true for β-D-galactose or related monosaccharide units which have a continuous non-polar surface from C-3 to C-6 and thus this face of the sugar is almost always observed to pack against aromatic side-chains from phenylalanine, histidine, or, most commonly, tryptophan. Hydrophobic interactions are believed to provide a significant contribution to the overall binding energy because apolar patches on the sugar and the aliphatic and aromatic side-chains of the protein are removed from the bulk solvent thus expelling unfavorably bound water molecules.

Divalent Cations

Mammalian C-type lectins are unique among structurally characterized lectins in that most require a calcium(II) ion to form direct coordinative bonds with the sugar ligands which have been found to be the primary determinants of affinity. Figure B.5.2 shows such an example [6] in which the full non-covalent binding potential of two vicinal OH groups is used. One lone pair of electrons from each OH group forms a coordinative bond with Ca^{2+}, the other lone pair accepts a hydrogen bond from a side-chain amino group, and the proton is donated to an acidic oxygen in a hydrogen bond. Calcium and other ions, for example manganese(II) ions, are, however, required for the activity of many families of carbohydrate-binding proteins, even if they do not interact directly with the ligands. In these instances the most common function of these ions is structural, in that the metal ion coordination shell orients important protein functional groups for optimum ligand binding.

Multivalency

The binding affinity of lectins is still surprisingly low – dissociation constants are usually approximately millimolar. Multivalency, i.e. the simulta-

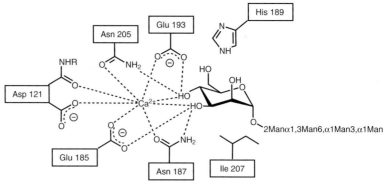

Fig. B.5.2. Crystal structure of an α-D-mannose unit bound to the C-type lectin *rat mannose-binding protein A* (MBP-A, PDB entry 2MSB [6]). Dashed lines indicate hydrogen and co-ordinative bonds.

neous association of several ligands of one biological unit (macromolecule, cell surface,...) with several receptors of another biological unit, provides a means of achieving high-affinity interactions applied by nature [7, 8]. Extension of carbohydrate binding sites to include direct or water-mediated contacts with multiple sugar units in oligosaccharide ligands ("subsite multivalence" [9]), however, results only in an increase in affinity to the micromolar range at best. High-affinity carbohydrate–protein interactions in the nanomolar range can be achieved by oligomerization of several lectin polypeptides each containing similar or identical simple binding sites ("subunit multivalence" [9]) or by clustering of several lectins on cell surfaces and interaction of these architectures with multiple carbohydrate epitopes presented in an appropriate manner on lipid or protein carriers [10].

Clustered binding sites are important because the free energy of binding of a multivalent ligand to multiple sites of an oligomeric lectin can be as large (or even larger) as the sum of the free energies of the individual binding interactions. This means that ideally the individual dissociation constants could potentially be multiplied to achieve high affinity for the complex ligand without the need to involve any additional molecular interactions beyond those seen with individual sugars in the single binding sites. Even if the requirements for ideal additivity are not often met in reality, because of additional geometric constraints, the affinity enhancement is still substantial. The strategy of employing structurally defined multivalent interactions is also very efficient in achieving maximum recognition of certain carbohydrate-coated surfaces – for example cell surfaces – while minimizing competitive binding to smaller saccharides or sugars attached to soluble glycoproteins at the same time. Structural studies revealed that many lectins that bind surfaces achieve the required planar array of sugar-binding sites by arrangement of polypeptide subunits in oligomers with cyclic symmetry, with all binding-sites located on one end of the oligomer.

References

1 S. H. BARONDES, *Trends Biochem. Sci.* **1988**, *13*, 480–482.
2 H.-J. GABIUS, *Eur. J. Biochem.* **1997**, *243*, 543–576; H.-J. GABIUS, ed., Animal Lectins, *Biochim. Biophys. Acta* **2002**, *1572*, 163–434.
3 H. LIS, N. SHARON, *Chem. Rev.* **1998**, *98*, 637–674.
4 D. C. KILPATRICK, *Handbook of Animal Lectins: Properties and Biomedical Applications*, Wiley, Chichester, **2000**.
5 F. A. QUIOCHO, N. K. VYAS, *Nature* **1984**, *310*, 381–386.
6 W. I. WEIS, K. DRICKAMER, W. A. HENDRICKSON, *Nature* **1992**, *360*, 127–134.
7 L. L. KIESSLING, T. YOUNG, K. H. MORTELL, in B. FRASER-REID, K. TATSUTA, J. THIEM, eds., *Glycoscience: Chemistry and Chemical Biology*, Vol. II, Springer, Heidelberg, **2001**, pp. 1817–1861.
8 M. MAMMEN, S.-K. CHOI, G. M. WHITESIDES, *Angew. Chem.* **1998**, *110*, 2908–2953; *Angew. Chem. Int. Ed.* **1998**, *37*, 2754–2794; J. J. LUNDQUIST, E. J. TOONE, *Chem. Rev.* **2002**, *102*, 555–578; C. F. BREWER, M. C. MICELI, L. G. BAUM, *Curr. Opin. Struct. Biol.* **2002**, *12*, 616–623.
9 J. M. RINI, *Annu. Rev. Biophys. Biomol. Struct.* **1995**, *24*, 551–577.
10 P. R. CROCKER, T. FEIZI, *Curr. Opin. Struct. Biol.* **1996**, *6*, 679–691.

2.2
Cyclopeptides as Macrocyclic Host Molecules for Charged Guests

Stefan Kubik

2.2.1
Introduction

Catalysis, transport, regulation, signal transduction, and the propagation of information are fundamental biochemical processes, all of which are essential for life on this planet. The precision and efficiency usually associated with these processes are a consequence of highly specific non-covalent interactions between the many compounds involved. This general principle has been termed *molecular recognition*. It is achieved in natural systems by inclusion of the substrate into a cavity formed by the appropriately folded peptide chain of an enzyme, a receptor, or an antibody. The shape of this cavity, the so-called active center, is either complementary to the shape of the substrate molecule or adapts to its shape very well. Additional contributions to binding selectivity come from functional groups of amino acid side-chains suitably placed inside the active center that interact with the included substrate via hydrogen bonding or electrostatic, van der Waals, or charge-transfer interactions. These groups also determine the function of the protein, for example, they participate in chemical modification of the substrate.

Many natural substrates are charged and Nature has therefore devised several special binding motifs in order to efficiently bind ionic compounds. In this chapter some natural ion receptors are presented and, by using the example of the cyclic peptides studied in my group, it is shown how the general principles of ion complexation can be used for the design of artificial receptors that bind charged substrates. (Another example is given in Chapter 2.3).

2.2.2
Cation Recognition

Important cation binding systems in Nature are, for example, a class of macrocyclic compounds termed *ionophores* and proteins that bind quaternary ammonium ions such as acetylcholine. The ionophores valinomycin **1**, nonactin, the enniatines, and baeuvericin are cation binders that are structurally quite diverse, yet

Highlights in Bioorganic Chemistry: Methods and Applications. Edited by Carsten Schmuck, Helma Wennemers
Copyright © 2004 WILEY-VCH Verlag GmbH & Co. KGaA, Weinheim
ISBN: 3-527-30656-0

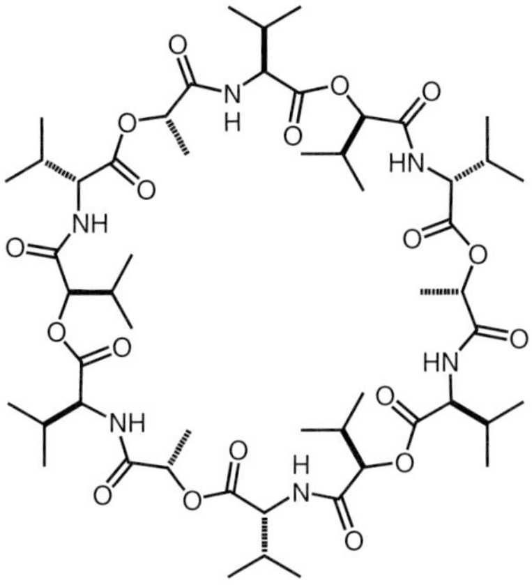

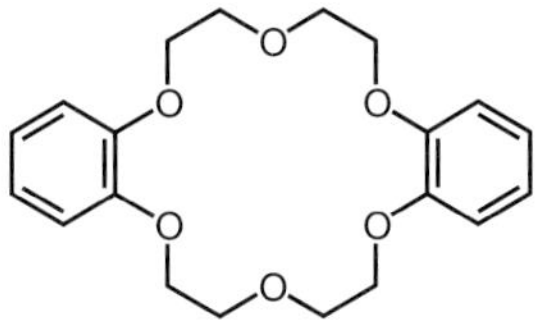

Fig. 2.2.1. Structures of valinomycin **1** (left), and a crown ether (right).

their mode of action is closely related [1]. All ionophores transport cations through biological membranes and thus cause a perturbation of the trans-membrane ionic balance. As a consequence, valinomycin and nonactin, for example, have high antibacterial activity (Figure 2.2.1).

Valinomycin is a cyclic depsipeptide containing a subunit sequence of L-valine, D-hydroxyisovaleric acid, D-valine, and L-lactic acid that is repeated three times in the ring. This structure enables **1** to adopt a conformation in which the six carbonyl groups of the valine residues are almost perfectly preorganized for the complexation of a K^+ ion. In the complex formed the isopropyl groups around the valinomycin ring all point outward, rendering the whole aggregate lipophilic, a property that accounts for the ability of **1** to transport potassium through a membrane. A cyclic arrangement of oxygen atoms is a common structural motif of all ionophores, and the cation affinity of these compounds therefore has a similar explanation.

Incidentally, C. J. Pedersen's first report on crown ethers and their complexes was published in the same year as the mechanism of the biological activity of valinomycin was clarified [2]. Crown ethers are cyclic derivatives of polyethylene glycol of varying ring size, an example of which is also depicted in Figure 2.2.1. The structural relationship with the ionophores is clearly visible. It is thus not surprising that crown ethers also bind metal cations by coordination with the oxygen atoms [1, 3].

Coordination by oxygen atoms is not the only mechanism with which cations can be bound in the cavity of a natural or non-natural receptor, however. The crystal structure of acetylcholinesterase, an enzyme that catalyzes the hydrolysis of the neurotransmitter acetylcholine into choline and acetate, with the inhibitor decamethonium ($Me_3N^+(CH_2)_{10}NMe_3^+$) included inside the active center showed an unexpected large number of aromatic amino acid side-chains in the vicinity of the cationic headgroups of the substrate [4]. The binding site responsible for the complexation of the quaternary ammonium group of acetylcholine is lined by 14 aromatic residues, for example, of which one tryptophan indole ring is in close contact with the bound cation. Similar contacts between aromatic groups and cations are quite common in natural cation-binding proteins, and they also contribute to the non-covalent stabilization of protein conformations [5]. This type of interaction has been termed cation–π interaction [6]. It is also responsible for the cation affinity of artificial cation receptors such as, for example, the calixarenes [7].

Calixarenes are formed by condensation of a *p*-substituted phenol with formaldehyde [8]. These macrocycles are conformationally quite flexible but, by introducing suitable substituents in the aromatic subunits, the so-called *cone* conformation, in which all aromatic subunits point into the same direction, can be stabilized. This conformation is usually best suited to complex guest molecules because it has a well defined hydrophobic cavity. An inclusion of cations such as ammonium ions or quaternary ammonium ions into this cavity can be demonstrated, for example, by the characteristic upfield shifts of guest signals in the NMR, an effect that is a consequence of the close proximity of the corresponding protons to the surfaces of the aromatic receptor subunits in the complex.

Calixarenes provided the main inspiration for the artificial receptors studied in my group. We wanted to develop a new class of macrocyclic host with binding properties and structural variability similar to calixarenes but a closer relationship to natural systems. The obvious choice was, of course, to base such receptors on cyclopeptides, macrocyclic compounds that are composed of the same subunits as the natural systems.

When we started our investigations, there were surprisingly few reports on the use of macrocyclic peptides as artificial receptors [9], most probably because cyclopeptides, especially larger ones, have a tendency to adopt conformations in solution unsuitable for inclusion of a guest molecule [10]. Moreover, prediction of the preferred solution conformations of such peptides is still difficult. These disadvantages can be overcome to some extent, however, by incorporating rigid amino acid subunits in the ring [11]. This strategy usually enables control of the conformational behavior of a cyclopeptide, and it should therefore also be suitable for stabilization of peptide conformations that have well defined cavities. In this respect we expected peptides containing 3-aminobenzoic acid as conformational constraint to have particularly interesting receptor properties. As Figure 2.2.2 illustrates, introduction of such aromatic subunits in every other position of a cyclopeptide ring affords structures that can be regarded as hybrids between conventional cyclopeptides and calixarenes. If these compounds adopt conformations in solution similar to those of calixarenes, one can also expect similar binding properties. The aromatic subunits of the cyclopeptides could, for example, induce cation affinity by means of cation–π interactions.

The starting point of our investigations was a cyclic hexapeptide containing 3-aminobenzoic acid and glutamic acid-5-isopropyl ester subunits **2** [12]. Our structural assignment showed that this cyclopeptide preferentially adopts conformations

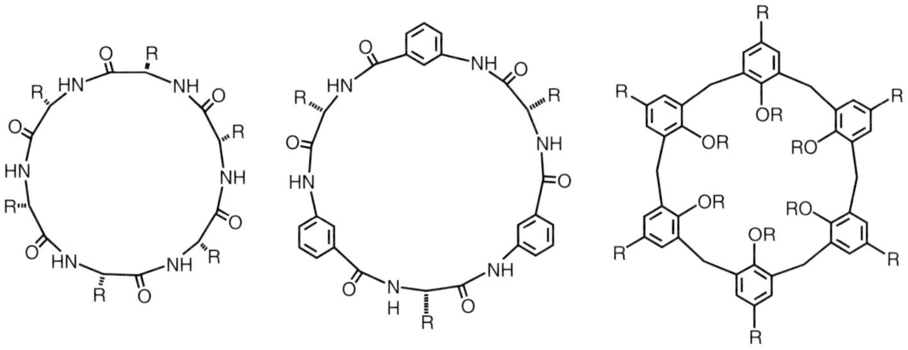

Fig. 2.2.2. Comparison of the structures of a cyclic hexapeptide (left), a calix[6]arene (right), and a cyclic hexapeptide composed of alternating natural amino acids and 3-aminobenzoic acid (center).

in chloroform stabilized by intramolecular hydrogen bonds between the aromatic NH and aromatic C=O groups. Because all aromatic peptide subunits point in the same direction, these conformations are closely related to the *cone* conformation of calixarenes. The cyclopeptide contains one aromatic subunit less than a calix[4]arene, however, and its cavity is shallower and has a slightly larger diameter.

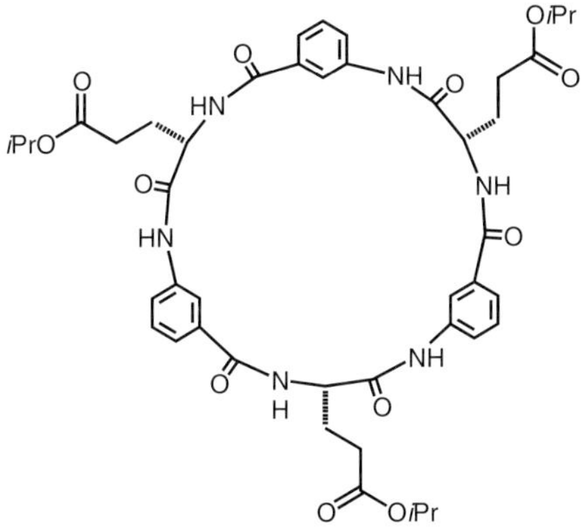

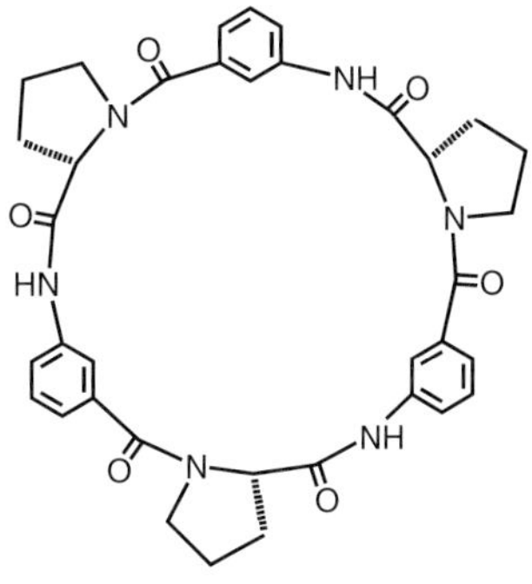

The ability of this peptide to bind cations could be demonstrated by the upfield shift observed for the cation signals in the NMR spectrum on addition of a salt of, for example, the *n*-butyltrimethylammonium ion (BTMA$^+$) to a solution of **2** in 0.2% d_6-DMSO/CDCl$_3$ [12]. This shift is a good indication of the interactions between the cation and **2**, and it enabled us to determine a stability constant of 300 M^{-1} for the complex formed. The maximum chemical shift, $\Delta\delta_{max}$, observed for, for example, the *N*-methyl signal of the cation amounts to only -0.05 ppm, however, and is thus significantly smaller than the shift usually associated with complexation of similar guests by calixarenes (typically −0.5 to −1 ppm). The effect of the aromatic subunits on the resonance of the guest protons is obviously weaker in **2**, most probably because the dimensions of the cyclopeptide cavity are not optimum for complexation of BTMA$^+$. To improve cation binding we therefore replaced the glutamic acid subunits of **2** with a more rigid amino acid, namely proline [13]. We expected the corresponding peptide **3** to be conformationally less flexible and, as a consequence, better preorganized for substrate binding. NMR spectroscopic investigations indeed confirmed this prediction. On complex formation between **3** and BTMA$^+$ picrate a significantly larger upfield shift of the BTMA$^+$ N-methyl signal of up to −0.70 ppm was observed; this could conveniently be used to determine a complex stability of 1260 M^{-1}. Thus, replacement of the glutamic acids by prolines led to an approximately fourfold increase in BTMA$^+$ complex stability, an effect that shows how sensitively the receptor properties of such peptides react toward structural modification. Currently we have no experimental evidence for participation of the peptide carbonyl groups in cation complexation, and therefore assume that complex formation is driven by cation–π interactions. A cooperative effect of the carbonyl groups on binding cannot be ruled out, however [14].

Further investigations revealed another interesting property of **3** [13]. On varying the anion in the BMTA$^+$ salt we found that anions other than picrate, for example iodide, tetrafluoroborate, or tosylate, bind strongly to the NH groups of **3** by hydrogen-bond formation. This interaction stabilizes a peptide conformation in which all NH groups converge to enable simultaneous complexation of the anion. The geometry of the tosylate complex of **3** was assigned by NOESY NMR spectroscopy. Its structural relationship to the iodide complex, of which a crystal structure was obtained (Figure 2.2.3), is high.

Figure 2.2.3 nicely illustrates that **3** simultaneously interacts with the anion and the cation. As expected, the cation is included in the shallow dish-shaped cavity formed by the aromatic peptide subunits. The cyclic arrangement of peptide NH

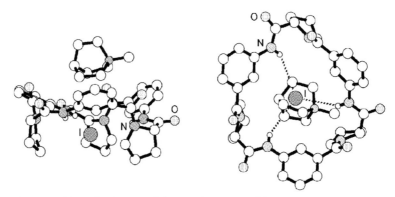

Fig. 2.2.3. Top and side views of the crystal structure of the N-methylquinuclidinium iodide complex of **3**.

Tab. 2.2.1. Stability constants K_a and maximum chemical shifts $\Delta\delta_{max}$ of the BTMA$^+$ complex of **3** in the presence of different anions (0.2% d_6-DMSO/ CDCl$_3$; $T = 298$ K; K_a in M^{-1}; $\Delta\delta_{max}$ maximum chemical shift of the BTMA$^+$ N-methyl protons in ppm; error limits for $K_a < 20$%).

Anion	K_a	$\Delta\delta_{max}$
Picrate	1260	0.70
Iodide	21100	1.11
Tetrafluoroborate	205000	0.99
Tosylate	5100000	1.16

groups enables additional anion complexation so that overall, **3** acts as a ditopic receptor in the presence of suitable anions with the ability to bind both components of an ion pair simultaneously. The aggregates formed are stabilized by strong electrostatic interactions between the oppositely charged ions, an effect reflected in the increase in cation complex stability with increasing stability of the anion complexes of **3** (Table 2.2.1) [15].

The corresponding dependence of cation complex stability on the anion differs profoundly from that of most other cation receptors such as cyclophanes or calixarenes [16]. For these cation complex stability decreases on changing the anion from picrate through iodide to tosylate, a dependence that has been attributed to ion-pair aggregation in non-polar solvents. Because the interaction of quaternary ammonium ions with tosylate or iodide in chloroform is considerably stronger than with picrate, cation complexes in the presence of the latter anion are usually more stable. Only when iodide or tosylate cooperatively contributes to cation binding, as in **3** or in some recently described calixarene derivatives [17], is reversal of this order observed.

Although peptide **3** binds cations significantly better than **2**, the preorganization of **3** for cation binding is still not optimum. Our structural investigations revealed that in the absence of anions that bind to NH groups, the secondary amide groups of **3** rotate relatively freely in solution [13]. This flexibility has no dramatic effect on the overall conformation of the peptide because is does not alter the orientation of the prolines with respect to the aromatic subunits, or the overall shape of the cavity available for cation complexation. Because it is nevertheless a disadvantage for complex formation, we looked for additional means of improving receptor preorganization.

Our next optimization step consisted in introduction of substituents in the 4-position of the aromatic peptide subunits of **3** [18]. These substituents should affect the rotation of neighboring amide groups by steric effects. More importantly, if they can form hydrogen bonds to the amide protons they should stabilize peptide conformations in which all NH groups point away from the cavity center. This effect would not only affect the flexibility but also the anion-binding ability of **3**, because, for the latter, a conformation with converging NH groups is required. To explore these possibilities we synthesized a series of cyclopeptides with different substituents in the 4-position of the aromatic rings. Here, only the effects of methoxy **4a** and methoxycarbonyl groups **4b** on cation and anion complexation will be summarized.

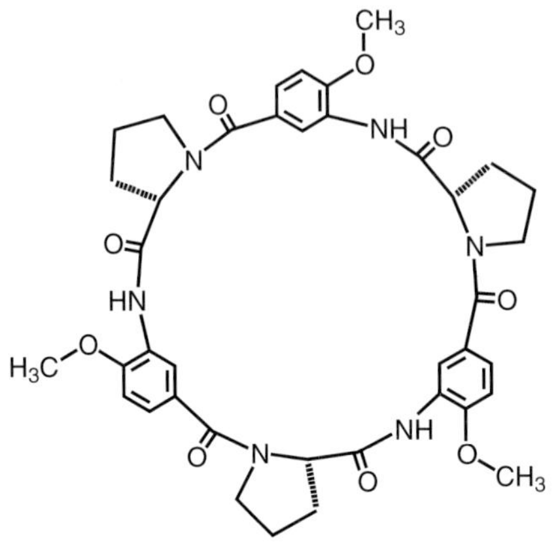

4a

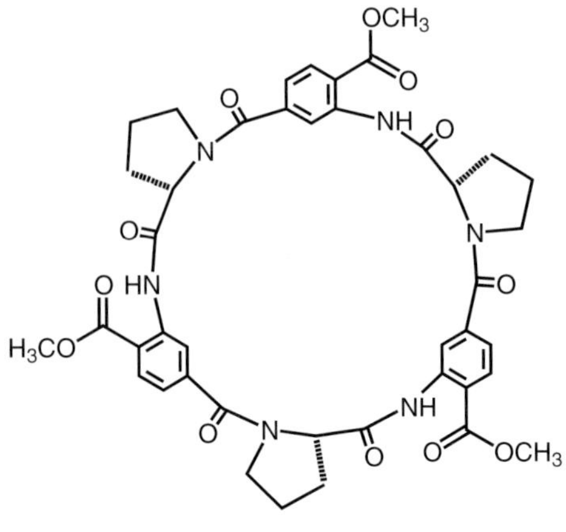

4b

Spectroscopic investigations confirmed that intramolecular hydrogen bonds between the oxygens of the aromatic substituents in **4a** and **4b** and the neighboring NH protons prevent amide rotation and cause the peptide NH groups to diverge [18]. Since the geometry of the cavity required for cation complexation is not significantly affected by this, both peptides are able to bind cations. The strong increase in BTMA$^+$ complex stability observed for **4a** and **4b** in comparison with **3** demonstrates the favorable effect of the reduced cyclopeptide flexibility on binding properties (Table 2.2.2). As predicted, the conformational control imposed by the aromatic subunits not only improves cation affinity, it also eliminates the ability of **4a** and **4b** to interact with anions. This loss is reflected, for example, in the de-

Tab. 2.2.2. Stability constants K_a and maximum chemical shifts $\Delta\delta_{max}$ of the BTMA$^+$ complex of **4a** and **4b** in the presence of different anions (0.2% d_6-DMSO/CDCl$_3$; $T = 298$ K; K_a in M^{-1}; $\Delta\delta_{max}$ maximum chemical shift of the BTMA$^+$ N-methyl protons in ppm; error limits for $K_a < 20\%$).

	Picrate		Iodide		Tosylate	
	K_a	$\Delta\delta_{max}$	K_a	$\Delta\delta_{max}$	K_a	$\Delta\delta_{max}$
3	1260	0.70	21100	1.11	5100000	1.16
4a	2700	0.68	850	0.73	340	0.56
4b	10800	0.54	3310	0.59	740	0.54

creasing stability constants of the BTMA$^+$ complexes when the counter-anion picrate is replaced by iodide or tosylate (Table 2.2.2), an order of complex stability now consistent with that observed for other cation receptors that cannot bind anions [16].

Systematic investigations revealed that anion binding is mainly prevented by the steric effects of the aromatic substituents. A cyclopeptide with methyl groups on the aromatic rings, for example, also binds no anions. The high cation affinity of **4a** and **4b** has to be attributed to the stabilizing effects of the aromatic substituents on peptide conformation, however, because the methyl-substituted peptide, in which the amide groups are not fixed in a certain orientation, has a cation affinity that is even lower than that of **3** [18].

2.2.3
Anion Recognition

Our success in improving the cation affinity of the cyclopeptides by conformational control motivated us to use a similar approach to increase anion affinity also. Considering that 70–75% of all enzyme substrates and cofactors are negatively charged, the efficient and selective molecular recognition of anions is probably even more important in Nature than that of cations [19]. The most important biologically relevant anions are phosphate (inorganic or in the form of esters such as DNA, RNA, ATP, etc.), sulfate, carboxylate, and chloride. In anion complexation by natural systems two basic types of interaction can be distinguished – electrostatic interaction and hydrogen-bond formation, with the latter usually inducing higher binding selectivity because of the directionality of hydrogen bonds. An important binding motif for anions that occurs, for example, in the phosphate binding protein (PBP) and carboxypeptidase A, namely the interaction between an anion and the protonated guanidinium group of an arginine side-chain, makes use of a combination of both effects (for more on this ion pair interaction see Chapter 2.3). Other peptides, such as the sulfate binding protein (SBP) require no strong electrostatic interactions for anion recognition, and the main contribution to substrate binding comes from a defined array of hydrogen bonds inside the active center between the anion and NH groups of the protein backbone, serine OH, or tryptophan NH groups [19].

Although Nature demonstrates that proteins efficiently bind anions in aqueous solution, differences between the properties of cations and anions make complexation of the latter generally more difficult. Anions are, for example, usually larger than cations and therefore need a larger cavity for complexation. In addition, the hydration energy of most anions is higher than that of cations. Therefore, anion receptors have to compete more efficiently with solvent molecules. Moreover, many anions are involved in protonation equilibria, making it necessary for receptors to be active in an appropriate pH window. Finally, anions occur in a range of geometries such as spherical (halides), linear (N_3^-), planar (NO_3^-), tetrahedral (SO_4^{2-}), or octahedral (PF_6^-), which requires more elaborate receptor design.

These problems were the reason why the development of artificial anion receptors took considerably longer than that of cation receptors. Today, however, many systems exist with which the difficulties of anion complexation can be overcome [20]. With the exception of receptors that bind anions by coordinative interactions (Lewis acid $\cdots$ anion), most systems make use of the same basic principles of anion complexation also found in Nature, namely electrostatic interactions and hydrogen-bond formation. Guanidinium-based receptors, for example, mimic the anion recognition in PBP or carboxypeptidase A (see Chapter 2.3), whereas neutral urea- or amide-based receptors mimic that of SBP. Although anion binding relies solely on hydrogen-bond formation in the latter example, some neutral anion receptors have high anion affinity even in polar solvents such as acetonitrile or DMSO [21]. Because of the high hydration energies of anions, however, complex formation in water often proved to be weak if present at all (cf also Box 7).

In the course of our investigations on the interactions of cyclopeptides such as **2** or **3** with cations we also encountered the anion binding capacity of these compounds. For **3**, we showed, for example, that to enable optimum interactions with anions, the peptide adopts a conformation with all NH groups pointing into the cavity center [13], and by stabilizing a peptide conformation with diverging NH groups we could completely eliminate this anion binding capacity [18]. We were therefore curious to discover whether the reversed effect could also be achieved, i.e. an increase in anion affinity by stabilization of a peptide conformation with converging NH groups.

5

6

Our approach to induce such a conformation is based on the orienting effects of pyridyl ring nitrogens on adjacent NH bonds, and consisted in replacement of the 3-aminobenzoic acids in the cyclopeptide ring of **3** by 6-aminopicolinic acid subunits [22]. NH bonds usually adopt a parallel orientation to the lone pair of the ring nitrogen, because of the antiparallel arrangement of the corresponding dipole

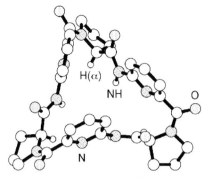

Fig. 2.2.4. Crystal structure of **5**·3 H$_2$O.

moments [23]. The picolinic acid subunits in **5** should therefore stabilize peptide conformations with all amide protons pointing into the cavity center, ideally pre-organized for anion binding. This arrangement was indeed found. The preferred conformation of **5** in solution and the solid state differs significantly from those observed for other peptides, such as, for example, **3** in its iodide complex (Figure 2.2.3), however [22].

The reason for this difference are cis amide conformations at the three proline subunits of **5**. Figure 2.2.4 shows that the presence of these amide conformations causes the three aromatic subunits of the peptide to be oriented almost parallel to the C_3 axis of the macrocycle, and the characteristic dish-shaped cavity observed for all other peptides to disappear. Despite this overall conformation, **5** is still able to interact with anions [22]. On addition of salts such as BTMA$^+$ tosylate to a solution of **5** in d_6-DMSO, for example, the NH signals of the peptides shift downfield in the NMR spectrum, a typical indication of hydrogen-bonding interactions between the NH protons and the anion. Almost no shift of the corresponding BTMA$^+$ cation signals could be detected, and the cation is thus not involved in anion binding. Complex formation strongly affects the resonance of the peptide H(α) protons, however. The corresponding downfield shift of the signal accounts for the close proximity of the peptide H(α) protons to the negative charge density of the guest in the complex, and is thus another indication of complex formation. It enabled us to investigate anion binding not only in aprotic solvents but also in polar, protic solvents such as water–methanol mixtures, in which the peptide NH protons are in rapid exchange with the protons of the solvent molecules.

To our surprise, we found that **5** binds anions such as halides or sulfate even in 80% D$_2$O–CD$_3$OD, which is remarkable considering that, despite the large excess of water molecules in these mixtures, the hydrogen-bonding interactions between the anions and the peptide are still efficient [22]. Assignment of the structure of the anion complexes of **5** revealed the reason for the unusual receptor properties of this peptide. We could show that **5** preferentially forms 2:1 sandwich-type complexes with suitable anions in which the guests are bound in a cavity located be-

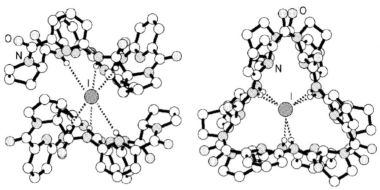

Fig. 2.2.5. Top and side views of the crystal structure of the iodide complex of **5**.

tween two almost perfectly interlocking cyclopeptide moieties. The crystal structure of the corresponding iodide complex is depicted in Figure 2.2.5.

This crystal structure shows that the iodide forms hydrogen bonds to all six NH of the two peptide moieties in the complex. It also demonstrates how effectively the anion is embedded between the cyclopeptides. Complex formation thus shields the guest from surrounding solvent molecules, an effect that strengthens receptor–substrate interactions; this might be one reason for the anion affinity of **5** in aqueous solution.

To improve the aqueous solubility of **5** we also synthesized a cyclopeptide **6** containing hydroxyproline subunits [24]. Although this compound is very water-soluble, and in solution adopts a similar conformation to **5**, it only forms 1:1 complexes with anions. A reason for the inability of **7** to form 2:1 complexes could be that the hydroxyproline subunits in **6** are better solvated than prolines in aqueous solution, and the desolvation required for aggregation of two cyclopeptide molecules thus occurs less readily. Steric hindrance of hydroxyl groups from different cyclopeptide moieties in the a dimeric complex of **6** could, moreover, also make aggregation difficult. Although peptide **6** cannot form sandwich-type complexes, it proved to be valuable for a quantitative determination of the anion affinity of **5**, the results of which are summarized in Table 2.2.3 [24].

This table shows that, for a given anion, the stability constants of the 1:1 complexes of both peptides are comparable, and that the stability of the 1:1 complexes increases in the order $Cl^- > Br^- > I^- > SO_4^{2-}$. This order can be rationalized in terms of the size of the ions, with larger ions forming more stable complexes because they fit better into the available peptide cavity. For sulfate, an additional contribution to complex stability from the higher charge of this anion must be considered.

The large stability constants K_2 of the 2:1 complexes of **5** indicate that, once formed, the 1:1 complexes of this peptide have a strong tendency to bind the sec-

Tab. 2.2.3. Stability constants K_a and maximum chemical shifts $\Delta\delta_{max}$ of some anion complexes of **5** and **6** (80% D_2O–CD_3OD; $T = 298$ K; K_1 and K_2 in M^{-1}; K_a in M^{-2}; $\Delta\delta_{max}$ maximum chemical shift of the peptide H(α) protons in ppm; error limits of the stability constants of the complexes of **6** < 20%, and **5** < 40%).

		5		6	
		K_a	$\Delta\delta_{max}$	K_a	$\Delta\delta_{max}$
NaCl	K_1	5	0.58	8	0.32
	K_2	6770	0.23		
	K_a	0.34×10^5			
NaBr	K_1	16	0.49	13	0.43
	K_2	6820	0.46		
	K_a	1.09×10^5			
NaI	K_1	22	0.55	19	0.58
	K_2	7380	0.84		
	K_a	1.62×10^5			
Na_2SO_4	K_1	96	0.44	95	0.51
	K_2	1270	0.77		
	K_a	1.22×10^5			

ond cyclopeptide moiety, which makes complex formation a highly cooperative process. The overall complex stabilities of the 2:1 complexes of **5** are remarkably high, ranging between 10^4–10^5 M^{-2}. Thus, by structural variation of our basic cyclopeptide we have identified a highly potent anion receptor whose anion affinity in aqueous solutions is, to the best of our knowledge, not surpassed by any other artificial receptor described so far that binds anions by hydrogen-bond formation.

In summary, our approach of using cyclopeptides with natural amino acids and 3-aminobenzoic acid subunits for the development of macrocyclic receptors has afforded remarkably efficient hosts. The cation affinity of **4b**, for example, exceeds that of many calixarene derivatives. Even more interesting is the high anion affinity of **5** in aqueous solution. By introducing additional functional groups such as carboxylates to the periphery of the cavity, we recently also obtained cyclopeptides that interact with neutral substrates, for example, carbohydrates [25]. Our peptides therefore represent a versatile class of artificial receptor that should prove useful in supramolecular and bioorganic chemistry.

Acknowledgment

I thank all co-workers involved in this project, in alphabetical order: Joachim Bitta, Guido Heinrichs, Daniela Kubik, and Susanne Pohl. Thanks are also due to Dr R. Goddard for solving many crystal structures, Professor Dr G. Wulff and Professor Dr H. Ritter for their continuing support during recent years, and the Deutsche Forschungsgemeinschaft for generous funding.

References

1 M. Dobler, *Ionophores and their structures*, Wiley, New York, **1981**; R. Hilgenfeld, W. Saenger, *Top. Curr. Chem.* **1982**, *101*, 1–82.
2 C. J. Pedersen, *J. Am. Chem. Soc.* **1967**, *89*, 7017–7036.
3 E. Weber, F. Vögtle, *Top. Curr. Chem.* **1981**, *98*, 1–41; G. W. Gokel, *Crown Ethers and Cryptands*, RSC, Cambridge, **1994**.
4 M. Harel, I. Schalk, L. Ehret-Sabatier, F. Bouet, M. Goeldner, C. Hirth, P. H. Axelsen, I. Silman, J. L. Sussman, *Proc. Natl. Acad. Sci. USA* **1993**, *90*, 9031–9035.
5 J. P. Gallivan, D. A. Dougherty, *Proc. Natl. Acad. Sci. USA* **1999**, *96*, 9459–9464.
6 J. C. Ma, D. A. Dougherty, *Chem. Rev.* **1997**, *97*, 1303–1324; G. W. Gokel, S. L. De Wall, E. S. Meadows, *Eur. J. Org. Chem.* **2000**, 2967–2978.
7 P. Lhoták, S. Shinkai, *J. Phys. Org. Chem.* **1997**, *10*, 273–285.
8 C. D. Gutsche, *Inclusion compounds*, Vol. 4, J. L. Atwood, J. E. Davies, D. D. MacNicol, eds., Oxford University Press, Oxford, **1991**, pp. 27–63; S. Shinkai, *Tetrahedron* **1993**, *49*, 8933–8968. V. Böhmer, *Angew. Chem.* **1995**, *107*, 785–818; *Angew. Chem., Int. Ed. Engl.* **1995**, *34*, 713–745.
9 C. M. Deber, E. R. Blout, *J. Am. Chem. Soc.* **1974**, *96*, 7566–7568; E. Ozeki, S. Kimura, Y. Imanishi, *Int. J. Peptide Protein Res.* **1989**, *34*, 111–117; T. Ishizu, J. Hirayama, S. Noguchi, *Chem. Pharm. Bull.* **1994**, *42*, 1146–1148; D. Leipert, D. Nopper, M. Bauser, G. Gauglitz, G. Jung, *Angew. Chem.* **1998**, *110*, 3503–3505; *Angew. Chem., Int. Ed. Engl.* **1998**, *37*, 3311–3314.
10 I. L. Karle, *The Peptides – Analysis, Synthesis, Biology*, Vol. 4, E. Gross, J. Meienhofer, eds., Academic Press, New York, **1981**, pp. 1–54.
11 R. Haubner, D. Finsinger, H. Kessler, *Angew. Chem.* **1997**, *109*, 1440–1456; *Angew. Chem., Int. Ed. Engl.* **1997**, *36*, 1374–1389; D. Ranganathan, *Acc. Chem. Res.* **2001**, *34*, 919–930.
12 S. Kubik, *J. Am. Chem. Soc.* **1999**, *121*, 5846–5855.
13 S. Kubik, R. Goddard, *J. Org. Chem.* **1999**, *64*, 9475–9486.
14 S. Roelens, R. Torriti, *J. Am. Chem. Soc.* **1998**, *120*, 12443–12452; C. E. Cannizzaro, K. N. Houk, *J. Am. Chem. Soc.* **2002**, *124*, 7163–7169.
15 The stability constants in Table 2 were calculated from saturation curves obtained by NMR titration under the assumption that the complexation-induced shift of the cation protons is independent on whether an anion is simultaneously bound or not.
16 V. Böhmer, A. Dalla Cort, L. Mandolini, *J. Org. Chem.* **2001**, *66*, 1900–1902; S. Bartoli, S. Roelens, *J. Am. Chem. Soc.* **2002**, *124*, 8307–8315.
17 A. Arduini, G. Giorgi, A. Pochini, A. Secchi, F. Ugozzoli, *J. Org. Chem.* **2001**, *66*, 8302–8308.
18 S. Kubik, R. Goddard, *Eur. J. Org. Chem.* **2001**, 311–322.
19 S. Mangani, M. Ferraroni, *Supramolecular Chemistry of Anions*, A. Bianchi, K. Bowman-James, E. García-España, eds., Wiley–VCH, New York, **1997**, pp. 63–78.
20 A. Bianchi, K. Bowman-James, E. García-España, *Supramolecular Chemistry of Anions*, Wiley–VCH, New York, **1997**; F. P. Schmidtchen, M. Berger, *Chem. Rev.* **1997**, *97*, 1609–1646; P. D. Beer, P. A. Gale, *Angew. Chem.* **2001**, *113*, 502–532; *Angew. Chem., Int. Ed. Engl.* **2001**, *40*, 486–516.
21 Ishida et al. were, in fact, the first to report on the high phosphate affinity in DMSO of a cyclic peptide containing alanine and 3-aminobenzoic acid subunits in an alternating sequence: H. Ishida, M. Suga, K. Donowaki, K. Ohkubo, *J. Org. Chem.* **1995**, *60*, 5374–5375.
22 S. Kubik, R. Goddard, R. Kirchner, D. Nolting, J. Seidel, *Angew. Chem.* **2001**, *113*, 2722–2725; *Angew. Chem., Int. Ed. Engl.* **2001**, *40*, 2648–2651.
23 V. Berl, I. Huc, R. G. Khoury, M. J. Krische, J.-M. Lehn, *Nature* **2000**,

407, 720–723; V. BERL, I. HUC, R. G. KHOURY, J.-M. LEHN, *Chem. Eur. J.* **2001**, *7*, 2798–2809; V. BERL, I. HUC, R. G. KHOURY, J.-M. LEHN, *Chem. Eur. J.* **2001**, *7*, 2810–2820.

24 S. KUBIK, R. GODDARD, *Proc. Natl. Acad. Sci. USA* **2002**, *99*, 5127–5132.

25 J. BITTA, S. KUBIK, *Org. Lett.* **2001**, *3*, 2637–2640.

B.6
Ion Transport Across Biological Membranes

Stefan Kubik

Biological membranes are effective barriers protecting the cytoplasm of a cell and all its functional components from the surrounding medium. Only small uncharged molecules such as water or urea, or larger hydrophobic compounds such as benzene can diffuse freely through the nonpolar double-layer of a membrane. Passage of larger polar molecules, to provide the cell with nutrients, for example carbohydrates or amino acids, or of ions, to maintain the membrane potential, is impossible without special transport mechanisms. The different types of transport mechanism found in biological systems can be divided into passive transport that follows the electrochemical gradient across the membrane, and active transport that proceeds in the opposite direction. The latter requires energy often supplied by hydrolysis of ATP.

An effective means by which ions can passively penetrate a biological membrane is passage through a pore created by a transmembrane protein, an example of which is gramicidin A. This linear helical peptide is composed of 15 alternating L- and D-amino acids. Self-assembly of two gramicidin A molecules in the interior of a cell membrane results in formation of a channel through which up to 10^7 monocations can pass in per second (Figure B.6.1). Another class of transmembrane protein, termed porines, form ion channels consisting of 16 or 18 cyclically arranged β-sheets, the so-called β-barrels. An interesting property of porines is that ion transport depends on the membrane potential – above or below a certain potential the porine channels close. The opening or closing of an ion channel can be triggered by the binding of a suitable substrate to the extracellular part of the transmembrane protein. A typical example is the nicotinic acetylcholine receptor responsible for transfer of nerve pulses across synapses. This transfer starts with release of acetylcholine from the pre-synaptic membrane into the synaptic gap. Binding of the neurotransmitter to the acetylcholine receptor located in the post-synaptic membrane causes the opening of an ion channel and, because Na^+ ions now permeate the membrane, a change in membrane potential.

Ionophores such as valinomycin also transport charged substrates across membranes. These often cyclic, low-molecular weight compounds do not

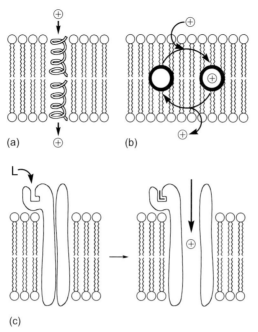

Fig. B.6.1. Schematic representation of the action of gramicidin A (a), an ionophore (b), and a ligand-controlled ion channel (c) in the transport of ions across a biological membrane.

form ion channels, however, but use a carrier mechanism for ion transport. As a consequence, the transport rate is much slower than a flow of ions through a channel. Transport selectivity is achieved as a result of the different affinities of the ionophores for different ions.

The action of transmembrane proteins capable of an active ion transport is more complex. The Na^+/K^+-ATP-ase, for example, the enzyme responsible for maintaining the high K^+ concentration inside cells, and the high Na^+ concentration outside, is made up of two pairs of large (α) and small (β) proteins. The α-proteins span the membrane and occur in different conformations in the presence of Na^+ or K^+. Phosphorylation of the Na^+/enzyme complex by ATP leads to conformational reorganization of the protein, whereupon sodium ions are transported to the extracellular side of the membrane where they are released. Subsequently, binding of K^+ in conjunction with dephosphorylation of the enzyme triggers a reversal of the conformational change and a transport of potassium into the cell interior. The overall process has a turnover of ca 150 s^{-1} and consists of a transport of three Na^+ ions out of the cell and two K^+ ions into the cell with simultaneous consumption of one molecule of ATP. Light is another energy source used for active membrane transport of ions by, for example, bacteriorhodopsin in halobacteria.

Model systems have been developed for many of these ion-transport mechanisms in the context of bioorganic chemistry. Examples are the cyclic peptides, described by M. R. Ghadiri et al., that have antibiotic activity similar to that of ionophores, a property that is most probably caused by the ability of these peptides to self-assemble inside biological membranes into channels [1]. Other compounds able to induce the formation of membrane pores are the bouquet-molecules introduced by J.-M. Lehn [2]. Artificial β-barrels have been developed by S. Matile's group [3]. Many host molecules used in bioorganic chemistry can serve as carriers for ions across membranes and have even made possible the development of systems with which active ion transport can be achieved [4].

References

1 S. Fernandez-Lopez, H.-S. Kim, E. C. Choi, M. Delgado, J. R. Granja, A. Khasanov, K. Kraehenbuehl, G. Long, D. A. Weinberger, K. M. Wilcoxen, M. R. Ghadiri, *Nature* **2001**, *412*, 452–455.

2 M. J. Pregel, L. Jullien, J.-M. Lehn, *Angew. Chem.* **1992**, *104*, 1695–1697; *Angew. Chem. Int. Ed. Engl.* **1992**, *31*, 1637–1640; J. Canceill, L. Jullien, L. Lacombe, J.-M. Lehn, *Helv. Chim. Acta* **1992**, *75*, 791–812; L. Jullien, T. Lazrak, J. Canceill, L. Lacombe, J.-M. Lehn, *J. Chem. Soc., Perkin Trans 2* **1993**, 1011–1020.

3 S. Matile, *Chem. Soc. Rev.* **2001**, *30*, 158–167.

4 M. Okahara, Y. Nakatsuji, *Top. Curr. Chem.* **1985**, *128*, 37–59.

2.3
Bioorganic Receptors for Amino Acids and Peptides: Combining Rational Design with Combinatorial Chemistry

Carsten Schmuck, Wolfgang Wienand, and Lars Geiger

2.3.1
Concept

One major research project in our group is currently the design, synthesis, and evaluation of artificial bioorganic receptors for binding amino acids and small oligopeptides in aqueous solvents. Peptides as a substrate are interesting for two reasons – their biological significance and their chemical structure. First, there are many biochemical or medicinal processes, for example enzymatic activity, bacterial infections, or neurodegenerative diseases, in which a selective molecular interaction of a peptide with another molecule or with itself (self-association) plays a decisive role [1]. Hence study of artificial receptor systems capable of selective binding to a specific biological peptide can help us understand better such peptide–molecule or peptide–peptide interactions on a detailed molecular basis. In contrast to most native proteins or natural receptors, bioorganic models are rather small and can be varied structurally deliberately. Therefore, they can be subjected to detailed physical–organic studies which cannot be performed as easily with many natural systems [2] (see also Chapter 1 for work on structural models for biological systems). The results obtained from such bioorganic model studies are not only useful for explaining and enabling better understanding of the underlying natural process, they might in the long term also facilitate the design of biosensors, the targeting of cellular processes by chemical tools, or the discovery of new medicinal therapeutics for a variety of diseases such as Alzheimer's, bacterial infections, or cancer. Second, beside this biological significance, peptides have a large variety of different potential binding sites for a receptor molecule to bind to – an amide backbone with H-bond donors and acceptors and side-chains with polar and non-polar groups. This is a great advantage if one is interested in designing peptide receptors as it enables exploitation of the whole spectrum of non-covalent interactions for complex formation [3]. In addition to the H-bond network to the peptide backbone, salt bridges to ionic amino acid residues (as in arginine, lysine, aspartic acid, etc.) or at the termini of the peptide (carboxylate or ammonium) and hydrophobic interactions with apolar side-chains (as in phenylalanine, valine, leucine, etc.) can further stabilize the receptor–substrate complex. The challenge of

Highlights in Bioorganic Chemistry: Methods and Applications. Edited by Carsten Schmuck, Helma Wennemers.
Copyright © 2004 WILEY-VCH Verlag GmbH & Co. KGaA, Weinheim
ISBN: 3-527-30656-0

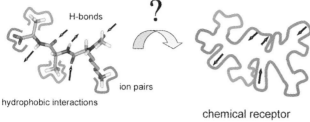

Fig. 2.3.1. Potential binding sites of a peptide.

designing a peptide receptor is therefore to translate the structural features of a given peptidic substrate into a suitable receptor molecule with complementary binding sites, as shown schematically in Figure 2.3.1 [4] (for a similar problem with carbohydrates see Chapter 2.1).

We will demonstrate in this chapter our approach towards modular receptors for complexation of biologically relevant peptides in water. A new binding motif for carboxylates, the guanidiniocarbonyl pyrroles, has been designed; this enables the formation of stable ion pairs even in aqueous solvents. By a stepwise elongation of this binding motif with additional interaction sites receptors are obtained that bind not only carboxylates but also single amino acids, both side-chain- and stereoselectively, and even tetrapeptides.

Because many biologically important small peptides contain a free C-terminus, which under physiological conditions is an anionic carboxylate, the first step of our approach to modular peptide receptors was to develop an efficient carboxylate binding site (CBS) which also functions in highly polar solvents up to aqueous solutions (for work on artificial hosts for spherical anions see Chapter 2.2). To get an idea of what is necessary for strong complexation of carboxylates under such challenging conditions one can take a look at Nature: For example, in carboxypeptidase A [5], an enzyme that hydrolytically cleaves an amino acid from the free C-terminus of a peptide chain, the peptidic carboxylate is essentially bound by an ion pair with the guanidinium group of arginine 145 and two additional H-bonds from asparagine 144 and tyrosine 248 (Figure 2.3.2).

The enzyme–ligand interaction is, furthermore, significantly facilitated by the overall hydrophobic character of the binding pocket which reduces competitive solvation of the binding sites by water molecules. As such ion pair formation between arginine and a carboxylate is widely found in Nature, it is not surprising that this binding motif has already been much used by supramolecular chemists over recent decades [6]. Unfortunately, without the hydrophobic shielding of an enzyme pocket the guanidinium–carboxylate ion pair is only stable in solvents of low polarity such as chloroform or acetonitrile (Box 7). Even the smallest amounts of more polar solvents such as DMSO, methanol, or even water cause immediate dissociation of these ion pairs. This is a general problem in supramolecular chem-

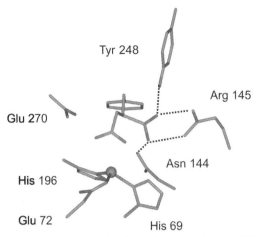

Fig. 2.3.2. Carboxylate binding within the active site of Carboxypeptidase A.

istry [7]. Molecular recognition is based on non-covalent interactions but, in contrast with covalent bonds, their strength is highly dependent on external conditions such as solvent composition, polarity, or even temperature. In this context water as a solvent is the most challenging. On the one hand, electrostatic interactions (H-bonds or ion pairs), which are quite well understood and have been extensively used in artificial supramolecular systems due to their complementarity and directionality, are rather weak in this solvent whereas, on the other hand, until now the stronger hydrophobic interactions [8] are much more difficult to design and use in artificial receptors.

The best solution is therefore to use not only one but several non-covalent interactions simultaneously to achieve strong ligand complexation, even in water. Although every individual contact between host and substrate by itself might be rather weak, their combined effect can still lead to high association constants ("Gulliver effect") [9]. Our idea for the design of an efficient CBS was, therefore, to improve the binding affinity of the guanidinium cation by use of suitable additional binding sites (how efficient arginine receptors can be designed is explained in Chapter 2.4). Based on theoretical calculations we therefore introduced cationic guanidiniocarbonyl pyrroles of type **1** as a new and easily accessible binding motif for carboxylate anions [10].

These acyl guanidinium receptors combine several advantages compared with simple guanidines; this makes them attractive candidates for the binding of carboxylates:

- Acyl guanidines have pK_a values in the order of 7 whereas simple guanidines have pK_a values of approximately 13. This increased acidity favors the formation of hydrogen-bonded ion pairs and hence increases the binding affinity.
- Additional prospective hydrogen-bond donors such as the amide NH can further enhance the stability of the complex.
- The binding motif is planar and rather rigid and therefore ideally preorganized for binding of planar anions such as carboxylates.

2.3.2
Structural and Thermodynamic Characterization of the New Binding Motif

As expected from the modeling, guanidiniocarbonyl pyrrole receptors indeed strongly bind carboxylates even in aqueous solvents, by a combination of ion pairing and multiple hydrogen bonding, as can be seen in the NMR spectrum. For example, addition of acetate to a solution of the ethylamide-substituted guanidiniocarbonyl pyrrole 2 in dimethyl sulfoxide causes significant complexation-induced shifts (CIS) of the various protons of 2 in the ^{1}H NMR spectrum (Figure 2.3.3).

The ^{1}H NMR spectrum of the receptor itself contains the "normal" signals expected for an acyl guanidinium cation (Figure 2.3.3, front) [11]. On addition of the carboxylate, large downfield complexation-induced shifts of some of the NH protons of the guanidiniocarbonyl pyrrole are observed (Figure 2.3.3, back). Such shift changes indicate that these protons form hydrogen bonds to the carboxylate. On formation of an H-bond the electron density around the proton is further diminished and this deshielding causes a downfield shift of the corresponding signal in

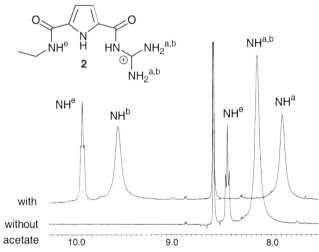

Fig. 2.3.3. ^{1}H NMR spectrum of receptor 2 (picrate salt) with (back) and without (front) acetate (NMe$_4^+$ salt) in [D$_6$]DMSO, showing the CIS of the guanidinium NH protons and the amide NH.

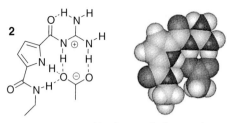

Fig. 2.3.4. Proposed binding mode for complexation of carboxylates by guanidiniocarbonyl pyrrole receptors like **2**.

the NMR. The observed shift changes can therefore be used to deduce information on the structure of the complex formed. For guanidiniocarbonyl pyrroles all the experimental data are consistent with the general binding mode depicted above (Figure 2.3.4). The guanidinium cation forms an ion pair with the carboxylate which is simultaneously hydrogen bonded by the pyrrole NH and the amide NH.

The quantitative dependence of the complexation-induced shift changes on the ratio of host to substrate can, furthermore, be used to calculate the binding constant [12]. With increasing amounts of the guest added to a solution of the host (or vice versa), the NMR host signals will be shifted depending on the relative amount of complex present in solution. This again depends on the ratio of host to guest, their concentration in solution and of course the thermodynamic dissociation constant of the complex formed. For our guanidiniocarbonyl pyrroles such ^{1}H NMR titration experiments showed that even in 50% water in dimethyl sulfoxide the association constants for the binding of carboxylates are still in the order of $K_{ass} \approx 10^3$ mol^{-1} whereas no complexation is observed for simple guanidinium cations under these conditions [10]. Our goal of designing an improved version of a guanidinium cation that also enables binding of carboxylates in solvents of high polarity was therefore achieved.

But which additional feature of our binding motif is responsible for this improved complex stability? The increased acidity of the acyl guanidinium cation or the additional H-bonds, or both? Unfortunately, it is impossible to determine experimentally the binding energy of an individual bond or type of interaction within an array of several non-covalent interactions, because only the overall stability can be measured. To estimate the individual energetic contributions of the different non-covalent interactions within this binding motif a systematically varied series of receptors **2–6** was therefore studied [10]. As each new receptor contains only one more potential binding site, comparison of their relative binding properties enables estimation, at least, of the energetic contribution of this interaction.

Within this series the binding constants were experimentally determined by NMR titration experiments in 40% water in dimethyl sulfoxide. The corresponding binding isotherms are shown in Figure 2.3.5, with the calculated binding constants. These data show that the simple guanidinium cation **3** does not bind carboxylates under these conditions. The more acidic acetyl guanidinium **4**, although much better, still has no more than a rather weak association. Only the guanidiniocarbonyl pyrroles form sufficiently strong complexes. But in the guanidiniocarbonyl pyrroles the different weak interactions do not contribute equally to the binding process. Still the interplay of ion pairing and hydrogen bonding between the acyl guanidinium moiety and the carboxylate turns out to be the most important driving force for stable complexation in aqueous solvents. But this ion pair is significantly stabilized especially by an H-bond from the amide NH in position 5 of the pyrrole whereas the pyrrole NH itself is less important.

2.3.3
Selective Binding of Amino Acids

So far we now have available a very efficient binding motif for carboxylates. For use of this binding motif as a building block in modular peptide receptors it is also

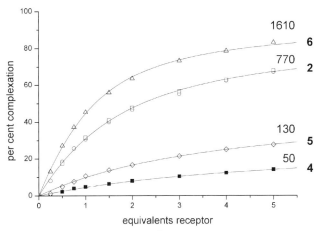

Fig. 2.3.5. NMR titration curves of receptors **2–6** (picrate salts) with N-acetylalanyl carboxylate (NMe$_4^+$ salt, 1 mM) in 40% water in [D6]DMSO; the calculated binding constants (in M^{-1}) are shown.

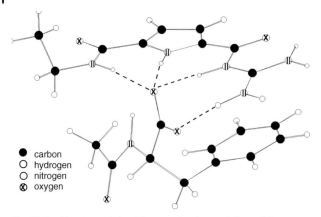

Fig. 2.3.6. Structure, derived from molecular modeling, of **2** with N-acetylphenylalanine in water (intermolecular hydrogen bonds are shown as broken lines).

necessary to achieve substrate selectivity. Additional interactions between the substrate and the receptor are therefore needed to make the recognition process selective. Surprisingly, this is already true in the simple guanidiniocarbonyl pyrroles presented so far. The binding strength for complexation of different N-acetyl amino acid carboxylates by the ethylamide-substituted receptor **2** in aqueous dimethyl sulfoxide depends significantly on the type of amino acid [13]. The association constant for the binding of D/L-phenylalanine ($K = 1700$ M^{-1}) is more than twice as strong as for D/L-alanine ($K = 770$ M^{-1}); whereas the association constant for D/L-lysine ($K = 360$ M^{-1}) is approximately half as strong. The differences in complex stability among the various amino acids must result from secondary interactions of their side-chains with the receptor. For phenylalanine the aromatic ring probably π-stacks with the acyl guanidinium unit of **2**, at least according to molecular modeling studies. This cation–π interaction [14] further stabilizes the complex (Figure 2.3.6). In contrast with this, the positively charged ω-ammonium group in lysine reduces the binding affinity relative to alanine, because of unfavorable electrostatic interactions with the positively charged guanidinium group.

The chiral, L-valine derived receptor **6** not only had side-chain selectivity but also moderate stereoselectivity [10]. For example, for alanine the L enantiomer is bound by a factor of three better than the D enantiomer. Because experimental structural information was not available, molecular modeling was used to help rationalize the observed binding selectivity. In the complex with the L enantiomer the binding constant ($K = 1610$ mol^{-1}) is larger than for binding by the ethylamide receptor **2** ($K = 770$ mol^{-1}), probably because of an additional hydrogen bond from the terminal carbamoyl group. This is also indicated by a corresponding complexation-induced shift change of this proton in the NMR. With D-alanine there is an unfavorable steric repulsion between the methyl group of the amino acid and the isopropyl side-chain of the receptor which is not present in the complex with the L enantiomer, in which the methyl group points away from the isopropyl group.

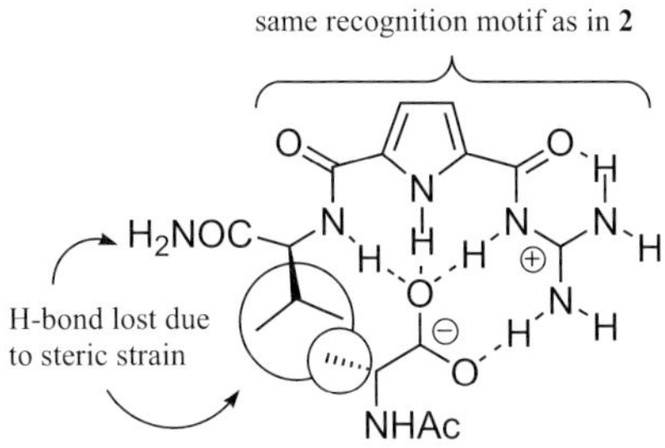

Fig. 2.3.7. An unfavorable steric interaction between the two side-chains in the complex between D-alanine and receptor **6** reduces the binding affinity.

Because the H-bond from the terminal carbamoyl group to the carboxylate is not strong enough to compensate for this increased steric strain, this H-bond is most probably lost in the complex with D-alanine (Figure 2.3.7). The remaining binding motif therefore resembles the simple amide-substituted guanidiniocarbonyl pyrrole **2**, which lacks the sterically demanding isopropyl group. In accordance with this, the binding constant for the D enantiomer is also approximately the same ($K = 730 \text{ mol}^{-1}$).

2.3.4
Binding of Small Oligopeptides

Going now from single amino acids to even larger substrates requires further binding sites attached to the guanidiniocarbonyl pyrroles for additional interactions with the peptides' other binding sites besides the carboxylate (Figure 2.3.1). How should such a peptide receptor look like? In principle there are two distinct approaches [15]. One can try to design a complete receptor de novo, rationally with the help of theoretical calculations. The larger the substrate, the more difficult this becomes, however, because theoretical calculations are not yet sufficiently reliable for complete design of a tailor-made artificial host for a large substrate. Another possibility is to use a random trial and error approach and to identify suitable receptors with the help of combinatorial chemistry [16, 17] (for more work on combinatorial chemistry see Chapters 3.1, 3.3, or 5.4). The best guarantee for success is, of course, to integrate both methods – to use binding motifs designed for a specific target as building blocks for a combinatorial synthesis. In such a *focused* combinatorial library [18] the chances of finding a hit are much higher than in a completely random library, because the structural diversity is already positively biased for a given problem, e.g. binding of a specific target. It is, therefore, sufficient to use even small libraries with only a couple of hundred different members.

Following this concept, we set out to find efficient receptors for tetrapeptidic substrates. As a first target the hydrophobic tetrapeptide, Ac–Val–Val–Ile–Ala–OH was chosen. This tetrapeptide represents the C-terminal sequence of the amyloid-β-

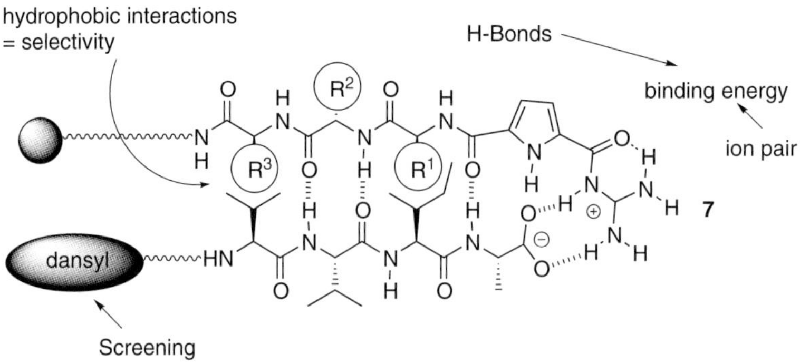

Fig. 2.3.8. A tripeptide-based library of cationic guanidiniocarbonyl pyrrole receptors **7** designed for the binding of Val–Val–Ile–Ala, a tetrapeptide representing the C-terminus of Aβ.

peptide (Aβ) which is responsible for the formation of protein plaques within the brain of patients suffering from Alzheimer's disease [19]. This specific peptide sequence is thought to promote the formation of self-aggregated β-sheets of Aβ stabilized through a combination of H-bonds and hydrophobic interactions [20]. An artificial receptor which effectively binds to the model tetrapeptide Ac–Val–Val–Ile–Ala–OH can therefore enable us to learn more about the molecular basis of the self-aggregation of the amyloid-peptide.

Our general design of a potential receptor **7** for this substrate is shown in Figure 2.3.8. The ion pair between the carboxylate and the guanidiniocarbonyl pyrrole serves as a starting point for complex formation. An additional tripeptide unit attached to the pyrrole provides further binding sites for the formation of a hydrogen-bonded antiparallel β-sheet with the backbone of the tetrapeptide substrate (Chapter 2.4 contains more about the use of such β-sheet mimics). In addition to these multiple electrostatic interactions, hydrophobic contacts between the amino acid side-chains both in the substrate and the receptor should, especially in aqueous solvents, further stabilize the complex and also guarantee the necessary substrate selectivity. To identify which amino side-chains in the receptor will be most efficient for this purpose a combinatorial approach was used.

A solid phase bound library of 512 different but structurally related receptors **7** was therefore synthesized using a standard Fmoc-procedure and a split–mix approach [21] (Boxes 11 and 25). In each of the three coupling steps eight different amino acids were used; these were selectively chosen to provide a range of structurally varying hydrophobic or steric interactions. In the second step each of the various 512 different tripeptides thus obtained was coupled with the guanidiniocarbonyl pyrrole binding motif. The advantage of such a solid-phase-bound combinatorial receptor library is, besides the fast and time-saving synthesis, that the whole library can be tested for a specific feature, in this case its binding properties towards the tetrapeptide substrate, in a single experiment [22]. For this purpose a fluorescence label in form of a dansyl group was attached via a water-soluble spacer

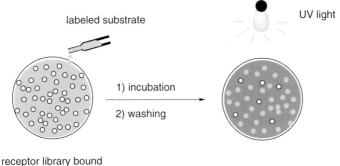

Fig. 2.3.9. Fluorescence binding assay.

to the N-terminus of the tetrapeptide substrate. After incubation of the library with this labeled substrate, a simple UV-assay can be used to identify efficient receptors (Figure 2.3.9). Only those beads on which the attached receptor is capable of bind the peptide can show the characteristic fluorescence of the dansyl group. All the other receptors which do not bind the peptide under the specific experimental conditions remain dark.

Such binding assays showed that our one-armed cationic receptors **7** are indeed capable of efficient binding of the tetrapeptide in polar solutions, even in water [23]. Interestingly, the uncharged methyl ester of the substrate binds only weakly and rather unselectively to the receptor library, suggesting that side-chain interactions alone are not strong enough to form a stable complex. The negatively charged carboxylate substrate is, however, selectively bound only by some and not all of the receptors, although ion pairing with the guanidiniocarbonyl pyrrole unit is the same for all the different receptors within the library. Hence, the binding of the tetrapeptide by the receptors requires both electrostatic and hydrophobic interactions. Neither of these interactions alone is sufficiently strong to ensure complex formation in polar solution (Figure 2.3.10).

One must, of course ensure that the fluorescence activity observed in the assay is really due to a selective complexation of the tetrapeptide substrate by the receptor. This was done by appropriate control experiments:

1. Because the labeled tetrapeptide does not bind to the unmodified solid support Amino-TentaGel, the observed fluorescence activity is not because of unspecific interaction with the solid support itself.
2. The dansylated spacer alone does not bind to the receptor library, showing that the binding indeed occurs between the peptide part of the substrate and the receptor.
3. The percentage of receptors that bind the substrate is concentration-dependent – at high concentrations nearly all of the library members bind the substrate, which shows that the observed binding specificity is not a result of selective quenching of the dansyl fluorescence rather than selective binding.

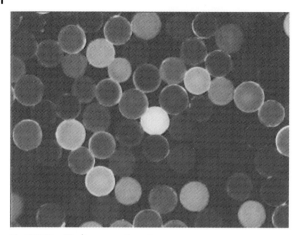

Fig. 2.3.10. Binding assay in water.

These qualitative assays show that one-armed cationic guanidiniocarbonyl pyrrole receptors can indeed effectively bind tetrapeptides even in water. Molecular modeling studies suggest a complex structure as shown for one specific example, the receptor Val–Val–Val–CBS, in Figure 2.3.11. Receptor and substrate form a hydrogen bonded β-sheet which is further stabilized by additional hydrophobic interactions between the apolar groups in the side-chains. Recognition of the tetrapeptide thus seems to be controlled by a fine balanced interplay between electrostatic and hydrophobic interactions.

Another advantage of such a combinatorial binding assay is that it can also be performed quantitatively, enabling direct determination of the binding constants of the different receptors [24]. The association constants for each receptor can be calculated from the fluorescence intensity of the substrate in solution before and after incubation and the loading of the resin. Even though such binding constants determined on a solid support are not the same, and less accurate than data obtained in solution (for example from NMR or UV titration experiments), a comparison of relative data within a series of related receptors can, at least, help rationalize aspects such as complex structure, stability and selectivity on a molecular basis. One can identify structural features that are associated with strong or weak bind-

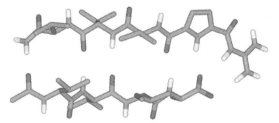

Fig. 2.3.11. Proposed structure for the complex between receptor (top) and tetrapeptide (below).

ing. Which parts of our modular receptors are most important for binding or selectivity? What kind of binding sites, electrostatic or hydrophobic, in the various positions of the receptor are needed? In other words a supramolecular structure–binding relationship can be derived from binding data obtained on a solid support.

We have so far performed a detailed thermodynamic analysis for binding of the tetrapeptide in methanol [23]. As these data show the binding is exceptionally strong with approximate association constants of 10^4 M^{-1} for the best receptors. In this assay the binding is measured relative to the formate counter-ion of the receptors, which also binds to the guanidiniocarbonyl pyrrole motif, although to a lesser extent than other carboxylates. Therefore, the interaction between the tetrapeptide and the receptors is actually even much stronger than suggested by these numbers. Hence, these one-armed hosts are among the most efficient peptide receptors in polar solvents yet reported. The selectivity of the receptors towards the tetrapeptide substrate is, furthermore, surprisingly high. The association constants for the various receptors differ by a factor of more than 100 among the library! Even small changes in the structure of the receptor have obviously pronounced effects on the binding properties. This also proves that even within such a small combinatorial library of only limited structural diversity the binding selectivity can be rather high – a necessary prerequisite for also achieving selective binding of different tetrapeptides by this general receptor class.

A closer look at the binding data enables correlation of complex stability and receptor structure. For example, the quantitative binding constants suggest that hydrophobic interactions of the receptor with the first amino acid residue of the substrate (Val) are most important. Exchange of valine in the position opposite of this residue in the receptor for N-Boc-protected lysine reduces the binding affinity by a factor of 10. The side-chain of lysine is probably too small to provide enough hydrophobic shielding of the isopropyl group of the substrate in the complex. This is excellent confirmation of previous studies of the Aβ self-aggregation which have shown that hydrophobic interactions with Val 39, which corresponds to the first amino acid of our tetrapeptide substrate, are especially important [25]. This again emphasizes the feasibility of using carefully chosen small bioorganic models for the analysis of more complex natural systems. First screening experiments with a larger library have already shown that, even in water, receptors of this general type efficiently bind the model tetrapeptide with association constants of the order of 10^4 M^{-1} (in tris-buffer of pH 6).

2.3.5
Conclusion

In conclusion, we have demonstrated that modular bioorganic receptors can be used for effective binding of amino acids and C-terminal oligopeptides in aqueous solutions. The central building block of our receptors is a guanidiniocarbonyl pyrrole moiety which was designed to be a very efficient carboxylate binding site even in aqueous solvents. This was achieved by rationally improving the binding strength of a guanidinium cation by forming additional hydrogen-bonds to the bound carboxylate. By extending this binding motif with additional interaction sites larger substrates such as amino acids and tetrapeptides can also be bound. In this context, the combination of target-oriented rational design with the power of combinatorial chemistry proved to be the method of choice for identification of suitable receptor structures for a given substrate.

References

1 N. Sewald, H.-D. Jakubke, *Peptides: Chemistry and Biology*, Wiley–VCH, Weinheim **2002**.
2 H. Dugas, *Bioorganic Chemistry*. Springer, New York, **1996**.
3 (a) G. A. Jeffrey, *An Introduction to Hydrogen Bonding*. Oxford University Press, New York, **1997**; (b) D. H. Williams, M. S. Westwell, *Chem. Soc. Rev.* **1998**, *27*, 57–64.
4 (a) M. W. Peczuh, A. D. Hamilton, *Chem. Rev.* **2000**, *100*, 2479–2494; (b) H. J. Schneider, *Angew. Chem.* **1993**, *105*, 890–892; (c) T. H. Webb, C. S. Wilcox, *Chem. Soc. Rev.* **1993**, *22*, 383–395.
5 S. Mangani, M. Ferraroni, *Supramolecular Chemistry of Anions*, A. Bianchi, K. Bowman-James, E. Garcia-Espana, eds., Wiley–VCH, New York, **1997**, pp. 63–78.
6 Review articles on the binding of carboxylates by artificial hosts including guanidinium-based systems: (a) R. J. Fitzmaurice, G. M. Kyne, D. Douheret, J. D. Kilburn, *J. Chem. Soc., Perkin Trans. 1* **2002**, 841–864; (b) F. P. Schmidtchen, M. Berger, *Chem. Rev.* **1997**, *97*, 1609–1646.
7 J.-M. Lehn; *Supramolecular Chemistry; Concepts and Perspectives*, VCH, Weinheim **1995**.
8 For a review on aromatic interactions see: (a) C. A. Hunter, K. R. Lawson, J. Perkins, C. J. Urch, *J. Chem. Soc., Perkin Trans. 2* **2001**, 651–669. (b) E. A. Meyer, R. K. Castellano, F. Diederich, *Angew. Chem.* **2003**, *115*, 1244–1287.
9 L. J. Prins, D. N. Reinhoudt, P. Timmerman, *Angew. Chem.* **2001**, *113*, 2446–2492; *Angew. Chem. Int. Ed.* **2001**, *40*, 2382–2426.
10 C. Schmuck, *Chem. Eur. J.* **2000**, *6*, 709–718.
11 R. D. Dixon, S. J. Geib, A. D. Hamilton, *J. Am. Chem. Soc.* **1992**, *114*, 365–366.
12 (a) C. S. Wilcox, *Frontiers in Supramolecular Chemistry and Photochemistry*, H. J. Schneider, H. Dürr, eds., VCH, Weinheim, **1990**, pp. 123–144. (b) K. A. Connors, *Binding Constants*, Wiley, New York, **1987**.
13 C. Schmuck, *Chem. Commun.* **1999**, 843–844.
14 (a) J. C. Ma, D. A. Dougherty, *Chem. Rev.* **1997**, *97*, 1303–1324; (b) D. A. Dougherty, *Science* **1996**, *271*, 163–168.
15 M. H. V. Van Regenmortel, *J. Mol. Recognit.* **2000**, *13*, 1–4.
16 (a) L. A. Thompson, J. A. Ellman, *Chem. Rev.* **1996**, *96*, 555–600; (b) F. Balkenhohl, C. von dem Busche-Hünnefeld, A. Lansky, C. Zechel, *Angew. Chem.* **1996**, *108*, 2436–2487; *Angew. Chem. Int. Ed.* **1996**, *36*, 2288–2337; (c) G. Lowe, *Chem. Soc. Rev.* **1995**, 309–317.
17 For related work on other peptide substrates see, e.g.: (a) K. Jensen, T. M. Braxmeier, M. Demarcus, J. G. Frey, J. D. Kilburn, *Chem. Eur. J.* **2002**, *8*, 1300–1309. (b) R. Xuo, G. Greiveldinger, L. E. Marenus, A. Cooper, J. A. Ellman, *J. Am. Chem. Soc.* **1999**, *121*, 4898–4899. For a review see: (c) M. W. Peczuh, A. D. Hamilton, *Chem. Rev.* **2000**, *100*, 2479–2493.

18 R. Breinbauer, I. R. Vetter, H. Waldmann, *Angew. Chem.* **2002**, *114*, 3002–3015.
19 (a) E. Zerovnik, *Eur. J. Biochem.* **2002**, *269*, 3362–3371; (b) J. Hardy, D. J. Selkoe, *Science* **2002**, *297*, 353–356; (c) B. Austen, M. Manca, *Chem. Brit.* **2000**, 28–31; (d) L. Gopinath, *Chem. Brit.* **1998**, 38–40; (e) P. T. Lansbury Jr, *Acc. Chem. Res.* **1996**, *29*, 317–321; (f) B. A. Yankner, *Neuron* **1996**, *16*, 921–932.
20 J. T. Jarret, E. P. Berger, P. T. Lansbury Jr, *Biochemistry* **1993**, *32*, 4693–4697.
21 K. S. Lam, M. Lebl, V. Krchnak, *Chem. Rev.* **1997**, *97*, 411–448.
22 W. C. Still, *Acc. Chem. Res.* **1996**, *29*, 155–163.
23 C. Schmuck, M. Heil, *Org. Biomol. Chem.* **2003**, *1*, 633–636.
24 S. S. Yoon, W. C. Still, *Tetrahedron* **1995**, *51*, 567–578.
25 P. T. Lansbury Jr, P. R. Costa, J. M. Griffiths, E. J. Simon, M. Auger, K. J. Halverson, D. A. Kocisko, Z. S. Hendsch, T. T. Ashbury, R. G. S. Spencer, B. Tidor, R. G. Griffin, *Nat. Struct. Biol.* **1995**, *2*, 990–998.

B.7
The Effect of Solvents on the Strength of Hydrogen Bonds

Carsten Schmuck

Because of their specificity and directionality hydrogen bonds play an important role in determining the three dimensional structure of chemical and biological systems, especially in combination with other non-covalent forces such as ionic or hydrophobic interactions. The main draw back of hydrogen bonds is their limited strength. The more polar a solvent, the weaker are the hydrogen bonds. A single hydrogen bond thus has a substantial binding energy only in non-polar aprotic solvents, and not in water.

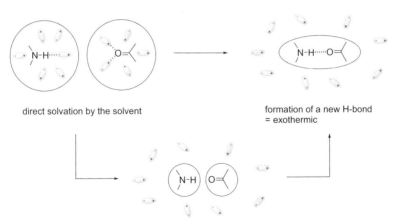

Fig. B.7.1.

Even for the most stable hydrogen bond in FHF$^-$ the binding energy decreases from 160 kJ mol^{-1} in the gas phase to only 3–4 kJ mol^{-1} in water! The effect of the polar solvent on the strength of H-bonds is twofold:

On a macroscopic level, the bulk solvent properties such as the dielectric constant are changed. According to the Coulomb law this weakens all electrostatic interactions and hence also hydrogen bonds.

On a molecular level, specific direct solvation of the donor and acceptor sites by individual solvent molecules occurs. In other words, the solvent functions as a competitive binding partner. Therefore, desolvation of both donor and acceptor sites must occur before a new hydrogen bond can be formed. The energetic price which must be paid for this desolvation, reduces the binding energy of the H-bond. In very polar solvents, the net energy change might then even be endothermic, but the association still can be favorable because of the release of ordered solvent molecules into the bulk solution upon desolvation, which increases the entropy of the system (Figure B.7.1).

2.4
Artificial Receptors for the Stabilization of β-Sheet Structures

Thomas Schrader, Markus Wehner, and Petra Rzepecki

2.4.1
β-Sheet Recognition in Nature

The β-sheet structure is one of the fundamental secondary structures of proteins (Box 8). It greatly stabilizes protein domains by forming an array of several parallel peptide backbone pieces five to fifteen amino acids long, which are interconnected purely by multiple hydrogen bonds. β-Sheets can, however, also be exposed to solvent; they then have another important role, namely as binding sites for external ligands. Therefore the active center of many enzymes contains a β-sheet fragment which binds to a target peptide delivering its critical functionality to the catalytically active "hot spot". A prominent example is the discovery that peptidal substrates and inhibitors bind to diverse proteolytic enzymes as β-strands [1]. In other instances enzymes are only active as dimers, which by themselves represent small β-sheet structures. Thus two identical domains of the HIV-I protease form a four-stranded β-sheet at their N-termini. This in turn leads to formation of the catalytically active cleavage channel which cuts the predecessor peptide into the correct pieces which automatically self-assemble to the complete virus. Such "functional" β-sheet structures are intrinsic targets for the development of new drugs aimed at blocking the active site of enzymes [2].

β-Sheet structures can, on the other hand, pose a severe problem for the organism if they are formed spontaneously by circumventing the body's strict control mechanisms. Because of their mutual saturation of all hydrogen-bond donors and acceptors with non-polar side-chains extended to the solvent, they readily precipitate from aqueous solution, and form large aggregates. In recent years, protein folding diseases have become increasingly important; the percentage of elderly people suffering from Alzheimer's dementia, e.g., has exceeded 0.1% in the western hemisphere. European countries have been haunted by epidemics like BSE threatening humans with the related new variant of CJD. Such pathological β-sheets represent another medicinal challenge, calling for the development of small ligands capable of keeping these proteins in solution or even reversing plaque formation [3].

Highlights in Bioorganic Chemistry: Methods and Applications. Edited by Carsten Schmuck, Helma Wennemers.
Copyright © 2004 WILEY-VCH Verlag GmbH & Co. KGaA, Weinheim
ISBN: 3-527-30656-0

2.4.2
Artificial β-Sheets and Recognition Motifs

Some twenty years ago chemists began their contributions on β-sheet research by creating small soluble β-sheets which could be examined structurally. Rigid templates and artificial β-turns were adorned with growing peptide chains, and the formation of two- and three-stranded artificial β-sheets was proven by NMR techniques and CD spectroscopy [4]. Kelly even postulated that the combination of his non-polar benzofuran template with the first lipophilic amino acids mimicked the hydrophobic cluster, which has been suggested as nucleation site for the formation of biological β-sheets in the protein-folding process [5]. The propensity of the twenty proteinogenic amino acids for adopting the β-sheet conformation in a peptide strand has been studied by various techniques and has even been extended to discrimination between parallel and antiparallel β-sheets [6]. Recipes for the construction of larger β-sheet peptides consisting entirely of natural amino acids have been put forward by Gellman and others, leading to the design of extended β-sheets which remain in solution [7].

In contrast, very few external ligands have been found that are capable of docking on to an existing peptide and thereby stabilize its β-sheet conformation. The first example was presented by our group in 1996, when we introduced aminopyrazole derivatives for the backbone recognition of dipeptides in organic solution [8]. Later (1998) the Hamilton group truncated the diaminoquinolone motif used by Kemp et al. in their epindolidione receptors [9] for the intramolecular stabilization of short peptides in the β-sheet conformation [10]. Bartlett recently (2002) presented the azacyclohexenone fragment for incorporation in drugs targeting β-sheets [11]. All these recognition motifs can be characterized by a general scheme – they recognize the acceptor–donor–acceptor pattern (ADA) found in natural β-sheets by a complementary arrangement of hydrogen-bond donors and acceptors (DAD) in a linear fashion, with optimized distances of 2.7 Å and 3.7 Å (Figures 2.4.1 and 2.4.2).

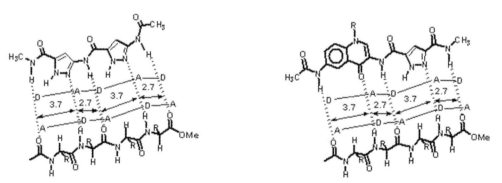

Fig. 2.4.1. Two illustrations for the general DAD pattern of artificial β-sheet ligands with optimized geometry.

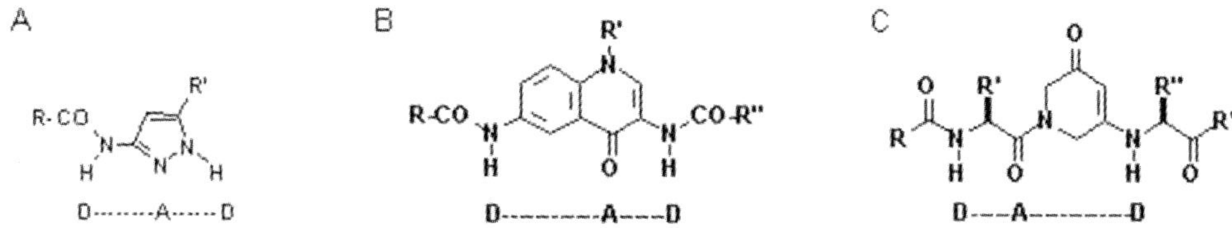

Fig. 2.4.2. Recognition motifs following the general rule outlined in Figure 2.4.1 (A) 3-aminopyrazoles, (B) 3,6-diaminoquinolones; (C) 3-azacyclohexenones.

To avoid self-association the overall shape of larger structures has been chosen to be slightly curved, or ionic groups have been incorporated for recognition of the N- or C-terminus.

2.4.3
Sequence-selective Recognition of Peptides by Aminopyrazoles

A formidable challenge in the artificial recognition of peptides is the achievement of sequence selectivity. The contributions of Schmuck (Chapter 2.3), Nestler (Chapter 3.1), or Wennemers (Chapter 5.4) in this book demonstrate the power of the combinatorial approach, originally introduced by Still et al. [12]. With our aminopyrazoles available we chose an alternative route, i.e. rational design of modular peptide receptors tailored for recognition of the main classes of amino acids [13]. The concept is that because complexation of a peptide in its β-sheet conformation brings all side-chains into a horizontal orientation, their specific recognition can only be achieved by introduction of an additional binding site coming from the top. To this end we connected to the aminopyrazole a U-turn in the form of Kemp's triacid, and attached to its opposite end an interchangeable tip with binding sites for polar, aromatic or basic side amino acid chains. As soon as the aminopyrazole docks on to the peptide's backbone, the additional binding site is lowered from above on to the respective side-chain and forms an additional interaction specific for the class to which this amino acid belongs (Figure 2.4.3).

Monotrifluoroacetylated diaminopyrazole was first reacted with the free Kemp's triacid to produce the imide, followed by *N*-Boc protection and amide-coupling with a *m*-substituted aniline derivative. Final Boc-deprotection occurred on the chromatography column leading directly to the new receptor modules. The recognition site X was chosen to be ethyl as a neutral reference, acetyl for polar side-chains, nitro for electron-rich aromatic residues and carboxylate for basic amino acids (Figure 2.4.4).

The structures of **3–6** show a high degree of preorientation, a prerequisite for the highly selective recognition of amino acid side-chains. Intramolecular hydrogen bonding locks the imide and the neighboring aminopyrazole in the same plane, whereas the sub-van der Waals distance between both aromatics keeps them parallel to each other. This was suggested by Monte Carlo conformational searches and proven by NOE measurements [14].

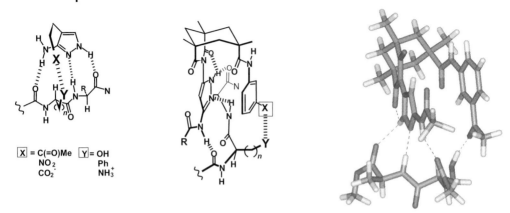

Fig. 2.4.3. Rational design of modular building blocks for the specific recognition of amino acid side-chains in the β-sheet conformation. Right: force-field calculation of the serine recognition event.

No self-association occurs between the receptor molecules of **4** in a concentration range between 10^{-2} M and 10^{-3} M, as was proven with a dilution experiment. Hence, all complexation studies were performed with alanine-containing dipeptides in this concentration range. As usual, the peptide's top face NH proton shows a downfield shift on complexation with the aminopyrazole. In almost every exam-

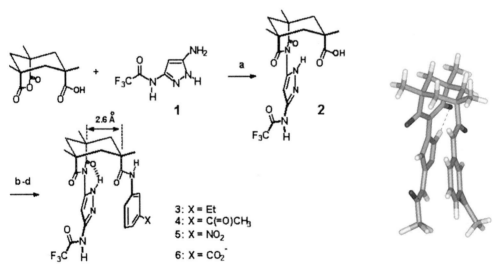

Fig. 2.4.4. Synthesis of hosts **3–6** from Kemp's triacid and monotrifluoroacetylated diaminpyrazole **1**; (a) 110 °C; (b) Boc$_2$O; (c) *m*-substituted aniline derivative, PyCloP; (d) silica gel. Right: Productive conformation of the receptor modules found in Monte Carlo simulations (MacroModel 7.0).

Tab. 2.4.1. Association constants K_a [M^{-1}] for complex formation between hosts **3–6** and various dipeptides, determined by NMR titrations in CDCl$_3$.

Host	Standard	Selective binding	Special
triflAMP[a]	Ala–Ala: 50	–	Phe–Ala: 40
Reference **3**	Orn–Ala: 280	Ala–Ala: 70	Phe–Ala: ≤40
XH-binder **4**	Ala–Ala: 80	Ser–Ala: 900	Ser–Val: no shifts
Arene-binder **5**	–	Phe–Ala: 350	Ala–Phe: no shifts
Cation-binder **6**	Propylamine: 490	Orn–Ala: 2360	–

[a] triflAMP = 3-trifluoroacetylamino-5-methylpyrazole

ple we obtained a clear 1:1 stoichiometry, assured by means of a Job plot [15]. NMR titrations in CDCl$_3$ furnished association constants below 100 M^{-1} (Ac–Ala–Ala–OMe/Ac–Phe–Ala–OMe) [16]. These relatively small numbers are explained by the low propensity of alanine-containing peptides to adopt the β-sheet conformation and were also found with the best binders developed by us earlier for dipeptides (Table 2.4.1).

The recognition module **3** with an acetyl group for polar side-chains furnished K_a values of 80 M^{-1} and 900 M^{-1} for Ac–Ala–Ala–OMe and Ac–Ser–Ala–OMe, respectively. Thus, a more than tenfold increase was observed when the additional hydrogen bond was formed between the hydroxymethyl group of serine and the acetyl group of the receptor. The reference compound **3** formed a 2:1 complex with the serine-containing dipeptide whose association constant for each step could be estimated at 180 M^{-1}. Obviously, the introduction of an additional hydrogen-bond acceptor in the rigid framework of the receptor module resulted in a substantial increase in free binding enthalpy for a serine-containing dipeptide, but not for dialanine nor for the reference compound **3**.

With our nitroarene tip in host **5**, we tried to establish additional π–π interactions with electron-rich residues in aromatic amino acids. To this end, we titrated Ac–Phe–Ala–OMe first with reference compound **3** and then with host **5** and obtained 1:1-association constants of ≤40 M^{-1} and 350 M^{-1}, respectively. Again, a ninefold increase from the steric repulsion by the reference compound and the π-stacking attraction by host **5** was found. Substantial upfield shifts in the aromatic regions of host and guest ^{1}H NMR spectra confirmed the additional stabilizing interaction.

For electrostatic interaction with basic amino acids we investigated complex formation between ornithine-containing Ac–Orn–Ala–OMe and anionic host **6**. To avoid suppression of backbone recognition by the superior electrostatics we chose acetate as the counter-ion for the ornithine ammonium ion. Preliminary NMR experiments revealed marked shifts in both host and guest signals, especially those close to the hydrogen-bond donors and acceptors. Even the ammonium signal shifted by ~1 ppm indicating the formation of the new ion pair. A high K_a value of ~2400 M^{-1} demonstrated the efficient recognition process. The electrostatic interaction alone was probed with **6** and n-propylammonium acetate, and gave a bind-

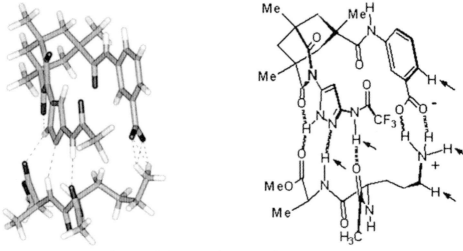

Fig. 2.4.5. Left: Monte Carlo-optimized structure of 6@Ac–Orn–Ala–OMe in chloroform (1000 steps). Right: Arrows indicate complexation-induced shifts in host and guest during the titration.

ing constant of 490 M^{-1}. The difference between these values must represent the backbone recognition operating simultaneously. The reference compound **3** gave only a relatively low number of 280 M^{-1} in its complex with Ac–Orn–Ala–OMe. Thus, introduction of a properly placed benzoate anion into the receptor module led to an additional ion-pair-reinforced hydrogen bond with an exemplary basic amino acid (Figure 2.4.5). Again, a selectivity of ~9:1 was established for the selective complexation of a representative basic dipeptide by the new anionic host **6**.

Karplus analyses of the NH-a-CH coupling constants which correlate with the characteristic torsion angle θ, offer valuable information about the conformation of the peptide [17]. We compared the ^3J values of the free peptides with those in the complex with their optimized binders. Usually signals became much sharper and the coupling constants markedly increased, approaching the calculated values (MacroModel).

In detailed comparative experiments we checked the binding behavior of our new hosts for dipeptides with the inverted sequence. Ac–Ala–Phe–OMe was completely rejected, Ac–Ala–Ser–OMe was bound with only 60 M^{-1}. Replacement of alanine with valine completely prevents complex formation, as does exchange of the acetyl protecting group for the sterically more demanding Boc group. The modification of the aminopyrazole binding site with the Kemp's triacid derivatives leads to highly selective hosts discriminating mainly as a result of steric factors.

We are currently designing special recognition tips in our new hosts for nonpolar and acidic amino acids in peptides. We are also covalently connecting two or more of the modules to achieve predictable sequence-selective recognition of larger peptides (for another approach to peptide receptors see Chapters 2.3 and 3.1). This

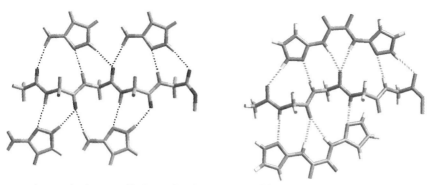

Fig. 2.4.6. Multiplication of hydrogen bonds – lining up of four aminopyrazole molecules along a tetrapeptide. Entropy gain: dimeric aminopyrazoles.

should in turn enable docking to characteristic peptide sequences on protein surfaces and might ultimately be used for protein-tagging, protection, or allosteric inhibition of enzymes.

2.4.4
Recognition of Larger Peptides with Oligomeric Aminopyrazoles

Binding experiments with tripeptides revealed that aminopyrazole derivatives can bind to the top and bottom faces simultaneously; both are active in the efficient three-point binding mode [18]. According to modeling studies, the aminopyrazoles can even be lined up along both faces of peptide strands and thereby stabilize their β-sheet conformation (Figure 2.4.6). After being capped by external ligands, however, the tendency of these peptides to form β-sheets should be greatly reduced, because the aminopyrazoles carry no hydrogen-bond donors on their outer side. Pathological protein aggregation could be prevented or even reversed with this new concept; in cooperation with the Riesner group (physical biology, Düsseldorf) the new β-sheet ligands have been tested to determine their effect on the aggregation behavior of the Prion and the Alzheimer's protein (for model studies of interactions with this peptide see Chapter 2.3).

The second generation of β-sheet ligands consists of aminopyrazole dimers, interconnected by rigid diacyl bridges [19]. They have the two DAD sequences in a defined geometrical relationship – with varying distances and torsion angles between both heterocyclic planes. Their synthesis can be conducted in a straightforward fashion – after Boc-protection of the reactive pyrazole-NH, the amino groups are reacted with diacylchlorides, followed by deprotection with trifluoroacetic acid. According to this strategy many dimeric aminopyrazoles can be synthesized rapidly (Figure 2.4.7).

Despite their large structural variation, all dimeric ligands bind to the model

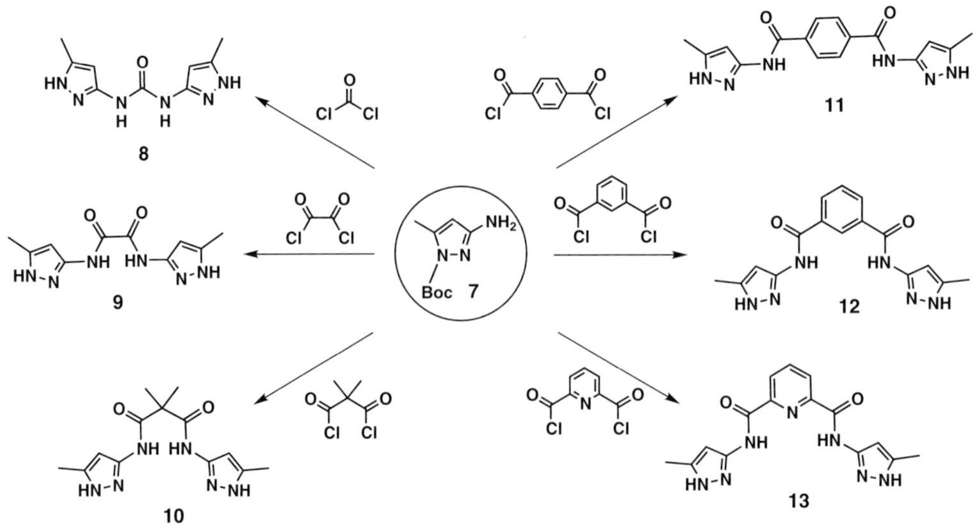

Fig. 2.4.7. Synthesis of the new dimeric aminopyrazole derivatives.

peptide Ac–Ala$_4$–OMe – in a clear 1:1-fashion (Job plots); significant downfield shifts of at least two peptidic *NH* protons in the ^{1}H NMR spectrum point to a specific peptide–ligand interaction. NMR titrations were conducted in CDCl$_3$ with 6% DMSO and produced association constants over a wide range between 0 and 1000 M^{-1} (Table 2.4.2). In dilution experiments ligands **9** and **13**, which had the strongest affinity for peptide **14**, were shown to undergo only a negligible self-association in the concentration range of 2×10^{-2} to 2×10^{-3} M.

Two head-to-head dimers have high affinity for tetraalanine – the oxalyl-bridged **9** and the pyridine-bridged **13** (Figure 2.4.7). On the other hand, the urea and the terephthaloyl spacer in **8** and **11** lead to no or negligible complex formation. Favorable conformations for multipoint binding were found for **9** and **13** by Monte Carlo simulations; although the carbonyl groups of the oxalyl bridge in **9** avoid mutual repulsion, they can easily adopt a staggered conformation, which still

Tab. 2.4.2. Association constants K_a from NMR titrations of Ac–Ala$_4$–OMe **14** with dimeric aminopyrazole ligands **8–13** in CDCl$_3$ with 6% DMSO at 20 °C.

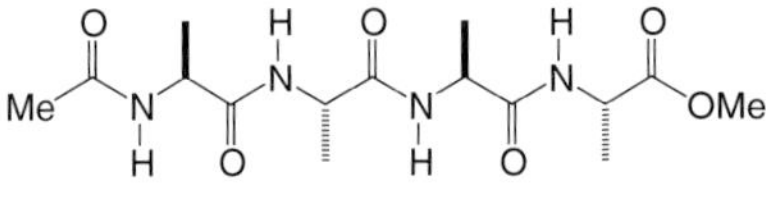

Ligand	K_a [M^{-1}]	ΔG [kcal mol^{-1}]
8	80	2.5
9[a]	570	3.7
10	220	3.1
11	0	0
12	170	3.0
13[a]	970	4.0

[a] 1:1 stoichiometry proven by Job plots.

Ac-Ala$_4$-OMe **14**

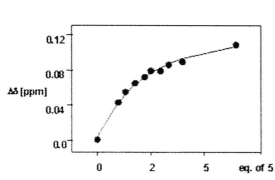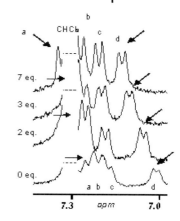

Fig. 2.4.8. NMR titration of tetraalanine derivative **14** with dimeric aminopyrazole **13** (Amp–Pyr).

brings both aminopyrazole DAD sites roughly in one line (Figure 2.4.6). In **13** two intramolecular hydrogen bonds between the pyridine N-atom and the amide NH-groups in their vicinity fix both aminopyrazole nuclei in a kinked orientation. Although the other ligands can also adopt flat geometries with both heterocyclic moieties pointing in one direction, their distance seems not well suited to interaction with model peptide **14**. Especially in the terephthaloyl-bridged dimer the second aminopyrazole cannot be fitted on to the ADA pattern of the peptide. It is, nevertheless, remarkable that dimeric aminopyrazoles are capable of forming multiple hydrogen bonds even with a tetrapeptide containing only alanines, which are known for their relatively low β-sheet propensity. In both successful examples more than four hydrogen bonds are formed simultaneously, leading to superior interactions compared with all other aminopyrazole dimers.

The dimeric aminopyrazoles have also been tested on the Prion and Alzheimer's proteins (for details, see Section 2.4.5). Preliminary experiments showed at an early stage of the investigation that even covalent aminopyrazole dimerization produced peptide ligands which could significantly retard protein aggregation under physiological conditions. These ligands have some promising features in common – they are easily synthesized, contain non-toxic components, have low molecular weights of approximately 300, and are neutral, stable molecules in their biologically active form.

To greatly expand the number of possible hydrogen-bond contacts between aminopyrazoles and peptides we created a third generation of hosts relying on oligomerization of a hitherto unknown unnatural amino acid, 3-aminopyrazole-5-carboxylic acid. This building block can be coupled with itself or other natural and unnatural amino acids by standard peptide synthesis procedures and leads to oligomers with a rigid planar geometry. They have a hydrogen-bond donor and acceptor pattern which can be fine-tuned to be perfectly complementary to that of larger peptides in the β-sheet conformation. Because of the five-membered ring,

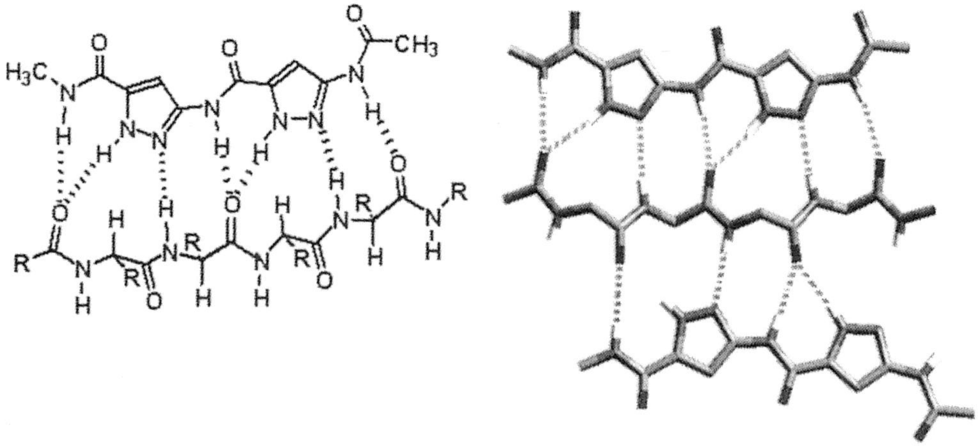

Fig. 2.4.9. Dimeric 3-aminopyrazole-5-carboxylic acid in its 1:1- and 2:1-complex with a tetrapeptide.

the oligomers have the form of a wide bow, preventing extensive self-association. Modeling experiments suggest that even a tetrapeptide can be bound by two dimeric aminopyrazolecarboxylic acids with a total of 13 hydrogen bonds (Figure 2.4.9). With this increasing number of non-covalent contacts, strong binding should become possible in polar solution, especially with higher aminopyrazole oligomers.

The parent heterocycle can be prepared from the nitro derivative **15** by a variety of reduction techniques. This nitro group, however, can also be conveniently used as a transient protecting group and subjected to standard peptide coupling conditions. It is highly critical to chose the correct protecting group for the pyrazole ring-NH; the *p*-methoxybenzyl moiety is stable under all the coupling procedures and can be cleaved in the final step with trifluoroacetic acid without causing racemization of α-amino acids [20].

Hybrid compounds with natural amino acids have a dual advantage over the pure aminopyrazole oligomers. Introduction of amino acids greatly increases the solubility of the whole host compound, and their sequence could serve as affinity tag for recognition of certain characteristic areas on protein surfaces, thereby rendering the β-sheet ligand specific for a certain protein of interest. A prominent feature for the Aβ-recognition is, e.g., an internal pentapeptide sequence KLVFF, which is also critical for the aggregation process [21]. For the synthesis of these hybrid systems we developed, in cooperation with the König group (Organic Chemistry, Regensburg), a solid phase procedure with Fmoc- and PMB-protected 3-aminopyrazole-5-carboxylic acid as key building block (for more on solid-phase chemistry see Chapters 3.6 and 6.1 and Box 25). Thus, a large variety of aminopyrazole-based oligomers is now accessible by convenient standard proce-

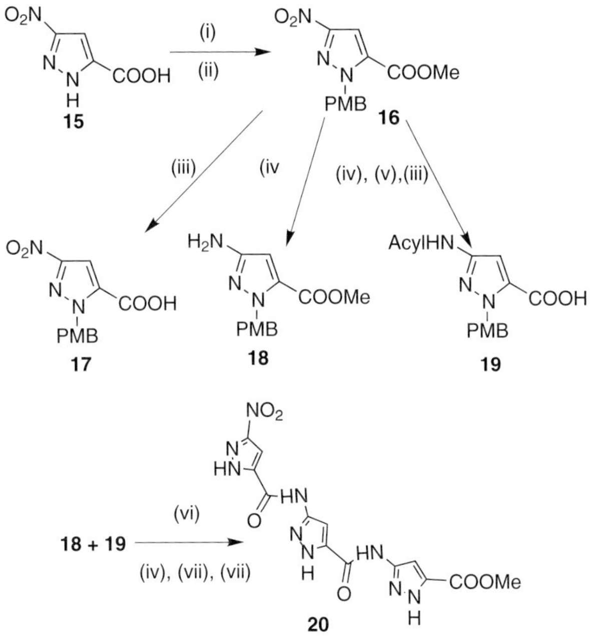

Fig. 2.4.10. Synthesis of oligomeric aminopyrazoles starting from 3-nitropyrazole-5-carboxylic acid **15**: (i) HCl, MeOH; (ii) PMB-Br, K_2CO_3; (iii) LiOH; (iv) Pd/C, H_2; (v) 2-chloro-1-pyridiniumiodide, DIEA; (vi) PyClop, DIEA; (vii) **18**, PyClop, DIEA; (viii) TFA, Δ.

dures, and will be tested in a broad screening program against a variety of proteins prone to pathological aggregation.

2.4.5
Recognition of Proteins with Aminopyrazoles

Protein-folding diseases are characterized by the formation of insoluble protein plaques in the brains' nerve cells. The mechanism of these misfolding events is often unclear. In BSE/CJD an infectious protein particle could be identified, leading to the famous prion hypothesis [22]. It is called prion protein scrapie (PrP^{Sc}). After refolding of the native protein newly formed β-sheet domains aggregate with those in other PrP^{Sc} molecules and lead to precipitation of the protein from physiological solution [23]. A similar process might be operating in the Alzheimer's protein, which also forms amyloid plaques in the brain consisting of very regular, stacked twisted β-sheets. Our concept intervenes at the beginning of the aggregation process – we aim at preventing dimerization of the pathological protein domains, by capping the solvent-exposed β-sheet region with our oligomeric aminopyrazole

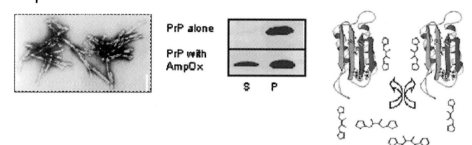

Fig. 2.4.11. Ultracentrifugation experiments with PrP alone (10 μM) and after incubation with AmpOx **9** (1 mM) under harsh aggregation conditions (37 °C, 1 day); S = solution, P = pellet. Left: typical prion rods (white bar ~ 100 nm) [25]. Right: schematic representation of the concept: capping the β-sheets [26].

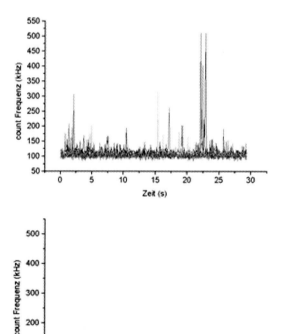

Fig. 2.4.12. FCS measurements of the aggregates formed from soluble Aβ (10 μM). Ten different 30 s runs are superimposed. Left: without additive; right: with dimeric aminopyrazole **9** (1 mM).

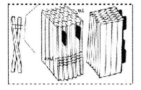 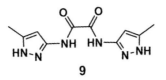 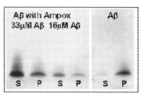

9

Fig. 2.4.13. Left: typical amyloid fibrils are twisted β-sheets [27]. Right: differential ultracentrifugation experiment with Aβ (1-42, 33 and 16 μM) in the absence (right) or presence (left) of dimeric aminopyrazole **9** (Ampox, 1 mM). S = solution, P = pellet.

ligands. Because the back face of these ligands contains no hydrogen-bond donors, the aggregation process should be completely stopped or even be reversed.

The biophysical experiments of our cooperation partners focus mainly on two different techniques – a rough estimate of aggregation prevention is found by differential ultracentrifugation (UC), but more detailed information about the kinetics and aggregate size comes from fluorescence correlation spectroscopy (FCS) [24].

The dimeric β-sheet-binders were first tested on the prion-protein by the Riesner group. An initial precipitation experiment showed that **9** was able to keep about one third of PrP^C in solution (Figure 2.4.11 – the detergent sodium dodecylsulfate was diluted out, a procedure which always leads to complete aggregation of the protein.). This was the starting point of our tests; monomeric aminopyrazole derivatives had no influence on the aggregation behavior of PrP. That one of the dimeric aminopyrazoles of the second generation was a promising candidate for a new therapeutic approach to protein folding diseases encouraged us to develop higher generations and generally test all new candidates in bioassays.

The dimeric aminopyrazole ligands were also tested to determine their effect on the aggregation behavior of the Alzheimer's protein. To this end we used the model compound Aβ (1-42), which was fluorescence-labeled with Oregon green. The FCS spectra demonstrate beautifully how Aβ spontaneously aggregates in sodium phosphate buffer containing 5% DMSO after incubation for 12–72 h. After preincubation of the same samples with selected aminopyrazole ligands, e.g. **9** and **20**, however, the aggregation could be completely suppressed (Figure 2.4.12).

The results of the respective differential ultracentrifugations correlate very well with the FCS measurements. Without addition of our ligands Aβ (1-42) almost completely precipitates and is found exclusively in the pellet (P). Addition of **9** or **20** before the start of the aggregation process keeps most of the protein in solution (S, Figure 2.4.13), however.

References

1 Review: R. Hirschmann, *Angew. Chem. Int. Ed. Eng.* **1991**, *30*, 1278; β-strand peptidomimetics: A. B. Smith III, R. Hirschmann, A. Pasternak, M. C. Guzman, A. Yokoyama, P. A. Sprengeler, P. L. Darke, E. A.

Emini, W. A. Schleif, *J. Am. Chem. Soc.* **1995**, *117*, 11113–11123.

2 R. Zutshi, J. Franciskovich, M. Shultz, B. Schweitzer, P. Bishop, M. Wilson, J. Chmielewski, *J. Am. Chem. Soc.* **1997**, *119*, 4841–4845.

3 Reviews: (a) T. Wisniewski, P. Aucouturier, C. Soto, B. Frangione, *Amyloid: Int. J. Exp. Clin. Invest.* **1998**, *5*, 212; (b) R. W. Carrell, D. A. Lomas, *Lancet* **1997**, *350*, 134–138.

4 Reviews: J. P. Schneider, J. W. Kelly, *Chem. Rev.* **1995**, *95*, 2169; J. S. Nowick, *Acc. Chem. Res.* **1999**, *32*, 287–296.

5 K. Y. Tsang, H. Diasz, N. Graciani, J. W. Kelly, *J. Am. Chem. Soc.* **1994**, *116*, 3988; P. Chitnumsub, W. R. Fiori, H. A. Lashuel, H. Diaz, J. W. Kelly, *Bioorg. Med. Chem.* **1999**, *7*, 39–59.

6 C. A. Kim, J. M. Berg, *Nature* **1993**, *362*, 267; J. S. Nowick, S. Insaf, *J. Am. Chem. Soc.* **1997**, *119*, 10903.

7 K. H. Mayo, E. Ilyina, H. Park, *Protein Science* **1996**, *5*, 1301–1315; H. L. Schenck, S. H. Gellman, *J. Am. Chem. Soc.* **1998**, *120*, 4869–4870; G. J. Sharman, M. S. Searle, *J. Am. Chem. Soc.* **1998**, *120*, 5291–5300.

8 T. Schrader, C. Kirsten, *J. Chem. Soc., Chem. Commun.* **1996**, 2089.

9 D. S. Kemp, B. R. Bowen, C. C. Muendel, *J. Org. Chem.* **1990**, *55*, 4650.

10 W. S. Weiner, A. D. Hamilton, *Bioorg. Med. Chem. Lett.* **1998**, *8*, 681–686.

11 S. T. Phillips, M. Rezac, U. Abel, M. Kossenjans, P. A. Bartlett, *J. Am. Chem. Soc.* **2002**, *124*, 58–66.

12 (a) W. C. Still, *Acc. Chem. Res.* **1996**, *29*, 155–163; (b) H. Wennemers, M. Conza, M. Nold, P. Krattiger, *Chem. Eur. J.* **2001**, *7*, 3342–3347; (c) T. Braxmeier, M. Demarcus, T. Fessmann, S. McAteer, J. D. Kilburn, *Chem. Eur. J.* **2001**, *7*, 1889–1898.

13 T. Schrader, M. Wehner, *Angew. Chem. Int Ed.* **2002**, *114*, 1827–1831.

14 MacroModel 7.0, Schrödinger, Inc., Monte Carlo simulation, chloroform, 3000 steps.

15 (a) P. Job, *Compt. Rend.* **1925**, *180*, 928; (b) M. T. Blanda, J. H. Horner, M. Newcomb, *J. Org. Chem.* **1989**, *54*, 4626.

16 (a) H. J. Schneider, R. Kramer, S. Simova, U. Schneider, *J. Am. Chem. Soc.* **1988**, *110*, 6442; (b) C. S. Wilcox, *Frontiers in Supramolecular Chemistry*, H. J. Schneider, ed., Verlag Chemie, Weinheim, **1991**, p. 123.

17 (a) M. Delepierre, C. M. Dobson, F. M. Poulsen, *Biochemistry* **1982**, *21*, 4756; (b) V. F. Bystrov, *Prog. Nucl. Magn. Reson. Spectrosc.* **1976**, *10*, 41.

18 C. Kirsten, T. Schrader, *J. Am. Chem. Soc.* **1997**, *118*, 10295–10299.

19 T. Schrader, M. Wehner, unpublished results.

20 C. Subramanyam, *Synth. Commun.* **1995**, *25*, 5761–5774.

21 C. Soto, E. M. Sigurdsson, L. Morelli, R. A. Kumar, E. M. Castano, B. Frangione, *Nat. Med.* **1998**, *4*, 822.

22 S. B. Prusiner, *Science* **1982**, *216*, 136–144; S. B. Prusiner, *Trends Biochem. Sci.* **1996**, *21*, 482–487; R. S. Hegde, P. Tremblay, D. Groth, S. J. DeArmond, S. B. Prusiner, V. R. Lingappa, *Nature* **1999**, *402*, 822–826.

23 K. Post, M. Pitschke, O. Schäfer, H. Wille, T. R. Appel, D. Kirsch, I. Mehlhorn, H. Serban, S. B. Prusiner, D. Riesner, *Biol. Chem.* **1998**, *379*, 1307; K. Jansen, O. Schäfer, E. Birkmann, K. Post, H. Serban, S. B. Prusiner, D. Riesner, *Biol. Chem.* **2001**, *382*, 683–691.

24 T. Schrader, D. Riesner, L. Nagel-Steger, K. Aschermann, C. Kirsten, P. Rzepecki, O. Molt, R. Zadmard, M. Wehner, *Patent Application DE* 102 21 052.7 of 5/10/**2002**.

25 Prion rod electron microscopy: D. Riesner, *Chemie in unserer Zeit*, **1996**, *30*, 66–74.

26 Scrapie protein model: R. Mestel, *Science* **1996**, *273*, 184–189.

27 Amyloid fibril schematic: G. G. Glenner, *N. Eng. J. Med.* **1980**, *302*, 1283–1292.

B.8
Secondary Structures of Proteins

Thomas Schrader

α-Helix and β-Sheet

Functional proteins are characterized not only by their amino acid sequence (called the primary structure) but also by their exact geometrical arrangement in space (called the secondary, tertiary, and quaternary structure) [1]. Only correctly folded proteins can perform their biological function correctly. This complicated process is currently under active investigation; a whole class of folding helper proteins, the chaperones [2], are used by Nature to prevent misfolding events with their often dramatic consequences (TSE, Alzheimer's disease, etc., vide infra).

Most proteins contain one or both of the fundamental regular peptide backbone arrays, i.e. the α-helix or the β-sheet. These not only confer mechanical stability on the protein but also protect active centers, create the correct microenvironment for the catalyzed reaction, or even participate in the catalytic process. In α-helices all NH groups point in one direction and hydrogen-bond to the respective C=O groups of the $i+4$ residues, all of which point into the opposite direction. By means of linear, intramolecular hydrogen bonds a right-handed helix is formed with a helical pitch of 0.54 nm (Figure B.8.1). Non-polar, aliphatic side-chains are compatible with the torsion angles necessary for an α-helix (only several areas of torsion angles θ and ϕ are allowed; these can be visualized in Ramachandran plots) [3]. Thus, helices are often found in transmembrane regions of proteins

Fig. B.8.1. Right-handed α-helix with intramolecular hydrogen bonds (dotted lines) [4].

responsible for signal transduction, e.g. G-protein-coupled receptors, ion channels, etc. In addition, α-helices are often found in fibrous (α-ceratin, collagen) and globular proteins (myoglobin), in the latter with an average length of 11 residues [5].

In a β-sheet, the peptide chain adopts an extended conformation with alternating pairs of NH and C=O groups pointing to the top and bottom face. Thus, β-sheets are sticky to the exterior. In proteins, a 5–15 amino acid strand is usually followed by a β-turn-motif (often containing proline residues), which reverses the direction of the main chain. In this antiparallel arrangement, the NH and C=O pairs can form multiple linear hydrogen bonds, leading to a regular sheet structure with remarkable stiffness (Figure B.8.2) [6]. Here, the side chains are placed horizontally to the left or the right in alternating order.

Certain combinations of secondary superstructures are often found in proteins and control their structure and function. The most frequent is the βαβ-unit, where an α-helix bridges two β-strands. This is the prevailing feature in most coenzyme-binding domains of dehydrogenases [7]. Other important superstructures include α,α-dimers, β-meanders and β-barrels.

Protein Structure Determination

Even with modern supercomputers it is impossible to predict the absolute thermodynamic minimum of an unknown folded protein chain. If the

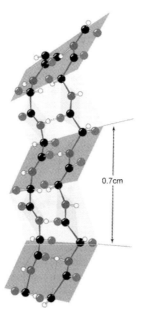

Fig. B.8.2. Double-stranded antiparallel β-sheet with intermolecular hydrogen bonds (dotted lines) [4].

structures of related proteins with similar sequences are known, homology modeling will produce very good estimates. In recent years conventional structure determination by X-ray crystallography has been complemented by the new NMR NOESY technique used for elucidation of solution structures (K. Wüthrich, Nobel Prize 2002). Although the results from both techniques are often perfectly superimposable, sometimes the structures differ substantially; in addition, flexible tails and protein dynamics can only be seen by NMR. This technique has hitherto been limited to a protein size of <200 kD, but progress is being made swiftly [8].

References

1 G. D. FASMAN, ed., *Prediction of Protein Structure and the Principles of Protein Conformation*. Plenum, New York, **1989**.
2 S. WALTER, J. BUCHNER, *Angew. Chem.* **2002**, *114*, 1142–1158.
3 G. N. RAMACHANDRAN, V. SASSIEKHARAN, *Adv. Protein Chem.* **1968**, *23*, 283–437.
4 D. VOET, J. G. VOET, Biochemie, VCH, Weinheim, **1994**.
5 E. Y. JONES, A. MILLER, *J. Mol. Biol.* **1991**, *218*, 209–219.
6 F. R. SALEMME, *Prog. Biophys. Mol. Biol.* **1983**, *42*, 95–133.
7 M. G. ROSSMANN, A. LILJAS, C.-I. BRÄNDEN, L. J. BANASZAK, in P. D. BOYER, ed., *The Enzymes*, Vol. 11, 3rd edn., Academic Press, **1975**.
8 K. WÜTHRICH, *Nature Struct. Biol.* **2000**, *7*, 188–189.

2.5
Evaluation of the DNA-binding Properties of Cationic Dyes by Absorption and Emission Spectroscopy

Heiko Ihmels, Katja Faulhaber, and Giampietro Viola

2.5.1
Introduction

The association of external molecules with DNA has attracted considerable interest in bioorganic and medicinal chemistry [1]. Such complex formation leads to significant modification of the structure of the DNA and therefore may have an important influence on the physiological function of the DNA, namely gene expression [2]. Most notably, the strong association of a molecule with DNA may lead to cell death and is not a desirable process in healthy tissue. Nevertheless, the suppression of the DNA replication and gene transcription by this association may also be used to destroy unwanted cells such as tumor cells or other infected tissue in living systems. Therefore, one of the most challenging goals in this area of research is the design of DNA-binding molecules which selectively bind to DNA in unwanted cells and lead to cell death without damaging the healthy cells. Along these lines, many investigations have been performed to gain more insight into different aspects of the association process of large and small molecules with DNA; and several classes of DNA-binding molecule have been established and investigated in detail [3]. For example, proteins represent a group of macromolecules whose association with DNA constitutes important steps in the gene-transcription process [4]. Consequently, one promising lead structure for functional DNA-binding drugs is based on polyamide derivatives [5]. Oligosaccharides have also been shown to be important DNA-binding moieties in DNA-targeting drugs such as Calicheamycin [6]. Organometallic complexes have been shown to associate with DNA [7], and also used to investigate charge-transfer through DNA [8] (for more information on this topic see Chapter 4.6). Cationic organic dyes are another important class of DNA-binding molecule. Representative examples are proflavine **1a** [9], acridine orange **1b** [10], methylene blue **2** [11], ethidium bromide **3** [12], thiazole orange **4** [13], anthraquinone derivatives [14], pyrido-annelated indolinium derivatives [15], acridizinium derivatives **5** [16], and protoberberines [17]. Such dyes have the advantage that their interaction with DNA can be easily evaluated, qualitatively and quantitatively, by absorption and emission spectroscopy, because of the significant change of their pronounced dye properties (absorption and

Highlights in Bioorganic Chemistry: Methods and Applications. Edited by Carsten Schmuck, Helma Wennemers.
Copyright © 2004 WILEY-VCH Verlag GmbH & Co. KGaA, Weinheim
ISBN: 3-527-30656-0

emission) on DNA binding. These simple and straightforward spectroscopic methods are especially advantageous because organic dyes absorb and emit at wavelengths which do not interfere with the absorption of the DNA bases ($\lambda_{max} \approx$ 260 nm).

1a: R = H; X = CH, Y = NH
1b: R = Me, X = CH, Y = NH
2 : R = Me, X = N, Y = NH

3

4

5a: R = NH_2
5b: R = H

2.5.2
Binding Modes

In general, two binding modes are possible between guest molecules and the host DNA: (a) minor or major groove binding, and (b) intercalation (Figure 2.5.1). A third binding mode, external binding, is also known; this results exclusively from attractive electrostatic interactions between a positively charged molecule and the negatively charged phosphate backbone of the DNA. In contrast, groove binding

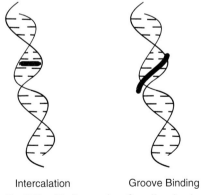

Intercalation   Groove Binding

Fig. 2.5.1. Binding modes of small molecules with DNA

and intercalation may be viewed as the formation of a supramolecular assembly based on associative interactions such as π stacking, hydrogen bonding, attractive van der Waals, or hydrophobic interactions. Each binding interaction is usually initiated by hydrophobic transfer of the DNA binder from the polar aqueous solution into the less polar environment of the DNA. If the DNA binder carries a positive charge, association with DNA also leads to a release of DNA counter cations from the DNA grooves.

2.5.2.1
Groove Binding

The DNA helix has two grooves of different size, the minor and the major grooves, which may serve as binding sites for guest molecules. Whereas relatively large molecules such as proteins bind preferentially to the major groove of DNA [4], the minor groove is the preferred binding site for small ligands [1c, 3]. The binding pocket of a DNA groove is defined by two different regions, the "bottom", formed by the edges of the nucleic bases that face into the groove, and the "walls", which are formed from the deoxyribose–phosphate backbone of the DNA. Groove binders usually consist of at least two aromatic or heteroaromatic rings whose connection allows conformational flexibility such that a crescent-shaped conformation may be achieved and the molecule fits perfect into the groove. In addition, functional groups are required to form hydrogen bonds with the nucleic bases at the bottom of the groove. Typical minor-groove binders are Hoechst 33258 (**6**) [18] and netropsin **7** [19].

Most groove binders have binding selectivity toward AT-rich areas, because grooves, which consist of GC base pairs, are sterically hindered by the guanine amino functionality at C-2 and its hydrogen bond with the C-2 carbonyl functionality of cytosine. It has also been observed that in AT-rich grooves the electrostatic potential is higher than in GC-rich regions. Thus positively polarized or charged ligands also have greater affinity for the AT-rich groove, because of favorable electrostatic attraction. Nevertheless, some groove binders are known that bind preferentially in GC-rich grooves, because they are substituted with functionalities that form strong attractive interactions with the guanine amino group [19].

2.5.2.2
Intercalation

In a DNA helix, the nucleic bases are located in an almost coplanar arrangement. In his pioneering work Lerman showed that this arrangement enables planar polycyclic aromatic molecules to intercalate between two base pairs [9, 20]. Important driving forces for this binding mode are dipole–dipole interactions and π stacking of the guest molecule with the aromatic nucleic bases. Other than groove binding, intercalation has a significant influence on the DNA structure, because the DNA needs to unwind so that the intercalator fits between the two base pairs. This unwinding leads to a lengthening of the helix of more than 340 pm, along with a significant change of the deoxyribose conformation [1c].

It has been observed that the binding of one intercalator between two base pairs hinders access of another intercalator to the binding site next to the neighboring intercalation pocket. This "neighbor exclusion principle" is well accepted and may be used as a general rule; it is, however, not fully understood. It has been proposed that structural changes of the DNA on intercalation lead to limited access to the neighboring binding pocket for steric reasons. Also, the intercalation process may reduce the negative electrostatic potential at the intercalation site, so that attractive electrostatic interactions no longer occur close to this site.

2.5.3
Evaluation of the Binding

In a host–guest complex, the physical properties of the DNA and the intercalator or groove binder are usually significantly different from those of the uncomplexed host and guest molecules. Complex formation may therefore be detected by monitoring a change of a particular physical property (see also Chapter 4.2). The extent of this change usually depends on the binding mode and, thus, enables intercalators and groove binders to be distinguished. If the focus is on the properties of the DNA molecule, hydrodynamic methods such as the viscosity or the sedimentation coefficient are helpful for monitoring the association process [21]. The thermodynamic stability of the DNA helix is also influenced by complex formation, so association with a guest molecule might also be determined by measuring a change of the melting temperature [21] (Box 17). Hydrodynamic methods are especially reliable and helpful for distinguishing the binding mode, because on intercalation the observed change of the DNA structure and thus the change of the viscosity or the sedimentation coefficient is relatively large whereas groove binding leads to marginal structural changes only.

Absorption and emission spectroscopy are also useful tools for monitoring DNA-binding processes. The interaction of dyes, especially, with DNA may be conveniently observed by these methods, because their absorption and emission properties change significantly on complex formation. Along with straightforward determination of absorption or emission intensity and wavelength, several variations and additional experiments are possible. For example, absorption of circularly- or linearly polarized light can be used in CD and LD spectroscopy to gain further knowledge of the orientation of the dye molecule relative to the DNA. Steady-state fluorescence polarization measurements and fluorescence energy transfer from the DNA bases to the bound dye have also been used as reliable criteria to elucidate the binding mode [21b], although a recent paper shows that the latter method should be used with caution [22].

It should be noted that determination of the binding mode needs to be performed with much care and that the use of only one method may lead to misinterpretation. In critical reviews it has been pointed out that only a combination of selected methods provides sufficient information to draw conclusions about the mode of binding [21].

In this paper we wish to report that spectrometric methods are useful tools for study of the DNA-binding properties of organic dyes. As a representative example, 9-aminoacridizinium bromide **5a** was chosen and, in some instances, compared with the parent system **5b** [16].

2.5.3.1
UV–Visible Spectroscopy

In a complex with DNA the guest molecule is positioned in an environment which is different from that of the uncomplexed molecule in solution. The guest molecules, especially solvatochromic compounds such as organic dyes [23], usually have different absorption properties in the complexed and uncomplexed forms. Thus, on addition of DNA to a solution of an intercalator or groove binder, a shift of the absorption maximum to longer wavelengths (bathochromic shift or red shift) and a decrease of the absorbance (hypochromicity) occurs. In practice the association process is monitored by a spectrophotometric titration, during which aliquots of DNA solution are added to a solution of the guest molecule. The absorption spectra at each dye-to-DNA ratio are determined and superimposed [24]. A representative spectrophotometric titration is shown for 9-aminoacridizinium bromide (**5a**) with calf thymus DNA (Figure 2.5.2). The long-wavelength absorption maximum of **5a** is shifted to longer wavelength by 12 nm on addition of DNA, and a significant decrease of the absorbance takes place; this is indicative of an associative interaction between the dye and DNA.

Further information might be extracted from spectrophotometric titrations if isosbestic points are observed. An isosbestic point appears when each absorption spectrum of the titration has the same absorbance at a particular wavelength, i.e. a

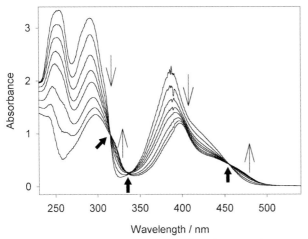

Fig. 2.5.2. Spectrophotometric titration of acridizinium salt **5a** with ct DNA in phosphate buffer (10 mM, pH 7.0); thin arrows indicate the increasing or decreasing absorption bands during the course of the titration; thick arrows indicate isosbestic points.

point of intersection in *all* superimposed absorption spectra (see Figure 2.5.2, thick arrow). Such isosbestic points reveal that each absorption spectrum arises from almost exclusively two different absorbing species (e.g. complexed and uncomplexed dye). Because it is likely that each binding mode results different absorption properties, an isosbestic point indicates that mainly one particular binding mode between the DNA and the guest molecule occurs. It should, nevertheless, be noted that a second, but minor, binding mode only contributes marginally to the overall absorption spectrum and might not have a significant influence on the isosbestic points.

Data from spectrophotometric titrations may also be used to determine the association constant (K) between the dye and DNA [25]. The data from spectrophotometric titrations, i.e. absorbance data (A_{obs}) at a fixed wavelength, are used to determine the concentration of bound dye (C_b), the concentration of uncomplexed dye (c), and the number of bound dye molecules per base pair (r) according to Eqs (1)–(3).

$$C_b = (A_f - A_{obs}/A_f - A_b)C_0 \tag{1}$$

$$c = C_0 - C_b = [1 - (C_b/C_0)] \times C_0 \tag{2}$$

$$r = C_b/C_{DNA} \tag{3}$$

where A_f is the absorbance of the uncomplexed dye, A_b is the absorbance of the bound dye, A_{obs} is the absorbance of a mixture of the free and bound compounds,

Tab. 2.5.1. Binding constants and binding-site size of acridizinium bromides **5a** and **5b** as determined from spectrophotometric titrations with st DNA, (poly[dA-dT]-poly[dA-dT]), and (poly[dG-dC]-poly[dG-dC]).

		5a	5b
st DNA:	K^a [M^{-1}] × 10^4	6.15 ± 0.21	1.24 ± 0.10
	n^a [bp]	2.01 ± 0.41	5.02 ± 0.21
(poly[dG-dC]-poly[dG-dC]):	K^a [M^{-1}] × 10^4	9.18 ± 0.43	5.74 ± 0.46
	n^a [bp]	2.00 ± 0.06	2.00 ± 0.05
(poly[dA-dT]-poly[dA-dT]):	K^a [M^{-1}] × 10^4	3.40 ± 0.20	0.80 ± 0.05
	n^a [bp]	4.21 ± 0.05	4.70 ± 0.05

[a] K is the binding constant (in bases) and n is the binding-site size (in base pairs), determined in ETN buffer solution (1 mM EDTA, 10 mM Tris buffer, 10 mM NaCl, pH 7.0)

and C_0 is the total concentration of the dye. The absorbance of the bound dye, A_b, is obtained from the absorption spectrum at full saturation. These data are used to represent the binding isotherms in a Scatchard plot [26], that is, a plot of r/c against r. Binding analysis of the experimental data is performed according the model developed by McGhee and von Hippel to determine the binding constant (K) and the binding-site size (n). Thus, the data may be fitted using Eq. (4) [27].

$$r/c = K(1 - nr)\{(1 - nr)/[1 - (n - 1)r]\}^{n-1} \quad (4)$$

For example, the binding constants and the binding-site size of complexes between acridizinium derivatives **5a** and **5b** and salmon testes DNA (st DNA) were determined by this method (Table 2.5.1). The binding constants reveal that the complexes between DNA and acridizinium derivatives **5a** and **5b** are reasonably stable. Moreover, the amino-substituted acridizinium salt **5a** has affinity for DNA which is approximately five times larger than that of the parent compound **5b**. The binding-site size, n, is a measure of the binding sites (i.e. base pairs) occupied by one guest molecule and gives additional information about the binding mode. Thus, a significant difference between the binding-site size of intercalators and groove binders is observed. According to the neighbor-exclusion principle an intercalator occupies two binding sites, so that at full saturation, an alternating sequence of occupied and free binding sites is observed. Thus perfect intercalators have a binding-site size of $n = 2$. A groove binder, however, covers several base pairs on binding to DNA, so n is significantly larger than 2. In **5a** and **5b**, binding-site sizes $n = 2$ and $n = 3$ were observed. Both values give evidence of intercalation of these compounds according to the neighbor-exclusion principle. Nevertheless, because the binding-site size for **5b** is slightly larger than the ideal value, it may be assumed that a binding mode other than intercalation contributes marginally to the overall binding.

In additional experiments, the base selectivity of the association may be investigated by spectrophotometric titrations with synthetic polynucleotides such as (poly[dA-dT]-poly[dA-dT]) and (poly[dG-dC]-poly[dG-dC]). The data (Table 2.5.1) reveal a significantly higher affinity of both acridizinium derivatives for binding to (poly[dG-dC]-poly[dG-dC]) compared with (poly[dA-dT]-poly[dA-dT]). The binding-site size for compounds **5a** and **5b** with (poly[dG-dC]-poly[dG-dC]) is, moreover, compatible with the neighbor-exclusion model, whereas no such fit was observed with (poly[dA-dT]-poly[dA-dT]). From these experiments it might be concluded that the acridizinium derivatives **5** intercalate preferentially in GC-rich regions of DNA.

2.5.3.2
Emission Spectroscopy

If a molecule has emission properties such as fluorescence or phosphorescence, addition of DNA may, as in the spectrophotometric titrations, result in a bathochromic shift of the emission maximum. More significantly, the emission intensity may change on complex formation. In the latter circumstances either an increase or a decrease of the emission intensity might be observed [28].

Dyes whose fluorescence intensity increases on binding to DNA (e.g. **3** and **4**) have especially high potential as DNA marker or detector molecules. In the absence of DNA the relatively low fluorescence quantum yield of these dyes results from a radiationless deactivation of the excited state by conformational changes or acid–base reactions with the solvent. On association with DNA, however, significant suppression of the conformational flexibility and a shielding of the dye from solvent molecules within the complex occurs, leading to an increase of the emission intensity.

In contrast, many dyes are known whose fluorescence intensity decreases on addition of DNA. The origin of this emission quenching is usually an electron-, energy- or hydrogen-transfer reaction between the excited dye and the DNA [28]. Organic dyes have absorption maxima at significantly longer wavelengths than the DNA bases, so energy transfer between the excited dye and DNA is energetically disfavored when the dye is excited at these wavelengths. Most often an electron transfer (ET) reaction between the excited dye and the DNA bases occurs, with guanine being the base with the highest propensity to be oxidized. Although, the actual oxidation potential of guanine within DNA is still under debate, the reduction potential of the guanosine radical (1.29 V relative to the NHE in H_2O, pH 7), which corresponds to the oxidation potential of guanine, can be used as reference [29]. Thus, according to the Rehm–Weller equation [30], the reduction potential of the dye in its excited state (E^*_{Red}) needs to be larger than ca. 1.3 V for an exergonic electron transfer ($\Delta G_{ET} < 0$) with guanine. The reduction potential of the excited fluorophore E^*_{Red} is estimated from the 0–0 transition, $E_{0,0}$ and the reduction potential of the ground state E^0_{Red} (Eq. 5).

$$E^*_{Red} = E^0_{Red} + E_{0,0} \tag{5}$$

The inverse ET mechanism has recently also been proposed for fluorescence quenching of ethidium bromide **3** by DNA, during which **3** is oxidized by particular DNA sequences [31]. The change of emission intensity on complex formation is followed by spectrofluorimetric titrations performed analogously to spectrophotometric titrations. It must, nevertheless, be considered that the absorption at a particular wavelength also changes on DNA addition (Figure 2.5.2), which also leads to a change of the emission intensity on excitation at this wavelength. To avoid this secondary effect on the emission spectrum, which does not reflect the direct influence of the DNA on the dye emission properties, the fluorophore should be excited at the isosbestic point, which is obtained from spectrophotometric titrations. A representative fluorimetric titration is shown for the dye **5a**. The aminoacridizinium **5a** emits in aqueous solution with a fluorescence quantum yield of $\phi_{fl} = 0.12$ [16a]. On addition of ct DNA the fluorescence is significantly quenched and the emission maximum is slightly red shifted (Figure 2.5.3A). It can be assumed that the fluorescence quenching is because of a photoinduced ET reaction (Scheme 2.5.1), because the excited acridizinium **5a*** has a reduction potential of ca. 1.9 V [32]. Consequently, fluorimetric titrations with synthetic polynucleotides (poly[dA-dT]-poly[dA-dT]) and (poly[dG-dC]-poly[dG-dC]) can be used to obtain evidence for this proposal. Depending on the oxidation potentials of the nucleic bases, the polynucleotides should have significantly different quenching abilities. Thus, addition of (poly[dA-dT]-poly[dA-dT]) to **5a** resulted in only a small decrease of the relative fluorescence quantum yield (Figure 2.5.3B). In contrast, addition of (poly[dG-dC]-poly[dG-dC]) to **5a** is accompanied by strong fluorescence quenching (Figure 2.5.3C). The different quenching of both polynucleotides might be because of the different binding constants with **5a** (Table 2.5.1). The difference between the two Stern–Volmer constants {(poly[dA-dT]-poly[dA-dT]) 5670 M^{-1}; (poly[dG-dC]-poly[dG-dC]) 51480 M^{-1}} obtained from the linear part of the Stern–Volmer plot is, nevertheless, significantly larger than that between the binding constants. Thus, the different quenching abilities of polynucleotides (poly[dA-dT]-poly[dA-dT]) and (poly[dG-dC]-poly[dG-dC]) are most probably a result of the different oxidation potentials of the nucleic bases (for more on nucleobase oxidation see Chapter 4.5) and a photoinduced ET reaction, as the main quenching mechanism, is evident.

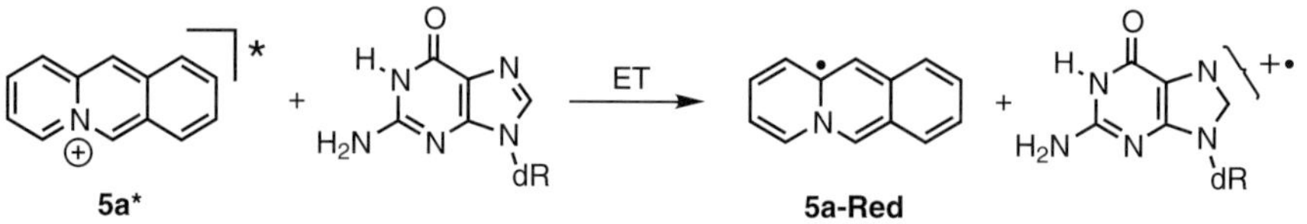

Scheme 2.5.1.

2.5.3.3
CD Spectroscopy

Although circular dichroism (CD) is not observed for achiral molecules, when they form complexes with DNA they are placed within a chiral environment and give

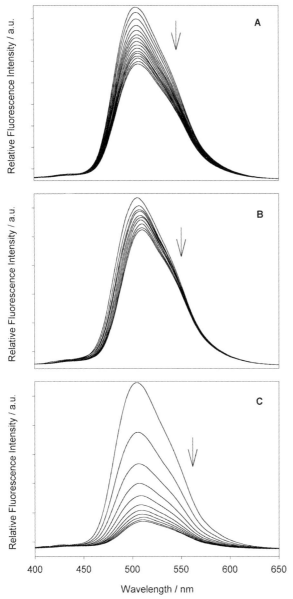

Fig. 2.5.3. Spectrofluorimetric titration of acridizinium salt **5a** with ct DNA (A), (poly[dA-dT]-poly[dA-dT]) (B), and (poly[dG-dC]-poly[dG-dC]) (C) in phosphate buffer (10 mM, pH 7.0); arrows indicate the decreasing emission intensity during the course of the titration.

an induced CD (ICD) signal [33], which results from non-degenerative coupling between the transition of the bound ligand and that of the nucleic-base transitions. The appearance of an ICD signal confirms the dye–DNA interaction and might provide further information about the position of a dye in its complex with DNA, because the intensity and the phase of the ICD signal depends on the position and the orientation of the chromophore relative to the DNA bases. An intercalator usually has a weak and negative ICD signal when its transition moment is polarized along the long axis of the binding pocket (i.e. parallel to the bisector of the base pairs). In contrast, relatively strong positive ICD bands appear when the transition moment is polarized perpendicular to the long axis of the binding pocket. Groove binders give even more intense ICD signals, usually with a positive band.

Thus aqueous solutions of the achiral acridizinium salts **5a** and **5b** alone have no CD activity, but an ICD is observed on addition of DNA to these salts; the spectra differ in phase and intensity, however (Figure 2.5.4). The shapes of the CD signals roughly resemble the broadened long-wavelength absorption spectrum of the corresponding acridizinium salts. The positive ICD of **5b** (Figure 2.5.4B, spectrum 2)

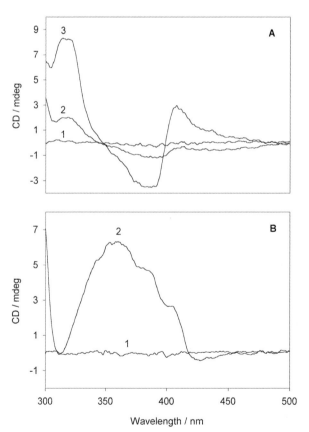

Fig. 2.5.4. CD spectra of salts **5a** (A) and **5b** (B) with ct DNA; 1. without DNA; 2. [dye]/[DNA] = 0.1; 3. [dye]/[DNA] = 0.05.

results from orientation of the short molecular axis of the dye parallel to the binding pocket [34]. In contrast, the CD spectrum of the amino-substituted acridizinium salt **5a** (Figure 2.5.4A, spectrum 2) in the presence of DNA gives negative signals for the S_0–S_1 transition at dye-to-DNA ratios smaller than 0.05; this is indicative of intercalative binding with the acridisinium long axis (which roughly resembles the polarization of the long-wavelength transition) oriented parallel to the long axis of the intercalation pocket. Nevertheless, at a higher dye-to-DNA ratio (0.1) a bi-signate signal pattern appears for the long-wavelength absorption of amine **5a** (Figure 2.5.4A, spectrum 3); this might result from exciton coupling [35]. This exciton band presumably arises because of aggregation of the dye and association of these aggregates with the DNA backbone. Such exciton CD signals are usually significantly stronger than those of intercalated molecules and can overlap with the latter.

2.5.3.4
LD Spectroscopy

Linear dichroism is defined as the differential absorption of linearly polarized light (Eq. 6).

$$LD = A_{\parallel} - A_{\perp} \tag{6}$$

$A_{\parallel}$ is the absorbance of the sample when the light is polarized parallel to a reference axis, and $A_{\perp}$ is the absorbance of light which is polarized perpendicular to this axis. The strength of the absorption depends on the orientation of the electric field vector of the light and the transition moment of the chromophore – parallel orientation results in maximum absorption whereas perpendicular orientation leads to zero absorption. By dividing the LD value by the absorbance of the unoriented sample under isotropic conditions (A_{iso}), the "reduced" linear dichroism (LD_r), i.e. the wavelength-dependent LD, is obtained (Eq. 7) [36].

$$LD_r = LD/A_{iso} = 3/2 S(3\cos^2 \alpha - 1) \tag{7}$$

The LD_r correlates with the orientation of the transition moment of the dye relative to the reference axis, as quantified by the angle α. LD_r is also proportional to an orientation factor S ($S = 1$ denotes perfect alignment of the dye, $S = 0$ random orientation). For an isolated, non-overlapping transition, Eq. (7) establishes the correlation between LD_r, α and S. These definitions lead to the qualitative rule that with an angle $\alpha > 55°$, a negative LD signal is observed, whereas with $\alpha < 55°$, a positive signal appears in the spectrum. Thus, with an appropriate set-up the orientation of a chromophore relative to a reference axis can be determined.

From Eq. (7) it is obvious that in an isotropic medium a LD signal cannot be detected because of the statistical orientation of the transition moments under these conditions. Thus, methods are required to promote preferential orientation of molecules. For LD experiments two general approaches have been established:

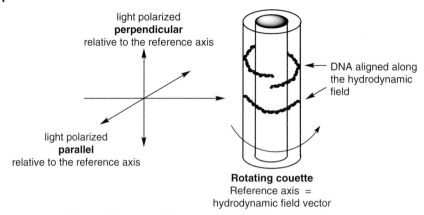

Fig. 2.5.5. Schematic illustration of the orientation of a cylindrical Couette flow cell relative to the linearly polarized light.

(a) the generation of an electric field ("electric *LD*") which forces the molecules to arrange in this field depending on their intrinsic dipole moment; and (b) the implementation of a hydrodynamic field in a rotating couette ("flow *LD*", Figure 2.5.5), in which molecules can align along the flow field. The latter method is only useful for macromolecules such as peptides or DNA, because small molecules cannot be oriented along the flow lines of the field. Nevertheless, for the determination of the binding mode of a dye–DNA complex flow-*LD* spectroscopy has been shown to be useful. In a hydrodynamic field most of the DNA molecules are partially arranged along the flow lines (with the flow lines as reference axis, $\alpha \approx 90°$), so that the DNA bases afford a clear negative *LD* signal (Figure 2.5.6A, solid line). Consequently, an intercalator should also give a negative *LD* signal, because the transition moment is almost coplanar to those of the nucleic acid bases. In contrast, for a groove binder the angle α is 45° to the helix axis and thus to the flow lines. Consequently, a groove binder should give a positive signal, which is relatively weak compared with that of an intercalator.

The flow *LD* spectrum of aminoacridizinium **5a** in the presence of st DNA at a molar dye:DNA ratio of 0.025 contains a negative *LD* signal in the long-wavelength absorption region of the acridisinium salt (Figure 2.5.6A, dashed line). The negative sign of the *LD* signals in this region is indicative of an intercalative binding mode between **5a** and DNA. The LD_r spectrum also provides information about the average orientation of the molecular plane of the aromatic dye relative to those of the DNA bases (Eq. 7). Typically, LD_r bands are of constant signal intensity, except for the region of overlap between different polarizations. A LD_r band with varying signal intensity usually results from heterogeneous binding. For the 9-amino derivative **5a**, a nearly constant LD_r value over the range 350–500 nm was

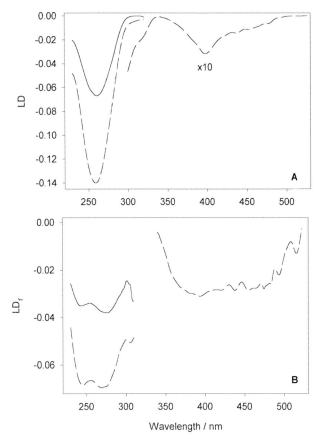

Fig. 2.5.6. Linear-flow-dichroism (A), and reduced linear-flow-dichroism (B) spectra of acridizinium salt **5a** in buffer solution (1 mM EDTA, 10 mM Tris buffer, 10 mM NaCl, pH 7.0); solid line, DNA without dye; dashed line, [**5c**]/[DNA] = 0.025.

observed (Figure 2.5.6B); this indicates that its orientation properties are fully consistent with intercalation into the DNA.

Further information can be gained from inspection of DNA base absorption. In the absence of acridizinium **5a** the DNA bases give a negative LD band in the absorption region 230–300 nm (Figure 2.5.6A, solid line). Most notably, a significant increase of this LD absorption occurs on addition of the 9-aminoacridizinium salt **5a** (Figure 2.5.6A, dashed line); this indicates better orientation of the macromolecule within the hydrodynamic field. It can this be concluded that the alignment of the DNA becomes more pronounced because of stiffening of the helix on intercalation of the ligand.

In summary, it has been demonstrated that complex formation between dyes and DNA may be conveniently monitored by absorption and emission spectroscopy and that these methods provide useful data for discussion of the binding strength and binding mode.

Acknowledgment

The authors thank the Bundesministerium für Bildung und Forschung, the Deutsche Forschungsgemeinschaft, the Deutscher Akademischer Austauschdienst, and CRUI (Vigoni program), the Fonds der Chemischen Industrie, and the Universitätsbund Würzburg for generous financial support.

References

1 See, for example, selected excellent reviews and references cited therein: (a) I. HAQ, J. LADBURY, *J. Mol. Recognit.* **2000**, *13*, 188–197; (b) P. B. DERVAN, R. W. BÜRLI, *Curr. Opin. Chem. Biol.* **1998**, *3*, 688–693; (c) W. D. WILSON, *Nucleic Acids in Chemistry and Biology*, G. M. BLACKBURN, M. J. GAIT, eds., IRL Press, Oxford, UK, **1996**, pp. 329–374; (d) B. H. GEIERSTANGER, D. E. WEMMER, *Annu. Rev. Biophys. Biomol. Struct.* **1995**, *24*, 463–493; (e) *Recent Advances in DNA Binding Agents. Symposia in print number 4, Biorg. Med. Chem.* R. S. COLEMAN, ed., **1995**, *3*, 611–872; (f) T. H. KRUGH, *Curr. Opin. Struct. Biol.* **1994**, *4*, 351–364; (g) E. M. TUITE, J. M. KELLY, *J. Photochem. Photobiol. B: Biology* **1993**, *21*, 103–124; (h) L. H. HURLEY, *J. Med. Chem.* **1989**, *32*, 2027–2033; (i) M. J. WARING, *Drugs Exptl. Clin. Res.* **1986**, *12*, 441–453.

2 (a) *Cancer Chemotherapeutic Reagents*, W. O. FOYE, ed., American Chemical Society, Washington, DC, **1995**; (b) *Nucleic Acid Targeted Drug Design*, C. L. PROBST, T. L. PERUN, eds., Marcel Dekker, New York, **1992**.

3 *DNA and RNA Binders*, M. DEMEUNYNCK, C. BAILLY, W. D. WILSON, eds., Wiley–VCH, Weinheim, **2002**.

4 W. D. WILSON, *Nucleic Acids in Chemistry and Biology*, G. M. BLACKBURN, M. J. GAIT, eds., IRL Press, Oxford, UK, **1996**, pp. 375–417.

5 J. M. GOTTESFELD, L. NEELY, J. W. TRAUGER, E. E. BAIRD, P. B. DERVAN, *Nature* **1997**, *387*, 202–205.

6 (a) T. TAKAHASHI, H. TANAKA, A. MATSUDA, T. DOI, H. YAMADA, T. MATSUMOTO, D. SASAKI, Y. SUGIURA, *Bioorg. Med. Chem. Lett.* **1998**, *8*, 3303–3306; (b) K. C. NICOLAOU, B. M. SMITH, J. PASTOR, Y. WATANABE, D. S. WEINSTEIN, *Synlett* **1997**, 401–440; (c) K. M. DEPEW, S. M. ZEMAN, S. H. BOYER, D. J. DENHARDT, N. IKEMOTO, S. J. DANISHEFSKY, D. M. CROTHERS, *Angew. Chem. Int. Ed. Engl.* **1996**, *35*, 2797–2801.

7 See for example: (a) Y. XIONG, L. N. JI, *Coord. Chem. Rev.* **1999**, *185/186*, 711–733; (b) K. E. ERKKILA, D. T. ODOM, J. K. BARTON, *Chem. Rev.* **1999**, *99*, 2777–2795.

8 H.-A. WAGENKNECHT, E. D. A. STEMP, J. K. BARTON, *J. Am. Chem. Soc.* **2000**, *122*, 1–7.

9 L. S. LERMANN, *J. Mol. Biol.* **1961**, *3*, 18–30.

10 (a) R. W. ARMSTRONG, T. KURUCSEV, U. P. STRAUSS, *J. Am. Chem. Soc.* **1970**, *92*, 3174–3181; (b) Y. KUBOTA, Y. FUJISAKI, *Bull. Chem. Soc. Jpn.* **1977**, *50*, 297–298.

11 E. TUITE, B. NORDÉN, *J. Am. Chem. Soc.* **1994**, *116*, 7548–7556.
12 J.-B. LEPECQ, C. PAOLETTI, *J. Mol. Biol.* **1967**, *27*, 87–106.
13 (a) T. L. NETZEL, K. NAFISI, M. ZHAO, J. R. LENHARD, I. JOHNSON, *J. Phys. Chem.* **1995**, *99*, 17936–17947; (b) H. S. RYE, S. YUE, D. E. WEMMER, M. A. QUAESADA, R. P. HAUGLAND, R. A. MATHIES, A. N. GLAZER, *Nucleic Acids Res.* **1992**, *20*, 2803–2812.
14 D. T. BRESLIN, C. YU, D. LY, G. B. SCHUSTER, *Biochemistry* **1997**, *36*, 10463–10473.
15 A. MOLINA, J. J. VAQUERO, J. L. GARCIA-NAVIO, J. ALVAREZ-BUILLA, B. DE PASCAL-TERESA, F. GADO, M. M. RODRIGO, *J. Org. Chem.* **1999**, *64*, 3907–3915.
16 (a) H. IHMELS, K. FAULHABER, C. STURM, G. BRINGMANN, K. MESSER, N. GABELLINI, D. VEDALDI, G. VIOLA, *Photochem. Photobiol.* **2001**, *74*, 505–512; (b) H. IHMELS, B. ENGELS, K. FAULHABER, C. LENNARTZ, *Chem. Eur. J.* **2000**, *6*, 2854–2864.
17 (a) D. S. PILCH, C. YU, D. MAKHEY, E. J. LaVOIE, A. R. SRINIVASAN, W. K. OLSON, R. S. SAUERS, K. J. BRESLAUER, N. E. GEACINTOV, L. F. LIU, *Biochemistry* **1997**, *36*, 12542–12553; (b) W. D. WILSON, A. N. GOUGH, J. J. DOYLE, M. W. DAVIDSON, *J. Med. Chem.* **1976**, *19*, 1261–1263.
18 P. PJURA, K. GRZESKOWIAK, E. R. DICKERSON, *J. Mol. Biol.* **1987**, *197*, 257–271.
19 C. BAILLY, J. B. CHAIRES, *Bioconjugate Chem.* **1998**, *9*, 513–538.
20 L. S. LERMAN, *Proc. Natl Acad. Sci. USA* **1963**, *49*, 94–102.
21 (a) E. C. LONG, J. K. BARTON, *Acc. Chem. Res.* **1990**, *23*, 273–279; (b) D. SUH, J. B. CHAIRES, *Biorg. Med. Chem.* **1995**, *3*, 723–728.
22 K.-M. HYUN, S.-D. CHOI, S. LEE, S. K. KIM, *Biochim. Biophys. Acta* **1997**, *1334*, 312–316.
23 Solvatochromism is the dependence of the absorption and emission properties (i.e. signal maximum, intensity, lifetime) of a compound on the solvent, or the surrounding medium in general: P. SUPPAN, N. GHONEIM, *Solvatochromism*, The Royal Society of Chemistry, London, **1997**.
24 Note: If large volumes of titrant solution are used or only small absorption changes occur upon complex formation, dilution effects must be considered. To avoid such an effect the DNA solution should contain the dye at the same concentration as in the dye solution to maintain a constant concentration of the dye during the titration.
25 For a detailed discussion see: C. R. CANTOR, P. R. SCHIMMEL, *Biophysical Chemistry, Part III*, W. H. FREEMAN, San Francisco, **1980**, pp. 1239–1262.
26 G. SCATCHARD, *Ann. N.Y. Acad. Sci.* **1949**, *51*, 660–672.
27 J. D. MCGHEE, P. H. VON HIPPEL, *J. Mol. Biol.* **1974**, *86*, 469–489.
28 G. LÖBER, *J. Luminescence* **1981**, *22*, 221–265.
29 S. STEENKEN, S. JOVANOVIC, *J. Am. Chem. Soc.* **1997**, *119*, 617–618.
30 D. REHM, A. WELLER, *Isr. J. Chem.* **1970**, *8*, 259–271.
31 A. I. KONONOV, E. B. MOROSHKINA, N. V. TKACHENKO, H. LEMMETYINNEN, *J. Phys. Chem. B* **2001**, *105*, 535–541.
32 $E_{0,0} = 2.59$ V; $E^0{}_{Red} = -0.72$ V relative to the NHE; D. DEMUTH, M. SCHMITTEL, H. IHMELS, unpublished results.
33 B. NORDÉN, T. KURUCSEV, *J. Mol. Recognit.* **1994**, *7*, 141–156.
34 A similar interpretation was presented for the ICD for complexes of 9-substituted anthracene derivatives with DNA: H. C. BECKER, B. NORDÉN, *J. Am. Chem. Soc.* **1999**, *121*, 11947–11952.
35 C. R. CANTOR, P. R. SCHIMMEL, *Biophysical Chemistry, Part II*, W. H. FREEMAN, San Francisco, **1980**, pp. 392–463.
36 (a) B. NORDÉN, M. KUBISTA, T. KURUCSEV, *Q. Rev. Biophys.* **1992**, *25*, 51–171; (b) B. NORDÉN, *Appl. Spectroscop. Rev.* **1978**, *14*, 157–248.

B.9
Binding of Small Molecules to DNA – Groove Binding and Intercalation

Heiko Ihmels, Carsten Schmuck

In general, two binding modes between guest molecules and the host DNA are possible – minor or major groove binding and intercalation (Figure B.9.1). A third binding mode, external binding, is also known; this results exclusively from attractive electrostatic interactions between a positively charged molecule and the negatively charged phosphate backbone of the DNA. In contrast, groove binding and intercalation can be viewed as the formation of a supramolecular assembly on the basis of associative interactions such as π stacking, hydrogen bonding, and attractive van der Waals or hydrophobic interactions. Each binding interaction is usually initiated by hydrophobic transfer of the DNA binder from the polar aqueous solution to the less polar environment of the DNA. If the DNA binder carries a positive charge, association with the DNA also leads to a release of DNA counter cations from the DNA grooves.

Groove Binding

The DNA helix has two grooves of different size, the minor and major grooves, which can serve as binding sites for guest molecules. Whereas relatively large molecules such as proteins bind preferentially to the major

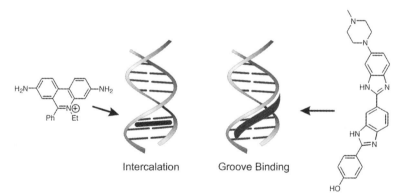

Fig. B.9.1. Molecules such as organic dyes, polycyclic aromatic compounds, organometallic complexes, saccharides, peptides, and polyamides bind to nucleic acids. Two major binding modes are possible – intercalation (left) and groove binding (right).

groove of DNA [1], the minor groove is the preferred binding site for small ligands [2, 3]. The binding pocket of a DNA groove can be defined by two different regions, the "bottom", formed by the edges of the nucleic bases that face into the groove, and the "walls", which are formed from the deoxyribose phosphate backbone of the DNA. Groove binders usually comprise at least two aromatic or heteroaromatic rings the connection of which enables conformational flexibility such that a crescent-shaped conformation can be achieved and the molecule fits perfectly into the groove. In addition, functional groups are required to form hydrogen bonds with the nucleic bases at the bottom of the groove. A typical minor-groove binder is Hoechst 33258 (shown to the right in Figure B.9.1). Most groove binders have binding selectivity towards AT-rich areas, because grooves which consist of GC base pairs are sterically hindered by the guanine amino functionality at C-2 and its hydrogen bond with the C-2 carbonyl functionality of cytosine. It has also been observed that in AT-rich grooves the electrostatic potential is larger than in GC-rich regions. Thus, positively polarized or charged ligands also have higher affinity for the AT-rich groove because of favorable electrostatic attraction. Despite these considerations groove binders are known that bind preferentially in GC-rich grooves, because they are substituted with functionality that forms strong attractive interactions with the guanine amino group [4].

Intercalation

In a DNA helix the nucleic bases are located in an almost coplanar arrangement. In his pioneering work, Lerman showed that this arrangement enables planar polycyclic aromatic molecules to intercalate between two base pairs [5, 6]. A typical intercalator is ethidium bromide (shown to the left in Figure B.9.1). Important driving forces for this mode of binding are dipole–dipole interactions and the π stacking of the guest molecule with the aromatic nucleic bases. Other than groove binding, intercalation has a significant effect on DNA structure, because the DNA must unwind to enable the intercalator to fit between the two base pairs. This unwinding leads to lengthening of the helix by more than 340 pm, with a significant change of the deoxyribose conformation [2].

It has been observed that the binding of one intercalator between two base pairs hinders access of another intercalator to the binding site next to the neighboring intercalation pocket. Although this "neighbor exclusion principle" is well accepted and can be used as a general rule, it is not fully understood. It has been proposed that changes in the structure of the DNA on intercalation result in limited access to the neighboring binding pocket for steric reasons. Intercalation might also reduce the negative electrostatic potential at the intercalation site, so attractive electrostatic interactions no longer occur close to this site.

References

1 W. D. Wilson, in G. M. Blackburn, M. J. Gait, eds., *Nucleic Acids in Chemistry and Biology*, IRL Press, Oxford, UK, **1996**, pp. 375–417.
2 W. D. Wilson, in G. M. Blackburn, M. J. Gait, eds., *Nucleic Acids in Chemistry and Biology*, IRL Press, Oxford, UK, **1996**, pp. 329–374.
3 M. Demeunynck, C. Bailly, W. D. Wilson, eds., *DNA and RNA Binders*, Wiley–VCH, Weinheim, **2002**.
4 C. Bailly, J. B. Chaires, *Bioconj. Chem.* **1998**, 9, 513–538.
5 L. S. Lermann, *J. Mol. Biol.* **1961**, 3, 18–30.
6 L. S. Lerman, *Proc. Natl Acad. Sci. USA* **1963**, 49, 94–102.

2.6
Interaction of Nitrogen Monoxide and Peroxynitrite with Hemoglobin and Myoglobin

Susanna Herold

2.6.1
Biosynthesis, Reactivity, and Physiological Functions of Nitrogen Monoxide

For a long time nitrogen monoxide (NO$^{\bullet}$) was primarily known as an air pollutant which was sometimes used in biochemical studies as a spectroscopic probe to examine the metal ligand environment in metalloproteins. However, the discovery that this simple diatomic inorganic molecule has diverse functions essential in physiology opened a completely new research area. Identification of the endothelium-derived relaxing factor (EDRF) as nitrogen monoxide was definitely one of the most exciting discoveries of biomedical research in the 1980s. Besides its potent vasodilatory effect, in certain circumstances NO$^{\bullet}$ was found to be responsible for the killing of microorganisms and tumor cells by activated macrophages, and to act as a novel, unconventional type of neurotransmitter [1].

In vivo NO$^{\bullet}$ is generated from the essential amino acid L-arginine by a family of enzymes called nitric oxide synthases (NOS) [1]. Three isoforms have been described and cloned – endothelial NOS (eNOS), brain or neuronal NOS (nNOS), and inducible macrophage-type NOS (iNOS). All isoforms of NOS make the same products, NO$^{\bullet}$ and L-citrulline, by incorporating an O-atom of dioxygen into L-arginine. Only the amount of NO$^{\bullet}$, the conditions under which it is made, and the location where it is synthesized differ among the three isoforms.

Many biological actions of NO$^{\bullet}$ are triggered by its interaction with guanylyl cyclase. This heme-containing enzyme is present in the cytosolic fraction of virtually all mammalian cells, with the highest concentrations found in the lungs and in the brain. Guanylyl cyclase, activated when NO$^{\bullet}$ binds to its reduced heme, catalyzes the conversion of guanosine triphosphate (GTP) into cyclic guanosine monophosphate (cGMP), an important intracellular signal molecule that is involved in the regulation of smooth muscle relaxation and causes blood vessels to dilate [2].

Nitrogen monoxide also undergoes several cGMP-independent reactions, mainly related to toxicological events. The two most important direct reactions of NO$^{\bullet}$ responsible for its toxicity in cells are interaction with metal centers of proteins (other than guanylyl cyclase) and reaction with radical intermediates of biological transformations, in particular with superoxide. Examples of cytotoxic effects of NO˙ are inhibition of key mitochondrial iron–sulfur enzymes such as NADH-ubiquinone oxidoreductase, NADH-succinate oxidoreductase, and aconitase, and inhibition of heme proteins such as cytochrome c oxidase and cytochrome P-450 enzymes [3].

2.6.1.1
The Biological Chemistry of Peroxynitrite

Simultaneous generation of NO˙ and superoxide favors the nearly diffusion-controlled ($k = (1.6 \pm 0.3) \times 10^{10}$ M^{-1} s^{-1}) production of the powerful oxidizing and nitrating agent peroxynitrite [systematic name: oxoperoxonitrate(1–)] [4, 5]. The peroxynitrite anion (ONOO$^-$) is stable, but the protonated form, peroxynitrous acid (HOONO, p$K_a = 6.8$), isomerizes to nitrate with a rate constant of 1.2 s^{-1} at 25 °C [5]. The reaction between peroxynitrite and CO_2, commonly present at mM concentrations in tissues, is a key route of peroxynitrite consumption in biological systems. The reaction between the peroxynitrite anion and CO_2 is quite fast (3×10^4 M^{-1} s^{-1} at 24 °C [6]) and yields the adduct 1-carboxylato-2-nitrosodioxidane (ONOOCO$_2^-$) [6, 7], a stronger nitrating agent than peroxynitrite [8]. In the absence of substrates ONOOCO$_2^-$ rapidly decays to nitrate and CO_2 ($t_{1/2} = 0.1$–3 ms [7, 9]). One of the effects of CO_2 is, therefore, to severely reduce the lifetime of peroxynitrite, partially preventing membrane crossing, and limiting its radius of action. Among the cellular components that can react directly with peroxynitrite in the presence of physiological CO_2 concentrations are thiols and metalloproteins, in particular hemoproteins [10].

Because of the instability of peroxynitrite under physiological conditions, the detection of 3-nitrotyrosine (NO$_2$-Tyr) has become a biochemical marker for the presence of peroxynitrite in pathophysiological processes. The biological significance of tyrosine nitration is a subject of great interest, because extensive evidence supports the formation of nitrotyrosine in vivo in diverse pathological conditions such as heart diseases, chronic inflammation and autoimmune diseases, cancer, Parkinson's disease, Alzheimer's disease, multiple sclerosis, amyotrophic lateral sclerosis, and ischemia–reperfusion injury [11].

2.6.2
Interaction of Nitrogen Monoxide and Peroxynitrite with Hemoglobin and Myoglobin

One of the most significant common aspects of the chemistry of NO˙ and peroxynitrite is their ability to react in a unique manner with the metal centers of numerous proteins, in particular hemoproteins [10]. We have used myoglobin (Mb) and hemoglobin (Hb) to investigate the diverse reactions that these simple inorganic biomolecules can undergo with different oxidation states of hemoproteins. Mutated forms of Mb and Hb are available [12, 13], and a large number of transition metals ions other than Fe ions have been successfully incorporated in these

2.6.2.1
The NO·-mediated Oxidation of Oxymyoglobin and Oxyhemoglobin

The rapid reaction of nitrogen monoxide with oxyhemoglobin (oxyHb) is of particular interest, because it significantly reduces the half-life of NO· in vivo and is the cause of increased blood pressure observed when extracellular hemoglobin-based blood substitutes are administered [16].

We have recently shown that, in analogy to the reaction between nitrogen monoxide and superoxide, the reactions of oxymyoglobin (oxyMb) and oxyHb with NO· generate intermediate iron(III)peroxynitrito complexes that were characterized by rapid-scan UV–visible spectroscopy [17, 18]. The intermediate complexes MbFe^{III}OONO and HbFe^{III}OONO can be observed at alkaline pH, but rapidly decay to nitrate and the aquoiron(III) form of the proteins (metMb and metHb, respectively) under neutral or acidic conditions. The best spectrum obtained for HbFe^{III}OONO in the visible region, shown as the first trace in Figure 2.6.1, has two absorption maxima at 504 nm (ε_{504} = 8.7 mM^{-1} cm^{-1}) and at 636 nm (ε_{636} = 5.4 mM^{-1} cm^{-1}) [18]. These maxima are characteristic of high-spin metHb and metMb derivatives with anionic ligands (Table 2.6.1). Almost identical absorption features were obtained for the myoglobin intermediate MbFe^{III}OONO (Table 2.6.1), despite its greater instability, that made it impossible to obtain an accurate spectrum.

The second-order rate constants, obtained from the linear plots of the observed pseudo-first-order rate constants against oxyMb or oxyHb concentrations, are pH

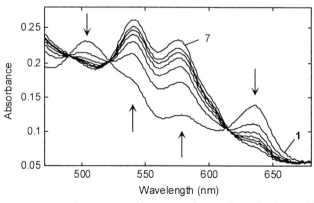

Fig. 2.6.1. Rapid-scan UV–visible spectra of the reaction of oxyHb (13.5 μM) with NO· (ca. 50 μM) in 0.1 M borate buffer at pH 9.5, 5 °C. Averaged spectra were collected (62 scans each s). The decay of the HbFe^{III}OONO (see arrows, spectrum 1) to HbFe^{III}OH (spectrum 7) is presented. Traces 1–7 were recorded 0, 160, 320, 480, 640, 800, and 1280 ms after mixing.

Tab. 2.6.1. Selected spectroscopic data for different methmyoglobin and methemoglobin complexes.

Complex	Soret $\lambda_{max}\ (\varepsilon)^a$	Visible $\lambda_{max}\ (\varepsilon)^a$	$\lambda_{max}\ (\varepsilon)^a$	Ref.
MbFe^{III}OONO	410 (138)	504 (8.0)	636 (3.2)	17
MbFe^{III}OH$_2$	408 (188)	502 (10.2)	630 (3.9)	9
MbFe^{III}ONO$_2$	409 (172)	502 (8.8)	629 (3.6)	19
MbFe^{III}ONO	410 (137)	504 (8.7)	631 (5.1)	19
MbFe^{III}NO$_2$	412 (137)	502 (8.4)	628 (4.2)	19
HbFe^{III}OONO	407 (165)	504 (8.7)	636 (5.4)	18
HbFe^{III}OH$_2$	405 (179)	500 (10.0)	631 (4.4)	9
HbFe^{III}ONO$_2$	408 (120)	527 (9.9)	628 (3.2)	19
HbFe^{III}ONO	410 (133)	537 (10.1)	617 (3.9)	20
HbFe^{III}NO$_2$	411 (132)	538 (10.0)	623 (3.6)	19
HbFe^{III}OC(O)CH$_3$	404 (178)	497 (10.5)	620 (5.5)	9
HbFe^{III}OC(O)H	404 (178)	496 (9.2)	620 (5.8)	9

$^a \lambda_{max}$ [nm] (ε [mM^{-1} cm^{-1}]).

dependent. At pH 7.0 we obtained $(43.6 \pm 0.5) \times 10^6$ M^{-1} s^{-1} and $(89 \pm 3) \times 10^6$ M^{-1} s^{-1} (per heme) for the NO$^{\cdot}$-mediated oxidation of oxyMb and oxyHb, respectively. Approximately the same rate constants were obtained when these reactions were performed in the pH range 5.0–7.0, whereas under alkaline conditions the second-order rate constants increased continuously up to $(97 \pm 3) \times 10^6$ M^{-1} s^{-1} and $(144 \pm 3) \times 10^6$ M^{-1} s^{-1} (per heme) at pH 9.5, for oxyMb and oxyHb, respectively.

As mentioned above, the rates of decay of the peroxynitrite complex of the two hemoglobin subunits (36 s^{-1} and 7 s^{-1} at pH 9.5) are significantly slower than that measured for MbFe^{III}OONO (205 s^{-1} at pH 9.5). However, these rates are significantly faster than that for the decay of free peroxynitrite under the same conditions (0.11 s^{-1} at pH 9.5 [25]). These differences can be rationalized by considering the influence of the distal histidine. As depicted in Scheme 2.6.1, the distal histidine (His64 in myoglobin) could facilitate the cleavage of the O–O bond by interacting with one of the two oxygen atoms. Acceleration of the decay of peroxynitrite when it is bound to the iron(III) center might therefore be a consequence of the presence of both the iron *and* this hydrogen bond that pull from two sides on the O–O bond.

Further mechanistic studies showed that no free peroxynitrite is formed during the reactions of NO$^{\cdot}$ with the oxy-forms of these proteins, and that nitrate is formed quantitatively, at both pH 7.0 and pH 9.0 [18]. Analysis of the proteins after ten cycles of oxidation by NO$^{\cdot}$ and reduction by ascorbic acid indicated that fewer than 1% of the tyrosine residues are nitrated. These results show that when peroxynitrite is coordinated to the heme of myoglobin or hemoglobin, it rapidly isomerizes to nitrate, and thus cannot nitrate the tyrosine residues of the globin.

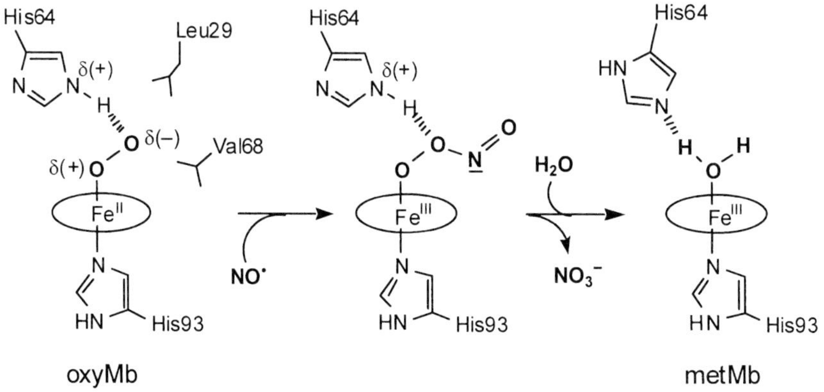

Scheme 2.6.1. The role of the distal histidine (His64) in the isomerization of peroxynitrite bound to the iron(III) center of myoglobin.

Interestingly, the iron(III)peroxynitrito complex is *not* detected when the aquo-iron(III) forms of Mb and Hb (metMb and metHb) are allowed to react with peroxynitrite under neutral or alkaline conditions [22]. Indeed, the reactivity of metMb towards peroxynitrite is regulated by the presence of the distal histidine (His64), which partly blocks the active site and stabilizes, via a strong hydrogen bond, the water ligand coordinated to the iron (see metMb structure in Scheme 2.6.1). In the presence of wild-type metMb the decay of peroxynitrite is only slightly accelerated ($k_{cat} = (1.4 \pm 0.1) \times 10^4$ M^{-1} s^{-1} at pH 7.4 and 20 °C [23]). In contrast, the myoglobin mutant in which His64 is replaced by an alanine (H64A) is an efficient catalyst for the isomerization of peroxynitrite ($k_{cat} = (6.0 \pm 0.1) \times 10^6$ M^{-1} s^{-1} at pH 7.4 and 20 °C [23]). Indeed, ion chromatographic analysis of the nitrogen-containing products showed that in the presence of 0.01 equivalents of H64A peroxynitrite decays quantitatively to nitrate. Moreover, HPLC analysis revealed that 0.05 equivalents of H64A prevent nitration of free tyrosine by peroxynitrite almost completely, both in the absence and presence of physiological amounts of carbon dioxide (1.2 mM).

2.6.2.2
The Peroxynitrite-mediated Oxidation of OxyMb and OxyHb

Despite the presence of a large concentration of carbon dioxide in the blood (ca. 1 mM), it has been reported that peroxynitrite can diffuse across the red-blood-cell membrane and react with oxyHb [24]. The anionic form (ONOO$^-$) crosses the erythrocyte membrane by using the anion channel band 3 whereas peroxynitrous acid crosses the lipid membranes by rapid passive diffusion [24].

We have recently shown that the peroxynitrite-mediated oxidation of oxyMb and oxyHb proceeds via intermediate oxoiron(IV) (ferryl) complexes, which, in a second step, react further with peroxynitrite to yield metMb and metHb, respectively [25]. The rate constants for the two steps of the reaction of peroxynitrite with oxyMb, at pH 7.3 and 20 °C, were determined as $(5.4 \pm 0.2) \times 10^4$ M^{-1} s^{-1} and $(2.2 \pm 0.1) \times 10^4$ M^{-1} s^{-1}, respectively [25]. The corresponding rates for the reaction with oxyHb, at pH 7.0 and 20 °C, are $(8.4 \pm 0.4) \times 10^4$ M^{-1} s^{-1} and $9.4 \pm 0.7 \times 10^4$ M^{-1} s^{-1}, respectively [25]. These rate constants suggest that oxyMb and oxyHb can successfully compete with CO_2 and, thus, might be involved in the detoxification of peroxynitrite under physiological conditions.

To act as efficient scavengers, however, it is important the proteins are not modified extensively during their reaction with peroxynitrite. To investigate this hypothesis, we have analyzed by HPLC peroxynitrite-treated proteins subjected either to acid hydrolysis or pronase digestion. Our data showed that only very low quantities of 3-nitrotyrosine are formed when peroxynitrite reacts with the oxy-form of these proteins. Comparable amounts of nitrated tyrosine residues are formed when metMb and metHb are treated with peroxynitrite under analogous conditions, but significantly larger yields are observed with apoMb and metMbCN (Figure 2.6.2). In apoMb the heme group is absent whereas in metMbCN the iron ion of the heme cannot interact directly with peroxynitrite because of the strong cyanide ligand. Taken together our data suggest that the heme center of Mb can act as an efficient scavenger of peroxynitrite, protecting the globin from nitration [26]. When the heme center is not available for direct reaction, peroxynitrite undergoes an unspecific reaction with the globin, which leads to significant tyrosine nitration.

Additional evidence for the role of oxyMb and oxyHb as scavengers of peroxynitrite is the observation that they effectively protect against peroxynitrite-mediated nitration of free tyrosine. Reaction of 200 μM peroxynitrite with 100 μM tyrosine leads to the formation of approximately 15% nitrotyrosine (at pH 7.4 and 20 °C). When the same reaction is carried out in the presence of 50 μM oxyMb no nitrotyrosine can be detected. In addition, the same amount of metMb reduces the

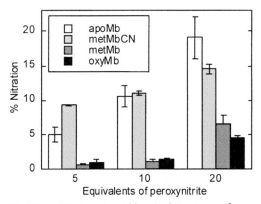

Fig. 2.6.2. Nitrotyrosine yields (% relative to the total tyrosine content of the protein, that is 2 tyrosine residues per heme), determined by HPLC after acid hydrolysis, from the reaction of apo-, met-, oxyMb, and metMbCN (100 μM) with different amounts of peroxynitrite (5, 10, and 20 equiv.) at 0 °C and pH 7.0.

nitration yield to 2.6%. However, 50 μM apoMb or metMbCN only slightly reduce the nitrotyrosine yield, to approximately 9 and 13%, respectively.

2.6.3
NO• as an Antioxidant

Interestingly, it has been proposed that NO• also has a protective role and, under certain conditions, acts as an antioxidant. Indeed, it has been shown that NO• can terminate a common pathological manifestation of oxidant overload to biological membranes, the deleterious process of lipid peroxidation [27], by reacting very rapidly with fatty alkoxyl (RO•) and peroxyl (ROO•) radical intermediates which would otherwise propagate the membrane damage. In addition, NO• can inhibit ferrylMb-induced oxidative damage by reducing ferrylMb to metMb [28].

2.6.3.1
The NO•-mediated Reduction of FerrylMb and FerrylHb

It has been suggested that heme-mediated redox reactions contribute to the organ dysfunction and/or tissue damage which occur in some pathological states characterized by release of Hb and Mb into the extracellular environment [29]. Autoxidation of oxyHb and oxyMb to their met-forms releases superoxide which dismutates to dioxygen and hydrogen peroxide, which in turn can cause tissue damage [30]. In addition, oxidation of these two proteins generates potentially cytotoxic products such as the ferryl species [31].

It has been proposed that the highly oxidizing species ferrylMb is, at least in part, responsible for the oxidative damage caused by the reperfusion of ischemic tissues. We have determined the rate constants for the reactions of ferrylMb and ferrylHb with NO• ($(17.1 \pm 0.3) \times 10^6$ M^{-1} s^{-1} and $(24 \pm 1) \times 10^6$ M^{-1} s^{-1} at pH 7.0 and 20 °C) [23, 24]. The large value of these rate constants implies that these reactions are very likely to occur in vivo and might represent a detoxifying pathway for ferrylMb and ferrylHb and, thus, an additional antioxidant function of NO•.

Interestingly, we have shown that also the reactions of ferrylMb and ferrylHb with NO• proceed via the rapid formation of an intermediate [19]. Because of the radical-like character of the oxo-ligand in the ferryl forms of the proteins [32], it is reasonable to assume that the first step of the reactions is rapid radical–radical recombination (Scheme 2.6.2), which leads to the O-nitrito intermediates MbFe^{III}ONO and HbFe^{III}ONO. As shown in Figure 2.6.3, the absorbance maxima in the spectrum of the Mb-intermediate are found at 504 nm ($\varepsilon_{504} = 8.7$ mM^{-1} cm^{-1}) and at 631 nm ($\varepsilon_{631} = 5.1$ mM^{-1} cm^{-1}), and are thus consistent with

$$\text{MbFe}^{IV}=O \longleftrightarrow \text{MbFe}^{III}-O^\bullet + NO^\bullet \longrightarrow \text{MbFe}^{III}-ONO \longrightarrow \text{MbFe}^{III}OH_2 + NO_2^-$$

Scheme 2.6.2. Mechanism of formation of the O-nitrito intermediate by reaction of ferrylMb with NO•.

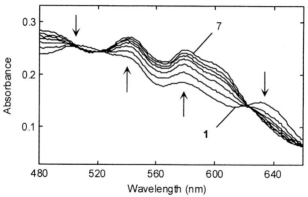

Fig. 2.6.3. Rapid-scan UV–visible spectra of the reactions of ferrylMb (14.7 μM) with NO· (50 μM in 0.1 M borate buffer at pH 9.5, 20 °C. Traces 1–7, recorded 0, 0.4, 0.8, 1.2, 1.6, 2.0, and 10 s after mixing, show the decay of the intermediate MbFe^{III}ONO (bold trace 1) to MbFe^{III}OH (trace 7).

this assumption (Table 2.6.1). The O-nitrito intermediates decay to nitrite and metMb and metHb, respectively, but are more stable than the corresponding peroxynitrito complexes.

Local nitrite concentrations in tissues are linked to the amounts of NO· produced. Indeed, except for nitrate generated from the reaction of NO· with oxyHb, nitrite is the major end product of NO· metabolism. Increased nitrite levels are thus found under pathophysiological conditions, for example inflammation, when NO· production is elevated. We have found that the rate constants for the reactions of the ferryl forms of Mb and Hb with nitrite are significantly lower than those for the corresponding reaction with NO· (16 ± 1 M^{-1} s^{-1} at pH 7.5 for Mb and $(7.5 \pm 0.1) \times 10^2$ M^{-1} s^{-1} at pH 7.0 for Hb, at 20 °C) [19, 20]. Thus, the reaction with nitrite probably plays a role only when NO· has been consumed completely and large concentrations of nitrite are still present.

In contrast with the protecting role of NO·, however, the reaction with nitrite generates nitrogen dioxide that can contribute to tyrosine nitration. Indeed, we have demonstrated that nitrite can cause nitration of added tyrosine in the presence of metMb (or metHb) and hydrogen peroxide [19]. As shown in Figure 2.6.4, the yield of nitrotyrosine increased with increasing nitrite concentration but reached a plateau (ca. 16% yield, relative to metMb) in the presence of 25–30 mM nitrite. On the basis of the reaction mechanism described in Scheme 2.6.3, the nitrite concentration dependence and, in particular, the plateau reached above 30 mM nitrite can be rationalized as follows. Reaction of metMb with hydrogen peroxide generates the one-electron oxidized form of ferrylMb which has an additional transient radical on the globin (·MbFe^{IV}=O) [33]. When nitrite is present in very high concentrations its reaction with either ·MbFe^{IV}=O or MbFe^{IV}=O might outcompete the reaction of tyrosine with ·MbFe^{IV}=O. Thus, Tyr· is generated in a concentration significantly lower than that of NO$_2$·. Consequently, the recombina-

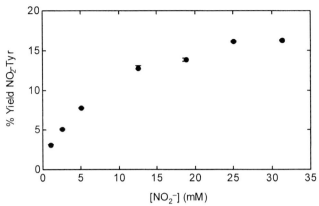

Fig. 2.6.4. Yield of nitrotyrosine, relative to metMb, generated by adding one equiv. H_2O_2 (relative to metMb) to a mixture of metMb (250–270 µM), 1 mM tyrosine, and different amounts of nitrite at pH 7.0.

$$MbFe^{III}OH_2 + H_2O_2 \longrightarrow {}^{\bullet}MbFe^{IV}=O + H_2O$$

$${}^{\bullet}MbFe^{IV}=O + Tyr \longrightarrow MbFe^{IV}=O + Tyr^{\bullet}$$

$${}^{\bullet}MbFe^{IV}=O + NO_2^- \longrightarrow MbFe^{IV}=O + NO_2^{\bullet}$$

$$MbFe^{IV}=O + NO_2^- \longrightarrow MbFe^{III}OH_2 + NO_2^{\bullet}$$

$$NO_2^{\bullet} + Tyr^{\bullet} \longrightarrow NO_2\text{-Tyr}$$

$$2\,NO_2^{\bullet} \rightleftharpoons N_2O_4 \quad \text{and} \quad N_2O_4 + H_2O \longrightarrow NO_2^- + NO_3^- + 2\,H^+$$

Scheme 2.6.3. Mechanism of formation of nitrotyrosine by reaction of metMb and nitrite in the presence of H_2O_2 and tyrosine.

tion of Tyr$^{\bullet}$ and NO$_2^{\bullet}$ to generate nitrotyrosine occurs to a lesser extent than the disproportionation of NO$_2^{\bullet}$ or its reaction with MbFe^{IV}=O to yield nitrate. This reaction mechanism might explain the observation that higher nitrite concentrations do not generate higher yields of nitrotyrosine.

2.6.4
Conclusion: A New Function of Myoglobin?

Although hemoglobin and myoglobin are probably the most thoroughly studied proteins, it has recently been proposed that these proteins might have additional biological functions; these remain a matter of continuing investigation.

The original role attributed to myoglobin (Box 10) has recently been challenged. The discovery that a mutant mouse devoid of myoglobin is capable of apparently normal muscle function raised questions about the real significance of this protein [34]. Our research in this field has shown that the reactions of NO· and peroxynitrite with oxyMb are rapid and are thus very likely to occur in vivo. We have also demonstrated that the globins are not modified significantly during these reactions.

NO· can inhibit mitochondrial respiration by reversibly binding to cytochrome c oxidase. Intracellular scavenging of NO· would therefore help to preserve respiration in the skeletal muscle and in the heart and, consequently, protect the energy-producing machinery. It might thus be possible that one of the essential functions of Mb is to scavenge peroxynitrite and excess NO·, generated under pathological conditions. This hypothesis is supported by the observation that in transgenic mice lacking myoglobin physiological properties of the heart are more severely affected by endogenously formed or exogenously applied nitrogen monoxide [35].

References

1 S. Moncada, R. M. J. Palmer, E. A. Higgs, Pharmacol. Rev. 1991, 43, 109–142.
2 A. J. Hobbs, Trends Pharmacol. Sci. 1997, 18, 484–491.
3 D. A. Wink, J. B. Mitchell, Free Radical Biol. Med. 1998, 25, 434–456.
4 J. S. Beckman, T. W. Beckman, J. Chen, P. A. Marshall, B. A. Freeman, Proc. Natl. Acad. Sci. USA 1990, 87, 1620–1624.
5 W. H. Koppenol, Redox Report 2001, 6, 339–341.
6 S. V. Lymar, J. K. Hurst, J. Am. Chem. Soc. 1995, 117, 8867–8868.
7 R. Meli, T. Nauser, W. H. Koppenol, Helv. Chim. Acta 1999, 82, 722–725.
8 A. Denicola, B. A. Freeman, M. Trujillo, R. Radi, Arch. Biochem. Biophys. 1996, 333, 49–58.
9 G. L. Squadrito, W. A. Pryor, Chem. Res. Toxicol. 2002, 15, 885–895.
10 R. Radi, Chem. Res. Toxicol. 1996, 9, 828–835.
11 H. Ischiropoulos, Arch. Biochem. Biophys. 1998, 356, 1–11.
12 E. E. Scott, Q. H. Gibson, J. S. Olson, J. Biol. Chem. 2001, 276, 5177–5188.
13 A. E. Miele, F. Draghi, A. Arcovito, A. Bellelli, M. Brunori, C. Travaglini-Allocatelli, B. Vallone, Biochemistry 2001, 40, 14449–14458.
14 B. M. Hoffman, The Porphyrins. Biochemistry, Part B, Vol. 7, D. Dolphin, ed., Academic Press, New York, 1979, pp. 403–444.
15 Y. Huang, T. Yonetani, A. Tsuneshige, B. M. Hoffman, G. K. Ackers, Proc. Natl. Acad. Sci. USA 1996, 93, 4425–4430.
16 D. H. Doherty, M. P. Doyle, S. R. Curry, R. J. Vali, T. J. Fattor, J. S. Olson, D. D. Lemon, Nature Biotechnol. 1998, 16, 672–676.
17 S. Herold, FEBS Lett. 1999, 443, 81–84.
18 S. Herold, M. Exner, T. Nauser, Biochemistry 2001, 40, 3385–3395.
19 S. Herold, F.-J. K. Rehmann, J. Biol. Inorg. Chem. 2001, 6, 543–555.
20 S. Herold, F.-J. K. Rehmann, Free Radical Biol. Med. 2003, 34, 531–545.
21 R. Kissner, T. Nauser, P. Bugnon, P. G. Lye, W. H. Koppenol, Chem. Res. Toxicol. 1997, 10, 1285–1292.
22 G. R. Martinez, P. Di Mascio, M. G. Bonini, O. Augusto, K. Briviba, H. Sies, P. Maurer, U. Röthlisberger, S. Herold, W. H. Koppenol, Proc. Natl. Acad. Sci. USA 2000, 97, 10307–10312.

23 S. Herold, T. Matsui, Y. Watanabe, *J. Am. Chem. Soc.* **2001**, *123*, 4085–4086.
24 A. Denicola, J. M. Souza, R. Radi, *Proc. Natl. Acad. Sci. USA* **1998**, *95*, 3566–3571.
25 M. Exner, S. Herold, *Chem. Res. Toxicol.* **2000**, *13*, 287–293.
26 S. Herold, K. Shivashankar, M. Mehl, *Biochemistry* **2002**, *41*, 13460–13472.
27 H. Rubbo, R. Radi, D. Anselmi, M. Kirk, S. Barnes, J. Butler, J. P. Eiserich, B. A. Freeman, *J. Biol. Chem.* **2000**, *275*, 10812–10818.
28 N. V. Gorbunov, A. N. Osipov, B. W. Day, B. Zayas-Rivera, V. E. Kagan, N. M. Elsayed, *Biochemistry* **1995**, *34*, 6689–6699.
29 D. Galaris, L. Eddy, A. Arduini, E. Cadenas, P. Hochstein, *Biochem. Biophys. Res. Commun.* **1989**, *160*, 1162–1168.
30 K. Shikama, *Chem. Rev.* **1998**, *98*, 1357–1373.
31 C. Giulivi, E. Cadenas, *Free Radical Biol. Med.* **1998**, *24*, 269–279.
32 D. T. Sawyer, *Comments Inorg. Chem.* **1987**, *6*, 103–121.
33 C. Giulivi, E. Cadenas, *Methods Enzymol.* **1994**, *233*, 189–202.
34 D. J. Garry, G. A. Ordway, J. N. Lorenz, N. B. Radford, E. R. Chin, R. W. Grange, R. Bassel-Duby, R. S. Williams, *Nature* **1998**, *395*, 905–908.
35 U. Flögel, M. W. Merx, A. Gödecke, U. K. M. Decking, J. Schrader, *Proc. Natl. Acad. Sci. USA* **2001**, *98*, 735–740.

B.10
Hemoglobin and Myoglobin

Susanna Herold

Hemoglobin, the constituent of red blood cells responsible for transport of dioxygen, and myoglobin, present in skeletal muscles to store dioxygen, are probably the most thoroughly studied proteins. Hemoglobins were first found in blood because of their very high concentration (15 g per 100 mL in normal human blood). Until recently it seemed that the oxygen carrier function was so specialized that it appeared to be the only role of these

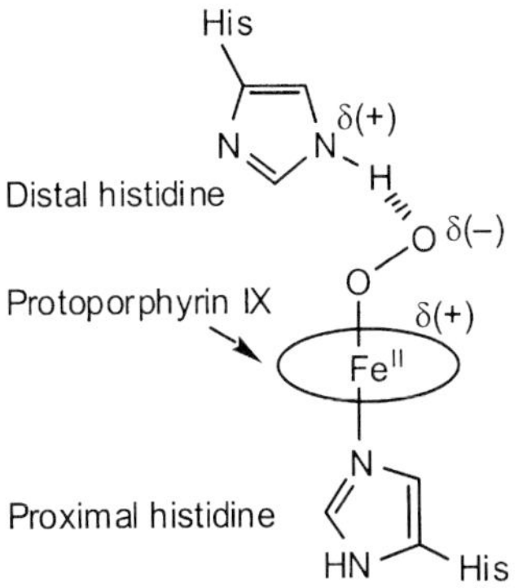

Scheme B.10.1. The active site of hemoglobin and myoglobin.

globins. However, recent discoveries of new hemoglobins in archea, bacteria, protozoa, plants, and invertebrate animals have revealed the broad diversity of these proteins and suggest that they are present in most organisms, if not all [1]. The hemoglobin that can be readily isolated from the blood of any vertebrate is a heterotetramer of two α and two β polypeptides, with a heme tightly bound to each monomer via the so-called proximal histidine residue (Scheme B.10.1) [2]. Despite the differences between their amino acid sequences, the α and β subunits have the same fold, which is made up of eight α-helices and called the globin fold. Movements and interactions between the α- and β-globin subunits lead to cooperative binding of oxygen, which enables hemoglobin to pick up oxygen readily in the lungs and to unload it efficiently in the peripheral respiring tissues. The coordinated dioxygen is stabilized by hydrogen bonding to an additional highly conserved histidine residue, the so-called distal histidine (Scheme B.10.1).

Myoglobin is a protein of conserved structure and function found in skeletal and heart muscles of vertebrates. It is believed to be used as a short-term reservoir of oxygen in working muscles and also to facilitate oxygen diffusion to mitochondria for respiration [3]. The myoglobin structure, the first protein structure to be revealed at the atomic level [4], has the typical globin fold formed by eight α-helices and is very similar to that of the two hemoglobin chains. The heme pocket is also analogous, with the proximal histidine coordinated to the iron center and the distal histidine stabilizing the coordinated dioxygen molecule (Scheme B.10.1).

References

1 R. HARDISON, *J. Exp. Biol.* **1998**, *201*, 1099–1117.
2 E. ANTONINI, M. BRUNORI, *Hemoglobin and Myoglobin in their Reactions with Ligands*, North–Holland, Amsterdam, **1971**.
3 B. A. WITTENBERG, J. B. WITTENBERG, *Annu. Rev. Physiol.* **1989**, *51*, 857–878.
4 J. C. KENDREW, R. E. DICKERSON, B. E. STRANDBERG, R. G. HART, D. R. DAVIES, D. C. PHILLIPS, V. C. SHORE, *Nature* **1960**, *185*, 422–427.

2.7
Synthetic Approaches to Study Multivalent Carbohydrate–Lectin Interactions

Valentin Wittmann

2.7.1
Introduction

The specific recognition of carbohydrate structures in biological systems (Box 5) by carbohydrate-binding proteins (lectins) is the basis of numerous intra- and intercellular events ranging from the control of protein folding to cell–cell communication during development, inflammation, and cancer metastasis [1]. Investigation of carbohydrate–lectin interactions can be approached from two directions. One is characterization of the protein part by molecular biology and structure determination (X-ray crystallography, NMR spectroscopy) [2]. In the other approach, which relies on synthetic organic chemistry, the specificity and affinity of modified or artificial lectin ligands and their effect on lectin function is studied [3, 4]. High-affinity lectin ligands are, furthermore, of considerable medicinal interest in the diagnosis and inhibition of carbohydrate-mediated processes such as inflammation or microbial adhesion [5]. The generation of high-affinity lectin ligands, however, is not trivial because most saccharide ligands bind to their protein receptors only weakly with dissociation constants typically in the milli- to micromolar range. Because many lectins have several binding sites or occur in oligomeric or clustered form on cell membranes, the creation of multivalent carbohydrate derivatives is a promising means of producing high-affinity ligands [3, 6–8].

2.7.2
Mechanistic Aspects of Multivalent Interactions

Multivalent interactions are characterized by the simultaneous binding of several ligands on one biological entity (surface, macromolecule) with several receptors on another entity (Figure 2.7.1) [8]. This type of interaction has unique collective properties that are qualitatively different from those of the corresponding monovalent systems. Not only it is possible to regulate the strength of an interaction by the number of receptor–ligand contacts, multivalent interactions also have different kinetic properties. As Whitesides et al. have demonstrated, it is possible to increase

Highlights in Bioorganic Chemistry: Methods and Applications. Edited by Carsten Schmuck, Helma Wennemers.
Copyright © 2004 WILEY-VCH Verlag GmbH & Co. KGaA, Weinheim
ISBN: 3-527-30656-0

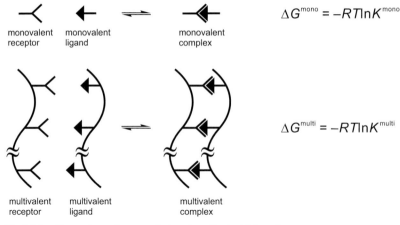

Fig. 2.7.1. Comparison of monovalent and multivalent interactions.

the rate of dissociation (k_{off}) of a multivalent complex by addition of a monovalent ligand [9]. A complete mechanistic description of multivalent binding is, however, difficult because of the complexity of such systems. The formation of a multivalent interaction involves many possible intermediates. Beside intramolecular binding of a multivalent receptor and a multivalent ligand, intermolecular binding may occur, leading to cross-linking and precipitation. The next section therefore focuses on the basic thermodynamics of the simplest multivalent system, the interaction of a bivalent ligand with a bivalent receptor, according to an analysis by Whitesides et al. [8].

The parameter ΔG^{multi} is made up of enthalpic (ΔH^{multi}) and entropic (ΔS^{multi}) components (Eq. 1) which have to be considered separately.

$$\Delta G^{multi} = \Delta H^{multi} - T\Delta S^{multi} \tag{1}$$

The enthalpy of binding (ΔH^{multi}) is, to a first approximation, the sum of the enthalpies of the individual monovalent interactions, i.e. for a bivalent system $\Delta H^{bi} = 2\Delta H^{mono}$ (Figure 2.7.2A, Case 1). This, however, only applies if the bivalent complex is unstrained and the binding events do not interfere with each other. If the binding of the first ligand interferes with binding of the second, the enthalpy of binding is less favorable (less negative) and $\Delta H^{bi} > 2\Delta H^{mono}$ (Figure 2.7.2A, Case 2). Such binding is enthalpically diminished and might occur if the bivalent complex is strained or if the first binding event exerts a negative allosteric effect on the second. Enthalpically enhanced binding is observed if the second binding event is more favorable than the first, because of a positive allosteric effect or because of favorable secondary interactions between the tether and the receptor and $\Delta H^{bi} < 2\Delta H^{mono}$ (Figure 2.7.2A, Case 3). A possible example of enthalpically enhanced binding is the interaction of the pentameric cholera toxin with five GM_1 molecules on cell surfaces [10].

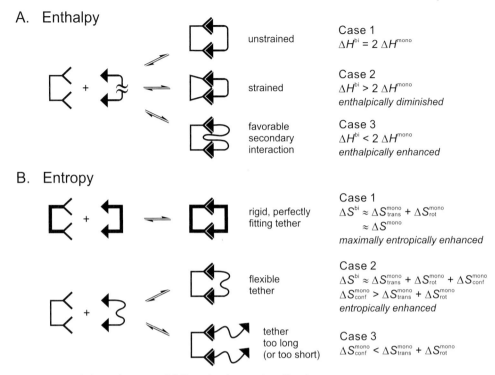

Fig. 2.7.2. Enthalpy and entropy of different binding modes of bivalent interactions.

The entropy of binding (ΔS^{multi}) of a multivalent interaction can be divided into contributions from changes in translational ($\Delta S_{\text{trans}}^{\text{multi}}$), rotational ($\Delta S_{\text{rot}}^{\text{multi}}$), conformational ($\Delta S_{\text{conf}}^{\text{multi}}$), and hydrational ($\Delta S_{\text{H}_2\text{O}}^{\text{multi}}$) entropy. The latter is assumed to be similar in each situation and is therefore ignored in this discussion. Also the weak logarithmic dependence of translational and rotational entropy on the mass and size of different molecules is ignored. For a bivalent interaction several cases can again be distinguished. If the two ligands and the two receptors are connected by rigid, perfectly fitting spacers, $\Delta S_{\text{conf}}^{\text{multi}} = 0$ and the interaction occurs with an entropy equivalent to a single monovalent interaction (Figure 2.7.2B, Case 1). This case of maximum entropic enhancement is, in general, unrealistic, because all tethers are somewhat flexible and $\Delta S_{\text{conf}}^{\text{multi}}$ is almost always unfavorable (less than zero). If this conformational cost is less than the total translational and rotational cost ($\Delta S_{\text{conf}}^{\text{multi}} > \Delta S_{\text{trans}}^{\text{multi}} + \Delta S_{\text{rot}}^{\text{multi}}$), the bivalent association is still entropically enhanced and favored over an intermolecular interaction (Figure 2.7.2B, Case 2). If $\Delta S_{\text{conf}}^{\text{multi}} < \Delta S_{\text{trans}}^{\text{multi}} + \Delta S_{\text{rot}}^{\text{multi}}$, bivalent binding is entropically diminished and a (1+2) association is favored (Figure 2.7.2B, Case 3).

According to this discussion, ΔG^{multi} for a (theoretical) bivalent system with rigid perfectly fitting spacers is given by Eqs (2) and (3).

$$\Delta G^{bi} = 2\Delta H^{mono} - T\Delta S^{mono} \tag{2}$$

$$\Delta G^{bi} = \Delta G^{mono} + \Delta H^{mono} \tag{3}$$

Because ΔS^{mono} is usually less than zero, in such a system ΔG^{bi} is even more favorable than $2\Delta G^{mono}$ and, therefore, $K^{bi} > (K^{mono})^2$. For systems with greater flexibility calculation of the binding enhancement is more complex. A suitable theoretical model has been published recently [11].

The discussion of a bivalent interaction in the general sense is applicable to N-valent interactions and illustrates that the design of the tethers connecting individual binding sites within a multivalent ligand is critical to obtaining high-affinity ligands. Usually, however, the optimization of multivalent ligands is performed by trial and error. In these cases the mechanisms by which binding occurs are less critical. To describe the binding enhancement in such uncharacterized multivalent systems compared with the corresponding monovalent system, Whitesides et al. proposed the empirical parameter β, which is the ratio of the association constants K^{multi} and K^{mono} as defined by Eqs (4) and (5) [8].

$$\beta = K^{multi}/K^{mono} \tag{4}$$

$$\Delta G^{multi} = \Delta G^{mono} - RT \ln \beta \tag{5}$$

Systems with high values of β are useful, irrespective of their mechanism of action. In fact, many multivalent systems with large values of β do not reach the binding enhancement which would be possible in the case of maximum entropic enhancement and enthalpic additivity.

Many examples of multivalent ligands employing numerous scaffolds have been described; these differ in size, carbohydrate content, and flexibility [3, 6–8]. Glycopolymers, for example, are able to cover large areas of cell surfaces and bridge several membrane-located lectins ("statistical" multivalency). Low-valent glycoclusters (miniclusters), on the other hand, bind preferentially to several binding sites of a single (oligomeric) lectin proximate in space and may be tailored to lectins with known 3D structure ("directed" multivalency).

2.7.3
Low-valent Glycoclusters for "Directed Multivalency"

Low-valent glycoclusters have been important for defining the structural features required for high-affinity binding to multivalent receptors. As discussed above, rigid miniclusters are particularly affine ligands – if the carbohydrates are properly oriented, enabling unstrained multidentate binding. Rigid miniclusters are, in principle, moreover, able to differentiate between various multivalent lectins with the same carbohydrate specificity but varying orientation of their binding sites. If, however, the 3D structure of the targeted lectin is unknown, large numbers of potential ligands have to be synthesized and screened to identify the required presentation of the sugar residues.

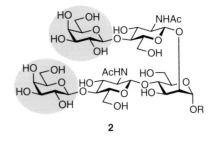
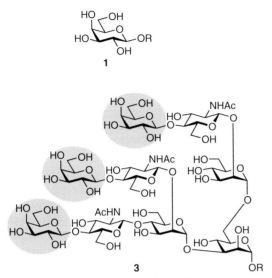

Compound	IC$_{50}$	β
1	0.3–1 mM	1
2	0.3 μM	1000–3333
3	7.4 nM	40540–135135

Fig. 2.7.3. Cluster glycosides synthesized by Lee et al. and their binding to ASGPR on isolated rabbit hepatocytes [7]. Carbohydrate residues involved in multivalent binding are highlighted in gray. The residual carbohydrates are assumed to function as a scaffold.

In pioneering investigations, Lee and coworkers have synthesized many glycoclusters employing different scaffolds during the development of high-affinity ligands for the asialoglycoprotein receptor (ASGPR) on intact hepatocytes and in the soluble form [4, 7, 12]. A remarkable series of compounds with increasing valency is shown in Figure 2.7.3. The IC$_{50}$ values for inhibition of a radiolabeled ligand for ASGPR binding to hepatocytes, which under the conditions used approach dissociation constants, were 0.3–1 mM for β-galactosides **1**, 0.3 μM for divalent oligosaccharide **2**, and 7.4 nM for trivalent oligosaccharide **3**. The phenomenon that the binding affinity of an oligovalent glycocluster increases geometrically with a linear increase in the number of sugar residues was termed the "glycoside cluster effect" by Lee et al.

Two research groups used the known X-ray structures of the heat-labile enterotoxin and the shiga-like toxin from *E. coli* to design inhibitors for these members of the AB$_5$ family of bacterial toxins. The most potent ligand of the heat-labile enterotoxin **7** prepared by Fan et al. had an IC$_{50}$ value of 560 nM, corresponding to an enhancement of $\beta = 10^5$ compared with monomeric galactose (Figure 2.7.4) [13]. Bundle and coworkers designed the decavalent glycocluster **8**, named Starfish, with an IC$_{50}$ value of 0.4 nM being comparable with the estimated affinity of the native ganglioside–pentamer interaction (Figure 2.7.5) [14]. Interestingly, the crystal structure of the toxin–inhibitor complex showed that each Starfish molecule was complexed by two toxin pentamers; this was, however, not planned.

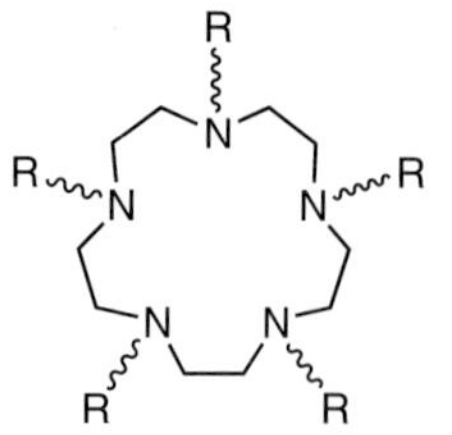

Compound	n	IC$_{50}$ (µM)	β
galactose		58000	1
4	1	242	240
5	2	16	3600
6	3	6	10000
7	4	0.56	104000

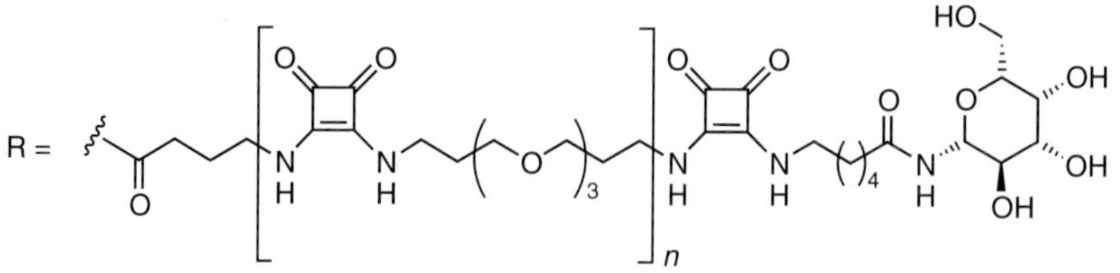

Fig. 2.7.4. Pentameric inhibitors of the heat-labile enterotoxin from *E. Coli* [13].

2.7.4
Spatial Screening of Lectin Ligands

To accelerate the process of finding the required presentation of carbohydrates when no structural information on the receptor is available, we developed a screening procedure for multivalent lectin ligands comprising four steps [15–17]:

1. split-mix synthesis (Box 11) of a library of scaffold molecules containing side-chain amino groups in varying amounts and spatial orientation;
2. attachment of several copies of a carbohydrate ligand to the amino groups;

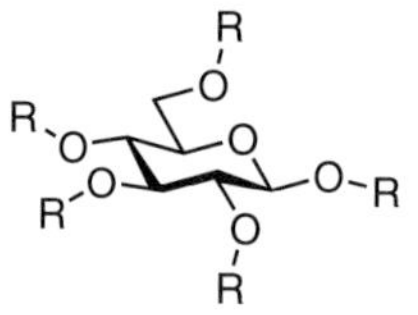

8: STARFISH

IC$_{50}$ = 0.4 nM

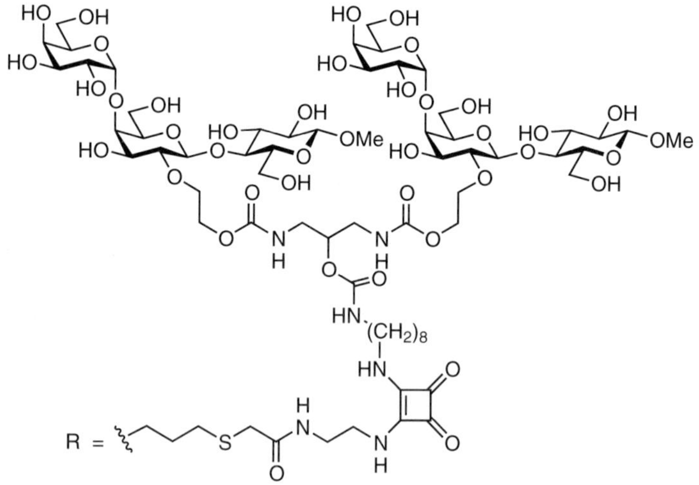

Fig. 2.7.5. Potent decameric inhibitor of the shiga-like toxin from *E. Coli* [14].

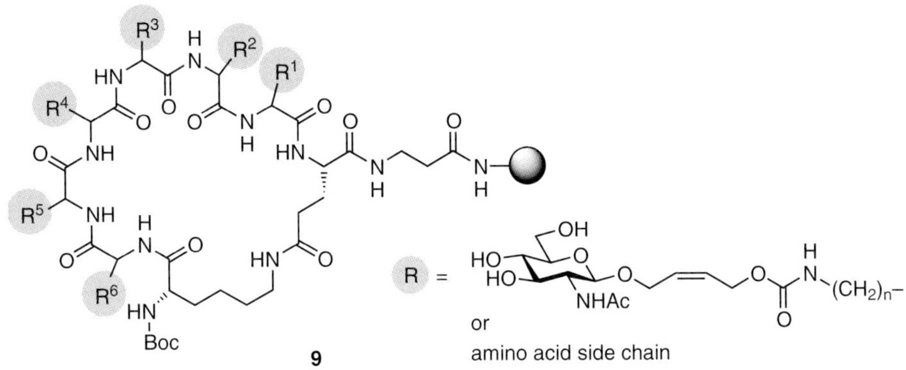

Fig. 2.7.6. Structure of multivalent neoglycopeptides used for spatial screening of lectin ligands [15, 17].

3. on-bead screening of the library for lectin-binding properties; and
4. identification of potent ligands by single-bead analysis.

2.7.4.1
Design and Synthesis of a Library of Cyclic Neoglycopeptides

As scaffolds for the multivalent presentation of carbohydrate ligands we chose cyclic peptides of general type **9** (Figure 2.7.6) [15]. At the combinatorially varied positions indicated by gray circles, D- and L-amino acids without side-chain functionality and D- and L-diamino acids such as lysine, diaminobutyric acid, or diaminopropionic acid are incorporated. The latter represent the points of attachment of the carbohydrates. This library design enables generation of spatial diversity in two dimensions. Positional diversity generates different carbohydrate patterns displayed on the scaffolds. Varying the stereochemistry of the amino acids increases spatial diversity by generating different backbone folds [18].

For attachment of the carbohydrates a new urethane-type linker based on the Aloc protecting group has been developed (Scheme 2.7.1) [15, 19] (for other linkers used in solid-phase synthesis see Chapter 6.1). In contrast with glycosylation reactions employing solid phase-bound peptides [20], the formation of an urethane bond proceeds in virtually quantitative yield. Scheme 2.7.2 shows the convergent solid-phase peptide synthesis of the 19,440 compounds containing library **15** of cyclic neoglycopeptides [17]. N-acetylglucosamine (GlcNAc) residues were attached to side-chain amino groups by employing active carbonate **12**. The carbohydrate content of the library members ranges from 0 (2.6% of all compounds) through 1 (14.5%), 2 (30.3%), 3 (30.9%), 4 (16.6%), 5 (4.5%) and 6 (0.5%).

2.7.4.2
On-bead Screening and Ligand Identification

Library **15** was screened for binding properties to wheat germ agglutinin (WGA) [17]. WGA is a 36 kDa lectin composed of two glycine- and cysteine-rich subunits.

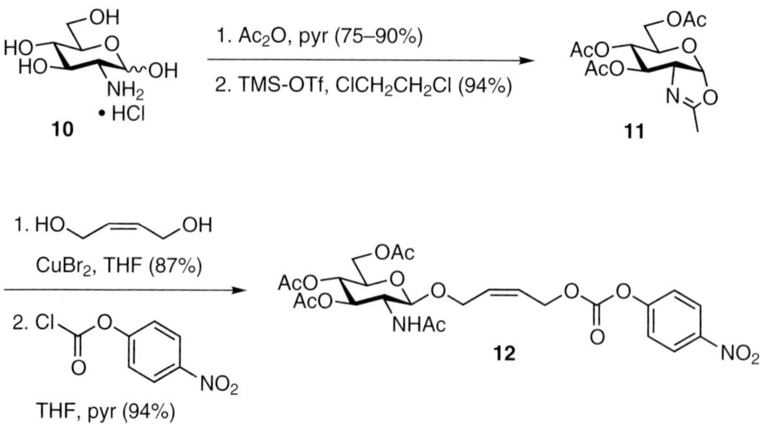

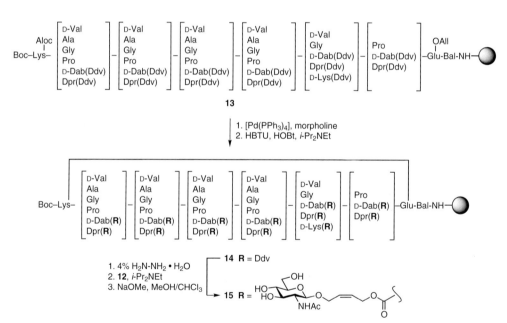

Scheme 2.7.1. Synthesis of the activated carbonate **12** for high-yield modification of side-chain amino groups of diamino acids [15, 19].

Scheme 2.7.2. Convergent solid-phase peptide synthesis of the library **15** of cyclic neoglycopeptides on amino-functionalized TentaGel without employing a linker [17]. (All = allyl, Aloc = allyloxycarbonyl, Bal = β-alanine, D-Dab = D-2,4-diaminobutyric acid, Ddv = 1-(4,4-dimethyl-2,6-dioxocyclohexylidene) isovaleryl, Dpr = L-2,3-diamino propionic acid, HBTU = O-benzotriazol-1-yl-N,N,N′,N′-tetramethyluronium hexafluorophosphate, HOBt = 1-hydroxybenzotriazole.)

2.7.4 Spatial Screening of Lectin Ligands

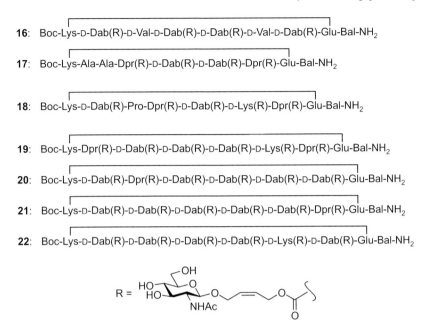

Fig. 2.7.7. Identified WGA ligands from spatial screening [17].

Each subunit contains four carbohydrate binding sites for N-acetylglucosamine and oligomers thereof, thus WGA is a promising candidate for multivalent interaction. Briefly, the resin-bound neoglycopeptides were incubated with biotinylated WGA followed by addition of an anti-biotin alkaline-phosphatase conjugate. Beads with bound lectin were detected by means of an alkaline phosphatase-catalyzed color reaction. When the assay was conducted in the presence of a competing monovalent ligand (GlcNAc), a small part (approx. 0.1%) of the beads stained very darkly. These beads were manually selected under a microscope and treated with [Pd(PPh$_3$)$_4$]/morpholine to remove the carbohydrates. After cleavage of the N-terminal Boc protecting group, "hit" structures were identified by automated single-bead Edman degradation.

Figure 2.7.7 shows the WGA ligands **16–22** identified from the screening process. If the binding assay responds to multivalency, it is expected to find glycoclusters with a large amount of GlcNAc residues. For compounds **16–22** with four to six sugars each, this is clearly the case. Interestingly, not all hexavalent glycopeptides contained in library **15** (0.5%) led to stained beads. Furthermore, beside the hexavalent compounds, one pentavalent and two tetravalent compounds which caused similar staining were identified.

To quantify their binding to WGA in solution, glycopeptides **16**, **18**, and **22** were re-synthesized as single compounds. IC$_{50}$ values for inhibition of the binding of porcine stomach mucin to peroxidase-labeled WGA (Table 2.7.1) were determined by an enzyme-linked lectin assay (ELLA) as described by Zanini and Roy [21]. The IC$_{50}$ values shown in Table 2.7.1 confirm that the binding behavior of **16**, **18**, and

Tab. 2.7.1. Inhibition of binding of porcine stomach mucin to peroxidase-labeled WGA by neoglycopeptides **16**, **18**, and **22**.

Compound	IC$_{50}$ (mM)	β
GlcNAc	83	1
16	0.381	218
18	0.134	619
22	0.146	568

22 in solution is similar to that on the solid phase, so the result of the on-bead screening was not a surface effect, as has been observed by Kahne et al. with an immobilized oligosaccharide library [22].

The fact that **18** and **22** have similar IC$_{50}$ values illustrates that the binding affinity is not only determined by the number of GlcNAc residues but also by the ligand architecture. The β values in Table 2.7.1 are the highest ever reported for oligovalent GlcNAc clusters of that size binding to WGA. With the presented convergent synthetic strategy it is possible to attach any desired carbohydrate ligand to a once-prepared cyclopeptide library, enabling rapid screening of different lectins.

2.7.5
Conclusion

Multivalency provides a means of enhancing the affinity of weak carbohydrate–lectin interactions. High-affinity lectin ligands can function as inhibitors or effectors of carbohydrate-mediated biological processes such as the inflammatory cascade or microbial adhesion to host cells. If the three-dimensional structure of the targeted lectin is known, multivalent carbohydrate ligands may be obtained by rational drug design. Spatial screening of multivalent lectin ligands on the other hand enables rapid identification of ligands with enhanced binding affinity without knowledge of the lectin structure.

References

1 R. A. DWEK, *Chem. Rev.* **1996**, *96*, 683–720; A. VARKI, R. CUMMINGS, J. ESKO, H. FREEZE, G. HART, J. MARTH, eds., *Essentials of Glycobiology*, Cold Spring Harbor Laboratory Press, Cold Spring Harbor, **1999**; R. S. HALTIWANGER, *Curr. Opin. Struct. Biol.* **2002**, *12*, 593–598.

2 J. M. RINI, *Annu. Rev. Biophys. Biomol. Struct.* **1995**, *24*, 551–577; K. DRICKAMER, *Curr. Opin. Struct. Biol.* **1999**, *9*, 585–590; R. LORIS, *Biochim. Biophys. Acta* **2002**, *1572*, 198–208.

3 B. T. HOUSEMAN, M. MRKSICH, *Top. Curr. Chem.* **2002**, *218*, 1–44.

4 Y. C. LEE, R. T. LEE, *Acc. Chem. Res.* **1995**, *28*, 321–327.

5 J. ALPER, *Science* **2001**, *291*, 2338–2343; H. RÜDIGER, H.-C. SIEBERT, D. SOLIS, J. JIMENEZ-BARBERO, A.

Romero, C.-W. von der Lieth, T. Diaz-Maurino, H.-J. Gabius, *Curr. Med. Chem.* **2000**, *7*, 389–416.

6 J. J. Lundquist, E. J. Toone, *Chem. Rev.* **2002**, *102*, 555–578; T. K. Lindhorst, *Top. Curr. Chem.* **2002**, *218*, 201–235; L. L. Kiessling, T. Young, K. H. Mortell, *Glycoscience: Chemistry and Chemical Biology*, Vol. II, B. Fraser-Reid, K. Tatsuta, J. Thiem, eds., Springer, Heidelberg, **2001**, pp. 1817–1861; R. Roy, *Top. Curr. Chem.* **1997**, *187*, 241–274.

7 R. T. Lee, Y. C. Lee, *Glycoconjugate J.* **2001**, *17*, 543–551.

8 M. Mammen, S.-K. Choi, G. M. Whitesides, *Angew. Chem.* **1998**, *110*, 2908–2953.

9 J. Rao, J. Lahiri, L. Isaacs, R. M. Weis, G. M. Whitesides, *Science* **1998**, *280*, 708–711; J. Rao, J. Lahiri, R. M. Weis, G. M. Whitesides, *J. Am. Chem. Soc.* **2000**, *122*, 2698–2710.

10 A. Schon, E. Freire, *Biochemistry* **1989**, *28*, 5019–5024.

11 J. M. Gargano, T. Ngo, J. Y. Kim, D. W. K. Acheson, W. J. Lees, *J. Am. Chem. Soc.* **2001**, *123*, 12909–12910.

12 Y. C. Lee, R. T. Lee, eds., *Neoglycoconjugates. Preparation and Applications*, Academic Press, San Diego, **1994**.

13 E. Fan, Z. Zhang, W. E. Minke, Z. Hou, C. L. M. J. Verlinde, W. G. J. Hol, *J. Am. Chem. Soc.* **2000**, *122*, 2663–2664.

14 P. I. Kitov, J. M. Sadowska, G. Mulvery, G. D. Armstrong, H. Ling, N. S. Pannu, R. J. Read, D. R. Bundle, *Nature* **2000**, *403*, 669–672.

15 V. Wittmann, S. Seeberger, *Angew. Chem.* **2000**, *112*, 4508–4512.

16 V. Wittmann, S. Seeberger, H. Schägger, *Tetrahedron Lett.* **2003**, *44*, in press.

17 V. Wittmann, S. Seeberger, *Angew. Chem.* **2003**, *115*, in press.

18 R. Haubner, D. Finsinger, H. Kessler, *Angew. Chem.* **1997**, *109*, 1440–1456.

19 V. Wittmann, D. Lennartz, *Eur. J. Org. Chem.* **2002**, 1363–1367.

20 A. Schleyer, M. Meldal, M. Renil, H. Paulsen, K. Bock, *Angew. Chem. Int. Ed. Engl.* **1997**, *36*, 1976–1978.

21 D. Zanini, R. Roy, *Bioconjugate Chem.* **1997**, *8*, 187–192.

22 R. Liang, L. Yan, J. Loebach, M. Ge, Y. Uozumi, K. Sekanina, N. Horan, J. Gildersleeve, C. Thompson, A. Smith, K. Biswas, W. C. Still, D. Kahne, *Science* **1996**, *274*, 1520–1522.

Part 3
Studies in Drug Developments

3.1
Building a Bridge Between Chemistry and Biology – Molecular Forceps that Inhibit the Farnesylation of RAS

Hans Peter Nestler

3.1.1
Prolog

To rationalize how two molecules communicate with each other is one of the most intriguing challenges of today's chemical and biochemical science. What do intermolecular interfaces look like? How were the receptor–ligand pairs selected and how do I create novel molecules to bind to a target protein? Today's understanding of molecular recognition does not really allow for the de-novo design of such pairs, but very often involves a lengthy and frustrating process of trials and errors. On the other hand, today's complex and delicately balanced biological pathways have not been designed in an ingenious master plan for the set-up of intermolecular interactions – the driving force throughout eons of evolution has been (and still is) the selection of the fittest entities from a variety of options.

Molecular biologists are accelerating the rate-limiting step of generating these options using highly diverse pools of proteins artificially. They rely on mechanisms that have been present through evolution – random change of present information (mutation), testing the new options for better performance (selection), and amplifying suitable information to feed it into a new cycle of evolution (inheritance) (this approach is exemplified in Chapters 4.3 and 4.4). The mutation step could be vastly accelerated by introducing high error rates for DNA replication (Box 20) or even by synthesizing the genetic material de novo. The diverse libraries are compartmentalized in cells and growing clones of selected cells then enables elucidation of the structures of selected library members. The "yeast-two-hybrid" system and the "phage display" technique are examples of how to study protein–protein and protein–peptide interactions in vivo, and "Selex" technology (Box 23) allows for the in-vitro evolution of short oligonucleotides that interact with small molecules and proteins or have ribozyme-like catalytic activity.

As mentioned above, the generation of new molecular entities by mutation is the rate-limiting step in accessing new biological options. To make efficient use of viable solutions, nature has adopted a modular domain-based approach to building complex proteins. Such an approach is featured in our immune system, that has to be capable of responding to external insults by selecting receptors (antibodies) to

Highlights in Bioorganic Chemistry: Methods and Applications. Edited by Carsten Schmuck, Helma Wennemers.
Copyright © 2004 WILEY-VCH Verlag GmbH & Co. KGaA, Weinheim
ISBN: 3-527-30656-0

hostile intruders (antigens) quickly. After stimulation, the immune system builds millions of different antibodies by combinatorial assembly from a fairly small number of protein modules. Combinatorial chemistry follows the same strategy to build vast collections of molecules from small sets of building blocks. Synthetic molecules cannot, however, be amplified by growing clones and chemists handling libraries faced two major challenges – isolation of active library members and the determination of their molecular structure. In recent years combinatorial chemistry has opened the way to the use of powerful library strategies for purposes where small molecules are involved. The development of new methodologies (such those as described in Chapters 3.6, 6.1, or 6.2 for example) has spurred much excitement and intense research activity has provided results that could only be dreamt of with conventional techniques. Although combinatorial chemistry has found its primary use in drug discovery and optimization, the potential of the technique is widespread and it is used in fields as diverse as material science, catalyst development (this is discussed in further detail in Chapter 5.4), and biochemistry to identify the substrates of novel enzymes [1–3]. Despite all their potential, combinatorial libraries have found only limited use for basic studies of intermolecular interactions [3, 4] (one example is discussed in Chapter 2.3). Thus one question remains – combinatorial libraries and molecular recognition – a match or a mismatch?

In our initial studies we developed libraries of small molecules with branched peptidic structure ("molecular forceps") and showed examples of highly specific interactions between these molecular forceps and peptide ligands in organic solvents [5–7]. We have employed these systems to derive artificial binding pockets that bind proteins of physiological importance at specific functional sites in aqueous media and we have used such molecules to inhibit the processing of RAS proteins at their carboxy termini. The approach we describe here outlines a new paradigm for chemical biology, as we block a biological process not by inhibiting the enzyme connected with this process (as shown in Chapter 3.2 in the case of new asthma therapeutics) but by masking the substrate with a small synthetic molecule. This account also tells the story of the challenges involved in crossing the frontier between chemistry and biology.

3.1.2
RAS – The Good, The Bad and The Ugly

We chose H-RAS as the first target protein for several reasons. RAS proteins are, perhaps, the most intensively studied and best understood signal transduction proteins from the perspective of structure and function [8]. The H-RAS crystal structure has been determined to the level of 1.5 Å, and the domains involved in guanine nucleotide binding and hydrolysis, effector interactions, interactions with regulating proteins, and processing have been extensively characterized. RAS proteins are initially synthesized in the cytoplasm where they undergo a series of post-translational modifications at their carboxy-terminal sequence, the CaaX-box (where "C" is cysteine, "a" an aliphatic amino acid, and "X" either serine or methionine), resulting in their farnesylation, and subsequent cleavage of the terminal tripeptide and carboxy-methylation [9, 10]. The processed proteins become localized on the cell membrane, a step that is essential to their functioning [11–13]. They are required for the transduction of signals from many membrane receptors, including tyrosine kinases and some G-protein linked receptors and the RAF protein kinase from mammalian cells, leading to activation of the MAP kinase cascade [14–16]. RAS proteins play a central role in normal cellular physiology, and point mutations that activate the oncogenic potential of RAS genes are commonly found in human tumors [17, 18].

Oncogenic RAS proteins have been considered to be a logical target for the development of cancer therapeutics. One approach has been to find agents that block RAS processing by inhibiting farnesyl transferase (FTase) [19]. Such agents can, in fact, induce morphological reversion of cells transformed by RAS at concentrations that do not arrest normal cell growth. The results with FTase inhibitors suggest that, in general, the level of an inhibitor needed to block the transforming activity of RAS will not necessarily reduce normal RAS activity below levels sufficient for normal cell growth [20]. Non-farnesylated oncogenic [^{12}Val]-RAS has, moreover, been shown to sequester RAF to the cytosol, thus leading to a cytostatic effect in transformed cells whereas growth and proliferation are slower in untransformed cells [21]. FTase inhibitors (Figure 3.1.1) might, therefore, be generally useful as anti-cancer agents. One potential obstacle to using such agents as drugs is that they can block the farnesylation of other critical proteins in humans, for example the γ-subunit of transducin and nuclear lamins.

On the other hand, a small molecule that binds to the carboxy terminus of a particular RAS might prevent FTase from acting upon this RAS without inhibiting the activity of the transferase on other substrates. Indeed, the precise CaaX-sequences and the two amino acid residues preceding them distinguish the mem-

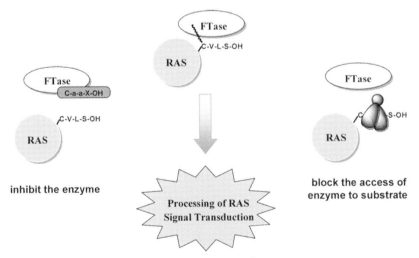

Fig. 3.1.1. Approaches to prevention of farnesylation of RAS.

Tab. 3.1.1. Various CaaX-boxes on proteins.

Protein	Sequence
H-RAS	... CKCVLS
K-RAS A	... KKCIIM
K-RAS B	... TKCVIM
N-RAS	... LPCVVM
R-RAS	... CPCVLL
RAL A	... ERCCIL
RAP 1A	... KSCLLL
RAP 2	... SACNIQ
Lamin B	... RSCAIM
Transducin, γ subunit	... GGCVIS

bers of the RAS family from each other, and these from other substrates for farnesyl and geranyl geranyl transferases. (Table 3.1.1) Thus a ligand that selectively binds to the CaaX-box of one particular RAS protein should be capable of specifically preventing processing of RAS.

3.1.3
Bridging the Gap

From our initial studies we had learned that recognition of peptides and their discrimination with single amino acid resolution does not require complex macromolecular structures but can be achieved with molecules that have two short peptide chains on a scaffold and can be selected from combinatorial libraries [5–7]. Although our initial experiments on molecular forceps were performed in organic solvents, we expected similar interactions should occur in aqueous surroundings and that molecular forceps should be able to recognize epitopes on physiologically important proteins and interfere with their biological functions. As it was not obvious whether a flexible or rigid structure would be advantageous for interactions with a particular target, we used both the rigid chenodeoxycholic acid scaffold with a glycine spacer **L1** or a flexible lysine **L2** as cores to generate an encoded combinatorial library of molecular forceps (MF) on Tentagel beads [22–24]. We also used a trilysine core **L3** to study the impact of increased molecular surface on the strength of the interactions. This library contained approximately 150 000 members.

We used this library to find receptors for the CaaX-box of the H-RAS protein. Because selecting molecular forceps from libraries for H-RAS binding does not ensure these forceps will recognize the carboxy terminus of H-RAS, however, we decided to take a two-pronged screening approach – in addition to screening the library with the protein, we sought to screen with the isolated carboxy-terminal peptide from H-RAS to select molecular forceps that bind this epitope specifically.

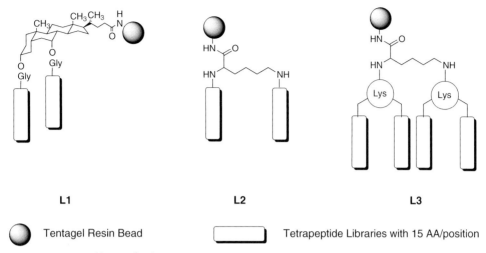

Fig. 3.1.2. Forceps libraries for the H-RAS receptors.

In the protein screen we employed fluorescein-5-maleimide (FM) to conjugate fluorescein to purified glutathione S-methyl transferase (FM-GST) and a GST-H-[^{12}Val]RAS fusion (FM-RAS) protein. First, about 7000 library beads were incubated extensively with FM-GST and the green fluorescent beads were removed to eliminate false positive binders. The remaining unstained beads were then incubated with FM-RAS. Among the approximately 7000 library beads used in the screening we found 65 relatively strong green beads. All positive molecules have either **L2** or **L3** as core template, and no molecules based on the **L1**-scaffold were found (Figure 3.1.2). From the thirty sequences we selected four molecular forceps, "MF4" through "MF7", for further studies. We decided to use one molecular forceps from a bead which failed to bind to FM-GST and FM-RAS ("MF8") as a negative control.

In the epitope approach, we used the His-tagged octapeptide derived from the CaaX-box of H-RAS. In non-exhaustive screening of the library we identified 15 molecules binding to the CaaX-box octapeptide; molecules with four peptide arms **L3** predominated as binders. We have re-synthesized three examples of peptide binders, "MF1" through "MF3". Each of the re-synthesized molecules that had been selected against the peptide binds with micromolar affinities selectively to the H-RAS protein in the presence of other proteins such as ovalbumin or BSA, supporting our assumption that we can screen libraries with isolated peptide epitopes to identify forceps binding to the parent full-length protein.

The re-synthesized molecular forceps (MF1 through MF8, except MF5 which was not soluble) were tested for inhibition of farnesylation by yeast FTase [25]. As expected, MF8, our negative selection, did not inhibit farnesylation of H-RAS protein. MF6 and MF7 from the protein screening also had no effect; neither did MF1 that we had obtained from peptide screening. MF4 had a weak impact on the far-

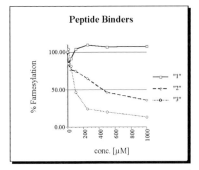 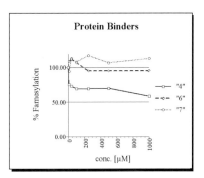

MF1: v-f-E-e-L3
MF2: V-E-F-E-L3
MF3: f-G-F-E-L3
MF4: p-E-K-S-L3
MF5: Q-E-k-p-L3
MF6: S-k-K-E-L3
MF7: f-p-K-s-L3
MF8: F-k-G-F-L3

Fig. 3.1.3. Sequences and inhibitions.

nesylation, with an IC$_{50}$ higher than 1 mM. The two molecular forceps MF2 and MF3 from the screening against the peptide epitope resulted in the best inhibition (Figure 3.1.3).

As discussed above, FTase processes many cellular proteins. We reasoned that if our forceps inhibited the farnesylation of RAS by binding to its carboxy terminus, they might not impede the processing of other proteins by FTase. Satisfyingly, molecular forceps MF3 had different effects on the farnesylation by FTase of four CaaX-containing peptides derived from Lamin B, K-RAS B, and H-RAS, and a chimera of H-RAS and K-RAS sequences ("H-RAS + K-RAS") (Figure 3.1.4a). Whereas MF3 had no effect on the modification of the Lamin B peptide at concentrations up to 1 mM, it had a weak effect on the K-RAS B peptide and a stronger impact on the H-RAS/K-RAS B chimera, correlating with the binding strengths to these substrates and the sequence similarities to the H-RAS sequence.

To confirm further that inhibition of farnesylation of GFP-CaaX-proteins is because of the binding of the molecular forceps to their CaaX-sequence, and not because of inhibition of FTase, we performed in-vitro binding assays with FTase. Whereas MF3 clearly interacts with H-RAS at a concentration of 250 μM, two and a half times the observed IC$_{50}$, we did not observe binding to Ftase (Figure 3.1.4b). Because the forceps bind to RAS and fail to bind FTase, we concluded that MF3 could not act as an enzyme inhibitor, but that its activity is caused by its interaction with the substrate [26].

3.1.4
Epilog

In recent years the use of combinatorial libraries has provided novel ways to generate artificial receptors. Our studies started from three assumptions:

- first, we should be able to find forceps that selectively recognize peptides in aqueous solution;

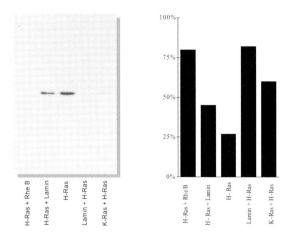

(a) Correlation of BInding Strength of MF3 to Various Proteins with the capacity to prevent the farnesylation of RAS

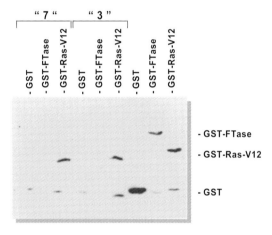

(b) Investigation of the interaction of MF3 and MF7 with FTase and H-RAS

Fig. 3.1.4. Selectivity and binding.

- second, these forceps should recognize the peptide as an epitope embedded in a protein; and
- finally, epitope-binding forceps should be capable of modulating the biological function of the target proteins.

By using a simple receptor design we generated a library of molecular forceps and screened them against the farnesylation site of the RAS protein. Our results prove the validity of our assumptions. While the actual numbers of molecules tested might be too low for final assessment of the most efficient screening strategy, we clearly demonstrated that artificial receptor molecules from libraries can specifically bind the epitopes of proteins and can modulate biochemical processes associated with these proteins.

The findings gave proof of our new paradigm, because we interfered with the enzymatic process not by inhibiting the enzyme but by masking the substrate epitope with a small molecule. In this work we used the carboxy-terminal CaaX-sequence of H-RAS. It is unlikely this is the only peptide epitope that will yield to this type of approach. Many other epitopes are subject to biochemical processing and could be targeted in the same fashion. Examples are glycosylation sites, myristylation motifs, or even proteolysis sites [27]. Many applications can be conceived for artificial receptor molecules that bind such epitopes. Besides the potential therapeutic application of these molecules, they can, with the appropriate modifications, be useful tools for the purification or diagnostic detection of proteins, in Western blots or in the physiological environment.

References

1 M. LEBL, Z. LEBLOVA, *Dynamic Database of References in Molecular Diversity*, World Wide Web: http://www.5z.com/divinfo/
2 D. J. MALY, L. HUANG, J. A. ELLMAN, *ChemBioChem* **2002**, *3*, 16–37.
3 H. P. NESTLER, R. LIU, *Combin. Chem. High Throughput Screen.* **1998**, *1*, 113–126.
4 H. P. NESTLER, *Curr. Org. Chem.* **2000**, *4*, 397–410.
5 R. BOYCE, G. LI, H. P. NESTLER, T. SUENAGA, W. C. STILL, *J. Am. Chem. Soc.* **1994**, *116*, 7955–7956.
6 C. GENNARI, H. P. NESTLER, B. SALOM, W. C. STILL, *Angew. Chem. Int. Ed. Engl.* **1995**, *34*, 1765–1768.
7 H. P. NESTLER, *Mol. Diversity* **1996**, *2*, 35–40.
8 J. C. LACAL, F. MCCORMICK, *The ras Superfamily of GTPases*, CRC Press, Boca Raton, **1993**.
9 S. CLARKE, J. P. VOGEL, R. J. DESCHENES, J. STOCK, *Proc. Natl. Acad. Sci. USA* **1988**, *85*, 4643–4647.
10 L. GUTIERREZ, A. I. MAGEE, C. J. MARSHALL, J. F. HANCOCK, *EMBO J.* **1989**, *8*, 1093–1098.
11 S. POWERS, S. MICHAELIS, D. BROEK, S. SANTA ANNA-A., J. FIELD, I. HERSKOWITZ, M. H. WIGLER, *Cell* **1986**, *47*, 413–422.
12 J. F. HANCOCK, A. I. MAGEE, J. E. CHILDS, C. J. MARSHALL, *Cell* **1989**, *57*, 1167–1177.
13 K. KATO, A. D. COX, M. M. HISAKA, S. M. GRAHAM, J. E. BUSS, C. J. DER, *Proc. Natl. Acad. Sci. USA* **1992**, *89*, 6403–6407.
14 A. MOODIE, B. M. WILLUMSEN, M. J. WEBER, A. WOLFMAN, *Science* **1993**, *260*, 1658–1661.
15 X.-F. ZHANG, J. SETTLEMAN, J. M. KYRIAKIS, E. TAKEUCHI-SUZUKI, S. J. ELLEDGE, M. S. MARSHALL, J. T. BRUDER, U. R. RAPP, J. AVRUCH, *Nature* **1993**, *364*, 308–313.
16 A. B. VOJTEK, S. M. HOLLENBERG, J. A. COOPER, *Cell* **1993**, *74*, 205–214.
17 M. BARBACID, *Ann. Rev. Biochem.* **1987**, *56*, 779–827.
18 D. R. LOWY, B. M. WILLUMSEN, *Ann. Rev. Biochem.* **1993**, *62*, 851–891.
19 S. AYRAL-KALOUSTIAN, E. J. SALASKI, *Current Medicinal Chemistry* **2002**, *9*, 1003–1032.
20 N. E. KOHL, F. R. WILSON, S. D. MOSSER, E. A. GIULIANI, S. J. DESOLMS, M. W. CONNER, N. J. ANTHONY, W. J. HOLTZ, R. P. GOMEZ,

T.-J. Lee, R. L. Smith, S. L. Graham, G. D. Hartman, J. B. Gibbs, A. I. Oliff, *Proc. Natl. Acad. Sci. USA* **1994**, *91*, 9141–9145.

21 M. Miyake, S. Mizutani, H. Koide, Y. Kaziro, *FEBS Lett.* **1996**, *378*, 15–18.

22 K. S. Lam, S. E. Salmon, E. M. Hersh, V. J. Hruby, W. M. Kazmierski, R. J. Knapp, *Nature* **1991**, *354*, 82–86.

23 Á. Furka, F. Sebestyén, M. Asgedom, G. Dibô, *Int. J. Pept. Protein Res.* **1991**, *37*, 487–493.

24 H. P. Nestler, P. A. Bartlett, W. C. Still, *J. Org. Chem.* **1994**, *59*, 4723–4724.

25 F. Tamanoi, H. Mitsuzawa, *Methods Enzymol.* **1995**, *255*, 82–91.

26 D. L.-Y. Dong, R. Liu, R. Sherlock, M. H. Wigler, H. P. Nestler, *Chem. Biol.* **1999**, *6*, 133–141.

27 Z. Zhang, T. Ly, T. Kodadek, *Chem. Biol.* **2001**, *8*, 391–397.

B.11
Split-and-mix Libraries

Hans-Peter Nestler and Helma Wennemers

Split synthesis enables the rapid generation of large numbers of related compounds (a "library") on a solid support. The procedure entails three steps that can be repeated many times. First, the resin is divided into equal portions (step 1). Each resin portion is then subjected separately to different reactions (step 2). The modified resin portions are combined and mixed (step 3) and are ready for the next cycle of the synthesis (Figure B.11.1). Successive cycles of distributing, reacting, and mixing of the beads lead to a combinatorial increase of the diversity of products – for example four cycles with 15 different reactions per cycle lead to $15^4 = 50625$ different com-

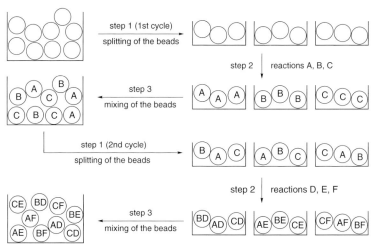

Fig. B.11.1. Basic concept of split synthesis.

pounds. Because each bead reacts with only one set of reagents in each cycle, each bead carries only a single compound ("one-bead-one-compound"); in other words: each compound is located on a different bead [1, 2].

Split-and-mix libraries have found numerous applications in the search for a substrate for a given receptor and vice versa. A typical affinity assay encompasses labeling of the host of interest with, e.g., a dye, a fluorophore, or radioactivity and equilibrating the labeled host with the bead-supported library of potential guests. The labeled host will be concentrated by those beads that carry molecules with affinity for the receptor. These beads are easily identified by visual inspection of the assay under a low-power microscope. Isolation and structural analysis reveals the structure of the active compound [3]. Recently, split-and-mix libraries are also successfully applied in the search for catalysts (see Chapter 5.4).

Because each resin bead carries, typically, only 100 pmol compound the structure of the compound cannot be elucidated by conventional IR, NMR, etc. Although MS is a convenient analytical tool, it fails if the library contains isomers with identical mass. An elegant method for determining the structure of each library compound is molecular encoding. Easily detectable molecular tags are attached to the beads as they proceed through the split-and-mix synthesis, thereby recording the reaction history of each individual bead. After the screening of the library, these molecular tags can be cleaved from each of the selected beads and analyzed to reveal the structures of the library members on these beads. The encoding methods most commonly used employ either microsequenceable oligonucleotides (PCR analysis), oligopeptides (Edman degradation), or small molecules that can be analyzed by gas chromatography, high-performance liquid chromatography, or mass spectrometry [4].

References

1 (a) Á. FURKA, F. SEBESTYÉN, M. ASGEDOM, G. DIBÔ, *Int. J. Pept. Protein Res.* **1991**, *37*, 487–493; (b) K. S. LAM, S. E. SALMON, E. M. HERSH, V. J. HRUBY, W. M. KAZMIERSKI, R. J. KNAPP, *Nature*, **1991**, *354*, 82–84.

2 K. S. LAM, M. LEBL, V. KRCHŇÁK, *Chem. Rev.* **1997**, *97*, 411–448.

3 W. C. STILL, *Acc. Chem. Res.* **1996**, *29*, 155–163.

4 P. SENECI, *J. Receptor Signal Transduct. Res.* **2001**, *21*, 409–445.

3.2
Inhibitors Against Human Mast Cell Tryptase: A Potential Approach to Attack Asthma?

Thomas J. Martin

3.2.1
Introduction

Asthma, one of the most serious of allergic and atopic diseases with great clinical and economic impact, has become an epidemic, affecting 155 million individuals throughout the world [1, 2]. At the current rate of growth it is expected that the number of patients suffering from asthma will increase to 22 million in the United States by the end of this decade. The annual death rate of more than 5000 is expected to increase, with corresponding cost to society [3]. The increasing cost of treating the disease is substantial. In the United States the total estimated cost of asthma in 1987 was $5.8 billion, in 1994 the cost of emergency treatment of asthma reached $10.7 billion; this has continued to rise. Hospitalization is the single greatest cost category, accounting for 48% of the total cost; this is followed by missed days from work and school and restricted activities at work [4, 5].

3.2.1.1
Asthma – Definition

Asthma is a generally chronic inflammatory pulmonary disorder characterized by tracheo-bronchial hyper-reactivity leading to paroxysmal airway narrowing, which may reverse spontaneously or as a result of treatment. Asthma (from a Greek word meaning "panting") is symptomatically accompanied by recurring attacks of wheezing, coughing, and labored breathing; it can be caused by allergies, environmental exposure, physical exertion, chemical irritation, or emotional stress [6].

Asthma is characterized by three key features:

1. airway obstruction (or airway narrowing) that is reversible (but not completely) either spontaneously or with treatment;
2. airway inflammation (mainly lymphocytes, neutrophils and degranulated mast cells); and
3. airway hyper-responsiveness (AHR) to a variety of stimuli (e.g. allergens such as pollen, animal secretions, molds, house dust mites) [1, 7–9].

In the context of atopic asthma AHR equates to an exaggerated bronchoconstrictor response not only to allergens, but also to a range of non-specific stimuli, including agents such as cold air or respiratory viral infections (in children) [2, 7]. The presence of active symptoms from airway obstruction and/or inflammation such as cough, wheeze, shortness of breath or exercise intolerance and chest tightness are typical characteristics of acute asthma. Although therapies for asthma have been developed which are effective in most patients, the disease is persistent and currently incurable [7, 8].

Asthma drugs attack from three different fronts [3, 9–12]:

1. by relaxing the smooth muscles surrounding the airways (theophylline and β_2-agonists which acts as bronchodilators);
2. by reducing the inflammation inside the airways (inhaled corticosteroids); and
3. by breaking the cycle of allergic/inflammatory response (leukotriene modifiers).

In 1987 the xanthine derivative theophylline was the most commonly used medication in the treatment of asthma, followed by β_2-agonists, such as albuterol or terbutaline, and inhaled corticosteroids, for example budesonide or flunisolide. Today, the most frequently used medication is inhaled β_2-agonists and it is expected that these therapy patterns have shifted toward greater use of inhaled corticosteroids [4, 5]. Nevertheless, whether used alone or in combination with other therapies, corticosteroids do not consistently abrogate airway inflammation in patients with asthma; common side effects associated with this type of drug are increased heart rate, nervousness, tremors, nasal irritation, nausea, and headaches [8, 10, 11].

An intensive search for anti-inflammatory drugs that are as effective as inhaled steroids but with fewer side effects has focused on the development of highly potent and selective tryptase inhibitors for treatment of allergic and inflammatory diseases, including asthma, interstitial lung diseases, psoriasis, and rheumatoid arthritis [3, 13, 14]. Human β-tryptase, a trypsin-like serine protease (Box 12) with a tetrameric structure, is the predominant protein of human lung mast cells [15] (together with basophils the most important cells in IgE (immunoglobulin E)-mediated reactions) and is thought to be involved in the pathogenesis of asthma and other allergic and inflammatory disorders [16, 17]. Several closely related isoforms of human tryptase ($\alpha 1$, $\alpha 2$, $\beta 1a$, $\beta 1b$, $\beta 2$, $\beta 3$) exist, but the β-tryptases are the only forms stored preformed in the mast cell granules, comprising 20–50% of the protein content of the mast cells [14, 18]. Stimulation of mast cells with antigens, mediated by IgE or other stimuli, results in the release of active tryptase into the extracellular environment, with other inflammatory mediators [18, 19]. In the airways of asthmatic patients tryptase levels are clearly elevated [20]. In vivo, inhalation of tryptase into the lung results in bronchoconstriction and development of airway hyper-responsiveness through mast cell activation [21]. Tryptase has trypsin-like endopeptidase activity [15], is resistant to all known endogenous proteinase inhibitors, and is enzymatically active only as a heparin-stabilized, non-covalent tetramer. In contrast with the tetrameric tryptase, all other serine proteases (trypsin, thrombin, factor Xa, chymase, elastase, etc.) seem to be expressed as monomeric units [16].

In tryptase the Asp-189 residue, located in the S1 pocket of the active site, can form a salt bridge to at least one basic head group (of an inhibitor molecule); this is required for binding in the active site and thus for the inactivation of the enzyme. The basic head groups range from amines to amidines to guanidines; heterocycles with some basicity should also be able to interact with the Asp-189 residue [14]. Several concepts in the design of potent tryptase inhibitors are currently under investigation [13, 14, 22]:

1. monofunctional inhibitors – molecules which reversibly or irreversibly interact at a single enzyme active site;
2. bifunctional inhibitors – molecules which reversibly interact at two enzyme active sites simultaneously (these inhibitors normally have a higher affinity for tryptase than other inhibitors that bind only one active site, thus conferring the selectivity for tryptase); and
3. tetramer disrupters – molecules which irreversibly inactivate the enzyme by enabling disassociation of the tetramers in its monomeric units.

Figure 3.2.1 shows the crystal structure of natively purified human β2-tryptase with its unique tetrameric architecture [20]. The distances between neighboring active sites are $\sim$20 Å and $\sim$40 Å (A and D, A and B) offering the chance of designing bifunctional inhibitors to span the gap between the active sites (symbolized bifunctional inhibitor which blocks two active sites of the enzyme).

The monofunctional tryptase inhibitor APC-366 (Axys Pharmaceuticals) reduces the acute airway response and histamine release to allergen in a pig model of allergen-induced asthma [19]. APC-366 is also effective in a sheep model of allergen-induced asthma but was only poorly effective in asthma patients (proof-of-principle) [8]. The compound was in clinical development phase II for asthma (inhalative). Although highly selective for tryptase over plasmin and plasma kallikrein, APC-366 was not selective against thrombin and trypsin [13]. Another monofunctional tryptase inhibitor is bis(5-amidino-2-benzimidazol-yl)methane (BABIM) which has been shown to be effective in the sheep. Further development of the compound was, however, discontinued, maybe because of the lack of selectivity over trypsin [13, 16, 17] (Figure 3.2.2).

3.2.2
Chemistry

Because of their high selectivity against other serine proteases, for example thrombin, trypsin, or factor Xa, the development of bifunctional tryptase inhibitors has attracted much attention during recent years. Several classes of effective dibasic tryptase inhibitor have been reported, recently [13, 14, 23]. Herein, we describe syn-

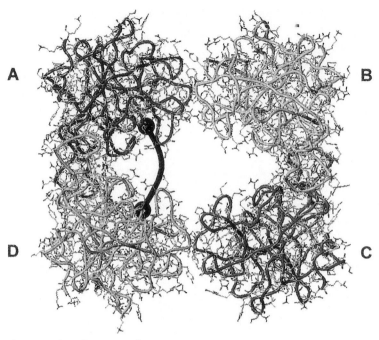

Fig. 3.2.1. Crystal structure of tetrameric β-tryptase, consisting of monomers A, B, C, and D, with a symbolized bifunctional inhibitor interacting with two active sites simultaneously. The distances between neighboring active centers are ~20 Å and ~40 Å (A–D and A–B) [16].

theses of two generations of dibasic tryptase inhibitor, including structure–activity relationships (SAR) and typical selectivity data. Initially, target inhibitors containing pyran rings as central scaffolds (Scheme 3.2.1, **1** and **2**), derived from tri-*O*-acetyl glucal **10**, were created, leading to more polar compounds of type I. With the introduction of aryl diynes as central templates (Scheme 3.2.1, **3** [24], **4** [25], **5** [26],

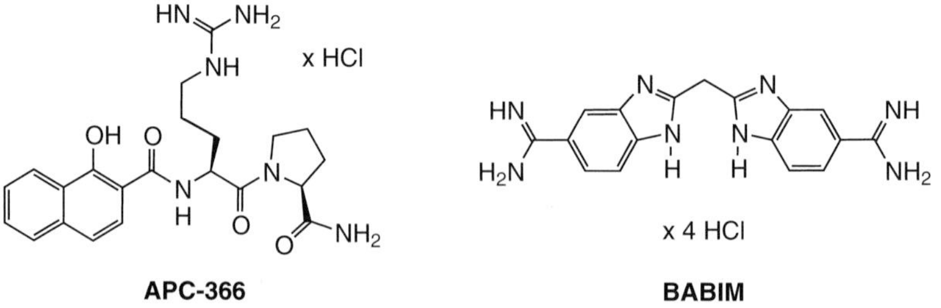

APC-366   **BABIM**

Fig. 3.2.2. Two monofunctional tryptase inhibitors for inhalative application – APC 366 (Axys Pharmaceuticals) and BABIM.

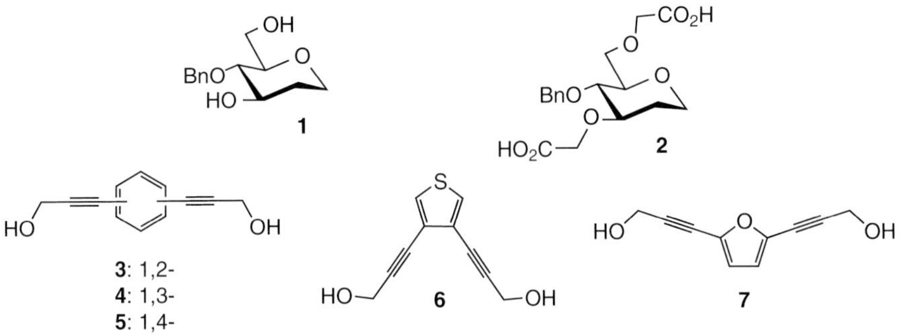

Scheme 3.2.1. Central core templates: pyran derivatives **1**, **2**, type I, and aryl diynes **3–7**, type II.

6, **7**) smaller and more "drug-like" target inhibitors (type II) could be achieved for the first time.

Typically the symmetrical inhibitors were prepared, in accordance with the strategies symbolized in Scheme 3.2.2, by coupling reactions of a central template and a one-sided protected linker-head intermediate (i). Subsequent cleavage of the protecting groups (ii), normally Boc groups, gave the desired target compounds of type I (e.g. **8**). The linker-head intermediates were obtained by condensation of a Boc-protected head group and a one-sided protected linker moiety. Alternatively, a template of type II can also be coupled directly with a partially protected head group (iii) which led after deprotection (iv) to the corresponding small-molecule inhibitors of type II (e.g. **9**).

Basically, the synthesized target molecules were achieved in amounts of 200–500 mg with a standard on purity of ≥95%. The purities of the compounds were routinely checked by analytical HPLC. The core template **1** was prepared by palladium-catalyzed hydrogenation of commercially available glucal **10** followed by deacetylation, according to the literature, which afforded derivative **11** [27] (Scheme 3.2.3). Free hydroxy compound **11** was selectively silylated with two equivalents of TBDMSCl in the presence of imidazole **12**, followed by benzylation and subsequent cleavage of the TBDMS groups **1**.

Scheme 3.2.4 shows the synthetic route employed for the synthesis of linker-head intermediate **14** from half-sided Boc-protected amine **13** [28], via urea formation, mediated by 4-nitrophenyl chloroformate with 1-Cbz-piperazine, followed by hydrogenation of the Cbz group with palladium on carbon.

The synthesis of linker-head intermediate **16**, illustrated in Scheme 3.2.5, began with Boc-tranexamic acid **15** [29] which, on amide coupling with Cbz-piperazine and EDC, again followed by palladium-catalyzed hydrogenation of the Cbz group as in the latter case, furnished derivative **16**.

Synthesis of the target inhibitors **8** and **17** is exemplified in Scheme 3.2.6. Treatment of template **1** with CDI (N,N-carbonyldiimidazole) and further reaction with building blocks **14** and **16**, respectively, followed by subsequent deprotection of the Boc groups, led to the final compounds **8** and **17** as HCl salts.

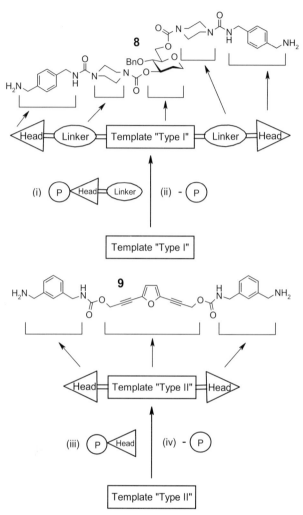

Scheme 3.2.2. (i) Coupling of template type I and linker-head intermediate. (ii) Deprotection, leading to compounds such as **8**. (iii) Coupling of template type II and head group. (iv) Deprotection, leading to compounds such as **9**.

Scheme 3.2.7 shows the formation of dicarboxy template **2**, starting by double alkylation of **1** with methyl bromoacetate and sodium hydride, followed by saponification. Target compound **18** was achieved by amide coupling of **2** with intermediate **16** in the presence of EDC and again cleavage of the Boc groups.

To create target inhibitors with more flexible side chains and reduced molecular weights we attached protected aminopentane tethers to both hydroxy groups of scaffold **1**, via double substitution reaction, leading to the formation of ether

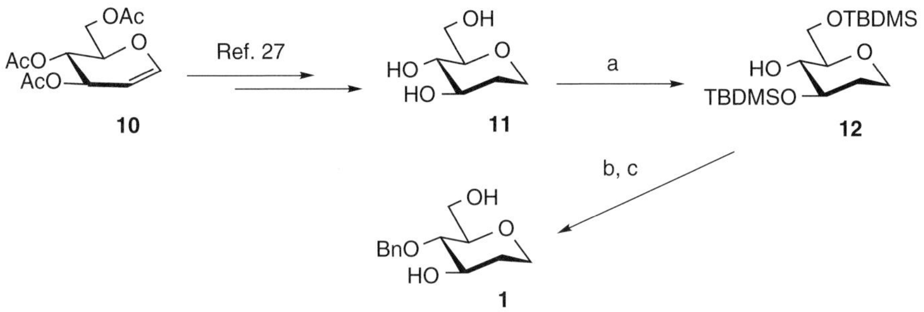

Scheme 3.2.3. (a) TBDMSCl (2 equiv.), Im, DMF, 98%; (b) BnBr, NaH, DMF, 73%; (c) TBAF, THF, 96%.

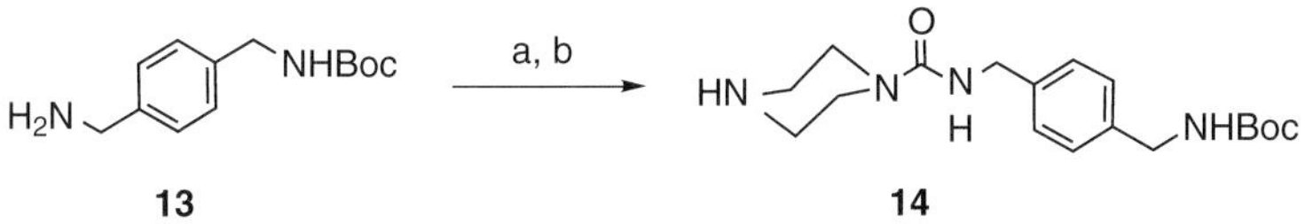

Scheme 3.2.4. (a) 1-Cbz-piperazine, 4-nitrophenyl chloroformate, Et$_3$N, CH$_2$Cl$_2$, 76%, (b) Pd/C, H$_2$, MeOH, 89%.

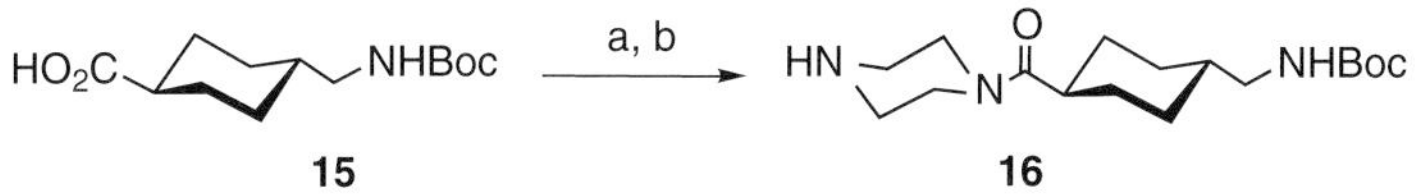

Scheme 3.2.5. (a) 1-Cbz-piperazine, EDC, HOBT, Et$_3$N, CH$_2$Cl$_2$, 94%; (b) Pd/C, H$_2$, MeOH, 91%.

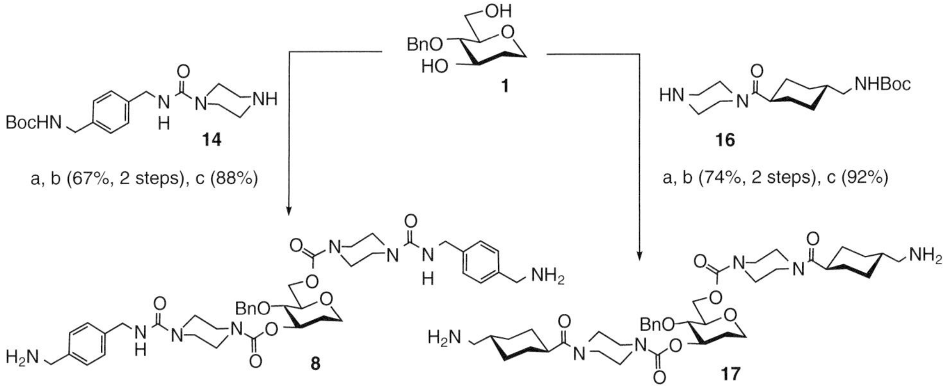

Scheme 3.2.6. (a) CDI, CH$_2$Cl$_2$; (b) **14** or **16**, Et$_3$N, CH$_2$Cl$_2$; (c) HCl/dioxane, dioxane.

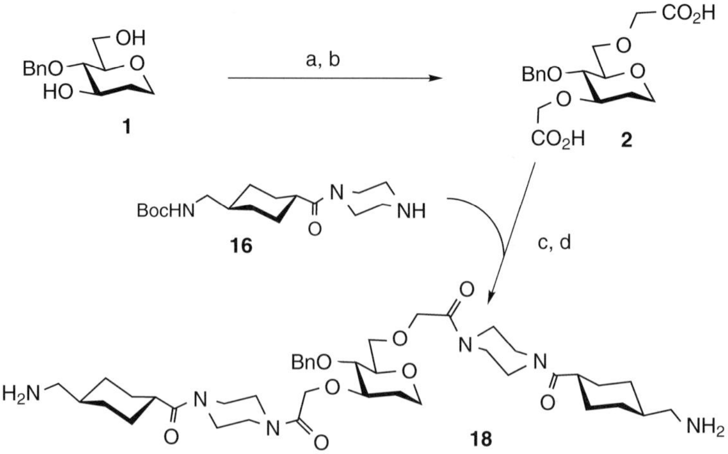

Scheme 3.2.7. (a) BrCH$_2$CO$_2$Me, NaH, dioxane, reflux, 38%; (b) NaOH (2 M), MeOH, 1 h, 92%; (c) **16**, EDC, HOBt, Et$_3$N, CH$_2$Cl$_2$, 85%; (d) HCl/dioxane, dioxane, 98%.

bonds (Scheme 3.2.8). Reaction of a twofold excess of phthalimido-protected aminopentane triflate **19** [30] with dihydroxy template **1**, under deprotonating conditions in the presence sodium hydride, resulted, after removal of the phthalimido groups with hydrazine hydrate, in isolation of the diamino intermediate **20**. Both steps provided nearly quantitative yields. Amide coupling of **20** with Boc-protected cyclohexanecarboxylic acid derivative **15** in the presence of EDC and HOBT, followed by the usual Boc cleavage with HCl in dioxane, afforded inhibitor **21** as the HCl salt.

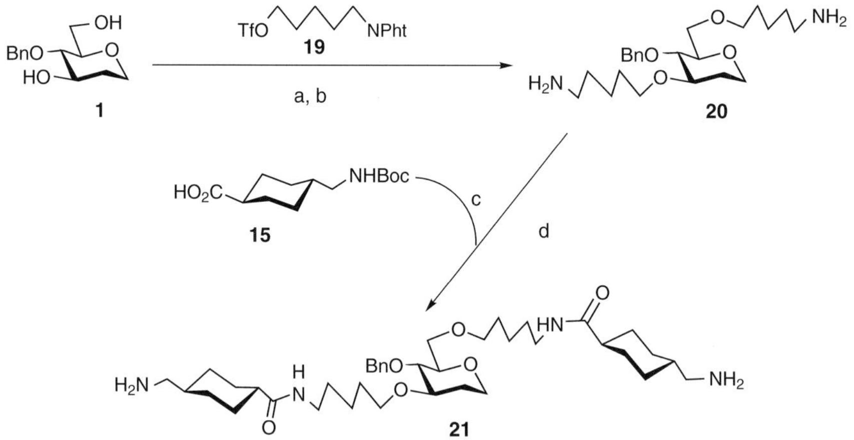

Scheme 3.2.8. (a) NaH, 15-Crown-5, CH$_2$Cl$_2$, quant.; (b) H$_2$N–NH$_2$·xH$_2$O, EtOH, 97%; (c) **15**, EDC, HOBT, Et$_3$N, CH$_2$Cl$_2$, 24%; (d) HCl/dioxane, dioxane, 89%.

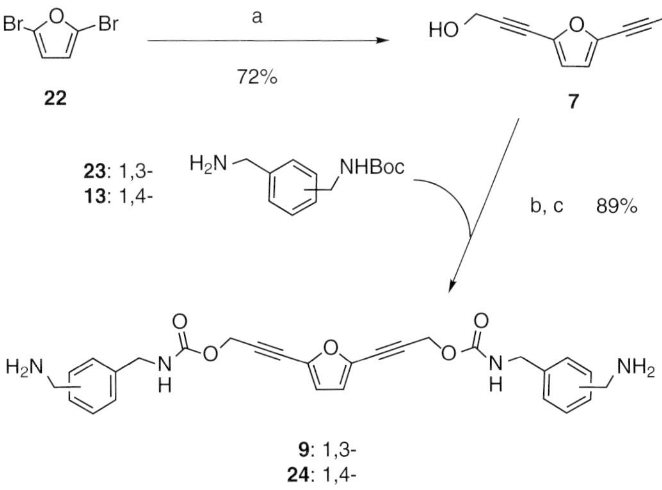

Scheme 3.2.9. (a) HC≡CCH₂OH, Pd[PPh₃]₄, CuBr·SMe₂, Et₃N, 70 °C, 10–12 h; (b) CDI, CH₂Cl₂; (c) **13** or **23**, Et₃N, CH₂Cl₂; (d) HCl/dioxane, dioxane.

An alternative series of dibasic tryptase inhibitors (type II) containing aryl diyne moieties as core template was designed to reach smaller and more "drug-like" target molecules. With the introduction of the aryl diyne motif highly active and selective inhibitors with molecular weights of approximately 500 were synthesized. The range of molecular weights in the previous pyran series (type I) was substantial – anything up to 750. An efficient preparative route in only three steps was employed for synthesis of aryl diyne-containing target inhibitors exemplified by compounds **9** and **24** (Scheme 3.2.9). Starting with a palladium-catalyzed double Heck reaction of dibromofuran **22** and propagyl alcohol, the corresponding core template **7** was obtained. Subsequent coupling reaction of **7**, mediated by CDI, with one-sided Boc-protected amines **13** and **23** [31], and further removal of the Boc groups, provided the required target compounds **9** and **24**. The synthetic approach for preparation of inhibitors **25–29** (Table 3.2.1) was analogous.

3.2.3
Biological Results and Discussion

The in-vitro inhibition activity of the synthesized target compounds against human mast cell tryptase is summarized in Tables 3.2.1 and 3.2.2 Compound **8**, which contains aminomethyl benzyl moieties as head groups, was the most potent of the pyran series (type I), with a K_i value of 1.3 nM. Compared with **8**, the tryptase inhibitors with aminomethylcyclohexyl substituents (**17**, **18**, and **21**), were 100-fold ($K_i = 130$ nM), 60-fold ($K_i = 75$ nM), and approximately 90-fold ($K_i = 120$ nM) less

Tab. 3.2.1. The synthesized type II target inhibitors **9**, and **24–29**, and their K_i values.

Compd	Structure	K_i [µmol]
9		0.0075
24		0.08
25		0.15
26		0.3
27		0.032
28		1.0
29		0.009

effective, whereas the tryptase inhibition activity increased slightly (approx. two-fold) from compound **17** to the lengthened derivative **18**. The more flexible compound **21** had almost the same activity as **17**, but was slightly less potent than compound **18**.

The most potent compound of the aryl diyne series (type II) was furan derivative **9** with a K_i value of 7.5 nM. An approximately tenfold decrease of potency was observed when the 1,3-substitution pattern of the head groups of **9** was replaced by 1,4-substitution to give **24** with a K_i value of 80 nM. Substitution of the 2,5-furan ring in **9** by the 3,4-thiophene ring gave **25**, with a twentyfold loss of tryptase inhibition activity (K_i = 150 nM). Replacement of the furanyl diyne template with the 1,3-substituted phenyl diyne template **29** was well tolerated (K_i = 9 nM), whereas 1,4-substituted head groups resulted in a fourfold loss of activity (**27**, K_i = 32 nM) compared with **29**. Combination of a 1,2-substituted phenyldiyne

Tab. 3.2.2. The synthesized type I target inhibitors **8**, **17**, **18**, and **21**, and their K_i values.

Compd	Structure	K_i [μmol]
8		0.0013
17		0.130
18		0.075
21		0.120

moiety with 1,4-substituted head groups **26** resulted in potency (K_i = 300 nM) ten- and thirtyfold less than that of **7** and **29**, respectively. A substantial loss of potency (110-fold), compared with **29**, was observed when the 1,3-phenyldiyne core and the 1,3-substituted head groups were replaced by the corresponding 1,4-substitution pattern (**28**, K_i = 1 μM).

3.2.4 Conclusion

We have created efficient syntheses of remarkably potent and selective bifunctional tryptase inhibitors, which are also competitive and reversible, containing pyran moieties and hetero and non-hetero aryl diynes as scaffolds. Several modifications at the core templates and the linker moieties are well tolerated without significant loss of inhibition activity. In contrast with previous results published recently [32], it was also apparent from the aryl diyne inhibitors that the distance between the two terminal amino groups can be considerably less than 30 bonds in highly potent target compounds (e.g. **9** and **29** with 26 bonds each). The in-vitro potencies of the compounds were between 1 μM for **26** and 1.3 nM for **15** with high selectivity against other serine proteases (trypsin, thrombin, and factor Xa, respectively) in

Tab. 3.2.3. Selectivity data for inhibitor **9** compared with other serine proteases.

Compd	Tryptase [μmol]	Trypsin [μmol]	Thrombin [μmol]	Factor Xa [μmol]
9	0.0075	32.4	13.5	95.5

the range 1000 to 100 000, exemplified by compound **9** (Table 3.2.3). Compounds **9** and **29** are both valuable biological tool substances, useful for further optimization with regard to bioavailability and metabolism studies. Continuing biological in-vivo screening will be necessary to prove the relevance of tryptase in connection with allergic and inflammatory disorders such as asthma.

Acknowledgment

The author thanks his colleagues T. Bär and W.-R. Ulrich for many helpful discussions and valuable suggestions. Special thanks to C. P. Sommerhoff for performing the in-vitro screening measurements.

References

1 W. Cookson, *Nature* **1999**, *402* (Suppl), B5–B11.
2 W. W. Busse, R. F. Lemanske Jr, *N. Engl. J. Med.* **2001**, *344*(5), 350–362 and references cited therein.
3 *Nat. Biotechnol.* **2000**, *18* (Suppl.), IT10–11 (http://www.biotech.nature.com).
4 D. H. Smith, D. C. Malone, K. A. Lawson, L. J. Okamoto, C. Battista, W. B. Saunders, *Am. J. Respir. Crit. Care. Med.* **1997**, *156*, 787–793.
5 P. J. Gergen, *J. Allergy Clin. Immunol.* **2001**, *107*(5), S445–S448.
6 *Academic Press Dictionary of Science Technology* (http://www.academicpress.com/inscight/02091998/asthma1.htm).
7 P. G. Holt, C. Macaubas, P. A. Stumbles, P. D. Sly, *Nature* **1999**, *402* (Suppl.), B12–B17.
8 P. J. Barnes, *Nature* **1999**, *402* (Suppl.), B31–B38.
9 G. Krihnaswamy, *Hospital Practice*, Aug 15, **2001**, (http://www.hosppract.com/issues/2001/08/krish.htm).
10 D. Spina, *Drugs and the Lung*, C. P. Page, W. J. Metzger, eds., Raven Press, New York, **1994**, p. 101.
11 R. Brattsand, O. Seroos, *Drugs and the Lung*, C. P. Page, W. J. Metzger, eds., Raven Press, New York, **1994**, p. 101.
12 British asthma guidelines coordinating committee. British guidelines on management: 1995 review and position statement. *Thorax* **1997**, *52* (Suppl. 1), 51–52; I. Ziment, UCLA School of Medicine, Los Angeles (http://www.imhotep.net/ziment.html).
13 L. E. Burgess, *Drug News Pers.* **2000**, *13*(3), 147–157 and references cited therein.
14 B. J. Newhouse, *IDrugs* **2002**, *5*(7), 682–688 and references cited therein.
15 M. Castells, *The Internet J. Asthma, Allergy and Immunol.* **1999**, *1*(1) and references cited therein (http://www.ispub.com/journals/ijaai.htm).
16 P. J. B. Pereira, A. Bergner, S. Macedo-Rebeiro, R. Huber, G. Matschiner, H. Fritz, C. P. Sommerhoff, W. R. Bode, *Nature* **1998**, *392*, 306–311; C. P. Sommerhoff, W. Bode, P. J. Pereira, M. T.

Stubbs, J. Stürzebecher, G. P. Piechottka, G. Matschiner, A. Bergner, *Proc. Natl. Acad. Sci. USA* **1999**, *96*(20), 10984–10991.

17 G. H. Caughey, *Am. J. Respir. Cell. Mol. Biol.* **1997**, *16*, 621–628.

18 C. D. Wright, A. M. Havill, S. C. Middleton, M. A. Kashem, D. J. Dripps, W. M. Abraham, D. S. Thomson, L. E. Burgess, *Biochem. Pharm.* **1999**, *58*, 1989–1996.

19 H. Sylvin, M. Stensdotter, I. van der Ploeg, K. Alving, *ATS* **2001**, D31(Poster K33), San Francisco, USA.

20 S. E. Wenzel, A. A. Fowler III, L. B. Schwartz, *Am. Rev. Respir. Dis.* **1988**, *137*, 1002–1008.

21 S. He, M. D. A. Gaca, A. F. Walls, *J. Pharm. Exp. Ther.* **1998**, *286*, 289–297.

22 K. D. Rice, R. D. Tanaka, B. A. Katz, R. P. Numerof, W. R. Moore, *Current Pharmaceutical Design* **1998**, *4*, 381–396.

23 T. J. Martin, *5th Int. Electronic Conf. Synth. Org. Chem. (ECSOC-5)*, 1–30 Sep **2001**, C0011; T. J. Martin, *10. Nachwuchswissenschaftler Symposium Bioorg. Chem.*, 10–12 Sept. **2001**, Cologne.

24 G. Just, R. Singh, *Tetrahedron Lett.* **1987**, *28*, 5981–5984; A. Basak, K. R. Rudra, *Tetrahedron Lett.* **2000**, *41*, 7231–7234.

25 G. J. Bodwell, T. J. Houghton, D. Miller, *Tetrahedron Lett.* **1998**, *39*, 2231–2234.

26 H. Hopf, P. G. Jones, P. Bubenitschek, C. Werner, *Angew. Chem.* **1995**, *107*(21), 2592–2594.

27 M. J. Kelley, M. Roberts, *J. Chem. Soc. Perkin Trans 1* **1991**, 787–797; M. Kanai, Y. Hamashima, M. Shibasaki, *Tetrahedron Lett.* **2000**, *41*(14), 2405–2410.

28 J. F. Callahan, D. Ashton-Shue, H. G. Bryan, W. M. Bryan, G. D. Heckman, et al. *J. Med. Chem.* **1989**, *32*(2), 391–396.

29 C. M. Svahn, F. Merenyi, L. Karlson, L. Widlund, M. Graells, *J. Med. Chem.*, **1986**, *29*(4), 448–453.

30 R. Hirschmann, K. C. Nicolaou, S. Pietranico, E. M. Leahy, J. Salvino, et al. *J. Am. Chem. Soc.* **1993**, *115*(26), 12550–12568.

31 D. W. P. M. Loewik, C. R. Lowe, *Eur. J. Org. Chem.* **2001**, *15*, 2825–2840.

32 N. Schaschke, A. Dominik, G. Matschiner, C. P. Sommerhoff, *Biorg. Med. Chem. Lett.* **2000**, *11*, 2361–2366.

B.12
Serine Proteases

Thomas J. Martin

The serine proteases are a class of proteolytic enzyme (they catalyze the hydrolysis of either ester or peptide bonds in proteins) that require an active site residue for covalent catalysis. The active site residue, the catalytic Ser-195, is particularly activated by hydrogen-bonding interactions with His-57 and Asp-102. Crystal structures show that Ser-195, His-57, and Asp-102 are close in space. Together these three residues, which are located in the substrate binding (S1) pocket, form the famed catalytic triad of the serine proteases. In humans and mammals serine proteases perform many important functions, especially the digestion of dietary protein, in the blood-clotting cascade, and in the complement system:

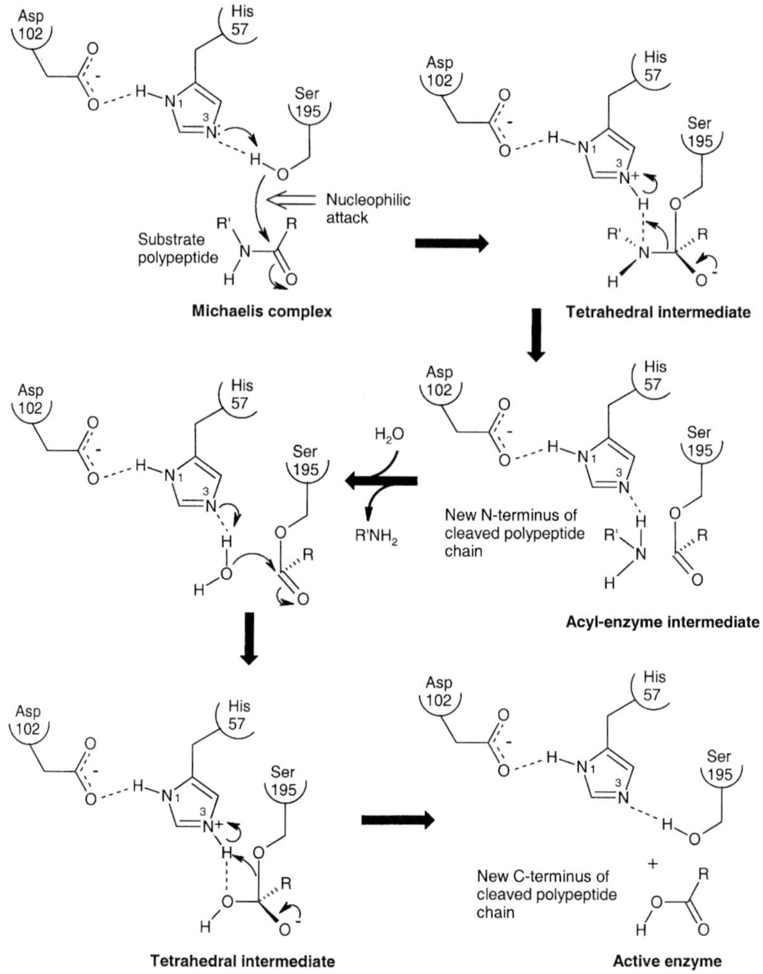

Fig. B.12.1. Schematic diagram of the mechanism of reaction of serine proteases.

- Trypsin, chymotrypsin, and elastase are three of the most important protein-digesting enzymes secreted by the pancreas. Despite their similarities they have different substrate specificity, that is, they cleave different peptide bonds during protein digestion.
- Several activated clotting factors are serine proteases, including thrombin, plasmin, factor 10 (X), and factor 11 (XI).
- Several proteins involved in the complement cascade are serine proteases, for example C1r and C1s, and the C3 convertases C4b, 2a and C3b, Bb.

The mechanism of reaction of the serine proteases is depicted schematically in Figure B.12.1; it is divided in an acylation step and a deacylation step.

Acylation Step

The nucleophilic OH group of Ser 195 attacks the carbonyl carbon of the scissile bond (Michaelis complex), forming the tetrahedral transition state. Here, the C=O bond becomes a single bond, leaving a negative charge on the O atom (an *oxyanion*) while the fourth valence of the carbon atom is occupied by a bond with the serine O-gamma.

The proton donated from the OH group of Ser 195 to His 57 is then donated to the N atom of the scissile bond, cleaving the C–N peptide bond (or the C–O ester bond) to produce the amine and the acyl–enzyme intermediate. The amine is that part of the substrate which follows the scissile bond in the sequence; the acyl–enzyme intermediate is the remaining fragment covalently bound to Ser 195.

Deacylation Step

The acylation step essentially occurs in reverse, but involving a water molecule instead of the amine (which diffuses away). The water loses a proton to His 57 and the resulting OH nucleophile attacks the acyl–enzyme intermediate forming another tetrahedral transition state. The proton is then donated to the O-atom of Ser 195, releasing the acid product.

General Reference

D. Voet, J. Voet, *Biochemistry*, 2nd edn., J. Wiley and Sons, **1995**.

3.3
Preparation of Novel Steroids by Microbiological and Combinatorial Chemistry

Christoph Huwe, Hermann Künzer and Ludwig Zorn

3.3.1
Introduction

Steroids are important pharmacologically active scaffolds [1]. The relevance of microbiological transformations for the production of steroidal drugs and hormones was first recognized when the 11α-hydroxylation of progesterone by *Rhizopus arrhizus* and *R. nigricans* was developed by Murray and Petterson [2]; this resulted in a tremendous improvement of the synthesis of corticoids. Starting from deoxycholic acid 31 steps had been needed for chemical synthesis of cortisone, whereas the microbiological procedure starting from diosgenin shortened the synthesis to 13 steps [3]. Since then, microbiological transformations have found broad application in the synthesis of pharmaceuticals and natural products [4].

In microbiological chemistry whole cells are used instead of isolated enzymes. This approach frequently proves to be beneficial, because isolation and purification of enzymes [5] are tedious and often uneconomical procedures which can also result in significant loss of enzyme activity. The main applications of microbiological reactions, which can be performed by use of growing, resting [6], or immobilized cells [7], include hydroxylation of non-activated C–H bonds [8], oxidation reactions [9], dehydrogenation of saturated to α,β-unsaturated ketones [10], reduction of keto groups [11], and partial degradation of complex molecules [12] (two such examples are discussed in Chapters 6.3 and 6.4). In these examples, the oxidizing or reducing enzyme is regenerated by other enzymes in the cell. Although isolated enzymes are often used for hydrolyzing reactions, because no coenzyme needs to be regenerated [5, 13], it is also possible to hydrolyze esters selectively using whole cells [14]. Similar to enzymatic reactions, microbiological transformations can be rather substrate-specific, thus limiting the flexibility of the approach. Similar to chemical catalytic reactions, moreover, microbiological transformations often require optimization of the reaction conditions before high yields and high enantiomeric purities can be obtained. Once a microorganism performing the desired transformation is found, there are several ways of improving the yield of a reaction. For example, addition of water-soluble organic solvents [15], cyclodextrins, surfac-

Highlights in Bioorganic Chemistry: Methods and Applications. Edited by Carsten Schmuck, Helma Wennemers.
Copyright © 2004 WILEY-VCH Verlag GmbH & Co. KGaA, Weinheim
ISBN: 3-527-30656-0

tants, or organic carbon or nitrogen sources [16], and optimization of pH [17] and incubation conditions often lead to higher yields. Transformations can also be improved by mutation and selection [16].

Combinatorial chemistry [18], on the other hand, is a well established means of synthesizing relatively large numbers of analogs relatively quickly (see also Chapters 2.3, 3.1 or 5.4 for examples). Use of laboratory automation [19], in conjunction with the advanced chemoinformatics systems needed to handle the enormous amounts of related data [20], enables full exploitation of this principle. These technologies have recently been widely applied to drug discovery programs [21] and to chemical development and material sciences [22]. In drug discovery early chemistry-driven approaches typically directed at large numbers of crude products have been widely supplemented or replaced by more focussed, structure-based and property-biased strategies [23]. In these approaches only a subset of a virtual library most likely to show the desired properties is synthesized, preferably selected by means of virtual screening, pharmacokinetic property calculations, and medicinal chemistry know-how, thus leading to high-quality compounds.

Combination of microbiological chemistry, often yielding scaffolds not easily obtained by purely chemical means, and combinatorial chemistry, enabling rapid and efficient synthesis of analogs, provides a valuable tool for generation of novel test compounds. As an example [24] we describe here the application of our lipoic acid-derived thioketal linker [25] to the solid-phase synthesis of Δ^4-3-keto steroidal ureas from β-sitosterol.

3.3.2
Results

The phytosterol β-sitosterol **1** is readily available from soy oil and is therefore, next to diosgenin, the most important re-growing starting material for the partial synthesis of novel steroids [26]. This compound was subjected to microbiological oxidation (*Mycobacterium sp.*) yielding the $\Delta^{1,4}$-3-keto intermediate **2**, followed by selective hydrogenation (Wilkinson's catalyst) to give the Δ^4-3-keto compound **3**, which was chosen as the scaffold for the model library described herein. Introduction of our thioketal linker, protection of the hydroxyl group of **4** as a TBS ether, and selective hydrolysis of the methyl ester **5** gave carboxylic acid intermediate **6**, ready to be attached to the solid support (Scheme 3.3.1).

Immobilization of **6** on a commercially available aminomethylated polystyrene resin [27] was readily tracked by means of solid-phase infrared spectroscopy (IR $\sim$ 1660 cm^{-1}) and yielded material with a resin loading of $\sim$0.75 mmol g^{-1} [28]. Removal of the silyl protecting group of **7** (IR $\sim$ 3410 cm^{-1}) and activation of the resulting hydroxyl group of **8** as a mesylate (IR $\sim$ 1175 cm^{-1}) gave intermediate **9**, ready for diversification (Scheme 3.3.2).

This material **9** was then placed in the 40 reactors of an automated synthesizer [29] and, in the sense of a matrix synthesis, sequentially treated with a set of five

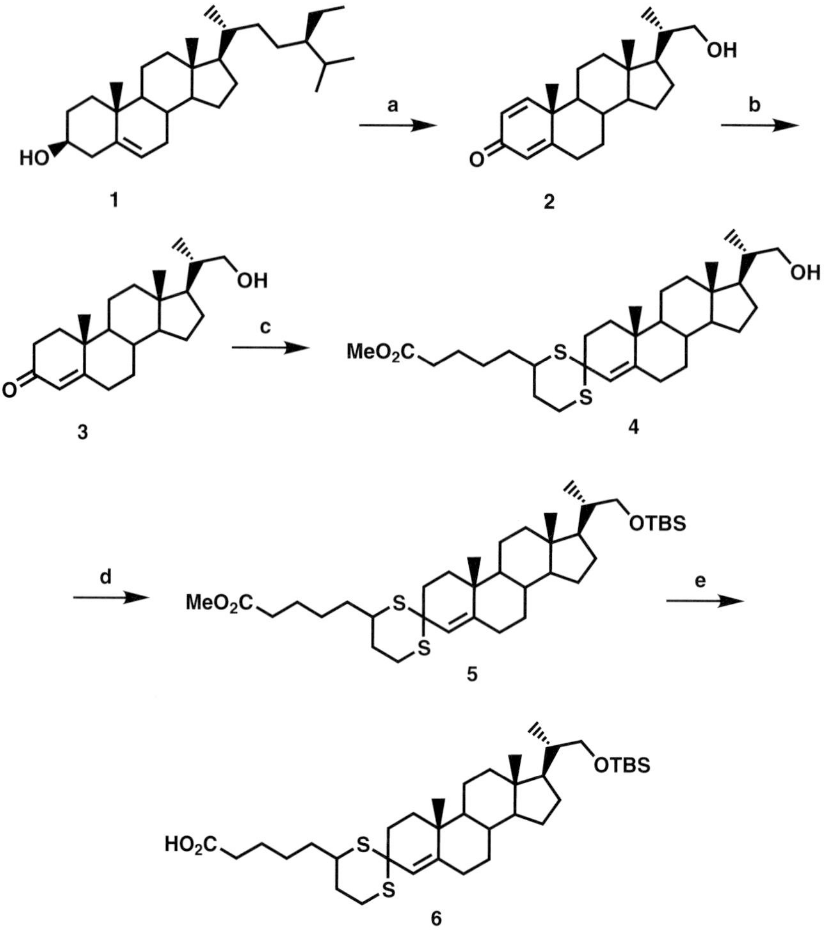

Scheme 3.3.1. Scaffold synthesis from re-growing starting material. Reagents and conditions: (a) *Mycobacterium sp.*; (b) H$_2$, (PPh$_3$)$_3$RhCl; (c) 1. (±)-α-lipoic acid methyl ester, NaBH$_4$, MeOH, r.t., 2 h, then aqueous work-up, CH$_2$Cl$_2$; 2. add **3**, BF$_3$·Et$_2$O, r.t., 5 days (90%); (d) TBSCl, imidazole, THF, r.t., 2 h (∼quant.); (e) NaOH, H$_2$O, THF, r.t., 16 h (76%).

primary amines (R^{1}NH$_2$), to give immobilized secondary amines **10**, and a set of eight isocyanates (R^{2}NCO), to generate a 5 × 8 matrix of 40 immobilized steroidal ureas **11**. Cleavage of the products from the solid support [25], then aqueous work-up, finally gave the desired compounds **12** (Scheme 3.3.3) [30].

In summary, microbiological chemistry provides a means of rapid and often selective modification of a complex structure in one step, thus furnishing compounds difficult to obtain by purely chemical means. While not being as broadly

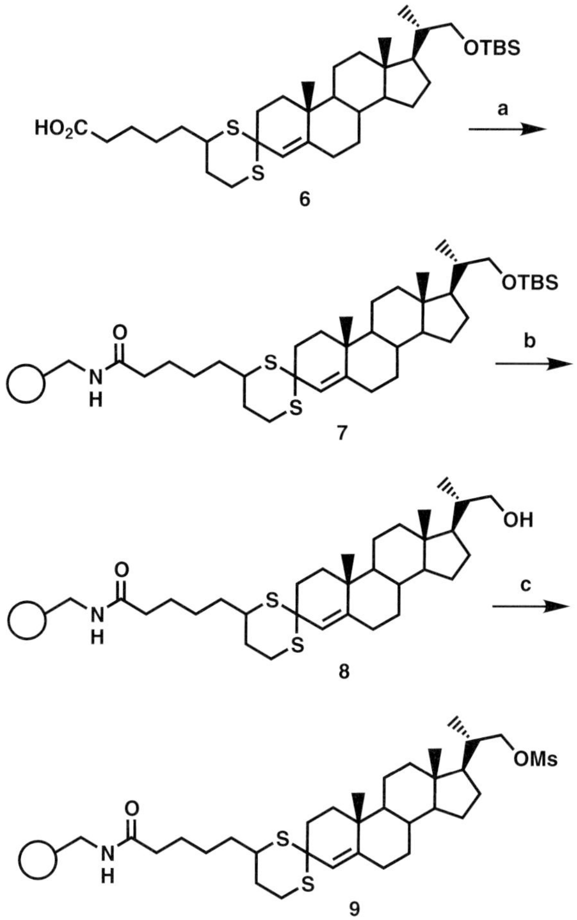

Scheme 3.3.2. Immobilization and preparation for derivatization. Reagents and conditions: (a) aminomethylpolystyrene, DIC, HOBT, DMF, r.t., o/n; (b) Bu$_4$N$^+$F$^-$·H$_2$O, DMF, r.t., o/n; (c) MsCl, pyridine, CH$_2$Cl$_2$, r.t., o/n.

applicable as chemical synthesis, microbiological chemistry often proves to be very successful if a microorganism is found that performs a transformation of interest. Further refinement by optimization of incubation conditions and nutrition media, and mutation and selection approaches, frequently leads to highly effective syntheses. Second generation combinatorial chemistry, on the other hand, enables rapid synthesis of large numbers of high-quality compounds. Combination of microbiological and combinatorial chemistry is, therefore, a powerful strategy for generation of novel, interesting test compounds.

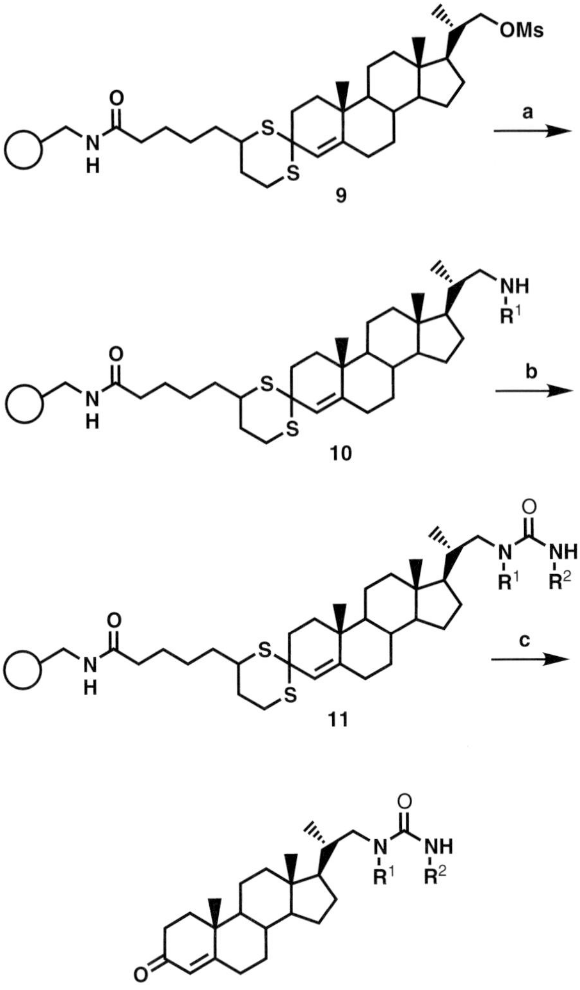

Scheme 3.3.3. Combinatorial diversification by automated solid-phase synthesis. Reagents and conditions: (a) R¹NH₂, DMF, 80 °C, 10 h (set of five amines); (b) R²NCO, DMF, r.t., 4 h (set of eight isocyanates); (c) PhI(Tfa)₂, CH₂Cl₂, EtOH, H₂O, r.t., 30 min. (∼10% overall from **7**).

References

1 FIESER, L. F.; FIESER, M. *Steroide*, Verlag Chemie, Weinheim, **1961**.
2 (a) CHARNEY, W.; HERZOG, W. *Microbial Transformations of Steroids*, Academic Press, New York, **1967**, p. 5; (b) SMITH, K. E.; AHMED, F.; ANTONIOU, T. *Biochem. Soc. Trans.* **1993**, *21*, 1077.

3 Fieser, L. F.; Fieser, M. *Steroide*, Verlag Chemie, Weinheim, **1961**, p. 705.
4 (a) Rufer, C.; Kosmol, H.; Schröder, E.; Kieslich, K.; Gibian, H. *Liebigs Ann. Chem.* **1967**, *702*, 141; (b) Hofmeister, H.; Annen, K.; Laurent, H.; Petzoldt, K.; Wiechert, R. *Drug Res.* **1986**, *36*, 781; (c) Roberts, S. M. *Biocatalysts for Fine Chemicals Synthesis*, John Wiley and Sons, Chichester, **1999**.
5 Drauz, K.; Waldmann, H. *Enzyme Catalysis in Organic Synthesis*, VCH, Weinheim, **1995**, 45.
6 Drauz, K.; Waldmann, H. *Enzyme Catalysis in Organic Synthesis*, VCH, Weinheim, **1995**, 157.
7 Drauz, K.; Waldmann, H. *Enzyme Catalysis in Organic Synthesis*, VCH, Weinheim, **1995**, 481.
8 Petzoldt, K. *Methods of Organic Chemistry (Houben–Weyl)*, Vol. E21e, Georg Thieme, Stuttgart, **1995**, p. 4857.
9 Li, Z.; van Beilen, J. B.; Deutz, W. A.; Schmid, A.; de Raadt, A.; Griengl, H.; Witholt, B. *Curr. Opin. Chem. Biol.* **2002**, *6*, 136.
10 (a) Charney, W.; Herzog, H. L. *Microbial Transformations of Steroids*, Academic Press, New York, **1967**, p. 5; (b) Pinheiro, H. M.; Cabral, J. M. S. *Enzyme Microb. Technol.* **1992**, *14*, 619.
11 Gottwald, M. *Methods of Organic Chemistry (Houben–Weyl)*, Vol. E21d, Georg Thieme, Stuttgart, **1995**, p. 4143.
12 Lo, C.-K.; Pan, C.-P.; Liu, W.-H. *J. Ind. Microbiol. Biotechnol.* **2002**, *28*, 280.
13 Wong, C.-H.; Whitesides, G. M. *Enzymes in Synthetic Organic Chemistry*, Elsevier, Amsterdam, **1994**.
14 Petzoldt, K.; Dahl, H.; Skuballa, W.; Gottwald, M. *Liebigs Ann. Chem.* **1990**, 1087.
15 Bortolini, O.; Medici, A.; Poli, S. *Steroids*, **1997**, *62*, 564.
16 Mahato, S. B.; Majumdar, I. *Phytochemistry* **1993**, *34*, 883.
17 Lo, C.-K.; Wu, K.-L.; Liu, W.-H. *Food Sci. Agric. Chem.* **2001**, *3*, 30.
18 (a) Lazo, J. S.; Wipf, P. *J. Pharm. Exp. Ther.* **2000**, *293*, 705; (b) Appell, K.; Baldwin, J. J.; Egan, W. J. *Sep. Sci. Technol.* **2001**, *3*, 23.
19 (a) Coates, W. J.; Hunter, D. J.; MacLachlan, W. S. *Drug Disc. Today* **2000**, *5*, 521; (b) Zechel, C. *Methods Princip. Med. Chem.* **2000**, *9*, 243.
20 (a) Gedeck, P.; Willett, P. *Curr. Opin. Chem. Biol.* **2001**, *5*, 389; (b) Manly, C. J.; Louise-May, S.; Hammer, J. D. *Drug Disc. Today* **2001**, *6*, 1101.
21 Seneci, P.; Miertus, S. *Mol. Div.* **2000**, *5*, 75.
22 Hagemeyer, A.; Jandeleit, B.; Liu, Y.; Poojary, D. M.; Turner, H. W.; Volpe, A. F.; Weinberg, H. W. *Appl. Catal. A: General* **2001**, *221*, 23.
23 (a) Böhm, H.-J.; Stahl, M. *Curr. Opin. Chem. Biol.* **2000**, *4*, 283; (b) Matter, H.; Baringhaus, K.-H.; Naumann, T.; Klabunde, T.; Pirard, B. *Comb. Chem. High Throughput Screen.* **2001**, *4*, 453.
24 Huwe, C. M.; Künzer, H.; Schlicht, C. *Fifth Int. Electronic Conf. on Synthetic Organic Chemistry*, **2001**, Poster C0020.
25 Huwe, C. M.; Künzer, H. *Tetrahedron Lett.* **1999**, *40*, 683.
26 *Römpp Lexikon Chemie*, Georg Thieme, Stuttgart, **1998**.
27 Aminomethylated polystyrene resin was obtained from Rapp Polymere GmbH, Germany.
28 The polymer loading was determined gravimetrically, i.e., calculated from the observed change of resin mass.
29 A Chemspeed ASW2000 synthesizer from Chemspeed, Switzerland, was used.
30 Yields were determined gravimetrically, products were characterized by HPLC–MS.
31 The authors wish to thank Ms. C. Schlicht for skilled technical assistance.

3.4
Enantiomeric Nucleic Acids – Spiegelmers

Sven Klussmann

Abstract

Spiegelmers are enantiomeric nucleic acids – either RNA or DNA – with high biological stability and high affinity and specificity for a given target molecule. The target binding properties can be designed by an in-vitro selection process called SELEX (systematic evolution of ligands by exponential enrichment; Box 23) starting from a random library of 10^{15} different molecules. Each molecule represents a different epitope, which is determined by the primary structure of the oligonucleotide.

Spiegelmers are a new class of substance for a wide field of applications, especially drug design. Their functional properties can be compared with the action of monoclonal antibodies. The production of spiegelmers is as simple as the synthesis of other oligonucleotides because standard chemical methods, for example common phosphoramidite chemistry, can be employed.

3.4.1
Towards Nucleic Acid Shape Libraries

After the discovery of ribozymes in the early 80s [1, 2] (Box 22) it became clear that nucleic acid molecules, especially RNA molecules, have more potential than being solely the passive conveyor of genetic information or the "glue" between ribosomal proteins. In several studies it was shown that even small RNA oligonucleotides, such as the so-called hammerhead ribozyme, can function as real enzymes [4]. These findings could be explained only by assuming that the complexity of the structure of these catalytically active molecules must be comparable with that of real protein enzymes.

Indeed, the discovery of catalytic RNA revived the discussion about an "RNA world" and the question whether polynucleotides or polypeptides were first to evolve [4]. What makes nucleic-acid-like molecules so interesting and unique is their inherent capacity to store and to replicate information. These properties, in combination with their structural diversity, makes nucleic acid-like molecules the

Highlights in Bioorganic Chemistry: Methods and Applications. Edited by Carsten Schmuck, Helma Wennemers
Copyright © 2004 WILEY-VCH Verlag GmbH & Co. KGaA, Weinheim
ISBN: 3-527-30656-0

ideal substance class for creating huge combinatorial shape libraries (the structural diversity of RNA is discussed in Chapter 1.1, for example). Thus, geno- and phenotype are combined in each single compound of a nucleic acid library, so that replication enables Darwinian evolution to occur at the molecular level [5].

Combinatorial libraries of nucleic acids can be synthesized very easily by use of standard chemistry methods. By having four different nucleotides and randomizing each position in a 25-mer oligonucleotide a library of 1.1×10^{15} different sequences can be generated [6]. A randomization over N nucleotides with the four nucleotides leads to a sequence space of 4^N. By assuming that the different sequences in these libraries determine different three dimensional structures a shape library of more than 10^{15} molecules can be created. Although longer random regions than 25 nucleotide are possible and in fact widely used, the theoretical sequence space of 4^N is limited to approx. 10^{15} different sequences because of experimental restrictions.

3.4.2
In-vitro Selection or SELEX Technology

Systematic evolution of ligands by exponential enrichment (SELEX) is a process used to screen libraries of single-stranded oligonucleotides for desired activity, e.g. binding to a target molecule or enzymatic activity [7] (Box 23). In a first step, a library of 10^{14} or 10^{15} different sequences is synthesized by standard solid-phase chemistry. The library can be either RNA or DNA. The random region is often between two fixed regions which serve as primer binding sites for the amplification step. An initial library of DNA can easily be transcribed into an RNA library with an RNA polymerase.

If the task is to screen for binding activity, the library is brought into contact with the target molecule. In an appropriate partitioning step, that can be either a kind of filtration or affinity chromatography, a very tiny fraction of the total library that binds to the target molecule is isolated. The isolated molecules are amplified by polymerase chain reaction (PCR); prior PCR RNA sequences are reverse-transcribed into cDNA. After the amplification step the slightly enriched library of the first selection cycle is then brought into contact again with the target molecule and after several partitioning–selection cycles and amplification the enriched sequences are cloned into an appropriate vector system and sequenced. Isolated sequences that bind to the target molecule are called aptamers [8] (lat. *aptus*, suited).

By use of the SELEX procedure aptamers have been isolated against a huge variety of targets. Although most aptamers reported so far are directed against proteins (excellent reviews are given by Gold et al. [9] and James [10]) aptamers can also be generated against other substance classes. Aptamers have been reported to bind also small molecules such as organic dyes, nucleosides, amino acids, and antibiotics [11].

Because it is necessary to employ enzymes during in-vitro selection, natural nucleotides must be used in the process to isolate aptamers. Consequently, the resulting first-generation aptamers are highly susceptible to attack from abundant degrading enzymes in the general environment and are, in particular, extremely unstable in biological media. Any prospect of pharmaceutical application would therefore require a complex procedure to exchange most standard nucleotides by modified nucleotides that are more stable against metabolizing activity (resulting in second-generation aptamers) [12, 13]. However:

- not all nucleotides in a selected oligonucleotide can be exchanged without losing the required three-dimensional structure of the aptamer;
- these post-selection modifications are time-consuming and expensive; and
- the degradation products of altered oligonucleotides might be toxic

Addressing the problem of the biological instability of aptamers by simultaneously taking advantage of the properties of oligonucleotides and the SELEX process, a mirror approach and the so-called spiegelmer technology was developed. In spiegelmer technology, the in-vitro selection process is combined with the principles of symmetry and chiral inversion.

3.4.3
Aspects of Chirality

The essential step during in-vitro selection is amplification of the rare species represented in the combinatorial oligonucleotide library that bind to the target molecule. Amplification of nucleic acids can be accomplished very easily by employing enzymes – DNA polymerases (Box 16) – in a process called PCR. On the other hand, the degradation of aptamers occurs as a result of abundant nucleases which attack specific sites within the naturally configured nucleic acids.

Most enzymes are proteins composed of chiral building blocks, the L-amino acids (with the exception of achiral glycine). Because of the chirality of its building blocks the enzymes themselves are inherently chiral. The word chiral is derived from the Greek word *"cheir"*, meaning hand. The phenomenon of chirality as an inherent property of biological products was first described by Louis Pasteur [14]. A satisfying definition was given by Lord Kelvin in 1904, in his Baltimore Lectures on molecular dynamics and the wave theory of light: "I call any geometrical figure, or group of points, chiral, and say it has chirality, if its image in a plane mirror, ideally realized, cannot be brought to coincide with itself."

The basic concepts of stereochemistry were independently developed by van't Hoff and Le Bel in 1874. Based on their findings Emil Fischer suggested that biological macromolecules are composed of chiral L amino acids and D sugars and in 1894 proposed his famous "key and lock" hypothesis that two chiral molecules interact through shape complementarity with each other and are therefore stereospecific [15].

Thus, enantiomeric nucleic acids which are composed of mirror image nucleotides should be resistant against nucleolytic degradation by naturally occurring enzymes. In mirror-image nucleic acids all nucleotides are converted to the synthetic, enantiomeric form which means inversion of each chiral center (1′, 2′, 3′, 4′ in RNA and 1′, 3′, 4′ in DNA). In the two-dimensional representation of mirror-image nucleic acids only the nucleobase of each nucleotide is switched from the right to the left hand side.

Unfortunately, enantiomeric or mirror-image nucleic acids cannot be used directly in the SELEX process because of the lack of (mirror-image) enzymes which would be needed to amplify them.

Most interesting targets for aptamers are peptides or proteins which are chiral targets; both interacting partners are of the same natural chirality. Following the principles of stereochemistry, the mirror-image configurations of identified aptamers should interact with the corresponding enantiomers of the targets (first line of Figure 3.4.1). In this circumstance both partners in the complex have the un-

aptamer • target

enantio-target • enantio-aptamer

enantio-target • aptamer

enantio-aptamer • enantio-target
(Spiegelmer)

Fig. 3.4.1. Biological proteins or peptides are composed of naturally occurring L amino acids whereas all naturally occurring nucleic acids are composed of D sugars. If an isolated aptamer interacts with a protein target by building a stable complex, both structures composed of the synthetic mirror image form (enantio-target and enantio-aptamer) should form a complex having the same characteristics. If an aptamer recognizes a synthetic mirror-image protein, the same aptamer sequence synthesized in the synthetic mirror-image form should consequently bind to the mirror form of the mirror-image protein, i.e. the naturally occurring protein.

natural chirality. By having two different chiralities in two substance classes – proteins and nucleic acids – in principle two additional pairings or complexes are possible, Figure 3.4.1, bottom line – an aptamer that would bind with high affinity and specificity to the synthetic mirror image form of a target molecule should code the sequence information for a shape, a mirror-image oligonucleotide, that would bind to the naturally occurring form of that target molecule. If these chiral principles are coupled to the powerful screening technology of SELEX, oligonucleotide ligands should be generated that would have the desired binding properties for a given target and at the same time these ligands should be unsusceptible to nucleolytic degradation by naturally occurring enzymes.

3.4.4
Spiegelmer Technology

First, a target molecule for which an antagonist is desired is identified. If this target molecule is a peptide, the sequence is synthesized in the synthetic mirror image configuration. Standard methods of synthesis, for example Fmoc or Boc chemistry (Box 25), can be employed using D amino acid building blocks [16]. If the target is a bigger protein it is useful to define epitopes or domains of that protein which can be synthesized in the mirror image configuration. By also using peptide ligation techniques, peptide sequences more than 100 amino acids long can be synthesized [17]. The targets are not limited to peptides or proteins; other substance classes are also accessible if the enantiomer of the intended target is synthesizable.

If the synthetic mirror-image form of the target is prepared, the standard in-vitro selection scheme is run to identify aptamers that bind to the mirror image target. After cloning and sequencing single sequences are identified and tested for their binding properties. The binding oligonucleotides or aptamers can be converted very easily to the mirror image forms by synthesizing the corresponding enantiomeric oligonucleotides. To discriminate aptamers from "mirror-image aptamers" the name spiegelmers for the latter substance class was coined (mirror, German "*Spiegel*"), Figure 3.4.2.

3.4.5
Examples and Properties of Mirror-image Oligonucleotides

3.4.5.1
Spiegelmers Binding to Small Molecules

The first functional mirror-image oligonucleotides, spiegelmers, described in 1996 were generated to bind to small molecules, i.e. the amino acid arginine and the nucleoside adenosine [18, 19]. For in-vitro selection against adenosine, the target in the enantiomeric configuration was covalently coupled to a matrix. A standard

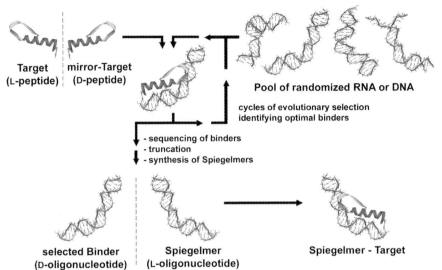

Fig. 3.4.2. Schematic representation of spiegelmer technology. A library of 10^{15} different oligonucleotides is synthesized and incubated with the mirror-image (or enantiomeric) form of a naturally occurring target. After several cycles of selection and amplification the enriched library is cloned in an appropriate vector and sequenced. Individualized binding oligonucleotides are further characterized and tested for binding to the mirror-image form of the naturally occurring target. The sequences of identified binders are then synthesized in the mirror-image form to create mirror-image oligonucleotides, the so called spiegelmers. Because of the rules of stereochemistry the spiegelmers bind to the naturally occurring form of the intended target molecule.

library containing 60 random positions flanked by primer binding sites was synthesized. A typical scheme showing how to create an RNA library is depicted in Figure 3.4.3.

Isolation of L-adenosine binders was essentially as outlined in Figure 3.4.2 and described by Sassanfar and Szostak [20]. To improve stereospecificity counter-selection using the natural target D-adenosine was introduced in later stages of selection [21]. After ten cycles of in-vitro selection and subsequent cloning and sequencing, high-affinity binding motifs (see appendix) against adenosine were identified that had the expected reciprocal chiral specificity; Figure 3.4.4. Competition binding experiments using equilibrium dialysis and radioactively labeled L- and D-adenosine were performed to estimate binding constants. Figure 3.4.4A shows the competition curves with the cognate and the non-cognate binding partners. The estimated dissociation constants for the binding complexes are, on average, 2.3 µM whereas the non-binding complexes result in significantly weaker binding of approx. 20 mM, if at all (Figure 3.4.4B). The mirror-inverted CD spectra of the adenosine binding aptamer (D-RNA) and spiegelmer (L-RNA) indicate that

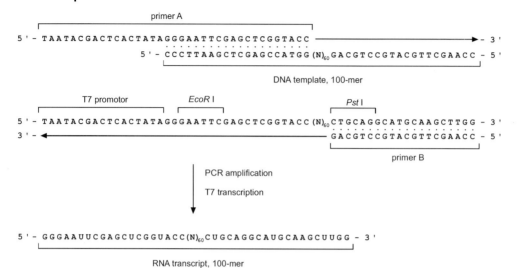

Fig. 3.4.3. Construction of a combinatorial RNA library. First, a DNA template having a random region of N positions between two fixed primer binding sites (primer A and B) is synthesized. A T7 promoter can be introduced by using an extended primer. After PCR the DNA template is double-stranded and can be transcribed into the RNA library by T7 RNA polymerase. *Eco*R I and *Pst* I are cleaving sites for restriction enzymes.

the tertiary structures of both oligonucleotides are also mirror-inverted (Figure 3.4.4C); thus the chiral inversion is responsible for the reciprocal chiral specificity. The biological stability of spiegelmers in comparison with aptamers was analyzed by incubating similar amounts of both oligonucleotides in human serum. The spiegelmer showed the expected stability against nucleolytic degradation for several days whereas the corresponding aptamer was completely digested within seconds (Figure 3.4.5). Similar results were obtained for the arginine binding motifs (see Appendix).

3.4.5.2
Mirror-image DNA Inhibiting Vasopressin in Cell Culture

The first bioactivity of a functional mirror image DNA oligonucleotide directed against the hormone peptide vasopressin was shown by Williams et al. in 1997 [22]. An initial in-vitro selection experiment using a library containing 60 random positions that was brought into contact with pre-immobilized D-vasopressin revealed only one binding sequence after 13 cycles. The sequence was truncated to a functional 68-mer. In a second in-vitro selection experiment a new pool containing the original 68-mer was doped at 30% so that at each doped position the parental base was present at a frequency of 70% and the other three bases at 10%. After four more cycles of in-vitro selection mainly variants of the dominant binder of the first selection experiment were isolated. An extremely large amount of conserva-

A

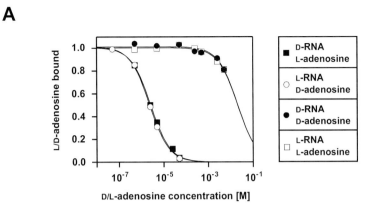

B

RNA	K_d (L-adenosine)	K_d (D-adenosine)	discrimination factor	
D-RNA (aptamer)	2.4 0.1 µM	20.1 ± 2.7 mM	8400	1500
L-RNA (Spiegelmer)	20.2 1.7 mM	2.2 ± 0.1 µM	9200	1200

C

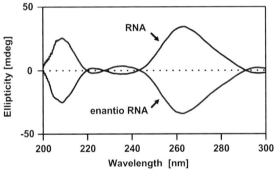

Fig. 3.4.4. Reciprocal chiral specificity. (A) The selected D-RNA sequence (aptamer) binds to the unnatural synthetic L-adenosine but not to D-adenosine and the corresponding L-RNA sequence (spiegelmer) recognizes naturally occurring D-adenosine but not L-adenosine. (B) The non-cognate interactions are approx. 9000-fold weaker than the intended interactions. (C) These results imply that the functional L-RNA and D-RNA sequences are folding into mirror-image shapes; this is consistent with the observed circular dichroism (CD) spectra of the functional 58-mer RNA sequences. CD spectra were recorded in 0.1 M NaCl and 10 mM sodium phosphate, pH 7.0, at 4 °C on a Jasco J-600 spectropolarimeter.

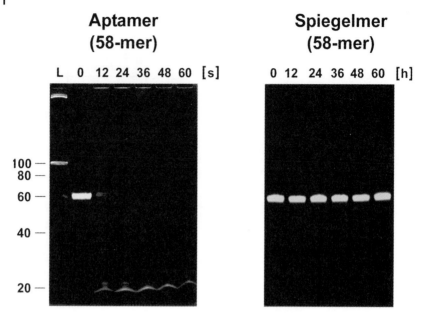

Fig. 3.4.5. Biological stability of an aptamer (D-oligonucleotide) and a spiegelmer (L-oligonucleotide) in human serum. Both oligonucleotides were incubated in buffered human serum and after the times indicated samples were taken and analyzed on a polyacrylamide gel [18]. After 24 s no full length aptamer could be detected whereas the spiegelmer was fully stable even after 60 h. (Reprinted with permission from the Nature Publishing Group, Nat. Biotechnol. 1996, 14, 1112–1115).

tion was observed for 31 of the 68 doped positions. On the basis of this information a 55-mer truncated vasopressin binder (see Appendix) was designed and synthesized in both enantiomeric configurations. The dissociation constants for both complexes D-oligonucleotide/D-vasopressin and L-oligonucleotide/L-vasopressin were determined by equilibrium dialysis to be approximately 1 µM. In a cell based assay antagonism was demonstrated for the L-oligonucleotide (spiegelmer) whereas a control sequence with the same base composition and similar secondary structure was almost inactive.

3.4.5.3
RNA and DNA Spiegelmers Binding to GnRH

More biological data for mirror-image oligonucleotides were generated by evaluating the properties of GnRH (gonadotropin-releasing hormone) binding spiegelmers [23]. GnRH is a ten-residue peptide released from the hypothalamus which acts as a key hormone of mammalian reproduction. The hormone binds to a single class of high affinity G-protein-coupled receptors on the plasma membranes of pituitary gonadotrophs and regulates the secretion of LH (luteinizing hormone) and FSH (follicle stimulating hormone). Inhibition of the GnRH–GnRH receptor

3.4.5 Examples and Properties of Mirror-image Oligonucleotides

system seems to be an appropriate form of treatment for certain sex hormone-dependent cancer indications such as prostate or breast cancer.

RNA and DNA libraries containing each 60 random positions and approx. 10^{15} different sequences were generated to isolate both RNA and DNA aptamers against D-GnRH. For partitioning, the oligonucleotide libraries were applied to sepharose columns which were derivatized with modified D-GnRH-Cys. The binding oligonucleotides were eluted by affinity competition with selection buffer containing unmodified D-GnRH. After six and eight rounds of in-vitro selection for the RNA and DNA libraries, respectively, the enriched libraries were cloned and sequenced. The sequences of the enriched RNA library were dominated by one main sequence and some related sequences which differed only in point mutations. The K_D of the main RNA sequence was found to be 92 ± 12 nM. The enriched DNA library revealed three clones that occurred between 10 and 15 times, as well as several orphan sequences. One of the orphans had the best K_D, 55 ± 7 nM, and was therefore chosen for further analysis. The RNA sequence was readily truncated to a 50-mer (see Appendix). Both RNA and mirror-image RNA had binding properties similar to those of their respective targets, although the dissociation constants rose slightly to 190–260 nM as a result of the truncation. The DNA could be truncated and modified to a 60-mer (see Appendix); both DNA and mirror-image DNA bound to GnRH with equilibrium dissociation constants of approximately 46 ± 7 nM (Figure 3.4.6). The RNA and DNA spiegelmers showed specificity by not binding to other peptides, even if they were related to GnRH, like buserelin and chicken LHRH. The functional activity of the spiegelmers was shown in a cell based assay employing Chinese hamster ovary cells expressing the GnRH receptor. GnRH receptor binding results in intracellular Ca^{2+} release that can be detected by

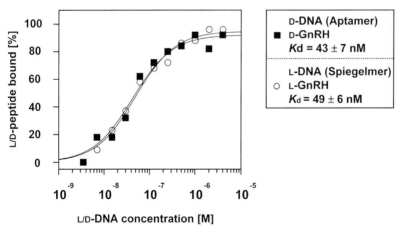

Fig. 3.4.6. Affinity of the GnRH-binding DNA sequence. The dissociation constants (K_D) were determined by equilibrium dialysis assays. Increasing aptamer (black squares) and spiegelmer (circles) concentrations were equilibrated against constant concentrations of tritiated D-GnRH and L-GnRH, respectively. The maximum bound fraction was normalized to 100%.

use of a pre-applied intracellular fluorescence marker a few seconds after stimulation with GnRH. Both spiegelmers were able to inhibit binding of L-GnRH to its cell surface receptor, with IC_{50} values of 200 and 50 nM, respectively, for the RNA and DNA spiegelmers.

3.4.5.4
In-vivo Data of GnRH Binding Spiegelmers

To increase the affinity and improve the temperature-stability of the anti-GnRH DNA sequence the original in-vitro selection was re-initiated after the first selection stage. The selection procedure was therefore performed at 37 °C and with greater stringency. After nine additional cycles of selection and amplification the fraction that bound to D-GnRH was increased to 13%. Enriched molecules of the tenth cycle were cloned and sequenced. Thirty-eight out of 48 sequences contained a clone that had significant binding to the selection target. The clone was truncated to a 67-mer, synthesized in its mirror-image configuration, and named NOX 1255. A 40-kDa poly(ethylene glycol) moiety was attached at the 5' end of NOX 1255 to give NOX 1257. NOX 1255 and NOX 1257 were subjected to extensive studies to determine their binding characteristics in cell culture, their immunogenic potential, and their activity in an animal model [24].

To evaluate binding characteristics the Chinese hamster ovary cell system was chosen as described above. NOX 1255 and the PEGylated spiegelmer inhibited GnRH response with an IC_{50} of approximately 20 nM, which corresponds to the dissociation constants. Hence, the terminal PEG-modification did not influence GnRH binding. Neither spiegelmer inhibited action of Buserelin, a GnRH receptor agonist structurally closely related to GnRH. These experiments showed again the extremely high specificity of oligonucleotide binders.

The activity in vivo of NOX 1255 and NOX 1257 was shown in a rat model. Male rats were castrated, resulting in an elevated GnRH level. GnRH itself regulates the serum LH concentration which is measured in the model. After castration the elevated serum LH concentration can be down-regulated by antagonizing GnRH action. This pharmacological effect can be achieved with the well known peptide analog Cetrorelix within 30–60 min after administration. Cetrorelix acts as a receptor antagonist and prevents GnRH from binding so that the LH level is as low as in non-castrated rats. The administration of the spiegelmers should result in the formation of GnRH–spiegelmer complexes and would therefore antagonize the hormone action directly. In fact NOX 1255 had a significant effect between 60 and 90 min but the GnRH antagonism leveled off after three to six hours; this might be because of rapid clearance of oligonucleotides. A different result was observed after administration of PEGylated NOX 1255. Although the effective molarity was only half as high as for NOX 1255, NOX 1257 antagonized GnRH action to an extent comparable with that of the GnRH receptor antagonist Cetrorelix over 24 h. This was the first demonstration of the pharmacological potency of spiegelmers.

Immunogenicity is one major concern with macromolecular drugs. For this reason NOX 1255 and 1257 were analyzed for their immunogenic potential. Both

Tab. 3.4.1. Comparison of spiegelmers NOX 1255 and NOX 1257.

	NOX 1255	NOX 1257
Molecular weight (g mol^{-1})	~21 000	~61 000
Dissociation constant, K_D	~20 nM	~20 nM
Cell culture, IC$_{50}$	~20 nM	~20 nM
Applied dosage (rats)	~4.1 µmol kg^{-1}	~2.4 µmol kg^{-1}
Activity in rat model, LH level	Significant effect between 1 and 6 h; no basis level of LH	Decrease of LH level to baseline after 1 h; activity stable for 24 h
Immunogenicity, rabbits		
– Spiegelmer	No effect	No effect
– Spiegelmer + adjuvant	No effect	No effect
– Spiegelmer conjugated	Low serum titers after 35 days	Low serum titers after 63 days

molecules were administered to rabbits over six weeks using a procedure that generally leads to the formation of specific antibodies for the administered antigen. Apart from the standard immunization, a second group of animals received the spiegelmers with adjuvant and a third group received the spiegelmers conjugated to the immunogen bSA (bovine serum albumin) as a haptenization approach. Neither NOX 1255 nor NOX 1257 triggered the formation of specific antibodies; only in the third group which received the conjugated spiegelmers could very low serum titers of antibodies be observed. The results of the biological studies for NOX 1255 and NOX 1257 are summarized in Table 3.4.1:

3.4.6
Conclusion

It has been shown that mirror-image oligonucleotides are usually suitable for development as potential drugs. The combination of the powerful screening platform SELEX with the chiral principles of spiegelmer technology leads to molecules that can antagonize pharmacologically meaningful targets; at the same time these so-called spiegelmers, although being macromolecular substances, have extremely high stability against enzymatic degradation and very low, if any, immunogenic potential. Several studies are in progress to further explore the potential of mirror-image oligonucleotides.

Acknowledgments

The author would like to thank Britta Wlotzka and Robert McLeod for helpful discussions and critical reading of the manuscript.

1a

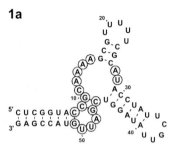

1b

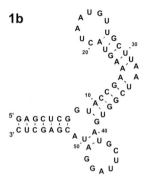

1c

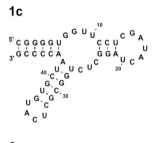

1d

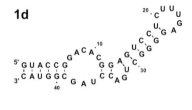

2a

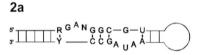

2b

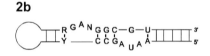

3

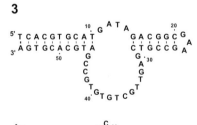

4a

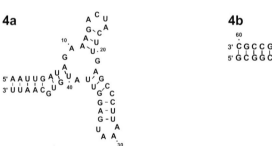

4b

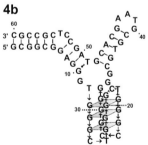

Appendix

Proposed secondary structures of identified oligonucleotides that bind in their mirror image configuration to naturally occurring target molecules. **1a** through **1d** represent different types of adenosine binding sequences; the circularized nucleotides in **1a** indicate an identified motif. **2a** and **2b** are arginine-specific binders. The vasopressin binding DNA sequence is presented in No. **3**. **4a** is a GnRH-specific sequence based on RNA and **4b** is a proposed G-quartet containing DNA specific oligonucleotide that binds to GnRH also.

References

1 K. Kruger, P. J. Grabowski, A. J. Zaug, J. Sands, D. E. Gottschling, T. R. Cech, *Cell* **1982**, *31*, 147–157.
2 C. Guerrier-Takada, K. Gardiner, T. Marsh, N. Pace, S. Altman, *Cell* **1983**, *35*, 849–857.
3 R. H. Symons, *Annu. Rev. Biochem.* **1992**, *61*, 641–671.
4 R. Lewin, *Science* **1986**, *231*, 545–546.
5 G. F. Joyce, *Nature* **1989**, *338*, 217–224.
6 D. J. Kenan, D. E. Tsai, J. D. Keene, *Trends Biochem. Sci.* **1994**, *19*, 57–64.
7 C. Tuerk, L. Gold, *Science* **1990**, *249*, 505–510.
8 A. D. Ellington, J. W. Szostak, *Nature* **1990**, *346*, 818–822.
9 L. Gold, B. Polisky, O. Uhlenbeck, M. Yarus, *Annu. Rev. Biochem.* **1995**, *64*, 763–797.
10 W. James, *Encyclopedia of Analytical Chemistry*, R. Meyers, ed., John Wiley and Sons, Chichester, **2000**, pp. 4848–4871.
11 M. Famulok, *Curr. Opin. Struct. Biol.* **1999**, *9*, 324–329.
12 J. M. Bacher, A. D. Ellington, *Drug Discovery Today* **1998**, *3*, 265–273.
13 C. E. Tucker, L. S. Chen, M. B. Judkins, J. A. Farmer, S. C. Gill, D. W. Drolet, *J. Chromatogr. B Biomed. Sci. Appl.* **1999**, *732*, 203–212.
14 M. Pasteur, *Compt. Rend. Hebt. Acad. Sci. Paris* **1848**, *26*, 534–538.
15 E. Fischer, *Chem. Ber.* **1894**, *27*, 2985–2993.
16 R. C. Milton, S. C. Milton, S. B. Kent, *Science* **1992**, *256*, 1445–1448.
17 J. A. Borgia, G. B. Fields, *Trends Biotechnol.* **2000**, *18*, 243–251.
18 S. Klussmann, A. Nolte, R. Bald, V. A. Erdmann, J. P. Furste, *Nat. Biotechnol.* **1996**, *14*, 1112–1115.
19 A. Nolte, S. Klußmann, R. Bald, V. A. Erdmann, J. P. Fürste, *Nat. Biotechnol.* **1996**, *14*, 1116–1119.
20 M. Sassanfar, J. W. Szostak, *Nature* **1993**, *364*, 550–553.
21 R. D. Jenison, S. C. Gill, A. Pardi, B. Polisky, *Science* **1994**, *263*, 1425–1429.
22 K. P. Williams, X. H. Liu, T. N. Schumacher, H. Y. Lin, D. A. Ausiello, P. S. Kim, D. P. Bartel, *Proc. Natl. Acad. Sci. USA* **1997**, *94*, 11285–11290.
23 S. Leva, A. Lichte, J. Burmeister, P. Muhn, B. Jahnke, D. Fesser, J. Erfurth, P. Burgstaller, S. Klussmann, *Chem. Biol.* **2002**, *9*, 351–359.
24 B. Wlotzka, S. Leva, B. Eschgfaller, J. Burmeister, F. Kleinjung, C. Kaduk, P. Muhn, H. Hess-Stumpp, S. Klussmann, *Proc. Natl. Acad. Sci. USA* **2002**, *99*, 8898–902.

3.5
Aspartic Proteases Involved in Alzheimer's Disease

Boris Schmidt and Alexander Siegler

3.5.1
Introduction

"Alzheimer's disease (AD) is a devastating illness that robs humans of their ability to remember, to think and to understand all the things we cherish most about being human." (P.F. Chapman)

Alzheimer's disease (AD) is an epidemic neurodegenerative disorder claiming millions of victims per year. The aging of the world population will be accompanied by an even higher toll. According to the World Health Organization the prevalence of Alzheimer's disease is 5.5% above 60 years of age and increases for elderly people (clinical AD: 16% 85 years, 22% 90 years) [1, 2]. "The onset of Alzheimer's disease is usually after 65 years of age, though earlier onset is not uncommon. As age advances, the incidence increases rapidly (it roughly doubles every 5 years)." Thus age is the dominant risk factor overruling even the positive impacts of nutrition and education. The socio-economic impact of Alzheimer's disease, the care needed for disabled and chronically wasting patients, the consequences for patients, relatives, and caretakers alike will be a major social and financial issue for the coming decades. "The direct and total costs of this disorder in the United States have been estimated to be US$536 million and US$1.75 billion, respectively, for the year 2000."

The exact cause of Alzheimer's disease remains unknown, although a number of factors have been suggested. These include metabolism and regulation of amyloid precursor protein, plaque-related proteins, tau proteins, zinc, copper, and aluminum [1].

Improvements in medication by use of acetyl cholinesterase inhibitors and general therapy significantly reduce symptoms at the onset of the disease [3, 4] but do not address the severe mortality in the final stages. A causal therapy is, therefore, still very much in demand, because no existing therapy effectively stops or even cures the disease. Identification of gene mutations linked to Alzheimer's disease-afflicted families in London and Sweden and additional polymorphisms that either cause or promote Alzheimer's disease have provided some insight into the biological pathways and the involvement of the amyloid precursor protein (APP) [5–8].

Highlights in Bioorganic Chemistry: Methods and Applications. Edited by Carsten Schmuck, Helma Wennemers
Copyright © 2004 WILEY-VCH Verlag GmbH & Co. KGaA, Weinheim
ISBN: 3-527-30656-0

Processing of APP by Secretases

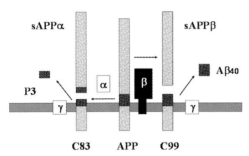

Scheme 3.5.1. Amyloid precursor protein (APP) degradation.
$\alpha = \alpha$-secretase, $\beta = \beta$-secretase, $\gamma = \gamma$-secretase, C83 = 83 C-terminal amino acid, C99 = 99 C-terminal amino acids.

A simplified diagram of amyloid precursor protein (APP) processing is depicted in Scheme 3.5.1. The up to 771 amino acid long APP, which occurs in three isoforms: APP695, APP751, and APP771, includes a signaling sequence, a large extramembranous sequence, and the crucial membrane-spanning domain, which is followed by a short cytoplasmic tail. Non-pathological cleavage occurs between Lys^{687} and Leu^{688} (K^{16}L^{17} in Scheme 3.5.2) under the action of α-secretase (also called TACE) producing α-APP and ultimately the fragments p3 and C83. α-Secretase belongs to the ADAM family and is sensitive to membrane cholesterol levels and can thus be modulated by cholesterol-reducing diets or drugs [9]. The most relevant point mutations for Aβ formation are K^{670}M^{671} → NL and V^{717} → Phe (Stockholm or Indiana), which cause familial Alzheimer's dementia (FAD). The molecular consequence of these point mutations is a different modulation of the three secretases, which act in concert to degrade APP. Usually 90% of APP is degraded by the α-secretase pathway, and despite years of intense research, the purpose of this degradation is still obscure. Rate-limiting β-secretase usually cleaves just 10% of all APP between the Met^{671}-Asp^{672} residues, but prefers the preceding amino acids Asn^{670}Leu^{671} of the Swedish mutation over Lys^{670}Met^{671}. The V^{717} → Phe mutation results in enhanced cleavage after Ala^{714}, leading to the notorious Aβ_{42}.

The genetic background of Alzheimer's disease is quite heterogeneous, and many associations have been made with locations on almost every other chromosome. Replicated or confirmed associations are, however, few. Late-onset Alzheimer's disease is linked to the ε4-allele of ApoE (which is the 4th identified functional mutation of an important transporter protein coded by multiple genes on different chromosomes). Mouse models expressing mutated human APP and presenilin 1 display many symptoms of Alzheimer's disease, although no model represents the full range of pathologies of the human disease. In particular the inflammation processes in humans and mice do not adequately relate to each

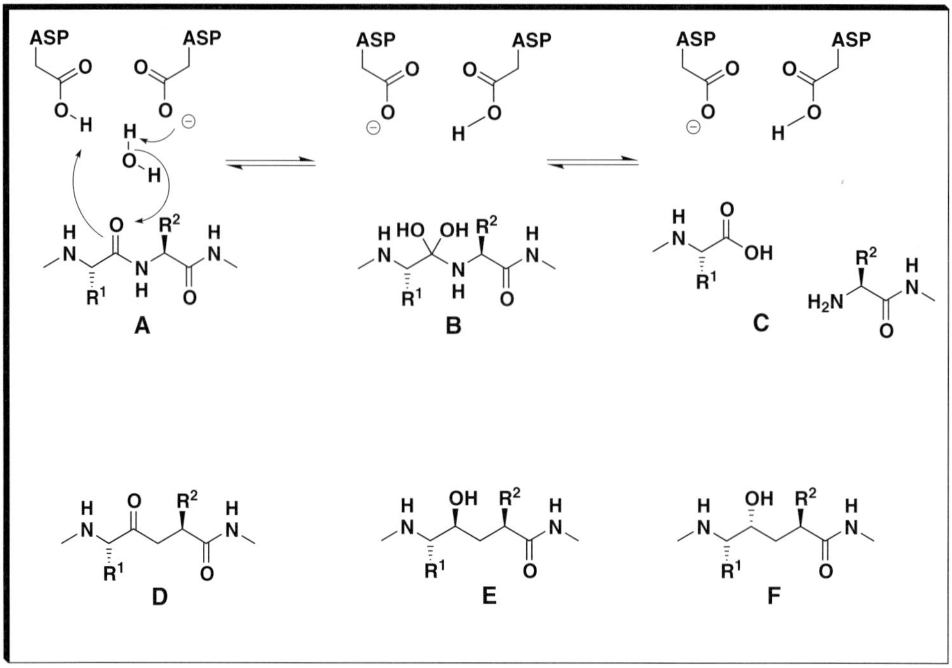

Scheme 3.5.2. General hydrolysis by aspartic proteases.

other. The observed loss of neurons is accompanied by the formation of plaques consisting of amyloid β-peptide (Aβ). A rational approach to a successful, causal therapy is based on a detailed understanding of Aβ formation, deposition and the inflammatory consequences. Decisive functions were assigned to the amyloid precursor protein (APP) and its degrading aspartic proteases β-secretase (also called β-amyloid-converting enzyme: BACE) and the presenilins. Presenilin is sometimes called γ-secretase, although there is a significant difference – presenilin is a mandatory component of γ-secretase activity, but it serves also as a functional domain for other proteins such as Notch. This is very important, because the Notch protein complex regulates cell differentiation at the embryonic stage and has functions in adult tissue.

The C-terminal transmembrane domain of β-secretase is not strictly required for activity, but location of enzyme and substrate in the same membrane enhances kinetics and specificity.

Only after subsequent proteolysis by the innermembrane protease, γ-secretase, between Val[711] and Ile[712] or Ala[713] and Thr[714] is the Aβ protein finally released, 40 or 42 amino acids long, resulting in Aβ40 and Aβ42 and the C-terminal fragment C99. Despite being the minor cleavage product, Aβ42 is the dominant factor in Aβ deposition and plaque formation, serving as a deposition nucleus. Aβ formation, deposition, and clearance are, therefore, highly attractive targets for drug develop-

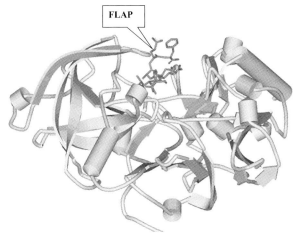

Fig. 3.5.1. BACE complexed to Glu-Val-Asn-Ψ(Leu-Ala)-Ala-Glu-Phe (3, OM99-2) (1FKN).

ment, although their normal regulatory function in healthy tissue is not fully understood (Aβ or peptide models thereof are also dealt with in Chapters 2.3 and 2.4).

Although several reviews on secretase inhibition have been published [10–12], the rapid progress in the field demands continuous survey. And despite this significant progress potent non-peptidic inhibitors of β-secretase are still unknown. Several peptide-based inhibitors were patented or reported immediately after J. Tang's disclosure of the BACE–inhibitor complex X-ray structure in 2000. Figures 3.5.1 and 3.5.2 show fragments of the homodimeric structures, which were reviewed recently [13]. Non-peptidic inhibitors of presenilin are known from patents

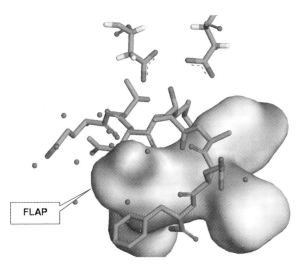

Fig. 3.5.2. BACE complexed to Glu-Val-Asn-Ψ(Leu-Ala)-Ala-Glu-Phe (3, OM99-2) (1FKN).

by Elan/Eli Lilly, Bristol Myers Squibb, and DuPont. A single original publication appeared for DAPT (**16**, difluorophenylacetylaminopropionylaminophenylacetic acid *tert*-butyl ester) and its drug-metabolism and pharmacokinetics (DMPK) outside that literature, but thanks to the commercialization of DAPT, which is a phase II candidate by Eli Lilly, it will turn into the standard for other compounds to come. Peptidic presenilin inhibitors [14, 15], for example Merck's L-685 458, which is still the most potent inhibitor, were patented prior to publication in scientific journals [16].

The well known, beneficial influence of non-steroidal anti-inflammatory drugs (NSAID) on the progress of Alzheimer's disease has been confirmed for some NSAID subtypes. The work by Weggen et al. indicates the potential of COX 1 inhibitors (e.g. Diclofenac, Sulindac, Indomethacin, Ibuprofen, but not the most prominent, Aspirin) in PS inhibition [17].

The reports of Nicastrin, which is a protein linked to familial dementia in the Italian town Nicastro, and its co-precipitation with presenilin by presenilin-specific antibodies [18] stimulated the ongoing debate about the identity of γ-secretase/PS. C. Haass has suggested that mature nicastrin plays a crucial role in PS1 trafficking from the endoplasmic reticulum to the plasma membrane [19].

β-Secretase (BACE) was established as an aspartic protease (Box 13) by molecular biology despite the initial lack of selective inhibitors. It bears all the hallmarks of a typical aspartic protease including the flexible flap region, which is crucial for substrate docking. The two states, *open* and *closed*, contribute to the selectivity and activity of the enzyme [20]. BACE1 is anchored to the membrane via its transmembrane domain (455–480); the catalytic domain is stabilized by three cystines in analogy to other aspartic proteases. The fully active BACE1 used by Tang for co-crystallization lacks the transmembrane and intracellular domains and some flexible N-terminal regions were not resolved by X-ray structure determination. The inhibitor is placed in the active site as intended by design (Figure 3.5.2): the transition state analog hydroxyethylene is coordinated through four hydrogen bonds to the two catalytic aspartic acids. Another ten hydrogen bonds are established between inhibitor, binding pocket, and flap region. Despite analogies with other aspartic proteases, there are significant differences in side-chain preferences. S4, S3′ are hydrophilic and readily accessible by water, the hydrophilic S4′, which holds the phenylalanine, is located at the surface and contributes less to binding. Therefore shortened peptidomimetics **4–6** retain activity. The S1′ position has space for more than just an alanine, as in the co-crystallized inhibitor **3**. This was realized by ethyl substitution of the hydroxyurea **5**. The importance of the flap region for structural reorganization and activity modulation was concluded from kinetics of statine-based peptides (hydroxyethylene) [21]. A detailed analysis of BACE distribution, structure, species variation, and properties was published recently [13]. BACE2, which is very similar, leads to additional hydrolysis close to Phe[20] (Scheme 3.5.3).

Aspartic proteases hydrolyze the amide bond as a result of concerted effort by an aspartic acid and an aspartate (Box 13). The aspartic acid protonates and activates the peptide **A** (Scheme 3.5.2) towards nucleophilic attack, and the aspartate is re-

ISEVKMD¹AEFRHDSGYEVHHQKLVFFAEDVGSNKGAIIGLMVGGVV⁴⁰IA⁴²TVI⁴⁵VTLVMLKK
NL Swedish   G²² arctic   F⁴⁵ Stockholm
                                                                      I⁴⁵
                                                                      G⁴⁵ London

Scheme 3.5.3. APP amino acid sequence close to the cleavage sites and point mutations in Aβ annotation.

quired to coordinate and deprotonate water to supply the nucleophilic hydroxyl anion (Scheme 3.5.2), which leads to the tetrahedral intermediate **B**. Collapse of the hydrated amide releases both a new C-terminal acid and a new N-terminal amine, as in **C**. Any precursor which gives way to more stable hydrates, e.g. ketones as in **D**, will interfere with hydrolysis. Electron deficiency of the ketones will stabilize the hydrate even further, therefore α,α-difluoroketones are well established aspartic protease inhibitors. A wide range of isosteric amide replacements was identified from natural protease inhibitors (**1**, pepstatin, Scheme 3.5.4) or resulted from insightful design. These dipeptide mimetics can be cited in a shorthand version using the three-letter amino acid code to indicate the junction: -Ψ(aa₁–aa₂)-. The hydroxyethylene isosters **E** and **F** were inspired by naturally occurring peptide mimetics, which are known to inhibit aspartic proteases. The specific placement of this pseudo-intermediate between the aspartic acid and the hydrolytic water is reflected by the different stereochemistry of the alcohols in **E** and **F**. The neighboring amino acids have to be adapted to the individual requirements of the targeted aspartic protease, be it renin, HIV-protease, plasmepsin, or β-secretase.

The high affinity complex of Glu-Val-Asn-Ψ(Leu-Ala)-Ala-Glu-Phe (**3**, OM99-2) with β-secretase results in complete inhibition of β-secretase activity and enabled crystallization and structure determination at 1.9 Å resolution (Figures 3.5.1 and 3.5.2) [22, 23]. The subsite specificity was established by determination of cleavage rates of combinatorial substrate mixtures and resulted in the discovery that Glu-Leu-Asp-Ψ(Leu-Ala)-Val-Glu-Phe was the most potent inhibitor (K_i 0.31 nM) of β-secretase. Recently a second member of the BACE family with high similarity was identified – BACE2 (also called memapsin 1), which causes additional cleavage reminiscent of α-secretase activity [24].

The pharmacological evidence compiled for γ-secretase is indicative of the activity of an aspartic protease requiring at least one additional cofactor. The location of the active site within the membrane makes γ-secretase quite unique. Currently, there is only one precedent for a similar, tricky enzyme, signal peptidase, which shares several features and most of the problems associated with inner-membrane location [25]. It will not, unfortunately, be easy to isolate and purify the membrane-stabilized protease while retaining its activity. It has, therefore, so far escaped crystallization and X-ray structure determination. Mutation analysis of the two conserved aspartic acids of all presenilins supports their key role in γ-secretase

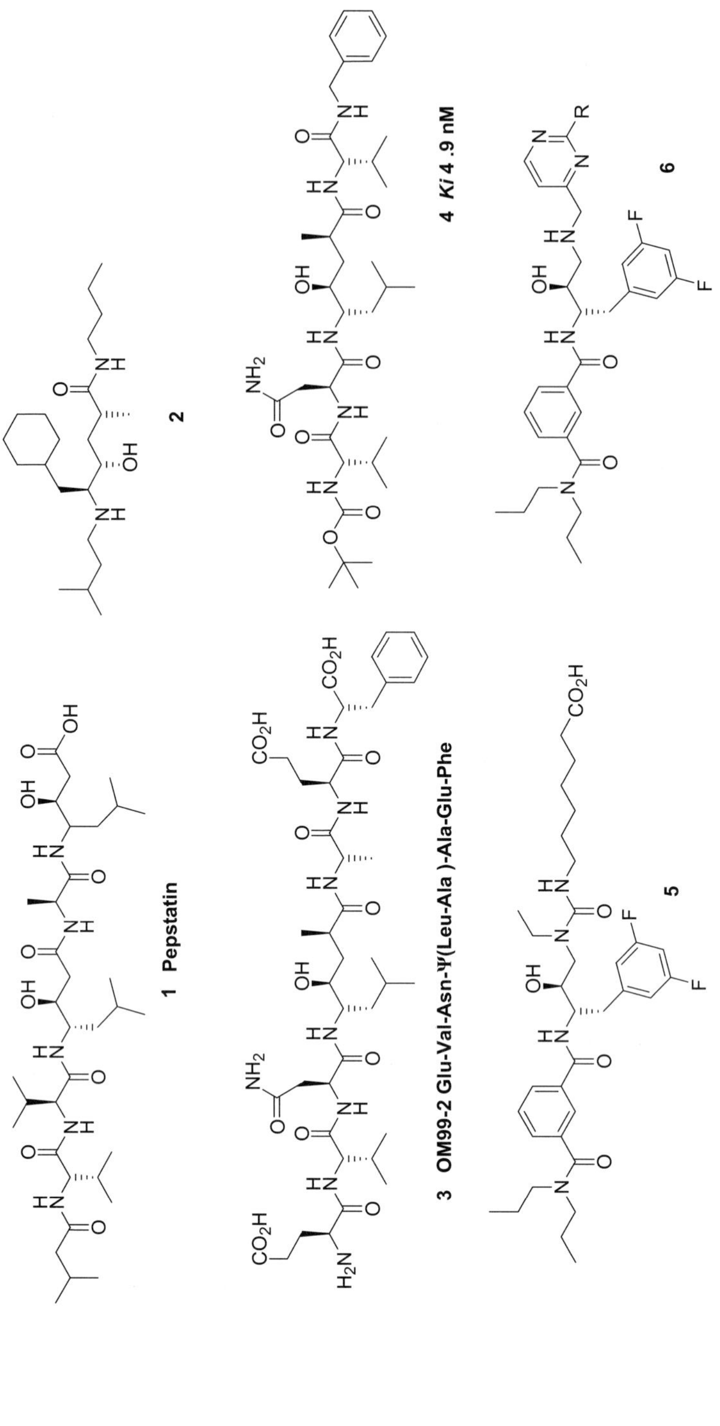

268 | 3.5 Aspartic Proteases Involved in Alzheimer's Disease

1 Pepstatin

2

3 OM99-2 Glu-Val-Asn-Ψ(Leu-Ala)-Ala-Glu-Phe

4 K_i 4.9 nM

5

6

Scheme 3.5.4.

activity, however. There is, moreover, evidence for an autoproteolytic mechanism, which is required to deliver active presenilin. The presenilin sheds the exon E9 (Scheme 3.5.7, below) in this cleavage and simultaneous maturation towards active γ-secretase. A proposal for the arrangement of the transmembrane helices has been made, but it does not fully explain the observed cleavage pattern [26]. A rudimentary scheme for the eight membrane-spanning domains is depicted below.

3.5.2
β-Secretase Inhibitors

Several inhibitors of β-secretase have been identified in cellular assays but, more often than not, the true nature of the inhibition mechanism was not reported. Broad-spectrum protease inhibitors, for example pepstatin 1, known aspartic protease inhibitors from renin, and HIV protease programs and cocktails thereof, had little inhibitory effect and gave misleading results. But the consequent utilization of the Swedish mutation and structure–activity relationships for an early Bristol–Myers Squibb compound, 2, resulted in OM99-2 (3) and successful co-crystallization with BACE. Activities have been reported for Leu–Ala hydroxyethylene isosters such as 4, which provide insight into the mode of binding. These compounds do not really invite drug development, because the obstacles in Alzheimer therapy will be even greater than inhibition of renin and HIV protease. Significant efforts have been made to reduce the molecular weight and the flexibility of the lead structure. The improved binding to the pocket P2' by valine instead of alanines, and the omission of the small interactions of P4 Glu and P4' Phe were first steps taken on a bumpy road. The Elan compounds 5, 6 have lost a good part of their peptide heritage, which is mandatory to obtain sufficient oral absorption and blood–brain barrier penetration.

Despite all the efforts of the pharmaceutical companies and academic groups, non-peptidic lead structures for BACE inhibition are very scarce (Scheme 3.5.5). Takeda reported the tetraline 7, which is not an obvious scaffold for protease inhibition, and therefore likely to originate from high-throughput screening efforts [27]. The activity is poor (IC$_{50}$ ≥ 1 µM) and the mode of action is uncertain. Latifolin 8, isolated from the heartwood of *Dalbergia sissoo*, was found to inhibit Aβ formation with an IC$_{50}$ of 180 µM [28].

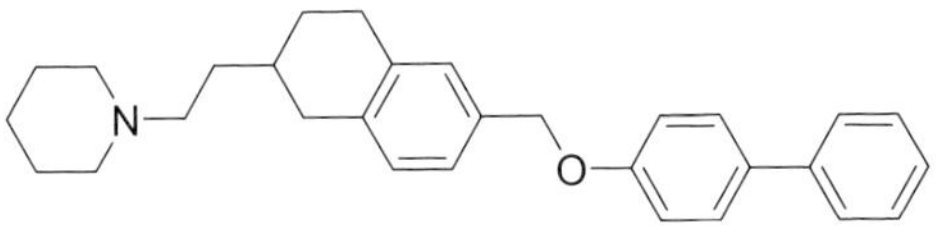

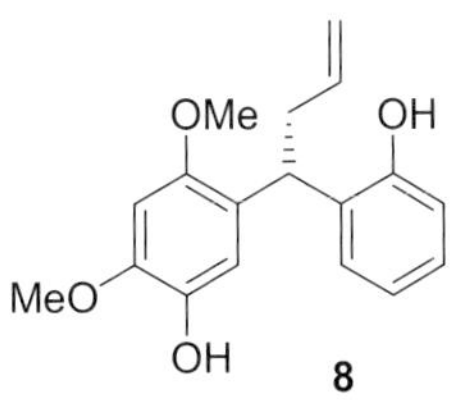

Scheme 3.5.5. Non-peptidic BACE inhibitors.

3.5.3
γ-Secretase Inhibitors

Several peptidic aldehydes have been reported to be inhibitors for either γ-secretase or β-secretase or both. Common to both series are lipophilic di- and tripeptides with bulky N-terminal protection, e.g. Z-LLL-CHO (MG132), Z-YIL-CHO, and Boc-GVV-CHO. The general lack of specificity of these aldehydes and their simultaneous inhibition of serine and cysteine proteases makes interpretation of data rather cumbersome. Indirect mechanisms through general protease inhibition result in complex concentration activity observations. Z-LLL-CHO (MG132), in fact, blocks maturation of the amyloid precursor protein. Some of these drawbacks were avoided by difluoro ketones as pioneered by Merryl Dow (Scheme 3.5.6), which

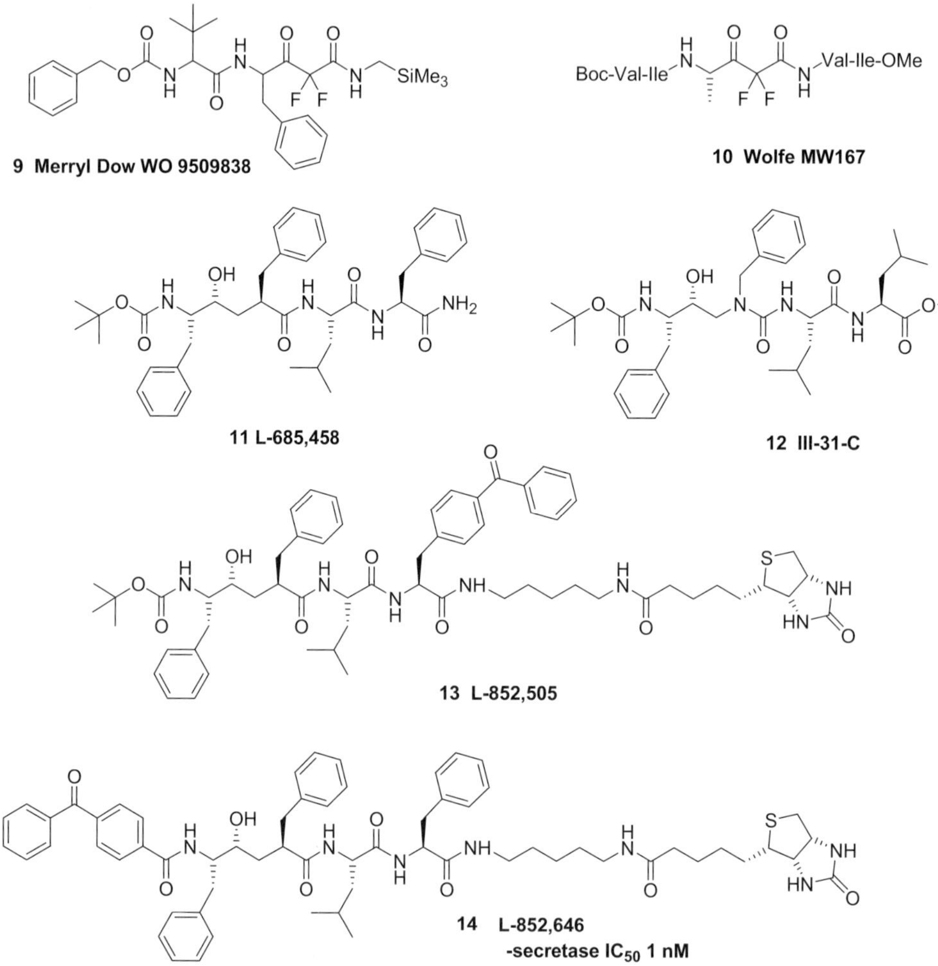
9 Merryl Dow WO 9509838
10 Wolfe MW167
Boc-Val-Ile
Val-Ile-OMe
SiMe3
11 L-685,458
12 III-31-C
13 L-852,505
14 L-852,646
-secretase IC50 1 nM

Scheme 3.5.6. γ-Secretase inhibitors.

again highlight the promiscuous nature of the enzyme. This results in weak inhibitory activity and the small impact of the amino acid variation as reported by M. S. Wolfe for derivatives of his MW167 **10** [29]. Thus this structural motif is limited to tools for assay development and labeling, e.g. difluoro ketones were used to distinguish γ-secretase and the Notch receptor, which share cleavage in the transmembrane region and several other features. The small impact of different difluoro ketones on the Aβ40/42 ratio supports the results from point mutation analysis and the very important phenylalanine scan of the transmembrane region of amyloid precursor protein [26]. The observed pattern of these phenylalanine introductions and their outcome on the Aβ40/42 ratio strongly supports α-helical presentation of the C99 fragment to the γ-secretase. If confirmed, this will make γ-secretase quite unique amongst the aspartic proteases, because there is only one report of proteolysis of such an α-helical substrate. All others prefer non-helical orientation or unfold their substrate to an extended geometry prior to proteolysis [30]. But α-helix mimetics of the scission site are not convincingly active.

A giant leap forward was achieved by the serendipitous identification of Merck's L-685 458, **11**, which was taken from a previous protease program, including the diastereomeric hydroxyethylene, the corresponding ketone and the parent tetrapeptide, the latter readily being cleaved. The all-lipophilic sequence with three phenylalanines was somewhat expected, because several studies had indicated the proximity of large lipophilic binding pockets (P2–P1, P1′–P2′, even P4′ and P7′) to the cleavage site. The inversion of the hydroxyethylene moiety reduces the inhibition 270-fold. This established the preference of β- and γ-secretase for hydroxyethylene isosters and enables their differentiation by control of the absolute stereochemistry. Labeling studies were conducted with different non-radioactive approaches, linking biotin and photoreactive fragments N- or C-terminally to L-852 505 **13** and L-852 646 **14**. The biotin was introduced to facilitate isolation and identification of the irreversibly labeled adducts via their streptavidin-enzyme-linked conjugates. Both attachments of photoreactive benzophenones (L-852 646 **14**, L-852 505 **13**) retained potent inhibition ($IC_{50} < 1$ nM for γ-secretase). Photolysis in the presence of solubilized γ-secretase provided a 20 kD protein (L-852 505 **13**) after isolation on a biotin-specific streptavidin-agarose gel followed by partial digestion. This 20 kD fragment was shown to be the C-terminal fragment of Presenilin 1 (PS1-CTF) by use of specific PS1-CTF antibodies. Binding to wild-type PS1 was negative in a control experiment, yet binding to the deletion construct PS1ΔE9, which lacks the cytosolic E9 loop, was positive (Scheme 3.5.7). Additional experiments support that PS1ΔE9 is part of the catalytically active complex, but lacks activity on its own. Additional information resulted from photolysis of L-852 646 **14** in the presence of solubilized γ-secretase; this led to isolation of a 34 kD fragment, which was found to be an N-terminal fragment. A similar transition-state motif, the hydroxyethylurea, was used for activity-based affinity purification. The immobilization of III-31-C (**12**, $IC_{50} < 300$ nM) on affi-gel 102 by exchanging the methyl ester for an hydrophilic amide linkage enabled isolation and identification of PS1-CTF, PS1-NTF, and Nicastrin from solubilized γ-secretase preparations [31]. Disappointingly, all strategies used to free active γ-secretase from the affinity

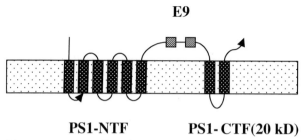

Scheme 3.5.7. Presenilin architecture.

gel failed. This is partially due to the strong binding affinity of the III-31-C **12** core for the target protein complex and partially due to the deep and narrow binding site, which requires strong denaturing conditions to break up the binding interactions. Co-precipitation of the inhibited γ-secretase with its substrates C83 and C99 gave rise to speculation about additional binding sites, where the substrate is recognized prior to transfer to the active site. These speculations are in accordance with the observed promiscuous nature of the cleavage, because they assign the specific recognition to other complex domains.

In comparison with β-secretase, for which detailed structural information and the enzyme kinetics are available, very little is known about the more complex γ-secretase. The diversity of selective, non-peptidic γ-secretase inhibitors is, therefore, something of a surprise. Elan's development of DAPT (**16**, difluorophenylacetyl-aminopropionylaminophenylacetic acid *tert*-butyl ester; Scheme 3.5.8) originated from an *N*-dichlorophenylalanine ester **15**, which had an $IC_{50} > 30$ μM in cellular screening. After several cycles of refinement and hundreds of compounds, activity peaked for the dipeptide mimetic DAPT **16** featuring the non-natural amino acid phenylglycine, which is crucial for activity (human embryonic kidney cells $IC_{50} = 20$ nM). Another key contribution stems from the difluorophenylacetic acid, resulting in a very steep structure–activity relationship; broad variation of the difluorophenyl moiety confirmed the demand for small electron-withdrawing substituents. Branched esters had similar activity and, despite speculations about the labile nature of tertiary butyl ester, which might be cleaved at the low pH of the gut, DAPT is not a prodrug.

Subcutaneous application of 100 mg kg^{-1} to mice resulted in a 50% reduction of Aβ brain levels after 3 h. The reduction was 40% 3 h after an oral dose of 100 mg kg^{-1}. Further development of the compound by Eli Lilly included stereoselective placement of the hydroxyl group and locking of the spatial arrangement of two phenyl rings in a seven-membered lactam to yield **17** (LY 411575, HEK $IC_{50} < 1$ nM).

After several failures with peptidic structures such as **2**, all of which suffered from toxicity problems during development, Bristol Myers Squibb and Merck [32] published details of almost 1000 derivatives of 4-chloro-*N*-(2,5-difluorophenyl)-benzenesulfonamides. Five-hundred of these were reported to be very good in-

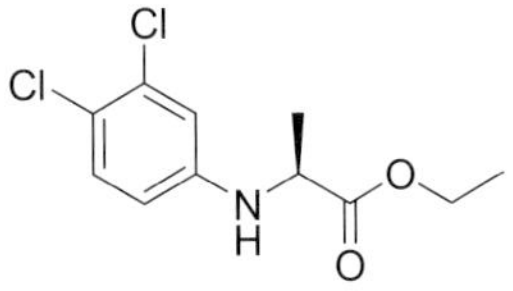

15

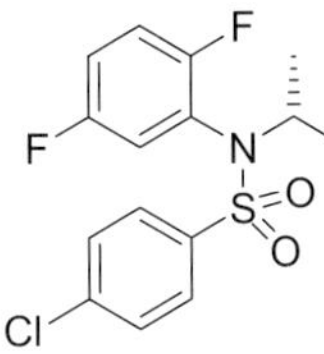

18

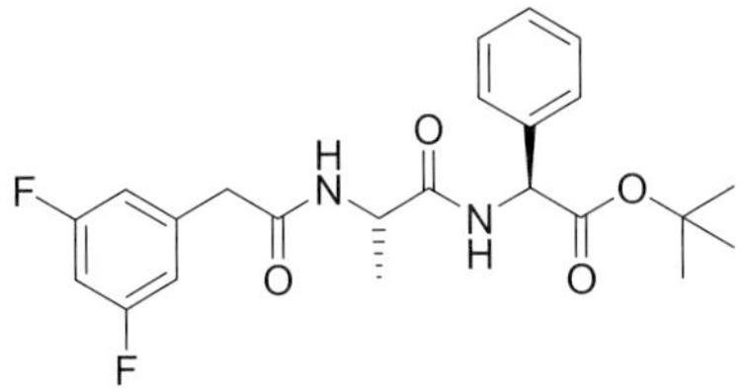

16 DAPT LY374973

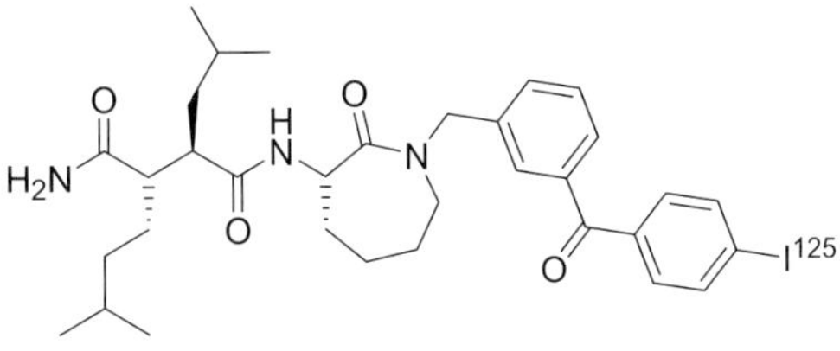

19

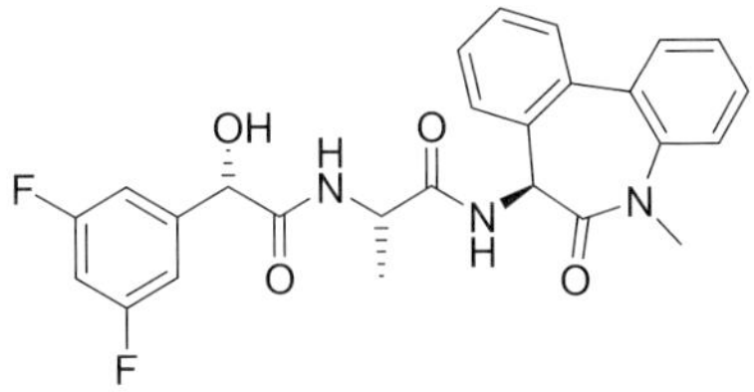

17 LY 411575

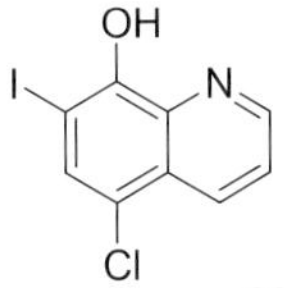

20

Scheme 3.5.8.

hibitors of γ-secretase activity. One of the activity clusters centers around the core structure **18**, with wide variation of substituent R to modulate bioavailability.

DuPont's hybrid structure **19** [33] bears the signature of a dipeptide-based SAR on γ-secretase and the reminiscent lead, which was synthesized in a matrix metalloproteinase (MMP) program. Removal of the central amide bond of the parent dipeptide, replacement of the hydroxamic acid by an amide, and introduction of a seven-membered lactam resulted in high activity and removed some of the problems associated with dipeptide lead structures. "Hot" labeling by photoactivation of I^{125}-benzophenone specifically cross-linked the inhibitor to three cell-membrane proteins.

3.5.4
Outlook

Despite tremendous progress in the field, drugs that cure Alzheimer's disease are still years away. Advances in β-secretase assay technology, combined with the availability of structural information, will profit from previous lessons learned on renin and HIV-protease. β-Secretase inhibitors are, therefore, likely candidates for clinical trials in the near future. Despite the lack of robust and cell-free assays and the simultaneous absence of detailed structural information about γ-secretase there are several development candidates, although all of these will have to pass their acid test on Notch. A cell-free assay and structural elucidation will have large impact on this selectivity issue. Substrate optimization is, furthermore, usually required for proteases to enhance the kinetics, and feasibility of high-throughput assays. This optimization was reported for β-secretase recently [34], but is still needed for γ-secretase. Unfortunately, this will not to be trivial, because of the specificities of the inner-membrane location of a rather promiscuous enzyme. Alternative approaches might be based on the cofactors of the presenilin complex or α-secretase modulation, because enhanced cleavage by α-secretase might result in reduced Aβ production. Cholesterol reducing drugs of the statine class have been shown to reduce Alzheimer's dementia; this might be explained by the sensitivity of α-secretase for cholesterol membrane concentrations [35]. Very promising immunotherapy with Aβ suffered from a severe setback because of negative clinical trials recently, but second-generation antigens are under investigation [36]. Metal chelators have frequently been investigated to reduce copper levels in brain tissue, which is thought to be partially responsible for Aβ toxicity. In most in-vitro studies, however, the copper concentrations required to observe the effect were magnitudes higher than those found in vivo. Recent studies, with 20 patients, of clioquinol **20**, a metal chelator that crosses the blood–brain barrier readily and has high affinity for zinc and copper ions, furnished interesting results, although lack of a control group in the study leaves ample room for other explanations, e.g. inflammation stimulus. Considering the multitude of approaches and the tremendous efforts under way, chances are high a cure will be found before the authors and the reader reach the crucial decades.

Acknowledgments

B.S. thanks the Fonds der Chemischen Industrie and the Deutsche Forschungsgemeinschaft (SCHM1012-3-1) for support of this work.

References

1 WHO, *The World Health Report 2001 Mental Health: New Understanding, New Hope*, **2001**.
2 T. POLVIKOSKI, R. SULKAVA, L. MYLLYKANGAS, I. L. NOTKOLA, L. NIINISTO, A. VERKKONIEMI, K. KAINULAINEN, K. KONTULA, J. PEREZ-TUR, J. HARDY, M. HALTIA, *Neurology* **2001**, 56, 1690.
3 J. S. JACOBSEN, *Curr. Top. Med. Chem.* **2002**, 2, 343.
4 S. IKEDA, Y. YAMADA, N. IKEGAMI, *Dementia Geriatr. Cognit. Disord.* **2002**, 13, 33.
5 A. M. SAUNDERS, *Pharmacogenomics* **2001**, 2, 239.
6 C. HOLMES, *Br. J. Psychiatry* **2002**, 180, 131.
7 R. E. TANZI, L. BERTRAM, *Neuron* **2001**, 32, 181.
8 J. HARDY, *PNAS* **1997**, 94, 2095.
9 M. L. MOSS, J. M. WHITE, M. H.

Lambert, R. C. Andrews, *Drug Discovery Today* **2001**, 6, 417.
10 M. S. Wolfe, C. Haass, *J. Biol. Chem.* **2001**, 276, 5413.
11 H. Steiner, C. Haass, *Nat. Rev. Mol. Cell Biol.* **2000**, 1, 217.
12 S. S. Sisodia, W. Annaert, S.-H. Kim, B. De Strooper, *Trends Neurosci.* **2001**, 24, S2.
13 S. Roggo, *Curr. Top. Med. Chem.* **2002**, 2, 359.
14 M. S. Shearman, D. Beher, E. E. Clarke, H. D. Lewis, T. Harrison, P. Hunt, A. Nadin, A. L. Smith, G. Stevenson, J. L. Castro, *Biochemistry* **2000**, 39, 8698.
15 Y.-M. Li, M. Xu, M.-T. Lai, Q. Huang, J. L. Castro, J. Dimizio-Mower, T. Harrison, C. Lellis, A. Nadin, J. G. Neduvell, R. B. Register, M. K. Sardana, M. S. Shearman, A. L. Smith, X.-P. Shi, K.-C. Yin, J. A. Shafer, S. J. Gardell, *Nature* **2000**, 405, 689.
16 A. Nadin, J. M. S. López, J. G. Neduvelil, S. R. Thomas, *Tetrahedron* **2001**, 57, 1861.
17 S. Weggen, J. L. Eriksen, P. Das, S. A. Sagi, R. Wang, C. U. Pietrizik, K. A. Findlay, T. E. Smith, M. P. Murphy, T. Bulter, D. E. Kang, N. Marquez-Sterling, T. E. Golde, E. H. Koo, *Nature* **2001**, 414, 212.
18 F. Chen, G. Yu, S. Arawaka, M. Nishimura, T. Kawarai, H. Yu, A. Tandon, A. Supala, Y. Q. Song, E. Rogaeva, P. Milman, C. Sato, C. Yu, C. Janus, J. Lee, L. Song, L. Zhang, P. E. Fraser, P. H. St George-Hyslop, *Nat. Cell Biol.* **2001**, 3, 751.
19 D. Edbauer, E. Winkler, C. Haass, H. Steiner, *PNAS* **2002**, 99, 8666.
20 D. Leung, G. Abbenante, D. P. Fairlie, *J. Med. Chem.* **2000**, 43, 305.
21 J. Marcinkeviciene, Y. Luo, N. R. Graciani, A. P. Combs, R. A. Copeland, *J. Biol. Chem.* **2001**, 276, 23790.
22 L. Hong, G. Koelsch, X. Lin, S. Wu, S. Terzyan, A. K. Ghosh, X. C. Zhang, J. Tang, *Science* **2000**, 290, 150.
23 R. T. Turner III, G. Koelsch, L. Hong, P. Castenheira, A. Ghosh, J. Tang, *Biochemistry* **2001**, 10001.
24 M. Farzan, C. E. Schnitzler, N. Vasilieva, D. Leung, H. Choe, *Proc. Natl. Acad. Sci. USA* **2000**, 97, 9712.
25 A. Weihofen, K. Binns, M. K. Lemberg, K. Ashman, B. Martoglio, *Science* **2002**, 296, 2215.
26 S. F. Lichtenthaler, R. Wang, H. Grimm, S. U. Uljon, C. L. Masters, K. Beyreuther, *PNAS* **1999**, 96, 3053.
27 M. Miyamoto, J. Matsui, H. Fukumoto, N. Tarui, (Takeda Chemical Industries, Ltd., Japan), WO 0187293, **2001**, 86 pp.
28 N. V. S. Ramakrishna, E. K. S. V. Kumar, A. S. Kulkarni, A. K. Jain, R. G. Bhat, S. Parikh, A. Quadros, N. Deuskar, B. S. Kalakoti, *Indian J. Chem. Sect. B* **2001**, 40B, 539.
29 C. L. Moore, D. D. Leatherwood, T. S. Diehl, D. J. Selkoe, M. S. Wolfe, *J. Med. Chem.* **2000**, 43, 3434.
30 D. P. Fairlie, J. D. A. Tyndall, R. C. Reid, A. K. Wong, G. Abbenante, M. J. Scanlon, D. R. March, D. A. Bergman, C. L. L. Chai, B. A. Burkett, *J. Med. Chem.* **2000**, 1271.
31 W. P. Esler, W. T. Kimberly, B. L. Ostaszewski, W. Ye, T. S. Diehl, D. J. Selkoe, M. S. Wolfe, *PNAS* **2002**, 99, 2720.
32 D. W. Smith, B. Munoz, K. Srinivasan, C. P. Bergstrom, P. V. Chaturvedula, M. S. Deshpande, D. J. Keavy, W. Y. Lau, M. F. Parker, C. P. Sloan, O. B. Wallace, H. H. Wang, (Merck and Co., Inc., USA; Bristol-Myers Squibb Company), WO 0050391, **2000**, 377 pp.
33 D. Seiffert, J. D. Bradley, C. M. Rominger, D. H. Rominger, F. Yang, J. E. Meredith Jr., Q. Wang, A. H. Roach, L. A. Thompson, S. M. Spitz, J. N. Higaki, S. R. Prakash, A. P. Combs, R. A. Copeland, S. P. Arneric, P. R. Hartig, D. W. Robertson, B. Cordell, A. M. Stern, R. E. Olson, R. Zaczek, *J. Biol. Chem.* **2000**, 275, 34086.
34 R. T. Turner 3rd, J. A. Loy, C. Nguyen, T. Devasamudram, A. K. Ghosh, G. Koelsch, J. Tang, *Biochemistry* **2002**, 41, 8742.
35 T. Hartmann, *Trends Neurosci.* **2001**, 24, S45.
36 D. Schenk, D. Games, P. Seubert, *Neurosci. News* **2000**, 3, 46.

B.13
Aspartic Proteases

Boris Schmidt

Aspartic proteases hydrolyze the amide bond by a concerted effort with aspartic acid and aspartate. The aspartic acid protonates and activates the peptide A toward nucleophilic attack and the aspartate is required to coordinate and deprotonate water to supply the nucleophilic hydroxy anion which leads to the tetrahedral intermediate B (Figure B.13.1). Collapse of the hydrated amide releases both a new C-terminal acid and a new N-terminal amine, as in C. Any precursor which forms more stable hydrates, e.g. ketones as in D, will interfere with hydrolysis. Electron deficiency on the ketones stabilizes the hydrate even further, so α,α-difluoroketones are well established aspartic protease inhibitors. A wide range of isosteric amide replacements has been identified from natural protease inhibitors (1, pepstatin, Scheme 4) or has resulted from insightful design. These dipeptide mimetics can be cited in a shorthand version by using the three-letter amino acid code to indicate the junction: $-\Psi(aa_1-aa_2)-$. The depicted hydroxyethylene isosteres (E, F), are inspired by naturally occurring peptide mimetics which are known to inhibit aspartic proteases. The specific placement of this pseudo-intermediate between the aspartic acid and the hydrolytic water is reflected by the different stereochemistry of the alcohols in E and F. The neighboring amino acids have to be adapted to the individual requirements of the targeted aspartic protease, be it renin, HIV-protease, plasmepsin or β-secretase.

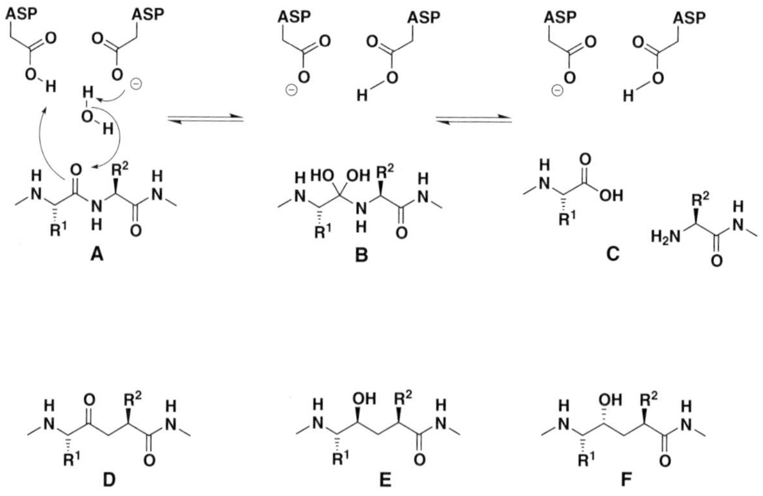

Fig. B.13.1.

3.6
Novel Polymer and Linker Reagents for the Preparation of Protease-inhibitor Libraries

Jörg Rademann

3.6.1
A Concept for Advanced Polymer Reagents

Despite the current success and popularity of polymer reagents the severe limitations of the available resins have become obvious during recent years. For many synthetically important transformations reliable reagents are not available and polymer-assisted synthesis (Box 14) was usually restricted to small scale applications and suffered from the inherent limitations of the standard support material, cross-linked polystyrene, for example solvent intolerance, undesired adsorption of reagents, or the chemical reactivity of the resin backbone [1].

Advanced polymer reagents might extend polymer-assisted conversion to more demanding transformations. The novel reagents should replace hazardous, toxic, and other undesirable chemicals by clean and reliable polymer-supported alternatives. Ideally, these polymer reagents should be recyclable or active in catalytic amounts. Advanced polymer reagents will also enable efficient use in large-scale applications which can be realized by novel, high-loaded polymer supports and/or by catalytic use of the reagents.

Generation and investigation of activated reactants and reactive intermediates in polymer gels have substantially extended the field of polymer-assisted conversions and has opened access to advanced polymer-supported reagents. For example, carbenium ions have been generated and released on polymer supports leading to the novel concept of alkylating polymers [2]. Alkylating polymers comprise resin-bound 1,3-alkylaryl triazenes which function as diazoalkane analogs, enabling versatile alkylations of carboxylic acids and phenols with numerous aliphatic, benzylic, or allylic groups. Radical release is another example of polymer-bound reactive intermediates [3]. Released radicals can undergo reactions exploiting solid-phase-specific reactivities and can be used as a linker concept.

In this article we will focus on two types of novel polymer reagent useful for preparation of protease inhibitor libraries. Oxidizing polymers have been developed for synthesis of amino and peptide aldehydes (Chapter 3) which are an important class of protease inhibitors by themselves and can also be used as reactive electrophiles in subsequent transformations.

Highlights in Bioorganic Chemistry: Methods and Applications. Edited by Carsten Schmuck, Helma Wennemers.
Copyright © 2004 WILEY-VCH Verlag GmbH & Co. KGaA, Weinheim
ISBN: 3-527-30656-0

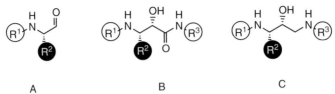

Fig. 3.6.1. Variability in the synthesis of protease inhibitors. N-terminal and C-terminal positions (R_1, R_3) of protease inhibitors are easily varied by combinatorial methods. In contrast, the substituent of the isosteric building block (R^2) and its stereochemistry are not accessible to diversity-oriented synthetic methods. Prominent protease inhibitor structures include peptide aldehydes A, norstatines B, and 1,3-diamino-propanols C.

To gain access to a broader spectrum of natural products including protease inhibitors, C–C-bond formation is often required in central steps. For this purpose polymer-supported carbanion equivalents have been investigated (Chapter 4). Several strategies leading to supported acyl anion equivalents are presented; these have been employed for general synthesis of protease inhibitors containing the α-hydroxy-β-amino motif.

3.6.2
Protease-inhibitor Synthesis – A Demanding Test Case for Polymer Reagents

The development of protease inhibitors faces several serious challenges. First, high affinity is demanded. Addressing a single protease in the presence of many others expressed in the human organism is, however, an even more difficult task (Box 15).

The selectivity of inhibitors is usually achieved by fine-tuning of substituents around an inhibitory element (Figure 3.6.1). For efficiency, this fine-tuning requires a synthetic strategy that is robust, tolerates diverse functional groups, enables parallel synthesis on a small scale, automation, facilitated work-up, and, finally, isolation of the products.

For many of the prominent inhibitor families (Fig. 3.6.1, Boxes 12 and 13), the available synthetic strategies do not fulfil these requirements. In the synthesis of peptide aldehydes, e.g., the parallel synthesis of variable structures without racemization is an unsolved problem. The same applies to the synthesis of many peptide isosteres such as the norstatines or the 1,3-diamino-2-propanols. Typically, combinatorial variation of the N-terminal and C-terminal positions can be easily attained. The side chain of the isosteric building block itself and its stereochemistry can, however, currently be varied only by a synthetic effort; this excludes versatile variation of the central inhibitory element.

Thus, developing diversity-oriented methods for construction of isosteric building blocks will make an important contribution to propelling protease inhibitor research.

3.6.3
The Development of Advanced Oxidizing Polymers

The oxidation of alcohols to carbonyl compounds is one of the most relevant transformations in organic synthesis, owing to the large diversity of products that can be obtained from aldehyde and ketone precursors. A variety of methods has been described for solving this task [4]. Common oxidizing agents for this transformation include a variety of heavy metal reagents, usually based on either chromium or ruthenium oxides, pyridine–SO_3, and dimethyl sulfoxide (DMSO) in combination with acetic anhydride, carbodiimide, or oxalyl chloride for activation. One of the most prominent methods for reliable conversion of sensitive compounds is the Dess–Martin reagent or its non-acetylated equivalent, 1-hydroxy-(1H)-benzo-1,2-iodoxol-3-one-1-oxide (2-iodoxybenzoic acid, IBX).

3.6.3.1
Polymer-supported Heavy-metal Oxides

There are several examples of polymer-supported oxidation reagents, including heavy-metal oxides bound to ion-exchange resins. Perruthenate resin [5], the immobilized analog of tetrapropylammonium perruthenate (TPAP), can be employed both stoichiometrically and catalytically. For the latter application additional co-oxidants are required; N-methylmorpholino-N-oxide (NMO) is usually used. The use of elemental oxygen has also been described. Perruthenate resin has recently been employed in a reaction sequence leading to heterocycles (Figure 3.6.2) [6]. Published examples indicate limited reactivity with non-benzylic alcohols. In general, ion-exchange resins suffer from potential leaching of heavy metals into the product solution, because they can be exchanged for any other anions present in the solution.

3.6.3.2
Oxidation with Immobilized Oxoammonium Salts

The 2,2,6,6-tetramethylpiperidinoxyl (TEMPO) radical was first prepared in 1960 by Lebedev and Kazarnovskii by oxidation of its piperidine precursor. TEMPO is a highly persistent radical, resistant to air and moisture, which is stabilized primarily by the steric hindrance of the NO-bond. Paramagnetic TEMPO radicals can be used as powerful spin probes for investigating the structure and dynamics of biopolymers such as proteins, DNA, and synthetic polymers by ESR spectroscopy [7]. A versatile redox chemistry has been reported for TEMPO radicals. The radical species can be transformed by two-electron reduction into the respective hydroxylamine or by two-electron oxidation into the oxoammonium salt [8]. One-electron oxidations involving oxoammonium salts have also been postulated [9]. The TEMPO radical is usually employed under phase-transfer conditions with, e.g., sodium hypochlorite as activating oxidant in the aqueous phase. In oxidations of primary alcohols carboxylic acids are often formed by over-oxidation, in addition to the de-

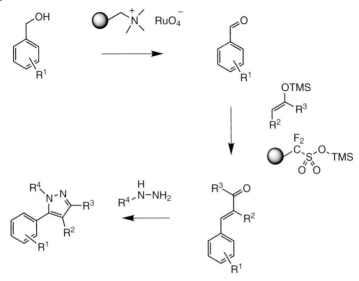

Fig. 3.6.2. Perruthenate resin, an oxidizing resin based on ion exchange of heavy metal oxides, has been successfully employed for preparation of heterocycle libraries. In this example the benzaldehydes were generated and reacted in an aldol reaction with Nafion-TMS as Lewis acid.

sired aldehydes. For catalytic oxidation the oxoammonium salt was postulated as the active intermediate [10].

Thus isolation of oxoammonium salts on insoluble, cross-linked polymer supports was investigated together with integration of the active resins in polymer-assisted solution phase synthesis [11]. These isolated oxoammonium salts could be employed in a water-free system with the intention of generating a highly reactive oxidizing agent that avoids over-oxidation to the acid, because of the absence of water.

The 4-hydroxy-TEMPO radical was coupled to 1% divinylbenzene polystyrene resin, using sodium hydride as base, yielding resin **1** with a loading of 0.93 mmol g^{-1} (Figure 3.6.3). ESR spectroscopy proved the presence of the free radical electron, because of the characteristic triplet signal as a result of coupling with the ^{14}N nucleus. Oxidation of the radical **1** to the oxoammonium resin **2** was best performed with N-chlorosuccinimide. The oxidation is accompanied by a distinct color change from colorless to a bright orange–red for chloride as counter-ion and brown–red for bromide. Chloride proved to be a superior counter-ion, because it was more reactive and led to fewer by-products.

The versatility of the novel reagent was investigated with a diverse selection of alcohols at room temperature with 3 equiv. of the reagent for 1 h. Results from the oxidations can be summarized as follows. Clean and rapid quantitative conversion to the respective aldehyde or ketone product was observed for all benzylic, allylic,

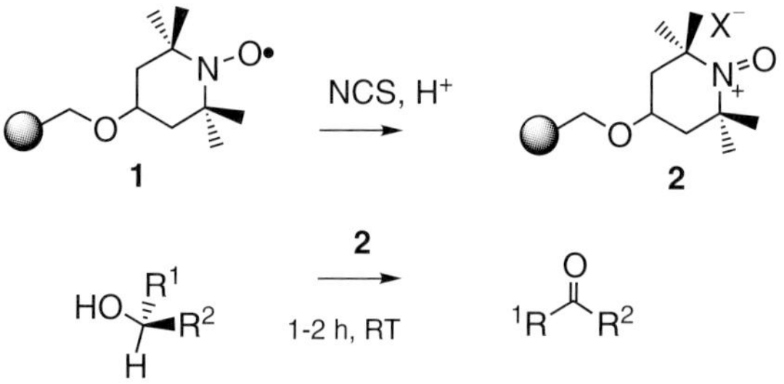

Fig. 3.6.3. Polymer-supported oxoammonium salts are highly reactive oxidants generated in situ by oxidation of the TEMPO radical with N-chlorosuccinimide.

and primary aliphatic alcohols and for most secondary aliphatic alcohols; yields were approximately 90%. As expected, diols yielded lactones in the secondary oxidation step. Easily enolizable primary ketones obtained from cyclohexanol, 1-phenylpropan-2-ol, and cholesterol could be further converted to the respective 1,2-diones. In this reaction the primary oxidation product, e.g. cyclohexanone, is transformed to the final diketone via an enolized intermediate. In addition, oxidizing resin **2** was effective in the conversion of a compound collection comprising 15 chemically diverse alcohols. Under the non-aqueous conditions described, however, the oxoammonium resin failed in the conversion of nitrogen-containing substrates such as protected amino alcohols, presumably because of to a single-electron oxidation reported earlier [9].

In summary, the polymer-bound oxoammonium reagent was highly efficient in polymer-supported oxidation of different alcohols and was capable of cleanly converting complex compound collections. No overoxidation to carboxylic acids was observed. It is obvious that the reagent will be of great value in polymer-supported transformations in solution, in automated parallel synthesis operations, and in flow-through reactors in up-scaled production processes.

Catalytic applications employing TEMPO resin **1** are particularly desirable for preparations on larger scale, increasing the efficiency of the polymer-supported reagent. As additional work-up is required for removal of co-oxidants or mediators, catalytic applications do not fit well into the synthetic format of polymer-assisted synthesis. Supported TEMPO radical can be employed with sodium hypochlorite as a stoichiometric amount of co-oxidant for oxidation of alcohols. An interesting alternative is the use of oxone as insoluble co-oxidant together with tetrabutylammonium bromide as a transfer reagent [12]. This reagent system can easily be extended to a three-phase system, employing immobilized TEMPO radical, dissolved transfer reagent, and insoluble oxone as co-oxidant (Figure 3.6.4) [13]. Elemental oxygen can also be employed as co-oxidant for oxidations mediated by TEMPO-resin **1**. Although soluble copper(II) salts can be employed as mediators, the rate of the oxidation is considerably slower than with oxone as co-oxidant.

Immobilized TEMPO radical has also been used for one-pot oxidation of alcohols to carboxylic acids [14]. For this purpose TEMPO-resin **1** was combined with two

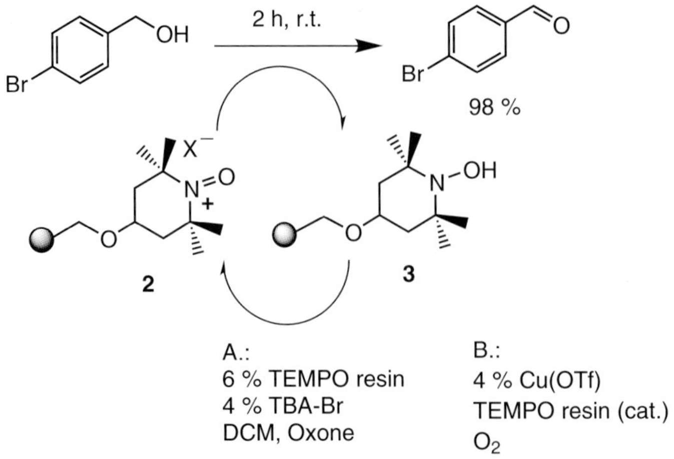

Fig. 3.6.4. Reactive oxoammonium resins can be regenerated catalytically in a three-phase system. With only 6% of TEMPO resin, gram amounts of alcohols can be converted efficiently.

ion-exchange resins loaded with chlorite anions and hydrogen phosphate and in the presence of catalytic amounts of potassium bromide and sodium hypochlorite in solution. The reaction requires work-up to remove salts, but tolerates several protecting schemes and affords pure products in good to excellent yields. The reaction can be rationalized by catalytic TEMPO-oxidation to aldehydes driven by dissolved hypochlorite followed by oxidation to the carboxylic acids effected by chlorite.

3.6.3.3
Oxidations with Immobilized Periodinanes

In recent years hypervalent iodine compounds have experienced extensive investigations yielding many results of practical synthetic importance. Supported iodinanes, i.e. iodoso or iodine(III) reagents, have been prepared by several groups, mainly as the bis-acetoxyiodoso derivatives [15–18] or as the respective dihalogeno compounds [19]. Iodoso reagents are employed in the oxidation of hydroquinones and phenols that have been exploited in the formation of spiroketals from a variety of tyrosines.

In contrast, periodinanes (i.e. iodoxo or iodine(V) reagents), preferably as the 1-hydroxy-(1H)-benzo-1,2-iodoxol-3-one-1-oxide (2-iodoxybenzoic acid, IBX) [20, 21] or its acetylation product, the Dess–Martin reagent [22], have been widely used for oxidation of sensitive and complex alcohols. Periodinanes have not yet been prepared on a polymer support; a silica-supported IBX has been reported recently [23].

The limitations of oxoammonium resins in the oxidation of nitrogen-containing moieties prompted investigation of polymer-supported periodinanes as poten-

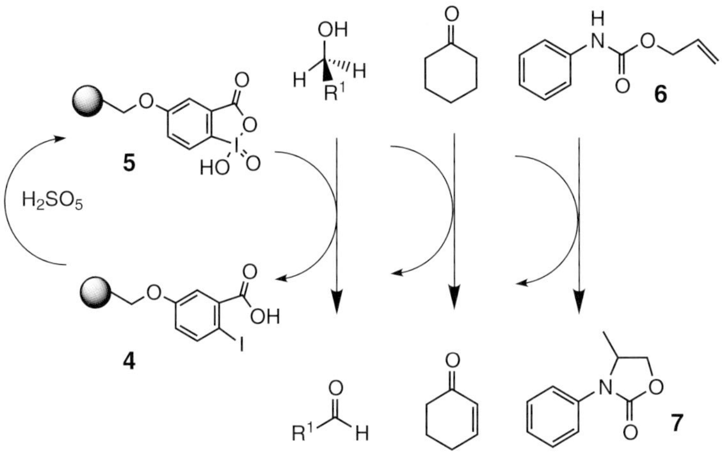

Fig. 3.6.5. Polymer-supported IBX can be activated and recycled with monoperoxy sulfonic acid (Caro's acid). The reagent is capable of alcohol oxidations, dehydrogenations, and radical cyclization reactions.

tial alternatives [15]. To obtain a functional iodine(V) reagent, a derivative of 2-iodobenzoic acid was required which would be suitable for immobilization yet retain oxidation properties similar to those of the parent compound. 5-Hydroxy-2-iodobenzoic acid esters can be efficiently immobilized on chloromethyl polystyrene via the phenoxide. The alkoxy derivatives have been prepared and oxidized to the iodoso derivative, but not to the iodoxo derivative. Methyl 5-hydroxy-2-iodobenzoate was obtained in two steps from 3-hydroxyanthranilic acid by a Sandmeyer reaction followed by esterification with thionyl chloride in methanol. It was coupled to chloromethyl polystyrene crosslinked with 1% divinylbenzene (1.20 mmol g^{-1}) by using cesium carbonate as base (Figure 3.6.5). The loading of the resin was determined by elemental analysis and was close to the theoretical value (98%). Saponification was effected by treatment with potassium trimethylsilanoxide in THF, yielding resin **4**.

Oxidation of **4** to resin **5** has been investigated under different conditions. Initial screening of oxidizing activity was conducted by HPLC analysis of the reaction with piperonyl alcohol as test substrate. Generation of monoperoxysulfonic acid (i.e. Caro's acid) was identified as the best oxidation conditions. By using an equimolar mixture of tetrabutylammonium oxone with methylsulfonic acid (DCM, RT, 3 h) resin **5** was furnished with a high oxidizing activity of 0.8 mmol g^{-1}. Resin **5** was characterized by IR spectroscopy, elemental analysis, and MAS NMR. Elemental analysis indicated a loading of 0.84 mmol g^{-1}, corresponding to a yield of 94% relative to the initial loading and taking into account the mass increase of the resin. No loss of iodine was observed under the strongly acidic reaction conditions. The oxidizing polymer **5** was stable towards air and moisture and could be stored without loss of activity.

The oxidation properties of periodinane reagent 5 (1.75 equiv., DCM, RT, 3 h) were investigated by reaction with a collection of a variety of alcohols including benzylic, allylic, and primary and secondary alcohols, including the unsaturated terpene alcohols citronellol and geraniol, and the carbamate-protected aminoalcohols Fmoc-Phe-ol and Fmoc-Ile-ol. All reactions were followed by GC–MS or by HPLC (215 and 280 nm). Products were identified by NMR spectroscopy and by mass spectrometry (EI, 70 eV); isolated yields were determined by weight. Most alcohols were converted to the respective aldehyde or ketone products in good to excellent yields and purities. After extensive washing resin 5 that had not been used at elevated temperatures could be recycled by repeated oxidation. In addition to the oxidation of alcohols, further important transformations effected by IBX were investigated with resin 5. Cyclohexanol reacted with 5 in a closed vessel (2.3 equiv., DCM, 2 h, 65 °C) yielding the α,β-unsaturated cyclohexenone via cyclohexanone and a postulated iodine–enol ether intermediate [24]. The unsaturated carbamate 6 was treated with reagent 5 (4 equiv., THF/DMSO 10:1, 90 °C, 16 h) to effect radical cyclization affording product 7 in 30% yield [25]. It should be noted that IBX at elevated temperatures can oxidize benzylic positions, which are abundant in the polystyrene backbone of 4; this might account for a competing reaction pathway.

Resin 5 was prepared as the first polymer-supported periodinane reagent. The resin was obtained with high loading (0.8 mmol g^{-1}) and was capable of converting a collection of diverse alcohols including complex and sensitive structures efficiently and in good to excellent yields to the respective carbonyl compounds. The α,β-desaturation of carbonyl compounds and the radical cyclization of an unsaturated carbamate were also demonstrated. The novel reagent is likely to find broad application in polymer-assisted solution phase synthesis. The new oxidizing resin should also be well suited to integration into parallel polymer-supported reaction sequences and to conversion of compound libraries.

3.6.3.4
Preparation of Peptide Aldehyde Collections

Peptide aldehydes are important biochemical tools for specific and reversible inhibition of serine, cysteine, and aspartate proteases in-vitro and in-vivo. Whereas several methods for the preparation of peptide aldehydes have been described [26–28], their reliable parallel synthesis remained a tedious endeavor. Peptide aldehydes suffer from configurational lability in the α-carbonyl position and from high chemical reactivity.

A polymer-supported approach employing oxidizing IBX-resin could be a valuable contribution to solving the task [29]. Thus, a collection of 24 C-terminal peptide alcohols was synthesized on trityl resin. Cleavage of the fully protected alcohols could be effected by treatment with hexafluoroisopropanol. The lyophilized pure products were then treated with 2 equiv. IBX-resin for 2 h. Conversions could be followed by NMR spectroscopy (Figure 3.6.6). A single aldehyde signal indicated conversion to the peptide aldehydes without racemization. Prolonged storage in

3.6.4 Polymer-supported Acylanion Equivalents | 285

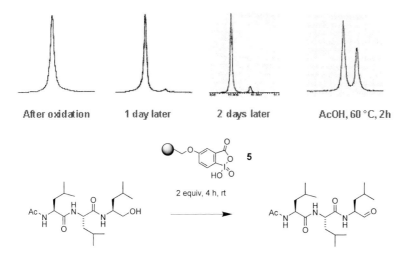

Fig. 3.6.6. Smooth and racemization-free polymer-assisted preparation of peptide aldehydes with oxidizing polymer **5**. Racemization of the peptide aldehydes was effected by acid treatment and monitored by ^{1}H-NMR at 9.3–9.5 ppm.

CDCl$_3$ solution led to the slow formation of a second diastereomer. Heating in dilute acetic acid (60 °C) effected rapid racemization.

Prolonged storage and release of pure peptide aldehydes could be realized by a threonine-scavenging resin. Peptide aldehydes were scavenged directly from the oxidation solution after washing, drying, and storage of the peptide aldehyde as C-terminal 1,3-oxazolidine derivative. Release of pure peptide aldehydes was effected by treatment with 1% AcOH at room temperature (Figure 3.6.7).

3.6.4
Polymer-supported Acylanion Equivalents [30]

Polymer-supported carbanion equivalents are the obvious supplement to polymer-supported access to reactive electrophiles either by oxidizing polymers or by release of carbenium ions. The combination of an oxidizing resin with a support carrying carbanion equivalents will be especially rewarding, enabling reaction sequences with C–C coupling steps and thus opening access to a wealth of potentially relevant products.

Bestatin **6a** (Figure 3.6.8) is the lead structure of a family of highly potent protease inhibitors containing the 3-amino-2-hydroxy-4-phenylbutanoic acid as the active, isosteric motif, mimicking the tetrahedral intermediate formed during amide bond cleavage [31]. Several clinically employed HIV-protease inhibitors are from this class of compound. The oxidized α-keto-β-amino derivatives have also

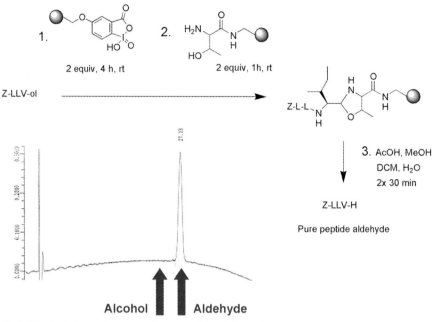

Fig. 3.6.7. Isolation and storage of pure peptide aldehydes by using an improved catch-and-release procedure [39]. Purity of peptide aldehydes was determined by RP-HPLC (214 nm).

been described as powerful inhibitors [32, 33]. Several bestatin syntheses have been described, although syntheses of variations of the general structure **6** are tedious and have been limited so far, especially in respect of the 4-substituent [34, 35]. Thus, a method enabling efficient variation of the target structure in all relevant positions would be highly desirable. Such a diversity-oriented procedure based on novel polymeric reagents was the objective of our work.

In Nature the ubiquitous α-hydroxy carbonyl motif is provided by acyl anion equivalents derived from thiamine pyrophosphate (vitamin B_1) as enzymatic cofactor. Thus our synthesis of compounds **8** was based on the construction and in-

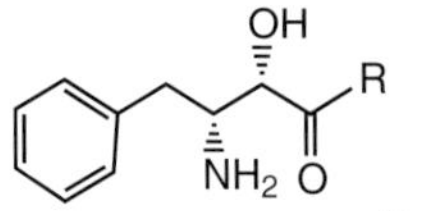

Bestatin (2S,3R): R = Leu
Probestin: R = Leu-Pro-Pro
Phebestin: R = Leu-Phe

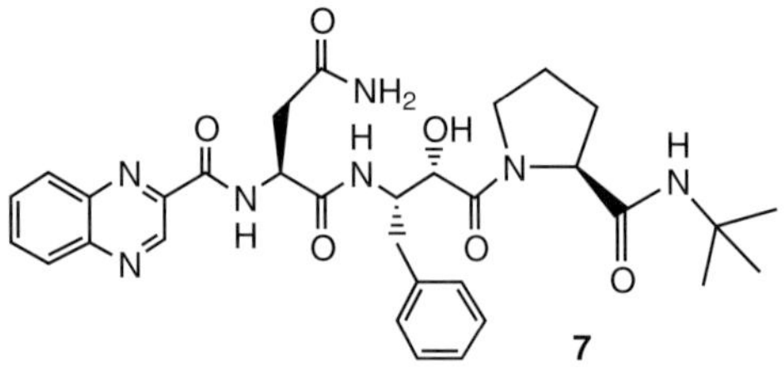

R-87366 (2S,3S)

Fig. 3.6.8. Structures of protease inhibitors containing the β-amino-α-hydroxy carbonyl motif.

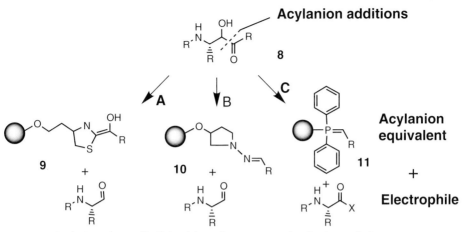

Fig. 3.6.9. Synthetic pathways (A–C) involving polymer-supported acyl anion equivalents.

vestigation of polymer-supported acyl anion equivalents. Ideally, the successful method should enable reaction with a readily available building block, under mild neutral conditions, to enable further derivatization after the central CC-coupling reaction. On polymer supports, dithioacetals have been suggested as acyl anion equivalents; these require strongly basic conditions for addition and mercury for release, however, thus failing in efficient preparation of complex and diverse functionality.

Three approaches leading to **8** were considered (Figure 3.6.9): Following the biosynthetic pathway directly, polymer-supported thiamine **9** was constructed (path A) and could lead via crossed acyloin couplings to the target structure. Polymer-supported hydrazones **10** were reported to add directly to aldehydes in a non-catalyzed Umpolung reaction (path B) with results reported in due course. Finally, phosphine ylides **11** were investigated as polymer-supported acyl anion equivalents (path C).

Pathway C seemed to be especially attractive, because it should enable addition of acyl anion equivalents to a large number of readily accessible activated carboxylic acids (Figure 3.6.10). Thus diversity in all relevant positions should be readily attainable. High-loaded triphenyl phosphine resin **12** (1.6 mmol g^{-1}) was alkylated with bromoacetonitrile under the action of microwave irradiation yielding phosphonium salt **13** quantitatively. **13** was converted into stable ylide **14** by treatment with tertiary amine. Carboxylic acids were activated in the presence of N-(3-dimethylaminopropyl)-N′-ethylcarbodiimide hydrochloride (EDC) and reacted with **14** yielding acyl cyanophosphoranes **15**. The reaction was monitored by ATR-IR; coupling yields could be determined by spectrophotometric Fmoc-determination and were 90% for Fmoc-phenylalanine as reference amino acid.

To investigate possible variation in the R^2 position different amino acids were investigated in the acylation reaction. Acyl cyanophosphoranes **15** could be deprotected and varied at the N-terminus, enabling access to diverse product variations.

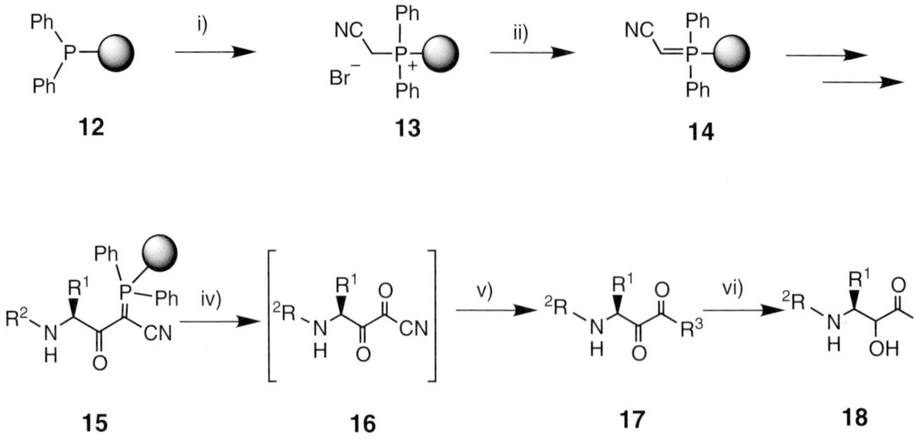

Fig. 3.6.10. General access to β-amino-α-hydroxy carbonyl compounds: (i) 5 equiv. BrCH$_2$CN, toluene, 150 °C, MW, 15 min; (ii) 5 equiv. TEA, DCM, RT, 2 h; (iii) 5 equiv. protected amino acid (R^1 = FMOC, BOC), 5 equiv. EDCI, 0.5 equiv. DMAP, DCM, RT, 4 h; (iv) O$_3$, DCM, −78 °C, 10 min; (v) O-, N-, and S-nucleophiles; (vi) (polystyrylmethyl)trimethylammonium borohydride, MeOH, RT, 16 h.

Cleavage was effected by ozonolysis at −78 °C, the intermediate α,β-diketo nitriles **16** are activated amino acids that could be converted in situ with O-, N-, and S-nucleophiles, respectively. Excess nucleophile employed in this reaction can be removed by a scavenger resin. Direct conversion of compounds **17** to the target structure **18** was accomplished with the reducing resin (polystyrylmethyl)-trimethylammonium borohydride (Novabiochem). Under the conditions selected both diastereomers were obtained in roughly equal amounts and could be fully assigned by NMR after purification by reversed-phase HPLC (Figure 3.6.11). Asymmetric reduction of α-ketocarbonyl compounds has been reported in the literature [20].

In summary, we have established a powerful novel concept of polymer-supported acyl anion equivalents that have been demonstrated to enable addition of a large variety of readily accessible activated carboxylic acids. Diversity in all relevant positions of the norstatine core structure (Figure 3.6.1B) could be achieved smoothly.

The supported acyl anion reagents are a novel class of reagent linker combining the intermediate immobilization of a substrate with a C–C-coupling as transformation step.

3.6.5
Conclusions

Generation and release of polymer-supported reactive intermediates or activated reactants have been demonstrated as a powerful concept leading to several novel

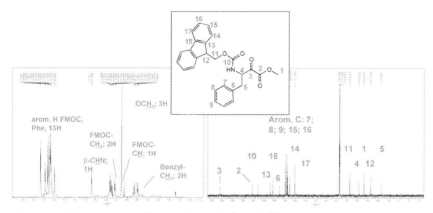

Fig. 3.6.11. Characterization of raw-product **17** by ^{1}H and ^{13}C NMR spectroscopy.

advanced polymer reagents for important transformations. The reagents are capable of cleanly converting sensitive single compounds and complex mixtures in solution. They are recyclable and/or can be employed catalytically.

Oxidizing polymers and polymer-supported carbanion equivalents, in particular, have been employed successfully for preparation of protease inhibitor collections as a practically relevant example. With peptide aldehydes and differently substituted α-hydroxy-β-amino carbonyl moieties two inhibitory motifs have been prepared with hitherto unprecedented ease and efficiency. Now it remains to be demonstrated that the synthetic methodology as introduced here is capable of assisting the identification of novel biologically active structures.

References

1 J. RADEMANN, M. BARTH, R. BROCK, H.-J. EGELHAAF, G. JUNG, *Chem. Eur. J.* **2001**, *7*, 3884–3889.

2 J. RADEMANN, J. SMERDKA, G. JUNG, P. GROSCHE, D. SCHMID, *Angew. Chem.* **2001**, *113*, 390–393; *Angew. Chem. Int. Ed.* **2001**, *40*, 381–385.

3 J. RADEMANN, P. SCHULER, G. NICHOLSON, unpublished work.

4 L. A. PAQUETTE, ed., *Encyclopedia of Reagents for Organic Synthesis*, Wiley, Chichester, **1995**.

5 B. HINZEN, S. V. LEY, *J. Chem. Soc. Perkin Trans. 1* **1997**, 1907–1910.

6 F. HAUNERT, M. H. BOLLI, B. HINZEN, S. V. LEY, *J. Chem. Soc. Perkin Trans. 1* **1998**, 2235–2237.

7 S. L. REGEN, *J. Am. Chem. Soc.* **1974**, *96*, 5175–5276.

8 A. E. J. DE NOOY, A. C. BESEMER, H. VAN BEKKUM, *Synthesis* **1996**, 1153–1174.

9 M. F. SEMMELHACK, C. R. SCHMID, D. A. CORTES, *Tetrahedron Lett.* **1986**, *27*, 1119–1122.

10 P. L. ANELLI, C. BIFFI, F. MONTANARI, S. QUICI, *J. Org. Chem.* **1987**, *52*, 2559–2562.

11 S. WEIK, G. NICHOLSON, G. JUNG, J. RADEMANN, *Angew. Chem.* **2001**, *113*, 1489–1492; *Angew. Chem. Int. Ed.* **2001**, *40*, 1436–1439.

12 C. BOLM, A. S. MAGNUS, J. P. HILDEBRAND, *Org. Lett.* **2000**, *2*, 1173–1175.

13 S. Barthelemy, S. Weik, J. Rademann, unpublished results, 2001.
14 K. Yasuda, S. V. Ley, *J. Chem. Soc. Perkin Trans. 1* **2002**, 1024–1025.
15 G. Sorg, A. Mengel, G. Jung, J. Rademann, *Angew. Chem.* **2001**, *113*, 4532–4535; *Angew. Chem. Int. Ed.* **2001**, *40*, 4395–4397.
16 M. L. Hallensleben, *Angew. Makromol. Chem.* **1972**, *27*, 223–227.
17 S. V. Ley, A. W. Thomas, H. Finch, *J. Chem. Soc. Perkin Trans. 1* **1999**, 669–671.
18 G.-P. Wang, Z.-C. Chen, *Synth. Commun.* **1999**, *29*, 2859–2866.
19 M. Zupan, A. Pollak, *J. Chem. Soc., Chem. Commun.* **1975**, 715–716.
20 M. Frigerio, M. Santagostino, *Tetrahedron Lett.* **1994**, *35*, 8019–8022.
21 C. Hartmann, V. Meyer, *Chem. Ber.* **1893**, *26*, 1727–1732.
22 D. B. Dess, J. C. Martin, *J. Org. Chem.* **1983**, *48*, 4156–4158.
23 M. Mülbaier, A. Giannis, *Angew. Chem.* **2001**, *113*, 4530–4532; *Angew. Chem. Int. Ed.* **2001**, *40*, 4393–4395.
24 K. C. Nicolaou, Y.-L. Zhong, P. S. Baran, *J. Am. Chem. Soc.* **2000**, *122*, 7596–7597.
25 K. C. Nicolaou, Y.-L. Zhong, P. S. Baran, *Angew. Chem.* **2000**, *112*, 639–642; *Angew. Chem. Int. Ed.* **2000**, *39*, 625–628.
26 J. A. Fehrentz, M. P. A. Heitz, J. Velek, F. Winternitz, J. Martinez, *J. Org. Chem.* **1997**, *62*, 6792–6796.
27 B. J. Hall, J. D. Sutherland, *Tetrahedron Lett.* **1998**, *39*, 6593–6596.
28 N. J. Ede, A. M. Bray, *Tetrahedron Lett.* **1997**, *38*, 7119–7122.
29 G. Sorg, B. Thern, G. Jung, J. Rademann, manuscript in preparation, **2002**.
30 S. Weik, J. Rademann, *Angew. Chem.* **2003**, *115*, 2595–8; *Angew. Chem. Int. Ed.* **2003**, *42*, 2491–2494.
31 H.-H. Otto, T. Schirrmeister, *Chem. Rev.* **1997**, *97*, 133–171.
32 Z. Li, A.-C. Ortega-Vilain, G. S. Patil, D.-L. Chu, J. E. Foreman, D. D. Eveleth, J. C. Powers, *J. Med. Chem.* **1996**, *39*, 4089–4098.
33 B. Munoz, C.-Z. Giam, C.-H. Wong, *Bioorg. Med. Chem.* **1994**, *2*, 1085–1090.
34 H. H. Wasserman, M. Xia, A. K. Petersen, M. R. Jorgensen, E. A. Curtis, *Tetrahedron Lett.* **1999**, *40*, 6163–6166.
35 M.-K. Wong, C.-W. Yu, W.-H. Yuen, D. Yang, *J. Org. Chem.* **2001**, *66*, 3606–3609.

B.14
Polymer-supported Synthetic Methods – Solid-phase Synthesis (SPS) and Polymer-assisted Solution-phase (PASP) Synthesis

Jörg Rademann

Although complex organic molecules can be constructed in homogeneous solution, for diversity-oriented purposes it is often advantageous to employ a multiple-phase system. This might facilitate isolation and separation procedures, the removal of excess reagents, and the completion of reactions. Solid-phase synthesis, the most widely applied example of use of multiple-phase systems in combinatorial chemistry, has significant advantages over use of homogeneous single phases. For SPS synthesis an insoluble polymer is employed as a point of attachment for surfaces; soluble polymers, fluorous biphase systems, or supercritical carbon dioxide are examples of other multiple phase systems employed in synthesis or in screening.

I. Solid phase synthesis (SPS)

II. Polymer-assisted solution phase synthesis (PASP synthesis)

a) Polymer reagents

b) Scavengers

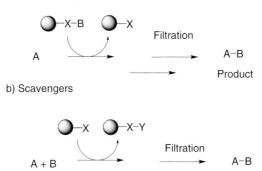

Fig. B.14.1. Classical solid-phase synthesis (SPS) and polymer-assisted solution phase (PASP). In SPS (I) products are constructed step-wise, covalently attached to an insoluble polymer support. PASP synthesis combines the merits of solution-phase chemistry with the advantages of facilitated phase separation by using polymer reagents (a) or scavenger resins (b).

Synthesis in solution, however, has indisputable advantages with regard to the versatility of the reactions applicable, the ease of analytical monitoring, and the accumulated knowledge of synthetic procedures. An ideal synthetic strategy would be one which combined these merits with the advantages of solid-phase synthesis – the possibilities of using reagents in high excess, to remove them by filtration, and to employ automated multiple synthesizers. This combination is realized in polymer-assisted solution-phase (PASP) synthesis, either by using scavenger resins or by implementation of polymer reagents (Figure B.14.1) [1–4]. Polymer reagents can be used in high excess and are removed by filtration, the products can be easily analyzed and further transformed in solution. They are especially suited to the transformation of products obtained by parallel and split–recombine combinatorial synthesis. Furthermore, polymer reagents enable the preparation of complex libraries by multi-step syntheses in solution. They can be used in automated and flow-through systems, and, finally, can be used to transform single compounds and complex mixtures.

The first polymer-supported reagents were derived from ion-exchange resins by immobilizing ionic reagents on macroporous polystyrene resins [5]. This approach enables easy access to many reagents. For preparation, a

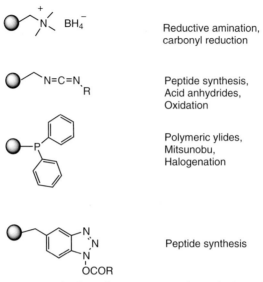

Fig. B.14.2. The first polymer reagents, either in the form of ion-exchange resins or covalently attached, were employed for a variety of simple transformations in the 1960s and 70s.

solution of the respective salt is added to the resin in excess, and removed after equilibration by washing with non-ionic solvents. Leaching of the reactive ions is, however, a general problem of this type of support. In principle, the immobilized ions can be exchanged by other competing ions in the solution. Today several important polymer reagents are still based on ion-exchange resins; examples include the borohydride resins [6, 7] for reductive amination and the perruthenate resin [8] for oxidation.

The next generation of polymer-supported reagents was based on covalently linked reactants. The concept was initiated in the 1960s and 1970s with the introduction of peptide coupling reagents such as supported carbodiimides [3, 9], and active esters [10], of supported phosphines [11], and of polymeric bases (Figure B.14.2). A major advantage of covalent linking is to avoid leaching of the immobilized reagents from the resin. At that time polymer-supported chemistry was, however, still limited to few fundamental conversions and was not accepted as a useful synthetic method beyond the areas of peptide and oligonucleotide chemistry. Only when solid-phase chemistry became a powerful tool of combinatorial chemistry in the 1990s did polymer-supported reagents also gain acceptance and were employed for generation of compound libraries and for multi-step syntheses of natural products.

For these reasons research on polymer reagents has soared during recent years. Today, increasingly demanding reactions can be performed with the help of polymer reagents based on the successful isolation of reactive intermediates [2]. Further progress must be made in the development of support materials tailored as carriers of polymer reagents [12].

References

1 A. Akelah, D. C. Sherrington, Chem. Rev. **1981**, *81*, 557–587.
2 This volume, Chapter 3.6.
3 A. Kirschning, H. Monenschein, R. Wittenberg, Angew. Chem. **2001**, *113*, 670–701; Angew. Chem. Int. Ed. **2001**, *40*, 650–679.
4 S. V. Ley, I. R. Baxendale, R. M. Bream, P. S. Jackson, A. G. Leach, D. A. Longbottom, M. Nesi, J. S. Scott, R. I. Storer, S. J. Taylor, J. Chem. Soc. Perkin Trans. 1 **2000**, *23*, 3815–4195.
5 F. Helfferich, Ion Exchange, McGraw–Hill, New York, 1962.
6 B. Sansoni, O. Sigmund, Naturwissenschaften **1961**, *48*, 598–599.
7 H. W. Gibson, F. C. Bailey, J. Chem. Soc. Chem. Commun. **1977**, 815.
8 B. Hinzen, S. V. Ley, J. Chem. Soc. Perkin Trans. 1 **1997**, 1907–1910.
9 Y. Wolman, S. Kivity, M. Frankel, J. Chem. Soc. Chem. Commun. **1967**, 629–630.
10 R. Kalir, A. Warshawsky, M. Fridkin, A. Patchornik, Eur. J. Biochem. **1975**, *59*, 55–61.
11 W. Heitz, R. Michels, Angew. Chem. **1972**, *84*, 296–297; Angew. Chem. Int. Ed. **1972**, *12*, 298–299.
12 J. Rademann, M. Barth, Angew. Chem. **2002**, *114*, 3087–3090; Angew. Chem. Int. Ed. **2002**, *41*, 2975–2978.

B.15
Inhibition of Proteases

Jörg Rademann

Protease inhibitors are of interest as tools to model and understand the binding of ligands to proteins and the function of hydrolases. They also constitute one of the major target groups of drugs in development and in clinical use. Proteases are ubiquitous and multifunctional in the human body. Deciphering the human genome revealed approximately 500 protease genes of the four major protease families – serine-, aspartate-, cysteine- [1], and metalloproteases [2]. In addition, many proteases of pathogens or viruses are also interesting drug targets [3]. Thus, to avoid side-effects selectivity is a major issue in the development of protease inhibitors. Because of the complexity and pharmaceutical relevance of protease inhibition, scientific work on the subject is overwhelmingly rich and has been documented in excellent reviews [3–5].

Essentially, all known protease inhibitors are substrate analogs binding to the active site of the enzyme (Figure B.15.1). Because the active site of all

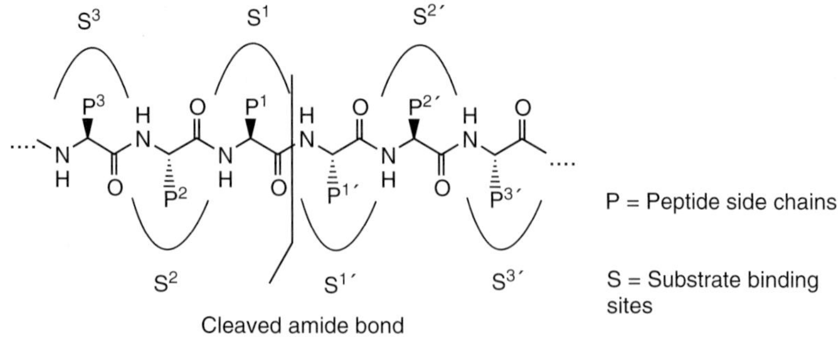

Fig. B.15.1. Most protease inhibitors are substrate analogs. Binding to the substrate binding sites (S) of the enzyme is established by interactions with the peptide side-chains (P) and the substrate backbone. The P-site of the substrate is located on the left hand side of the cleaved amide bond; the P'-site on the right hand side.

proteases has been selected for peptide or protein recognition during evolution, most inhibitors contain one or several amino acids and amide bonds. To constitute an inhibitor, the recognized ligand must contain a non-cleavable element at the active site that interacts with the catalytically active protease side chain.

Many of these inhibitory elements are either electrophiles or nucleophiles characteristic of one type of protease.

Peptide aldehydes constitute a rather general example of protease inhibitors. The electrophilic carbonyl group is attacked reversibly by the cleaving nucleophile, forming a covalent acetal or thioacetal intermediate. With cysteine proteases the preferred inhibitors are strong electrophiles, for example ketones, chloromethyl ketones, epoxides, or vinyl sulfones. Many cysteine protease inhibitors form an enzyme–inhibitor complex irreversibly; these are therefore denoted "suicide-inhibitors".

Members of a second prominent group of mechanism-based inhibitors contain a peptide isostere, an element mimicking an intermediate formed during amide bond cleavage. Inhibitors of metalloproteases often contain a metal-chelating unit such as hydroxamate or phosphonic acids, which act as a bidentate ligand.

Understanding and development of protease inhibitors faces several serious challenges. First, high affinity of the inhibitors is demanded. Often, however, it is even more difficult to target selectively only one of the many proteases that are expressed simultaneously by an organism. The selectivity of inhibitors for one or few proteases usually requires extended fine-tuning of substituents around the isosteric element to scan the active site of the targeted protease. In many instances the bioavailability of the amide-rich and flexible structures poses a third serious challenge to the medicinal chemist.

Tab. B.15.1. Principal types of protease inhibitor, representative structures, and preferred target proteases.

Entry	Inhibitor	Structure	Type of inhibition	Target proteases
1	Peptide aldehydes		Covalent, reversible	Serine-, cystein-, aspartate-
2	Suicide inhibitors		Covalent, irreversible	Cystein-
3	Peptide isostere		Transition state mimic	Aspartate-, serine-
4	Hydroxamates		Metal chelator	Metallo-

High-throughput synthesis and combinatorial chemistry employing polymer supports, with computer-based rational design, are important tools for systematic optimization of the affinity, selectivity, and bio-availability of protease inhibitors. It is very likely that currently available isosteres will not suffice for addressing the natural selectivity of proteases. Thus, synthetic methodology must be devised not only to decorate a predefined isosteric core but to define novel active site binders.

References

1 A. A. Hernandez, W. R. Roush, Curr. Opin. Chem. Biol. 2002, 6, 459–465.
2 D. H. Kim, S. Mobashery, Curr. Med. Chem. 2001, 8, 959–965.
3 L. Tong, Chem. Rev. 2002, 102, 4609–4626.
4 H.-H. Otto, T. Schirrmeister, Chem. Rev. 1997, 97, 133–171.
5 H. M. Abdel-Rahman, G. S. Al Karamany, N. A. El Koussi, A. F. Youssef, Y. Kiso, Curr. Med. Chem. 2002, 9, 1905–1922.

Part 4
Studies in Diagnostic Developments

4.1
Selectivity of DNA Replication

Andreas Marx, Daniel Summerer, and Michael Strerath

4.1.1
Introduction

All DNA synthesis required for DNA repair, recombination, and replication depends on the ability of DNA polymerases to recognize the template and correctly insert the complementary nucleotide. The mechanisms whereby these enzymes achieve this tremendous task has been a central topic of interest since the discovery of the first DNA polymerase, *E. coli* DNA polymerase I, by Arthur Kornberg approximately half a century ago [1]. Since then enormous efforts from scientists in many disciplines have been undertaken to gain insights into the complex mechanisms and functions of these molecular machines.

DNA polymerases catalyze DNA synthesis in a template-directed manner (Box 16). For most known DNA polymerases a short DNA strand hybridized to the template strand is required to serve as a primer for initiation of DNA synthesis. Nascent DNA synthesis is promoted by DNA polymerases by catalysis of nucleophilic attack of the 3'-hydroxyl group of the 3'-terminal nucleotide of the primer strand on the α-phosphate of an incoming nucleoside triphosphate (dNTP), leading to substitution of pyrophosphate. This phosphoryl transfer step is promoted by two magnesium ions that stabilize a pentacoordinated transition state by complexation of the phosphate groups and essential carboxylate moieties in the active site (Figure 4.1.1) [2].

Some DNA polymerases achieve selective information transfer according to the Watson–Crick rule with error rates as low as one mistake in one million synthesized nucleotides [3]. What are the underlying mechanisms accounting for this strikingly high selectivity? At first glance the formation of distinct hydrogen-bonding patterns between the nucleobases of the coding template strand and the incoming nucleoside triphosphate might be responsible for the selective information transfer. Nevertheless, as been suggested by Goodman and coworkers on the basis of thermal denaturing studies of matched and mismatched DNA complexes, these interactions alone are not sufficient to explain the extent of selectivity commonly observed for enzymatic DNA synthesis [3g,h, 4]. Thus, further more complex interactions, for example geometric constraints acting on the substrates in the

Highlights in Bioorganic Chemistry: Methods and Applications. Edited by Carsten Schmuck, Helma Wennemers.
Copyright © 2004 WILEY-VCH Verlag GmbH & Co. KGaA, Weinheim
ISBN: 3-527-30656-0

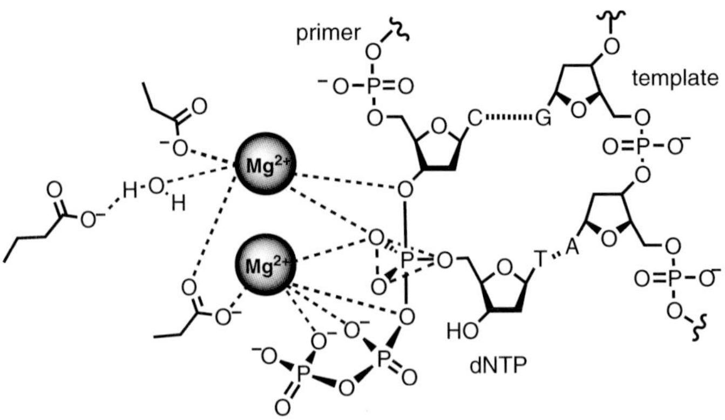

Fig. 4.1.1. DNA polymerase-catalyzed nucleotide insertion.

enzyme's active site, might significantly contribute to the observed high fidelity. Studies devoted to gaining insight into mechanisms that prevent nucleotide misinsertion have received most attention and are the major focus of this chapter. It should, nevertheless, be mentioned that beside misinsertion of non-canonical nucleotides, errors that occur as a result of misalignment of the template primer complex are also major sources of mutations such as single nucleotide deletions and insertions [3c]. Here, we will briefly summarize recent efforts and concepts dedicated to elucidation of the origin of DNA polymerase nucleotide insertion selectivity. The aim of this chapter is to highlight how a multitude of efforts from scientists of many disciplines ranging from biochemistry and structural biology to (bio)organic chemistry have contributed to gain valuable insights into a complex biological process such as enzymatic DNA replication.

4.1.2
Biochemical and Structural Studies

Valuable insights into how DNA polymerases process their substrates were obtained as a result of detailed kinetic studies of the enzymes. Benkovic and coworkers employed rapid quenching techniques to study the kinetics of transient intermediates in the reaction pathway of DNA polymerases [5]. Intensive studies revealed that *E. coli* DNA polymerase I follows an ordered sequential reaction pathway when promoting DNA synthesis. Important aspects of these results for DNA polymerase fidelity are conformational changes before and after the chemical step and the occurrence of different rate-limiting steps for insertion of canonical and non-canonical nucleotides. *E. coli* DNA polymerase I discriminates between canonical and non-canonical nucleotide insertion by formation of the chemical bond. Bond formation proceeds at a rate more than several thousand times slower when an incorrect dNTP is processed compared with canonical nucleotide insertion.

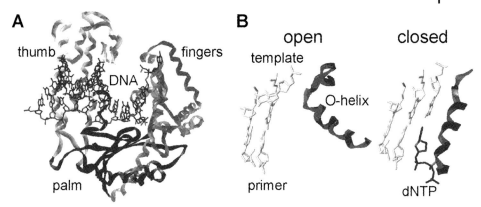

Fig. 4.1.2. A. Structure of *Thermus aquaticus* (*Taq*) DNA polymerase bound to the DNA primer–template complex. B. "Open" and "closed" conformation after dNTP binding of *Taq* DNA polymerase. Structures were built on PDB entries 3KTQ and 4KTQ.

Recently determined crystal structures of several DNA polymerases in complexes with their DNA and dNTP substrates have contributed significantly to our understanding of structure and substrate recognition by these complex enzymes [6]. Most DNA polymerases with known structures have a large cleft in which the primer template complex is embedded (Box 16). By analogy of this conformation with a half open right hand the enzyme domains are termed thumb, palm, and fingers (Figure 4.1.2A).

The palm domain harbors the catalytic center comprising the essential carboxylates involved in the phosphoryl transfer reaction. The high conservation of this domain throughout distinct DNA polymerase families, e.g. eukaryotic, prokaryotic, and viral DNA polymerases, is striking. In contrast, the finger and thumb domains, which make extensive contact with the primer template complex and the incoming dNTPs, differ significantly among DNA polymerases. Results from structural investigation of DNA polymerases strongly support the occurrence of large conformational changes from an "open" to a "closed" conformation before phosphodiester bond formation, triggered by dNTP binding (Figure 4.1.2B). Editing of nascent nucleotide base-pair geometry during these transitions is believed to be a crucial determinant of DNA polymerase selectivity.

Data available from crystal structures of DNA polymerases suggest the formation of nucleotide binding pockets which preferentially accommodate Watson–Crick base pairs. Nevertheless, DNA polymerase selectivity often varies significantly depending on the DNA polymerase [7]. By comparison of crystal structures derived from high- and low-fidelity enzymes valuable insights were gained into structural differences which might be the origin of the often considerably different selectivity of these enzymes. Error-prone *Sulfolobus solfataricus* P2 DNA polymerase IV (Dpo4) was found to adopt the familiar overall shape of a half-open right hand, found in several DNA polymerases, in which parts of the primer-template are embedded [6d]. The palm domain is structurally similar to those of other DNA polymerases and the essential carboxylates in Dpo4 are in identical positions as found in high-fidelity DNA polymerases and in HIV-1 reverse transcriptase. The finger and thumb domains of Dpo4 which surround the incoming triphosphate and template nucleotide are, however, unusually small. The "O-helix" (Figure 4.1.2B) that is believed to be involved in editing of correct nucleobase geometry and present in all high-fidelity DNA polymerases is absent in Dpo4. In total, the nascent base pair between the template and the incoming nucleotide in Dpo4 is less tightly surrounded in the vicinity of this enzyme in comparison to high-fidelity DNA polymerases. The open and solvent-accessible active site might be one structural reason for the error-prone replication of DNA by this kind of DNA polymerase. Another structure of this enzyme had a complexed non-canonical dNTP bound to the active site. From the structure it is apparent that conformations of the sugar phosphate moieties of the primer, template and nucleoside triphosphate in the active site differ significantly from that found when a canonical nucleotide is bound [6d]. Because of translocation of the template without replication of the first template base (G), the incoming ddGTP forms a canonical base pair with the next template base (C). Such an alignment might be the origin of the apparent faulty DNA synthesis of this kind of enzyme leading to frameshifts (Figure 4.1.3).

Taking together, DNA polymerase structural data indicate a high degree of shape complementary between the active sites of the enzymes and the nucleotide substrates, suggesting that geometrical constraints are at least one cause of DNA polymerase fidelity.

These assumptions are further supported by the finding that mutations that are believed to alter the geometry of the binding pocket or the conformational changes

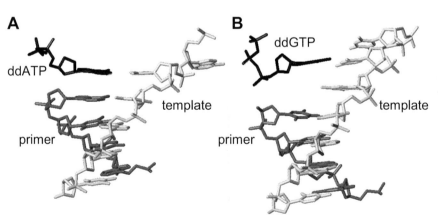

Fig. 4.1.3. DNA bound in the active site of Dpo4. A. Primer-template and incoming canonical triphosphate ddATP. B. Primer-template and incoming non-canonical triphosphate (ddGTP). Structures were built on PDB entry code 1JX4 and 1JXL.

needed to trigger catalysis effect the fidelity of the DNA polymerase [3a,b]. One of the most striking examples of this is the Arg283Ala mutation in DNA polymerase β. Arg283 is part of the nucleotide-binding pocket and its substitution with a sterically less demanding alanine moiety results in a marked decrease in fidelity. Similar results were obtained by mutation of *E. coli* DNA polymerase I and HIV-1 reverse transcriptase, strongly suggesting the participation of steric constraints in DNA polymerase selectivity mechanisms.

4.1.3
Use of Tailored Nucleotide Analogs to Probe DNA Polymerases

4.1.3.1
Non-polar Nucleobase Surrogates

Many nucleotide analogs have been synthesized, and their action on DNA polymerases has been studied. Most of this work has been summarized in recent excellent reviews [3a,d,f,g]. In this short summary we focus on selected examples used recently in endeavors devoted to a understanding DNA polymerase selectivity. Most of this work was undertaken to elucidate the non-covalent interactions that govern the formation of the nucleobase pairs. The impact of interstrand base pair hydrogen bonding on overall enzyme selectivity has been the main object of intensive study. As already mentioned, for a long time hydrogen bonding according to the Watson–Crick rule was viewed as "informational" [3a,d,g]. Consequently, it is a common perception that these interactions are primarily responsible for the ability of DNA polymerases to drive formation of the canonical nucleobase pairs.

To address the participation of hydrogen bonding of the coding nucleotide in the template strand with the nucleobase of the incoming dNTP in DNA polymerase selectivity mechanisms several new nucleotide analogs were designed and subsequently employed in in-vitro replication experiments.

To evaluate the participation of hydrogen bonding in DNA replication selectivity mechanisms Kool described a seminal functional strategy based on chemically modified DNA polymerase substrates [3a,d,g]. He developed nucleotide analogs in which the polar natural DNA nucleobases are replaced by non-polar aromatic molecules, which closely mimic the shape and size of the natural nucleobases but have at least significantly diminished ability to form stable hydrogen bonds (Figure 4.1.4).

These non-polar nucleotide isosteres were used as functional probes to elucidate the impact of hydrogen bonding on DNA polymerase selectivity. In preliminary experiments Kool et al. studied the insertion of dNTP opposite **F**, the non-polar isostere of thymidine (Figure 4.1.4), in the template strand [8]. If purely hydrogen-bonding drives selective nucleotide incorporation one would expect that incorporation opposite **F** is very inefficient and unselective. Interestingly, they observed the contrary when studying the Klenow fragment of *E. coli* DNA polymerase I. This

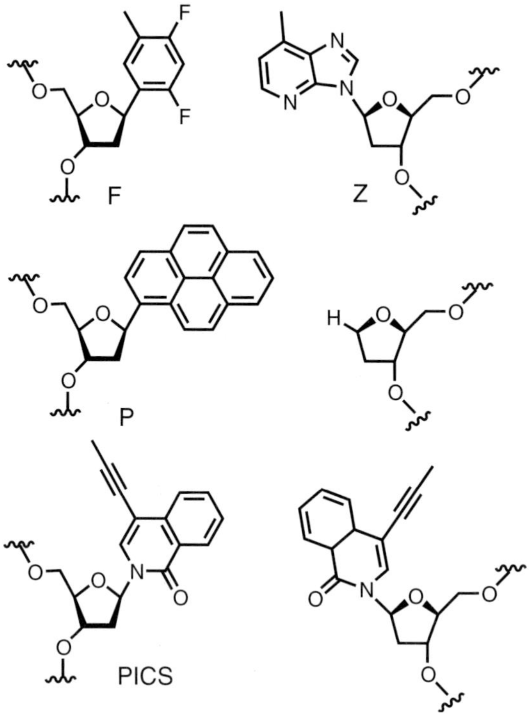

Fig. 4.1.4. Non-polar DNA polymerase substrates.

enzyme was able to promote nucleotide insertion opposite **F** efficiently with remarkable selectivity. Steady-state analysis revealed that the formation of the artificial **dA–F** pair was only a factor of four lower than that of the natural pair [8]. In the next set of experiments Kool et al. turned the base pair around. They found that **dFTP** was still inserted highly efficiently by the polymerase, and that the selectivity was as high as for the natural substrates [9]. On the basis of these results Kool proposed that hydrogen-bonding in enzymatic DNA synthesis was not as important as commonly believed. This assumption was further supported by recent findings that a completely artificial **F–Z** pair was processed by polymerases efficiently and with considerable selectivity (Figure 4.1.4), indicating that steric effects play at least a significant role in DNA polymerase selectivity processes [10]. Further evidence of such a steric model was obtained by studying **dPTP** insertion opposite abasic sites in the template strand (Figure 4.1.4) [11]. The pyrene group is nearly as large as a natural base pair and obviously has no significant hydrogen-bonding capacity. The space occupied by the pyrene moiety fills in the blank of the removed base in the template strand. Remarkably, the Klenow fragment and T7 DNA polymerase were found to insert **dPTP** opposite abasic sites more efficiently than opposite natural bases or another pyrene-bearing moiety in the template. Kool et al. concluded from these results that hydrogen bonding is not required to achieve high incorporation efficiencies and that significant levels of selectivity can be achieved without hydrogen bonds. DNA polymerase properties such as selectivity of DNA synthesis often vary significantly among different DNA polymerases [3, 7]. To gain insight into the origin of these processes Kool et al. performed a survey of protein–DNA interactions with seven different DNA polymerases and reverse transcriptases conducted with non-polar isosteres [12]. These studies indicated three different classes among the enzymes studied depending on the minor groove interactions and steric fit around a nascent base pair. For the majority of enzymes studied it seemed that no hydrogen bonds between the nascent base pairs are required if correct geometry and specific interactions through the minor groove are fulfilled. Kool concludes that close fitting to Watson–Crick geometry and satisfaction of specific minor groove interactions are among the most important factors in DNA replication.

That hydrogen bonding is not required for faithful nucleotide incorporation is further supported by recent impressive work from Schultz, Romesberg, and coworkers [13]. With the aim of expanding the genetic alphabet they synthesized an array of nucleotides bearing aromatic surrogates for the natural bases and subsequently tested them for their action on DNA polymerases. Interestingly, they could show that the Klenow fragment of E. coli DNA polymerase I was able to insert the propynylisocarbostyril (PICS) analogs opposite another PICS moiety in the template strand with high efficiency. These results confirmed that DNA polymerases can promote efficient base-pair formation in the absence of hydrogen bonding.

4.1.3.2
Analogs with Modified Sugar Moieties

Most functional studies have focused on nucleobase-recognition processes. Little is known about the impact of DNA polymerase interactions with the 2′-deoxyribose moiety and their participation in processes which contribute to fidelity. Crystal structures of DNA polymerases together with enzyme mutation studies suggest that the sugar moiety of the incoming triphosphate is fully embedded in the nucleotide binding pocket and undergoes essential interactions with the enzyme [3, 6]. This issue has been addressed and a functional strategy was developed to monitor steric constraints in DNA polymerases within the nucleotide binding pocket acting on the sugar moiety of an incoming nucleoside triphosphate [14, 15]. To sense interactions of DNA polymerases with the sugar moiety of incoming triphosphates alkyl labels at the 4′-position of the 2′-deoxyribose were introduced in such a way that they did not interfere with hydrogen bonding, nucleobase pairing, and stacking. The steric probes T^RTP were designed by substituting the 4′-hydrogen atom of thymidine triphosphate (TTP) with alkyl groups of continually increasing steric bulk (Figure 4.1.5).

If sugar recognition processes are involved in DNA polymerase fidelity mecha-

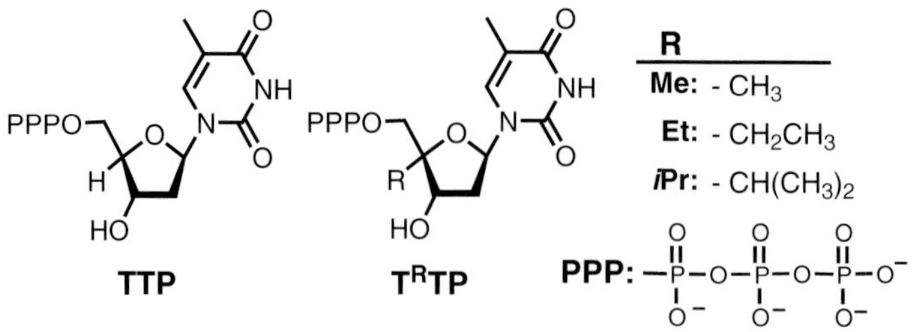

4.1 Selectivity of DNA Replication

Fig. 4.1.5. Steric probes with altered sugar moieties.

nisms 4′-alkyl modifications should significantly alter their substrate properties. On the basis of the steric model for DNA replication selectivity steric constraints within the active site during mismatch formation should be even more pronounced when processing the size-augmented nucleotide analogs. Consequently, this should lead to increased selectivity for nucleotide incorporation. Interestingly, that was indeed observed when the Klenow fragment was studied [14]. On the basis of steady-state kinetic analysis the enzyme is capable of inserting $T^{Me}TP$ and $T^{Et}TP$ with high efficiency opposite a canonical template base. On the other hand, misinsertion opposite non-canonical bases is approximately 100-fold less efficient than for the natural substrate T^HTP. These results further support the model that steric constraints are at least one crucial determinant of DNA polymerase selectivity.

As already mentioned, DNA polymerase selectivity often varies significantly depending on the DNA polymerase [3, 7]. The origin of this varying error propensity is elusive. It is assumed that DNA polymerases form nucleotide binding pockets that differ in properties such as shape and tightness (see also Chapter 4.3). Thus, high-fidelity DNA polymerases are believed to form more rigid binding pockets tolerating less geometric deviation whereas low-fidelity enzymes have more flexibility, leading to reduced fidelity. If varied active site tightness is, indeed, a crucial determinant of varied DNA polymerase selectivity, steric probes such as T^RTP should reflect this in differential action on DNA polymerases known for their error propensity, for example human immunodeficiency virus type 1 reverse transcriptase (HIV-1 RT). One would expect HIV-1 RT to process the bulkier thymidines more efficiently than the more selective Klenow fragment. Thus, the effects of thymidines T^RTP on wild-type HIV-1 RT and on the M184V mutant of the enzyme were tested [16, 17]. M184V mutation was used to introduce a β-methyl side-chain, present in valine, that is believed to contact the sugar ring of the incoming triphosphate [61]. Interestingly, this mutation has been shown to result in increased nucleotide insertion selectivity, which is attributed to increased steric constraints within the active site. On the basis of steady-state and transient kinetic investigations it was found that the M184V mutant incorporates the 4′-methylated thymidine $T^{Me}TP$ with significantly higher fidelity than the wild-type enzyme [16, 17].

This property can be attributed to the additional β-side-chain, present in valine and absent in methionine. This size-augmentation is monitored by the increased bulk of the steric probe $T^{Me}TP$ resulting in significantly lower misinsertion efficiency by the M184V mutant compared with the natural substrate.

Furthermore, with regard to "correct" insertion of the different T^RTP used, there is little difference between HIV-1 RT and the Klenow fragment on the basis of steady-state analysis. Analysis of misinsertion shows, however, that the two enzymes behave differently. Whereas 4'-methylation has little effect on the selectivity of HIV-1 RT, significant effects are observed for the Klenow fragment [14, 16]. On the basis of the concept of active-site tightness these results suggest that the enzymes differ most significantly when promoting misinsertion rather than insertion opposite canonical template bases. This might be the result of differential active site conformations causing different steric constraints while promoting "incorrect" nucleotide insertion. These studies provide the first conclusive experimental evidence that variations of steric constraints within the nucleotide binding pocket of at least two DNA polymerases cause differences in nucleotide incorporation selectivity.

4.1.4
Conclusions and Perspectives

The combined efforts of research groups from different disciplines, described briefly have yielded a more precise picture of the complex mechanisms which lead to faithful insertion of nucleotides according to Watson–Crick. It is now believed that a mere hydrogen bonding model as the sole driving force for fidelity of DNA polymerases cannot, alone, account for the observed selectivity. Geometric and steric effects seem to have much greater effect than originally assumed. We are, nevertheless, far from a complete understanding of these enzymes, which can be regarded as complex machines. To worsen the case, DNA polymerases participate in complex interactions with the primer-template strands during catalysis of DNA polymerization [3, 7]. Enzyme contacts with the primer-template are manifold and reach up to several nucleotide pairs beyond the catalytic center, usually leading to severe bending of the DNA near the primer 3'-end. The implication of these contacts on enzyme selectivity are far from being well understood. As a result of recent studies valuable new insight has been gained into these processes; that indicate the participation of these contacts in DNA polymerase fidelity mechanisms [18–20]. Interestingly, DNA-enzyme interactions might be exploited for development of new or improved highly selective DNA amplification approaches needed for efficient and reliable genome analysis [21]. Nevertheless, further detailed studies in a similar vein employing a multitude of multidisciplinary strategies are required to come even close to an understanding of these complex enzymes.

References

1. A. KORNBERG, T. A. BAKER, *DNA Replication*, 2nd edn., W. H. FREEMAN, New York, **1991**.
2. (a) C. M. JOYCE, T. A. STEITZ, *Annu. Rev. Biochem.* **1994**, *63*, 777–822; (b) T. A. STEITZ, S. J. SMERDON, J. JÄGER, C. M. JOYCE, *Science* **1994**, *266*, 2022–2025.
3. Reviews: (a) E. T. KOOL, *Annu. Rev. Biochem.* **2002**, *71*, 191–219; (b) P. H. PATEL, L. A. LOEB, *Nat. Struc. Biol.* **2001**, *8*, 656–659; (c) T. A. KUNKEL, K. BEBENEK, *Annu. Rev. Biochem.* **2000**, *69*, 497–529; (d) E. T. KOOL, J. C. MORALES, K. M. GUCKIAN, *Angew. Chem.* **2000**, *112*, 1046–1068; *Angew. Chem. Int. Ed.* **2000**, *39*, 991–1009; (e) T. A. KUNKEL, S. H. WILSON, *Nat. Struc. Biol.* **1998**, *5*, 95–99; (f) U. DIEDERICHSEN, *Angew. Chem.* **1998**, *110*, 1745–1747; *Angew. Chem. Int. Ed.* **1998**, *37*, 1655–1657; (g) M. F. GOODMAN, *Proc. Natl. Acad. Sci. USA* **1997**, *94*, 10493–10495; (h) H. ECHOLS, M. F. GOODMAN, *Annu. Rev. Biochem.* **1991**, *60*, 477–511.
4. (a) I. WONG, S. S. PATEL, K. A. JOHNSON, *Biochemistry* **1991**, *30*, 526–537; (b) D. L. SLOANE, M. F. GOODMAN, H. ECHOLS, *Nucleic Acids Res.* **1988**, *16*, 6465–6475; (c) J. PETRUSKA, M. F. GOODMAN, M. S. BOOSALIS, L. C. SOWERS, C. CHEONG, I. TINOCO JR, *Proc. Natl. Acad. Sci. USA* **1988**, *85*, 6252–6256.
5. (a) S. J. BENKOVIC, C. E. CAMERON, *Methods Enzymol.* **1995**, *262*, 257–269; (b) M. E. DAHLBERG, S. J. BENKOVIC, *Biochemistry* **1991**, *30*, 4835–4843; (c) R. D. KUCHTA, P. BENKOVIC, S. J. BENKOVIC, *Biochemistry* **1988**, *27*, 6716–6725; (d) R. D. KUCHTA, V. MIZRAHI, P. A. BENKOVIC, K. A. JOHNSON, S. J. BENKOVIC, *Biochemistry* **1987**, *26*, 8410–8417.
6. (a) L. F. SILVIAN, E. A. TOTH, P. PHAM, M. F. GOODMAN, T. ELLENBERGER, *Nat. Struct. Biol.* **2001**, *8*, 984–989; (b) B.-L. ZHOU, J. D. PATA, T. A. STEITZ, *Mol. Cell* **2001**, *8*, 427–437; (c) J. TRINCAO, R. E. JOHNSON, C. R. ESCALANTE, S. PRAKASH, L. PRAKASH, A. K. AGGARWAL, *Mol. Cell* **2001**, *8*, 417–426; (d) H. LING, F. BOUDSOCQ, R. WOODGATE, W. YANG, *Cell* **2001**, *107*, 91–102. (e) M. C. FRANKLIN, J. WANG, T. A. STEITZ, *Cell* **2001**, *98*, 413–416; (f) S. DOUBLIÉ, S. TABOR, A. M. LONG, C. C. RICHARDSON, T. ELLENBERGER, *Nature* **1998**, *391*, 251–258; (g) Y. LI, S. KOROLEV, G. WAKSMAN, *EMBO J.* **1998**, *17*, 7514–7525; (h) J. R. KIEFER, C. MAO, J. C. BRAMAN, L. S. BEESE, *Nature* **1998**, *391*, 304–307; (i) H. F. HUANG, R. CHOPRA, G. L. VERDINE, S. C. HARRISON, *Science* **1998**, *282*, 1669–1675; (j) D. L. OLLIS, P. BRICK, R. HAMLIN, N. G. XUONG, T. A. STEITZ, *Nature* **1985**, *313*, 762–766.
7. (a) M. F. GOODMAN, *Annu. Rev. Biochem.* **2002**, *71*, 17–50; (b) A. MARX, D. SUMMERER, *ChemBioChem* **2002**, *3*, 405–407; (c) E. C. FRIEDBERG, P. L. FISCHHABER, C. KISKER, *Cell* **2001**, *107*, 9–12; (d) Z. LIVNEH, *J. Biol. Chem.* **2001**, *276*, 25639–25642; (e) M. F. GOODMAN, B. TIPPIN, *Nat. Rev. Mol. Cell. Biol.* **2000**, *1*, 101–109; (f) U. HÜBSCHER, H. P. NASHEUER, J. E. SYVAOJA, *Trends Biochem. Sci.* **2000**, *25*, 143–147; (g) E. C. FRIEDBERG, W. J. FEAVER, V. L. GERLACH, *Proc. Natl. Acad. Sci. USA* **2000**, *97*, 5681–5683; (h) E. C. FRIEDBERG, V. L. GERLACH, *Cell*, **1999**, *98*, 413–416; (i) R. E. JOHNSON, M. T. WASHINGTON, S. PRAKASH, L. PRAKASH, *Proc. Natl. Acad. Sci. USA* **1999**, *96*, 12224–12226.
8. S. MORAN, R. X.-F. REN, S. RUMNEY IV, E. T. KOOL, *J. Am. Chem. Soc.* **1997**, *119*, 2056–2057.
9. S. MORAN, R. X.-F. REN, E. T. KOOL, *Proc. Natl. Acad, Sci. USA* **1997**, *94*, 10506–10511.
10. J. C. MORALES, E. T. KOOL, *Nat. Struc. Biol.* **1998**, *5*, 950–954.
11. T. J. MATRAY, E. T. KOOL, *Nature* **1999**, *399*, 704–708.
12. J. C. MORALES, E. T. KOOL, *J. Am. Chem. Soc.* **2000**, *122*, 1001–1007.
13. (a) D. L. MCMINN, A. K. OGAWA, Y.

Wu, J. Liu, P. G. Schultz, F. E. Romesberg, *J. Am. Chem. Soc.* **1999**, *121*, 11585–11586; (b) A. K. Ogawa, Y. Wu, D. L. McMinn, J. Liu, P. G. Schultz, F. E. Romesberg, *J. Am. Chem. Soc.* **2000**, *122*, 3274–3287; (c) Y. Wu, A. K. Ogawa, M. Berger, D. L. McMinn, J. Liu, P. G. Schultz, F. E. Romesberg, *J. Am. Chem. Soc.* **2000**, *122*, 3274–3287; (d) E. L. Tae, Y. Wu, G. Xia, P. G. Schultz, F. E. Romesberg, *J. Am. Chem. Soc.* **2001**, *123*, 7439–7440.

14 D. Summerer, A. Marx, *Angew. Chem.* **2001**, *113*, 3806–3808; *Angew. Chem. Int. Ed.* **2001**, *40*, 3693–3695.

15 M. Strerath, D. Summerer, A. Marx, *ChemBioChem* **2002**, *3*, 578–580.

16 M. Strerath, J. Cramer, T. Restle, A. Marx, *J. Am. Chem. Soc.* **2002**, *124*, 11230–11231.

17 J. Cramer, M. Strerath, A. Marx, T. Restle, *J. Biol. Chem.* **2002**, *277*, 43593–43598.

18 (a) W. A. Beard, K. Bebenek, T. A. Darden, L. Li, R. Prasad, T. A. Kunkel, S. H. Wilson, *J. Biol. Chem.* **1998**, *273*, 30435–30442; (b) K. Bebenek, W. A. Beard, T. A. Darden, L. Li, R. Prasad, B. A. Luton, D. G. Gorenstein, S. H. Wilson, T. A. Kunkel, *Nat. Struc. Biol.* **1997**, *4*, 194–197.

19 D. T. Minnick, K. Bebenek, W. P. Osheroff, R. M. Turner Jr, M. Astatke, L. Liu, T. A. Kunkel, C. M. Joyce, *J. Biol. Chem.* **1999**, *274*, 3067–3075.

20 D. Summerer, A. Marx, *J. Am. Chem. Soc.* **2002**, *124*, 910–911.

21 M. Strerath, A. Marx, *Angew. Chem.* **2002**, *114*, 4961–4963; *Angew. Chem. Int. Ed.* **2002**, *41*, 4766–4769.

B.16
Polynucleotide Polymerases

Susanne Brakmann

Polynucleotide polymerases, or nucleotidyl transferases, are enzymes that catalyze the template-instructed polymerization of deoxyribo- or ribonucleoside triphosphates into polymeric nucleic acid – DNA or RNA. Depending on their substrate specificity, polymerases are classed as RNA- or DNA-dependent polymerases which copy their templates into RNA or DNA (all combinations of substrates are possible). Polymerization, or nucleotidyl transfer, involves formation of a phosphodiester bond that results from nucleophilic attack of the 3′-OH of primer-template on the α-phosphate group of the incoming nucleoside triphosphate. Although substantial diversity of sequence and function is observed for natural polymerases, there is evidence that many employ the same mechanism for DNA or RNA synthesis. On the basis of the crystal structures of polymerase replication complexes, a "two-metal-ion mechanism" of nucleotide addition was proposed [1]; during this two divalent metal ions stabilize the structure and charge of the expected pentacovalent transition state (Figure B.16.1).

Since the discovery of *Escherichia coli* DNA polymerase I in 1957 [2], many polymerases have been identified in prokaryotes and eukaryotes, including the recent discovery of several error-prone DNA polymerases [3]. Primary sequence alignments revealed that these polymerases can be cate-

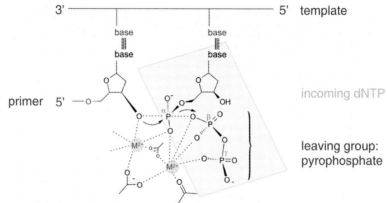

Fig. B.16.1. Suggested mechanism of DNA polymerization (modified from Ref. [1]).

gorized into families within which a large amount of amino acid sequence conservation is found. Currently, there are six known polymerase families (Table B.16.1) – A, B, C, X, reverse transcriptase (RT), and Y (UmuC/DinB) [4–7]. Among these, families A, B, and C are classified according to homologies with *E. coli* polymerases I, II, and III.

Tab. B.16.1. Representative members of families of DNA polymerases (modified from Ref. [6]).

Family	Prokaryotic	Eukaryotic	Archaea	Viral
A	Pol I	Pol γ, θ		T3, T5, T7
B	Pol II	Pol α, δ, ε, ζ	Pol BI, BII	HSV, RB69, T4
C	Pol III			
D			Pol D	
X		Pol β, λ, μ, T dt		
RT		Telomerase		HIV
Y	Pol IV, V	Pol η, ι, κ		

References

1 T. A. STEITZ, *Nature*, **1998**, *391*, 231–232.
2 A. KORNBERG, *Science*, **1960**, *131*, 1503–1508.
3 M. F. GOODMAN, B. TIPPIN, *Nat. Rev.*, **2000**, *1*, 101–109.
4 D. K. BRAITHWAITE, J. ITO, *Nucl. Acids Res.*, **1993**, *21*, 787–802.
5 R. SOUSA, *TIBS*, **1996**, *21*, 186–190.
6 P. H. PATEL, L. A. LOEB, *Nat. Struct. Biol.*, **2001**, *8*, 656–659.
7 H. OHMORI, E. C. FRIEDBERG, R. P. P. FUCHS, M. F. GOODMAN, F. HANAOKA et al., *Mol. Cell*, **2001**, *8*, 7–8.

4.2
Homogeneous DNA Detection

Oliver Seitz

4.2.1
Introduction

Many diseases with different appearances, for example cystic fibrosis, Tay Sachs disease, Huntington disease, familial hypercholesterolemia, and cancer are genetic disorders [1]. It is one of the chief aims of molecular diagnostics to detect a developing disease before symptoms appear. DNA-targeted analyses play a very important role and are used in a variety of clinical settings. The binding of a probe molecule to the complementary nucleic acid target is the molecular basis for most current methods in DNA-based diagnostics. In principal there are two different approaches, heterogeneous and homogeneous assays. Heterogeneous assays rely on immobilization of either the analyte or the probe molecule by a solid or gel phase; this facilitates removal of unbound binding partners (Figure 4.2.1a). Areas in which binding occurred are detectable by means of a reporter-group that is usually appended to the soluble binder. In contrast, homogeneous assays comprise a solution phase only and separation of unbound from bound molecules is not possible (Figure 4.2.1b). The design of a homogenous assay is conceptually more demanding, because the hybridization event must be coupled with alteration of a detectable variable. An advantage of homogeneous DNA-detection is that nucleic acid hybridization can be monitored in real-time even within a living cell. Single closed-tube assays are also feasible, which reduces the risk of contamination and speeds analysis. In the following discussion most important homogeneous DNA-based assays will be presented focusing on fluorescent detection techniques. The examples selected serve illustrative purposes and are not intended as a comprehensive overview. For more detailed information the reader is guided to some excellent review articles [2–5].

4.2.2
Non-specific Detection Systems

A host of detection principles can be used to sense the presence of double-stranded DNA. Fluorescent detection techniques employ dyes that change fluorescence properties upon binding to DNA (this is also discussed in detail in Chapter 2.5).

Highlights in Bioorganic Chemistry: Methods and Applications. Edited by Carsten Schmuck, Helma Wennemers.
Copyright © 2004 WILEY-VCH Verlag GmbH & Co. KGaA, Weinheim
ISBN: 3-527-30656-0

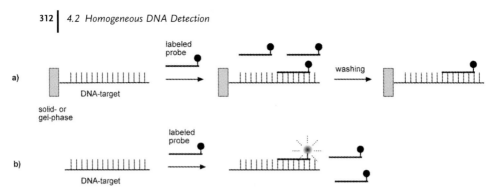

Fig. 4.2.1. Heterogeneous and homogenous DNA detection.

Interaction of molecules with duplex DNA follows three different binding modes governed by electrostatic interactions, groove binding interactions, or intercalation of planar ring systems between the Watson–Crick base pairs (Box 9). The dyes used for homogeneous DNA-detection typically bind to the minor groove or by intercalation (Figure 4.2.2). Often minor groove-binding dyes have cationic sites that are oriented in space by a crescent-shaped scaffold to fit into the helical curve. A twentyfold increase in fluorescence is observed for the commonly used DAPI dye on binding to duplex DNA [6]. The benzimidazole ring system of the so-called Hoechst dyes enables fluorescence enhancement by a factor of 90 [7]. There are, in addition, numerous proprietary dyes from Molecular Probes.

Intercalator dyes such as ethidium bromide or, more recently, cyanine dyes such as thiazole orange [8] selectively bind to duplex DNA by intercalation between the base pairs (Figure 4.2.2). Dimerization of two fluorophore moieties is a frequently found motif in modern dyes such as TOTO and is performed to increase the binding affinity to DNA [9, 10]. The extended ring systems are constrained into a planar arrangement as a result of intercalation. As a result the π-electrons are effectively delocalized and vivid fluorescence enhancement occurs. This property and the fluorescence increase on minor-groove binding provides a simple means of monitoring the production of duplex DNA during the polymerase chain reaction (PCR). Several platforms are now available for performing real-time PCR analyses and typically a PCR-thermocycler is equipped with a fluorescence detection device [4]. It must be remarked that intercalator dyes bind non-specifically to all duplex-DNA species including primer-dimers and non-specific amplification products. This is a major disadvantage of non-specific detection methods. Melting curve analysis can, however, help in distinguishing specific targets from primer dimers.

4.2.3
Specific Detection Systems

4.2.3.1
Single Label Interactions

For achieving a specific detection most assays rely on use of labeled probe molecules. Only one label is necessary when the hybridization strongly affects the

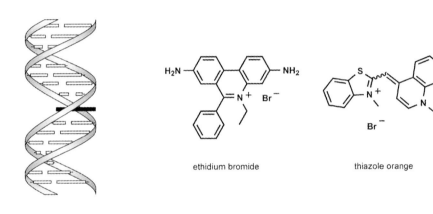

Fig. 4.2.2. Binding modes of typical stains used for homogenous detection of double stranded DNA. Binding to the minor groove or intercalation between base pairs can lead to fluorescence enhancement.

fluorescence properties. Fluorescence polarization anisotropy measurements can detect the hybridization-induced alteration of molecular weight. The underlying principle is that a fluorophore that is excited by polarized light also emits polarized light if molecular rotations are slow compared with the fluorescence lifetime. It was, for example, shown that the fluorescence anisotropy of the labeled probe **1** is enhanced by binding to the complementary DNA strand (Figure 4.2.3) [11]. The observed anisotropy differences are, however, not pronounced, even when high molecular weight PCR DNA is targeted (see **2**) [12].

The fluorescence of many fluorophores is environmentally sensitive and is, therefore, in principle, suitable for reporting a hybridization event. Meehan and coworkers attached a pyrene chromophore to the 5′-end of an oligonucleotide (Figure

4.2 Homogeneous DNA Detection

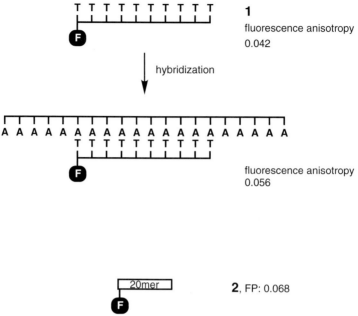

Fig. 4.2.3. Fluorescence anisotropy is increased on duplex formation.

4.2.4) [13]. The pyrene fluorescence of **3a** was quenched by a factor of 8 as a result of intercalation. A more recent example was reported by Fontecave and co-workers [14]. Double strand formation led to a 20-fold decrease of the fluorescence of the 5′-terminally appended deazaflavine moiety in **3b**. Although the measured fluorescence decays are significant, hybridization-induced fluorescence increases are clearly more desirable. For example, dansyl or pyrenyl groups were introduced at the 2′-position [15, 16]. The fluorescence of a pyrenyl oligoribonucleotide (see **4b**) increased by a factor of 16 to 60 when hybridized with an RNA-complement. The hybridization with a DNA strand led, however, to rather poor fluorescence enhancement. The structural basis of these oppositional fluorescence properties is unclear. In RNA duplexes that contained a mismatched base pair in the vicinity of the pyrene group the fluorescence intensity was dramatically reduced. It is remarkable that match/mismatch discrimination was accomplished at temperatures below the melting temperature, T_M, (Box 17) of the nucleotide.

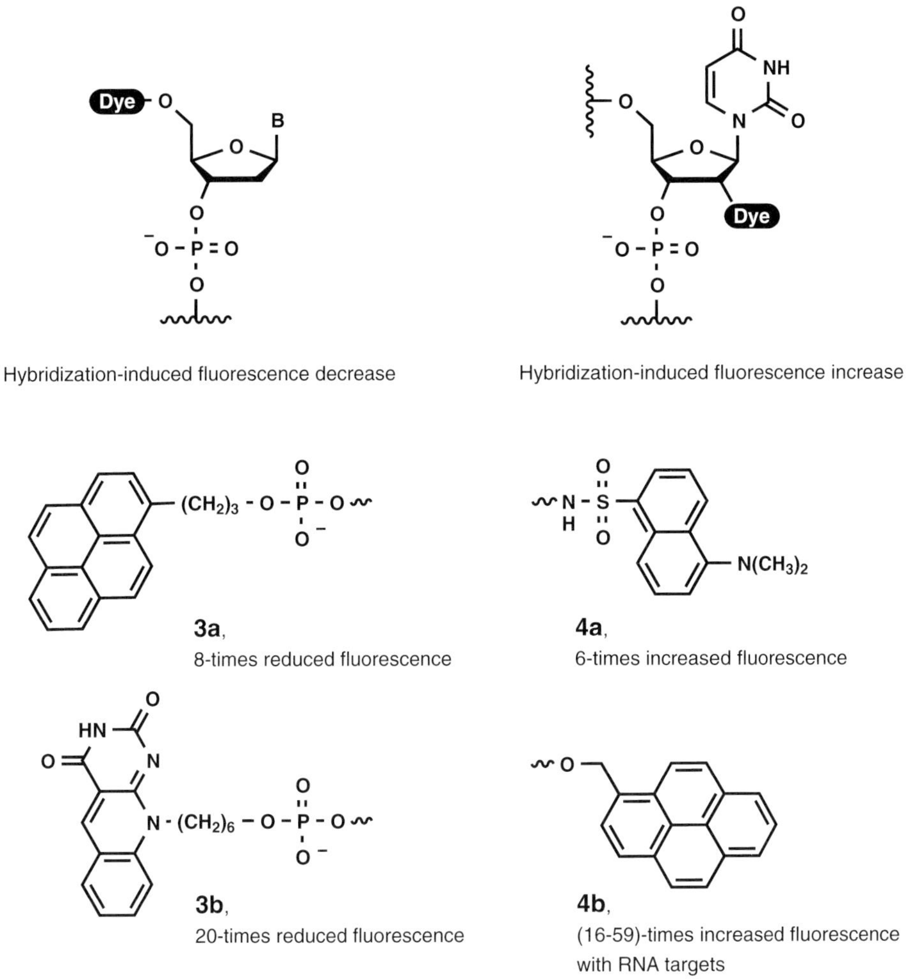

Hybridization-induced fluorescence decrease

Hybridization-induced fluorescence increase

3a, 8-times reduced fluorescence

4a, 6-times increased fluorescence

3b, 20-times reduced fluorescence

4b, (16-59)-times increased fluorescence with RNA targets

Fig. 4.2.4. Environmentally sensitive fluorophores can sense duplex formation.

A form of environmentally sensitive fluorescence is observed for intercalator dyes such as ethidium bromide. Typically the fluorescence intensity increases on intercalation. Intercalator dyes were conjugated to probe molecules through a flexible linker, which has to be long enough to enable intercalation between the base pairs of the formed duplex (Figure 4.2.5). Barton introduced the use of inorganic intercalators such as in **5** [17] (which can also be used to study charge transfer in DNA; see Chapter 4.6). Organic dyes were applied by Ishiguro and co-workers [18]. Hybridization of the oxazole yellow conjugate **6** increased the fluorescence intensity by a factor of two. The cyanine dye thiazole orange was recently appended to the terminus of the DNA analog PNA (peptide nucleic acid; Box 19) [19, 20]. The fluorescence of mixed sequence duplexes containing PNA–thiazole orange con-

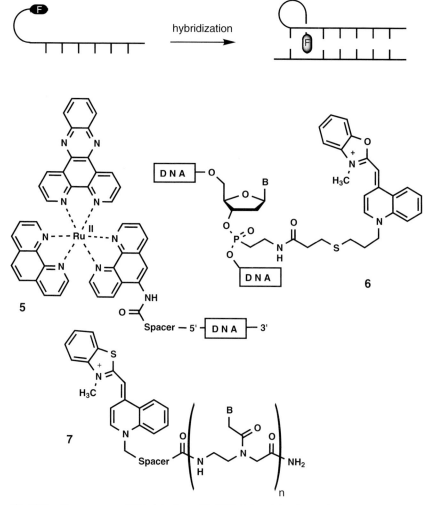

Fig. 4.2.5. Homogeneous DNA detection with DNA–intercalator dye conjugates.

jugates **7** was up to 16 times higher than single-strand fluorescence. In one instance a hybridization-induced fluorescence decrease was observed. Similar fluorescence enhancements were determined on binding to single mismatched DNA as long as measurements were performed at temperatures below the melting temperature of the corresponding PNA–DNA duplex. DNA targets and their respective single-base mutants were distinguishable at temperatures that led to dissociation of mismatched complexes.

More selective detection of single-base mutations can be achieved by replacing a nucleobase with a fluorophore (Figure 4.2.6). We envisaged that this *forced intercalation* of fluorophores or fluorescent base surrogates should enable the detection

Fig. 4.2.6. A fluorophore that serves as fluorescent base surrogate is forced to intercalate adjacent to the expected mutation site. This forced intercalation enables detection of mismatched base pairs even within a formed duplex.

of structural changes caused by single base mutations or other DNA-modifying events [21]. For example, thiazole orange (TO) was attached to the backbone of the DNA-analogous peptide nucleic acid (Scheme 4.2.1). The TO was equipped with carboxyalkyl spacers of different length to enable coupling with the PNA backbone. To facilitate the screening of suitable TO derivatives we developed an on-resin procedure for gaining divergent access to a variety of PNA–dye conjugates which relied on the use of the internal modifier in **8**. Combinatorial approaches uncovered the TO derivative in **9** as suitable fluorophore and revealed its preferred sequence context.

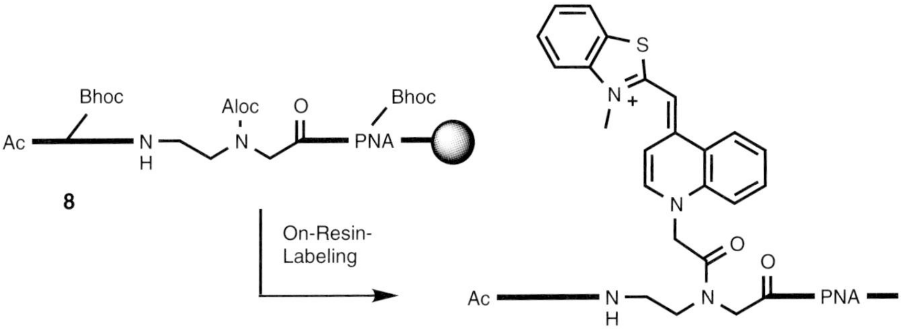

Scheme 4.2.1. Divergent solid-phase synthesis of FIT-probes.

For example, the thiazole orange-containing PNA-probe **10** was hybridized with oligonucleotides **11A** and **11C** (Figure 4.2.7) [O. Köhler, O. Seitz, Chem. Commun. 2003, DOI: 10.1039/B308299G]. Remarkably, the hybridization of **10** with **11A** led to a 19-fold fluorescence increase. The fluorescence intensity of the single mismatched duplex **17·11C** was considerably lower (by a factor of 4) than the emission of matched duplex **17·11C**. These data suggest that FIT probes should become of high utility in diagnostic assays particularly in single-base-mutation analysis.

4.2.3.2
Dual Label Interactions

Most homogenous detection methods make use of distance-dependent interactions between two labels. For example, a fluorescence donor can pass its energy by a ra-

10 GCCGTA-HN ... TAGCCG-Gly

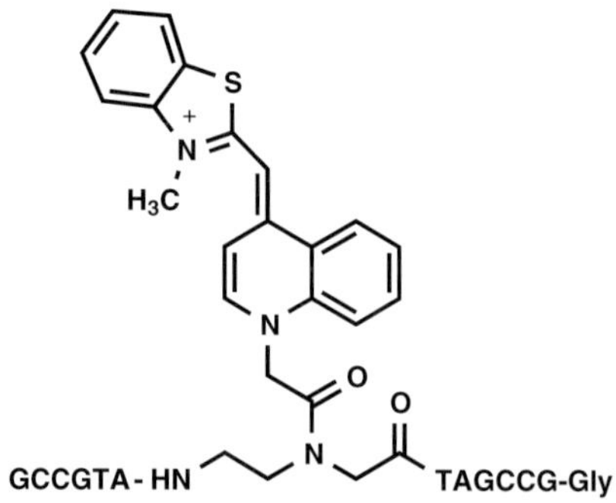

19-times increased fluorescence upon binding of **11A**: 5'-CGGCTATTACGGC-3'
5-times increased fluorescence upon binding of **11C**: 5'-CGGCTCTTACGGC-3

Fig. 4.2.7. Homogeneous DNA detection with FIT-probes.

diationless fluorescence resonance energy transfer (FRET) to an acceptor group. Single-labeled probes, termed ResonSense probes, bind to target DNA (Figure 4.2.8a) [24]. A DNA-intercalator such as SYBR Gold can bind to the double-stranded segment serving as fluorescence donor that sensitizes the fluorescence of the probe-bound acceptor dye. More commonly, both fluorescence donor and a fluorescence acceptor group are appended to probe molecules. In the adjacent probe format two probes are designed to bind adjacent to one another on the target strand, which brings a 5'-label in close proximity to a 3'-label (Figure 4.2.8b) [25, 26]. As a result the fluorescence of the donor dye becomes quenched whereas acceptor fluorescence is enhanced. An alternative approach utilizes double-stranded probes that dissociate on competitive hybridization with the target DNA [27]. This leads to disruption of the energy transfer between the 5'-label and the 3'-label, thereby increasing the fluorescence of the donor dye. The molecular beacons (Box 18) are an intramolecular version of this approach (Figure 4.2.8c) [28]. Molecular beacons are designed to form a hairpin structure with a target unrelated double-stranded stem sequence that holds two reporter groups in close proximity. Accordingly, the fluorescence is quenched as a result of FRET and static and dynamic (collisional) quenching. When the single-stranded loop segment anneals to the target sequence, structural reorganization increases the donor–quencher distance within the formed duplexes and fluorescence can occur. PCR-primers equipped with a molecular beacon structure are known as Scorpion probes [29, 30]. Molecular beacons incorporated in the DNA-template strand enable real-time monitoring of the DNA polymerase reaction [31]. It has been discovered that hybridization of linear dual-labeled oligonucleotides devoid of secondary structure-forming segments also leads to fluorescence enhancement [32]. This has been attributed to hybridization-induced stiffening that reduces quenching processes that operate in the flexible single-stranded probes. For example, oligonucleotide probes in which

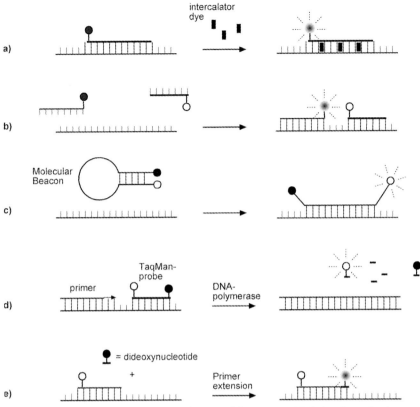

Fig. 4.2.8. Homogeneous DNA detection by employing fluorescence resonance energy transfer between two labels. Dual label interaction between (a) dyes that bind to dsDNA and fluorescence-labeled probe molecules; (b) adjacent probes and (c) in Molecular Beacons; (d) in TaqMan probes, and upon (e) template-directed dye-terminator incorporation.

terminal FAM donors and TAMRA acceptors were spaced by 20 to 27 bases afforded fluorescence increases of a factor of seven when hybridized to complementary nucleic acids [33].

In our work we explored the utility of dual-labeled PNA probes [34–36]. It was observed that thermal denaturation of single-stranded PNA, in contrast with DNA, resulted in a phase transition and considerable hyperchromicity, which indicated that base stacking might be a favorable process even in unhybridized PNA [37]. We reckoned that a suitably appended fluorescence donor could be located in close proximity to the fluorescence quencher because of a possible inter- or intramolecular association of PNA single-strands (Figure 4.2.9). As a result collisional quenching and fluorescence resonance energy transfer (FRET) would reduce the fluorescence of the single-strand. It was expected that quenching should be more

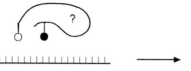 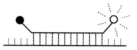

Fig. 4.2.9. Linear PNA-beacons: in unhybridized peptide nucleic acids (PNA) the averaged distance between appropriately appended fluorescence donor and fluorescence quencher groups is smaller than in the duplex structure. The fluorescence is quenched by collisional quenching and fluorescence resonance energy-transfer (FRET). When the probe sequence anneals to the target sequence fluorescence occurs.

efficient in dual-labeled PNA-probes than in dual-labeled DNA-probes. Hybridization to a complementary nucleic acid would, however, induce a structural reorganization that would lead to an increase of the averaged donor–quencher-distance. Thus, in the duplex, fluorescence would occur.

It was unclear how the hybridization-induced change of the distance between the donor and the quencher group could be maximized. We developed a highly flexible and automatable strategy which enabled us to perform all reactions, including the labeling steps, on the solid phase (Scheme 4.2.2). The use of the HYCRON-linker enabled the solid phase synthesis of unprotected PNA-resins such as **14** (for more information on linkers see Chapter 6.1). The resin-bound PNA **14** was subjected to a set of orthogonal ligation reactions. A Pd(0)-catalyzed allyl transfer accomplished the final detachment furnishing dual-labeled PNA conjugates such as **16**. The hybridization experiments revealed that duplex formation was accompanied by fluorescence enhancements reaching factors of 6. Use of mixed-sequence PNA oligomers that contain terminally appended fluorescent labels suggested that the fluorescence behavior of linear dual-labeled PNA oligomers is a general phenomenon that is independent of the sequence context [35]. Recent work compared the performance of Molecular Beacons, linear dual-labeled DNA, and linear dual-labeled PNA [38]. Briefly, dual-labeled PNA was found to hybridize rapidly and to provide a fluorescence response that was independent of the salt concentration.

Dual-labeled oligonucleotide probes known as TaqMan probes are amongst the most widely used type of probe in real-time PCR analysis (Figure 4.2.8d) [33, 39]. This is because of the 5′–3′ exonuclease activity of *Taq*-polymerase which provides an effective means of separating the donor from the acceptor fluorophore. During PCR the dual-labeled probe binds to the template. When the *Taq*-polymerase extends the primer it eventually reaches the 5′-end of the hybridized probe, which is then degraded by the exonuclease activity. This leads to an irreversible donor fluorescence increase that accumulates with each round of PCR-amplification. Irreversible cleavage of triple-stranded regions forms the basis of the Invader assay [29]. The opposite, namely the template-directed coupling of a fluorescence-labeled dye-terminator or oligonucleotide segment, forms the basis of the homogeneous variants of the Primer Extension [40, 41] (Figure 4.2.8e) and Oligonucleotide Ligation assays [42], respectively.

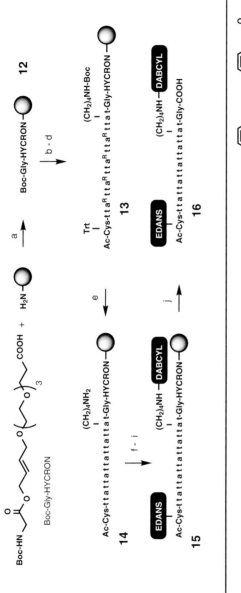
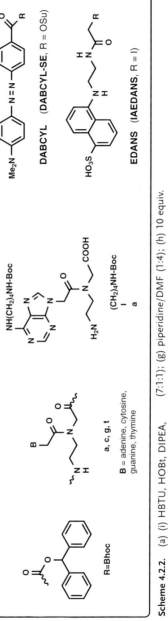

Scheme 4.2.2. (a) (i) HBTU, HOBt, DIPEA, DMF; (ii) Ac₂O, Pyr (1:10); (b) (i) TFA; (ii) DIPEA:DMF (1:9); (c) iterative cycles of: (i) piperidine/DMF (1:4); (ii) Fmoc-B^Bhoc-OH, HATU, iPr₂NEt, Pyr, DMF; (iii) Ac₂O, Pyr, DMF; (d) (i) piperidine/DMF (1:4); (ii) Ac₂O, Pyr, DMF; (e) TFA:ethanedithiol:H₂O (95:2.5:2.5); (f) 10 equiv. DABCYL-SE, DMF:Pyr:NMM (7:1:1); (g) piperidine/DMF (1:4); (h) 10 equiv. DTT, DMF:H₂O:NMM (9:3:1); (i) 10 eq IAEDANS, DMF:H₂O:NMM (9:3:1); (j) [Pd(PPh₃)₄], morpholine, DMSO, DMF, 10% (based on **12**). (DTT = dithiothreitol, HBTU = O-(benzotriazol-1-yl)-N,N,N′,N′-tetramethyl-uronium hexafluorophosphate, HOBt = 1-hydroxybenzotriazol)

4.2.4
Conclusion

From work in the early 1980s until today homogeneous DNA detection has matured into an invaluable tool of modern molecular diagnostics. Homogenous hybridization assays avoid the need to separate unbound from bound probe molecules and provide significant advantages such as convenience, high speed, and accuracy of DNA analysis. Most of the current probe systems such as adjacent probes, Molecular Beacons, TaqMan or Scorpion probes draw on the energy transfer between two interacting reporter groups. These methods distinguish between the bound and the unbound state of probe molecules and report the formation of the probe–analyte duplex. The sequence specificity of DNA detection is thus determined by the selectivity of probe hybridization.

Progress in homogeneous DNA detection has been driven by the achievements of DNA synthesis and labeling techniques which have provided the opportunity to equip DNA with virtually any kind of reporter group at various positions. There is, however, a lack of methods as far as base modification is concerned. The new developments particularly in the field of DNA analogs have made it possible to increase the repertoire of base surrogates. For example, we have shown that forced intercalation of thiazole orange affords large fluorescence enhancement on hybridization and marked sensitivity to the presence of adjacent base mismatches. It can be expected that suitably labeled DNA analogs such as the FIT probes proposed by us will provide valuable tools for homogenous nucleic acid detection in real-time PCR and in living cells.

References

1 www.ncbi.nlm.nih.gov/disease
2 L. E. Morrison, *J. Fluoresc.* **1999**, *9*, 187–196.
3 V. V. Didenko, *Biotechniques* **2001**, *31*, 1106–1121.
4 C. A. Foy, H. C. Parkes, *Clin. Chem.* **2001**, *47*, 990–1000.
5 L. J. Kricka, *Ann. Clin. Biochem.* **2002**, *39*, 114–129.
6 M. Kubista, B. Akerman, B. Norden, *Biochemistry* **1987**, *26*, 4545–4553.
7 T. Stokke, H. B. Steen, *J. Histochem. Cytochem.* **1985**, *33*, 333–338.
8 L. G. Lee, C. H. Chen, L. A. Chiu, *Cytometry* **1986**, *7*, 508–517.
9 H. S. Rye, J. M. Dabora, M. A. Quesada, R. A. Mathies, A. N. Glazer, *Anal. Biochem.* **1993**, *208*, 144–150.
10 H. S. Rye, S. Yue, D. E. Wemmer, M. A. Quesada, R. P. Haugland, R. A. Mathies, A. N. Glazer, *Nucleic Acids Res.* **1992**, *20*, 2803–2812.
11 A. Murakami, M. Nakaura, Y. Nakatsuji, S. Nagahara, Q. Tran-Cong, K. Makino, *Nucleic Acids Res.* **1991**, *19*, 4097–4102.
12 N. J. Gibson, H. L. Gillard, D. Whitcombe, R. M. Ferrie, C. R. Newton, S. Little, *Clin. Chem.* **1997**, *43*, 1336–1341.
13 J. S. Mann, Y. Shibata, T. Meehan, *Bioconj. Chem.* **1992**, *3*, 554–558.
14 C. Dueymes, J. L. Decout, P. Peltie, M. Fontecave, *Angew. Chem. Int. Ed.* **2002**, *41*, 486–489.
15 K. Yamana, Y. Ohashi, K. Nunota, H. Nakano, *Tetrahedron* **1997**, *53*, 4265–4270.

16 K. Yamana, H. Zako, K. Asazuma, R. Iwase, H. Nakano, A. Murakami, *Angew. Chem.* **2001**, *113*, 1143–1145; *Angew. Chem. Int. Ed.* **2001**, *40*, 1104–1106.

17 Y. Jenkins, J. K. Barton, *J. Am. Chem. Soc.* **1992**, *114*, 8736–8738.

18 T. Ishiguro, J. Saitoh, H. Yawata, M. Otsuka, T. Inoue, Y. Sugiura, *Nucleic Acids Res.* **1996**, *24*, 4992–4997.

19 N. Svanvik, A. Stahlberg, U. Sehlstedt, R. Sjoback, M. Kubista, *Anal. Biochem.* **2000**, *287*, 179–182.

20 N. Svanvik, G. Westman, D. Y. Wang, M. Kubista, *Anal. Biochem.* **2000**, *281*, 26–35.

21 O. Seitz, F. Bergmann, D. Heindl, *Angew. Chem.* **1999**, *111*, 2340–2343; *Angew. Chem. Int. Ed.* **1999**, *38*, 2203–2206.

22 J. Coste, D. Le-Nguyen, B. Castro, *Tetrahedron Lett.* **1990**, *31*, 205–208.

23 M. Plass, E. Hartmann, O. Müller, J. Kuhlmann, AF 116346, www.ncbi.nlm.nih.gov

24 M. A. Lee, A. L. Siddle, R. H. Page, *Anal. Chim. Acta* **2002**, *457*, 61–70.

25 M. J. Heller, L. E. Morrison, *Rapid Detection and Identification of Infectious Agents*, D. T. Kingsbury, S. Falkow, eds., Academic Press, New York, **1985**, pp. 245–256.

26 R. A. Cardullo, S. Agrawal, C. Flores, P. C. Zamecnik, D. E. Wolf, *Proc. Natl. Acad. Sci. USA* **1988**, *85*, 8790–8794.

27 L. E. Morrison, T. C. Halder, L. M. Stols, *Anal. Biochem.* **1989**, *183*, 231–244.

28 S. Tyagi, F. R. Kramer, *Nat. Biotechnol.* **1996**, *14*, 303–308.

29 D. Whitcombe, J. Theaker, S. P. Guy, T. Brown, S. Little, *Nat. Biotechnol.* **1999**, *17*, 804–807.

30 A. Solinas, L. J. Brown, C. McKeen, J. M. Mellor, J. T. G. Nicol, N. Thelwell, T. Brown, *Nucleic Acids Res.* **2001**, *29*, E96–U14.

31 D. Summerer, A. Marx, *Angew. Chem. Int. Ed.* **2002**, *41*, 3620–3622.

32 K. J. Livak, S. J. A. Flood, J. Marmaro, W. Giusti, K. Deetz, *Genome Res.* **1995**, *4*, 357–362.

33 K. J. Livak, S. J. A. Flood, J. Marmaro, W. Giusti, K. Deetz, *PCR-Methods Appl.* **1995**, *4*, 357–362.

34 O. Seitz, *Angew. Chem.* **2000**, *112*, 3389–3352; *Angew. Chem. Int. Ed.* **2000**, *39*, 3249–3252.

35 O. Seitz, O. Köhler, *Chem. Eur. J.* **2001**, *7*, 3911–3925.

36 J. M. Coull, B. D. Gildea, J. Hyldig-Nielsen (Boston Probes), WO A-9922018, **1999**.

37 K. L. Duholm, P. E. Nielsen, *New. J. Chem.* **1997**, *21*, 19–31.

38 H. Kuhn, V. V. Demidov, J. M. Coull, M. J. Fiandaca, B. D. Gildea, M. D. Frank-Kamenetskii, *J. Am. Chem. Soc.* **2002**, *124*, 1097–1103.

39 P. M. Holland, R. D. Abramson, R. Watson, D. H. Gelfand, *Proc. Natl. Acad. Sci. USA* **1991**, *88*, 7276–7280.

40 X. N. Chen, B. Zehnbauer, A. Gnirke, P. Y. Kwok, *Proc. Natl. Acad. Sci. USA* **1997**, *94*, 10756–10761.

41 X. Chen, P. Y. Kwok, *Nucleic Acids Res.* **1997**, *25*, 347–353.

42 V. V. Didenko, P. J. Hornsby, *J. Cell Biol.* **1996**, *135*, 1369–1376.

B.17
Melting Temperature T_M of Nucleic Acid Duplexes

Oliver Seitz

Watson–Crick base pairing mediates the mutual recognition of two complementary nucleic acid strands [1]. It is the molecular basis of life and

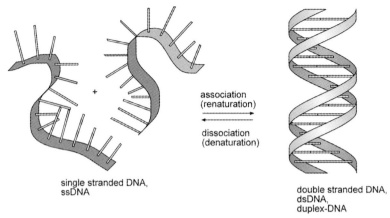

Fig. B.17.1. Hybridization leads to the formation of the PNA double helix.

of most current approaches in oligonucleotide-based diagnostics. The hybridization event anneals two single-stranded nucleic acids to form a duplex molecule with helically stacked nucleobases (Figure B.17.1). Nucleic acid hybridization is reversible and can be regarded as an association–dissociation or renaturation–denaturation process. The stability of a duplex molecule toward thermal denaturation depends on (a) base composition, (b) base sequence, (c) oligomer length, and (d) cation concentration.

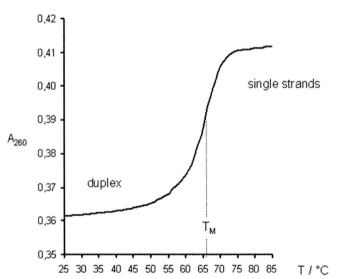

Fig. B.17.2. Real melting curve of a duplex with $T_M = 65.8\ °C$. The duplex comprised 5′-GGGCGCTGGAGGTGTG-3′ and the complementary strand.

At high temperatures duplex dissociation occurs. The temperature at which 50% of the duplexes are dissociated is called the melting temperature T_M. The T_M value is, hence, a measure of the thermal stability of a duplex. How are T_M values determined?

Nucleic acids absorb UV light with an absorption maximum at approximately 260 nm. This absorption mainly results from the nucleobases. Importantly, double-stranded DNA has a lower molar absorption coefficient, ε, than calculated for the sum of the molar absorption coefficients of the bases. Accordingly, duplex denaturation is accompanied by increased UV absorption [2]. This effect is termed "hyperchromicity" and originates from the disruption of base stacking.

The thermal denaturation of a DNA duplex is also called "DNA melting". The melting behavior of a duplex can be conveniently studied by measuring the temperature dependence of UV absorption. The resulting plot is known as the "melting curve" (Figure B.17.2). Typically, melting curves are sigmoid in form, which is indicative of co-operative base stacking. The inflection point is known as the "melting temperature, T_M".

References

1 J. D. WATSON, F. H. C. CRICK, Nature, **1953**, *171*, 737–738.
2 W. SAENGER, *Principles of Nucleic Acid Structure*, Springer, New York, **1983**.

B.18
Molecular Beacons

Oliver Seitz

A growing number of fluorescence-based assays makes use of distance-dependent interactions between a fluorophore and a chromophore [1]. For example, once excited by absorption of light a fluorescent group can pass its energy by radiationless fluorescence resonance energy-transfer (FRET) to an acceptor group, assuming donor emission overlaps (at least partially) with acceptor absorption. The acceptor group dissipates the energy as heat, or emits light if the acceptor is fluorescent itself. As a result donor emission becomes quenched. At very short distances a second quenching mechanism can operate – energy is transferred by contact. This so-called contact-mediated quenching can occur even with spectrally non-matched fluorophores. Irrespective of the quenching mechanism, donor fluorescence increases with the spatial separation of the two interacting fluorophores.

In molecular beacons fluorescence dequenching serves for detection of specific DNA segments in homogeneous solution, for example in real-time

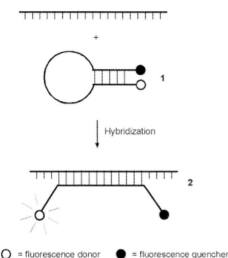

○ = fluorescence donor ● = fluorescence quencher

Fig. B.18.1. Molecular beacons are virtually non-fluorescent in the unhybridized form **1**. Target binding opens the molecular beacon to form **2** and fluorescence occurs.

PCR analysis and RNA detection in living cells [2]. Molecular beacons **1** are oligonucleotides designed to form hairpin structures with a target-unrelated double-stranded stem sequence that holds two reporter groups in close proximity (Figure B.18.1). In the closed form, **1**, the fluorescence is quenched because of FRET and contact-mediated quenching. The loop portion is a probe sequence that is complementary to a segment of the nucleic acid target. When the target-probe duplex is longer and more stable than the double-stranded stem, structural reorganization occurs. Target annealing rigidifies the loop segment, and, because stem duplex and target-probe duplex cannot co-exist, the hairpin is opened to form hybrid **2**. The disruption of the double-stranded stem separates the fluorescence donor from the acceptor, quenching becomes less likely, and fluorescence can occur. Hybridization of the target sequence to the probe is, therefore, accompanied by strong fluorescence enhancement (up to fiftyfold). A host of reporter groups can be used. Commonly, a fluorophore is combined with a non-fluorescent acceptor. It seems that contact-mediated quenching is the predominant quenching mechanism in the closed form **1** which facilitates the selection of fluorophores, particularly in multiplex formats [3].

References

1 (a) L. E. MORRISON, *J. Fluoresc.* **1999**, *9*, 187–196; (b) V. V. DIDENKO, *Biotechniques* **2001**, *31*, 1106–1121.
2 S. TYAGI, F. R. KRAMER, *Nat. Biotechnol.* **1996**, *14*, 303–308.
3 (a) S. BERNACCHI, Y. MÉLY, *Nucleic Acids Res.* **2001**, *29*, e62; (b) S. TYAGI, S. A. E. MARRAS, F. R. KRAMER, *Nat. Biotechnol.* **2000**, *18*, 1191–1196.

B.19
Peptide Nucleic Acids, PNA

Oliver Seitz

Nucleic acids have been called the "molecules of life" and it is widely believed that duplex formation by base pairing is a necessary requirement for the evolution of life as we know it. This has stimulated many chemists to synthesize nucleic acid analogs to gain an understanding of the essential features that govern nucleic acid recognition and hybridization. Interest in these studies has been further fueled by the demands of current approaches toward gene-targeted diagnosis and therapy which call, e.g., for improved base pairing properties and enhanced biostability. Synthetic compounds that sequence-specifically recognize and bind to a DNA or RNA segment are, therefore, of great value.

Peptide nucleic acids (PNA) are a particular successful example of DNA analogs (Figure B.19.1) [1]. Nielsen et al. replaced the entire ribose phosphate backbone by an uncharged, achiral pseudopeptide backbone to which the nucleobases were attached by an acetyl linkage [2]. This dramatic alteration resulted in remarkably improved binding to complementary nucleic acid sequences; this occurred with both high affinity and high selectivity. The thermal stability of identical sequences follows the order PNA–PNA > PNA–RNA > PNA–DNA. The structure of PNA–DNA duplexes resemble a B-form whereas PNA–RNA duplexes adopt an A-form. PNA–PNA duplexes form both left- and right-handed helices that are wide (28 Å), large pitched (18 bp) with the base pairs perpendicular to the helix axis. Despite

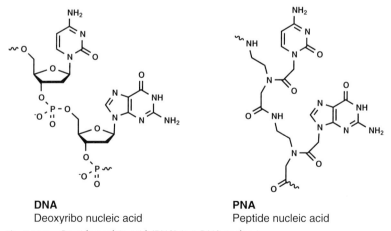

DNA
Deoxyribo nucleic acid

PNA
Peptide nucleic acid

Fig. B.19.1. Peptide nucleic acid (PNA) is a DNA analog in which the (deoxy)ribosephosphate backbone has been replaced by an artificial pseudopeptide.

their name, PNA are not acids rendering the hybridization event largely independent of salt concentration. PNA oligomers are, furthermore, capable of displacing one strand of duplex DNA to form a triplex containing one DNA and two PNA strands. A particularly attractive feature is that PNA are not subject to nuclease or protease-mediated degradation and are thus highly stable even in living cells. PNA have been advantageously employed in a variety of gene-targeting applications [3].

References

1 (a) B. Hyrup, P. E. Nielsen, Bioorg. Med. Chem. **1996**, 4, 5–23; (b) K. L. Dueholm, P. E. Nielsen, N. J. Chem. **1997**, 21, 19–31; (c) E. Uhlmann, A. Peyman, G. Breipohl, D. W. Will, Angew. Chem. **1998**, 110, 2954–2983; Angew. Chem. Int. Ed. **1998**, 37, 2797–2823.

2 P. E. Nielsen, M. Egholm, R. H. Berg, O. Buchardt, Science **1991**, 254, 1497–1500.

3 (a) D. A. Dean, Adv. Drug Deliv. Rev. **2000**, 44, 81–95; (b) P. E. Nielsen, Curr. Opin. Biotechnol. **2001**, 12, 16–20.

4.3
Exploring the Capabilities of Nucleic Acid Polymerases by Use of Directed Evolution

Susanne Brakmann and Marina Schlicke

4.3.1
Introduction

Polynucleotide polymerases are central players in the game of life that enable maintenance, transmission, and expression of genetic information. Polymerases also play another important role in molecular biology, however, being employed in catalyzing key processes such as DNA amplification by *polymerase chain reaction* (PCR), RNA production by in vitro transcription, cDNA synthesis, sequencing, or mutagenesis. In nature, *faithful* replication of DNA is crucial for long term survival of species – although mutations provide the variability that drives natural selection, they frequently result in loss of function and genetic instability. Thus, a balanced level of replication fidelity must be established by all living organisms. This is achieved by a sophisticated machinery of multiple nucleic acid polymerases. These enzymes seem to interact perfectly, meeting the requirements for repair of DNA damage and synthesis of DNA during cell replication.

The crystal structures of polymerases usually have a shape resembling a partially closed right hand with fingers, palm, and thumb subdomains [1] (Box 16). The palm subdomain comprises the floor of the polymerase active site and harbors the highly conserved amino acids that are likely to chelate the two catalytically important, divalent metal ions. The fingers bind the incoming nucleotide and interact with the primer-template whereas the thumb binds double-stranded DNA. Recent crystal structures of two polymerases of the Y family showed that the fingers and thumb of these highly error-prone enzymes are small in comparison to those of other DNA polymerases – a fact that suggests fewer interactions between polymerase and DNA that might result in reduced checking for correct base pairing [2–5]. The solution structure of a viral DNA polymerase X indicated for the first time that DNA polymerization might be catalyzed by an enzyme that does not have a hand-like architecture [6–8].

With regard to fidelity, the crystal structures of family A and B polymerases suggest that binding of the incoming nucleotide is associated with a conformational transition characterized by closing of the enzyme's nucleotide binding site and produced by the rotation of helices found in the fingers subdomain [9]. The

Highlights in Bioorganic Chemistry: Methods and Applications. Edited by Carsten Schmuck, Helma Wennemers.
Copyright © 2004 WILEY-VCH Verlag GmbH & Co. KGaA, Weinheim
ISBN: 3-527-30656-0

"closed conformation" is stabilized by precise positioning of all interactions with nucleotide, divalent cations, 3'-primer hydroxyl group, and Watson–Crick base pairing. It is assumed that the sum of these events restricts the conformations and structures of the incoming nucleotide and promotes the efficiency of correct nucleotide incorporation (this question is also discussed in Chapter 4.1).

4.3.2
Directed Evolution of Nucleic Acid Polymerases

Polynucleotide polymerases attract much attention – not only because of their central role in DNA metabolism, which suggests an important link to various diseases like tumor growth, defects of the immune system, stress-associated mutagenesis, or viral infections. Several polymerases are indispensable tools for molecular biotechnology, and could be even more valuable if the range of substrates accepted, or their stability and activity, could be "tuned" to specific requirements.

A major goal of polymerase research is to locate the lynchpins of fidelity, that is, to elucidate structural elements and conformational transitions that control this essential feature. The "classical" approach involved amino acid alignments of the polymerase domains from diverse family A enzymes and led to the identification of conserved motifs [10]. These investigations were accompanied by several site-directed mutagenesis studies that revealed the highly conserved amino acids critical for catalysis; that enabled location of some essential interactions between the enzyme and its substrates. The emerging "evolutionary techniques" that were successfully applied to the selection of functional nucleic acids using the SELEX procedure [11, 12] (Box 23) also led to a paradigm shift in investigating (and exploiting) the structure–function relationships of proteins: Directed evolution in the laboratory is highly attractive because its principles are simple and do not require detailed knowledge of structure, function, or mechanism (Box 20). Essentially, like natural evolution, this methodology involves iterative implementation of:

- generation of a *library* of mutated genes,
- its functional expression, and
- a sensitive assay to identify individuals with the desired properties.

Lawrence Loeb and coworkers performed pioneering directed evolution experiments to detect polymerase activity beyond their natural characteristics. These researchers were thus able to recognize structure–function links that had been hidden when conventional attempts were used.

In pursuing these efforts to unravel the complex mechanism that determines fidelity, we now address the following questions:

1. Can we "create" new mutant enzymes with altered fidelity of catalysis and tolerance on non-natural substrates?
2. Can we exploit these mutant enzymes for medical and technical purposes?

4.3.3
Practical Approaches to the Directed Evolution of Polymerase Function: Selection or Screening?

During the past few years a sound method repertoire has been developed to generate diversity in populations of genes [13]. Depending on the experimental problem to be solved the mutagenesis strategies applied might range from insertion of degenerate oligonucleotide cassettes [14] and error-prone DNA synthesis [15, 16] to diverse methods for the recombination (*shuffling*) of mutant DNA fragments [17–19]. The "sorting step", i.e. the identification of desirable members of the mutant population, is, however, most critical in directed evolution experiments. This step might be accomplished either by applying conditions that allow only variants of interest to appear (*selection*), or by assaying all members individually (*screening*).

4.3.3.1
Selection of Polymerases with Altered Activity and Fidelity

Selection strategies exploit conditions that favor the exclusive survival of desired variants. Most often this is achieved by *genetic complementation* of a host that is deficient in a certain pathway or activity (Figure 4.3.1a). Inventing a selection scheme for "low-fidelity DNA synthesis" is, however, a difficult task. Loeb and coworker Joann Sweasy made a major breakthrough in this field when they employed an *E. coli* host strain that encoded a temperature-sensitive mutant of DNA polymerase I (genotype: *recA718polA12*). At elevated temperatures this mutant strain fails to form colonies unless complemented by a DNA polymerase that can effectively substitute for DNA polymerase I [20]. Using strain *E. coli recA718polA12*, the researchers first identified active mutants of rat DNA polymerase β [21, 22], of *Thermus aquaticus* DNA polymerase I [23], and of HIV reverse transcriptase [24–26]. Shortly thereafter, Sweasy and coworkers extended this genetic selection system to identification of different DNA polymerase β mutator mutants, that is, error-prone variants of the enzyme (Figure 4.3.1b) [27, 28].

4.3.3.2
Screening Polymerase Libraries for Altered Activity

Obviously, an adaptation of Loeb's scheme to the selection of RNA polymerases (DNA-dependent transcriptases, or RNA-dependent replicases), reverse transcriptases, or DNA polymerases with altered substrate tolerance is difficult because these enzymes hardly enable the conferring of a growth advantage to host cells. In an exceptional approach to this problem, Andrew Ellington and colleagues focused

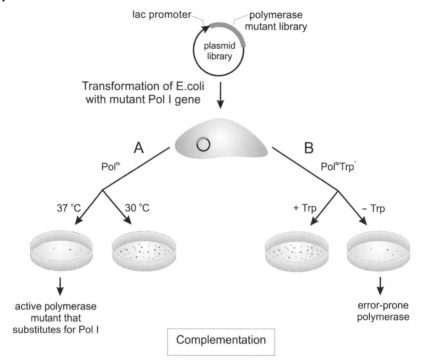

Fig. 4.3.1. Schemes for genetic selection of DNA polymerase function based on complementation. (a) Host cells of E. coli recA718polA12 that encodes a temperature-sensitive variant of DNA polymerase I (PolI^{ts}) are transformed with a polymerase mutant library. *Active* polymerase mutants substitute for DNA polymerase I at the non-permissive temperature (37 °C). (b) The host strain E. coli recA718polA12trpE65 has an additional chromosomal ochre mutation (that is, a stop codon). As a consequence, this strain is unable to synthesize anthranilate synthase, an enzyme needed to produce tryptophan (Trp). Transformants surviving on tryptophan-deficient media might occur by direct reversion of the stop codon because of an *active* and *error-prone* polymerase variant [27].

on a method opposite to that of Loeb – and simply picked clone duplicates that did *not* grow if the expressed polymerase variant fulfilled their requirements [29]. In searching for T7 RNA polymerase variants with altered promoter specificities they used the so-called autogene [30], that is, a T7 RNA polymerase (T7 RNAP) gene linked to a T7 promoter. The activity of the autogene can be initiated by basal level expression of T7 RNAP in the cell, and can lead to cell death if induced by IPTG.

Usually, however, identifying active polymerases requires a strategy that involves active searching of the library. Again, Loeb and coworkers pioneered this field by starting with a combination of selection and screening [31, 32]. A *Taq* DNA polymerase library was preselected for activity by genetic complementation of the above-mentioned selection scheme. Mutants that supported bacterial growth at the non-permissive temperature were isolated and screened for other interesting activity, for example RNA polymerase activity. They identified mutants with low efficiency of RNA synthesis only, however, because their two-step procedure aimed at substituting for a DNA polymerase in the first place. Floyd Romesberg, Peter Schultz, and coworkers therefore invented a phage display-based approach for converting the Stoffel fragment of *Taq* DNA polymerase I (SF) into an RNA polymerase [33]. They displayed polymerase variants and primer–template duplexes on phage surfaces, assayed these constructs for incorporation of ribonucleoside triphosphates, including biotinylated UTP, and captured successful candidates on streptavidine-coated magnetic beads. Thus, the researchers were able to identify three SF variants that polymerized ribonucleoside triphosphates as efficiently as the wild-type incorporates dNTP substrates.

Another challenging approach for screening pools of polymerase variants was presented by Philipp Holliger and coworkers [34]. They constructed a simple feedback loop consisting of a polymerase that replicates only its own encoding gene within compartmentalized individual self-replication reactions. To co-localize gene and encoded protein the researchers emulsified *E. coli* cells, the "carriers" of polymerase variants, within droplets of a water-in-oil emulsion. Using this technique, Holliger was able to identify variants of *Taq* DNA polymerase with 11-fold greater thermostability or with 130-fold increased resistance to the potent inhibitor heparin.

In principle, the screening concepts introduced here could be applied to other polymerases, and other activity. As yet, however, the contexts of low polymerase fidelity or of tolerance versus non-natural substrates have not been sufficiently targeted. In the course of our studies, we tackled these problems and presented solutions for both, selecting, or screening, polymerase libraries. Thereby, we detected an error-prone polymerase variant and polymerase activity in the sole presence of sterically demanding substrates.

4.3.4
Genetic Selection of an Error-prone Variant of Bacteriophage T7 RNA Polymerase

Studies with several RNA viruses have demonstrated their exceptionally high replication error rates that confer genetic flexibility, while resulting in substantial decreases of infectivity [35, 36]. Enhanced mutagenesis might even result in virus extinction, as predicted from virus entry into an "error catastrophe" [37, 38]. We are interested in studying and utilizing the link between error-prone polymerases and virus viability and started our search for an error-prone variant with the RNA polymerase of bacteriophage T7 (that is, the transcriptase of a DNA virus). We therefore developed a stringent positive selection scheme that rewarded inaccurate transcription by a T7 RNAP mutant with the survival of bacteria [39]. This genetic selection scheme involved a system of two compatible plasmids which coupled mutant polymerase genes to the essential, but inactivated, tetracycline resistance gene (Figure 4.3.2).

Feedback coupling was achieved by introducing a T7 promoter upstream of the tetracycline resistance gene (*tet*). This resulted in the exclusive dependence of re-

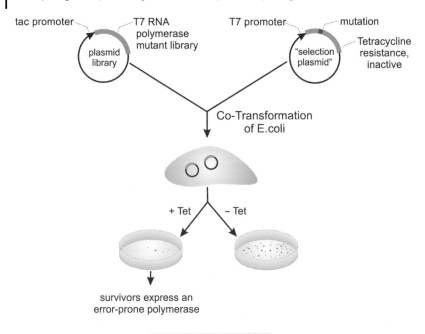

Fig. 4.3.2. Scheme for the genetic selection of an error-prone RNA polymerase variant based on the reversion of *antibiotic resistance*. The "selection plasmid" carries a mutant tetracycline resistance gene under the control of a T7 promoter. This construct renders the host cell sensitive to tetracycline (*tet*) unless an *active* and *error-prone* variant of T7 RNA polymerase is present. This enzyme would produce randomly altered transcripts among which one or more might exist that encode restored resistance activity (for example, as a consequence of random second-site mutation) [39].

sistance expression on the presence of T7 RNA polymerase. Substitution of a single amino acid of the *tet* fully inactivated the resistance which is mediated by a hydrophobic, membrane-associated efflux pump [40]. The mutation (Y100 → P) is located within a periplasmic region of the membrane-bound protein which previously had been identified to be critical for the performance of resistance [41]. Our positive selection scheme was based on the finding that functional interactions exists between the N- and C-terminal domains of the tetracycline efflux pump [42]. As a consequence of this interaction a deleterious mutation in one domain of the protein might be suppressed by some second-site mutation within the other domain. For selection, *E. coli* cells harboring the "selection plasmid", were first co-transformed with a library of $\sim 10^6$ plasmid-encoded, randomly mutated T7 RNAP gene variants, and then grown in liquid media containing 40% of the standard tetracycline dose. Surviving bacteria emerged immediately and were shown to express a single variant of T7 RNAP with three amino acid substitutions, F11 → L, C515 → Y, and T613 → A. In contrast to the wild-type enzyme that catalyzes transcription of the T7 genome with a nucleotide substitution error rate of $\leq 6 \times 10^{-5}$ only, a 20-fold-increased error rate of 1.25×10^{-3} (no preference for a certain type of mutation, or a certain sequence context) was observed for the isolated variant. Beyond that we observed that bacteria which constitutively express the selected polymerase variant are less efficiently infected and lysed by bacteriophage T7 – suggesting that the infecting phage utilizes the host-encoded error-prone enzyme.

Regarding the efficiency of our selection scheme, we can conclude that our expectations have been completely fulfilled, because the theoretical error rate of $\sim 10^{-3}$ (compensation of one missense mutation within ca. 1300 base pairs of *tet*) was reached. Attempts to obtain a further increase of the transcriptional error rate by compensating two missense mutations failed, however – surviving bacteria represented artifacts harboring either no T7 RNAP gene or non-functional polymerase fragments.

4.3.5
Screening for Polymerases with Altered Substrate Tolerance

Over the past 15 years a group of chemists worked to develop a novel strategy for DNA sequencing that is based on fluorescence detection at the single molecule level [43–47]. The proposed "single molecule sequencing" requires the complete and faithful synthesis of DNA copies exclusively from labeled analogs of the four types of nucleobase, A, G, C, and T, and thus unusual polymerase activity. Most natural DNA polymerases are known to discriminate against bulky fluorescent nucleotide analogs which often also have a net charge differing from that of the natural substrates [48–51]. Until recently, only few mutant bacteriophage T4 DNA polymerases were known to have an increased capacity to incorporate modified monomers [52].

To gain access to one or more DNA polymerases with the abilities to:

1. incorporate a fluorophore-labeled nucleotide,
2. extend the terminus by addition of the next fluorophore-labeled substrate, and
3. retain sufficient incorporation fidelity

we developed a functional screening approach and started the search by using the Klenow fragment (KF) of DNA polymerase I (Pol I) of *Escherichia coli*. Because elongation of a primer-template is not a single-step reaction that can be assessed with simple yes/no decisions, we needed a detection technique that yields quantitative information on molecular sizes, or fluorescence intensities, or both. Fluorescence correlation spectroscopy (FCS) [45, 53] is, in principle, a suitable technique for analysis of minute spontaneous fluctuations in the fluorescence emission behavior of small molecular ensembles that reflect inter- or intramolecular dynamics. FCS can, for example, give information about the velocity of translationally diffusing molecules. Because this value is related to the molecular size, we assumed that, ideally, chain lengths of polymerization products can be deter-

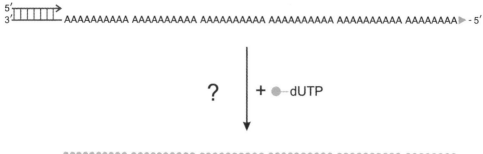

Fig. 4.3.3. Elongation of the primer hybridized to the 5'-biotinylated template requires polymerization along a homopolymeric (dA)$_{58}$ stretch. In the sole presence of fluorescence-labeled deoxynucleoside triphosphates (dye = tetramethylrhodamine, TAMRA), this selective constraint forces the multiple successive incorporation of sterically and electronically demanding substrates. Prior to fluorimetric assessment, reaction products can be immobilized on streptavidin-coated surfaces and purified from excessive fluorescent monomer [55].

mined. We thus applied series of primer extension reactions that required elongation of a primer-template along a homopolymeric (dA)$_{58}$ stretch of the a template (Figure 4.3.3).

While setting up our screening system with the wild-type KF, we observed the prevalence of a single molecular species that was identified to be the full length reaction product. The surprising result suggested that the wild-type enzyme was at least capable of catalyzing the template-instructed polymerization of 58 tetramethylrhodamine-(TAMRA)-labeled deoxyuridylic acid residues. In pursuance of labeling long, natural DNA molecules, we then utilized KF in further primer extension reactions and substituted:

1. TTP by its analog TAMRA-dUTP
2. dCTP by rhodamine110-(R110)-dCTP, and
3. both TTP and dCTP by the respective analogs.

Diverse analyses demonstrated that we prepared heteroduplex DNA fragments of 7000–9000 base pairs (bp) with complete substitution of both pyrimidine nucleotides [54]. Although the labeled DNA differs very much from natural DNA (for example, with respect to solubility, melting behavior, and circular dichroism), we were also able to show that the high degree of label incorporation so far did not alter the fidelity of KF.

It should be mentioned that our screening approach, i.e. comparative evaluation of individual polymerase variants, required two purification steps [55]. On the one hand, polymerase mutants cloned and expressed by use of *E. coli* had to be separated from competing host polymerases; on the other hand, reaction products had to be separated from excess fluorescent monomers that could interfere during the FCS measurement. Although all of these purification steps were performed with one-step procedures using commercially available microwell formats, the throughput was comparatively low. Future improvements should therefore aim at circumventing as many purification steps as possible, for example by alteration of the expression system or by employing detection techniques that enable more efficient distinction between educts and products.

4.3.6
Alternative Scenarios for Assaying Polymerase Activity

Several approaches have been reported for the screening of polymerase activity, for example radioisotope assays such as *scintillation proximity assays* [56, 57] or fluorescence-based assays [58–61]. Most of these assays, however, suffer from use of tedious procedures, the use of radioisotopes, or use of expensive reagents for fluorescence signal generation. A convenient means of online monitoring of DNA polymerase activity has recently been presented by Andreas Marx and Daniel Summerer [62]. Their technique involves a DNA template that forms a stable hairpin structure labeled at two positions:

1. internally at the origin of the hairpin stem (fluorescent dye: carboxyfluorescein, FAM), and
2. at the 5′-terminus (fluorescent dye: TAMRA).

Because of the hairpin formation, these dyes are in such a close proximity that their fluorescence is quenched (*molecular beacon*; Box 18) unless the structure is unfolded in the course of second-strand synthesis (Figure 4.3.4b). Thus, detection of a fluorescence signal from one of both dyes is a direct measure of the progress of the reaction. These researchers also showed that primer extension reactions can be monitored directly in cleared lysates of cells overexpressing the Klenow fragment of *E. coli* DNA polymerase I. Thus, the "molecular beacon assay" might supersede extensive purification.

Further promising polymerase assays employ fluorescence polarization (a property that evaluates the rotational diffusion of fluorophores, which is related to the molecular volume), or the detection of fluorescence intensity during incorporation of fluorescence-labeled substrates (Figure 4.3.4a).

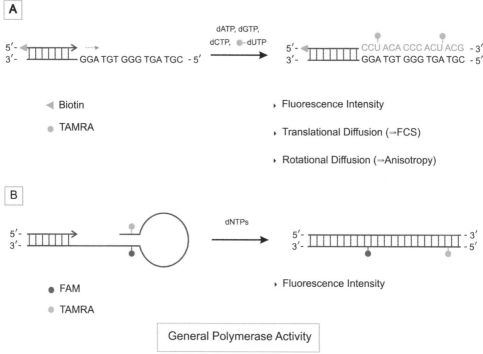

Fig. 4.3.4. Alternative schemes for (general) assessment of polymerase activity. (a) *Active* polymerases elongate the primer-template, thereby randomly incorporating fluorescence-labeled nucleotides (at a low density). The reaction progress can be monitored either by determination of the fluorescence intensity, or by using fluorescence techniques that also evaluate the product lengths (for example, by determining the velocity of translational or rotational diffusion; these values are related to the molecular volume, or mass). (b) A fluorescence signal arises if the dually labeled probe, a molecular beacon, opens during elongation of the primer-template [62].

4.3.7
Concluding Remarks

Directed evolution turns out to be a potent tool not only for improvement of polymerase function, or for detecting unexpected new activity but also for collecting information on polymerase structure and function with respect to fidelity. The work summarized in this chapter shows that polymerases with altered fidelity or altered substrate tolerance can be generated in vivo by using genetic selection, in vitro by employing a variety of assay schemes, or by combinations of these approaches. More efforts are desirable including:

- development of selection schemes that aim at generating novel RNA polymerases or reverse transcriptases;
- implementation of expression systems that circumvent polymerase purification in massively parallel screening approaches; and
- improvement of detection techniques with regard to sensitivity and information about the focused polymerase activity.

Directed evolution will contribute substantially to our understanding of substrate discrimination by polymerases and, thus, will enable use of these enzymes as important tools in a variety of synthetic and diagnostic applications.

References

1 D. L. OLLIS, P. BRICK, R. HAMLIN, N. G. XUONG, T. A. STEITZ, *Nature* **1985**, *313*, 762–766.
2 J. TRINCAO, R. E. JOHNSON, C. R. ESCALANTE, S. PRAKASH, L. PRAKASH, A. K. AGGARWAL, *Mol. Cell.* **2001**, *8*, 417–426.
3 L. F. SILVIAN, E. A. TOTH, P. PHAM, M. F. GOODMAN, T. ELLENBERGER, *Nat. Struct. Biol.* **2001**, *8*, 984–989.
4 T. ELLENBERGER, L. F. SILVIAN, *Nat. Struct. Biol.* **2001**, *8*, 827–828.
5 A. MARX, D. SUMMERER, *ChemBioChem* **2002**, *3*, 405–407.
6 W. A. BEARD, S. H. WILSON, *Nat. Struct. Biol.* **2001**, *8*, 915–917.
7 M. W. MACIEJEWSKI, R. SHIN, B. PAN, A. MARINTCHEV, A. DENNINGER, M. A. MULLEN, K. CHEN, M. R. GRYK, G. P. MULLEN, *Nat. Struct. Biol.* **2001**, *8*, 936–941.
8 A. K. SHOWALTER, I. L. BYEON, M. SU, M. D. TSAI, *Nat. Struct. Biol.* **2001**, *8*, 942–944.
9 P. H. PATEL, L. A. LOEB, *Nat. Struct. Biol.* **2001**, *8*, 656–659.
10 J. F. DAVIDSON, J. ANDERSON, H. GUO, D. LANDIS, L. A. LOEB, *Directed Molecular Evolution of Proteins*, S. BRAKMANN, K. JOHNSSON, eds., Wiley–VCH, Weinheim, **2002**, pp. 281–307.
11 D. IRVINE, C. TUERK, L. GOLD, *J. Mol. Biol.* **1991**, *222*, 739–761.
12 D. S. WILSON, J. W. SZOSTAK, *Ann. Rev. Biochem.* **1999**, *68*, 611–647.
13 S. BRAKMANN, *ChemBiochem* **2001**, *2*, 865–871.
14 M. S. HORWITZ, L. A. LOEB, *Proc. Natl. Acad. Sci. USA* **1986**, *83*, 7405–7409.
15 D. W. LEUNG, E. CHEN, D. V. GOEDDEL, *Technique* **1989**, *1*, 11–15.
16 R. C. CADWELL, G. F. JOYCE, *PCR Meth. Appl.* **1992**, *2*, 28–33.
17 W. P. C. STEMMER, *Proc. Natl. Acad. Sci. USA* **1994**, *91*, 10747–10751.
18 H. ZHAO, L. GIVER, Z. SHAO, J. A. AFFHOLTER, F. H. ARNOLD, *Nat. Biotechnol.* **1998**, *16*, 258–261.
19 Z. SHAO, H. ZHAO, L. GIVER, F. H. ARNOLD, *Nucl. Acids Res.* **1998**, *26*, 681–683.
20 J. B. SWEASY, L. A. LOEB, *J. Biol. Chem.* **1992**, *267*, 1407–1410.
21 J. B. SWEASY, L. A. LOEB, *Proc. Natl. Acad. Sci. USA* **1993**, *90*, 4626–4630.
22 J. B. SWEASY, M. CHEN, L. A. LOEB, *J. Bacteriol.* **1995**, *177*, 2923–2925.
23 M. SUZUKI, D. BASKIN, L. HOOD, L. A. LOEB, *Proc. Natl. Acad. Sci. USA* **1996**, *93*, 9670–9675.
24 B. KIM, L. A. LOEB, *Proc. Natl. Acad. Sci. USA* **1995**, *92*, 684–688.
25 B. KIM, T. R. HATHAWAY, L. A. LOEB, *J. Biol. Chem.* **1996**, *271*, 4872–4878.
26 B. KIM, *Methods* **1997**, *12*, 318–324.
27 S. L. WASHINGTON, M. S. YOON, A. M. CHAGOVETZ, S. X. LI, C. A. CLAIRMONT, B. D. PRESTON, K. A. ECKERT, J. B. SWEASY, *Proc. Natl. Acad. Sci. USA* **1997**, *94*, 1321–1326.

28 A. M. Shah, D. A. Conn, S. X. Li, A. Capaldi, J. Jäger, J. B. Sweasy, *Biochemistry* **2001**, *40*, 11372–11381.
29 J. Chelliserrykattil, G. Cai, A. D. Ellington, *BMC Biotechnol.* **2001**, *1*, 13; http://www.biomedcentral.com/1472-6750/1/13
30 J. W. Dubendorff, F. W. Studier, *J. Mol. Biol.* **1991**, *219*, 61–68.
31 M. Suzuki, A. K. Avicola, L. Hood, L. A. Loeb, *J. Biol. Chem.* **1997**, *272*, 11228–11235.
32 P. H. Patel, L. A. Loeb, *J. Biol. Chem.* **2000**, *275*, 40266–40272.
33 G. Xia, L. Chen, T. Sera, M. Fa, P. G. Schultz, F. E. Romesberg, *Proc. Natl. Acad. Sci. USA* **2002**, *99*, 6567–6602.
34 F. J. Ghadessy, J. L. Ong, P. Holliger, *Proc. Natl. Acad. Sci. USA* **2001**, *98*, 4552–4557.
35 C. K. Biebricher, E.-M. Düker, *J. Gen. Microbiol.* **1984**, *130*, 941–949.
36 E. Domingo, C. K. Biebricher, M. Eigen, J. J. Holland, eds., *Quasispecies and RNA Virus Evolution: Principles and Consequences*, Landes Bioscience, Georgetown, Texas, **2001**.
37 M. Eigen, *Naturwissenschaften* **1971**, *58*, 465–523.
38 S. Brakmann, *Biophys. Chem.* **1997**, *66*, 133–144.
39 S. Brakmann, S. Grzeszik, *ChemBioChem* **2001**, *2*, 212–219.
40 D. Schnappinger, W. Hillen, *Arch. Microbiol.* **1996**, *165*, 359–369.
41 P. McNicholas, I. Chopra, D. M. Rothstein, *J. Bacteriol.* **1992**, *174*, 7926–7933.
42 P. McNicholas, M. McGlynn, G. G. Guay, D. M. Rothstein, *J. Bacteriol.* **1995**, *177*, 5355–5357.
43 D. C. Nguyen, R. A. Keller, J. H. Jett, J. C. Martin, *Anal. Chem.* **1987**, *59*, 2158–2161.
44 W. P. Ambrose, P. M. Goodwin, J. H. Jett, M. E. Johnson, J. C. Martin, B. L. Marrone, J. A. Schecker, C. W. Wilkerson, R. A. Keller, *Ber. Bunsenges. Phys. Chem.* **1993**, *97*, 1535–1542.

45 M. Eigen, R. Rigler, *Proc. Natl. Acad. Sci. USA* **1994**, *91*, 5740–5747.
46 R. F. Service, *Science* **1999**, *283*, 1669.
47 J. Stephan, K. Dörre, S. Brakmann, T. Winkler, T. Wetzel, M. Lapczyna, M. Stuke, B. Angerer, W. Ankenbauer, Z. Földes-Papp, R. Rigler, M. Eigen, *J. Biotechnol.* **2001**, *86*, 255–267.
48 V. Folsom, M. J. Hunkeler, A. Haces, J. D. Harding, *Anal. Biochem.* **1989**, *182*, 309–314.
49 H. Yu, J. Chao, D. Patek, R. Mujumdar, S. Mujumdar, A. S. Waggoner, *Nucl. Acids Res.* **1994**, *22*, 3226–3232.
50 Z. Zhu, J. Chao, H. Yu, A. S. Waggoner, *Nucl. Acids Res.* **1994**, *22*, 3418–3422.
51 Molecular Probes, Eugene, Oregon, USA, *Handbook of Fluorescent Probes and Research Chemicals*, **2000**.
52 M. Goodman, L. Reha-Krantz, *Synthesis of Fluorophore-labeled DNA*, International Patent WO 97/39150 (PCT/US97/06493) **1997**.
53 D. Magde, E. L. Elson, W. W. Webb, *Phys. Rev. Lett.* **1972**, *29*, 705–711.
54 S. Brakmann, S. Löbermann, *Angew. Chemie Int. Ed. Engl.* **2001**, *40*, 1427–1429.
55 S. Brakmann, P. Nieckchen, *ChemBioChem* **2001**, *2*, 773–777.
56 S. Lutz, P. Burgstaller, S. A. Benner, *Nucl. Acids Res.* **1999**, *27*, 2792–2798.
57 D. L. Earnshaw, A. J. Pope, *J. Biomol. Screen.* **2001**, *6*, 39–46.
58 M. A. Griep, *Anal. Biochem.* **1995**, *232*, 180–189.
59 M. Seville, A. B. West, M. G. Cull, C. S. McHenry, *Biotechniques* **1996**, *21*, 664–672.
60 H. Tveit, T. Kristensen, *Anal. Biochem.* **2000**, *289*, 96–98.
61 J. H. Zhang, T. Chen, S. H. Nguyen, K. R. Oldenburg, *Anal. Biochem.* **2000**, *281*, 182–186.
62 D. Summerer, A. Marx, *Angew. Chemie Int. Ed.* **2002**, *41*, 3620–3622.

B.20
Directed Molecular Evolution of Proteins

Petra Tafelmeyer, and Kai Johnsson

Directed molecular evolution adopts the Darwinian approach to the evolution of proteins or peptides and, in contrast to rational approaches, does not require information about the sequence and the structure of the protein. In short, directed evolution consists in repetitive cycles of random mutagenesis of the protein/peptide sequence followed by screening or selection for candidates with the desired properties (Figure B.20.1).

Many different approaches are used to introduce mutations into a gene; most of those currently used are based on the polymerase chain reaction (PCR). Error-prone PCR, for example, uses a low-fidelity DNA polymerase and reaction conditions likely to introduce random base changes into the polynucleotide chain [1]. Care must be taken regarding the mutation rate – although a low level of mutagenesis will cover only a small fraction of the accessible sequence space, higher mutation rates also increase the chances of accumulating unfavorable or deleterious mutations, resulting in non-active protein. A combinatorial approach for enzyme evolution, called DNA shuffling, consists in random fragmentation of a gene then reassembly by self-priming PCR; this enables recombination of mutations from different

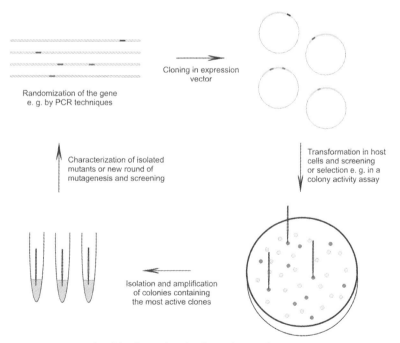

Fig. B.20.1. Example of the directed molecular evolution of a protein.

clones [2]. To further increase the accessible sequence space, DNA shuffling even enables recombination of families of homologous genes to generate protein chimeras [3].

In contrast with error-prone PCR and DNA shuffling, saturation mutagenesis enables randomization of a particular residue (or several residues) in the protein of interest with any of the 20 natural amino acids [4]. In this process, however, additional information is required to determine interesting positions for randomization.

More critical than choosing a particular method for randomization is the choice of a suitable screening or selection scheme to identify the clones with the desired activity from a library of an immense number of inactive or less active mutants (current library sizes vary between 10^4 and 10^{13} individual members). In other words, a link between the desired property (i.e. phenotype) and the corresponding gene (i.e. genotype) must be created.

In general, one must distinguish between screening and selection methods. Screening relies on inspection of all individual members of a library and isolation of the interesting ones on the basis of specific properties of the active mutants (often by visual or spectroscopic detection). Selection, in contrast, is based on elimination of undesired variants. This can be achieved either on the basis of the significant growth advantage provided by the active protein to its host or, in vitro, as a result of ligand binding and temporary immobilization of the active variants while undesired mutants are washed away. Techniques used in screening or selection experiments range from facile colony activity screening to yeast two-hybrid systems or in-vitro selection display systems (phage display, mRNA display, ribosome display), to mention only some of the numerous possibilities [5–10].

All these approaches have been used to alter protein function, to increase the activity or solubility of proteins, or to adapt enzymes for industrial applications. The goal of artificial man-made proteins with tailor-made activities is, however, still far away and none of the currently existing approaches provides the ultimate solution to the directed evolution of proteins. Nevertheless, numerous examples of successfully altered and improved proteins clearly show the power of directed evolution for protein design.

References

1 I. A. LORIMER, I. PASTAN, *Nucleic Acids Res.* **1995**, *23*, 3067–3068.
2 W. P. C. STEMMER, *Nature* **1994**, *370*, 389–391.
3 A. CRAMERI, S.-A. RAILLARD, E. BERMUDEZ, W. P. C. STEMMER, *Nature* **1998**, *391*, 288–291.
4 A. N. VALLEJO et al., in: C. W. DIEFFENBACH, G. S. DVEKSLER, eds., Cold Spring Harbor Laboratory Press, Cold Spring Harbor, **1995**, pp. 603–612.
5 A. IFFLAND, P. TAFELMEYER, C. SAUDAN, K. JOHNSSON, *Biochemistry* **2000**, *39*, 10790–10798.
6 Z. LIN, T. THORSEN, F. H. ARNOLD, *Biotechnol. Prog.* **1999**, *15*, 467–471.

7 C. T. Chien, P. L. Bartel, R. Sternglanz, S. Fields, *Proc. Natl. Acad. Sci. USA* **1991**, *88*, 9578–9582.

8 I. Ponsard, M. Galleni, P. Soumillion, J. Fastrez, *ChemBioChem* **2001**, *2*, 253–259.

9 A. D. Keefe, J. W. Szostak, *Nature* **2001**, *410*, 715–718.

10 J. Hanes, A Plückthun, *Proc. Natl. Acad. Sci. USA* **1997**, *91*, 4937–4942.

4.4
Labeling of Fusion Proteins with Small Molecules in vivo

Susanne Gendreizig, Antje Keppler, Alexandre Juillerat, Thomas Gronemeyer, and Kai Johnsson

4.4.1
Introduction

Characterizing the movement, interactions, and chemical microenvironment of a protein inside the living cell is the key toward detailed understanding of its function. Most strategies that aim at realizing this objective are based on genetically fusing the protein of interest to a reporter protein that monitors changes in the environment of the coupled protein. Examples include fusion with fluorescent proteins, the yeast two-hybrid system, and Split-Ubiquitin [1–3]. All these techniques have limitations, however. For example, fluorescent proteins have the following drawbacks:

1. only a limited number of emission wavelengths is available, limiting their use in applications such as fluorescence-energy transfer (FRET);
2. they are relatively large and tend to oligomerize, thereby affecting the function of the protein of interest; and
3. they fold only slowly into their fluorescent form, making it difficult to follow dynamic processes in vivo [1].

The yeast two-hybrid and the Split-Ubiquitin system, two approaches to studying protein interactions in vivo, are either limited to interactions that can be reproduced within the nucleus of the yeast cell or critically depend on the geometry of the protein complex studied [2, 3]. As a consequence of the limitations of current methods used to study proteins in the living cells substantial efforts are being made to label proteins specifically, in vivo, with small synthetic molecules capable of probing and modulating protein function. Up to now, these approaches have been based on the non-covalent binding of a small molecule to a specific binding protein, the formation of stable complexes between biarsenical compounds and peptides containing cysteines, or on incorporation of unnatural amino acids using suppressor tRNA technology [4–9]. We present in this article a general method for covalent labeling of fusion proteins in vivo that complements the existing methods

Highlights in Bioorganic Chemistry: Methods and Applications. Edited by Carsten Schmuck, Helma Wennemers.
Copyright © 2004 WILEY-VCH Verlag GmbH & Co. KGaA, Weinheim
ISBN: 3-527-30656-0

for protein labeling in vivo and in vitro and might lead to new ways of studying and manipulating proteins in living cells [10].

Covalent attachment of the small molecule to the fusion protein is achieved by the unusual mechanism of the human DNA repair protein O^6-alkylguanine-DNA alkyltransferase (hAGT), which irreversibly transfers the alkyl group from its substrate, O^6-alkylguanine-DNA, to one of its cysteine residues (Figure 4.4.1A) [11]. The substrate specificity of hAGT is relatively low, because it also reacts readily with the nucleobase O^6-benzylguanine (BG) [11]. Oligonucleotides containing derivatives of O^6-benzylguanosine with substituted benzyl rings are also accepted as substrates of hAGT [12]. On the basis of these observations we reasoned that BG derivatives of the type **1** could be used for in-vivo labeling of hAGT fusion proteins (Figure 4.4.1B). To test the feasibility of the approach we investigated the labeling of hAGT fusion proteins with biotin, fluorescein, and digoxigenin using BG derivatives BGBT, BGAF, and BGDG (Figure 4.4.1C). BGAF was synthesized as the diacetate of fluorescein, to increase its membrane permeability. Within the cell, the diacetate of fluorescein is readily hydrolyzed to fluorescein by esterases, yielding BGFL (Figure 4.4.1C) [5]. These BG derivatives were chosen as prototypes of a new class of label, because their covalent attachment to proteins in vivo and in vitro can be easily detected, thereby enabling evaluation of the approach. BGBT, BGDG, BGAF, and BGFL can be synthesized in four steps from 6-chloroguanine (Figure 4.4.1D). The synthesis of the final precursor **2** of the BG derivatives of type **1** can be performed on a gram scale and it is readily derivatized with active esters. This is important because it allows preparation of BG derivatives of the vast number of reporter groups that are commercially available as active esters (see for example *Molecular Probes*). In addition, the labeling experiments were performed with the hAGT mutant G160W (^{W160}hAGT), which has been reported to have increased activity against BG [13]. To investigate the reactivity of substrates such as BGBT, BGDG or BGFL with ^{W160}hAGT, the kinetics of the reaction of BGBT with purified ^{W160}hAGT were measured in vitro (Figure 4.4.2A, B). In these experiments, the second-order rate constant of the reaction of ^{W160}hAGT with BGBT was determined to be 600 s^{-1} M^{-1} compared with 3000 s^{-1} M^{-1} for reaction of ^{W160}hAGT with BG, demonstrating that BG derivatives substituted at the 4-position of the benzyl ring are substrates of ^{W160}hAGT and can be used to label ^{W160}hAGT in vitro. We then tested whether BGBT can selectively label ^{W160}hAGT in vivo when it is expressed in *E. coli* as a fusion protein with an N-terminal 6xHis tag. In these experiments, BGBT was added directly to the medium of *E. coli* expressing ^{W160}hAGT at concentrations of 10 μM and after washing away excess substrate proteins labeled with biotin were detected by probing the Western blots of total cell extracts with streptavidin–peroxidase conjugates (Figure 4.4.2C). The results demonstrate specific labeling of ^{W160}hAGT in *E. coli* by BGBT within 60 min; the only other major band in the Western blot corresponds to a protein that is also biotinylated in the absence of BGBT. If a different globular protein of similar size, either cytochrome *c* peroxidase (CCP, 30 kDa, Figure 4.4.2C) or glutathione *S*-transferase (GST, 25 kDa, data not shown), was overexpressed at similar levels no labeling of the overexpressed protein by BGBT was observed. The efficiency of in-vivo labeling with

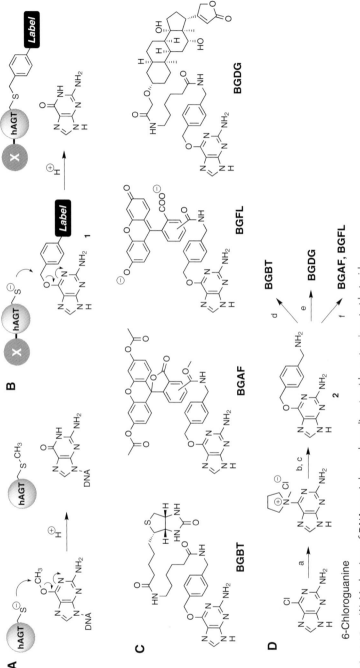

Fig. 4.4.1. (A) Mechanism of DNA repair by hAGT. (B) Covalent labeling of an X-hAGT fusion protein using O^6-benzylguanine (BG) derivatives of type 1. (C) Structures of the BG derivatives BGBT, BGAF, BGFL and BGDG used in this work. (D) Scheme of synthesis of BGBT, BGAF, BGFL, and BGDG; (a) 1-methyl-pyrrolidine, N,N-dimethylformamide (DMF) 66%; (b) 2,2,2-trifluoro-N-(4-hydroxymethyl-benzyl)acetamide, potassium tert-butoxide, DMF, 88%; (c) K_2CO_3, methanol, 85%; (d) N-(+)-biotinyl-6-aminocaproic acid N-succinimidyl ester, triethylamine, DMF, 69%; (e) Digoxigenin-3-O-methylcarbonyl-6-aminocaproic acid N-succinimidyl ester, DMF, triethylamine, 87%; (f) 5(6)-carboxyfluorescein diacetate N-succinimidyl ester (mixture of isomers), triethylamine, DMF, 8% (BGFL), 2% (BGAF).

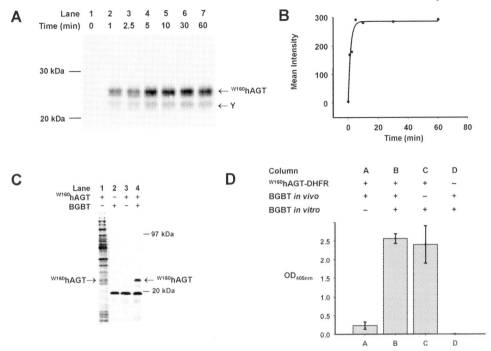

Fig. 4.4.2. Labeling of ^{W160}hAGT fusion proteins in vitro and in vivo. (A) Analysis of the in vitro reaction of ^{W160}hAGT (0.4 μM) with BGBT (15 μM). Aliquots were taken from the reaction mixture at the indicated time points and quenched with oligonucleotide containing BGBT (1 μM). Biotinylated ^{W160}hAGT was detected by Western blotting using a streptavidin–peroxidase conjugate. The band Y corresponds to a co-purified proteolytic degradation product of ^{W160}hAGT. (B) Fitting of the intensities obtained from the Western blot in (A) to a first-order reaction model, resulting in a second-order rate constant of 600 M^{-1} s^{-1} for the reaction of ^{W160}hAGT with BGBT. (C) Analysis of total cell extract of E. coli BL21(DE3) expressing ^{W160}hAGT with and without BGBT in the medium. Lane 1: SDS–PAGE of BL21(DE3) expressing ^{W160}hAGT and subsequent staining with Coomassie Blue; lanes 2–4: Western blot of total cell extracts of BL21(DE3) expressing or not expressing ^{W160}hAGT in the presence or absence of BGBT after 120 min. In lane 2, CCP was expressed instead of ^{W160}hAGT. A streptavidin–peroxidase conjugate was used to detect biotin-labeled proteins. The band at 20 kDa corresponds to a protein that is biotinylated in E. coli in the absence of BGBT. (D) Analysis of cell lysates of yeast S. cerevisiae expressing or not expressing a ^{W160}hAGT-DHFR fusion protein in the presence or absence of BGBT (10 μM, 2.5 h) using an ELISA. Samples in columns B–D were also incubated with an oligonucleotide containing the nucleobase BGBT after lysis. In column D a Ste14p-DHFR fusion protein was overexpressed instead of ^{W160}hAGT-DHFR. An anti-HA antibody was used as the primary antibody and an anti-mouse-HRP conjugate as a secondary antibody.

BGBT in E. coli was estimated to be 24% by comparing the signals resulting from the in-vivo labeling with those obtained from succeeding labeling of cells after lysis with oligonucleotides containing BGBT.

To demonstrate that the covalent labeling of hAGT fusion proteins in vivo can be applied to different hosts, we fused ^{W160}hAGT N-terminally to dihydrofolate reductase (DHFR) from mouse to yield ^{W160}hAGT-DHFR and expressed this fusion protein in yeast S. *cerevisiae*. In this construct DHFR also carried a C-terminal HA epitope tag. Analysis of the biotinylation of ^{W160}hAGT-DHFR in yeast using Western blots proved to be impossible as the streptavidin–peroxidase conjugate strongly cross-reacted with several yeast proteins, one of them of the same size as the ^{W160}hAGT-DHFR fusion protein. To demonstrate that the ^{W160}hAGT-DHFR fusion protein can be biotinylated in yeast, cells expressing ^{W160}hAGT-DHFR were incubated with BGBT (10 µM), then lysed and the cell extracts transferred into streptavidin-coated microtiter plates. After washing of the wells the immobilized ^{W160}hAGT-DHFR fusion protein was detected by ELISA using an anti-HA antibody as the primary antibody (Figure 4.4.2D). The signals in the ELISA obtained from such samples were significantly above background, which was defined as the signal obtained from yeast not incubated with BGBT. To quantify the efficiency of the labeling in yeast, the signal from lysates that were treated with oligonucleotides containing BGBT after labeling in vivo were compared with those that were labeled solely in vivo (Figure 4.4.2D). These data enabled us to estimate that 10% of ^{W160}hAGT-DHFR is biotinylated in vivo under these conditions. No signal above background was observed in this assay when a different DHFR fusion protein, a fusion protein with the S. *cerevisiae* Ste14p (Ste14p-DHFR), was expressed using the same vector and incubated in vivo and in cell extracts with BGBT (Figure 4.4.2D). The relatively low efficiency of the in-vivo labeling with BGBT in *E. coli* and yeast is most probably because of the low intracellular concentration of the substrate, because biotin derivatives with similar linker structures as BGBT have been shown to have relatively low cell permeability [5].

To demonstrate the feasibility of our approach in standard mammalian cell cultures, we investigated the labeling of a ^{W160}hAGT fusion protein with fluorescein in Chinese hamster ovarian (CHO) cells using BGAF. In contrast to AGTs from yeast and *E. coli* all mammalian AGTs characterized so far accept BG as substrate [11]. To achieve specific labeling of the ^{W160}hAGT fusion protein, we used a previously described AGT-deficient CHO cell line [14]. To facilitate evaluation of the fluorescence labeling, a ^{W160}hAGT fusion protein was constructed and targeted to the nucleus of the CHO cells by fusing ^{W160}hAGT to three consecutive SV40 large T antigen nuclear localization sequences (NLS), yielding ^{W160}hAGT-NLS$_3$ [15]. To demonstrate that ^{W160}hAGT-NLS$_3$ fusion proteins are targeted to the nucleus of the CHO cell, we transfected CHO cells with a vector expressing a ^{W160}hAGT-ECFP-NLS$_3$ fusion protein containing enhanced cyan fluorescent protein (ECFP). The location of ^{W160}hAGT-ECFP-NLS$_3$ in the nucleus of the cell was verified by confocal fluorescence microscopy via fluorescence of ECFP. CHO cells transiently transfected with a vector expressing ^{W160}hAGT-NLS$_3$ were incubated with BGAF (5 µM) for 5 min, washed, and monitored with confocal fluorescence microscopy (Figure 4.4.3A–C). During and after the washing step excess fluorophores leaked out of the cell and after 25 min only the nucleus of the cell had a strong fluorescence signal (Figure 4.4.3A–C). The background fluorescence in the cytosol was measured to be less than 10% under these conditions (Figure 4.4.3C). Two control experiments indicate that the observed fluorescence labeling of the nucleus is be-

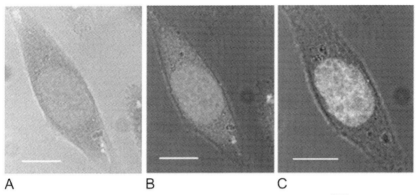

A B C

Fig. 4.4.3. Covalent labeling of nuclear targeted ^{W160}hAGT-NLS$_3$ in AGT-deficient CHO cells. Confocal micrographs A–C show overlays of transmission and fluorescence channels (exc 488 nm). The size bar in A–C corresponds to 10 μm. Confocal micrographs (A–C) illustrate the time course of the labeling of transiently expressed ^{W160}hAGT-NLS$_3$ with BGAF in AGT-deficient CHO cells. (A) AGT-deficient CHO cell transiently expressing ^{W160}hAGT-NLS$_3$ during incubation with BGAF (5 μM). (B) Same cell as in (A) after 5 min incubation with BGAF (5 μM) and three washes with PBS. (C) Same cell as in (B) after additional 25 min incubation in PBS, illustrating the specific labeling of hAGT-NLS$_3$ in the nucleus of a transiently transfected CHO cell.

cause of covalent labeling of ^{W160}hAGT-NLS$_3$. First, in AGT-deficient CHO cells not transfected with the gene of the hAGT fusion protein no fluorescence labeling of the nucleus was observed. Second, incubating CHO cells transiently expressing ^{W160}hAGT-NLS$_3$ with BG (10 μM) before incubation with BGAF to inactivate the hAGT fusion protein prevented the fluorescence labeling of the nucleus. The data thus show that hAGT can be covalently and specifically labeled in mammalian cell cultures. The concentration of fluorescence-labeled ^{W160}hAGT-NLS$_3$ in the nucleus was estimated to be 2 μM. By comparison, the concentration of the ^{W160}hAGT-ECFP-NLS$_3$ fusion protein in the nucleus after 24 h of transient expression was estimated by fluorescence of ECFP to be 3 μM, demonstrating that the in vivo concentration of the labeled hAGT fusion protein approaches that of the corresponding ECFP fusion protein.

To improve the speed and the efficiency of the in-vivo labeling and to obtain novel insights into the structure–function relationship of the protein, we also generated hAGT mutants with increased activity against BG derivatives. In these experiments, we chose to submit hAGT to directed evolution using phage display [16, 17] (Box 20). In the following phage display experiments hAGT was displayed as a fusion protein with the phage capsid protein pIII. Selections were based on incubating phages with BGDG and isolating those phages covalently labeled with digoxigenin, by use of immobilized anti-digoxigenin antibodies. To select for hAGT mutants with increased activity against BGDG, we randomized four amino acids that were in the proximity of either the benzyl ring (Pro140, Gly159, Gly160), or could make contact with the purine (Asn157). Using phage display, hAGT mutants were selected which had up to fifteenfold increased activity against BGBT than wild-type, the most active mutant with the sequence Pro140, Gly157, Glu159, Ala160 (PGEA). To demonstrate that these mutants can improve the efficiency of our in-vivo labeling technique, the mutant PGEA was transiently expressed as a PGEA-NLS$_3$ fusion protein in AGT-deficient CHO cells and its labeling with BGFL investigated by confocal fluorescence microscopy. In these experiments, the mutant PGEA led to a twofold higher fluorescence signal in the nucleus of the cell than the previously used mutant ^{W160}hAGT.

Besides its unusual mechanism, hAGT has a variety of properties that make it a suitable protein for specific and covalent labeling of fusion proteins in vivo. Most importantly, it has high reactivity against a substrate which is otherwise chemically inert and which can be derivatized by a wide variety of labels such as dyes, cross-linkers, or affinity tags without significantly affecting the rate of the reaction of hAGT with the substrate. None of the substrates tested so far had chemical toxicity during the in-vivo labeling and, because of the irreversibility of the labeling, excess substrate can be easily washed away. hAGT is, furthermore, a monomer of 207 residues, thereby reducing the likelihood that its fusion to other proteins will affect their oligomeric state, and the protein of interest can be fused either to the N or the C terminus of hAGT without affecting its reactivity. Experiments in mammalian cells should be performed in AGT-deficient cell lines, preventing the labeling of the endogenous AGT. For example, incubating HEK293 cells with BGAF leads to detectable fluorescence labeling of the endogenous AGT in the nucleus of the cell. The efficiency of the in vivo labeling for a given substrate will depend primarily on its cell permeability. For most applications, however, quantitative labeling of the protein of interest is not mandatory.

In conclusion, we have developed a general method enabling covalent and specific labeling of fusion proteins in vivo. Its applicability in different organisms and its independence of the nature of the label should make this method an important tool for functional studies of proteins in the living cell. AGT fusion proteins should, furthermore, become a powerful tool for all those in-vitro applications where the protein of interest must be either specifically labeled or immobilized.

Acknowledgment

Funding of this work was provided by the Swiss Science Foundation, the European Community and the EPFL. SG was supported by a fellowship from the Boehringer Ingelheim foundation.

References

1 R. Y. TSIEN, *Annu. Rev. Biochem.* **1998**, 67, 509.
2 S. FIELDS, O. SONG, *Nature* **1989**, 340, 245.
3 N. JOHNSSON, A. VARSHAVSKY, *Proc. Natl. Acad. Sci. USA* **1994**, 91, 10340.
4 J. FARINAS, A. S. VERKMAN, *J. Biol. Chem.* **1999**, 274, 7603.

5 M. M. Wu, J. Llopis, S. Adams, J. M. McCaffery, M. S. Kulomaa, T. E. Machen, H. P. Moore, R. Y. Tsien, *Chem. Biol.* **2000**, *7*, 197.

6 B. A. Griffin, S. R. Adams, R. Y. Tsien, *Science* **1998**, *281*, 269.

7 G. Gaietta, T. J. Deerinck, S. R. Adams, J. Bouwer, O. Tour, D. W. Laird, G. E. Sosinsky, R. Y. Tsien, M. H. Ellisman, *Science* **2002**, *296*, 503.

8 S. R. Adams, R. E. Campbell, L. A. Gross, B. R. Martin, G. K. Walkup, Y. Yao, J. Llopis, R. Y. Tsien, *J. Am. Chem. Soc.* **2002**, *124*, 6063.

9 L. Wang, P. G. Schultz, *Chem. Commun.* **2002**, 1.

10 A. Keppler, S. Gendreizig, T. Gronemeyer, H. Pick, H. Vogel, K. Johnsson, *Nature Biotechnology*, **2003**, *21*, 86.

11 A. E. Pegg, *Mutat. Res.* **2000**, *462*, 83.

12 R. Damoiseaux, A. Keppler, K. Johnsson, *ChemBioChem.* **2001**, *2*, 285.

13 M. Xu-Welliver, J. Leitao, S. Kanugula, W. J. Meehan, A. E. Pegg, *Biochem. Pharmacol.* **1999**, *58*, 1279.

14 B. Kaina, G. Fritz, S. Mitra, T. Coquerelle, *Carcinogenesis* **1991**, *12*, 1857.

15 D. Kalderon, B. L. Roberts, W. D. Richardson, A. E. Smith, *Cell* **1984**, *39*, 499.

16 R. Damoiseaux, P. G. Schultz, K. Johnsson, *ChemBioChem.* **2002**, *3*, 573.

17 A. Juillerat, T. Gronemeyer, H. Pick, H. Vogel, K. Johnsson, *Chemistry & Biology*, **2003**, *10*, 313.

4.5
Oxidative Splitting of Pyrimidine Cyclobutane Dimers

Uta Wille

4.5.1
Introduction

Thymidine cyclobutane dimers are important photoproducts formed by short-wave UV irradiation ($\lambda = 290$–320 nm) of DNA, by [2 + 2] cycloaddition between two adjacent thymine nucleobases in the same oligonucleotide strand (Scheme 4.5.1) [1]. They lead to profound biological effects in vivo, including mutation, cancer, and cell death [2] (Box 21). In a wide range of organisms the repair of these lesions in DNA is accomplished by enzymes (the *photolyases*), which regenerate undamaged thymidines by means of a photoinduced electron-transfer process [3].

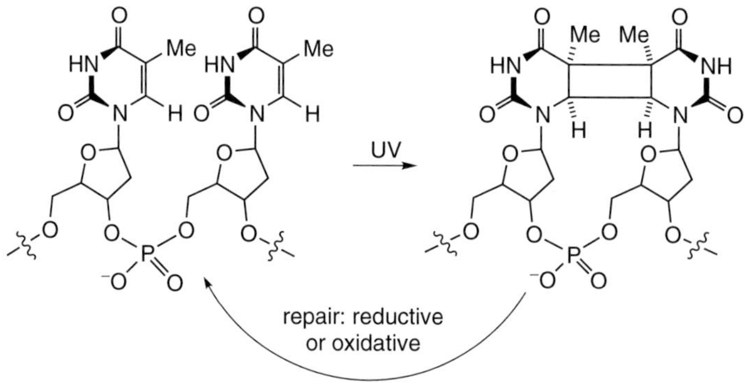

Scheme 4.5.1.

Several pathways have been pursued for the study of the splitting mechanism, often using simpler model systems than nucleosides, e.g. pyrimidine cyclobutane dimers (Pyr<>Pyr). Investigation of model systems has the advantage that spectroscopic techniques can be more readily applied and that they can be precisely controlled and manipulated in ways the natural system cannot. Whereas in DNA only pyrimidine cyclobutane dimers with a head-to-head (*syn*, s) orientation and *cis*

Highlights in Bioorganic Chemistry: Methods and Applications. Edited by Carsten Schmuck, Helma Wennemers.
Copyright © 2004 WILEY-VCH Verlag GmbH & Co. KGaA, Weinheim
ISBN: 3-527-30656-0

(c) configuration can be formed, because of the steric constraints in the double strand, artificial Pyr⟠Pyr can also be obtained with a head-to-tail (*anti, a*) orientation and trans (*t*)-configuration at the cyclobutane ring. Examples of the model systems most often used for Pyr⟠Pyr are shown in Scheme 4.5.2.

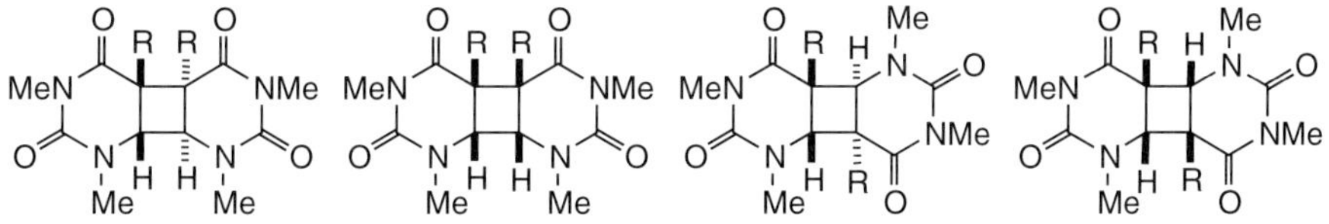

	trans,syn (t,s)	cis,syn (c,s)	trans,anti (t,a)	cis,anti (c,a)
R = H: **1**	1.34 V	1.46 V	1.86 V	1.88 V
R = Me: **2**	1.35 V	1.45 V	1.87 V	1.80 V

$E_{p/2}$ vs Ag/AgNO$_3$ (in acetonitrile); Ref [21a]

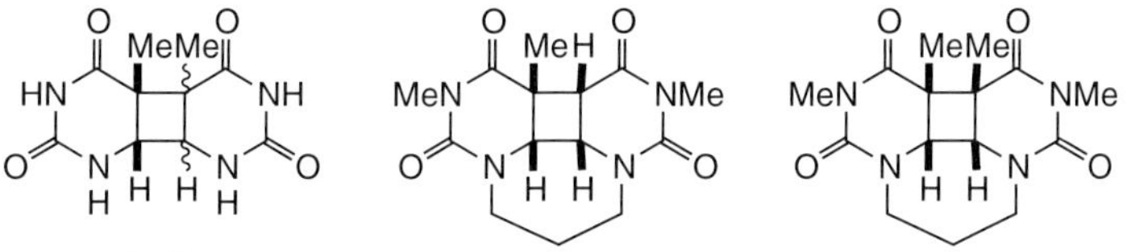

Scheme 4.5.2.

Numerous molecules and ions have been found to be capable of photochemically splitting Pyr⟠Pyr in solution. Because these sensitizers (S) use light of longer wavelengths than those absorbed by the dimer, direct energy transfer of the excited singlet state of the sensitizer to Pyr⟠Pyr has not been considered likely to be responsible for the splitting. Instead, these sensitizers have in common the fact that on excitation they become powerful oxidizing or reducing species. This has led to the formulation of two general mechanisms for dimer splitting, one proceeding via electron donation to the dimer, and one via electron abstraction from the dimer (Scheme 4.5.3). The dimer radical anion or radical cation formed as the key intermediate by this electron transfer then splits. In each case charge neutralization occurs in a subsequent electron transfer step.

$$S + Pyr\diamond Pyr \xrightarrow{h\nu} \begin{Bmatrix} S^{\cdot+}Pyr\diamond Pyr^{\cdot-} \longrightarrow S^{\cdot+}Pyr\,^{\cdot}Pyr \\ S^{\cdot-}Pyr\diamond Pyr^{\cdot+} \longrightarrow S^{\cdot-}Pyr\,^{\cdot+}Pyr \end{Bmatrix} \longrightarrow S + 2\,Pyr$$

Scheme 4.5.3.

Examination of the thermodynamics of the cleavage reaction of c,s-**2** (Scheme 4.5.2) revealed that splitting of the cyclobutane dimer is an exothermic process with an enthalpy of -79.3 kJ mol^{-1} [4], which reflects both loss of strain and formation of the conjugated C=C double bond. The enthalpy for the cleavage of the dimer cation radical and anion radical of c,s-**2** was determined to be -79.3 kJ mol^{-1} and -116.8 kJ mol^{-1}, respectively [4]. Although it is now established that the photolyase-assisted DNA repair of Pyr<>Pyr follows a reductive pathway [5], investigation of the mechanisms and intermediates of the oxidative dimer cleavage process offers insights into the alternative pathway, which is, apparently, avoided by nature in the evolution of photoenzymatic repair.

4.5.2
Mechanism of the Oxidative Splitting of Pyr<>Pyr

Many experimental and theoretical studies have been performed on the oxidative splitting of Pyr<>Pyr. It must, nevertheless, be stated that there remain uncertainties about mechanistic details of this process. Most experimental investigations were performed using redox photosensitizers to oxidize Pyr<>Pyr, but oxidizing radicals and radical ions were also employed. A general disadvantage of studies using photosensitizers as oxidants is the simultaneous formation of a radical anion of the sensitizer, which could serve as a convenient source of electrons for back transfer. This reduction could lead to significant difficulties in interpretation of the experimental data. The problem could be overcome by the use of oxidizing radicals and radical ions, which enable irreversible electron transfer. A compilation of the most common systems used in these studies is given in Table 4.5.1.

The assumed general mechanism for oxidative splitting of *syn*-configured Pyr<>Pyr is given in Scheme 4.5.4. Electron transfer from Pyr<>Pyr to the oxidant leads to formation of a dimer radical cation Pyr$^{\cdot+}$<>Pyr. The charge location in Pyr$^{\cdot+}$<>Pyr is shown on the nitrogen to emphasize the way in which an adjacent cyclobutyl bond is stabilized by electron abstraction; the actual charge distribution is expected to place significant charge density at other atoms in the radical ion.

Photo-CIDNP experiments using anthraquinones as photosensitizers for oxidation of a variety of uracil- and thymine-derived cyclobutane dimers, e.g. c,s-**1**, t,s-**1**, c,a-**1**, t,a-**1**, and c,s-**2**, **4**, and **5**, demonstrated the existence of both Pyr$^{\cdot+}$<>Pyr and its dissociation product, the monomer radical cation Pyr$^{\cdot+}$ [6, 7].

The lifetime of Pyr$^{\cdot+}$<>Pyr seemed to depend on the structure of the parent dimer. It was observed that the presence of a trimethylene bridge linking both pyrimidine units at N(1) and N(1′), as in the mixed thymine–uracil dimer **4** and in the thymine dimer **5**, leads to significant stabilization of the corresponding dimer radical cation, whereas in unbridged dimers a maximum lifetime of 10^{-10}–10^{-9} s was proposed for their dimer radical cations [6]. It was further shown that in bridged pyrimidines dimers of types **4** and **5** the interconversion of dimer and

Tab. 4.5.1. Most important redox-photosensitizers, radicals and radical ions used in the studies on the oxidative repair of pyrimidine cyclobutane dimers.

Redox-Photosensitizers		Radicals and Radical Ions	
$K_3Fe(CN)_6$	potassium hexacyanoferrate(III) Ref. [23]	$N_3^{\bullet}$	azide radical Ref. [19]
UO_2SO_4	uranyl(IV) sulfate Ref. [23]	$HO^{\bullet}$	hydroxyl radical Ref. [15]
		$NO_3^{\bullet}$	nitrate radical Ref. [8, 12, 22]
(anthraquinone-2-sulfonate structure)	anthraquinone-2-sulfonate (AQS) Ref. [6, 19, 24]	$Br_2^{\bullet -}$	bromine radical anion Ref. [15a]
		$SO_4^{\bullet -}$	sulfate radical anion Ref. [15a, 19]
(anthraquinone structure)	anthraquinone (AQ) Ref. [7a]	(phenanthrene radical cation structure)	phenanthrenyl radical cation Ref. [21]
(9,10-dicyanoanthracene structure)	9,10-dicyanoanthracene (DCA) Ref. [31]	(isoquinoline radical cation structure)	isoquinoline radical cation Ref. [25]
(DDQ structure)	2,3-dichloro-5,6-dicyano-1,4-quinone (DDQ) Ref. [11b]		
(tetraacetylriboflavin structure)	2′,3′,4′,5′-tetraacetylriboflavin, protonated (ac_4rfH^+) Ref. [13]		

monomer radical cation is reversible, possibly because of the proximity of the pyrimidine moieties, although the equilibrium between these two radical cations lies far towards the monomer radical cation intermediate [6a,b].

There are, nevertheless, experimental indications that a dimer radical cation $Pyr^{\bullet +}\!\!\diamond\!Pyr$, even without a linking bridge, must have a certain lifetime. The splitting of **t,s-1** initiated by the oxidizing nitrate radicals ($NO_3^{\bullet}$), generated by photolysis of cerium(IV) ammonium nitrate in acetonitrile, leads besides the "repaired" uracil **9**, also to a cyclobutane dimer species **8**, presumably with a hy-

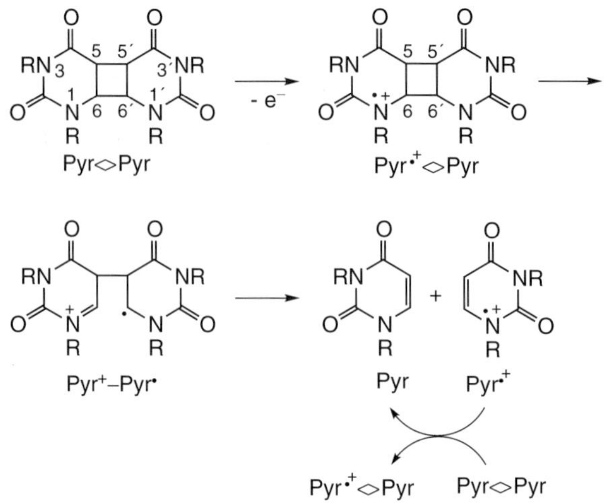

4.5 Oxidative Splitting of Pyrimidine Cyclobutane Dimers

Scheme 4.5.4.

droxymethylene group at N(1), as by-product (Scheme 4.5.5) [8]. The pathway leading to this compound is not yet known, but it can be proposed that it is formed first through deprotonation at N(1) in the dimer radical cation **6a**, and the resulting benzylic-type radical intermediate **7** is then stepwise converted to **8**. In addition, an interesting by-product is also formed during the NO$_3$·-induced oxidative cleavage of c,s-**1**, also presumably at the stage of the respective radical cation **6b**. The unusual structure of compound **10** with a pyrimidine ring being destroyed but the cyclo-

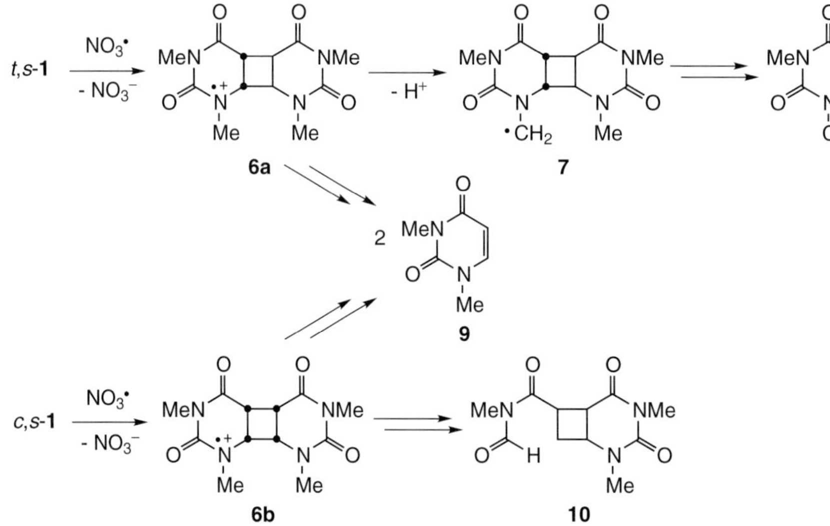

Scheme 4.5.5.

4.5.2 Mechanism of the Oxidative Splitting of Pyr◇Pyr

butane unit remaining intact was tentatively assigned from its fragmentation pattern in the mass spectrum (Scheme 4.5.5) [8].

The cycloreversion of the cyclobutane radical cation Pyr$^{•+}$◇Pyr could proceed in either a concerted or stepwise manner, and many attempts were made to determine the mechanism of this cleavage step. Because the radical cation is delocalized, it is not unreasonable that both the C(5)–C(5′) and the C(6)–C(6′) bonds are weakened by oxidation of Pyr◇Pyr. The observation of a substantial secondary deuterium isotope effect for the cleavage of the first bond [C(6)–C(6′)] and a small isotope effect for the cleavage of the second bond [C(5)–C(5′)] in various deuterated uracil-derived cyclobutane dimers was, however, taken as an indication of a stepwise splitting mechanism via the distonic radical cation Pyr$^+$–Pyr$^•$ [9]. Theoretical studies performed by Rösch, Michel-Beyerle et al. also strongly support the assumption of a successive cycloreversion [10].

Additional experimental evidence for stepwise cycloreversion with the C(6)–C(6′) bond being broken first was obtained from the photosensitized [11] or radical-induced [8] oxidative cleavage of the c,s-1, and t,s-1, which leads not only to the "repaired" **9**, but also to the C(5)–C(5′) linked 1,3-dimethyluracil dimer **13** (Scheme 4.5.6). Remarkably, in the case of the NO$_3^•$-induced splitting of t,s-1, the dimer **13** was found to be the major cleavage product and not the monomer **9** [8]. Compound **13** might be formed at the stage of the distonic radical cation **11** by elimination of a proton, followed by further oxidation and deprotonation of the radical intermediate **12**. The assumed competition between formation of the entirely cleaved product **9** and the semi-cleaved dimer **13** is supported by the finding that higher concentrations of the oxidizing NO$_3^•$ lead to an increasing ratio of 13/9 [12].

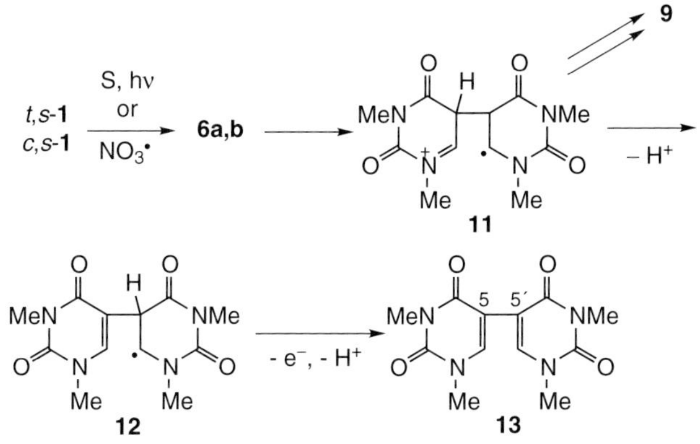

Scheme 4.5.6.

The process of cleavage of Pyr◇Pyr ends with formation of a monomeric Pyr and its radical cation Pyr$^{•+}$. The latter was found to initiate a radical chain by oxidation of further Pyr◇Pyr [8, 13]. In DNA, however, an analog chain reaction which would amplify the oxidative repair of pyrimidine dimer-derived damages could only be possible by long-range electron/hole transfer, because the oxidized nucleobase Pyr$^{·+}$ is located at a fixed position in an oligonucleotide strand [14].

Besides electron transfer, alternative pathways have also been proposed in literature for the initial step in the cleavage of Pyr$\diamond$Pyr. Heelis et al. [15a] and Grossweiner et al. [15b] observed that hydroxyl radicals (HO$^·$) and sulfate radical anions (SO$_4^{·-}$), both generated by pulse radiolysis in aqueous solution, were able to cleave c,s-3 to the monomer. Despite the high oxidizing strength of HO$^·$ (E^0 HO$^·$/OH$^-$ = 2.7 V relative to the NHE) [16], which lead Grossweiner et al. to the assumption of electron transfer as initial reaction step [15b] the reactions of with organic substrates in aqueous solutions rarely proceed by direct electron transfer but generally occur either by addition to π systems or by hydrogen abstraction. Although the detailed mechanism for thymine production from c,s-3 by HO$^·$ is not yet known, hydrogen atom transfer, presumably from C(6) or C(6'), was assumed to be the primary step.

The magnitude of the rate constant of the SO$_4^{·-}$-induced splitting of c,s-3, determined by Heelis et al. [14] from computer simulation of the build-up of thymine, also indicated hydrogen abstraction as the initial reaction step, even though SO$_4^{·-}$ is a very strong one-electron oxidant (E^0 SO$_4^{·-}$/SO$_4^{2-}$ = 2.4 V relative to the NHE) [16] and hydrogen abstractions by SO$_4^{·-}$ are usually several orders of magnitude slower than electron transfer [17]. Actually, the SO$_4^{·-}$- or N$_3^·$- (E^0 N$_3^·$/N$_3^-$ = 2.7 V relative to the NHE) [16, 18] initiated splitting of the stereoisomeric C(5)–C(5') linked dihydrothymine dimers **14** into the thymines **15** and 5,6-dihydrothymines **16** in aqueous solution has been suggested to proceed through initial electron transfer (Scheme 4.5.7) [19].

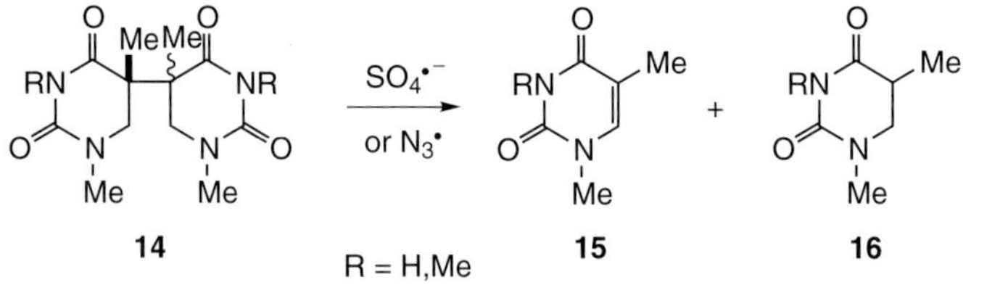

Scheme 4.5.7.

4.5.3
Stereoselectivity of the Oxidative Splitting of Pyr$\diamond$Pyr

The kinetics and energetics of reductive electron transfer in the enzyme–substrate complex of DNA photolyase indicates strong structural dependence of the photorepair process in different Pyr$\diamond$Pyr; this might arise from the recognition of the substrates in the binding pocket and/or from different capabilities of splitting in the photochemical process [20]. Likewise, for oxidative repair of Pyr$\diamond$Pyr it seems that not only the nature of Pyr, but also the constitution and configuration at the

Tab. 4.5.2. Stereoselectivity of the oxidative splitting of pyrimidine cyclobutane dimers.

Entry	Splitting efficiency[a]	Experimental conditions	Ref.
1	syn-1 > anti-1	in ethanol; unsensitized	11a
2	t,s-1 > c,s-1 ≫ c,a-1 > t,a-1	in acetonitrile; photosensitized (DDQ)	11b
3	t,s-1 > c,s-1 > c,s-2 > t,a-1 > c,a-1 > c,a-2	in acetonitrile; radical-induced ($NO_3^·$)	8, 22
4	c,s-3 > t,s-1 > c,s-1 > c,s-2 ≈ c,a-2 ≫ t,a-1 ≈ c,a-1	in water; photosensitized [$K_3Fe(CN)_6$]	23
5	c,s-1 > t,s-1 > c,s-3 > c,a-2 > c,s-2 ≫ c,a-1 ≈ t,a-1	in water; photosensitized (UO_2SO_4)	23
6	t,s-3 > c,s-3	in water; photosensitized (AQS)	24
7	t,s-1 > c,s-1 > 4 > 5	in acidified acetonitrile; photosensitized (protonated tetraacetylriboflavin [ac_4rfH^+])	13
8	c,s-2 > t,s-2	in dichloromethane; radical cation-induced (isoquinoline$^{·+}$)	25
9	c,s-2 > c,s-1 > t,s-2	in acetonitrile; radical cation-induced (phenanthrene$^{·+}$)	21a

[a] In order of decreasing reactivity.

cyclobutane ring has a significant effect on the rate of the cleavage process. Table 4.5.2 presents a compilation of the experimental results reported by various groups.

Because electron transfer is generally believed to be the initial step in the oxidative cleavage reaction, it might be expected that a correlation between reactivity and oxidation potential of Pyr◇Pyr should exist. In Scheme 4.5.2 are listed the irreversible half-peak anodic potentials $E_{p/2}$ for the stereoisomeric dimethyluracil- and dimethylthymine-derived cyclobutane dimers **1** and **2**, which were taken from Ref. [21a]. The *anti*-configured dimers clearly have higher oxidation potentials than the corresponding *syn* dimers. This might be because of considerable perturbation of the HOMO of the *syn*-configured dihydropyrimidine chromophore by the conjunction of the two chromophores forming the cyclobutane ring; this can be interpreted in terms of through-bond interactions between the *n* orbitals of N(1) and N(1′) involving the C(6)–C(6′) bond. In contrast, in the *anti* dimers no significant perturbation occurs, because the relevant *n* orbitals are separated by two C–C bonds [21a].

In fact, a faster oxidative splitting of the *syn*-configured dimers was usually observed. Elad et al. reported that both the unsensitized [11a] and sensitized splitting [11b] of the two *syn*-configured isomers of **1** is more effective than that of the *anti* compounds (entries 1 and 2). Analogous behavior was observed in the $NO_3^·$-induced splitting of a variety of dimethyluracil- and dimethylthymine-derived dimers (entry 3) [8, 22]. From the findings by Rosenthal et al. [23], who used transition metal salts as redox photosensitizers, it might, however, be concluded that *c,a*-**2** is more easily cleaved than *c,s*-**2** (entries 4 and 5). The reason for this discrepancy is not clear, but might, perhaps, be because of the different experimental conditions. Whereas the oxidation potentials $E_{p/2}$ given in Scheme 4.5.2 were measured in acetonitrile [21a], Rosenthal et al. [23] performed their studies in aqueous solution. No electrochemical data for Pyr<>Pyr measured under aqueous conditions are yet available in the literature.

Steric repulsion seem to affect the oxidation potential, because the more crowded c,s-1 and c,s-2 commonly have higher oxidation potentials, by ca. 0.1 V, than the corresponding trans,syn-configured isomers (Scheme 4.5.2) [21a]. On the basis of this, of the different isomeric dimethyluracil- and dimethylthymine-derived dimers t,s-1 and t,s-2 are assumed to be oxidatively cleaved more rapidly than their respective stereoisomers. This behavior was indeed observed in several instances, e.g. for the AQS-sensitized splitting of t,s-3 (entry 6) and the $Fe(CN)_6^{3+}$-sensitized cleavage of syn-configured 1 (entry 4), both in aqueous solution [23, 24]. Likewise, sensitizer- (entries 2 and 7) or radical-induced oxidative splitting (entry 3) of different Pyr<>Pyr in organic solvents was occasionally observed to correlate principally with the oxidation potentials $E_{p/2}$ of Pyr<>Pyr [8, 11b, 13, 22].

Despite this, the apparent discrepancies between reactivity and oxidation potential observed for many dimers indicates that the overall rate of the oxidative repair of Pyr<>Pyr cannot be simply equated with the rate of the initial electron transfer. When isoquinoline radical cations, generated in situ by irradiation of N-ethoxyisoquinolinium hexafluorophosphate, were used as oxidants, c,s-2 was observed to be cleaved faster than t,s-2 (entry 8) [25]. This finding was explained by dipole–dipole repulsion of the two carbonyl groups on the cyclobutane ring in the dimer radical cation of c,s-2, which lead to an enhancement of the splitting efficiency compared with the radical cation of t,s-2, thus exceeding the difference in the rate of the electron transfer between cis,syn and trans,syn dimers. Pac et al. [21a] explained the observed "reversed" reactivity with c,s-2 being faster cleaved than c,s-1 and t,s-2 using phenanthrenyl radical cations as oxidants, which were generated in situ by 1,4-dicyanobenzene-sensitized oxidation of phenanthrene, with steric repulsion in an excited state charge-transfer complex between the oxidant and the dimer (entry 9). Of the three dimers under investigation, c,s-2 is the most strained molecule, whereas t,s-2 is the less strained. The reactivity of c,s-1 is intermediate, because of the lack of the C(5) and C(5′) methyl groups.

In contrast with this, $NO_3^{\cdot}$-induced splitting reveals completely different behavior. It was observed from competition experiments that c,s-1 is cleaved significantly more rapidly than c,s-2 (entry 3) [22]. Because the oxidation potentials of both compounds are virtually identical (see Scheme 4.5.2), the difference in splitting efficiency must be because of the presence or absence of the C(5) and C(5′) methyl groups on the cyclobutane ring. $NO_3^{\cdot}$ is a strong oxidizing radical (E^0 $NO_3^{\cdot}$/$NO_3^- = 2.0$ V relative to the SCE; in acetonitrile) [26], and it is therefore believed that the electron abstraction should be fast and irreversible (whether this electron transfer proceeds through an inner- or outer-sphere process, is not yet clear). Because both $NO_3^{\cdot}$ and the reduced NO_3^- are small and planar molecules [27], it seems unlikely that steric interactions between the dimer or its radical cation and $NO_3^{\cdot}$ or NO_3^-, respectively, are responsible for the difference in splitting efficiency.

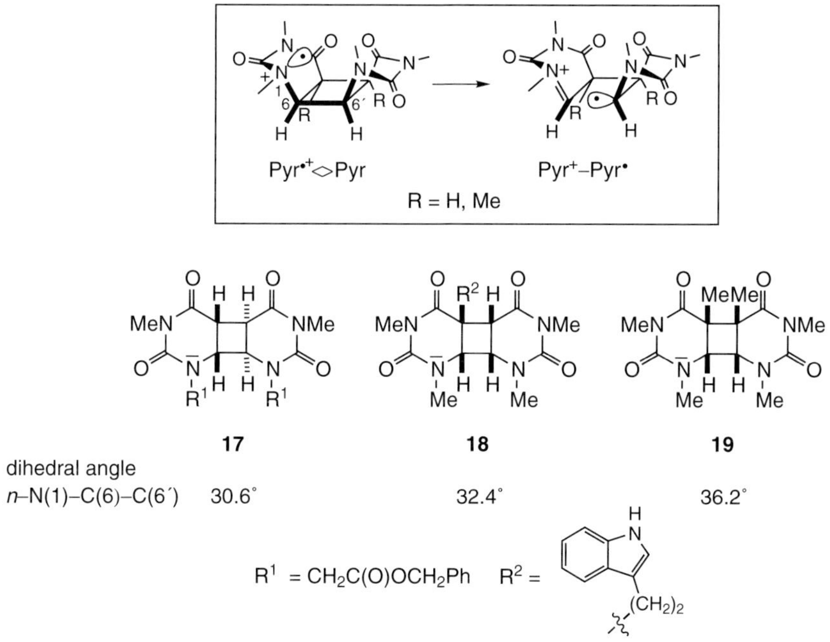

4.5.3 Stereoselectivity of the Oxidative Splitting of Pyr<>Pyr

Scheme 4.5.8.

	17	18	19
dihedral angle n–N(1)–C(6)–C(6′)	30.6°	32.4°	36.2°

$R^1 = CH_2C(O)OCH_2Ph$ $R^2 =$ (indole)(CH$_2$)$_2$

Considering the actual cyclobutane ring opening step Pyr$^{·+}$<>Pyr → Pyr$^+$–Pyr$^·$ (Scheme 4.5.4), however, where the C(6)–C(6′) σ bond is broken and a N(1)=C(6) π bond is simultaneously formed, this cleavage should be slower the more the involved orbitals, e.g. the SOMO and the σ orbitals of the N(1)–C(6) and C(6)–C(6′) bonds, deviate from coplanarity (Scheme 4.5.8).

No geometrical data calculated for dimethylthymine- and dimethyluracil-derived radical cations are, unfortunately, available in the literature. As an approximation, because it is not expected that removal of an electron would result in dramatic geometrical alterations, the X-ray data of the uncharged Pyr<>Pyr should tentatively reveal the influence of substituents at C(5) and C(5′) on the geometry of the dimer radical cation and, especially, of the cyclobutane ring. Assuming that the n orbital at N(1) in Pyr<>Pyr has a similar orientation as the SOMO in Pyr$^{·+}$<>Pyr, the dihedral angles along n–N(1)–C(6)–C(6′) in the various pyrimidine dimers 17 [28], 18 [29], and 19 [30], show that stepwise introduction of substituents at C(5) and C(5′) leads to an increase of the torsion at the cyclobutane ring (Scheme 4.5.8). The resulting decreasing overlap of the relevant orbitals could result in a slowdown of the cycloreversion, thus leading to a decrease of the overall splitting rate of Pyr<>Pyr. It might, therefore, be concluded that the ease of oxidative dimer splitting is governed by both oxidation potential of the dimer and the stereoelectronic requirements of its radical cation [22]. In contrast with this, the second step of the cycloreversion according to Pyr$^+$–Pyr$^·$ → Pyr + Pyr$^{·+}$ (Scheme 4.5.4) is not believed to be affected by comparable stereoelectronic effects, because any conformation favorable for splitting could be achieved by simple rotation around the C(5)–C(5′) σ bond.

An analogous stereoelectronic influence on the rate of the reductive repair process was also observed by Carell et al. [28]. The ring opening of the dimer radical anion also proceeds stepwise, but with the C(5)–C(5′) bond being broken first. The C(5) and C(5′)-methyl groups of thymine-derived dimers, which were found to be repaired more slowly than the uracil-derived dimers, lead to distortion of the geometry, which results in deceased overlap of the π^* C(4)–O(4) orbital with the σ^* C(5)–C(5′) orbital.

Rose et al. [13] also observed slower cleavage of thymine-derived bridged dimers **4** and **5**, compared with *syn*-**1** (entry 7). They explained this finding not by stereoelectronic effects but by the possible inability of a thymine radical cation to propagate the chain reaction by oxidizing the corresponding dimer. Because dimers **4** and **5** are linked by a trimethylene bridge, however, interference of the latter on the splitting efficiency of these compounds could not be excluded.

4.5.4
Conclusions

Although many experimental and theoretical studies on the oxidative repair of Pyr◇Pyr have been reported, it cannot yet be stated that the mechanism of this process is fully understood. This applies especially to conflicting data on the stereoselectivity of splitting efficiency. This inconsistency might, however, be partly because of the different reaction conditions. Most of the studies was performed using redox photosensitizers, and the possibility of back-electron transfer and formation of charge-transfer complexes in these systems might be one reason for the obvious discrepancies. It seems, therefore, that further investigations are required in which these possible interferences could be excluded, e.g. by using oxidants, which are:

- strong enough to exclude back-electron transfer, and
- relatively small, so that discriminations of the stereoisomeric Pyr◇Pyr by the oxidant caused by steric interactions can be avoided.

There are, nevertheless, indications that the efficiency of the oxidative cleavage does not only depend on the rate of the initial electron transfer but also on stereoelectronic effects in the resulting dimer radical cation. Which of these factors prevails during cleavage of a particular dimer is not yet clear. As a consequence, determination of absolute rate constants and identification of reaction intermediates appearing during the oxidative repair of Pyr◇Pyr, e.g. by spectroscopic methods, seems to be inevitable for a proper understanding of the splitting mechanism.

4.5.5
Experimental

4.5.5.1
Oxidative Cleavage of the 1,3-Dimethyluracil-derived Cyclobutane Dimers 1 by Nitrate Radicals (NO$_3^\bullet$)

In a Duran reactor 25 µmol **1** [8, 11a] and 5 µmol cerium(IV) ammonium nitrate are dissolved in 6 mL acetonitrile, deaerated for 5 min in an ultrasound bath and irradiated under argon for 2 h using a medium-pressure mercury lamp. The reaction mixture is filtered (SiO$_2$, ethyl acetate), concentrated and analyzed by GC. (Varian 3400cx; SE 30 column, 50 m, temperature program $120_5 \rightarrow 250_{22}$, heating rate 10° min^{-1}). t_{Ret}: t,s-**1** = 29.1 min, c,s-**1** = 30.2 min, t,a-**1** = 27.0 min, c,a-**1** = 27.5 min, **9** = 11.7 min.

References

1 (a) J. C. Sutherland, *Photochem. Photobiol.* **1977**, *25*, 435–440; (b) J.-H. Yoon, C.-S. Lee, T. R. O'Connor, A. Yasui, G. P. Pfeifer, *J. Mol. Biol.* **2000**, *299*, 681–693.

2 W. Harm, *Biological Effects of Ultraviolet Radiation*, Cambridge University Press, London, **1980**.

3 (a) T. Okamura, A. Sancar, P. F. Heelis, T. P. Begley, Y. Hirata, N. Mataga, *J. Am. Chem. Soc.* **1991**, *113*, 3143–3145; (b) S.-T. Kim, A. Sancar, E. Essenmacher, G. T. Babcock, *J. Am. Chem. Soc.* **1992**, *114*, 4442–4443.

4 M. P. Scannell, S.-R. Yeh, D. E. Falvey, *Photochem. Photobiol.* **1996**, *64*, 764–768.

5 (a) P. F. Heelis, R. F. Hartman, S. D. Rose, *Chem. Soc. Rev.* **1995**, 289–297; (b) S.-T. Kim, A. Sancar, *Photochem. Photobiol.* **1993**, *57*, 895–904.

6 (a) P. J. W. Pouwels, R. F. Hartman, S. D. Rose, R. Kaptein, *Photochem. Photobiol.* **1995**, *61*, 563–574; (b) P. J. W. Pouwels, R. F. Hartman, S. D. Rose, R. Kaptein, *J. Am. Chem. Soc.* **1994**, *116*, 6967–6968; (c) T. Young, R. Nieman, S. D. Rose, *Photochem. Photobiol.* **1990**, *52*, 661–668.

7 (a) H. D. Roth, A. A. Lamola, *J. Am. Chem. Soc.* **1972**, *94*, 1013–1014; (b) A. A. Lamola, *Mol. Photochem.* **1972**, *4*, 107–133.

8 O. Krüger, U. Wille, *Org. Lett.* **2001**, *3*, 1455–1458.

9 (a) R. Austin, S. McMordie, T. P. Begley, *J. Am. Chem. Soc.* **1992**, *114*, 1886–1887; (b) M. R. Widmer, E. Altmann, H. Young, T. P. Begley, *J. Am. Chem. Soc.* **1989**, *111*, 9264–9265.

10 (a) A. A. Voityuk, M.-E. Michel-Beyerle, N. Rösch, *J. Am. Chem. Soc.* **1996**, *118*, 9750–9758; (b) J. Rak, A. A. Voityuk, M.-E. Michel-Beyerle, N. Rösch, *J. Phys. Chem. A* **1999**, *103*, 3569–3574; (c) J. Rak, A. A. Voityuk, N. Rösch, *Theochem* **1999**, *488*, 163–168.

11 (a) D. Elad, I. Rosenthal, S. Sasson, *J. Chem. Soc. (C)* **1971**, 2053–2057; (b) S. Sasson, D. Elad, *J. Org. Chem.* **1972**, *37*, 3164–3167.

12 U. Wille, unpublished results.

13 R. F. Hartmann, S. D. Rose, *J. Org. Chem.* **1992**, *57*, 2302–2306.

14 P. J. Dandliker, R. E. Holmlin, J. K. Barton, *Science* **1997**, *275*, 1465–1468.

15 (a) P. F. Heelis, D. J. Deeble, S.-T. Kim, A. Sancar, *Int. J. Radiat. Biol.* **1992**, *62*, 137–143; (b) L. I. Grossweiner, A. G. Kepka, R. Santus, J. A. Vigil, *Int. J. Radiat. Biol.* **1974**, *25*, 521–523.

16 P. Wardmann, *J. Phys. Chem. Ref. Data* **1989**, *18*, 1637–1755.

17 (a) *Gmelin Handbuch der Anorga-*

nischen Chemie: Schwefel, Gmelin-Institut für Anorganische Chemie der Max-Planck-Gesellschaft zur Förderung der Wissenschaften, ed., 8th edn., Suppl. Vol. 3, Springer, Berlin, **1980**, p. 334; (b) D. SCHULTE-FROHLINDE, K. HILDENBRAND, *Free Radicals in Synthesis and Biology*, F. MINISCI, ed., NATO ASI Series, **1989**, p. 335; (c) B. C. GILBERT, J. R. LINDSAY SMITH, P. TAYLOR, S. WARD, A. C. WHITWOOD, *J. Chem. Soc., Perkin Trans. 2* **1999**, 1631–1637; and literature cited therein.

18 G. V. BUXTON, C. L. GREENSTOCK, W. P. HELMAN, A. B. ROSS, *Phys. Chem. Ref. Data* **1988**, *17*, 513–886.

19 T. ITO, H. SHINOHARA, H. HATTA, S. NISHIMOTO, S. FUJITA, *J. Phys. Chem. A* **1999**, *103*, 8413–8420.

20 (a) T. LANGENBACHER, X. ZHAO, G. BIESER, P. F. HEELIS, A. SANCAR, M. E. MICHEL-BEYERLE, *J. Am. Chem. Soc.* **1997**, *119*, 10532–10536; (b) S.-T. KIM, K. MALHOTRA, C. A. SMITH, J.-S. TAYLOR, A. SANCAR, *Biochemistry* **1993**, *32*, 7065–7068; (c) S.-T. KIM, A. SANCAR, *Biochemistry* **1991**, *30*, 8623–8630.

21 (a) C. PAC, J. KUBO, T. MAJIMA, H. SAKURAI, *Photochem. Photobiol.* **1982**, *36*, 273–282; (b) T. MAJIMA, C. PAC, J. KUBO, H. SAKURAI, *Tetrahedron Lett.* **1980**, *21*, 377–380.

22 U. WILLE, O. KRÜGER, unpublished results.

23 I. ROSENTHAL, M. M. RAO, J. SALOMON, *Biochem. Biophys. Acta* **1975**, *378*, 165–168.

24 E. BEN-HUR, I. ROSENTHAL, *Photochem. Photobiol.* **1970**, *11*, 163–168.

25 J. J. A. HUNTLEY, R. A. NIEMAN, S. D. ROSE, *Photochem. Photobiol.* **1999**, *69*, 1–7.

26 E. BACIOCCHI, T. DEL GIACCO, S. M. MURGIA, G. V. SEBASTIANI, *J. Chem. Soc., Chem. Commun.* **1987**, 1246–1248.

27 R. P. WAYNE, I. BARNES, P. BIGGS, J. P. BURROWS, C. E. CANOSA-MAS, J. HJORTH, G. LE BRAS, G. K. MOORTGAT, D. PERNER, G. RESTELLI, H. SIDEBOTTOM, *Atmos. Environ.* **1991**, *A25*.

28 J. BUTENANDT, R. EPPLE, E.-U. WALLENBORN, A. P. M. EKER, V. GRAMLICH, T. CARELL, *Chem. Eur. J.* **2000**, *6*, 62–72.

29 T. C. GROY, S.-T. KIM, S. D. ROSE, *Acta Cryst.* **1991**, *C47*, 1287–1290.

30 N. CAMERMAN, A. CAMERMAN, *J. Am. Chem. Soc.* **1970**, *92*, 2523–2527.

31 D. J. FENICK, H. S. CARR, D. E. FALVEY, *J. Org. Chem.* **1995**, *60*, 624–631.

B.21
DNA Damage

Uta Wille

Because of the complex structure of DNA various damage pathways are possible. The most important lesions are caused by radical species or UV irradiation, which can affect the sugar backbone, the nucleobases, or both. These lesions, if not repaired, can contribute to mutagenesis, carcinogenesis, aging, inherited disease, and cell death.

Damage to the Sugar Backbone

Damage to the sugar moiety in nucleotides is usually initiated by hydrogen abstraction. A variety of hydrogen-abstracting radical species can be present in cells, for example highly reactive hydroxy radicals (HO·), which are formed in human oxygen metabolism by reduction of hydrogen peroxide or fatty acid hydroperoxides by either photochemical or metal-catalyzed decomposition, or by the antitumor antibiotics bleomycin and neocarzinostatin in their respective activated form. In a typical B-form duplex one of the two 5'-hydrogens and the 4'-hydrogen, which point into the minor groove, were found to be the atoms most accessible to external attack, although initial hydrogen abstraction at the other positions is also possible [1]. Of particular importance are the 4'-sugar radicals, which were observed to be precursors to immediate strand breaks. The proposed pathway for the HO·-induced degradation of an oligonucleotide **1** under aerobic conditions is outlined in Scheme B.21.1 [2]. The intermediate hydroperoxide **3** undergoes a Crigee rearrangement, and subsequent elimination and hydrolysis lead to scission of the sugar-phosphate linkage resulting in formation of the cleavage products **6–8**.

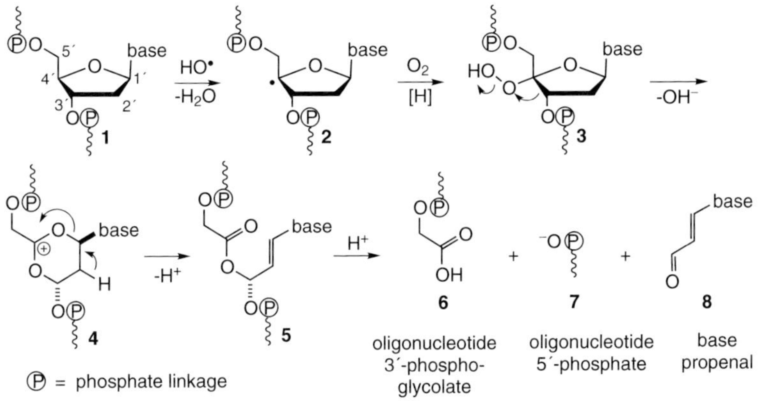

Scheme B.21.1.

Damage to the Nucleobases

Free radicals and radiation can both damage nucleobases. On the basis of its reactivity the most important radical in cells is HO·, which reacts with pyrimidine and purine nucleobases mainly by addition to the unsaturated bonds. In purines, HO· attack occurs preferentially at C4 and C8; this is shown as an example in Scheme B.21.2 for the reaction with guanosine (G) **9** [3a,b]. Elimination of water from the C4 adduct **10** leads to the comparatively stable radical **11** which can undergo oxidative degradation to the oxazolone **14** [3c]. The C8 adduct **12** can either undergo further oxidation to yield 8-oxoguanosine **15**, or ring opening followed by reduction to the formamide **16**.

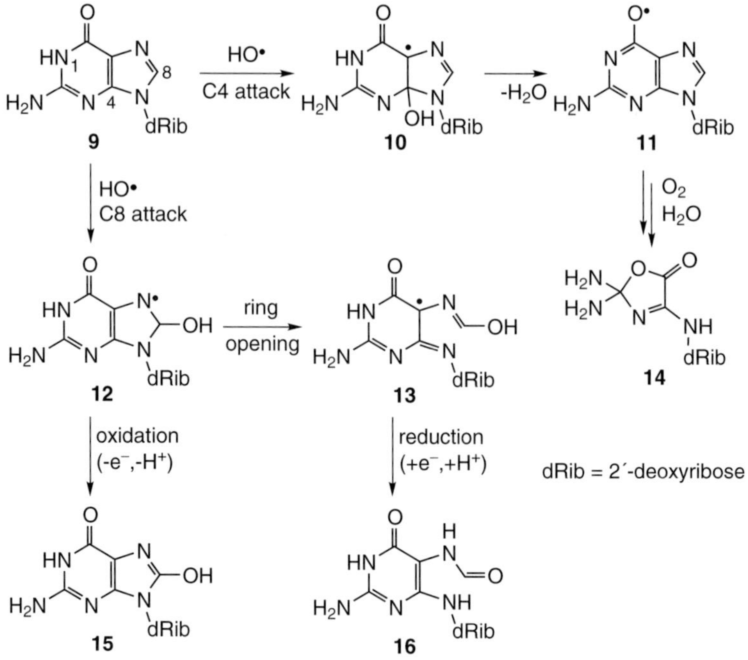

4.5 Oxidative Splitting of Pyrimidine Cyclobutane Dimers

Scheme B.21.2.

Radical attack on pyrimidines occurs at the C5=C6 double bond with a strong preference for addition to C5 [3b, 4]. Products formed by reaction of thymidine (T) **17** with HO˙ under aerobic conditions are outlined in Scheme B.21.3 [5]. Besides radical addition, hydrogen abstraction from the exomethyl group by HO˙ is also possible in T – in contrast with uridine (U). The initially formed radical intermediates **18**, **22**, and **25** can potentially also abstract hydrogen atoms from the adjacent ribosyl groups leading to direct strand scission (see above). The chemistry of cytidine (C) with HO˙ is complicated, because loss of the C5=C6 double bond makes the amino group at C4 more susceptible to hydrolytic deamination. U-derived compounds are, therefore, common products resulting from oxidation of C (not shown).

Direct UV irradiation of DNA leads to a variety of lesions at the nucleobase. The principle products are pyrimidine cyclobutane dimers T◇T, T◇C, and C◇C, formed in that order of preference, by [2+2] photocycloaddition of two adjacent pyrimidine nucleosides in the same oligonucleotide strand (Scheme B.21.4). With approximately 50% of the frequency of cyclobutane dimers, [2+2] dimerization involves a C=O double bond leading to an intermediate oxetane **29**, which later decomposes to the [6–4] photoproduct **30**.

It has been suggested that the C-containing dimers, e.g. T◇C, C◇C, and [4–6] photoproducts of type **30**, in particular, are significantly respon-

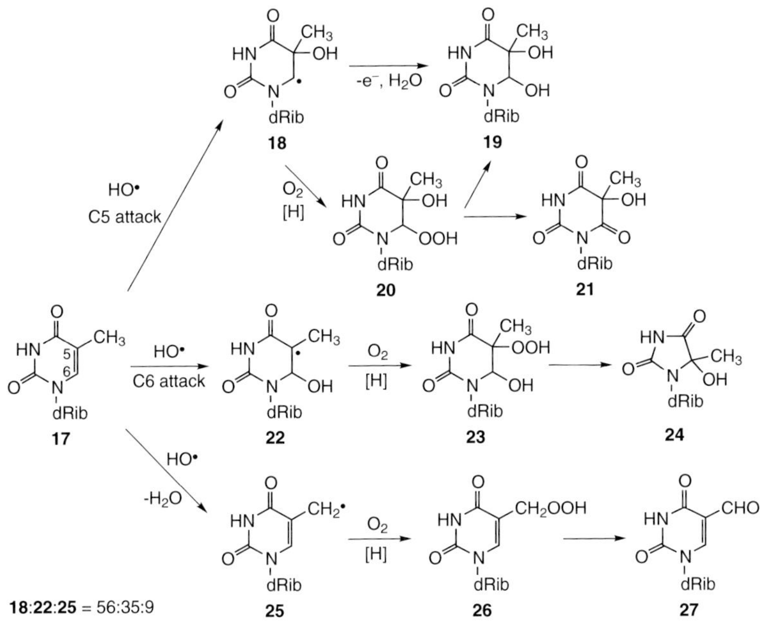
HO•
C5 attack
-e⁻, H2O
O2
[H]
HO•
C6 attack
HO•
-H2O
dRib
18
19
20
21
17
22
23
24
25
26
27

18:**22**:**25** = 56:35:9

dRib = 2´-deoxyribose

Scheme B.21.3.

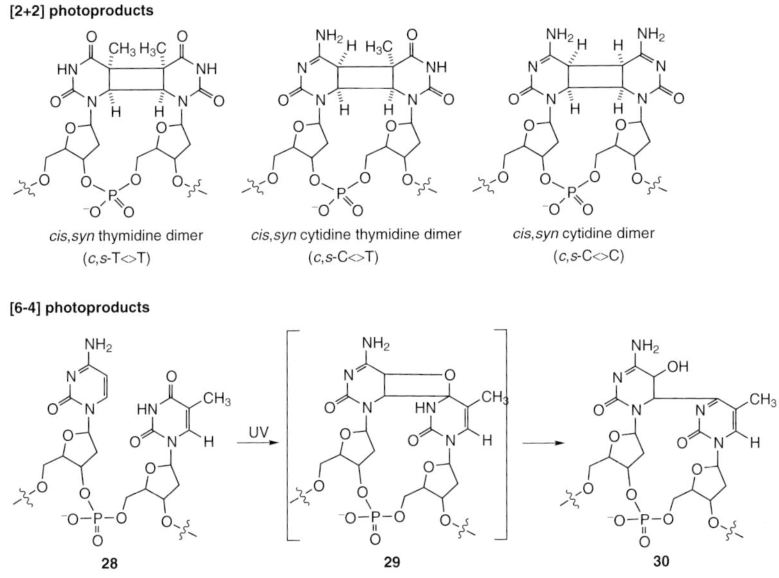
[2+2] photoproducts
cis,syn thymidine dimer
(c,s-T<>T)
cis,syn cytidine thymidine dimer
(c,s-C<>T)
cis,syn cytidine dimer
(c,s-C<>C)
[6-4] photoproducts
UV
28
29
30

Scheme B.21.4.

sible for the lethal effect of UV light. Because of the facilitated tautomerization of the exocyclic amino group at C4 in C, which is caused by loss of the C5=C6 double bond, the resulting imino tautomer has the same hydrogen bonding pattern as T. Consequently, this can lead to incorporation of adenosine (A) rather than G into the daughter strand during DNA replication. In addition, C-derived dimers are known to be very susceptible to hydrolysis. The net genetic result from either of these pathways is a C → T mutation [6].

References

1 Review: W. Knapp Pogozelski, T. D. Tullius, *Chem. Rev.* **1998**, *98*, 1098–1107.
2 A. Dussy, E. Meggers, B. Giese, *J. Am. Chem. Soc.* **1998**, *120*, 7399–7403.
3 (a) S. Steenken, *Chem. Rev.* **1989**, *89*, 503–520; (b) D. Schulte-Frohlinde, K. Hildenbrand, in: *Free Radicals in Synthesis and Biology*, F. Minisci, ed., NATO ASI Series, **1989**, p. 335; (c) W. Adam, S. Andler, W. M. Nau, C. R. Saha-Möller, *J. Am. Chem. Soc.* **1998**, *120*, 3549–3559.
4 (a) D. K. Hazra, S. Steenken, *J. Am. Chem. Soc.* **1983**, *105*, 4380–4386; (b) W. F. Ho, B. C. Gilbert, M. J. Davies, *J. Chem. Soc., Perkin Trans. 2* **1997**, 2533–2538.
5 Review: C. J. Burrows, J. G. Muller, *Chem. Rev.* **1998**, *98*, 1109–1151.
6 D. J. Fenick, H. S. Carr, D. E. Falvey, *J. Org. Chem.* **1995**, *60*, 624–631, and literature cited therein.

4.6
Charge Transfer in DNA

Hans-Achim Wagenknecht

4.6.1
Introduction

The possibility that the one-dimensional array of π-stacked base pairs in B-form DNA might serve as a pathway for charge migration was suggested over 40 years ago [1]. Since then, the fundamental question whether DNA serves as a medium for long-range charge transfer (CT) has been the subject of much controversy. Many different techniques have been used to explore DNA-mediated CT processes, among these are biochemical methods and spectroscopic measurements. Several mechanisms, for example superexchange and hopping mechanism, have been proposed and, at least in part, experimentally verified. CT chemistry through DNA is now widely accepted, and the discussion has moved to the question of its mechanism and, most recently, to its biological relevance and biotechnical application. It has become clear that DNA-mediated CT can occur on an ultrafast time scale and can result in reactions over long distances. Most experiments have also shown that CT in DNA is extremely sensitive to the π-stacking of the intervening DNA bases and to disruption and perturbation of the DNA structure or conformation.

4.6.2
Hole Transfer and Hole Hopping in DNA

Interest in DNA-mediated CT has been spurred by its relevance to oxidative damage which might cause mutagenesis and carcinogenesis [2]. Thus, most research groups have focused their work on the photochemically or photophysically induced oxidation of DNA, and furthermore, on the mobility of the created positively charged radical in the DNA which, in fact, is a hole transfer (HT). Most experiments have been performed according to the steps [1]:

- labeling of DNA with redox active probes through intercalation and/or covalent linkages;

Highlights in Bioorganic Chemistry: Methods and Applications. Edited by Carsten Schmuck, Helma Wennemers.
Copyright © 2004 WILEY-VCH Verlag GmbH & Co. KGaA, Weinheim
ISBN: 3-527-30656-0

- initiation of CT by photochemical or electrochemical techniques; and
- detection of CT processes by spectroscopic, electrochemical, or biochemical methods.

Using this approach, experiments are still ongoing to enable further understanding of the dynamics of DNA-mediated CT and its role in DNA damage and repair. Published experiments exploring CT reactions through DNA can be divided into spectroscopic and biochemical studies.

4.6.2.1
Spectroscopic Studies

When investigations of DNA-mediated CT were started all research groups interpreted their results according to the Marcus theory [3]. As a result, CT processes were described in terms of a superexchange mechanism. The charge tunnels in one coherent step from **D** (charge donor) to **A** (charge acceptor) and never resides on the intermediate DNA bridge (**B**) (Figure 4.6.1). In this case, the rate k_{CT} depends on the distance R between **A** and **D** and the exponential parameter β which itself is dependent on the nature of the bridge **B** and its coupling with **D** and **A** [4]. Values of β for CT through proteins lie in the range $1.0–1.4\,\text{Å}^{-1}$ [4]. In contrast, β

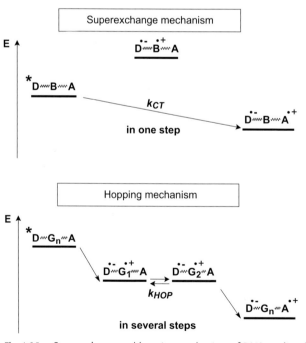

Fig. 4.6.1. Superexchange and hopping mechanism of DNA-mediated hole transfer.

Tab. 4.6.1. Summary of the spectroscopic studies of DNA-mediated HT. The rate k_{CT} and distance dependence β are given by $k_{CT} = k_0 e^{-\beta R}$.

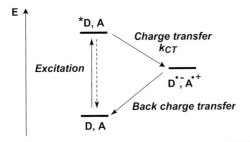

Charge donor/acceptor	Covalently attached?	β [Å^{-1}]	k_{CT} [s^{-1}]	Research group	Year
E/MV	No	–	10^5	Fromherz [1]	1986
E/MV, Ac/DAP	No	1.0	10^8	Harriman [1]	1992
Ru(II)/Rh(III) complexes	Yes	0.2	10^9	Barton [1]	1993
Ru(II)/Rh(III) complexes	Yes	1.0–1.5	10^6	Meade [1]	1995
St/G	Yes	0.64	10^8–10^{12}	Lewis [1]	1997
E/Rh(III) complex	Yes	–	10^{10}	Barton [1]	1997
Ac/G	Yes	1.4	10^5–10^{10}	Tanaka [1]	1998
E/Z	Yes	–	10^{12}	Barton/Zewail [1]	1999
N/A	Yes	–	10^9	Lewis [1]	1999
Ap/G	Yes	0.1–1.0	10^9–10^{10}	Barton [1]	1999
Ap/G	Yes	0.75	10^6–10^7	Shafirovich [6]	2000
Ap/G	Yes	0.6	10^9–10^{11}	Barton/Zewail [7]	2000
Ac/G, Ac/Z	Yes	–	10^7–10^{12}	Michel-Beyerle [8]	2001
Ru(II) complex/In	Yes	–	$\geq 10^7$	Barton [9]	2002

Abbreviations: A = adenine, Ac = acridine, Ap = 2-aminopurine, DAP = N,N′-dimethyl-2,7-diazapyrenium, E = ethidium, G = guanine, In = 4-methylindole, MV = methyl viologen, N = naphthalene, St = stilbenedicarboxamide, Z = 7-deazaguanine.

values determined for CT reactions in DNA can be found in a wide range from $\beta < 0.1$ Å^{-1} to $\beta = 1.5$ Å^{-1} (Table 4.6.1) [5].

Organic and inorganic intercalators not covalently attached to oligonucleotides were first reported for study of CT in DNA. These experiments provide little information, because of the lack of an accurate measurement of the distance between the **D** and **A** and the concern for pairing of the intercalators. A significant experimental improvement came with DNA assays bearing covalently linked intercalators. By use of these systems systematic measurement of distance dependence and base sequence dependence was possible. These experiments led to three important observations:

- Short-range CT reactions occur on a very fast time-scale ($k_{CT} = 10^9$–10^{12} s^{-1}).
- The typical β value for DNA-mediated CT is 0.6–0.8 Å^{-1}.
- Intercalation of **D** and **A** is crucial for rapid and efficient CT.

4.6 Charge Transfer in DNA

The occurrence of very small β values ($\leq 0.1\,\text{Å}^{-1}$) with very shallow distance dependence led to the description of an alternative mechanism – the hopping model. Among the four different DNA bases guanine (G) is the most easily oxidized. As a result, the G radical cation plays the role of the intermediate charge carrier during the hopping process. In contrast with the previously described superexchange mechanism, when the positive charge has been injected into the DNA base stack it hops from G to G and can finally be trapped at a suitable charge acceptor (Figure 4.6.1). Each hopping step itself is a tunneling process through the intervening adenine (A)–thymine (T) base pairs. In contrast with the superexchange mechanism the rate of HT by hopping does not depend on the overall distance between **D** and **A** but is only dependent on the number of hopping steps [10]. Lewis et al. were able to measure the rate for a single hopping step from G to GG, $k_{\text{HOP}} = 5 \times 10^7\,\text{s}^{-1}$ [11].

4.6.2.2
Biochemical Experiments

The G radical cation has been identified as the precursor of a variety of different oxidative G lesions which are normally described as G^{ox}. Some of these G oxidation products have been identified (Figure 4.6.2) [12]. Biochemical experiments explore

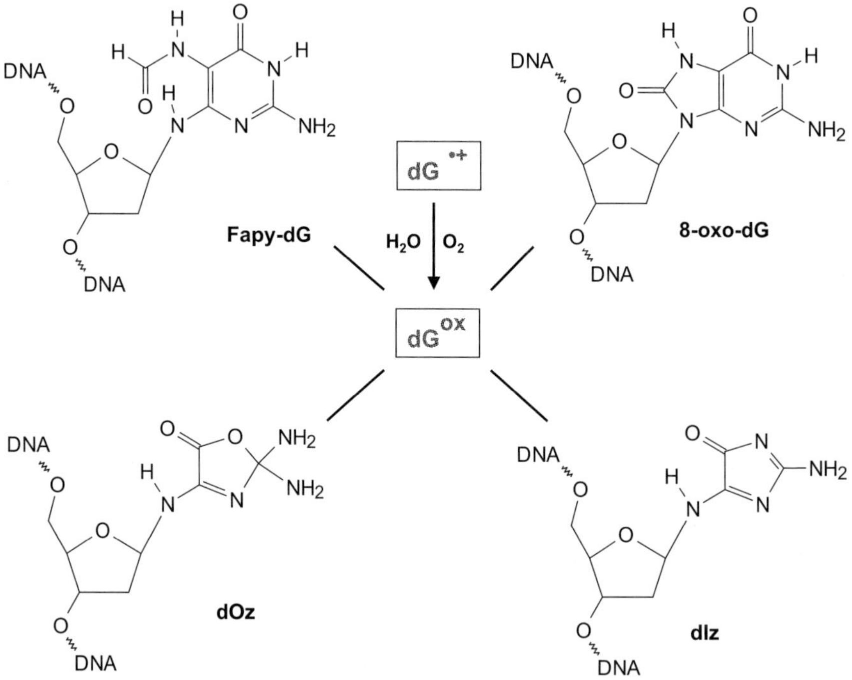

Fig. 4.6.2. Examples of oxidative G damage (G^{ox}) with formamidopyrimidine (Fapy-dG), 8-oxoguanine (8-oxo-dG), oxazolone (dOz), or imidazolone (dIz) structure.

HT reactions through DNA by an indirect method. As described above, after photochemical or photophysical oxidation of DNA with a suitable intercalator G is preferentially oxidized. The resulting G radical cation can react with H_2O and/or O_2 yielding oxidized G products G^{ox}. Such modified DNA strands can be cleaved at the site of G^{ox} by treatment with, e.g., piperidine at elevated temperature, then separated by gel electrophoresis and visualized by phoshorimagery using radioactive ^{32}P-labeling [13].

The most common photooxidants for DNA are metal complexes or organic intercalators such as Rh(III) complexes, Ru(II) complexes, ethidium derivatives, anthraquinone derivatives, uridine modified with cyanobenzoquinones, and modified 2'-deoxyribosides bearing a photoreactive group (Figure 4.6.3) [1]. These systems differ significantly in their structural properties, redox potentials, and absorbing wavelengths. It has, nevertheless, been observed in all systems that the positive charge can be transported with high efficiency over very long distances (up to 200 Å). The observed efficiency of HT seems to be strongly dependent on the distance and base sequence between each of the G:cytosine (C) pairs. In particular it has been shown that A can act as an intermediate charge carrier if G is not present [14].

4.6.3
Protein-dependent Charge Transfer in DNA

From spectroscopic and biochemical studies it has become clear that DNA-mediated CT is extremely sensitive to the π-stacking of the intervening DNA bases and to disruption and perturbation of the DNA structure or conformation. This indicates that sensing of DNA damage could be accomplished, at least in part, on the basis of CT chemistry. In considering these possibilities, it is important to discover whether DNA-mediated CT does occur within the cell. Charge transfer in HeLa cell nuclei has recently been probed by use of a rhodium photooxidant [15]. After incubation and irradiation the genomic DNA was isolated and analyzed. This revealed that base damage occurs preferentially at the 5'-G of GG sites. More importantly, oxidative G damage was found at protein-bound sites that were inaccessible to the rhodium photooxidant, as examined by footprinting. This clearly indicates that CT processes could occur in cells.

Specific DNA–protein interactions which either promote or inhibit CT processes through the protein–DNA interface would be the most crucial part of a biological system sensing DNA damage. Recent experiments have shown clearly that DNA-mediated CT processes are modulated both negatively and positively by DNA-binding proteins. Most importantly, each of the observed influences of the proteins can be explained by special structural features of the corresponding DNA–protein complexes. Thus, special DNA–protein interactions result in a characteristic modulation of the DNA-mediated CT.

Barton et al. developed an assay enabling the study of the influence of DNA-binding proteins simply by gel electrophoretic analysis of oxidative G damage

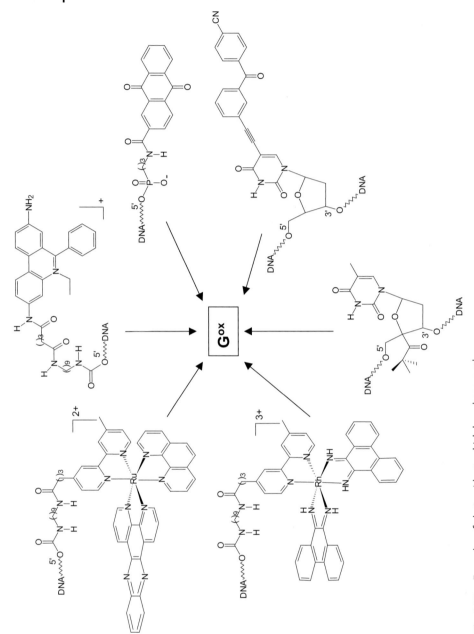

Fig. 4.6.3. Examples of photooxidants which have been used in biochemical studies of HT in DNA.

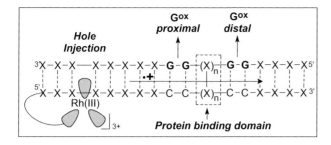

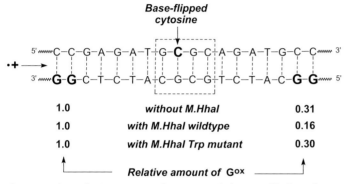

Fig. 4.6.4. Assay for investigation of protein-modulated CT (top). The effect of the protein is apparent from the relative amounts of oxidative G damage (G^{ox}) at the proximal and distal GG pairs. For example, base-flipping of the target C in the binding sequence of M.HhaI decreases the efficiency of HT through the DNA–protein complex. In wild-type M.HhaI the hole created by base-flipping is filled by the aliphatic side-chain of a Gln 237, interrupting the DNA base stack. In contrast, the M.HhaI Trp mutant intercalates the indole side-chain of Trp 237 which can $\pi-\pi$ interact with the neighboring bases. As a result HT efficiency is restored.

(Figure 4.6.4) [16]. The positively charged radical (hole) is injected by use of a Rh(III) complex which is covalently attached to the 5′-end of the oligonucleotide. Two GG pairs are located along the DNA duplex sequence (proximal and distal) and are the most preferable sites for trapping the radical. Oxidative G damage (G^{ox}), as the chemical result of HT, can be detected in different amounts at the distal and proximal GG pair. Because of their location G damage occurring at the distal GG pair reflects the effect of the DNA-binding protein whereas damage at the proximal GG pair is not affected by the protein. Hence, the proximal GG pair serves as an internal standard of radical damage.

Proteins that bind the major groove but do not perturb the normal B-DNA structure can enhance CT efficiency in DNA. This was demonstrated by using the restriction endonuclease PvuII and the transcription factor ANTP [17]. As a result of the binding of proteins, the DNA conformation is stiffened, the conformational movements are reduced, and, as a result, CT is facilitated. In contrast with R.PvuII and ANTP, the TATA-box binding protein induces two 90° bends in the DNA duplex. Because of this strong conformational change of the DNA structure CT efficiency decreases significantly [17]. Most recently, Saito et al. investigated the endonuclease BamHI–DNA complex [18]. In this complex direct contact of a positively charged guanidinium group of the protein with the recognition sequence of the DNA completely suppressed CT and dramatically reduced the efficiency of CT through the duplex.

One of the most interesting examples of how proteins can modulate CT in DNA is the cytosine methylase HhaI [19]. The protein recognizes the sequence 5'-GCGC-3'. En route to methylation of the internal C a base-flipped complex is formed. In this complex, the target C can be found in an extrahelical position and a glutamine (Gln) side-chain fills the space in the DNA duplex. Not surprisingly, M.HhaI binding profoundly inhibits long-distance CT, as revealed by diminution of distal GG damage (Figure 4.6.4). Base-flipping and insertion of the aliphatic side-chain of Gln clearly attenuated the HT as a result of interruption of the base stack. The situation is different for a M.HhaI mutant in which Gln 237 is substituted by tryptophan (Trp). Interestingly, long-range oxidation of the distal GG pair was restored on binding with this Trp mutant of M.HhaI (Figure 4.6.4). This is the result of electronic interaction of the flat aromatic indole heterocycle of Trp with the neighboring DNA bases. In conclusion, base-flipping interrupts CT but the indole side-chain of Trp can replace a normal DNA base electronically.

The goal of our experiments was spectroscopic verification that an intercalated Trp can participate in CT reactions in DNA. Aromatic amino acids such as tyrosine (Tyr) or Trp usually have lower oxidation potentials than G and can therefore be used as spectroscopic traps for holes [16]. It seemed reasonable for us to start spectroscopic investigations of protein-modulated CT in DNA by using the smallest possible DNA-binding peptides – Lys-Trp-Lys and Lys-Tyr-Lys, bearing only one DNA-base-interacting amino acid, Trp or Tyr, respectively. The assay (Figure 4.6.5) contained a Ru(II) complex covalently attached to the 5'-end of the DNA duplex and an AT alternating sequence as the potential binding site for the peptides [20]. The corresponding Ru(III) complex was generated in situ as a powerful ground-state oxidant capable of oxidizing G and, subsequently, DNA-bound Trp or Tyr as a part of the peptides Lys-Trp-Lys, or Lys-Tyr-Lys, respectively. The CT reaction was followed by emission and time-resolved transient absorption spectroscopy. The resulting oxidized products of the peptides (Trp^{ox}, Tyr^{ox}) and the DNA itself were analyzed by different biochemical methods. It was shown that the radicals of Trp and Tyr can be generated by DNA-mediated CT and occur on the microsecond time-scale and in high yields. CT from the Ru(III) complex to the tripeptides follows the hopping mechanism; this means the G radical cation occurs as an intermediate charge carrier. Interestingly, the peptide radicals generated have completely different reactivity. The DNA-bound Trp radical forms oxidized products in the presence of O_2 and the DNA-bound Tyr radical forms cross-links with the DNA bases at the peptide binding site (Figure 4.6.5).

Using this knowledge about spectroscopic observation of DNA-to-peptide CT, we applied the method to the M.HhaI-DNA complex, using the Trp 237 mutant men-

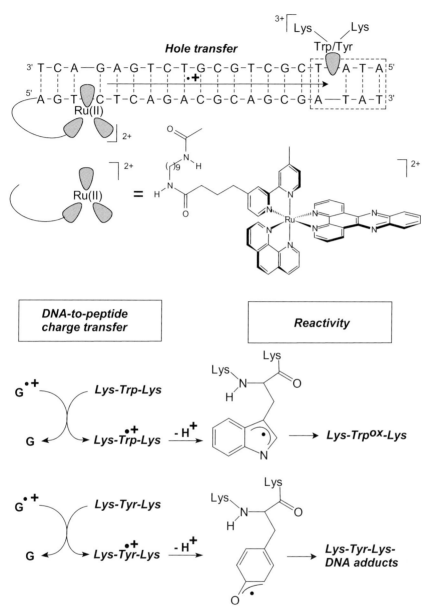

Fig. 4.6.5. Assay for spectroscopic investigation of how aromatic amino acids modulate DNA-mediated HT. The peptide radicals generated by DNA-mediated HT have different reactivity. The radical of Lys-Trp-Lys forms oxidized peptide products in the presence of O_2 whereas the radical Lys-Tyr-Lys forms crosslinks with the DNA bases at the binding site.

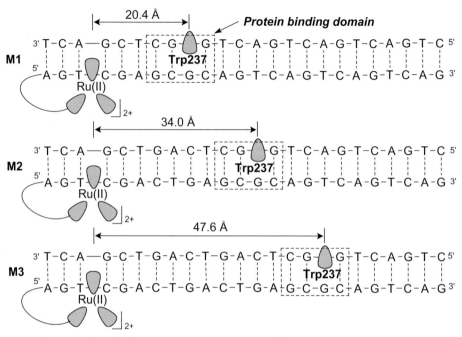

Fig. 4.6.6. Assay for the spectroscopic investigation of the distance-dependence of HT in DNA using the protein M.HhaI (Trp 237 mutant).

tioned above [19]. The rate of formation of the radical intermediates was measured as a function of the distance using three different DNA substrates (**M1**, **M2**, and **M3**; Figure 4.6.6). The CT reaction was observed by time-resolved transient absorption spectroscopy. The product radical was identified to be a mixture of the Trp and G radicals occurring in the DNA–protein contact area. The laser experiments revealed that the signals showed little variation in the rate of formation of the transient radical in the three different DNA substrates. Biexponential fitting yielded rate constants of $\sim 5 \times 10^6$ s^{-1} and $\sim 3 \times 10^5$ s^{-1} for all duplexes. This clearly establishes a lower limit for HT in DNA of $>10^6$ s^{-1} through 50 Å of the base stack. On the basis of the absence of significant distance dependence we concluded that HT through the DNA is not a rate-limiting step.

Given the results from protein-modulation described in this section, DNA-mediated charge-transfer chemistry requires consideration biologically and physiologically. It has, for example, been suggested that regions containing a large amount of G are typically found in CpG islands, introns, and telomeres and, therefore, that such areas are hot spots for G damage, and as a result, could prevent the genome from oxidative radical damage [21].

4.6.4
Reductive Electron Transfer in DNA

With regard to important biological consequences such as DNA damage, mutagenesis, and carcinogenesis, in most of the experiments described above only *oxidative* HT processes have been observed. As shown in the Sections 4.6.1 and 4.6.2, such HT results in oxidative G damage at remote sites on the nucleic acid. On the other hand, *reductive electron transfer* (ET) processes (Figure 4.6.7) are currently used in DNA chip technology [22] and DNA nanotechnology [23]. Despite broad knowledge about these bioanalytical and biomedical applications, little is known about the behavior of excess electrons in DNA.

Most knowledge about excess ET in DNA comes from γ-radiolysis studies, which suggest a thermally activated hopping process [14]. A new DNA assay recently published by Carell et al. enables photoinduced cleavage of T–T dimers by a flavin derivative from a distance (Figure 4.6.8) [24]. The flavin moiety was synthetically incorporated as an artificial nucleobase into oligonucleotides. Although spectroscopic measurements obtained by use of this system have not been published, T–T dimer splitting was interpreted as the chemical result of reductive ET through the DNA base stack. This interpretation is mainly based on:

1. the known redox properties of the flavin intercalator in its reduced and deprotonated state, and
2. the absence of typical DNA base-sequence-dependence which would be observed for a hole hopping process, as described above.

Zewail et al. recently reported femtosecond time-resolved studies on the reduction of T by photoexcited 2-aminopurine in DNA duplexes [7].

Lewis et al. have investigated photoinduced electron transfer in DNA hairpins synthetically capped by a stilbene diether derivative serving as an electron donor (Figure 4.6.9) [25]. The electron injection rates are larger when T is used as the electron acceptor ($>2 \times 10^{12}$ s^{-1}) than when C is used (3.3×10^{11} s^{-1}). This indicates that the reduction potential is lower for T than C in double-helical B-DNA. Interestingly, a small difference between electron-injection rate was detected when the C used as the electron acceptor was base-paired to G or to inosine (I). The hydrogen bonding in the C–G base pair is stronger than in the C–I base pair. As a result the electron injection rate into the C–G base pair is slower (3.3×10^{11} s^{-1}) than into the C–I base pair (1.4×10^{12} s^{-1}) indicating differences in reduction potentials of C as a result of different base-pairing.

The DNA bases most easily reduced are T and C, the reduction potentials of which are very similar [26]. It is therefore expected that excess electron migration through DNA occurs via a hopping mechanism involving all base pairs (C–G and T–A) and the radical anions C$^{\cdot -}$ and T$^{\cdot -}$ as stepping stones. We focused our work on 5-pyrenyl-2'-deoxyuridine (Py-dU) and 5-pyrenyl-2'-deoxycytidine (Py-dC) as nucleoside models for ET in DNA. Photoexcitation of the pyrenyl group results in

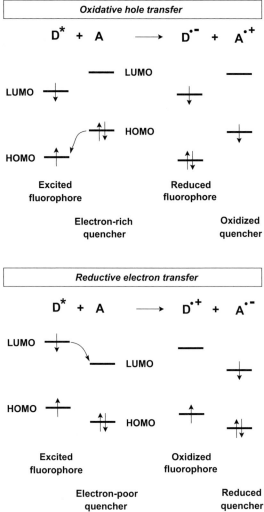

Fig. 4.6.7. Differences between oxidative HT and reductive ET. In both a photoexcited fluorophore is used to initiate the CT process. In oxidative HT an electron is removed from the HOMO of the DNA (**A**) into the HOMO of **D**. In reductive ET the photoexcited electron is shifted from the LUMO of **D** into the LUMO of the DNA (**A**). In the former DNA is oxidized whereas in the latter case DNA is reduced.

intramolecular ET yielding the corresponding pyrimidine radical anion and the pyrenyl radical cation (Py$^{\bullet+}$–dX$^{\bullet-}$). Py-dU and Py-dC were prepared by Suzuki–Miyaura-type cross-coupling reactions [27]. We subsequently characterized the properties and dynamics of the ET in Py-dU and Py-dC by steady-state fluorescence spectroscopy and femtosecond transient absorption spectroscopy (Figure 4.6.10) [28]. Our results showed that the pyrimidine radical anions generated are of dif-

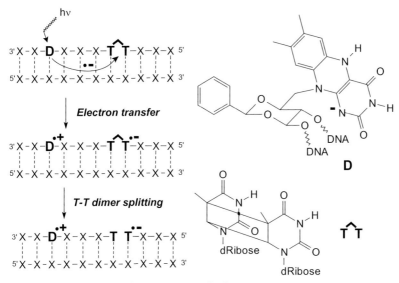

Fig. 4.6.8. Chemical assay for investigation of reductive ET in DNA. T–T dimer (T^T) splitting is the chemical result of photoinduced ET from a distant flavin derivative as the charge donor **D**.

ferent basicity, which is significant for understanding electron migration in DNA. The pK_a of the protonated Py$^{\cdot+}$–dU(H)$^{\cdot}$ biradical has been determined by steady-state fluorescence to be ~5.5. This shows clearly that neither water as the surrounding molecule or H-bonding donors such as the complementary DNA base A

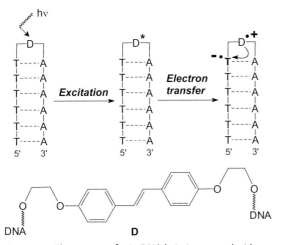

Fig. 4.6.9. Electron transfer in DNA hairpins capped with a stilbene diether derivative as the charge donor **D**.

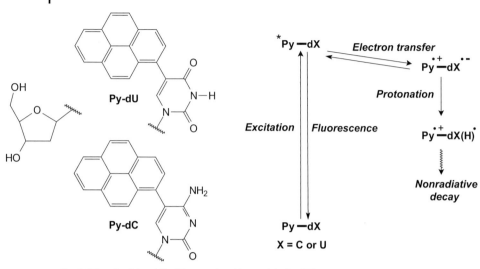

Fig. 4.6.10. Py-dU and Py-dC as nucleoside models for ET in DNA. When Py-dX (X = U or C) is excited at 340 nm intramolecular ET occurs. Subsequently, protonation can occur, yielding the biradical Py$^{\cdot+}$–dX(H)$^{\cdot}$.

can protonate the radical anions U$^{\cdot-}$ or T$^{\cdot-}$. In contrast, the non-protonated radical anion of dC (Py$^{\cdot+}$–dC$^{\cdot-}$) could not be observed in aqueous solutions after excitation. Although the situation in water cannot be directly compared with DNA, the results provide evidence that protonation of C$^{\cdot-}$ by the complementary DNA base G or the surrounding water molecules will occur rapidly. Although such protonation processes occur reversibly they might limit, or even terminate, electron migration in DNA, because of the separation of the spin from the charge. In conclusion, therefore, C$^{\cdot-}$ cannot play a major role as an intermediate charge carrier, whereas T$^{\cdot-}$ can act as a stepping stone for electron hopping in DNA (Figure 4.6.11).

By use of the nucleoside Py-dU we prepared a range of pyrene-modified duplexes by phosphoramidite chemistry [29]. The covalently attached pyrene group is located outside the DNA base stack. On excitation intramolecular ET in the Py-dU group represents injection of an excess electron into the DNA base stack. The differences between the prepared DNA duplexes are the bases located next to the Py-dU group.

Despite uncertainty related to irreversible electrochemistry, the trend for the reducibility of the nucleobases was established as: T,U ≈ C ≫ A > G [26]. Because of this trend, we expect that the U radical anion should only be able to reduce adjacent pyrimidine bases. In accordance with this assumption we expect different emission quantum yield, depending on the nature of the adjacent DNA bases (Figure 4.6.11). In fact, duplex **P1** has a high emission quantum yield, because ET from the Py-dU group to the adjacent A is not expected. In contrast, significant quenching of the emission can be observed when a T or C is placed adjacent to the

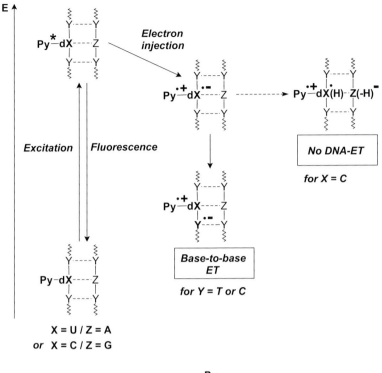

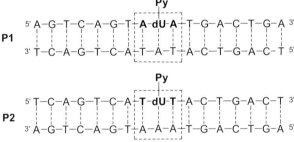

Fig. 4.6.11. Reductive ET in pyrenyl-modified DNA duplexes. Excitation of the Py-dX group at 340 nm results in ET, yielding the Py$^{•+}$–dX$^{•−}$ biradical. If ET to the adjacent DNA bases occurs as an alternative pathway, quenching of emission is observed.

Py-dU group, as in **P2**; this is indicative of ET from the Py-dU group to the adjacent T or C. This ET interpretation is supported by time-resolved pump-probe laser spectroscopy measurements.

Despite these efforts in understanding reductive ET in DNA, future experiments must focus on:

1. more detailed investigation of the dynamics of the electron injection process;
2. the base sequence dependence of ET;
3. the rate of base-to-base ET in terms of the hopping model; and
4. the chemical reactivity of radical anions in DNA yielding physiologically relevant DNA lesions.

Acknowledgments

Our work is supported by the Deutsche Forschungsgemeinschaft, the VolkswagenStiftung and the Fonds der Chemischen Industrie. I am grateful to my coworkers Nicole Amann, Elke Mayer, and Manuela Rist for their experimental contributions. I would also like to thank Dr Torsten Fiebig and his coworkers for fruitful collaboration and valuable discussions. I am also grateful to Professor Horst Kessler for generous support.

References

1 M. W. GRINSTAFF, Angew. Chem. Int. Ed. **1999**, *38*, 3629.
2 (a) P. O'NEILL, E. M. FRIEDEN, Adv. Radiat. Biol. **1993**, *17*, 53; (b) B. ARMITAGE, Chem. Rev. **1998**, *98*, 1171; (c) S. O. KELLEY, J. K. BARTON, Metal Ions Biol. **1999**, *36*, 211.
3 R. A. MARCUS, N. SUTIN, Biochim. Biophys. Acta **1985**, *811*, 265.
4 (a) J. R. WINKLER, H. B. GRAY, Chem. Rev. **1992**, *92*, 369; (b) M. R. WASIELEWSKI, Chem. Rev. **1992**, *92*, 435.
5 (a) M. E. NUNEZ, J. K. BARTON, Curr. Opin. Chem. Biol. **2000**, *4*, 199; (b) C. R. TREADWAY, M. G. HILL, J. K. BARTON, Chem. Phys. **2002**, *281*, 409.
6 V. SHAFIROVICH, A. DOURANDIN, W. HUANG, N. P. LUNEVA, N. E. GEACINTOV, Phys. Chem. Chem. Phys. **2000**, *2*, 4399.
7 (a) C. WAN, T. FIEBIG, O. SCHIEMANN, J. K. BARTON, A. H. ZEWAIL, Proc. Natl. Acad. Sci. USA **2000**, *97*, 14052; (b) T. FIEBIG, C. WAN, A. H. ZEWAIL, ChemPhysChem **2002**, *3*, 781.
8 S. HESS, M. GÖTZ, W. B. DAVIS, M. E. MICHEL-BEYERLE, J. Am. Chem. Soc. **2001**, *123*, 10046.
9 M. PASCALY, J. YOO, J. K. BARTON, J. Am. Chem. Soc. **2002**, *124*, 9083.
10 M. BIXON, B. GIESE, S. WESSELY, T. LANGENBACHER, M. E. MICHEL-BEYERLE, J. JORTNER, Proc. Natl. Acad. Sci. USA **1999**, *96*, 11713.
11 F. D. LEWIS, X. LIU, J. LIU, S. E. MILLER, R. T. HAYES, M. R. WASIELEWSKI, Nature **2000**, *406*, 51.
12 (a) C. J. BURROWS, J. G. MULLER, Chem. Rev. **1998**, *98*, 1109; (b) S. STEENKEN, Chem. Rev. **1989**, *89*, 503.
13 H.-A. WAGENKNECHT, Chem. in unserer Zeit **2002**, *36*, 318.
14 B. GIESE, Annu. Rev. Biochem. **2002**, *71*, 51.
15 M. E. NUNEZ, G. P. HOLMQUIST, J. K. BARTON, Biochemistry **2001**, *40*, 12465.
16 E. M. BOON, J. K. BARTON, Curr. Opin. Struct. Biol. **2002**, *12*, 320.
17 S. R. RAJSKI, J. K. BARTON, Biochemistry **2001**, *40*, 5556.
18 K. NAKATANI, C. DOHNO, A. OGAWA, I. SAITO, Chem. Biol. **2002**, *9*, 361.
19 (a) S. R. RAJSKI, S. KUMAR, R. J. ROBERTS, J. K. BARTON, J. Am. Chem. Soc. **1999**, *121*, 5615; (b) H.-A. WAGENKNECHT, S. R. RAJSKI, M. PASCALY, E. D. A. STEMP, J. K. BARTON, J. Am. Chem. Soc. **2001**, *123*, 4400.
20 (a) H.-A. WAGENKNECHT, E. D. A. STEMP, J. K. BARTON, J. Am. Chem.

Soc. **2000**, *122*, 1; (b) H.-A. WAGENKNECHT, E. D. A. STEMP, J. K. BARTON, *Biochemistry* **2000**, *39*, 5483.

21 (a) A. HELLER, *Faraday Discuss.* **2000**, *116*, 1; (b) K. A. FRIEDMAN, A. HELLER, *J. Phys. Chem. B* **2001**, *105*, 11859.

22 (a) E. M. BOON, J. E. SALAS, J. K. BARTON, *Nat. Biotechnol.* **2002**, *20*, 282; (b) G. HARTWICH, D. J. CARUANA, T. DE LUMLEY-WOODYEAR, Y. WU, C. N. CAMPBELL, A. HELLER, *J. Am. Chem. Soc.* **1999**, *121*, 10803; (c) N. M. JACKSON, M. G. HILL, *Curr. Opin. Chem. Biol.* **2001**, *5*, 209; (d) A. R. PIKE, L. H. LIE, R. A. EAGLING, L. RYDER, S. N. PATOLE, B. A. CONNOLLY, B. R. HORROCKS, A. HOULTON, *Angew. Chem. Int. Ed.* **2002**, *41*, 615; (e) F. PATOLSKY, Y. WEIZMAN, I. WILLNER, *J. Am. Chem. Soc.* **2002**, *124*, 770.

23 (a) C. MAO, W. SUN, Z. SHEN, N. C. SEEMAN, *Nature* **1999**, *397*, 144; (b) H.-W. FINK, C. SCHÖNENBERGER, *Nature* **1999**, *398*, 407; (c) D. PORATH, A. BEZRYADIN, S. DE VRIES, C. DEKKER, *Nature* **2000**, *403*, 635.

24 C. BEHRENS, L. T. BURGDORF, A. SCHWÖGLER, T. CARELL, *Angew. Chem. Int. Ed.* **2002**, *41*, 1763.

25 F. D. LEWIS, X. LIU, S. E. MILLER, R. T. HAYES, M. R. WASIELEWSKI, *J. Am. Chem. Soc.* **2002**, *124*, 11280.

26 (a) S. STEENKEN, J. P. TELO, H. M. NOVAIS, L. P. CANDEIAS, *J. Am. Chem. Soc.* **1992**, *114*, 4701. (b) C. A. M. SEIDEL, A. SCHULZ, M. H. M. SAUER, *J. Phys. Chem.* **1996**, *100*, 5541; (c) J. JORTNER, M. BIXON, A. V. VOITYUK, N. RÖSCH, *J. Phys. Chem. A* **2002**, *106*, 7599.

27 N. AMANN, H.-A. WAGENKNECHT, *Synlett* **2002**, 687.

28 (a) N. AMANN, E. PANDURSKI, T. FIEBIG, H.-A. WAGENKNECHT, *Angew. Chem. Int. Ed. Engl.* **2002**, *41*, 2978; (b) M. RAYTCHEV, E. MAYER, N. AMANN, H.-A. WAGENKNECHT, T. FIEBIG, manuscript submitted for publication.

29 N. AMANN, E. PANDURSKI, T. FIEBIG, H.-A. WAGENKNECHT, *Chem. Eur. J.* **2002**, *8*, 4877.

Part 5
Catalysis

5.1
Protease-catalyzed Formation of C–N Bonds

Frank Bordusa

5.1.1
Optimization of Proteases for Synthesis: Selection of Current Techniques

Proteases, or, following the recommendation of the NC-IUB, *peptidases*, belong to the hydrolase class of enzyme and are among the few enzymes active at the backbone of polypeptides. Several hundred of these enzymes are known and, in general, they all catalyze the same reaction – hydrolysis of peptide bonds (two specific classes are presented in Boxes 12 and 13). One might ask why such activity might be useful for organic chemists interested in a catalyst mediating exactly the reverse of peptide bond hydrolysis. As catalysts, true to the definition familiar in chemistry, proteases alter the rate at which the thermodynamic equilibrium of the reaction is reached, but do not change that equilibrium itself. This inevitably implies that these enzymes work reversibly in both directions of the reaction. The equilibrium constants for the reverse reaction are, however, in the range 10^{-3} to 10^{-4} L mol^{-1}. As a consequence, it is easy to recognize that even with the most efficient enzyme, proteases cannot act a priori as universal and perfect catalysts for C–N bond formation. Several serious drawbacks remain, mainly:

1. the formation of at least two synthesis products, i.e. that desired and the hydrolyzed acyl donor, because of the enzyme's native hydrolytic activity;
2. most important, the specificities of the available proteases do not enable all desired products to be assembled and, hence, only reactions with compounds closely related to preferred amino acid residues are of practical relevance whereas non-proteinogenic amino acid moieties are not usually acceptable substrates;
3. particularly in syntheses with longer peptides, there is a permanent risk of proteolytic side-reactions of both the starting compounds and the products formed; and
4. solvents, additives, and reaction conditions can strongly affect the enzyme's activity and stability.

Because of properties, proteases are far from being perfect tools for catalyzing a broad spectrum of organic syntheses, especially those based on the reverse activity of the enzyme. Further efforts are of decisive importance to overcoming this limitation – to suppress competitive acyl donor hydrolysis, to alter enzyme specificity, and to suppress undesired proteolytic side-reactions. Much attention has been devoted to approaches to the engineering of the synthesis medium in which the reaction is performed (see also Chapters 6.3 and 6.4). Historically these studies have mainly focused on investigation on the behavior of proteases towards organic solvents, initially used as co-solvents and later on as pure solvents practically without any water content. It should, however, be remembered that in addition to organic solvents other approaches are useful for manipulating enzyme properties. Although less popular, at least if the number of publications is considered to be the only criterion, these include studies in supercooled and frozen media, in supercritical fluids such as compressed carbon dioxide or propane, and in solid-to-solid reaction systems. A substantial number of reviews document this impressive development and highlight important aspects of non-aqueous biocatalysis in detail [1]. In addition, a very recent review covers the latest and most important findings in the whole field of proteases [2]. Because re-discussion of this topic, although exciting, is hindered by space constraints, this contribution is necessarily selective and focuses on two other efficient approaches based on engineering of the substrates (reactants) and of the enzyme itself.

5.1.2
Substrate Engineering

If undesired subsequent reactions are observed during protease-catalyzed syntheses it is of minor importance which bond is cleaved by the enzyme. These side-reactions merely show that the enzyme's specificity for the acyl donor does not lie sufficiently above its specificity for the peptide product. Because structural changes in the amino acid building blocks of the reactants are ruled out, the leaving group of the acyl donor, essential for mediating acceptance of the acyl moiety in the so-called kinetically controlled approach, remains the only variable for suppressing competitive reactions. Specific leaving-group manipulation at the carboxyl component is commonly used in practice. Efforts in this direction initially enabled reactions with cleavage-sensitive peptide reactants and finally led to the development of the "substrate mimetics" approach.

5.1.3
Classical Concept of Leaving-group Manipulation

The structural requirement for a protease substrate is the presence of specific amino acids at the C-terminus of the acyl moiety, for example Arg or Lys in the case of trypsin. Acyl donors lacking those site-specific amino acids are, therefore, usually not targets for protease-catalyzed syntheses. Also, because of the distinct specificity of proteases toward amino acid moieties recognized by the enzyme, only acyl donors with the most specific amino acid residue at the C-terminus can be coupled without proteolytic side-reactions. In contrast, coupling of the less specific amino acid moieties proceeds successfully only if the peptide reactants do not contain any of the more specific counterparts. In instances where differences between specificities are less pronounced, leaving group manipulation can be used to improve the specificity of the originally less specific acyl donor over that of the more specific and, thus, more cleavage-sensitive amino acid moiety located within the reactants. This general approach was demonstrated ten years ago by Jakubke et al. [3] and has since been confirmed several times [2]. Leaving group manipulation has also been found to be generally useful means of increasing enzyme activity toward originally less specific acyl donors and, thus, to accelerate the rate of reaction. In particular, preparative scale syntheses profit from the higher reaction rates of leaving group-engineered acyl donors. For example, approximately 90% of the expensive protease clostripain could be saved simply by using Z-Lys-SBzl instead of Z-Lys-OMe as the acyl donor in the first reaction of a stepwise synthesis of the highly trifunctional tetrapeptide H-Lys-Tyr-Arg-Ser-OH [4]. In contrast with the effect on the reaction rate, only a marginal effect of the leaving group on product yields are expected, because of the formation of identical acyl enzyme species which should undergo identical partitioning into the peptide product and the competitive hydrolyzed acyl donor. Leaving group-related differences in product yields can only result from incomplete acyl donor consumption or differences between the rates of spontaneous hydrolysis of the distinct acyl donors.

5.1.4
Substrate Mimetics-mediated Syntheses

In contrast with classical leaving group manipulation focusing on the adaptation of the leaving group to meet the specificity of the S' subsite of proteases, in substrate mimetics the leaving group is manipulated to bind to the enzyme's active site (S_1 position; subsite notation according to Ref. [5]; Box 15). Nature itself can be regarded as the prototype of this kind of substrate engineering. Indeed, the ribosomal peptidyl transferase, nature's expert for peptide bond formation, does not recognize the amino acid being coupled but the tRNA to which it is esterified. This strategy enables peptide synthesis in both a specific and universal manner which uses a single enzyme only to catalyze the C–N bond formation step. In substrate mimetics the shift in the location of the site-specific moiety from the peptide's C-terminus into the leaving group is also accompanied by a shift in enzyme activity, enabling serine and cysteine proteases to react with non-specific amino acid or peptide sequences (Scheme 5.1.1). Importantly, for this remarkable activity no further manipulation, of either the enzyme or the reaction medium, are, in principle, necessary. The first reported examples of such substrate mimetics were acyl-4-amidino and -4-guanidinophenyl esters (OGp) (Scheme 5.1.2, compounds **2** and **3**) which were found to be recognized by originally Arg-specific proteases almost independently of their individual peptide sequence [6]. Interestingly, this non-

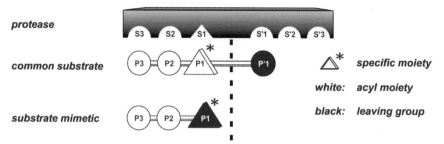

Scheme 5.1.1. Schematic comparison of the binding of substrate mimetics and common acyl donors to the active site of proteases, on the basis of the ideas of the conventional binding model of proteases.

specificity for the acyl moiety is equally true for coded L amino acids and their D-configured counterparts and even for sterically hindered α,α-dialkylated amino acids.

The generality of this approach has been proved by studies on the Glu-specific V8 protease; these established the first substrate mimetics, carboxymethyl thioesters (SCm), for non-arginine-specific proteases [7]. The SCm moiety **6** was selected empirically, on the basis of the close structural similarity to the side-chain of specific Asp and Glu residues. Interestingly, despite the restricted specificity of V8 protease, the SCm group was found to act as a suitable mimic mediating the acceptance of originally non-specific acyl moieties, as was found for the OGp esters and Arg-specific proteases. Finally, extension of the substrate mimetics approach to a third synthetically important class of proteases, those specific for aromatic amino acid moieties, has been described for chymotrypsin as example [8]. A computer-assisted docking approach has been used to predict the function of the OGp group as an artificial recognition site for this enzyme; this was finally confirmed by peptide synthesis reactions.

Application of substrate mimetics to the synthesis of longer polypeptides requires combination with chemical methods, primarily solid-phase peptide synthesis (SPPS; Box 25), to provide access to long-chain substrate mimetic reactants. This issue was initially addressed by using oxime resin strategy [9]. By use of this approach and Boc chemistry several peptide esters in the form of different substrate mimetics were synthesized and have subsequently served as carboxyl components for trypsin-, V8 protease-, and chymotrypsin-catalyzed fragment condensations. Longer-chain substrate mimetics can also be prepared by Fmoc SPPS by utilizing safety catch resin strategy. Aminolytic or thiolytic cleavage of the activated peptide–resin linkage yields substrate mimetics ready for enzymatic ligation. By using this approach and clostripain as catalyst the all-D amino acid version of the WW domain of the human peptidyl-prolyl-*cis*/*trans*-isomerase Pin1 (Box 1) has been synthesized in good yield and excellent purity (Scheme 5.1.3) [10].

Besides linear peptides, the substrate mimetics approach also enables synthesis of isopeptides [11]. The prerequisite for this activity is use of an iso-type of sub-

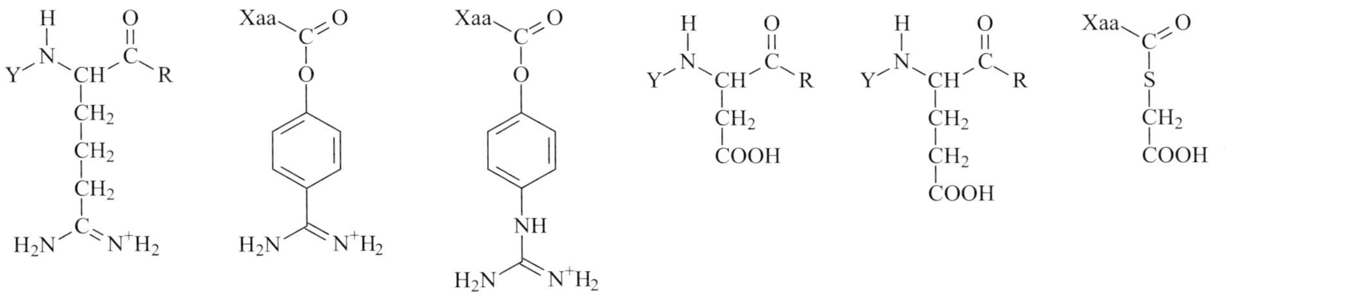

Scheme 5.1.2. Structures of substrates and substrate mimetic moieties for Arg- and Glu-specific proteases. 1. arginine; 2. 4-amidinophenyl ester; 3. 4-guanidinophenyl ester (OGp); 4. aspartic acid; 5. glutamic acid; 6. carboxymethyl thioester (SCm).

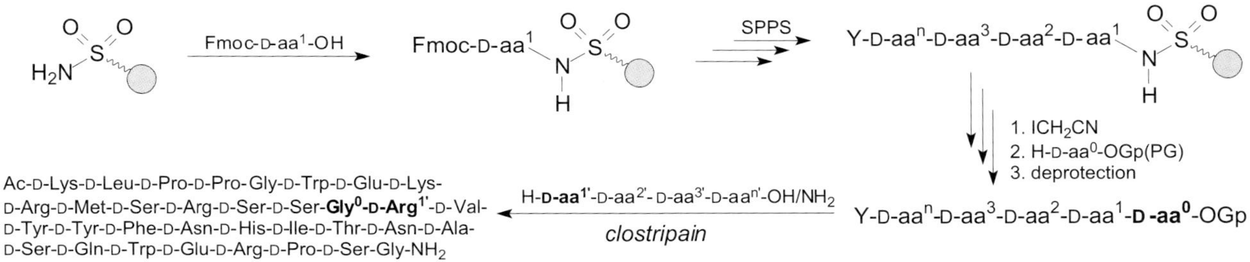

Fmoc-D-aa¹-OH ⟶ Fmoc-D-aa¹ ⟶ H-D-**aa¹'**-D-aa²'-D-aa³'-D-aaⁿ'-OH/NH₂ (clostripain)

Ac-D-Lys-D-Leu-D-Pro-D-Pro-Gly-D-Trp-D-Glu-D-Lys-D-Arg-D-Met-D-Ser-D-Arg-D-Ser-D-Ser-**Gly⁰-D-Arg¹'**-D-Val-D-Tyr-D-Tyr-D-Phe-D-Asn-D-His-D-Ile-D-Thr-D-Asn-D-Ala-D-Ser-D-Gln-D-Trp-D-Glu-D-Arg-D-Pro-D-Ser-Gly-NH₂

⟶ SPPS

Y-D-aaⁿ-D-aa³-D-aa²-D-aa¹

1. ICH₂CN
2. H-D-aa⁰-OGp(PG)
3. deprotection

⟶ Y-D-aaⁿ-D-aa³-D-aa²-D-aa¹-**D-aa⁰**-OGp

Scheme 5.1.3. Schematic representation of the synthesis of the all-D version of the WW domain of the human peptidyl-prolyl-*cis/trans*-isomerase Pin 1 by use of substrate mimetics ligation combined with the alkanesulfonamide safety-catch solid-phase peptide synthesis approach [10]. PG, Y, protecting group; OGp, 4-guanidinophenyl ester.

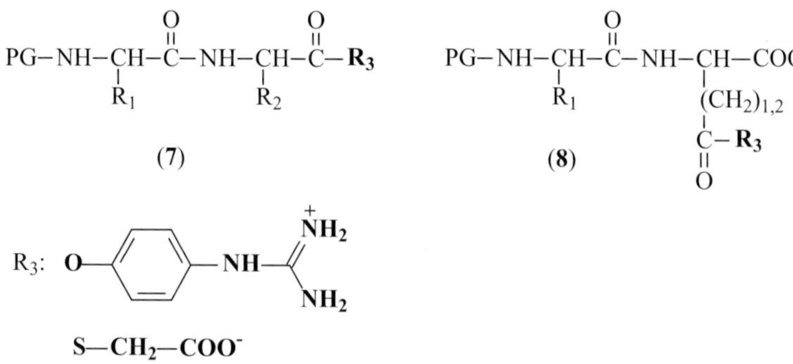

$$\text{PG-NH-CH(R}_1\text{)-C(=O)-NH-CH(R}_2\text{)-C(=O)-}\mathbf{R_3}$$

(7)

$$\text{PG-NH-CH(R}_1\text{)-C(=O)-NH-CH(COO}^-\text{)-(CH}_2\text{)}_{1,2}\text{-C(=O)-}\mathbf{R_3}$$

(8)

$$R_3: \text{O-C}_6\text{H}_4\text{-NH-C(=}\overset{+}{N}H_2\text{)-NH}_2$$

$$\text{S-CH}_2\text{-COO}^-$$

Scheme 5.1.4. General structures of linear **7** and iso-type **8** substrate mimetics. The site-specific ester leaving groups are emphasized by bold letters. PG, protecting group; R_1, R_2, individual side-chains; R_3, site-specific ester leaving group.

strate mimetic that directs the intrinsic synthesis activity of the protease to the side-chain carboxyl moiety of Asp and Glu (Scheme 5.1.4). Similar to linear substrate mimetics, the iso-type counterparts bear a site-specific ester leaving group which is, however, linked to the ω-carboxyl moiety of Asp and Glu instead of being at the C-terminus of the peptide donor. This different architecture leads to a shift in the synthetic activity of the enzyme to the side-chain of Asp and Glu; this finally results in the formation of isopeptides.

More efficient than other methods, substrate mimetics enable proteases to react not only with non-specific coded amino acids, but also with non-amino acid-derived acyl donors. By using OGp esters of 4-phenylbutyric acid (Pbu-OGp) and benzoic acid (Bz-OGp) as the donor, a variety of amino acid amides and peptides as acyl acceptors, and clostripain and chymotrypsin as the catalysts the appropriate isosteric peptide products could be obtained in excellent yields [8, 12]. By using the unique specificity of clostripain toward the amino component the approach could even be expanded to the synthesis of a wide variety of N-linked neo-peptidoglycans. The approach enables selective coupling of carboxylate moieties derived from the side chains of Asp, Glu, and the C-terminus of peptides both with simple monomeric and with highly complex carbohydrates, for example D-glucosamine, D-galactosamine, muramic acid, and moenomycin A (Scheme 5.1.5) [13]. Even C–N bonds completely outside peptidic structures have been synthetic targets of the approach. For example, non-amino acid-derived carboxylic components, for example Pbu-OGp and Bz-OGp, and a large number of non-peptidic amino components, for example aliphatic and aromatic amines, amino alcohols, non-α-amino carboxylic acids, and diamines, have been efficiently coupled [12]. To conclude, this remarkable activity opens up new possibilities of easy synthesis of a broad spectrum of linear peptides, isopeptides, all-D peptides, peptide isosteres, peptide–carbohydrate conjugates, and non-amino acid-derived carboxylic acid amides under extraordinary mild conditions unachievable with classical protease approaches.

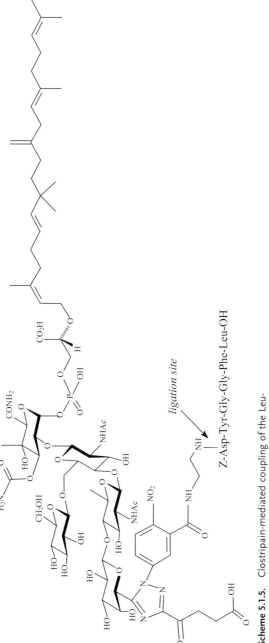

Scheme 5.1.5. Clostripain-mediated coupling of the Leu-enkephalin sequence Z-Asp-Tyr-Gly-Gly-Phe-Leu-OH with a moenomycin A analog.

5.1.5
Enzyme Engineering

Four primary goals are usually pursued when engineering a protease for synthesis. First, one needs to make the enzyme a more stable catalyst especially towards organic solvents. Second, and usually more difficult to accomplish, one needs to improve the enzyme's synthetic efficiency, mainly by reducing the rate of competitive hydrolysis reactions. Unfortunately, the molecular basis of control of the rate of hydrolysis of the acyl enzyme intermediate is only poorly understood so there is, currently, no general approach enabling direct suppression of competitive hydrolysis reactions by enzyme engineering and optimization of active-site specificity for better binding of the acyl acceptor is the only practical way of minimizing undesired hydrolysis of the acyl donor. Third, one needs to reduce the usually unwanted proteolytic activity of proteases to prevent competitive peptide cleavage during synthesis. Several approaches have been developed and a few have been found to be synthetically useful. Nevertheless, practically all the resulting enzyme variants suffer from significantly lower overall enzyme activity making it questionable whether they will reach industrial use in large-scale synthesis. Fourth, one needs to extend or alter the native specificity of the enzyme to fit the requirements of the synthesis. As for all other goals, chemical and genetic modification can both be used to create biocatalysts with the desired function different from that of the original parent enzyme. Selected examples related to proteases will be presented below.

5.1.6
Chemical Enzyme Modifications

Covalent chemical modification of enzymes can be regarded as the original method available for altering enzyme properties. This long history does not, however, mean that this kind of basic enzyme engineering has lost anything of its attraction, chemical modification has re-emerged as a powerful complementary technique to genetic approaches for tailoring enzymes. Several benefits contribute to this renaissance:

1. chemical modification is generally applicable;
2. it is usually inexpensive and easy to perform; and
3. it enables incorporation of non-coded amino acid moieties and, thus, leads to a variety of enzyme species which cannot be generated by routine genetic engineering.

Classical approaches to chemical enzyme modification, however, often suffer from lack of regio-selectivity, which can yield heterogeneous and irreproducible enzyme mixtures. For example, preparation of methyl-chymotrypsin, -subtilisin or -trypsin using methyl sulfonate reagents, originally used to methylate the histidine of the catalytic triad of the enzymes, yields enzyme mixtures in which the remaining histidines of the enzyme molecules are partly or completely methylated [2]. Because of the small size of the modifying moiety, however, undesired effects of this random modification on enzyme activity are only marginal. The methylated enzyme variants are now recognized as interesting biocatalysts for peptide synthesis, mainly because of loss of proteolytic activity with some of the esterase activity remaining. Although important, the very low synthetic activity of methylated proteases should, however, generally hinder their practical use. Oxidation of Met^{192} in chymotrypsin leads to a much more active enzyme variant which, however, has still proteolytic activity. Nevertheless, because of the improved esterase to amidase ratio and the higher stability of the enzyme towards basic conditions, the Met^{192}-sulfoxide-chymotrypsin was found to be a useful enzyme for peptide synthesis. The preparation of seleno-subtilisin and thiol-subtilisin can be regarded as further examples of small-size chemical modification [14]. The latter, which simultaneously marks the beginning of chemical enzyme engineering, entails conversion of the active-site serine 221 of subtilisin to cysteine. This first alteration remains one of the most useful. Similar to methylated proteases, subtilisin S221C is catalytically wounded to the point that it will barely hydrolyze peptide bonds yet is quite reactive with certain activated ester substrates. Thiol-subtilisin also benefits from the fact that thiol esters usually have higher aminolysis to hydrolysis ratios than regular oxo esters. This combination of properties has made it a useful tool for peptide synthesis and transesterification reactions [2]. It should be noted, however, that its catalytic activity is several orders of magnitude lower than that of the wild-type enzyme, although it is more active than the His-methylated species. Seleno-subtilisin is a much poorer enzyme than thiol-subtilisin. Its synthetic application for peptide synthesis essentially needs the use of highly activated esters as the acyl donors, but even then the reactions proceed very slowly. In addition, the enzyme is very sensitive to oxidants. Although less suitable for peptide synthesis, this behavior has made seleno-subtilisin a useful biocatalyst for mediation of peroxidase-like reactions [15].

Site-specific modifications essentially need the presence of a unique amino acid moiety which can undergo chemo-selective reactions. Having such a unique amino acid within an enzyme is, however, the exception rather than the rule. Subtilisin and carboxypeptidase Y can be regarded as such rare enzymes, because they contain no natural cysteines. Thus, incorporation of an artificial cysteine moiety by site-directed mutagenesis creates a unique reaction center that can be chemically modified to introduce an unnatural amino acid side-chain selectively (Scheme 5.1.6) [16]. By use of subtilisin a variety of structures differing in size and physicochemical properties have been covalently linked to artificial cysteines located at different positions within the enzyme. Interestingly, remarkable effects on both catalytic properties were observed. Although most chemical active-site modifications reduce the activity of an enzyme this was not always so for this type of active-site engineering. For example, incorporation of hydrophobic ligands at the enzyme's S_2 subsite increased the activity of the enzyme more than threefold compared with the parent enzyme [16c]. An improved ratio of esterase to amidase activity

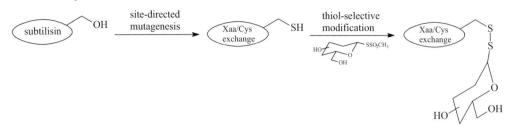

Scheme 5.1.6. Controlled site-selective modification of subtilisin by a combined site-directed mutagenesis chemical modification approach.

was also found for a number of enzyme variants; this occasionally correlated with broader substrate tolerance, as illustrated by the examples in Table 5.1.1 [17].

5.1.7
Genetic Enzyme Modifications

The history of genetic enzyme modification is closely connected with proteases and with subtilisin in particular. The first genetic modifications in this enzyme were conducted soon after the gene was cloned in the early 1980s [18]. Two decades later mutations in well over 50% of the 275 amino acids of subtilisin have been reported. Several other proteases also became targets of genetic modification and a variety of useful techniques now exists for introducing changes into the enzyme at the genetic level. Basic principles and selected examples related to proteases will be summarized below.

Site-directed mutagenesis can be regarded as one of the first genetic modification techniques; it has proved to be useful for engineering a protease for synthesis. Although simple to perform, it does, however, require that one have some idea of which residues are important. Having a three-dimensional structure of the enzyme in question is particularly helpful in selecting those important moieties. Because of the inability to predict long range structural changes, most protease engineering involves catalytic amino acids, substrate binding regions, and direct stabilizing mutations. With regard to peptide synthesis, subtilisin species with enhanced synthetic efficiency, modified specificity, improved stability, and altered pH profile have been designed by this method [2, 19]. Efforts in this field led, for example, to a double mutant ("subtiligase"), in which the catalytic Ser^{221} is exchanged with Cys, and Pro^{225} with Ala [20]. The enzyme variant was used to synthesize wild-type and mutant ribonuclease A in milligram quantities by stepwise ligation of six esterified peptide fragments [20b]. Single and multiple site-specific mutations have also been extensively used to obtain subtilisin variants with increased stability toward inactivation by organic solvents [19]. Although several aspects of the stability of the enzyme have been significantly improved, the successful rational design of

5.1.7 Genetic Enzyme Modifications

Tab. 5.1.1. Effect of active-site glycosylation at the artificial cysteine[166] at the base of the primary specificity S_1 pocket on the substrate tolerance of subtilisin (according to ref. [17b], with permission from the Royal Society of Chemistry).[a]

Acyl donor	Acyl acceptor	Product	Time (h)	Yield[b] (%) Wt-subtilisin	S166C-1	S166C-2	S166C-3	S166C-4
Z-L-Phe-OBn	H-Gly-NH$_2$	Z-L-Phe-Gly-NH$_2$	1	92	95	93	91	95
Z-L-Ala-OBn	H-Gly-NH$_2$	Z-L-Ala-Gly-NH$_2$	5	91	85	77	92	83
Z-L-Glu-OBn	H-Gly-NH$_2$	Z-L-Glu-Gly-NH$_2$	5	62	58	65	54	67
Z-L-Phe-OBn	H-L-Ala-NH$_2$	Z-L-Phe-L-Ala-NH$_2$	24[c]	57	28	34	31	32
Z-L-Ala-OBn	H-L-Ala-NH$_2$	Z-L-Ala-L-Ala-NH$_2$	24[c]	0	15	16	22	11
Z-L-Glu-OBn	H-L-Ala-NH$_2$	Z-L-Glu-L-Ala-NH$_2$	24[c]	0	48	50	51	55
Z-D-Phe-OBn	H-Gly-NH$_2$	Z-D-Phe-Gly-NH$_2$	48[d]	0	6	8	7	8
Z-D-Ala-OBn	H-Gly-NH$_2$	Z-D-Ala-Gly-NH$_2$	48[d]	0	80	77	72	70
Z-D-Glu-OBn	H-Gly-NH$_2$	Z-D-Glu-Gly-NH$_2$	48[d]	0	63	62	64	64

[a] Conditions: DMF–water, 1:1; 0.1 M donor; 0.3 M acceptor; 0.036 mol% enzyme.
[b] Isolated yields.
[c] 0.2 M acceptor.
[d] Further 0.036 mol% added after 24 h.
S166C-1, -(CH$_2$)$_2$-O-β-D-Glc(Ac)$_2$; S166C-2, -(CH$_2$)$_2$-O-β-D-Glc(Ac)$_3$; S166C-3, -(CH$_2$)$_2$-O-β-D-Gal(Ac)$_3$; S166C-4, -(CH$_2$)$_2$-O-β-D-Gal.

Tab. 5.1.2. Yields of intact hexapeptide products synthesized by trypsin D189E-catalyzed coupling of Bz-Xaa-OGp with specific amino acid-containing peptides (according to Ref. [21b], with permission from the American Chemical Society).[a]

Acyl donor	Acyl acceptor	Product	Yield (%)
Bz-Leu-OGp	Ala-Ala-*Lys*-Ala-Gly	Bz-Leu-Ala-Ala-*Lys*-Ala-Gly-OH	59.2
Bz-Leu-OGp	Ala-Ala-*Arg*-Ala-Gly	Bz-Leu-Ala-Ala-*Arg*-Ala-Gly-OH	68.0
Bz-Gln-OGp	Ala-Ala-*Lys*-Ala-Gly	Bz-Gln-Ala-Ala-*Lys*-Ala-Gly-OH	64.7
Bz-Gln-OGp	Ala-Ala-*Arg*-Ala-Gly	Bz-Gln-Ala-Ala-*Arg*-Ala-Gly-OH	69.1
Bz-Ala-OGp	Ala-Ala-*Lys*-Ala-Gly	Bz-Ala-Ala-Ala-*Lys*-Ala-Gly-OH	48.9
Bz-Ala-OGp	Ala-Ala-*Arg*-Ala-Gly	Bz-Ala-Ala-Ala-*Arg*-Ala-Gly-OH	45.7

[a] Conditions: 0.2 M Hepes buffer, pH 8.0, 0.2 M NaCl, 0.02 M $CaCl_2$, 25 °C, 10% methanol; [acyl donor]: 2 mM, [acyl acceptor]: 15 mM, [enzyme]: 3.1×10^{-6} M.

a completely organic solvent-stable protease remains an unsolved challenge. Although work in this field is dominated by subtilisin, several other proteases have become targets for site-directed genetic optimization. Recent examples are the design of: organic solvent-stabilized variants of α-lytic protease and thermolysin-type bacterial neutral protease; alteration of the P_1 and P_1' substrate preferences of carboxypeptidase Y; and the decrease of the proteolytic activity of *Streptomyces griseus* protease [2]. Site-directed mutagenesis had also led to several trypsin variants with improved synthetic properties for substrate mimetic-mediated peptide syntheses; the reduced proteolytic activity of these variants enables the coupling of specific amino acid-containing peptide reactants (Table 5.1.2) [21].

To conclude, enzyme engineering by site-directed mutagenesis continues to be a useful means of improving proteases for synthesis. Further improvements of this technology can be expected from better understanding of the relationship between the structure and function of enzymes and from sequence and structure alignments between related enzymes. The longstanding inability to predict the exact protein structure required for stereoselective reaction of a given substrate or which is necessary for high stability and activity of the enzyme in a given reaction medium remains, however, one of the main hindrances associated with this technology.

In-vitro-directed evolution using random genetic mutation and recombination has recently been explored as a more generally applicable approach to the modification of enzyme properties (Box 20 and Chapter 4.3). This technology has the advantage that is does not require a priori knowledge of the relationship between enzyme structure and function. Restrictions are given by the number of mutants which can be examined routinely while standard screenings are usually limited to a number of individual mutant species in the range 10^4–10^5 [19]. All combinations of double substitutions in subtilisin for example, would yield a total of 3×10^7 different variations. Accordingly, only the population of single mutations can be adequately searched for appropriate events. As a result, purely random evolution of a biocatalyst is especially useful for improving more global (cumulative) properties, such as stability, rather than the native activity or specificity of an enzyme, which usually depend on synergistic mutational events. For example, the preparation of randomly mutated subtilisin and its successful screening for enhanced thermostability has recently been reported [22] and the study of Zhao and Arnold [23] gave an impression of the number of accumulating point mutations that can contribute to an increase in stability. In total, eleven stabilizing mutations were identified, of which several could not have been predicted by use of rational design approaches. Further stabilization of the enzyme could be achieved by targeting random mutagenesis to positions at which stabilizing changes had already been found [24]. A similar combination of design and random mutagenesis has furnished a hyperstable calcium-free subtilisin that inactivates 250 000 times more slowly than the wild-type enzyme in 10 mM EDTA [25] and efforts directed at stabilizing the activity of subtilisin in aqueous–organic media continue to be successful. On the basis of earlier attempts a subtilisin variant has been created containing ten substitutions as a result of five sequential rounds of random mutagenesis and two additional substitutions by site-directed mutagenesis [26]. Subsequent hydrolytic studies verified a 256-fold increase in enzyme activity in 60% aqueous DMF and structural analysis showed that all ten substituents were located around the active site. Further improvements could be achieved by introducing three additional substitutions by random mutagenesis [27]. The resultant subtilisin variant with thirteen substitutions in total has been shown to have 471 times higher hydrolytic activity in 60% DMF than the original parent enzyme. Interestingly, in all instances the enzymes optimized towards DMF were less active in purely aqueous media.

Genetic modifications can also be introduced in the gene of interest by *phage display* methods. Commonly, more than 1×10^9 independent variants are covered by this technique which, for example, enables screening all combinations of amino acids at six specified positions of subtilisin. Although an obvious limitation of phage display is that selection is achieved as a result of binding activity, several publications report the use of this technology to improve the catalytic behavior of enzymes. For example, introduction of random mutations at 25 positions within the active site of the protease subtiligase resulted in increased peptide ligation activity [28]. In a second study fully active subtilisin was displayed on the phage surface with the aim of changing the S_4 specificity of the enzyme [29]. Uncontrolled proteolysis by the enzyme could be efficiently avoided by addition of a reversible inhibitor to the culture medium.

Another efficient way of creating new molecular diversity is by recombination of related genes by *DNA-shuffling* methods. An impressive example of application of the DNA-shuffling approach to proteases has recently been published [30]. A family of 25 subtilisin gene sequences was shuffled with a wild-type subtilisin gene and the resulting library of the corresponding enzymes was then tested for thermo- and solvent stability, activity, and pH-dependence. Multiple clones were identified that were significantly better than any of the parental enzymes for each property tested.

5.1.8
Conclusions

Both historically and currently proteases represent biocatalysts of outstanding interest in synthetic organic chemistry. It has been proved that after establishing the optimum conditions for synthesis multigram, kilogram, and even tons of natural products or complex synthetic products can be obtained by protease catalysis. Although research is broadened, the biocatalyst is still the most interesting target for improvements. Development of appropriate activity and specificity enables biocatalytic processes, and improvement of activity and stability may make a process economically feasible. It is, therefore, not surprising that research efforts have focused on improving enzyme properties such as substrate range and specificity, stability, and function. Whereas for resolution of racemic carboxylic acids, alcohols, and amines or the desymmetrization of prochiral and meso compounds proteases are now generally recognized as normal bench reagents, a general approach to enzymatic polypeptide synthesis remains to be formulated. By building on promising strategies, for example site-specific chemically modified enzymes and the substrate mimetics concept, significant improvements have been achieved and more can be expected in the near future. The final breakthrough might be achieved by combination of these strategies. Further input can be expected from the combination of enzyme engineering with directed evolution and gene-shuffling techniques, which currently seem to be the most fertile approaches to the design of proteases with tailored selectivity and synthetically relevant activity in essentially any suitable reaction medium. But rational enzyme design, particularly in combination with computational techniques is also expanding and might contribute significantly to improvement of proteases for synthesis. Lack of knowledge enabling prediction of long-range structural changes, which still remains as one of the main hindrances associated with this technology, might be partly compensated by the growing number of enzymes that are being studied systematically and can be used for structure–function alignments.

References

1 (a) A. M. KLIBANOV, Nature 2001, 409, 241–246; (b) Y. L. KHMELNITSKY, J. O. RICH, Curr. Opin. Chem. Biol. 1999, 3, 47–53; (c) G. COLOMBO, G. OTTOLINA, G. CARREA, Monatsh. Chem. 2000, 131, 527–547; (d) P. J. HALLING, Curr. Opin. Chem. Biol. 2000, 4, 74–80; (e) G. CARREA, S. RIVA, Angew. Chem. Int. Ed. 2000, 39, 2226–2254.
2 F. BORDUSA, Chem. Rev. 2002, 102, 4817–4867.
3 V. SCHELLENBERGER, A. GÖRNER, A. KÖNNECKE, H.-D. JAKUBKE, Peptide Res. 1991, 4, 265–269.
4 F. BORDUSA, D. ULLMANN, H.-D. JAKUBKE, Angew. Chem. Int. Ed. 1997, 36, 1099–1101.
5 I. SCHECHTER, A. BERGER, Biochem. Biophys. Res. Commun. 1967, 27, 157–162.
6 (a) V. SCHELLENBERGER, H.-D. JAKUBKE, N. P. ZAPEVALOVA, Y. V. MITIN, Biotechnol. Bioeng. 1991, 38, 104–108; (b) H. SEKIZAKI, K. ITOH, E.

Toyota, K. Tanizawa, *Chem. Pharm. Bull.* **1996**, *44*, 1577–1579; (c) 1585–1587; (d) F. Bordusa, D. Ullmann, C. Elsner, H.-D. Jakubke, *Angew. Chem. Int. Ed.* **1997**, *36*, 2473–2475; (e) H. Sekizaki, K. Itoh, E. Toyota, K. Tanizawa, *Tetrahedron Lett.* **1997**, *38*, 1770–1780; (f) H. Sekizaki, K. Itoh, E. Toyota, K. Tanizawa, *Chem. Pharm. Bull.* **1999**, *47*, 444–447; (g) M. Thormann, S. Thust, H.-J. Hofmann, F. Bordusa, *Biochemistry* **1999**, *38*, 6056–6062.

7 N. Wehofsky, F. Bordusa, *FEBS Lett.* **1998**, *443*, 220–224.

8 R. Günther, S. Thust, H.-J. Hofmann, F. Bordusa, *Eur. J. Biochem.* **2000**, *267*, 3496–3501.

9 V. Cerovsky, F. Bordusa, *J. Peptide Res.* **2000**, *55*, 325–329.

10 N. Wehofsky, S. Thust, J. Burmeister, S. Klussmann, F. Bordusa, *Angew. Chem. Int. Ed.*, in press.

11 N. Wehofsky, M. Alisch, F. Bordusa, *Chem. Commun.* **2001**, 1602–1603.

12 R. Günther, F. Bordusa, *Chem. Eur. J.* **2000**, *6*, 463–467.

13 N. Wehofsky, R. Löser, A. Buchynskyy, P. Weizel, F. Bordusa, *Angew. Chem. Int. Ed.* **2002**, *41*, 2735–2738.

14 (a) Z.-P. Wu, D. Hilvert, *J. Am. Chem. Soc.* **1989**, *111*, 4513–4514; (b) K. E. Neet, D. E. Koshland, *Proc. Natl. Acad. Sci. USA* **1966**, *56*, 1606–1611; (c) L. Polgar, M. L. Bender, *Biochemistry* **1967**, *6*, 610–620.

15 I. M. Bell, D. Hilvert, *Biochemistry* **1993**, *32*, 13969–139673.

16 (a) L. M. Bech, K. Breddam, *Carlsberg Res. Commun.* **1988**, *53*, 381–393; (b) H. Grøn, L. M. Bech, S. Branner, K. Breddam, *Eur. J. Biochem.* **1990**, *194*, 897–901; (c) K. Khumtaveeporn, A. Ullmann, K. Matsumoto, B. G. Davis, J. B. Jones, *Tetrahedron: Asymmetry* **2001**, *12*, 249–261; and references cited therein.

17 (a) G. DeSantis, J. B. Jones, *Curr. Opin. Biotechnol.* **1999**, *10*, 324–330; (b) K. Matsumoto, B. G. Davis, J. B. Jones, *Chem. Commun.* **2001**, 903–904.

18 (a) J. A. Wells, E. Ferrari, D. J. Henner, D. A. Estell, E. Y. Chen, *Nucleic Acids Res.* **1983**, *11*, 7911–7925.

19 P. N. Bryan, *Biochim. Biophys. Acta* **2000**, *1543*, 203–222.

20 (a) L. Abrahamsen, J. Tom, J. Burnier, K. A. Butcher, A. Kossiakoff, J. A. Wells, *Biochemistry* **1991**, *30*, 4151–4159; (b) D. Y. Jackson, J. Burnier, C. Quan, M. Stanley, J. Tom, J. A. Wells, *Science* **1994**, *117*, 819–820.

21 (a) R. Grünberg, I. Domgall, R. Günther, K. Rall, H.-J. Hofmann, F. Bordusa, *Eur. J. Biochem.* **2000**, *267*, 7024–7030; (b) S. Xu, K. Rall, F. Bordusa, *J. Org. Chem.* **2001**, *66*, 1627–1632; (c) K. Rall, F. Bordusa, *J. Org. Chem.* **2002**, *67*, 9103–9103.

22 A. Sattler, S. Kanka, K. H. Maurer, D. Riesner, *Electrophoresis* **1996**, *17*, 784–792.

23 H. Zhao, F. H. Arnold, *Protein Eng.* **1999**, *12*, 47–53.

24 K. Miyazaki, F. H. Arnold, *J. Mol. Evol.* **1999**, *49*, 389–391.

25 S. Straussberg, P. Alexander, D. T. Gallagher, G. Gilliland, B. L. Barnett, P. Bryan, *Biotechnology* **1995**, *13*, 669–673.

26 K. Chen, F. H. Arnold, *Proc. Natl. Acad. Sci. USA* **1993**, *90*, 5618–5622.

27 L. You, F. H. Arnold, *Protein. Eng.* **1996**, *9*, 77–93.

28 S. Atwell, J. A. Wells, *Proc. Natl. Acad. Sci. USA* **1999**, *96*, 9497–9502.

29 D. Legendre, N. Laraki, T. Graslund, M. E. Bjornvad, M. Bouchet, P. A. Nygren, T. V. Borchert, J. Fastrez, *J. Mol. Biol.* **2000**, *296*, 87–102.

30 J. E. Ness, M. Welch, L. Giver, M. Bueno, J. R. Cherry, T. V. Borchert, W. P. C. Stemmer, J. Minshull, *Nat. Biotechnol.* **1999**, *17*, 893–896.

5.2
Twin Ribozymes

Sabine Müller, Rüdiger Welz, Sergei A. Ivanov, and Katrin Bossmann

5.2.1
Introduction

Twenty years ago the catalytic activity of RNA molecules was discovered. The self-splicing pre-ribosomal RNA of the ciliate *Tetrahymena* was the first example of an RNA molecule that forms a catalytic active site for a series of precise biochemical reactions [1]. Shortly after, activity was discovered in the RNA component of Ribonuclease P, a ribonucleoprotein enzyme that performs catalysis in a multiturnover fashion [2]. These findings indicated that proteins do not have a monopoly on biological catalysis, but that RNA molecules can also serve as both genetic material and cellular enzymes. Catalytic RNA molecules (ribozymes; Box 22) are structural components of several naturally occurring RNAs. The hammerhead ribozyme, the hairpin ribozyme, the hepatitis delta virus ribozyme, the neurospora varkud satellite ribozyme, the group I and group II self-splicing introns, and the RNA subunit of RNase P, mentioned above, belong to this class of natural ribozymes which catalyze the cleavage and ligation of phosphodiester bonds [3].

In addition to natural ribozymes several synthetic ribozymes with new activity have been obtained by in-vitro selection techniques. Starting with a pool of random RNA sequences, molecules with a desired activity can be isolated by successive cycles of activity selection, reverse transcription into DNA, and amplification by polymerase chain reaction [4a,b].

5.2.2
Application of Ribozymes

Because of their ability to cleave a suitable RNA target in trans, ribozymes have great potential as gene inhibitors. Concepts for using ribozymes in gene therapy were developed over the past decade and have been transformed into a number of clinical trials, although success is still limited. In general, ribozyme-based strategies focus on suppression of pathological gene products to treat diseases caused by viral infections or malignancies. The major aim is to block gene expression by interfering with RNA transcription (Figure 5.2.1).

Highlights in Bioorganic Chemistry: Methods and Applications. Edited by Carsten Schmuck, Helma Wennemers.
Copyright © 2004 WILEY-VCH Verlag GmbH & Co. KGaA, Weinheim
ISBN: 3-527-30656-0

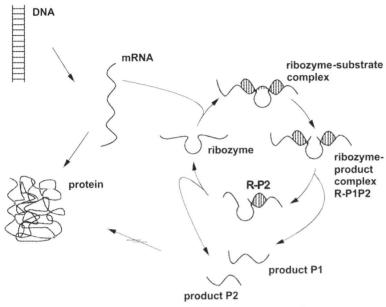

Fig. 5.2.1. Schematic representation of ribozyme based inhibition of gene expression.

Ribozymes recognize their substrates by Watson–Crick base pairing. By altering substrate recognition sequences target RNA molecules can therefore be cleaved in a highly sequence-specific manner. Among the natural RNA catalysts the hammerhead and the hairpin ribozyme, in particular, are used for this purpose. Because of their simple structure and small size both species are particularly well suited to be engineered for cleavage of a desired RNA (Figure 5.2.2).

In addition to the hammerhead and hairpin ribozyme, nature harbors several other catalytic RNAs with useful activities. Group I introns, in particular, have been studied as tools for therapeutic RNA manipulation. The activity of group I introns includes a two step transesterification with guanosine as co-factor (cis-splicing, Figure 5.2.3). Natural cis-splicing introns can be engineered to perform this reaction in trans and thus gain importance as therapeutic agents to correct genetic disorders (trans-splicing, Figure 5.2.3).

Trans-Splicing ribozymes have been generated initially to repair mutant lacZ transcripts in bacteria [5] and mammalian cells [6], and more recently to amend mutant transcripts associated with myotonic dystrophy [7] and many cancers in mammalian cell lines [8], and with sickle cell anemia in erythrocyte precursors isolated from patients with sickle cell disease [9]. Trans-Splicing group I introns can be designed virtually against any RNA target, because the only conserved nucleotide is a uridine 5' to the splice site. Because of the short recognition sequence (6–9 bases), however, non-specific splicing has been detected [6]. Furthermore, the whole 3' fragment which should replace the cleaved mutated sequence (Figure 5.2.3) has to be attached to the approximately 350-nts-long group I intron. De-

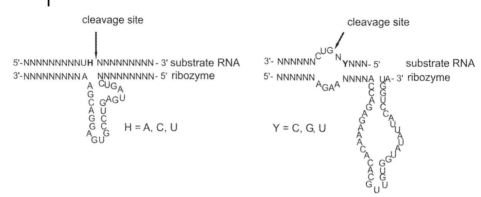

Fig. 5.2.2. Consensus sequence of (a) the hammerhead and (b) the hairpin ribozyme for therapeutic RNA cleavage.

pending on the lengths of the transcript and the location of the mutation very large constructs (~500 nts) have to be used.

We sought to develop a system for correction of genetic disorders based on small ribozymes. As discussed above, the endonucleolytic activity of hammerhead and hairpin ribozymes has been used in gene therapeutic approaches. Both ribozymes are small, synthetically available, and easy to handle. Whereas the hammerhead ribozyme cleaves its substrate 200 times faster than it ligates the formed products [10], the hairpin ribozyme is a twofold better ligase than an endonuclease [11]. Because repair requires a tool with both cleavage and ligation activity, the hairpin ribozyme is an appropriate basis for the design of an RNA repair ribozyme. Our strategy involved combination of two catalytic modules in one molecule (twin ribozyme) to generate a ribozyme with two processing sites at the target RNA. After removal of a mutated RNA sequence by a two-site cleavage in the first step, a new fragment carrying the correct sequence could be incorporated in the second step, now using the ligation activity of the ribozyme (Figure 5.2.4). If it is possible to combine and control cleavage and ligation activity, a hairpin-derived twin ribozyme might be a potential tool for RNA manipulation in vitro with the possibility in future of applying a similar method in vivo.

Construction of a hairpin ribozyme targeting more than one site within the substrate RNA also yields further insight into the structural limits of hairpin ribozyme-catalyzed RNA processing and can provide information on the structural and mechanistic requirements of catalysis.

5.2.3
Building Blocks for Twin Ribozymes

5.2.3.1
The Conventional Hairpin Ribozyme (HP-WT)

The conventional hairpin ribozyme is derived from the minus strand of *Tobacco ringspot virus* satellite RNA [12a,b]. It works in cis (intramolecularly) in nature, but

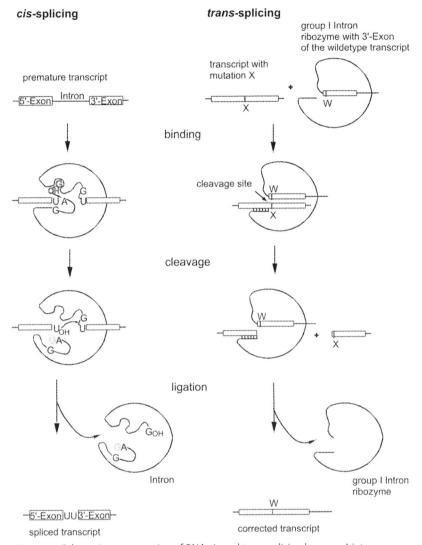

Fig. 5.2.3. Schematic representation of RNA cis and trans splicing by group I introns.

it has been engineered to work in trans (intermolecularly) and thus catalyze site-specific reversible cleavage at the 5' side of a GUC triplet (Figure 5.2.5), generating a 2',3'-cyclic phosphate and a free 5'-hydroxyl group (Figure 5.2.6) [13].

The ribozyme consists of two independently folded domains, A and B. Each of these contains an internal loop (A, B) flanked by two Watson–Crick helices (H-1, H-2 in domain A, H-3, H-4 in domain B). The RNA substrate is bound to the ribozyme to generate Helix 1 and 2 and is reversibly cleaved within loop A. Catalysis is supported by positively charged metal ions, typically Mg^{2+}. On the basis of several biochemical studies it has been suggested the metal ions are important for

5.2 Twin Ribozymes

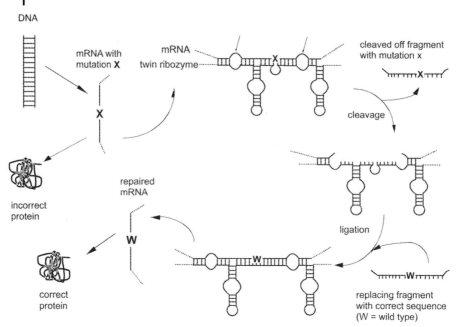

Fig. 5.2.4. Strategy for the repair of genetic disorders by use of twin ribozymes.

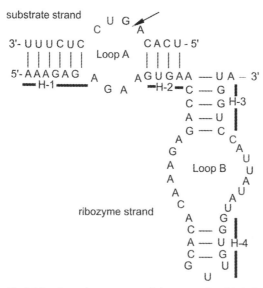

Fig. 5.2.5. Secondary structure of the conventional hairpin ribozyme HP-WT. The arrow denotes the cleavage site. The four helices (H-1 through H-4) are marked by bars.

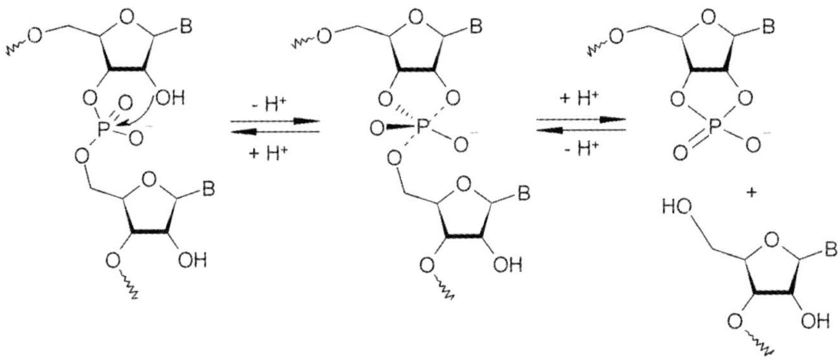

Fig. 5.2.6. Mechanism of the reversible cleavage of a phosphodiester bond catalyzed by the hairpin ribozyme.

stabilization of the ribozyme active conformation and not as thought for a long time, to participate in active site chemistry. Last, the crystal structure, which has recently been solved [14], showed that no metal ions are present in the active center, providing additional proof for the structural role of positively charged cofactors.

Folding into the active conformation requires a sharp bend between helix 2 and helix 3 to orient the two helical domains in an antiparallel fashion and to enable Loop A to interact with Loop B (Figure 5.2.7). Catalysis is achieved by specific positioning of the phosphodiester bond to be processed within the ribozyme architecture, such that nucleophilic *in line* attack of the 2′-OH at the phosphorus atom in an S_N2-like mechanism becomes possible (Figure 5.2.6).

5.2.3.2
The Reverse-joined Hairpin Ribozyme (HP-RJ)

Reverse-joined hairpin ribozymes have been introduced by Komatsu et al. [15a,b]. This type of ribozyme is derived from the conventional hairpin ribozyme by dissecting the two domains at the hinge between helix 2 and 3 and re-joining helix 4 to helix 1 via a linker comprising six unpaired bases (Figure 5.2.8).

The design of reverse-joined hairpin ribozymes is based on the assumption that the bent active conformation of the hairpin ribozyme can be reached from distinct

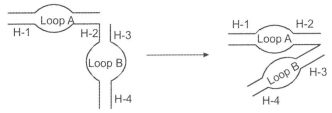

Fig. 5.2.7. Schematic representation of folding into the hairpin ribozyme active conformation.

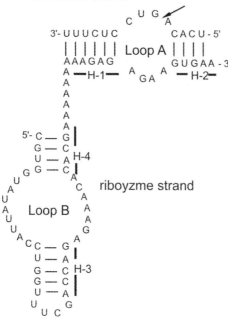

Fig. 5.2.8. Secondary structure of the reverse-joined hairpin ribozyme HP-RJ. The arrow denotes the cleavage site. The four helices (H-1 through H-4) are marked by bars.

starting points. Specific contacts between the nucleotides within Loops A and B keep the hairpin ribozyme in the active conformation. It is, therefore, likely that joining the two domains in reverse order also will create an RNA structure that is able to fold into a conformation with loop–loop contact, as long as the capacity of the loop nucleotides for specific interaction is preserved. On the basis of this principle we have designed and synthesized the reverse-joined hairpin ribozyme HP-RJ shown in Figure 5.2.8 [16]. The A_6-linker acts as a flexible hinge between helix 1 and helix 4 and is required to bridge the distance of 34 Å between the distal ends of helix 1 and helix 4, as observed in the crystal structure [14].

The reverse-joined hairpin ribozyme catalyzes the same specific reaction as the conventional hairpin ribozyme and has similar kinetic behavior [16].

Conventional and reverse-joined hairpin ribozymes as described above can be combined in one molecule to create a twin ribozyme. Their use is, however, limited by the specific structure. Because folding into the active conformation requires a bend between helix 2 and helix 3 in the conventional hairpin ribozyme, and between helix 1 and helix 4 in the reverse-joined hairpin ribozyme, these helices form a sort of hinge and are, therefore, not susceptible to arbitrary changes. Whereas helix 1 in HP-WT can be extended without distortion of tertiary folding, the situation in the reverse-joined hairpin ribozyme is the opposite, helix 2 can be extended, and helix 1 cannot. This enables combination of the two catalytic modules in a single mode as shown in Figure 5.2.11, namely the conventional hairpin ribozyme on the right hand side attached to the reverse-joined unit on the left hand side of the molecule.

To design hairpin ribozyme units that can be placed at any position in a twin ribozyme we have developed three-way junction hairpin ribozymes [17] and the branched reverse-joined hairpin ribozymes described below.

5.2.3.3
Three-way Junction Hairpin Ribozymes (HP-TJ)

To explore the catalytic potential of three-way junction hairpin ribozymes we have synthesized the ribozyme HP-TJ, with an additional helix at the branch-point of the two domains, and have studied the kinetic properties in comparison to those of the wild-type ribozyme HP-WT. The new type of hairpin ribozyme contains an extra sequence at the 3' side of the ribozyme strand to enable it to hybridize with the 5' side sequence of a 28-mer substrate RNA. Thus, in comparison to the natural hairpin ribozyme, HP-TJ consists of an additional helix 5 between the ribozyme and part of the substrate strand (Figure 5.2.9).

The linker of three unpaired cytidines at the hinge between helix 3 and helix 5 was introduced to preserve the necessary hinge flexibility to enable the ribozyme to bend and thus to bring loop A and loop B in close proximity to each other. Investigation of three-way junction ribozymes with a different number of cytidine residues (0–5) at the hinge showed that the construct containing three cytidines is most active [17].

5.2.3.4
Branched Reverse-joined Hairpin Ribozymes (HP-RJBR)

As mentioned above the specific structure of HP-RJ has a clear disadvantage for use as universal catalytic module. The substrate binding domain cannot be extended in the 5' direction, requiring HP-RJ to be placed exclusively at the 5' end of a multi-target ribozyme. To overcome this limitation we prepared the branched ribozyme HP-RJBR (Figure 5.2.10).

In this structure the catalytic domain is connected to the substrate binding domain via the 4-position of a thymidine analog, enabling extension of the substrate-binding domain in 5' direction by a natural phosphodiester bond at the 5'-OH of the sugar unit belonging to the same nucleoside. Before RNA synthesis we prepared the modified monomer 2'-deoxy-N^4-(n-6-hydroxyhexyl)-5-methylcytidine which, after suitable protection and on incorporation into the RNA, was used for assembly of the second dimension chain via the hexyl-OH group, resulting in the generation of a non-natural branch (Figure 5.2.10).

Activity tests showed that the chemically synthesized branched ribozyme catalyzed the cleavage of the substrate RNA shown in Figure 5.2.10.

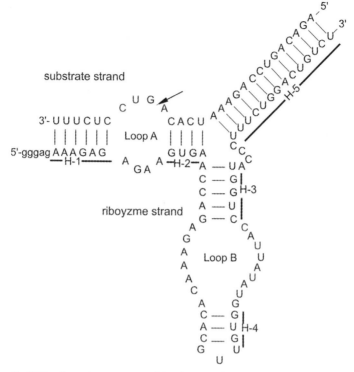

Fig. 5.2.9. Secondary structure of the three-way junction hairpin ribozyme HP-TJ. The arrow denotes the cleavage site. The five helices (H-1 through H-5) are marked by bars. 5′-dangling nucleosides result from in-vitro transcription.

5.2.4
Design, Synthesis and Characterization of Twin Ribozymes

With the described set of catalytic modules at hand we designed and synthesized several twin ribozymes, three of those are shown as examples in Figure 5.2.11.

For combination in a twin ribozyme both units should specifically interact with a unique substrate sequence. To this end we changed the substrate sequence in the single units and determined the kinetic constants of the cleavage reaction before constructing the twin ribozymes. All single catalytic modules under single turnover conditions had rate constants between 0.2 and 0.5 min^{-1}.

The preparation of twin ribozymes included both chemical and enzymatic procedures. Basically, in our laboratory RNA strands longer than 60 nucleotides are transcribed from synthetic DNA templates using T7 RNA-polymerase. By using 5-benzylmercapto-1H-tetrazole [18a,b] as activator in the phosphoramidite approach, however, we also succeeded in synthesizing a chemical version of HP-

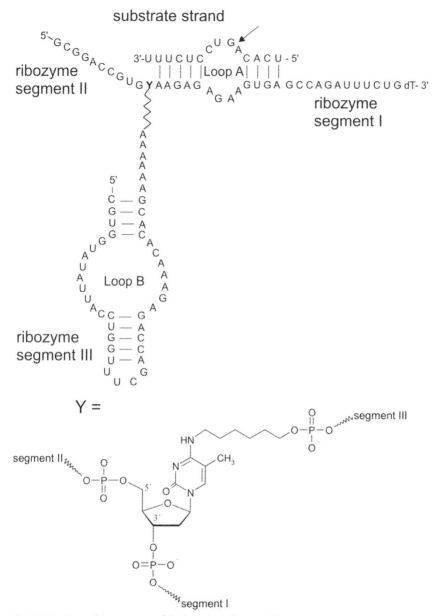

Fig. 5.2.10. Secondary structure of the branched ribozyme HP-RJBR and structure of the branching nucleoside. The arrow denotes the cleavage site. The single segments within the ribozyme strand are marked.

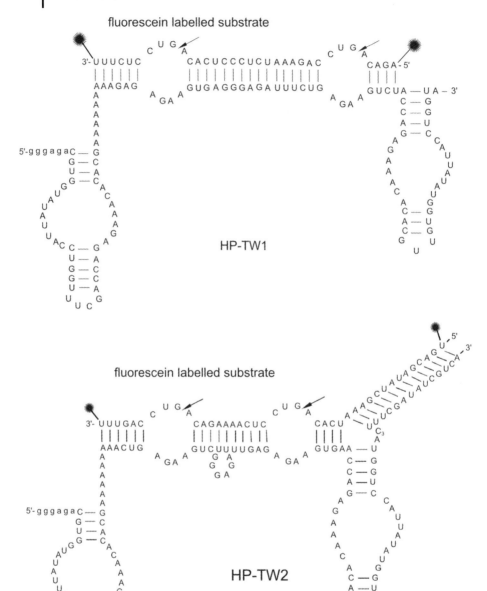

Fig. 5.2.11. Secondary structures of twin ribozymes. The arrows mark the cleavage sites. The substrates are fluorescein-labeled at both ends.

5.2.4 Design, Synthesis and Characterization of Twin Ribozymes

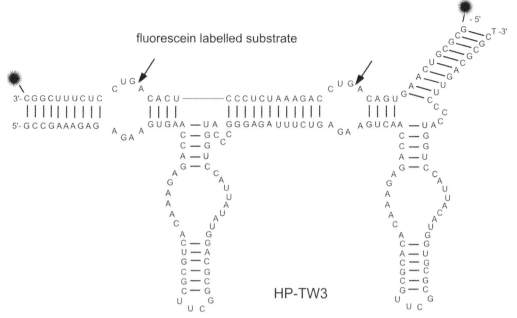

Fig. 5.2.11 *(continued)*

TW2. Fluorescein-labeled substrates were synthesized exclusively by solid-phase chemistry and the fluorescein residue was attached during this procedure [16].

HP-TW1 was assembled from the catalytic strand HP-TW1 and the substrate strand (Figure 5.2.11). The 34-mer substrate-RNA has a fluorescein-label at both the 3′- and the 5′-ends, to enable monitoring of the cleavage at both predicted sites by use of our previously developed quantitative assay of hairpin ribozyme activity [16]. Specific cleavage by the reverse-joined ribozyme unit within the twin ribozyme would produce a 9-mer and a 25-mer RNA; cleavage by the conventional hairpin ribozyme unit would produce a 5-mer and a 29-mer RNA. Both the 29-mer and the 25-mer fragments are still substrates for single cleavage at the second site. Formation of the 5-mer and 9-mer cleavage products is therefore expected to accelerate during reaction and these should be found as the main products (Figure 5.2.12).

Because the substrate is labeled at the 3′-end and at the 5′-end, short product RNAs and the longer product strands are both detectable. Analysis of the cleavage reaction shows that the substrate RNA is cleaved at two specific sites after incubation with the twin ribozyme, leading to the characteristic 5-mer and 9-mer fragments and to a 20-mer, which cannot be detected by laser-induced fluorescence, as main products. Whereas the fractions of the 5-mer and 9-mer produced increase with time, the fractions of the 25-mer and the 29-mer decrease after an initial increase, based on cleavage occurring at the second specific site (Figure 5.2.13). Similar behavior was observed for the cleavage reaction of HP-TW2 and HP-TW3.

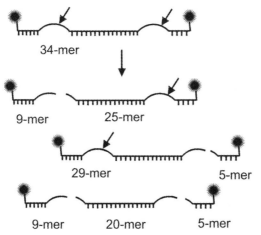

Fig. 5.2.12. Schematic representation of cleavage of the 34-mer substrate RNA by HP-TW1 carrying a fluorescein label at both ends.

5.2.5
Application of Twin Ribozymes

After having shown that combination of two catalytic modules in one molecule results in twin ribozymes with double cleavage activity we wondered whether twin ribozymes can also catalyze RNA ligation and thus be used for catalysis of RNA fragment-exchange reactions. If all fragments (substrates and cleavage products)

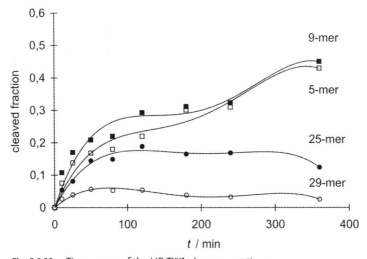

Fig. 5.2.13. Time course of the HP-TW1 cleavage reaction.

remain bound to the ribozyme, the hairpin ribozyme mentioned above is a better ligase than an endonuclease, it ligates twice as fast as it cleaves its substrate [11]. Because of rapid dissociation of cleavage products in the minimal construct as shown in Figure 5.2.5, however, the fragments are not available for re-ligation and substrate cleavage is the predominant reaction. This in turn, opens up control of the activity of the ribozyme by duplex stability. Substrates that are weakly bound to the ribozymes should be preferentially cleaved, whereas strongly bound substrates should be ligated.

On the basis of this assumption we designed the twin ribozyme HP-TW2 (Figure 5.2.11) for catalysis of an RNA exchange reaction. HP-TW2 consists of two catalytic modules, a reverse-joined (left unit) and a three-way junction hairpin ribozyme (right unit). On binding of the substrate a GGGAGA loop is formed in the central region of the ribozyme, because no complementary bases are provided in the substrate sequence. Thus after cleavage at both predicted sites a 14-mer RNA fragment is produced which, because of helix destabilization by the loop, should easily dissociate from the ribozyme. The incoming fragment to be ligated contains the six additional complementary bases and thus forms a more stable duplex with the ribozyme than the removed 14-mer (Figure 5.2.14). Preliminary results suggest that HP-TW2 catalyzes the exchange of those two fragments, although still with rather low yield.

5.2.6
Summary and Outlook

A Canadian study has shown that up to the age of 25, approximately 53 out of 1000 persons develop a disease with a genetic background. Most of those genetic diseases rely on changes in a single gene, which on the basis of ever-growing progress in the elucidation of the human genome, can be located more and more precisely.

Ribozymes can cleave and ligate phosphodiester bonds and thus have great potential for gene therapeutic applications that focus on the correction of genetic disorders rather than on destruction of the messenger RNA coding for a pathological gene product. First attempts have been made to use the large group I intron ribozyme for this purpose and have shown that repair of mutated RNA is possible.

Small ribozymes such as our twin ribozymes might have advantages over the large group I intron for a number of reasons. They are much smaller and, in principle, synthetically available. Because of their small size they are easy to study in the laboratory and their structure can be better optimized. With two processing sites, twin ribozymes target only a small area of the substrate RNA. They can be modified in the substrate binding domain, which enables construction of custom-designed ribozymes to target a desired RNA sequence.

Still working with arbitrary chosen sequences for proof of principle, in future we aim to apply selection procedures to develop twin ribozymes against RNA targets with genetic relevant mutations. In parallel, our studies on the mechanism of RNA catalysis and RNA structure–function relationships will help us to understand how

5.2 Twin Ribozymes

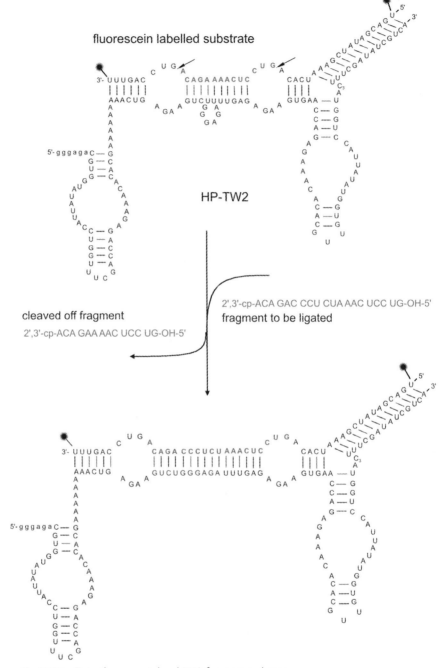

Fig. 5.2.14. Twin ribozyme-catalyzed RNA fragment exchange.

ribozymes work and will flow back into the de-novo design of optimized twin ribozymes.

Worldwide on-going work in the field of nucleic acid-based gene therapy targets the question of the delivery, transport, and in-vivo activity of nucleic acid drugs. The results obtained from these studies will be taken into consideration for the in-vivo design of twin ribozymes after the in-vitro assay has proven successful.

References

1 K. Kruger, P. J. Grabowski, A. J. Zaug, J. Sands, D. E. Gottschling, T. R. Cech, Cell 1982, 31, 147–157.
2 C. Guerrier-Takada, K. Gardiner, T. Marsh, N. Pace, S. Altman, Cell 1983, 35, 849–857.
3 J. A. Doudna, T. R. Cech, Nature 2002, 418, 222–228 (Review).
4 (a) A. Jäschke, B. Seelig, Curr. Opin. Chem. Biol. 2000, 4, 257–262; (b) G. A. Soukup, R. R. Breaker, Curr. Opin. Struct. Biol. 2000, 10, 318–325 (Reviews).
5 B. A. Sullenger, T. R. Cech, Nature 1994, 371, 619–622.
6 J. T. Jones, S. W. Lee, B. A. Sullenger, Nature Med. 1996, 2, 643–648.
7 L. A. Phylactou, C. Darrah, M. J. Wood, Nature Genet. 1998, 18, 378–381.
8 T. Watanabe, B. A. Sullenger, Proc. Natl. Acad. Sci. USA 2000, 97, 8490–8494.
9 N. Lan, R. P. Howrey, S. W. Lee, C. A. Smith, B. A. Sullenger, Science 1998, 280, 1593–1596.
10 K. J. Hertel, T. K. S. Zimmermann, G. Ammons, O. C. Uhlenbeck, Biochemistry 1994, 33, 3374–3385.
11 X. Zhuang, H. Kim, M. J. B. Pereira, H. P. Babcock, N. G. Walter, S. Chu, Science 2002, 296, 1473–1476.
12 (a) P. A. Feldstein, J. M. Buzayan, G. Bruening, Gene 1989, 82, 53–61; (b) A. Hampel, R. Tritz, Biochemistry 1989, 28, 4929–4933.
13 For recent review see D. M. J. Lilley, ChemBioChem 2001, 2, 729–733.
14 P. B. Rupert, A. R. Ferré-D'Amaré, Nature 2001, 410, 780–786.
15 (a) Y. Komatsu, I. Kanzaki, M. Koizumi, E. Ohtsuka, J. Mol. Biol. 1995, 252, 296–304; (b) Y. Komatsu, I. Kanzaki, E. Ohtsuka, Biochemistry 1996, 35, 9815–9820.
16 C. Schmidt, R. Welz, S. Müller, Nucleic Acids Res. 2000, 28, 886–894.
17 R. Welz, B. Scholz, C. Schmidt, S. Müller, Collection Symposium Series 2, A. Holy, M. Hocek, eds., Academy of Sciences of the Czech Republic, Prague, 1999, pp. 331–334.
18 (a) S. Pitsch, P. A. Weiss, L. Jenny, A. Stutz, X. Wu, Helv. Chim. Acta 2001, 84, 3773–3795; (b) R. Welz, S. Müller, Tetrahedron Lett. 2002, 43, 795–797.

B.22
Ribozymes

Sabine Müller

Ribozymes are RNA molecules with catalytic activity (*ribo-* like *ribo*nucleic acid, *-zyme* like en*zyme*). Despite having only four different chemical subunits (compare protein enzymes – 20 different amino acids) RNA molecules have amazing structural variety. They fold into complex tertiary structures analogous to highly structured proteins and catalyze a broad range of chemical reactions. RNA folding is a very hierarchical process – initially formed secondary structural elements undergo further folding into a three-dimensional complex that contains binding pockets for ligands or substrates and that provides catalytic centers for chemical reactions. Several ribozymes occur in nature, for example the hammerhead ribozyme, the hairpin ribozyme, the hepatitis delta virus ribozyme, and the neurospora VS ribozyme. These ribozymes are rather small and are derived from viruses, virusoides, viroids, and satellite RNA. Apart from these small ribozymes nature also harbors large RNA with catalytic properties, for example the self-splicing group I and group II RNA or the most prominent ribozyme, the ribosome. Although most naturally occurring ribozymes catalyze the cleavage or formation of phosphodiester bonds, peptide bond formation is catalyzed in the ribosome. Apart from this natural activity, ribozymes have been selected from large pools of RNA sequences to catalyze reactions such as aminoacylation, RNA polymerization, N-glycosidic bond-formation and cleavage, pyrophosphate bond-formation and cleavage, N-alkylation, S-alkylation, porphyrin metalation, biphenyl isomerization, and Diels–Alder and Michael reactions.

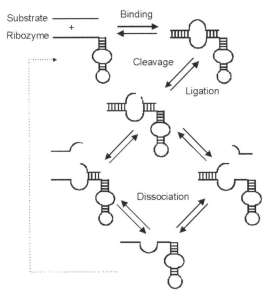

Fig. B.22.1. Catalytic cycle of a ribozyme. A suitable RNA substrate is bound to the ribozyme by Watson–Crick base pairing. On binding the catalytically active structure is formed and the substrate is cleaved at a specific position. Cleavage products dissociate from the ribozyme and the latter is free for the next round of catalysis.

Ribozymes are paradigms of structure–function relationship, folding into a catalytically active structure is mirrored in activity. Most ribozymes behave like true catalysts; they have multiple turnover kinetics that can be described by the Michaelis–Menten model (Figure B.22.1).

The remarkable structural and functional versatility of RNA has prompted conjecture on the possibility of a life-form based primarily on RNA and preceding our DNA- and protein-based life. In this "RNA world" genetic information might have resided in the sequence of RNA molecules, with the phenotype being derived from the catalytic properties of RNA.

Further Reading

J. A. DOUDNA, T. R. CECH, *Nature* **2002**, *418*, 222–228.

A. JÄSCHKE, *Curr. Opin. Struct. Biol.* **2001**, *11*, 321–326.

5.3
RNA as a Catalyst: the Diels–Alderase Ribozyme

Sonja Keiper, Dirk Bebenroth, Friedrich Stuhlmann, and Andres Jäschke

5.3.1
Introduction

The ability of biological macromolecules to catalyze chemical transformations in aqueous media with high specificity is a most fascinating phenomenon and the basis of our life. Although this catalysis is performed predominantly by proteins, for example enzymes and antibodies, nucleic acids can also act as catalysts. For both proteins and nucleic acids the catalytic function is determined by complex three-dimensional structures. Nucleic acids, especially RNA, have a variety of structural features as a result of base pairing. This can occur intermolecularly as in double-stranded DNA, or intramolecularly, the latter generating folding patterns, for example hairpin loops, bulges, mismatches, 3-way and 4-way junctions, and pseudoknots. A remarkable variety of different three-dimensional shapes is thereby formed (the folding properties of RNA are further discussed in Chapter 1.1, for example).

The so-called ribozymes (Box 22) were discovered in 1982 by T. Cech and S. Altman. The naturally occurring species catalyze predominantly one reaction type – hydrolysis or transesterification of phosphodiester bonds in RNA. A very important natural ribozyme is the ribosome. On the basis of X-ray crystallographic investigations it was recently shown that the active site for the peptide bond-formation reaction is composed exclusively of RNA.

By simulating evolution in vitro it has become possible to isolate artificial ribozymes from synthetic combinatorial RNA libraries [1, 2]. This approach has great potential for many reasons. First, this strategy enables generation of catalysts that accelerate a variety of chemical reactions, e.g. amide bond formation, N-glycosidic bond formation, or Michael reactions. This combinatorial approach is a powerful tool for catalysis research, because neither prior knowledge of structural prerequisites or reaction mechanisms nor laborious trial-and-error syntheses are necessary (also for non-enzymatic reactions, as discussed in Chapter 5.4). The iterative procedure of in-vitro selection enables handling of up to 10^{16} different compounds

Highlights in Bioorganic Chemistry: Methods and Applications. Edited by Carsten Schmuck, Helma Wennemers.
Copyright © 2004 WILEY-VCH Verlag GmbH & Co. KGaA, Weinheim
ISBN: 3-527-30656-0

in a single test tube, a number far beyond the capacity of high-throughput screening. Selective pressure enables fine-tuning of the properties of the catalyst [3, 4]. Their catalytic properties also make ribozymes attractive candidates for medical research. Numerous approaches have been used to apply ribozymes in the therapy of severe diseases, for example cancer and HIV infections. A more academic motive for studying RNA catalysis is to support the hypothesis of the existence of a prebiotic RNA world. If it were possible to find ribozymes for each crucial reaction step of our present protein enzyme-dominated biology, this would demonstrate that in the past RNA could have been the repository of genetic information and the executor of biochemical processes at the same time, and thus form the evolutionary precursor for DNA and proteins [5, 6]. Although several artificial ribozymes have been isolated by in-vitro selection, their reaction mechanisms have yet to be elucidated. Our laboratory has successfully isolated a ribozyme that accelerates Diels–Alder reactions. In this chapter, we will summarize our results in this project.

5.3.2
Diels–Alder Reaction

Diels–Alder cycloaddition is one of the central reactions in organic synthesis and, despite its minor biological significance, has often been used in innovative approaches in catalysis research. In the context of the "RNA world" hypothesis, the formation of carbon–carbon bonds would be essential in anabolic pathways.

We specifically chose the reaction between anthracene and maleimide (Figure 5.3.1) for a number of reasons. Most importantly, we assumed that the completely different overall geometry of reactants and products would facilitate enrichment of catalysts that are capable of multiple turnovers. Anthracene is planar, in contrast with the ~120° angles between the different rings in the Diels–Alder product. A ligand that can bind to anthracene should, therefore, not be able to bind to the product except after extensive refolding. The availability of sensitive UV absorbance and fluorescence assays for anthracene was another practical reason to choose this reaction.

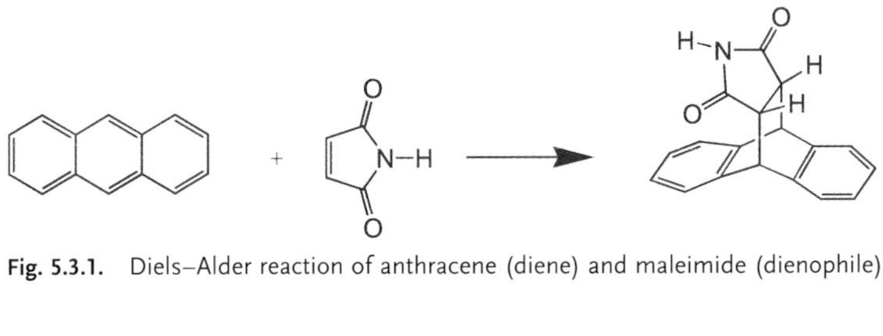

Fig. 5.3.1. Diels–Alder reaction of anthracene (diene) and maleimide (dienophile).

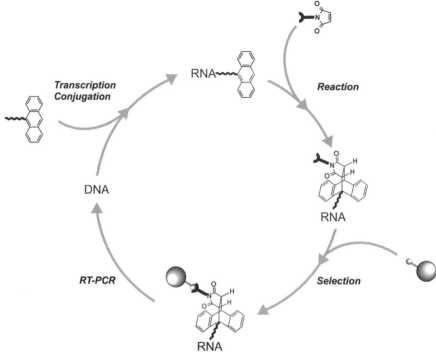

Fig. 5.3.2. Direct selection of catalytic RNA for a Diels–Alder reaction with tether-coupled anthracene.

5.3.3
In-vitro Selection

To select RNA catalysts for the Diels–Alder reaction between anthracene and maleimide (Figure 5.3.2), anthracene was covalently attached to each molecule of a combinatorial RNA library via a tether [7]. This spacer divides the molecules into a catalytic and a substrate domain. Poly(ethylene glycol) (PEG) was found to be suitable for this purpose, because it is flexible and water-soluble and also chemically and structurally inert. We generated a pool of RNA–PEG–anthracene conjugates with 120 randomized nucleotides. The starting library contained $\sim 2 \times 10^{14}$ sequence variants, and about 50 copies of each species were present. Chemically synthesized anthracene–PEG was attached to the 5′ end of the RNA molecules by transcription initiation of the chemically synthesized DNA pool, using the fact that T7 RNA polymerase tolerates modifications at the 5′ end of initiating guanosines. During the transcription reaction radioactively labeled nucleotides were incorporated into the RNA; this enabled monitoring of every step of a selection cycle by simple quantification.

This library of RNA–linker–substrate conjugates was then incubated with maleimide. The maleimide was derivatized with biotin as an anchor group for easy separation. If an RNA molecule catalyzes the reaction between the attached anthracene and biotin maleimide, it becomes tagged with the biotin group, and can easily be isolated by use of a solid support. After removal of all non-biotinylated members of the pool by means of immobilized streptavidin, a biotin binding protein, the bound RNAs were reverse-transcribed and the resulting DNA library was amplified by PCR. Transcription of the resulting DNA, again in the presence of anthracene–PEG-substituted initial guanosine, yielded an enriched pool of RNA conjugates which was used as input for the next round of selection. This iterative procedure enables isolation of extremely rare species from gigantic libraries (Box 23).

In the first five cycles of in-vitro selection, very little (0.1–0.2%) of the total RNA reacted with biotin, predominantly because of the uncatalyzed background reaction. Several potential cofactors, for example Lewis acids, metal ions, a dipeptide compound, and a dipyridyl compound, were present in the reaction mixture. From cycle 6 onwards significant acceleration of the reaction was observed. As a consequence, in the following cycles the selection pressure was increased by shortening the reaction time and reducing the concentration of maleimide. Compared with the starting library a 6500-fold acceleration of the reaction was observed for the enriched pool after cycle 10.

5.3.4
Sequence Analysis and Ribozyme Engineering

Even after ten cycles of in vitro selection, this was still a library. To separate the active sequences, the RNA was reverse-transcribed and PCR-amplified, and the resulting DNA was cloned, amplified, and sequenced. Thirty-five different catalytically active sequences were found, 32 of which shared a common secondary structure motif (Figure 5.3.3) – a hexanucleotide region AAUACU and a pentanucleotide region UGCCA. The neighborhood of these regions was pairwise complementary, indicating formation of a double strand with a bulge in the middle. Upstream of these sequence elements a nucleotide region complementary to the 5' constant primer region could be identified, indicating the presence of a third helix. These structural proposals were confirmed by use of RNAdraw and RNA structure software.

To verify experimentally that the catalytic activity was caused by this secondary structure motif, we prepared a 53-nucleotide truncated version of one clone, which contained a condensed version of the motif near the 5' end. The activity of this oligoribonucleotide was found to be unchanged compared with that of the parent sequence. Next, we rationally increased the thermodynamic stability by eliminating unpaired bases from the helices, increasing the GC content of the helices, and replacing loops 1 and 2 with the stabilizing tetraloop sequences UUCG. The

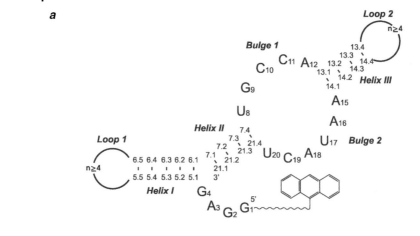

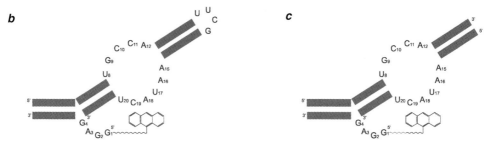

Fig. 5.3.3. (a) Proposed catalytic motif and numbering scheme of the Diels–Alderase ribozyme motif, and schematic depiction of (b) bipartite ribozyme version and (c) tripartite ribozyme version.

resulting sequence contains only 49 nucleotides and accelerates the Diels–Alder reaction approximately 18 500-fold in the intramolecular reaction. The reaction had Michaelis–Menten-type saturation kinetics with regard to biotin maleimide [8].

We then prepared a 38-mer ribozyme derived from the 49-mer by formally cutting loop 1. A substrate oligonucleotide–PEG–anthracene conjugate 11 nucleotides long was prepared. Samples were incubated under single-turnover conditions and aliquots withdrawn at specified times were analyzed by denaturing polyacrylamide gel electrophoresis (PAGE). From the ratio of the bands a 16 000-fold rate acceleration was calculated; this is only 15% slower than the self-modifying reaction. A tripartite system of 24-mer, 18-mer, and 11-mer–PEG–anthracene conjugate produced by formally cutting loop 2 is also very effective in catalysis. These constructs were found to be very useful for investigation of the relationship between structure and function, for example mutation analyses, because short oligonucleotides can be synthesized effectively by solid-phase chemical methods.

5.3.5
Mutation Analysis

Comparative sequence analysis suggested that the single-stranded regions comprising the bulge and the 5′-GGAG terminus are crucial for catalysis. Single-base substitutions were therefore inserted and the activities of the mutated ribozymes were measured by means of an electrophoretic gel mobility shift assay. Figure 5.3.4 shows the resulting activity profile on a logarithmic scale.

Nucleotides G1, G2, U8, C10, C11, and U20 were absolutely conserved and nucleotides A3, G4, G9, and A18 were highly conserved. Other positions, for example A15 and A16, were less sensitive to mutation. Clues about a direct involvement of conserved nucleotides in the catalytic function cannot yet be drawn. Certain essential nucleotides might be needed for correct three-dimensional folding of the catalytic structure without being part of a proposed binding pocket. At position 12 an A to G transition mutation is strongly preferred compared to transversions. Similarly, at position 17 a U to C transition is preferred.

To investigate possible base-pairing interactions between these essential nucleotides, two point mutations were inserted simultaneously at respective positions inside the single stranded regions. If the deleterious effect of a first mutation is attenuated or reversed by a second mutation ("rescue"), direct interaction between these positions is likely. Again the tripartite ribozyme version was used, because of the advantage of ready combination of mutated fragments by hybridization without the need to synthesize the complete RNA strand. No evidence of base pairing was found for some obvious proposals, namely between bases G1, G2, and C10, C11, or between G9 and C19. These bases could not be replaced by other complementary pairs, according to the standard Watson–Crick scheme, without complete loss of activity.

The picture changed when positions 3/20 and 4/19 were examined. The pair A3/U20 could be exchanged by a G–C pair in both orientations almost without loss of activity, whereas exchange by an A–U in the other orientation was not well tolerated. The two bases G4/C19 could, furthermore, be replaced by a G–C in the other orientation and by the two possible A–U pairs also, without great loss of activity. A G–U wobble was also very well tolerated. These observations strongly support the assumption of base pairing at those positions according to the standard Watson–Crick scheme. Non-standard hydrogen-bonding interactions between bases, as described for other RNA structures, could not be detected by this method [9].

5.3.6
True Catalysis

Whereas all experiments described so far used systems in which the diene was covalently tethered to the RNA, true catalysis with multiple turnovers was the final goal of this project. We therefore performed measurements in which the tether

5.3 RNA as a Catalyst: the Diels–Alderase Ribozyme

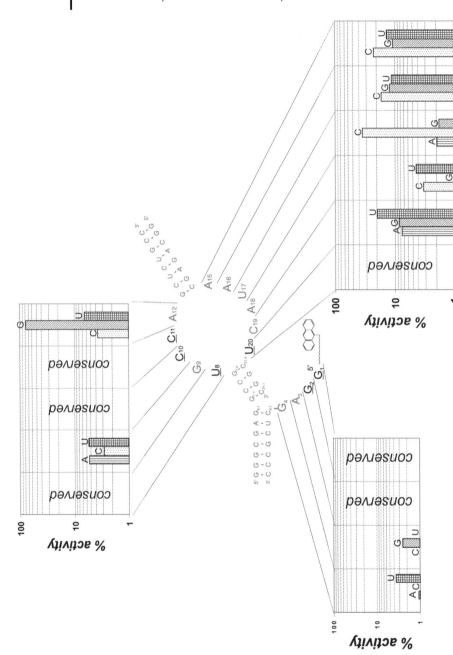

Fig. 5.3.4. Mutation analysis of the tripartite ribozyme version (absolutely conserved nucleotides are in black and underlined). The activity profile is shown on a logarithmic scale.

between RNA and substrate was omitted [10]. By employing UV–visible spectroscopy the progress of the reaction could be observed by following the decrease of the anthracene absorption at 365 nm. The catalytic action of this ribozyme on different anthracene and maleimide derivatives was explored and turned out to be successful. Thus, this Diels–Alderase ribozyme is among the first that catalyze chemical reaction between two small organic molecules and remains unchanged after multiple catalytic cycles.

5.3.7
Kinetics

The kinetics of the ribozyme-catalyzed reactions were determined by measuring the initial rates of the catalyzed reaction. For reasons of solubility, anthracene-hexaethyleneglycol, and maleimidocaproic acid were used for these measurements. The initial velocities were determined by monitoring the decrease of anthracene absorbance at 365 nm over ∼5% conversion. The initial rate of the ribozyme-catalyzed reaction was found to be proportional to the ribozyme concentration. The reaction shows saturation-type kinetics with respect to both substrates. Catalysis was examined as a completely random bireactant system.

The secondary double-reciprocal plot of the data from Figure 5.3.5 gives Michaelis constants of 0.37 mM for the diene and 8 mM for the dienophile. The calculated maximum rate V_{max} is 0.15 mM min^{-1}, which at a ribozyme concentration of 7 μM corresponds to a k_{cat} of 21 min^{-1}. The highest initial rate that was measured directly corresponds to a catalytic turnover of approximately 6 min^{-1}. With these catalytic properties the 49nt Diels–Alderase ribozyme is among the faster

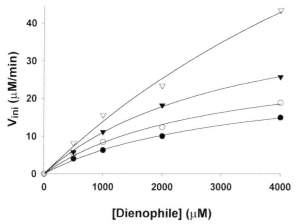

Fig. 5.3.5. Dependence of the rate of reaction on the concentration of dienophile at fixed diene concentrations ● 55, ○ 99, ▼ 190, ▽ 415 μM. Initial rates (V_{ini}) were obtained by monitoring the UV absorbance over the first 5% of conversion.

5.3 RNA as a Catalyst: the Diels–Alderase Ribozyme

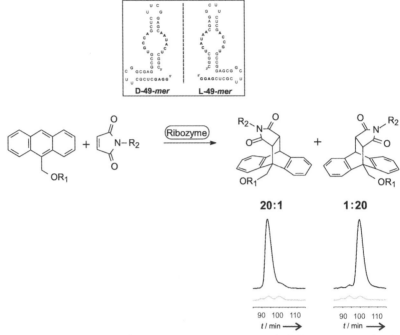

Fig. 5.3.6. HPLC analysis of Diels–Alder reactions catalyzed by 49nt D-RNA and the enantiomeric 49nt L-ribozyme. Samples were analyzed on a chiral column with UV detection at 210 nm. The light gray curve in the chromatograms corresponds to the uncatalyzed background reaction that occurs in the absence of ribozyme.

biomolecular catalysts for the Diels–Alder reaction. It compares favorably with catalytic antibodies raised against transition state analogs and with ribozymes that contain modified bases [11].

5.3.8
Stereoselectivity

RNA is a homochiral polymer. It occurs in nature only in the D configuration. We investigated how the chiral 49nt catalyst affects the stereochemistry of the reaction by analysis of the products by chiral HPLC (Figure 5.3.6). An enantiomeric excess (ee) of >95% was observed for the catalyzed reaction between anthracene-hexaethyleneglycol and N-pentylmaleimide.

For 9-anthrylmethanol, the ee was merely 16%, for anthracene-monoethyleneglycol it was 33%. Obviously, the size of the substituent in the 9-position plays an important role in determining the ee, whereas variation of the substituent on the maleimide does not lead to significant changes in enantioselectivity. For the 49nt Diels–Alderase it could be shown for the first time that an unnatural L-ribozyme has the opposite enantioselectivity to its natural mirror image counterpart. Other physical properties, e.g. rate acceleration, remain unchanged, of course [10].

5.3.9
Substrate Specificity and Inhibition

For the diene there are few structural requirements for acceptance by the 49nt ribozyme as a substrate [12]. The diene must contain three linearly annelated rings. Enlargement of the anthracene by introduction of methyl or hydroxymethyl groups at positions 1, 4, 9′, and 10 did not impede the action of the ribozyme. Substitutions at the 2- and 3-positions did, however, prevent the ribozyme from accelerating the Diels–Alder reaction. Thus, in a putative binding pocket there seems to be space in front and behind the anthracene but there is tight fit at the sides. The substituent in the 9-position is unnecessary for recognition, because ether, thioether, or alkyl residues are accepted; this also indicates no hydrogen bonds are formed between the ribozyme and the anthracene substrate.

For the dienophile some structural requirements are necessary for acceptance by the ribozyme as a substrate (Figure 5.3.7). It must be a five-membered ring and substituents on the reactive double bond are not permitted. The biotinyl residue is not necessary for recognition of maleimide by the 49nt RNA, because the initially used biotin maleimide can be truncated to a maleimide and an alkyl chain, both of

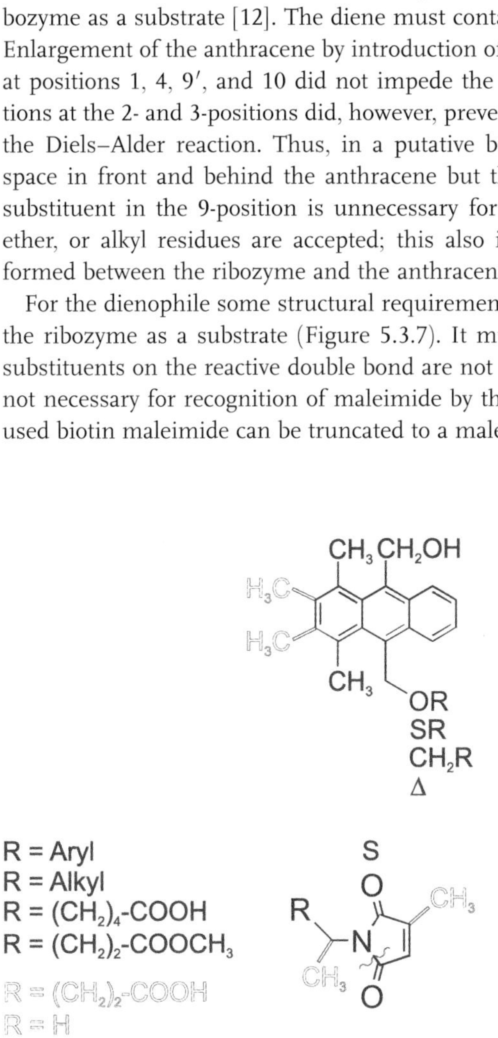
CH3 CH2OH
H3C
H3C
CH3
OR
SR
CH2R
Δ
R = Aryl
R = Alkyl
R = (CH2)4-COOH
R = (CH2)2-COOCH3
R = (CH2)2-COOH
R = H
S
O
CH3
R
N
CH3
O

Fig. 5.3.7. Summarized structural requirements of dienes and dienophiles for acceptance by 49nt ribozyme. Substitutions shown in black are tolerated whereas those denoted by empty letters are deleterious. Δ denotes removal of the whole substituent.

which seem to be essential for catalysis. The activity is maximum for N-pentyl-maleimide. No branching in the α-position is tolerated.

Because of the product-like nature of the transition state of the Diels–Alder reaction, interactions between product and ribozyme might provide valuable mechanistic information. Cycloaddition products and analogs were tested for their ability to inhibit the catalyzed reaction. The requirements for the effect of inhibitors are essentially the same as for substrates. The catalyzed reaction proceeds highly enantioselectively, and product inhibition was also found to be stereospecific. A R,R-product enantiomer inhibits the natural D-ribozyme approximately 20-fold more strongly than the S,S-product. For the mirror image L-ribozyme, the S,S product is the stronger inhibitor. Each ribozyme enantiomer is inhibited more effectively by its respective product. An anthracene with a strongly electron-withdrawing group in the 9-position is no substrate, yet is bound by the catalyst, as demonstrated by its inhibitory effect on the catalyzed reaction.

5.3.10
Conclusions

The mechanism by which the ribozyme achieves catalysis is currently under investigation. Aforementioned strategies, for example mutation analysis and variation of substrate properties, give some suggestions about the geometry of the ribozyme. According to the general mechanism of Diels–Alder reactions, the catalytic pocket must be wide enough to accommodate the two reactants in a stacked, coplanar arrangement. Intercalation of the diene is unlikely, because attack of the dienophile would be prevented. No heteroatom is necessary for recognition of the anthracene, making hydrogen bonding and metal ion coordination improbable. We assume the anthracene is bound primarily by stacking interactions. Behind the anthracene the binding site accommodates small substituents in the 4- and 10-positions. Recognition of the maleimide is also governed by hydrophobic interactions. The hydrophobic cleft quite unspecifically accommodates large substituents (benzyl, heptyl) unless there is no branching at the α-position. Thus, one possibility is that the rate of the reaction is accelerated by increasing the local concentration of both reactants in a hydrophobic pocket. Apart from this, the energy of the dienophile's LUMO could be reduced by withdrawal of electron density, as in Lewis-acid-catalyzed Diels–Alder reactions, which could easily be achieved via the maleimide's carbonyl oxygens. The RNA could, on the other hand, be increasing the energy of the diene's HOMO by destabilizing the diene and bending it toward the transition state.

There is only one other example of ribozyme-catalyzed carbon–carbon bond formation, and this also is a Diels–Alder reaction [13, 14]. The ribozymes isolated by the Eaton group accelerated the reaction of a tethered aliphatic diene with biotinylated maleimide approximately 800-fold, although no activity was observed toward the two free reactants. A second feature is that these ribozymes contain modified nucleotides and their activity depends on the presence of cupric ion.

The Houk group recently compared several artificial catalytic Diels–Alderase systems, including the ribozymes described here [11]. This study came to the conclusion that in none of these artificial systems there is a significant specific stabilization of the transition state. Acceleration arises predominantly from binding of the reactants, converting a second-order reaction of diene with dienophile into a first-order reaction of the termolecular complex of host, diene, and dienophile. The simultaneous presence of the two components of an intermolecular Diels–Alder reaction within the confined space of a cavity is the driving force that facilitates the reaction.

Our results, with further biochemical studies of the accessibility of certain regions of the RNA by use of enzymatic and chemical probing experiments or nucleotide analog interference mapping, could enable computation of a model structure. Results from initial NMR studies are indicative of structural changes that seem to depend on binding of the diene. The most important approach toward a comprehensive structure model is probably crystallization of the ribozyme as complex with substrate or inhibitor molecules. Because few structures of ribozymes have yet been determined, there is general interest in obtaining deeper insight into the binding geometry of small molecules to the RNA and possibly finding new elements of tertiary structure formation.

References

1 M. FAMULOK, A. JENNE, Curr. Chem. **1999**, *202*, 101–131.
2 A. JÄSCHKE, B. SEELIG, Curr. Opin. Chem. Biol. **2000**, 257–262.
3 J. R. LORSCH, J. W. SZOSTAK, Acc. Chem. Res. **1996**, *29*, 103–110.
4 G. F. JOYCE, Curr. Opin. Struct. Biol. **1994**, *4*, 331–336.
5 L. E. ORGEL, J. Theor. Biol. **1986**, *123*, 127–149.
6 R. F. GESTELAND, T. R. CECH, J. F. ATKINS, The RNA World, Cold Spring Harbor Laboratory Press, **1999**.
7 B. SEELIG, A. JÄSCHKE, Bioconjugate Chem. **1999**, *10*, 371–378.
8 B. SEELIG, A. JÄSCHKE, Chem. Biol. **1999**, *6*, 167–176.
9 D. BEBENROTH et al., manuscript in preparation.
10 B. SEELIG, S. KEIPER, F. STUHLMANN, A. JÄSCHKE, Angew. Chem. Int. Ed. **2000**, *39*, 4576–4579.
11 S. P. KIM, A. G. LEACH, K. N. HOUK, J. Org. Chem. **2002**, *67*, 4250–4260.
12 F. STUHLMANN, A. JÄSCHKE, J. Am. Chem. Soc. **2002**, *124*, 3238–3244.
13 T. M. TARASOW, S. L. TARASOW, B. E. EATON, Nature **1999**, *389*, 54–57.
14 A. JÄSCHKE, Biol. Chem. **2001**, *382*, 1321–1325.

B.23
SELEX: Systematic Evolution of Ligands by Exponential Enrichment

Andres Jäschke and Sonja Keiper

The method of in vitro-selection was developed in the early 1990s in the laboratories of Gold, Szostak, and Joyce. It is based on the repeated se-

5.3 RNA as a Catalyst: the Diels–Alderase Ribozyme

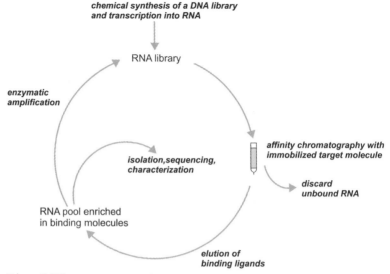

Scheme B.23.1.

lection of active species from nucleic acid libraries by separation from unreactive molecules and enzymatic amplification of the enriched library. Because of these two key steps the method is called SELEX (systematic evolution of ligands by exponential enrichment). Two aims can thus be attempted. The first is the selection of molecules with specific binding properties, called aptamers [1]. In this approach the target molecule of interest is immobilized on a solid matrix (Scheme B.23.1). The combinatorial nucleic acid library is then applied to the affinity matrix in a suitable buffer. Unbound nucleic acid molecules are washed away and the binding molecules are retained. These are later competitively eluted with a buffer containing the free target molecule. The eluted molecules undergo enzymatic amplification as a result of addition of a template-dependent polymerase. Thus, several thousand copies of each selected nucleic acid molecule are generated. This results in a library in which molecules with high affinity for the target are enriched. This process must be repeated several times until molecules with the desired properties dominate the enriched library. The members of this library are then identified by sequencing.

The second aim is selection of molecules with new catalytic function, called ribozymes. Two different approaches are used to find catalytic nucleic acids. One is to synthesize a transition-state analog (TSA) of the corresponding reaction [2]. The TSA is then used as target molecule in the affinity selection scheme described above. The selected aptamers are screened to find molecules that catalyze the respective reaction that proceeds via this transition state. This concept has been successfully used for catalytic anti-

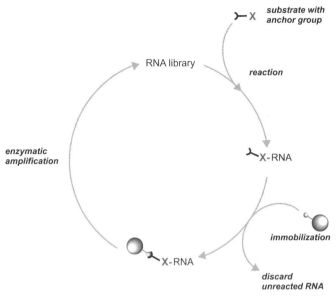

Scheme B.23.2.

bodies; unsuccessful attempts have usually been reported for nucleic acids. The most successful method for identification of nucleic acid catalysts is direct selection in which members of a combinatorial DNA or RNA library are isolated if reaction with a substrate X is accelerated (Scheme B.23.2). X carries an anchor group so that RNA molecules that react with the substrate acquire the anchor group and can subsequently be isolated by affinity chromatography on a suitably derivatized matrix [3]. Unreacted RNA does not bind to the matrix and can be removed by washing. Again this cycle is repeated until active molecules dominate the library. One limitation of this method is that the isolated molecules are self-modifiers (i.e. not true catalysts) and perform the reaction once only. In this book chapter a method is described that enables this limitation to be overcome, enabling generation of nucleic acid catalysts for true bimolecular organic reactions.

References

1 FAMULOK, M.; SZOSTAK, J. W. Angew. Chem. Int. Ed. Engl. **1992**, 31, 979–989.
2 PRUDENT, J. R.; UNO, T.; SCHULTZ; P. G. Science **1994**, 264, 1924–1927.
3 WILSON, C.; SZOSTAK, J. W. Nature **1995**, 374, 777–782.

5.4
Combinatorial Methods for the Discovery of Catalysts

Helma Wennemers

5.4.1
Introduction

The quest to produce compounds in ever better yields and higher enantiomeric purity makes the discovery of highly active and specific catalysts one of the most fundamental goals of chemistry. This task is a major challenge, because many catalytic reactions are governed by subtle mechanistic aspects that are difficult to rationalize. As a result, during the development of a new catalyst much time is often consumed by sequential testing of series of ligands with different yet similar substituents and by trials for the best reaction conditions, for example stoichiometry and concentration of the reactants, or choice of metal. Thus, a more empirical approach that mimics the natural catalyst development process of random mutations and selection of the fittest is needed to make the catalyst-discovery process more efficient. Combinatorial chemistry has the potential to fill this gap because it enables simultaneous generation of many compounds either by automated parallel synthesis or by the split-and-mix procedure (Box 11) (combinatorial chemistry can also be used to identify artificial receptors; see Chapters 2.3 and 3.1). The challenge is posed by the need for general and reliable screening methods that enable facile identification of catalytically active library members within a large library of potential catalysts. This challenge is even greater in testing of the members of split-and-mix libraries, because the intrinsic problem of the free diffusion of reactants and products must be met by smart screening methods.

This article summarizes methods that enable testing of parallel libraries and split-and-mix libraries for catalytic activity [1, 2]. It is designed to give an overview of screening methods that enable simultaneous testing *and* read-out of many potential catalysts. Testing of compound arrays that relies on offline analysis of a catalytic event by conventional analytical methods (HPLC, GC, MS, etc.) has been reviewed in several excellent articles and will be neglected here [1, 2]. Also, methods for testing of catalysts for gas-phase reactions (e.g. resonance-enhanced multiphoton ionization [3], scanning mass spectrometry [4], and photoacoustic detection [5]) will not be discussed here.

Highlights in Bioorganic Chemistry: Methods and Applications. Edited by Carsten Schmuck, Helma Wennemers.
Copyright © 2004 WILEY-VCH Verlag GmbH & Co. KGaA, Weinheim
ISBN: 3-527-30656-0

5.4.2
Testing of Parallel Libraries for Catalytic Activity

Parallel approaches have been successful in finding catalysts when the tested molecular diversity is limited to approximately one hundred compounds or when modern automation and miniaturization equipment enables rapid, high-throughput monitoring of many thousand reactions. Simultaneous testing of parallel libraries requires monitoring of a signal that is indicative of catalytic activity. Most methods rely either on measurement of the change in the reaction enthalpy by IR–thermography or on monitoring of the emergence or disappearance of color or fluorescence over the course of a reaction.

5.4.2.1
Colorimetric and Fluorescent Screening

Hydrosilation of Alkenes and Imines

Crabtree and coworkers used the disappearance of color to monitor the activity of 12 potential catalysts for hydrosilation reactions [6]. The group used alkenes and imines with ferrocenyl and pyrimidyl substituents at either end as colored reactants. On hydrosilation the color is bleached, because conjugation between the electron donor and acceptor groups is reduced. The bleaching is easily observed by eye and can be recorded quantitatively by means of a digital camera (Figure 5.4.1). By use of this method, the known Wilkinson catalyst was identified in a proof-of-concept experiment. A palladacycle, $[Pd\{(o\text{-tolyl})_2PC_6H_4\}OAc]_2$, usually used in Heck reactions, was also found to be catalytically active.

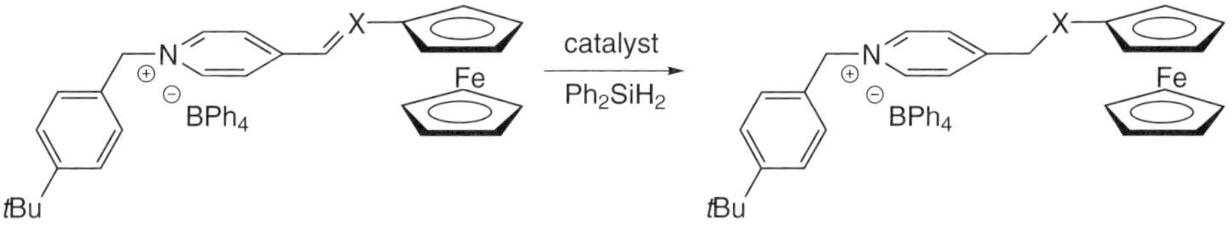

1a X = CH (deep purple)
1b X = N (dark blue)

2a X = CHSiHPh$_2$ (light yellow)
2b X = NHSiHPh$_2$ (light yellow)

Fig. 5.4.1. Bleaching of reactive dyes.

This method might suffer from lack in generality, because a reactive dye was used as a substrate. Control reactions proved that the relative order of reactivity of the identified hydrosilation catalysts is, however, identical when conventional unsaturated systems instead of the reactive dyes are employed.

Heck Reactions

The formation of a fluorescent product anchored to a solid support was used by Hartwig's group to monitor the capacity of 45 different phosphine and diphos-

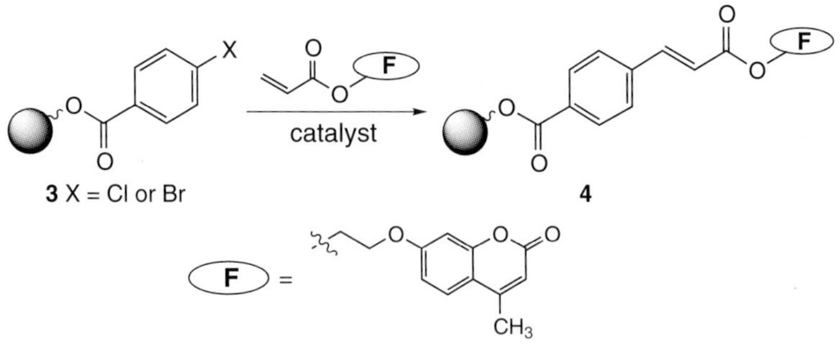

Fig. 5.4.2. Addition of fluorescent reaction partner.

phine ligands to form active catalysts with Pd(dba)$_2$ for Heck reactions [7]. The group used the coupling of a fluorescent alkene to a solid supported aryl halide as a test reaction (Figure 5.4.2). Unreacted fluorescent alkene is washed away and successful coupling is indicated by fluorescent beads.

Although reactivity on a solid support does not necessarily match that in homogeneous solution, the screening resulted in the identification of di(*tert*-butylphosphino)ferrocene and tri(*tert*-butyl)phosphine as highly effective ligands for Heck coupling of aryl chlorides and aryl bromides in homogeneous and heterogeneous media.

Allylic Alkylations

Morken and Lavastre used the formation of a colored side product to identify catalysts for the allylation of β-dicarbonyl compounds [8]. The researchers employed 1-naphthyl allyl carbonate **5** as an allyl source and the diazonium salt of fast red as an indicator. Formation of the active π-allyl complex furnishes CO_2 and 1-naphthoxide which deprotonates the 1,3-dicarbonyl compounds which can, in turn, react with the π-allyl metal complex. 1-Naphthol is the only species in the reaction mixture that can react with the diazonium salt **6** to generate the bright red azo dye fast red. Thus the red color is indicative of successful formation of the active π-allyl complex (Figure 5.4.3).

Combinations of eight different ligands and twelve different metal salts were screened for their efficiency to catalyze the allylation of β-dicarbonyl compounds. The assay identified not only the well known catalyst system Pd(OAc)$_2$ combined with a phosphine ligand but also the combination [{IrCl(cod)}$_2$] and *i*Pr-pybox or 1,10-phenanthroline as efficient catalysts. These are the first examples of non-phosphane iridium catalysts capable of allylic alkylations.

In contrast to the examples already discussed this method is based not on monitoring of the coloration or decoloration of the desired reaction product but on the formation of a colored product between a reagent-specific side product and a specific indicator.

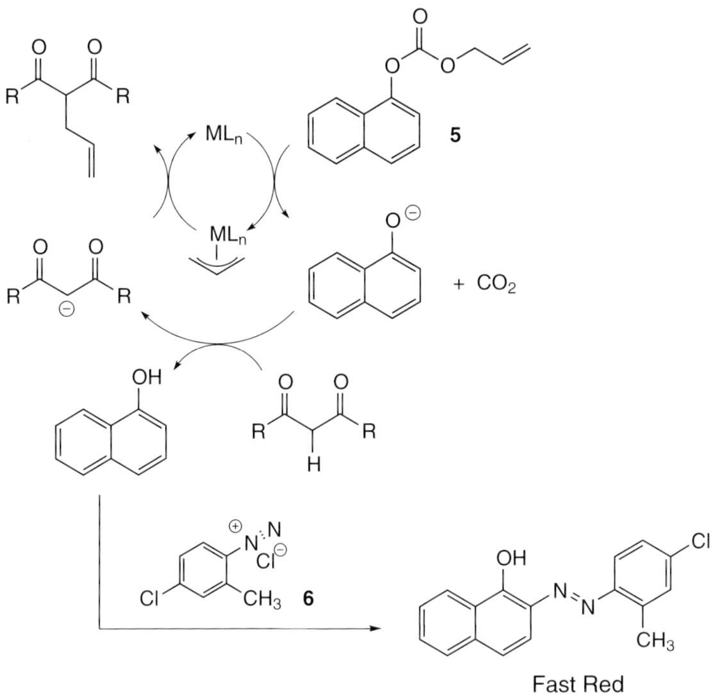

Fig. 5.4.3. Allylic alkylations.

5.4.2.2
IR–Thermography

Infrared cameras detect infrared radiation and enable recording of local spatial temperature changes with time. Because most chemical reactions have a measurable heat of reaction, $\Delta H_r°$, IR–thermography is a general tool for monitoring the progress of a catalytic reaction. Catalysts with the highest activity will evoke the largest temperature changes, because the temperature change is proportional to the turnover number of the catalyst and the heat of the reaction. Besides the testing of catalysts for gas-phase reactions [9, 10] and the testing of bead-supported catalysts, IR–thermography has been used to monitor the catalytic activity of parallel compound libraries in homogeneous liquid phases.

Reetz and coworkers tested catalysts for different reactions such as enantioselective acylation of a chiral secondary alcohol by lipases, the enantioselective ring opening of epoxides to non-racemic diols, and metathesis reactions [11, 12]. The two first examples are exothermic reactions and catalyst activity is revealed by "hot spots" in the IR image. The catalytic performance found by use of time-resolved IR–thermography correlated well with already known activity of the tested catalysts [11]. The metathesis reaction is particularly interesting, because it is the first example of the monitoring of endothermic reactions by means of an IR camera [12].

The rates of initiation of four known precatalysts for the metathesis reaction of 1,7-octadiene were examined by monitoring the temperature changes in the reaction wells. The IR thermographic pictures taken of three of the four reactions after one and two minutes revealed "cold spots" whereas the reaction compartment of the fourth, least active, precatalyst remained unchanged. The observation of cold spots implies the uptake of heat and indicates an endothermic effect. Although it was not possible to quantify the activity of the different catalysts, because the temperature decrease was also partially caused by evaporation of one of the reaction products, ethylene, relative catalyst activity was in good agreement with that from corresponding laboratory-scale reactions.

The general applicability of IR–thermography renders it a good tool for monitoring catalytic activity. The method is, however, limited by the lack of information on the structure of the reaction products formed.

5.4.3
Testing of Split-and-mix Libraries for Catalytic Activity

In parallel libraries each compound is located in a defined reaction container, for example, one well of a 96-well plate. Because the molecular diversity of split-and-mix libraries usually exceeds that of parallel libraries, many more potential catalysts can be tested in split-and-mix libraries. Their screening is, however, a much greater challenge if the ensemble of all the beads is tested simultaneously. In contrast to parallel libraries, in which each potential catalyst is located in a defined reaction container, reactants and products are free to diffuse in the solvent that surrounds all the beads and leave the catalyst – naturally – unchanged. The challenge has been met by:

1. IR–thermography;
2. the formation of insoluble colored reaction products;
3. fluorescent pH indicators;
4. the use of gels as reaction media; and
5. substrate-catalyst co-immobilization.

5.4.3.1
IR–thermography

Taylor and Morken extended the use of IR–thermography to the monitoring of the change in the heat of reaction on and in the surroundings of a bead carrying an active catalyst (Figure 5.4.4) [13]. In a search for acylation catalysts an encoded library of 3150 different potential nucleophilic catalysts was prepared by the split-and-mix procedure and tested for their acylation properties. The library beads were spread in a reaction solution of chloroform–ethanol–triethylamine–acetic anhydride, 40:6:6:3, and monitored with an IR camera. Whereas no detectable thermal

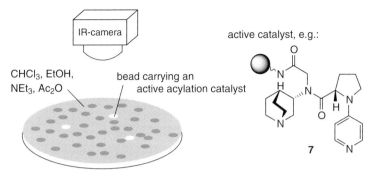

Fig. 5.4.4. On-bead IR–thermographic screening for acetylation activity.

change was observed for most of the beads, a few beads showed up as "hot spots". Separation and decoding of the sequences of the "hottest beads" revealed a diastereomeric pair of N-4-pyridylprolines to be the compounds on these beads. Kinetic experiments with the compounds underlined their high activity in comparison with other compounds which were present in the library but had no activity in the thermal screening.

5.4.3.2
Formation of Insoluble Reaction Products

In a trial to identify hydrolytically active members in a metal-complexed undecapeptide library, Berkessel and coworkers used the phosphate ester **8** as a substrate [14]. On hydrolysis the insoluble blue indigo dye **9** is formed and precipitates on beads that carry compounds with hydrolytic activity (Figure 5.4.5).

Catalytic activity was observed when a 625-membered undecapeptide library was complexed with Zr(IV). Beads with active catalysts were easily identified under a low-power microscope and the structures of the active peptides were analyzed by Edmann degradation. The activity of four Zr(IV)–peptide complexes was further proven and quantified in solution-phase experiments not only with the examined indigo derivative but also with the corresponding *para*-nitrophenyl phosphate. The group extended the screening for phosphodiesterase activity and enabled discovery of solid phase bound peptide–Eu^{3+} complexes as hydrolyzing agents [15].

5.4.3.3
Fluorescent pH Indicators

Miller and coworkers introduced pH-sensitive molecular sensors (Figure 5.4.6) for discovery of novel acylation catalysts [16]. Aminomethylanthracene was used as an acid sensor, because it undergoes photoinduced electron transfer (PET) when in

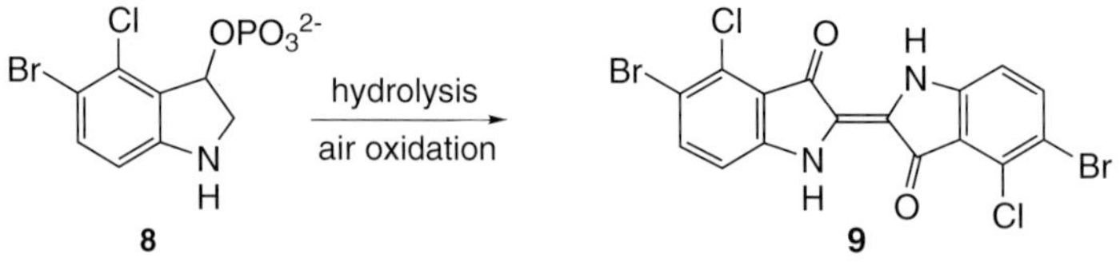

5.4 Combinatorial Methods for the Discovery of Catalysts

[Structures 8 and 9: hydrolysis, air oxidation]

9 turquoise, insoluble indigo dye

undecapeptide library: ⬤~~Phe–X–GlyGly–X–GlyGly–X–GlyGly–X

X = L-Arg, L-His, L-Tyr, L-Trp or L-Ser

active sequences:

H$_2$N-Ser-(Gly)$_2$-His-(Gly)$_2$-Arg-(Gly)$_2$-His-Phe-CO$_2$H
H$_2$N-Ser-(Gly)$_2$-Ser-(Gly)$_2$-Ser-(Gly)$_2$-His-Phe-CO$_2$H
H$_2$N-Ser-(Gly)$_2$-Arg-(Gly)$_2$-His-(Gly)$_2$-His-Phe-CO$_2$H

Fig. 5.4.5. Insoluble indigo as reaction product of hydrolysis reactions.

the form of the free amine but fluoresces when protonated. This fluorescence readout system can be applied both in parallel solution phase assays and in single-bead assays. In the later approach beads were co-functionalized with the pH sensor and the potential catalyst. The compounds were then screened for activity in acylation reactions of alcohols with acetic anhydride, a reaction that can easily be monitored, because acetic acid is released in the acyl transfer reaction and triggers the fluorescence response.

The testing of a 7.5×10^6 membered octapeptide library with fixed N-terminal π-(CH$_3$)-histidine (Pmh) and C-terminal alanine resulted in identification of catalysts with higher activity than DMAP for the acetylation of *sec*-phenylethanol [17]. The identified catalyst Boc-Pmh-L-Asn(Trt)-D-Val-L-His(Trt)-D-Phe-D-Val-D-Val-L-Ala-resin **10** then served as a parent compound for a second-generation library. This screening yielded catalysts with yet greater activity and specificity. Interestingly, all

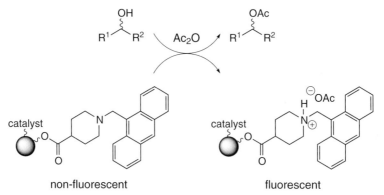

Fig. 5.4.6. pH-Sensitive fluorescence indicators.

catalysts were even more active and selective when tested in homogeneous solution. The catalysts with the highest activity also had the highest selectivity for one enantiomer of phenylethanol over the other. Thus although the assay was designed to screen for catalytic activity only it also identified, at the same time, catalysts with high selectivity.

5.4.3.4
Gels as Reaction Media

Gels enable slow diffusion of reactants and products and fix the beads in a defined position. The area around an embedded bead can thus be regarded as a mini-reaction compartment where reaction products remain long enough to be detectable. Miller proved the feasibility of the idea by preparing a PEGA-type polymer covalently linked to aminomethylanthracene, the fluorescent pH sensor [18]. Cross-linking of the polymer was accomplished around library beads functionalized with potential peptidic catalysts. When a library was tested for acetylation reactions (see above) fluorescent halos formed only on and around beads carrying acetylation catalysts, as sequencing of the peptides on the active beads revealed. Because the reaction matrix PEGA swells in different solvents (DMF, CH_2Cl_2, THF, H_2O, etc.) reactions can be performed in a wide variety of solvents.

Davis showed for aqueous reaction media that the principle can also be applied simply by using a low-melting-point agarose [19]. In proof-of-concept experiments the researchers embedded non-functionalized beads and beads functionalized with β-D-galactosidase in the agarose and left them to react with o-nitrophenyl-β-D-galactopyranoside, the substrate of the enzyme. The yellow color of the hydrolysis product, o-nitrophenolate, emerged only on and around beads functionalized with the hydrolytic enzyme. The group extended the study to the monitoring of a pH change, simply by adding a non-covalently bonded pH indicator (methyl red) to the agarose. Esterase activity, leading to the generation of acid, was monitored successfully by use of this method.

5.4.3.5
Substrate-Catalyst Co-immobilization

Our group has developed a method based on the immobilization of one reaction partner and each library member, the potential catalyst, on the same bead [20]. This can easily be accomplished by means of a bifunctional linker that carries on one end the reactant A and on the other end the library member (Figure 5.4.7). If the reaction partner B is labeled with, e.g., a dye, a fluorophore, or radioactivity a reaction between A and B leads to the covalent attachment of the marker to those beads that carry catalytically active library members. These beads are then readily detected with a low-power microscope.

The feasibility of the method was demonstrated by testing the members of a tripeptide library for their ability to catalyze the acylation of an alcohol (reactant A) by a dye-marked pentafluorophenyl (Pfp) ester (reactant B).

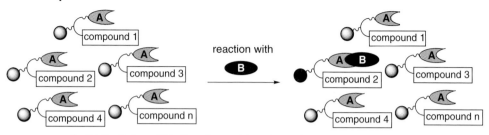

Fig. 5.4.7. Co-immobilization of potential catalyst and reactant (A) on the same bead. In the example only compound **2** is an active catalyst.

The library was designed not to contain amino acids that could be self-acylated, thus Ser, Thr, Lys, and Cys were not part of the library. His was incorporated as a known weak acylation catalyst and was expected to be selected. After reacting the library with the dye-marked Pfp ester, and thorough washing, several beads remained red, indicating covalent attachment of the dye to the resin beads. Analysis of the peptides on these red beads showed that each active bead carried at least one His. There is, therefore, no cross catalysis between different beads. The method can be used to search for catalysts of any bimolecular reaction in which one reaction partner can be attached to a solid support and the other labeled with a dye, a fluorophore, or radioactivity.

5.4.4
Conclusions

The rapid development of combinatorial screening methods has been accompanied by the development of ever more efficient high-throughput analysis technologies. These not only enable analysis of catalytic activity but also the determination of enantiomeric excess [2, 21]. Taking these developments together, research in this field can be expected to yield highly active and selective catalysts with structures that could have not been predicted by conventional means.

References

1 For reviews see: (a) S. DAHMEN, S. BRÄSE, *Synthesis* **2001**, 1431–1449; (b) H. WENNEMERS, *Comb. Chem. High Throughput Screening* **2001**, *4*, 273–285; (c) B. JANDELEIT, D. J. SCHAEFER, T. S. POWERS, H. W. TURNER, W. H. WEINBERG, *Angew. Chem.* **1999**, *111*, 2648–2689; *Angew. Chem. Int. Ed.* **1999**, *38*, 2494–2532; (d) R. H. CRABTREE, *Chem. Commun.* **1999**, 1611–1616; (e) K. D. SHIMIZU, M. L. SNAPPER, A. H. HOVEYDA, *Chem. Eur. J.* **1998**, *4*, 1885–1889.

2 M. T. REETZ, *Angew. Chem.* **2001**, *113*, 292–320; *Angew. Chem. Int. Ed.* **2001**, *40*, 312–329.

3 S. M. Senkan, *Nature* **1998**, *394*, 350–353.

4 (a) P. Cong, R. D. Doolen, Q. Fan, D. M. Giaquinta, S. Guan, E. W. McFarland, D. M. Poojary, K. Self, H. W. Turner, W. H. Weinberg, *Angew. Chem.* **1999**, *111*, 507–512; *Angew. Chem. Int. Ed.* **1999**, *38*, 483–488; (b) M. Orschel, J. Klein, H.-W. Schmidt, W. F. Maier, *Angew. Chem.* **1999**, *111*, 2961–2965; *Angew. Chem. Int. Ed.* **1999**, *38*, 2791–2794.

5 T. Johann, A. Brenner, M. Schwickardi, O. Busch, F. Marlow, S. Schunk, F. Schüth, *Angew. Chem.* **2002**, *114*, 3096–3100; *Angew. Chem. Int. Ed.* **2002**, *41*, 2966–2968.

6 A. C. Cooper, L. H. McAlexander, D.-H. Lee, M. T. Torres, R. H. Crabtree, *J. Am. Chem. Soc.* **1998**, *120*, 9971.

7 K. H. Shaughnessy, P. Kim, J. F. Hartwig, *J. Am. Chem. Soc.* **1999**, *121*, 2123.

8 O. Lavastre, J. P. Morken, *Angew. Chem.* **1999**, *111*, 3357–3359; *Angew. Chem. Int. Ed.* **1999**, *38*, 3163–3165.

9 F. C. Moates, M. Somani, J. Annamalai, J. T. Richardson, D. Luss, R. C. Willson, *Ind. Eng. Chem. Res.* **1996**, *35*, 4801–4803.

10 A. Holzwarth, H. W. Schmidt, W. F. Maier, *Angew. Chem.* **1998**, *110*, 2788–2792; *Angew. Chem. Int. Ed.* **1998**, *37*, 2644–2647.

11 M. T. Reetz, M. H. Becker, K. H. Kühling, A. Holzwarth, *Angew. Chem.* **1998**, *110*, 2792–2795; *Angew. Chem. Int. Ed.* **1998**, *37*, 2647–2650.

12 M. T. Reetz, M. H. Becker, M. Liebl, A. Fürstner, *Angew. Chem.* **2000**, *112*, 1294–1298; *Angew. Chem. Int. Ed.* **2000**, *39*, 1236–1239.

13 S. J. Taylor, J. P. Morken, *Science* **1998**, *280*, 267–270.

14 A. Berkessel, D. A. Hérault, *Angew. Chem.* **1999**, *111*, 99–102; *Angew. Chem. Int. Ed.* **1999**, *38*, 102–105.

15 A. Berkessel, R. Riedl, *J. Comb. Chem.* **2000**, *2*, 215–219.

16 (a) G. T. Copeland, S. J. Miller, *J. Am. Chem. Soc.* **1999**, *121*, 4306–4307; (b) E. R. Jarvo, C. A. Evans, G. T. Copeland, S. J. Miller, *J. Org. Chem.* **2001**, *66*, 5522–5527.

17 G. T. Copeland, S. J. Miller, *J. Am. Chem. Soc.* **2001**, *123*, 6496–6502.

18 R. F. Harris, A. J. Nation, G. T. Copeland, S. J. Miller, *J. Am. Chem. Soc.* **2000**, *122*, 11270–11271.

19 M. Müller, T. W. Mathers, A. P. Davis, *Angew. Chem.* **2001**, *111*, 3929–3931; *Angew. Chem. Int. Ed.* **2001**, *40*, 3813–3815.

20 P. Krattiger, C. McCarthy, A. Pfaltz, H. Wennemers, *Angew. Chem.* **2003**, *115*, 1763–1766; *Angew. Chem. Int. Ed.* **2003**, *42*, 1722–1724.

21 For excellent reviews see Ref. [2] and M. G. Finn, *Chirality* **2002**, *14*, 534–540.

Part 6
Methodology, Bioengineering and Bioinspired Assemblies

Highlights in Bioorganic Chemistry: Methods and Applications. Edited by Carsten Schmuck, Helma Wennemers.
Copyright © 2004 WILEY-VCH Verlag GmbH & Co. KGaA, Weinheim
ISBN: 3-527-30656-0

6.1
Linkers for Solid-phase Synthesis

Kerstin Knepper, Carmen Gil, and Stefan Bräse

6.1.1
Introduction

Polymers derived from polystyrene have been known since 1839 [1]. However, organic chemists started to regard polymers as valuable tools for organic synthesis only when Merrifield published the first solid-supported peptide synthesis [2]. Since then synthetic methods on polymeric supports have been employed for the preparation of synthetic oligonucleotides, oligosaccharides, and peptides. Particularly in the last decade the use of solid-phase synthesis has become vital in drug discovery. Large numbers of relatively small drug-like molecules with a molecular weight of 500 g mol^{-1}, which are potential new drugs, are readily available using solid-phase synthesis [3, 4].

The required elements of a solid-phase synthesis consist of three structural parts: the polymeric support (resin), a molecule or fragment, which eventually yields the final product, and the linker, which enables a suitable connection between the two parts [5]. Occasionally the linker can be located further from the polymeric support by use of a spacer (Figure 6.1.1).

Solid-phase synthesis always starts by attachment of the building block to the resin. The next step is a reaction which can be either transformation on the attached building block or coupling of another molecule. After this, a washing process enables simple removal of excess reagents or coupling partners. Reiteration of the sequence of reaction and washing is usually feasible. Finally, the compound is released by cleavage from the resin. Major advantages of solid phase organic synthesis (SPOS) are the easy isolation of the molecules at each step by simple filtration and the ease of automation (Figure 6.1.2). The pseudo dilution is also effective, because reactive functional groups are locally separated. The main disadvantages are the extra effort required to develop solid phase routes and limited analytical methods. Recently, however, new analytical methods have been developed, overriding this drawback [6].

Definition: The International Union of Pure and Applied Chemistry (IUPAC) defines a linker as a "Bifunctional chemical moiety attaching a compound to a solid support or soluble support which can be cleaved to release compounds from

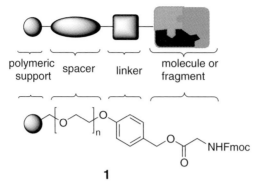

Fig. 6.1.1. Components of polymeric synthesis.

the support. A careful choice of the linker allows cleavage to be performed under appropriate conditions compatible with the stability of the compound and assay method" [7].

The starting material for a solid-phase synthesis is attached reversibly to a linker, which is bound either directly or via a spacer to the resin (usually with divinylbenzene cross-linked polystyrene). The choice of the linker is therefore very important in the design of a solid-phase chemical route. They must be stable in the presence of reagents and enable cleavage under mild conditions. Traditionally, linkers were designed to release molecules with one particular functional group and hence acted more or less as bulky protecting groups [8]. Initially, solid-phase

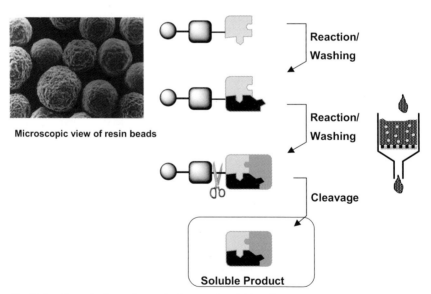

Fig. 6.1.2. General scheme for synthesis on an insoluble polymeric support.

synthesis was mainly used to obtain peptides and the release of carboxylic acids and amines, which are essential for peptide synthesis, has therefore been studied extensively. However, small drug-like molecule libraries require linkers that are more versatile [9].

A particular linker should resist the entire synthetic sequence without bias to the diversity or structure of the target compound library and without limiting the chemical methodology. None of the reaction conditions used for construction of the building blocks should lead to a premature cleavage (orthogonality principle). Cleavage from the resin must be as mild as possible to ensure the product will not be affected. In general, linkers for various kinds of building blocks are known and different cleavage strategies have been developed, for example photocleavage, safety catch, and traceless linkers [10].

6.1.2
General Linker Structures

6.1.2.1
Immobilization of Molecules

The anchoring of molecules to a resin can usually be realized by two different strategies [8]. Direct loading is clearly the most straightforward technique for setting up solid-phase organic synthesis. A molecule with a reactive or potentially reactive functional group is coupled directly to the preformed linker. This strategy is useful if the linker and the building block can be coupled efficiently. Coupling rates >90% are essential at this point. Successful examples are, for instance, the formation of amide bonds, reductive amination reactions, alkylation reactions (including Mitsunobu reactions), and olefin metathesis. Because building blocks can be used without an additional purification step, this method is especially suitable for anchoring libraries of starting materials and/or automated synthesis attachment of molecules to a particular resin is highly dependent on the nature of the linker. Whereas simple and rapid mixing of reagents is sometimes sufficient to drive the attachment to completion, occasionally tedious monitoring is necessary. Sometimes attachment proceeds under similar conditions to those of detachment, for example formation of ketals, for which excess reagents are required to drive the reaction to completion.

In a second general method for attachment, the building block can be coupled to give the preformed linker in solution. The thus formed fragment, which is called the "handle" (see, for example, Ref. [11]; Figure 6.1.3), can then be activated for attachment on to the resin using the coupling strategies outlined above. Although this method requires an additional step in solution, the purity of the building block can be enhanced. This handle approach can be particularly favorable when the activated linker tends to decompose (for example silyl chlorides) or can be formed only in moderate yields and purities. In automated synthesis, however, the required solution-phase step is clearly a drawback.

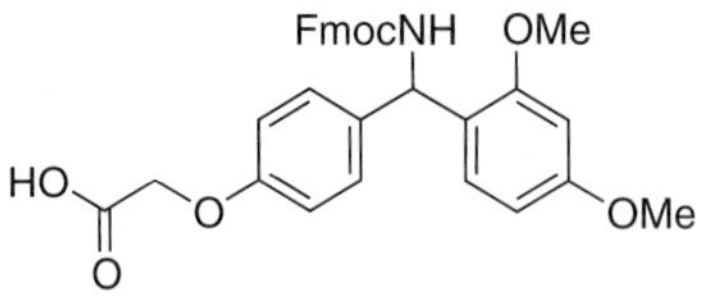

Fig. 6.1.3. The unloaded Knorr linker **2** as a handle.

6.1.2.2
Spacers

Linkers can be attached directly to the resin or might be located further from the polymeric backbone by use of spacers. These bifunctional constructs can either be built sequentially or attached via a handle approach. The spacer acts as a connection to give the building block more mobility; the kinetics for a given reaction might thus be superior to those for the corresponding resin without spacers [12]. Large spacers also substantially change the physical properties, for example swelling behavior, as demonstrated with Tentagel resin [13], which consists of poly(ethylene glycol)-grafted polystyrene. It must, however, be remembered that use of spacers not only requires an additional synthetic step, it usually results in a decrease of loading capacity, and the spacer must be as robust as the linker towards the reaction conditions applied to the bead. For example, Tentagel resin tends to lose attached building blocks, because of cleavage of the ethereal poly(ethylene glycol) (PEG) moiety.

For characterization by NMR spectroscopy of compounds attached to a polymeric support flexible spacers are advantageous, because of the increased mobility of the substrate and the reduction of line broadening usually observed in NMR spectra of polymers.

Usually, it is different to make a clear distinction between linker and spacer. The linker is the minimum part of the resin required for the functional cleavage (for silyl linkers it is the silyl group, for trityl linkers it is the triphenylmethyl moiety, and for the triazene linker it is the 1-aryltriazenyl group, etc.). The spacer is, therefore, the part between the linker and the resin as depicted in Figure 6.1.1.

6.1.3
Linker Families

Linkers are usually categorized according to the kind of functional group or substrate class they can selectively immobilize (linkers for carboxylic acids, alcohols, amines, etc.). Because a variety of types of linker is available for solid-phase synthesis, many belong to certain well-established classes of protecting group (Table 6.1.1) and can therefore be grouped into linker families. The members of each family have certain reactivity patterns in common.

Tab. 6.1.1. Common protecting groups and the analogous linker families.

Protecting group in the liquid phase	Functional group protected	Linkers or linker families
Benzyl	Alcohols, Esters	Benzyl-type linkers
Allyl	Amines	Allyl-type linkers
Cbz (Z)	Amines	Carbamate-based linkers [14]
Alloc	Amines	Allyl-type linkers
Boc	Alcohols	Boc-type linker [15]
Silyl ether	Alcohols	Silyl-type linkers [16, 17]
Alkyl ester	Carboxylic esters	
SEM (trimethylsilylethoxymethyl)	Alcohols	SEM linker [18]

Whereas a linker is the chemical structure essential for loading and cleavage of a particular functional group, a linker system provides the whole procedure for the attachment to and cleavage from the resin.

Because of an increased demand for flexible anchoring of molecules, other new families of linkers such as the sulfur linkers and triazene linkers have emerged. The most abundant type of linkers developed so far are based on benzylic-type groups.

6.1.3.1
Benzyl-type Linkers

Benzyl-type linkers are the most common anchoring groups for various kinds of functionality. Esters, amides, amines, alcohols, and thiols, in particular, can be immobilized by this linker family. This was demonstrated by Merrifield [2] and Wang [19] and is the starting point of modern linker development. Benzylic linkers are typically cleaved by strong acids (for example trifluoroacetic acid, TFA), which cause protonation and subsequent elimination. A nucleophilic scavenger usually quenches the resonance-stabilized cation thus formed.

The prototype of a functional group with an appropriate breakable bond is the Wang resin **3**, which contains a 4-hydroxybenzyl alcohol linker moiety. The benzyl alcohol hydroxyl group can be functionalized using either electrophilic or nucleophilic substrates (Scheme 6.1.1) to give a benzylic linkage. It is very stable in many reactions, but can be cleaved by acids such as trifluoroacetic acid or HF. Acids, alcohols, esters, and amides can be obtained as products after cleavage.

An increase of acid lability can thus be achieved by stabilization of this intermediate, for example by *o*- and *p*-substitution of methoxy groups on to the ring [20]. This has been demonstrated in the development of the SASRIN resin (**6**, super acid sensitive resin: 1% TFA cleavable) [21, 22] with one additional alkoxy group related to the Wang resin **3** and the HAL linker (**7**, hypersensitive acid-labile: 0.1% TFA cleavable) [23] having two additional alkoxy groups (Figure 6.1.4). In addition, benzyl-type linkers can be cleaved by ammonolysis [24], light, metal salts [25], and

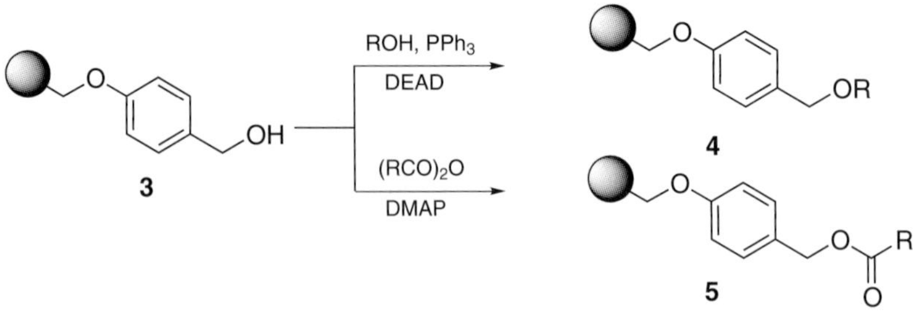

Scheme 6.1.1. Loading of the Wang resin **3**.

oxidizing agents (for example, Wang resin with H_2O_2 [26] or DDQ [27]). The introduction of nitro groups on to benzyl-type linkers leads to photolabile systems.

The Rink resin **8**, with an electron-rich benzhydryl moiety, is particularly useful for attachment of a variety of functionality, for example primary amines [28] (Scheme 6.1.2). Loading of the amine can be achieved via the corresponding chloride or triflate.

Two linkers particularly suitable for peptide amides and cyclopeptides are the peptide amide linker (PAL) and the backbone amide linker (BAL) [29] (Scheme 6.1.3). With these electronic and steric factors enable acylation of secondary benzylamines and relatively mild cleavage.

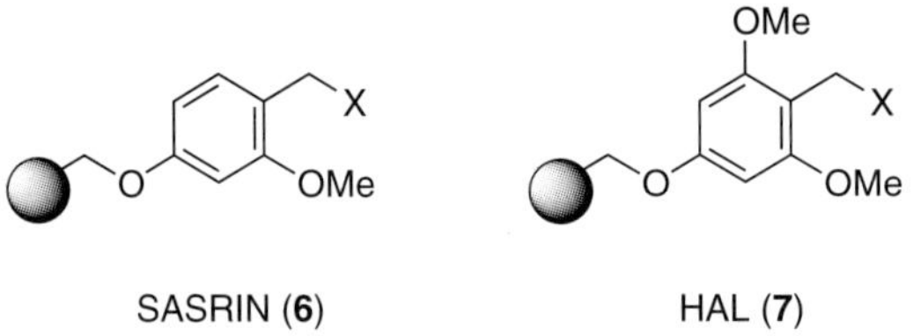

Fig. 6.1.4. The SASRIN and HAL linkers.

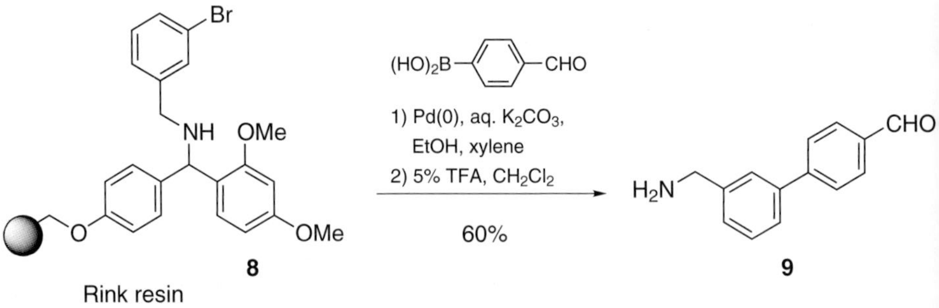

Scheme 6.1.2. The Rink resin as a linker for primary amines.

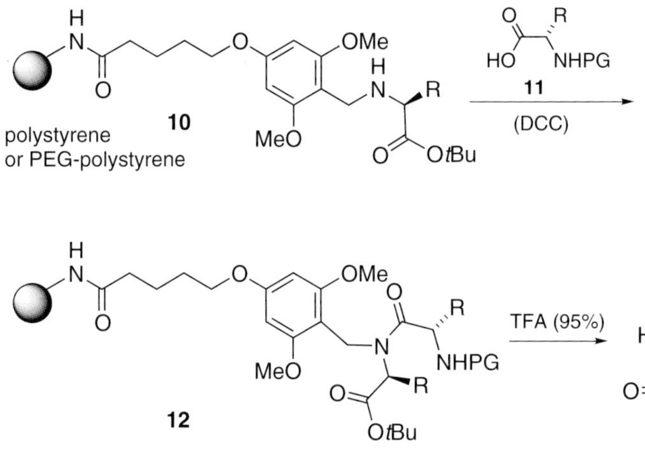

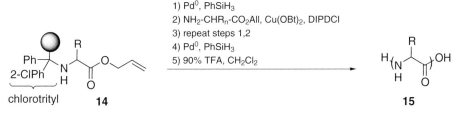

Scheme 6.1.3. Backbone amide linkage using the PAL linker.

6.1.3.2
Trityl Resins

Trityl resins are particularly suitable for immobilization of nucleophilic substrates such as acids, alcohols, thiols, and amines. They are quite acid-sensitive and are cleavable even with acetic acid; this is useful when acid-labile protecting groups are used. The stability of trityl resin can be tailored by use of substituted arene rings, as shown by chlorotrityl resin, which furnishes a more stable linker than the trityl resin itself. Steric hindrance also prohibits formation of diketopiperazines during the synthesis of peptides. Orthogonality toward allyl-based protective groups was demonstrated in the reverse solid-phase peptide synthesis of oligopeptides [30] (Scheme 6.1.4).

Scheme 6.1.4. Reverse (N → C) directed solid-phase peptide synthesis with the 2-chlorotrityl resin.

6.1.3.3
Allyl-based Linkers

Allyl-based linkers are particularly suitable for attachment of carboxylic acids, which can be detached by metal catalysis, in particular by means of a palladium

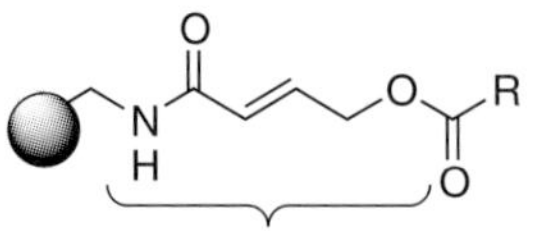

HYCRAM (16)

Fig. 6.1.5. The HYCRAM linker.

catalyst. The advantages of linker cleavage by palladium-catalysis are the mild reaction conditions and the orthogonality to a variety of protecting groups. Kunz et al. [31, 32] developed the first and most simple linker to use the π-allyl detachment strategy. Starting from 2-bromocrotonic acid, attachment to an amino group on a resin and further reaction with the cesium salt of an appropriate protected amino acid or peptidic structure yields HYCRAM resin (**16**, hydroxycrotonylamide) [33] (Figure 6.1.5). The allylic cleavage proceeds with Pd(PPh$_3$)$_4$ and morpholine or hydroxybenzotriazole as nucleophiles [34].

The readily available HYCRON (hydroxycrotyl-oligoethylene glycol-n-alkanoyl) linker [35] is based on a similar concept, except that a handle comprising an amino acid and a preformed linker has been used to minimize the risk of racemization on cleavage. Stability toward unwanted nucleophilic cleavage was greater than for the HYCRAM linker. Incorporation of a β-alanine moiety facilitates monitoring of the reaction. An example of the use of the allyl linkers is the synthesis of peptide nucleic acids (PNA) [35] (Scheme 6.1.5).

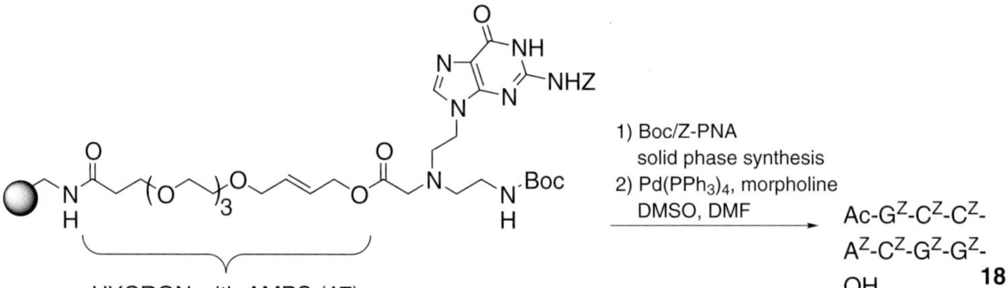

HYCRON with AMPS (17)

1) Boc/Z-PNA solid phase synthesis
2) Pd(PPh$_3$)$_4$, morpholine DMSO, DMF

Ac-G^Z-C^Z-C^Z-A^Z-C^Z-G^Z-G^Z-OH   **18**

Scheme 6.1.5. Synthesis of PNA using the HYCRON linker.

6.1.3.4
Ketal Linkers

Ketals and their corresponding sulfur analogs are well-established protecting groups in solution-phase synthesis. The most versatile ketal linker for solid-phase organic synthesis is the tetrahydropyranyl (THP) linker [36]. This linker enables the attachment and detachment of alcohols, phenols, and nitrogen functionality under acidic conditions. Ketal-type linkers are stable towards bases and organometallic reagents. An example of the synthesis of complex molecules using THP-

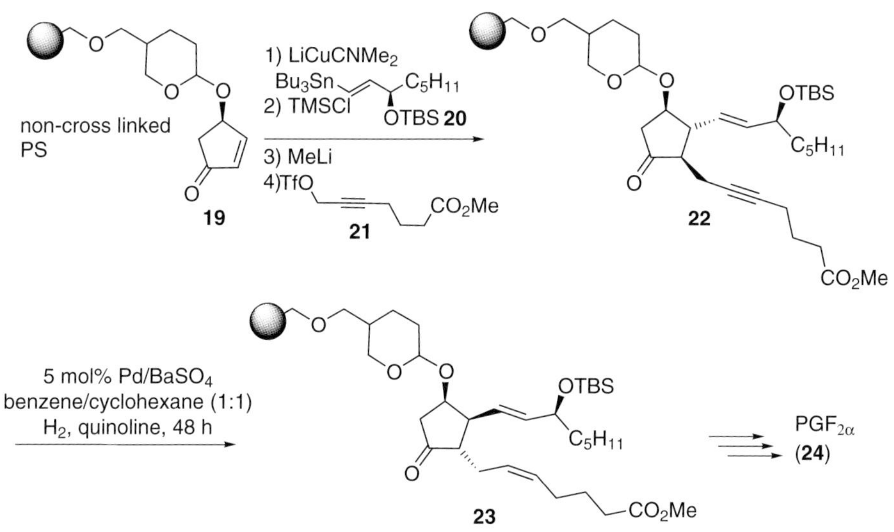

Scheme 6.1.6. Prostaglandin synthesis on a solid phase.

type linkers is the synthesis of prostaglandins on non-cross-linked polystyrene [37] (Scheme 6.1.6).

6.1.3.5
Ester and Amide Linkers

Ester and amide moieties are, apart from being used in the benzyl and allyl linker-type, are also suitable for attachment of building blocks (Figure 6.1.6).

Carboxylic esters have been released by the action of alkoxides on ester resins. Cleavage is usually performed by use of methoxide in methanol (for example Ref. [38], Scheme 6.1.7). A drawback is certainly the need to remove excess of metal salts and/or aqueous work-up. Alternatively, esters might be cleaved under acidic conditions.

Arylhydrazides can serve as safety-catch linkers for C-terminal carboxylic acids, amides, or esters. Cleavage proceeds via oxidation with copper(II) salts and subsequent cleavage of the diazenyl moiety by means of a nucleophile [39] (Scheme 6.1.8).

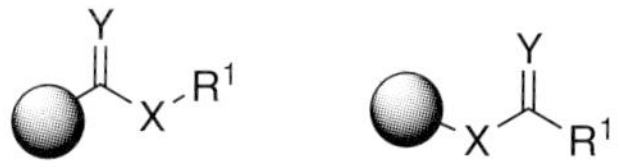

X = O, NR^2, N=N; Y = O, S, NR^2

Fig. 6.1.6. Ester linkers: general structures.

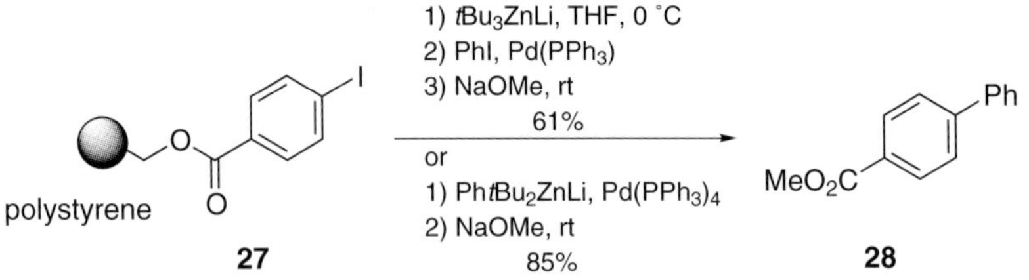

Scheme 6.1.7. The use of esters as linkers for benzoic acid derivatives.

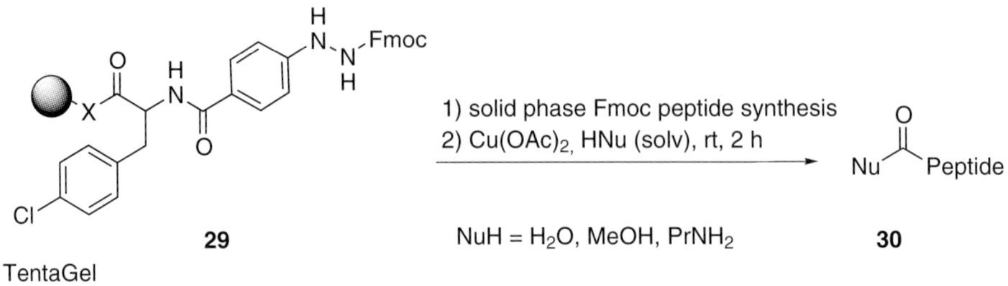

Scheme 6.1.8. Detachment of peptides from hydrazide resins.

6.1.3.6
Silicon- and Germanium-based Linkers

The electronic and steric properties of silicon compounds have been used in many applications for the design and use of new linker types. The robustness of silicon linkers against basic and organometallic reagents makes them especially suitable for solid-phase organic synthesis. Cleavage can be effected by means of electrophiles, preferably protons (TFA). A special feature of silyl linkers is their sensitivity to fluoride ions; this makes them ideally orthogonal to the variety of other functionality present in the molecule. Fine-tuning of their electronic and steric properties is possible by use of different substituents on the silicon atom (e.g. trimethylsilyl and *tert*-butyldimethylsilyl).

The first traceless linkers for arenes were silyl linkers. This linker type was used in the synthesis of a benzodiazepine library, a milestone in the solid-phase synthesis of small organic molecules (Scheme 6.1.9). Synthesis of the linker involves lithium–halogen exchange and, after chlorosilane attachment, coupling with an

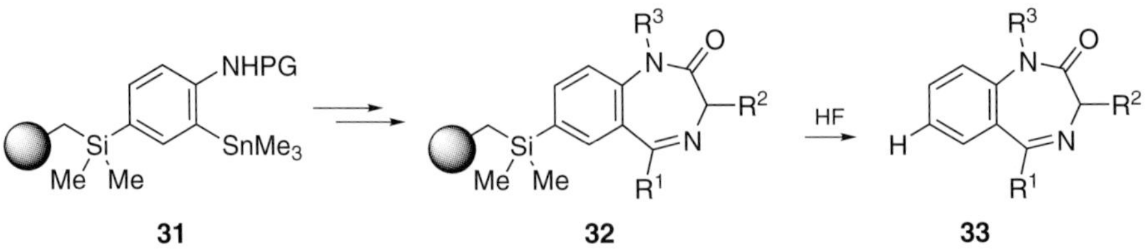

Scheme 6.1.9. Synthesis of a benzodiazepine library using a silyl linker.

aryl halide. Because the silyl arene can be cleaved in the unwanted direction to give silylated arenes, further improvement led to the development of a germanium linker [40]; this gives rise to the formation of pure material. In addition, cleavage of these linkers can be accomplished by electrophiles, for example iodine, bromine, and chlorine, to yield halogenated residues thus converting this linker into a multiple cleavage linker system [41].

6.1.3.7
Boron Linkers

Boronates have been used in a variety of linker types either as linkers for diols [42] or as precursors for metal-mediated cleavage. A boronic acid ester, which contains an aryl iodide moiety attached by an appropriate tether, can act as an intramolecular arylation agent. A polymer-bound precursor furnished a macrocyclic constrained β-turn peptide mimic *via* biaryl coupling, leading to cleavage [43] (Scheme 6.1.10).

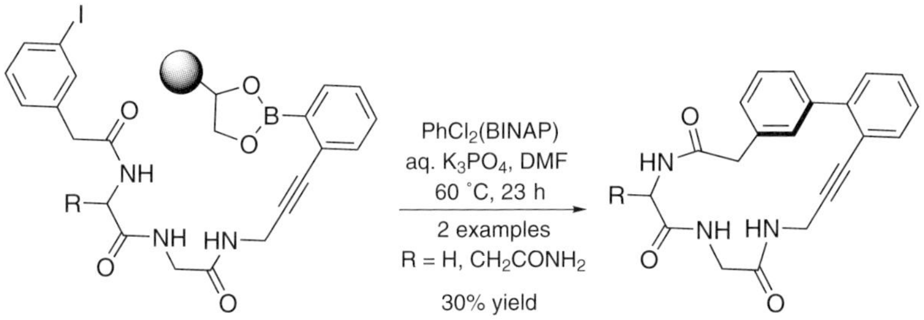

Scheme 6.1.10. Intramolecular cleavage Suzuki coupling.

A set of modern linkers is based on sulfur, stannane and selenium chemistry. A reason for this popularity is obviously because these elements can be tailored favorably for use as fragile points of attachment.

6.1.3.8
Sulfur Linkers

Sulfur has been used in linkers such as thioethers, sulfoxides, sulfones, sulfonic acids and their corresponding derivatives. A safety-catch linker for amines is based on 2-(thiobenzyl)ethylcarbamates [44]. Linkage is performed with preformed handles containing ethenyloxycarbonyl-protected amines **37**. Attachment to thiomethylated polystyrene **38** is performed under conditions involving radicals. Cleavage was performed with an oxidizing agent, which forms the retro Michael substrate (Scheme 6.1.11).

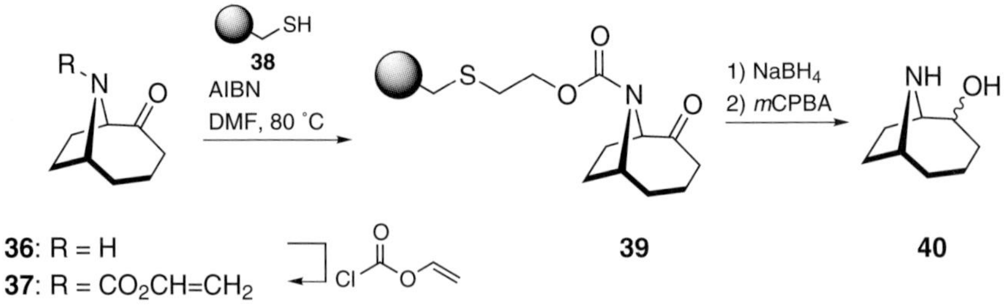

6.1 Linkers for Solid-phase Synthesis

36: R = H
37: R = CO$_2$CH=CH$_2$

Scheme 6.1.11. Thiobenzylethylcarbamates as linkers for amines.

6.1.3.9
Stannane Linkers

Stannanes have become prominent in multifunctional anchoring groups. A polymer-bound tin hydride **41** has been used to hydrostannylate alkynes under the action of palladium-catalysis to give polymer-bound alkenylstannanes **42**. These alkenyl stannanes have been employed in intermolecular [45] and intramolecular Stille reactions [46]. Alkenylstannanes can also undergo protonation to give alkenes **44** in a traceless fashion. This linker is therefore multifunctional (Scheme 6.1.12).

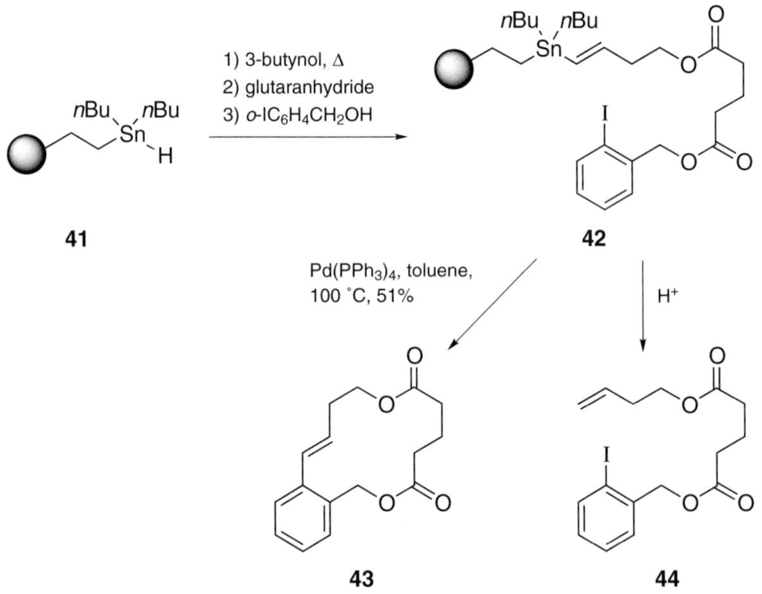

Scheme 6.1.12. The stannane linker for Stille reactions.

6.1.3.10
Selenium Linkers

The selenium–carbon bond is, because of its weakness ($E = 217$ kJ mol^{-1}), prone to homolytic cleavage to give radicals; this provides ample possibilities for design of new linkers [47, 48]. Starting from polystyrene a variety of steps including selenation with selenium powder or MeSeSeMe give rise to the formation of selenium resins which can be alkylated to give selenoethers **47**. Traceless cleavage yielding alkanes **48** can be conducted by reduction with tributyltin hydride, whereas formation of alkenes **49** can be observed after mild oxidation (Scheme 6.1.13). This linker holds promise for wide applicability, because the starting materials (alkenes, alkyl halides) are readily available, although toxicity of reagents and starting materials must be considered.

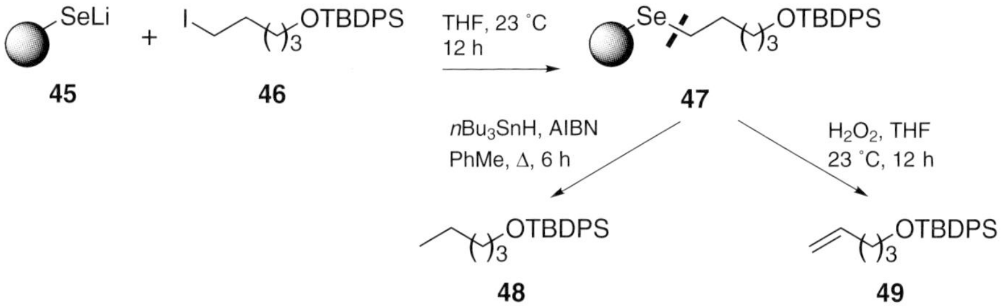

Scheme 6.1.13. The selenium linker for alkanes and alkenes.

6.1.3.11
Triazene Linkers

The chemistry of diazonium salts provides tremendous opportunities for construction of a wide range of aromatic compounds. Triazenes, which have been used as traceless linkers for arenes [49], not only provide interesting new possibilities for activation of the ortho position of the arenes, they are also ideal synthons for diazonium salts. Triazenes are stable towards daylight, moisture and bases including alkyl lithium reagents, reducing agents, and oxidizing reagents; they are, however, cleaved by Brønsted acids and certain Lewis acids into diazonium salts and amines. Two linkers based on triazene chemistry have been developed. Whereas the T1 linker system consists in 3,3-dialkyl-1-aryl triazenes bound to the support *via* the alkyl chain [49] (Scheme 6.1.14) the T2 linker family is based on immobilized aryl diazonium salts [50].

The triazene T1 linker has been successfully used as a linker for arenes. Approximately 100 different anilines **50** have been immobilized and functionalized on the beads. Mild acidic cleavage of the triazenes yields the amine resin **52** and the modified aryl diazonium salts **55**. The latter can be transformed into a variety of

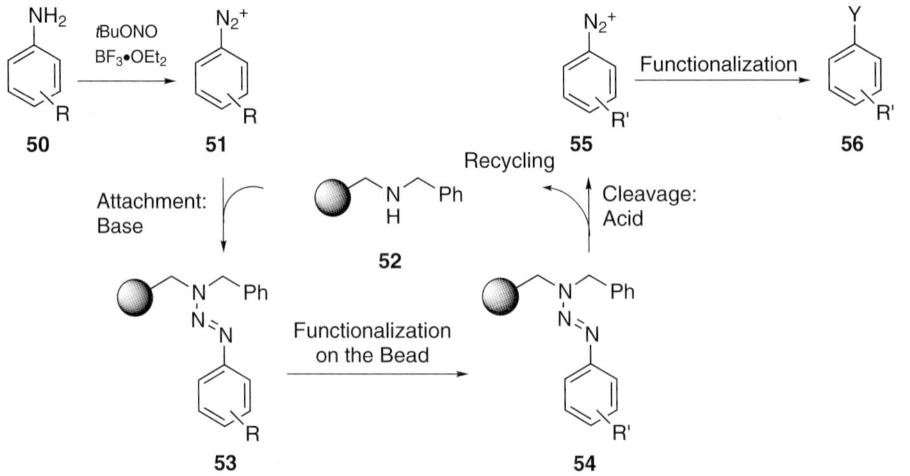

6.1 Linkers for Solid-phase Synthesis

Scheme 6.1.14. Concept of the T1 linker.

different products yielding modified arenes in high yields and purities directly at the cleavage step (Scheme 6.1.15).

Diazonium salts, for example, can be reduced to hydrocarbons **62-H** in THF with the aid of ultrasound, which facilitates this reduction because of a radical pathway. An alternative reagent for this reduction is trichlorosilane, which is not only a source of traces of hydrochloric acid, which cleaves the triazene moiety; as a hydride donor it can also reduce the diazonium ions cleanly. Addition of methyl iodide to a triazene resin at elevated temperature (110 °C) gives rise to aryliodides **62-I** (Nu = I), whereas arylhalides **62-X** (X = Cl, Br, I) are readily available by the action of lithium halides in the presence of an acidic ion-exchange resin or with the corresponding trimethylsilyl halide at room temperature. A mixture of acetic anhydride and acetic acid produces phenol acetates **62-OAc**. Although quite flexible in the range of possible electrophiles that can be employed, the most striking feature was the development of a cleavage–cross coupling strategy [51]. Starting from modified triazene resins, a one-pot cleavage–cross-coupling reaction was conducted with two equivalents of trifluoroacetic acid in MeOH at 0 °C to give a diazonium ion. In-situ coupling with a variety of alkenes **64** in the presence of catalytic amounts (5 mol%) palladium(II) acetate furnished the corresponding cross coupling products **65** in excellent yield and purity.

In the examples above, the diazonium group, on cleavage from the resin, is lost as dinitrogen. A suitable nucleophilic ortho substituent, however, favors cyclization to give heterocyclic structures. Benzotriazoles **61**, for example, are accessible from o-aminoaryl-substituted triazenes [52]. Other heterocyclic systems, for example cinnolines, are available, from o-alkynylaryl triazenes, by a Richter cleavage reaction strategy [53]. Cleavage was conducted with aqueous hydrogen chloride or hydrogen bromide in acetone or dioxane at room temperature to afford cinnolines **60** (Scheme 6.1.15).

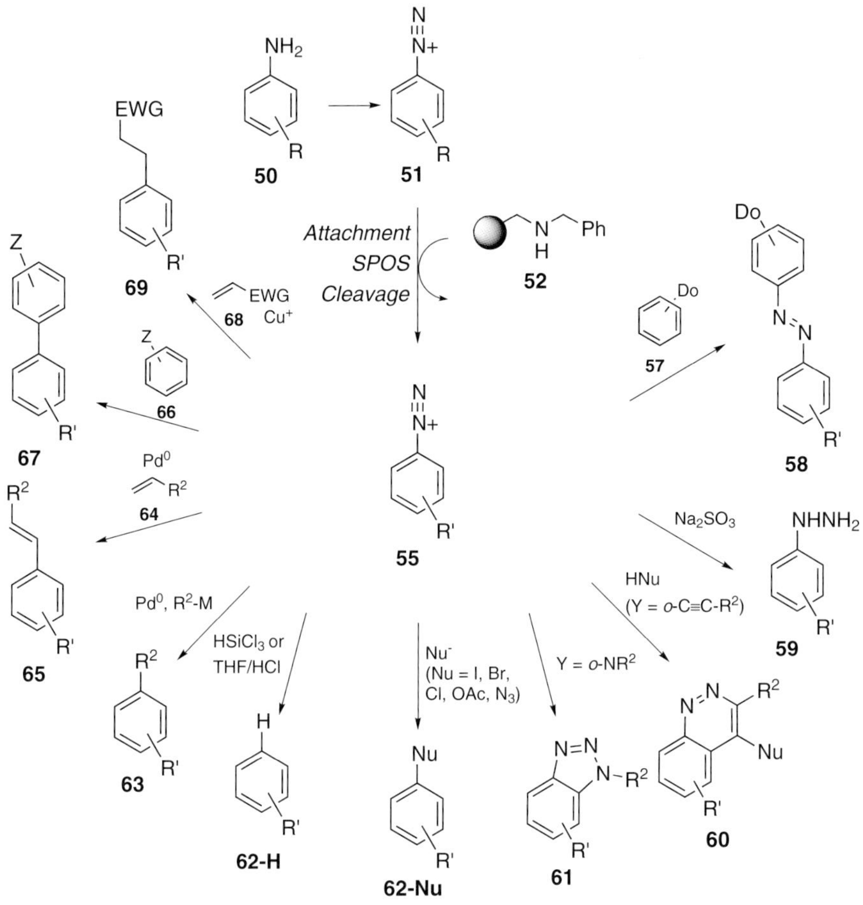

Scheme 6.1.15. Possibilities of the T1 triazene linker.

Whereas the T1 linker involves immobilization of a diazonium salt on an amine resin, the T2 linker is the reversal of this concept. An immobilized diazonium salt **73** was prepared from Merrifield resin **70** in two steps; subsequent addition of primary and secondary amines generated triazenes **74**. Attachment of hydroxylamine, hydrazines, sulfoximines, and phenols (to give azo coupling products) proceeds equally well (Scheme 6.1.16).

Secondary amines can be cleaved from the resin either directly or after modification. Primary amines can be derivatized on the free N–H function and can therefore be modified to an array of products. Thus, ureas **81** [54], thioureas **80**, guanidines **79**, and carboxamides **82** were prepared in excellent yield (Scheme 6.1.17).

Whereas cleavage of trisubstituted triazenes gives rise to secondary amines in excellent yields, cleavage of disubstituted triazene **77** gives rise to aliphatic diazonium salts. The aliphatic diazonium ion formed undergoes substitution with nu-

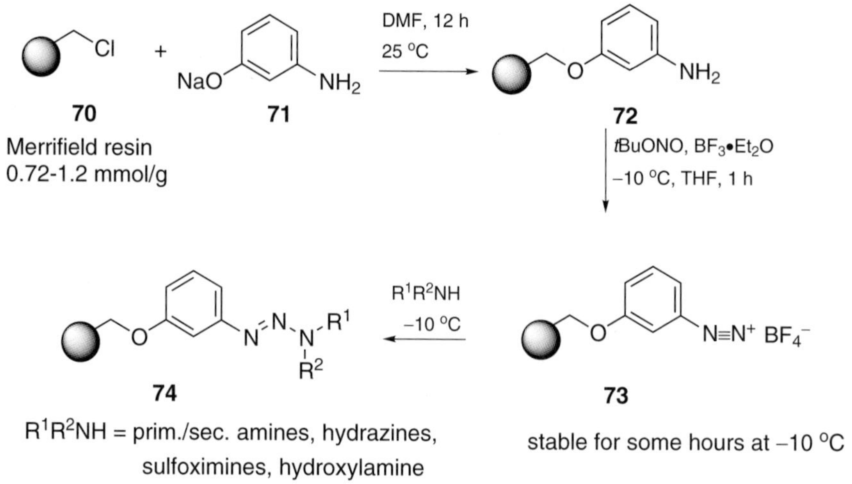

6.1 Linkers for Solid-phase Synthesis

Scheme 6.1.16. Synthesis of the T2 linker.

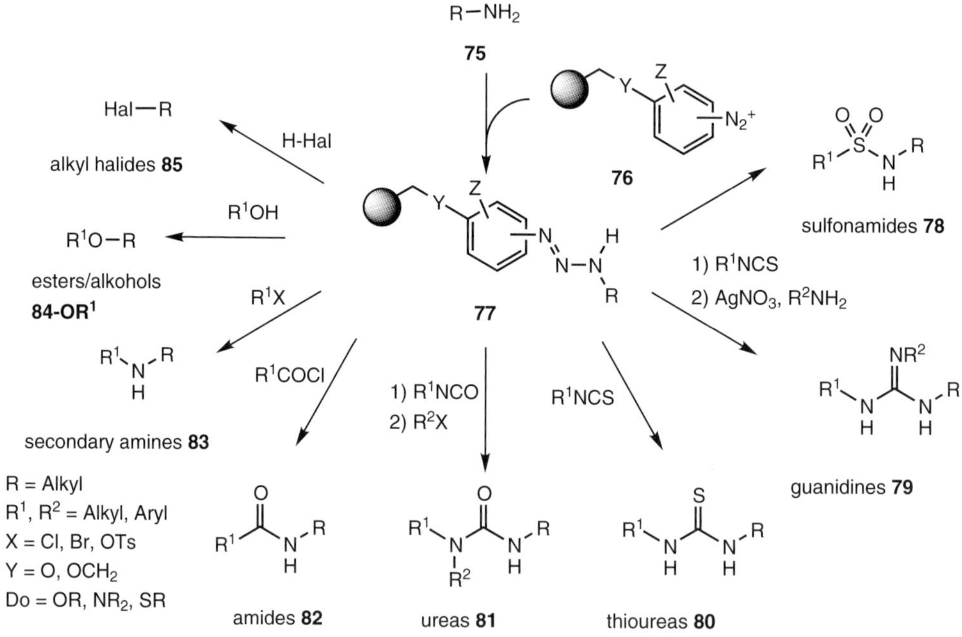

70 Merrifield resin 0.72-1.2 mmol/g

71

72

73 stable for some hours at –10 °C

74

R^1R^2NH = prim./sec. amines, hydrazines, sulfoximines, hydroxylamine

Scheme 6.1.17. Possibilities with the T2 linker.

- alkyl halides **85**
- esters/alkohols **84-OR¹**
- secondary amines **83**
- amides **82**
- ureas **81**
- thioureas **80**
- guanidines **79**
- sulfonamides **78**

R = Alkyl
R^1, R^2 = Alkyl, Aryl
X = Cl, Br, OTs
Y = O, OCH$_2$
Do = OR, NR$_2$, SR

cleophiles present in the reaction mixture. Therefore, alkyl halides **85**, alcohols **84-OH**, alkyl ester **84-OCOR**, and sulfonic esters **84-OSO₂R** can be formed by cleavage with trimethylsilyl halides (X = I, Br, Cl), carboxylic acids (X = for example, OAc, OTfa) and sulfonic acids, respectively. The regioselectivity of the cleavage can be explained by the presence of one tautomer of the triazene in which the hydrogen atom is next to the arene ring. Overall, this reaction sequence enables substitution of an amino group by oxygen functionality or a halogen (Cl, Br, I).

The structure of the diazonium salt clearly influences the stability of the diazonium moiety. The thermally stable diazonium ion **77a** ($Z = Cl$, $Y = CH_2O$) [$t^{1/2}$ (25 °C) > 100 days] is also capable of scavenging a variety of nucleophiles (amines, phenols, and anilines) [55].

6.1.4
Orthogonality Between Linkers

Orthogonality of linkers is important in the design and execution both of simple and complex reaction sequences performed on a solid support. A set of innovative linkers and cleavage strategies has been disclosed enabling a full set of orthogonality.

As discussed above, each linker family is sensitive toward a certain spectrum of cleavage conditions and is therefore stable to dissimilar conditions. Since most of the linkers are based on well-established protecting groups, table 6.1.2 can be used for the determination of orthogonality. For example, benzyl-type linkers, most of which are cleaved by electrophiles, and are stable towards nucleophiles, can be combined with ester-based protective groups.

Tab. 6.1.2. Orthogonality of linker families.[a]

Cleavage reagents	Benzyl-type linkers	Ketal/acetal linkers	Esters/amide linkers	Silyl linkers	Triazene linkers	Selenium/sulfur/stannane linkers
Electrophiles	++	++	++	++	++	++
Nucleophiles	0	0	++	0	0	0
Fluoride ions	0	0	0	++	0	0
$h\nu$ (light)	(++)	0	0	0	+	0
Oxidizing conditions	0	0	0	0	0/+	++
Reducing conditions	0	0	+	0	0/+	++

[a] (++): specially designed linker; ++: cleavage; 0: no cleavage; + partial cleavage.

6.1.5
Cleavage of Linkers

A vast array of different methods and reagents are suitable for cleavage of linkers. Such cleavage is usually conducted with acids. The most popular cleaving reagent is therefore trifluoroacetic acid in a variety of solvents and at different concen-

Tab. 6.1.3. Typical electrophiles and Lewis acids used for detachment.

Electrophile (concn)	Solvent	Additive	Example of a suitable linker	Product
HF			[56–58]	
HF			Wang linker	Thiols
HF		Anisole	PAM resin	Carboxylic acids
HF		Cresol	p-Acyloxy BHA resin [59]	Amides
CF$_3$SO$_3$H				
TFA (0.1%–neat)	CH$_2$Cl$_2$		Various linkers	
TFA		p-cresol	[58, 60, 61]	
TFA		p-cresol, Me$_2$S	[58]	
TFA		Anisole (PhOMe)		
TFA (25%)		Et$_3$SiH	[62]	
TFA		ethanedithiol (EDT)	[63]	
TFA		iPr$_3$SiH	[64]	
TFA		Et$_3$SiH	[65, 66]	
TFA		PhSMe	[67]	
HCl	Dioxane		Ketals [68]	Ketones
CF$_3$SO$_2$OSiMe$_3$			[69]	
HBr/Ac$_2$O			[70, 71]	
AcOH	CH$_2$Cl$_2$		[72]	
AlCl$_3$	CH$_2$Cl$_2$/MeNO$_2$		[73]	
Et$_2$AlCl			[74]	
Me$_3$SiCl			[54]	

trations. Because of its low boiling point, removal is readily achievable. A variety of other acids has also been used. Anhydrous HF, a rather toxic reagent, or triflic acid are required for more stable linkers. A mild reagent is trimethylsilyl chloride, which solvolyzes slowly into HCl and hexamethyldisiloxane. Electrophiles other than protons, photons, oxidizing and reducing reagents, and nucleophiles can often be used (Tables 6.1.3 and 6.1.4).

Typical nucleophilically cleavable linkers are the REM resin linker **86** [75] (polymer-bound benzyl acrylate: regenerated Michael acceptor) and the Dde group **87** (ADCC (4-Acetyl-3,5-dioxo-1-methylcyclohexane carboxylic acid) anchor) [76], which are linkers for tertiary and primary amines, respectively (Figure 6.1.7).

6.1.5.1
Oxidative/Reductive Methods

Besides electrophiles and nucleophiles, several linkers are designed to be cleaved by oxidative or reductive methods (Tables 6.1.5 and 6.1.6). Besides the feature of orthogonality with other cleaving methods, a drawback in the use of oxidative or reductive reagents is the need to remove excess reagents or by-products.

6.1.5 Cleavage of Linkers

Tab. 6.1.4. Typical nucleophiles used for detachment.

Nucleophile	Solvent	Additive	Suitable Linker	Product(s)
RMgX			Carboxyl linker	Ketones, alcohols
RMgX	THF		Thioester linker	Alcohols
R_2CuLi	THF		Thioester linker	Ketones
NaOH or KOH	MeOH		Carboxyl linker	Carboxylic acids
KOH	THF/H_2O		Reissert complex	Isoquinolines [77]
NaOMe			Carboxyl linker	Methyl esters
Fluoride ions			Silyl linkers	Hydrocarbons, alcohols
HSCH$_2$CH$_2$OH (2-mercaptoethanol)	DMF	NMM, AcOH	Dinitroaryl linker	Thiols [78]
nPrNH$_2$	DMF		Dde	Primary amines
N_2H_4	DMF		Dde [76]	Primary amines
NH_3	(vapor)			[79]

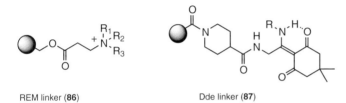

REM linker (86)   Dde linker (87)

Fig. 6.1.7. The REM and the Dde linkers.

Tab. 6.1.5. Some reducing agents used for detachment.

Reducing agent	Solvent	Suitable linker	Product
nBu$_3$SnH	Toluene	Selenium linker	Hydrocarbons
NaBH$_4$	THF/H_2O	Amide	Alcohols [80]
LiBH$_4$	THF	Thioester	Alcohols
Na/Hg	MeOH	Sulfone linker	Alkanes
Phosphines		Disulfide linker	Thiols [81–84]

Tab. 6.1.6. Some oxidizing agents used for detachment.

Oxidant	Solvent	Suitable linker	Product
mCPBA	CH$_2$Cl$_2$	Thiol-based safety-catch linkers	Secondary amines
mCPBA	CH$_2$Cl$_2$	Selenium linkers	Alkenes
[Bis(trifluoroacetoxy)iodo]benzene		Thioketal-based linker	Ketones
DDQ		Wang resin	Alcohols [27]
H_2O_2		Wang resin	Carboxylic acids [26]
Cu(OAc)$_2$	Pyr, MeCN	Hydrazide linker	Amides [85], Arenes [86]

6.1.5.2
Special Linkers

Photocleavable Linkers

Light-induced cleavage offers new possibilities for orthogonal use of linkers and acid- or base-labile protecting groups. The first photolabile linker, **88**, which was based on the *o*-nitrobenzyl protecting group, was developed by Rich and Gurwara [87] for synthesis of protected peptides (Scheme 6.1.18). This linker was developed further, because in the synthesis of the original linker a nitration of Merrifield resin was involved, leading to nitration of excess phenyl rings. Preformed handles have, therefore, usually been used. On UV-photolysis the photo-by-product, a nitrosobenzaldehyde, is also always photoactive and leads to reduction in cleavage yield from the support. To circumvent this problem an additional methyl group was introduced to the linkers; this led to the photoreactive nitrosoacetophenone system [88]. Introduction of methoxy groups *para* to the nitro groups (vanillin-type linker) [89] also improves cleavage properties and the compounds are typically released within 3 h in >90% yield and >95% purity under neutral conditions [90, 91]. These linkers are therefore biocompatible and suitable for Fmoc solid-phase peptide synthesis (Fmoc-SPPS) [92].

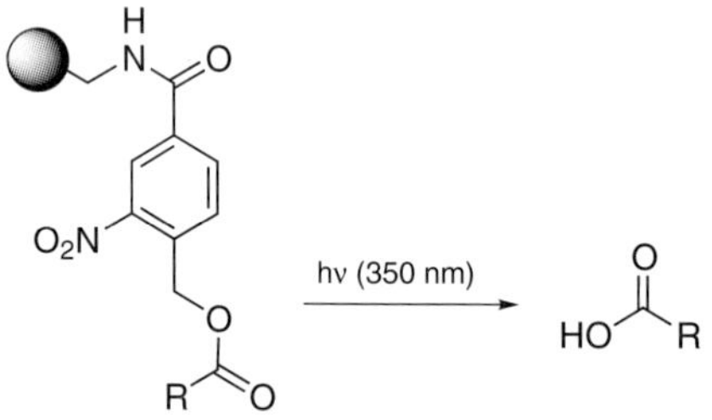

Scheme 6.1.18. The prototype photolabile linker.

6.1.5.3
Metal-assisted Cleavage

Cleavage mediated or catalyzed by (transition) metals is particularly interesting for several reasons. First, this type of cleavage is usually orthogonal to other procedures thus enabling various types of transformation. Second, reactive intermediate organometallic compounds can be suitable for further transformations.

In particular, cleavage of substrates from a solid support by use of palladium-promoted or -catalyzed reactions has some advantages over other cleavage methods. Because most protecting groups and functionalities are resistant toward palladium complexes, selective surgical removal is frequently possible. In addition, intermediate π-allyl- and σ-aryl-palladium complexes can be used in principle for further derivatization with the use of appropriate linker types.

6.1.5 Cleavage of Linkers | 469

The detachment of molecules with concomitant cross-coupling or allylic substitution is an elegant means of increasing diversity on cleavage. A common drawback of most methods is contamination with transition metal catalysts and organometallic by-products, although a variety of methods is available for the sequestering of transition metals from the cleaved products. The same applies to removal of other by-products.

π-Allyl-based Linkers

Cleavage of polymer-bound allyl esters with palladium catalysts provides general access to π-allyl complexes, which can react with a variety of nucleophiles. This has been used in the development of π-allyl-based linkers. Ene–yne cross metathesis and subsequent cleavage in the presence of different nucleophiles yields the corresponding functionalized dienes **93** [93] (Scheme 6.1.19).

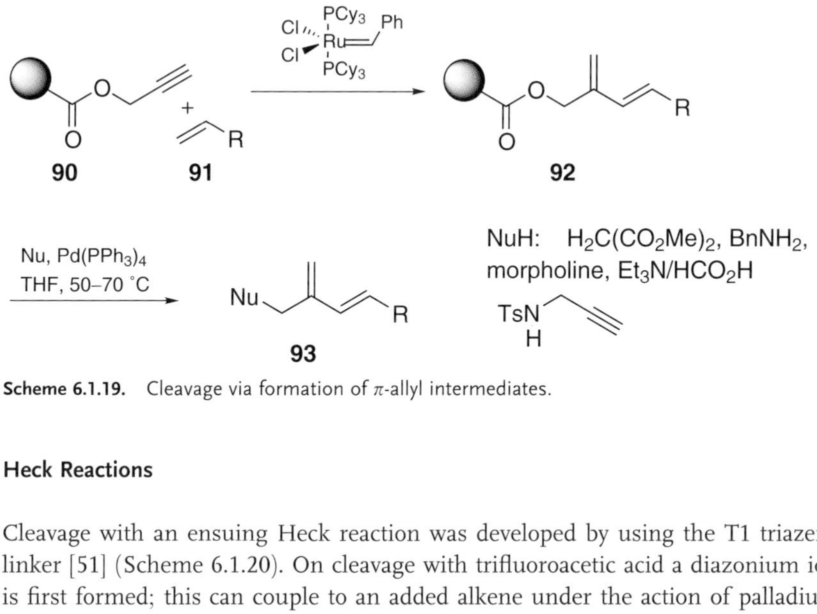

Scheme 6.1.19. Cleavage via formation of π-allyl intermediates.

Heck Reactions

Cleavage with an ensuing Heck reaction was developed by using the T1 triazene linker [51] (Scheme 6.1.20). On cleavage with trifluoroacetic acid a diazonium ion is first formed; this can couple to an added alkene under the action of palladium catalysis. The coupling proceeds well with simple terminal alkenes, styrenes, and di- and even trisubstituted alkenes. The advantage of this process is clearly the possibility of using volatile alkenes (and alkynes) without contamination by any salt or other less volatile by-product, particularly with the use of palladium on charcoal as the catalyst.

Stille Couplings

Intermolecular Stille reaction of aryl halides with immobilized stannanes (Scheme 6.1.21) provided the coupling products in good yield [45]. In addition, the stanny-

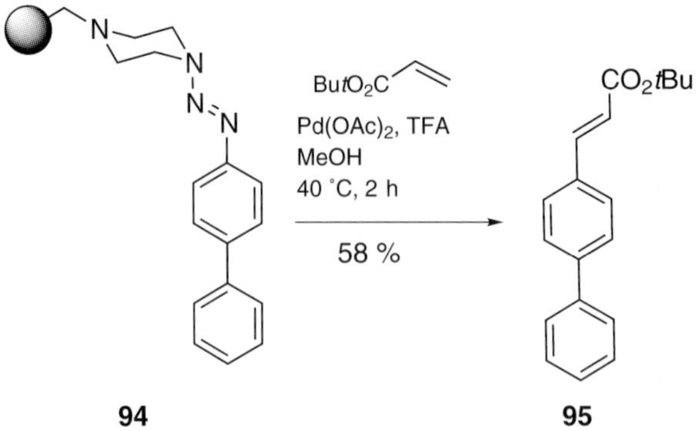

Scheme 6.1.20. Cleavage with ensuing Heck coupling using the triazene linker.

lated resin produced by cleavage of the coupling can be recycled. Although the products obtained were not contaminated by any stannane, they were separated from the excess of reactive electrophiles that had to be used in the cleavage–coupling step. The intramolecular version, which was used to produce macrocyclic ring systems such as the natural product (S)-zearalenone **97** does not have this drawback [45] (Scheme 6.1.21).

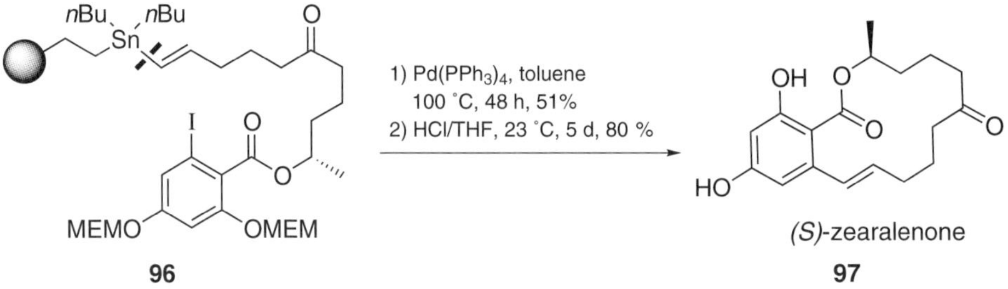

Scheme 6.1.21. A cleavage Stille strategy using a stannane linker for the synthesis of zearalenone **97**.

Suzuki Coupling

Suzuki coupling after a cleavage reaction is potentially applicable in a multifunctional sense. Because of the tendency of the boronic acid derivative to give homo-coupling products, the need to use additional ligands, and the low volatility of the boronic acid derivative, however, more or less tedious work-up is required after these types of transformation. A few studies have proven that some functionality, when generated during cleavage, can act as the leaving groups for a subsequent Suzuki reaction.

Arylmethyl(homobenzyl)ethylsulfonium salts are appropriate substrates for Suzuki-type coupling reactions. In this reaction, performed on a polymer-bound sulfonium tetrafluoroborate, the benzyl fragment on the sulfur atom was transferred to the boronic acid residue. The sulfonium salt was prepared from an alkylthiol resin by alkylation with a substituted benzyl halide to give thioether **98** and subsequent alkylation with triethyloxonium tetrafluoroborate. Reaction with a boronic acid derivative yielded diaryl methanes **99** [94] (Scheme 6.1.22).

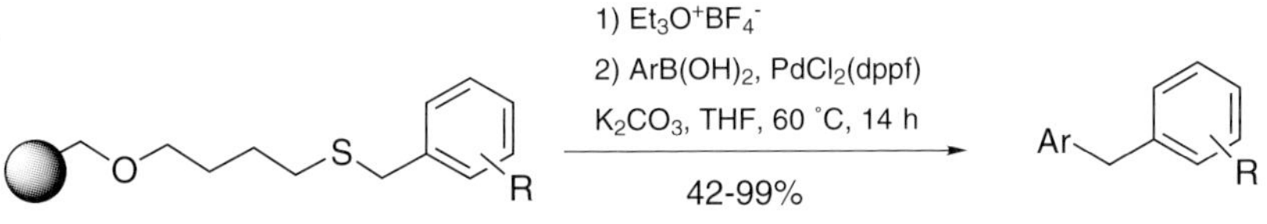

Scheme 6.1.22. Cleavage then Suzuki coupling using sulfonium salts.

Cleavage via Alkene Metathesis

Cleavage via alkene metathesis is particularly useful because clean and selective scissoring of molecules is possible. Cleavage by metathesis can be performed either by cyclization during cleavage [95–101] (ring closing metathesis, RCM), intermolecular metathesis [101] (cross metathesis), or intramolecular metathesis [95] (Scheme 6.1.23).

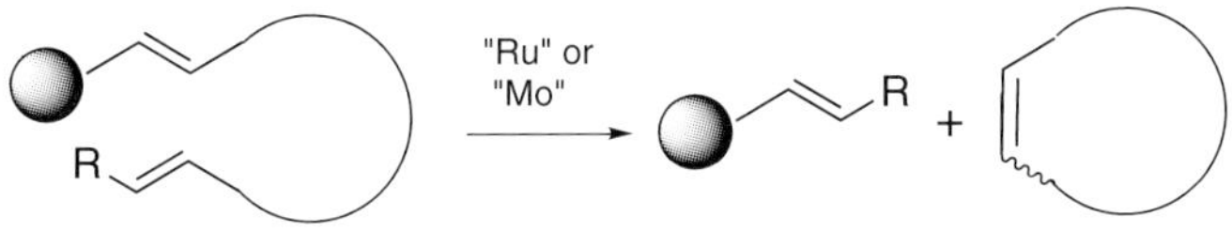

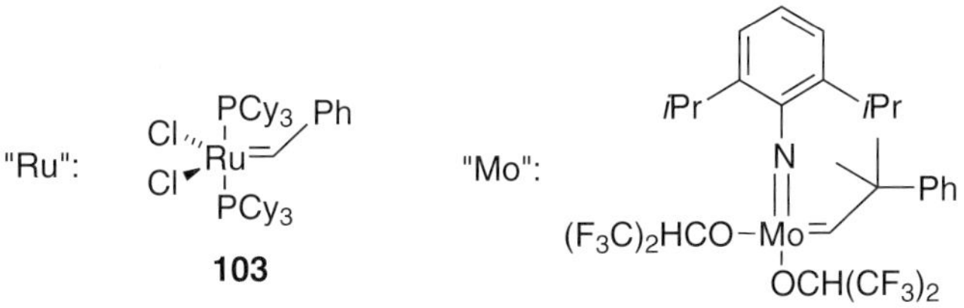

Scheme 6.1.23. The concept of metathetic cleavage.

An anchoring group for solid-phase synthesis of oligosaccharides has been based on metathetic cleavage [102]. Although alkenyl units are not suitable for glycosidation reactions in the presence of strong electrophilic activators, by use of the 4,5-dibromooctane (DBOD) anchor an iodide-mediated elimination reaction provides the active linker which can be cleaved under metathesis conditions [102] (Scheme 6.1.24).

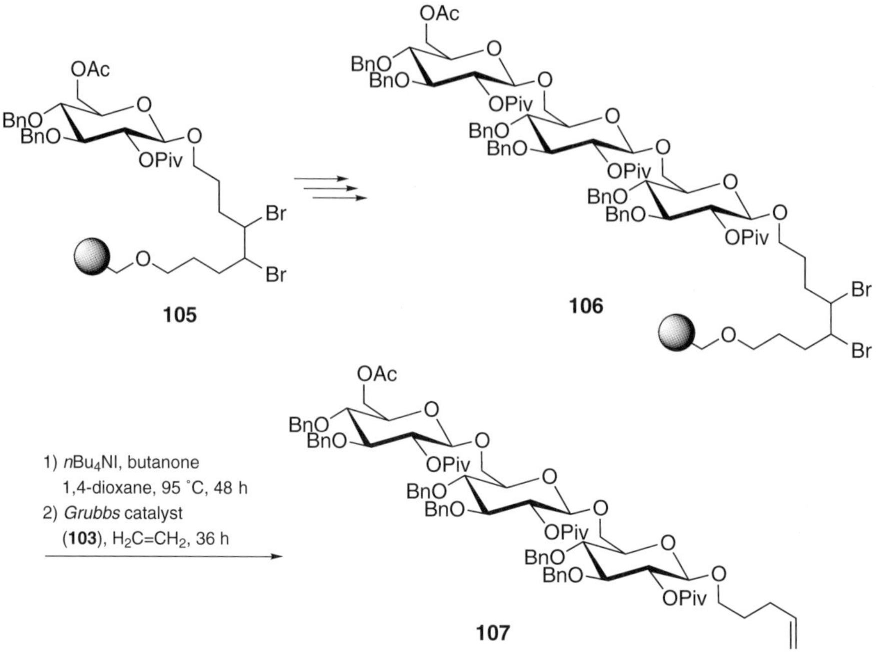

Scheme 6.1.24. Oligosaccharide synthesis on a solid support using the DBOD linker.

6.1.6
Linker and Cleavage Strategies

Apart from simple monofunctional cleavage, a variety of different linker strategies has been developed in recent years. In particular, new concepts based on safety-catch linkers, cyclative cleavage strategies, and fragmentation reactions have been presented. Cleavage of linkers can be monofunctional or with functionalization of the linking site, whatever is desired [103] (Figure 6.1.8). In the latter, also known as the multifunctional cleavage strategy, the number of library compounds is multiplied by the number of building blocks or functional groups that can be incorporated during the cleavage step. An anchoring group capable of functionalization and traceless linking is, therefore, a versatile tool for enhancing diversity in a given system.

6.1.6 Linker and Cleavage Strategies | 473

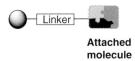

Attached molecule

'Safety catch' linkers:

Cyclization-cleavage linkers:

Fragmentation/ Cycloreversion-cleavage linkers:

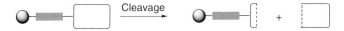

Monofunctional linkers:

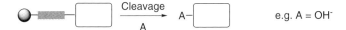

e.g. A = OH$^-$

Traceless linkers:

Cleavage "H" → H—☐   e.g. A = H$^+$

Multifunctional linkers:

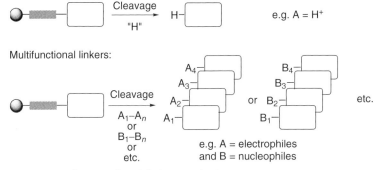

e.g. A = electrophiles and B = nucleophiles

Fig. 6.1.8. Linker types for solid-phase synthesis.

6.1.6.1
Safety-catch Linkers

A safety-catch linker is a linker which is cleaved by performing two different reactions instead of the normal single step, thus providing greater control over the timing of compound release [7]. The "safety catch" principle consists of a linker, which is inert towards the cleavage conditions during the synthesis and has to be activated prior to cleavage (Scheme 6.1.25). Because "safety catch" means the activation of the linker before cleavage, such a strategy can be applied to monodirectional, e.g. traceless, linkers or multifunctional linkers, and to cyclative cleavage strategies. An example of this strategy is the novel version [104] of the Kenner sulfonamide linker [105] as exemplified in the synthesis of vinylsulfones [106] (Scheme 6.1.26).

Scheme 6.1.25. General scheme for safety-catch linkers.

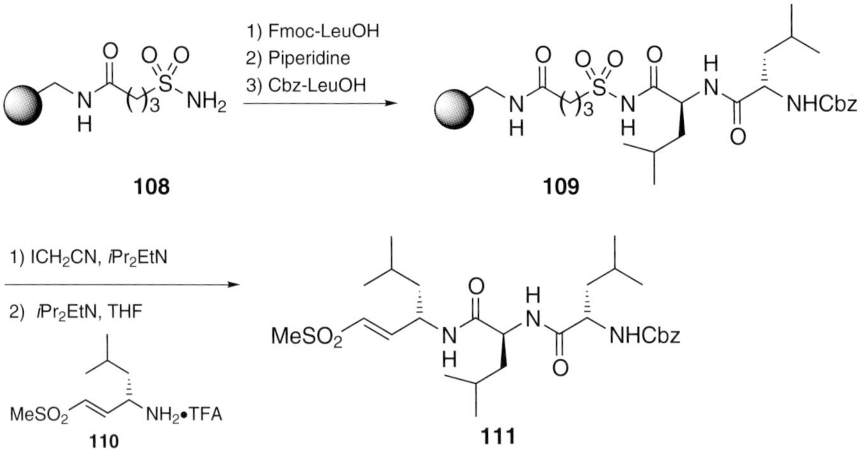

Scheme 6.1.26. The Kenner safety-catch linker.

6.1.6.2
Cyclative Cleavage (Cyclorelease Strategy)

Cyclative cleavage results from intramolecular reaction at the linker which results in a cyclized product. The cleavage can also act as a means of purification if resin-bound side-products are incapable of cyclizing, and thus remain attached to the solid support on release of the desired material [7]. The cyclization–cleavage strategy (cyclative cleavage or cyclorelease strategy) is typically used for synthesis of cyclic structures on a solid support. It uses the characteristics of quasi high-dilution kinetics on solid supports and thus has advantages which are not found in solution-phase chemistry.

In general, the starting material for cyclative cleavage is anchored to the resin *via* a leaving group (Scheme 6.1.27). An internal nucleophile affords ring closure by displacement of the leaving group either directly or after activation. Apart from nucleophilic attack, cyclative cleavage can be effected, for example, by the Stille (Scheme 6.1.21) or Wittig–Horner reactions [107, 108].

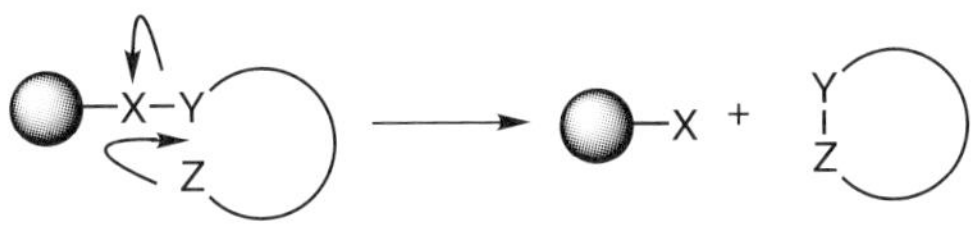

Scheme 6.1.27. General scheme for cyclative cleavage.

Because the intramolecular reaction is much faster than any intermolecular step, this strategy provides an additional purification step, because only the cyclized products are detached from the bead. Incomplete building blocks will remain attached to the solid support. However, cyclative cleavage must be distinguished from reactions in which the cyclization occurs in solution after cleavage, because unsuccessful cyclization precursors will remain in the liquid phase.

The precursor for cyclization cleavage is usually linked *via* an ester bond to the solid support whereas the nucleophile is based on amine functionality. Thus the product formed is a cyclic amide or an analog. Indeed, one of the first examples of this type of reaction was the pioneering benzodiazepine synthesis by Camps et al. in 1974 [109]. In this synthesis benzodiazepine ring formation proceeds by simultaneous cleavage from the bead. A vast number of examples has appeared since then and these provide various types of heterocyclic system, as demonstrated by an approach to 3,5-disubstituted 1,3-oxazolidinones **117** via a ring-opening–cyclization–cleavage step [110] (Scheme 6.1.28).

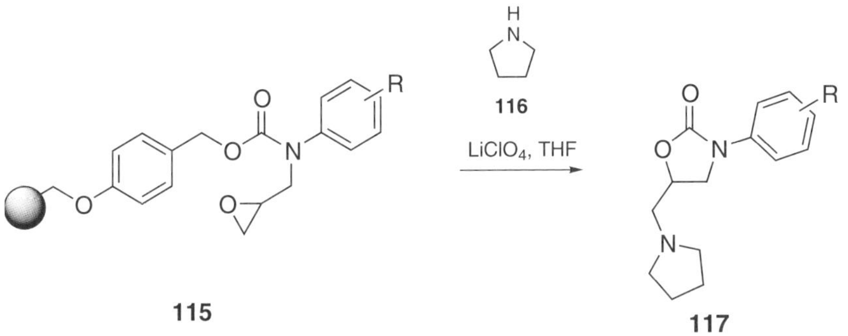

Scheme 6.1.28. Synthesis of oxazolidines **117** by cyclative cleavage.

6.1.6.3
Fragmentation Strategies

The fragmentation strategy is related to the traceless anchoring groups, as defined, and also includes strategies which can be regarded as retro-cycloaddition cleavage, cycloelimination, or cyclofragmentation reactions (Scheme 6.1.29).

Scheme 6.1.29. General scheme for fragmentation strategy.

In these strategies a double or triple bond results from 1,n-elimination processes. These are synthetically useful operations with a wide range of possibilities for construction of rigid templates of different ring sizes. Occasionally, a retro cycloaddition must also be regarded as a fragmentation cleavage.

Only one example of an attachment of heteroarenes by addition/elimination strategy has been devised [77, 111]. Although arenes are more or less resistant toward addition, heteroaromatic systems such as isoquinolines **118** are prone to addition of nucleophiles. Subsequent reaction with addition of electrophiles furnishes the so-called Reissert compounds **120**. These are stable compounds which can, for example, be alkylated. In solid-phase synthesis the electrophile chosen was a polymer-based acid chloride. Detachment can be achieved by simple addition of hydroxide ions (Scheme 6.1.30).

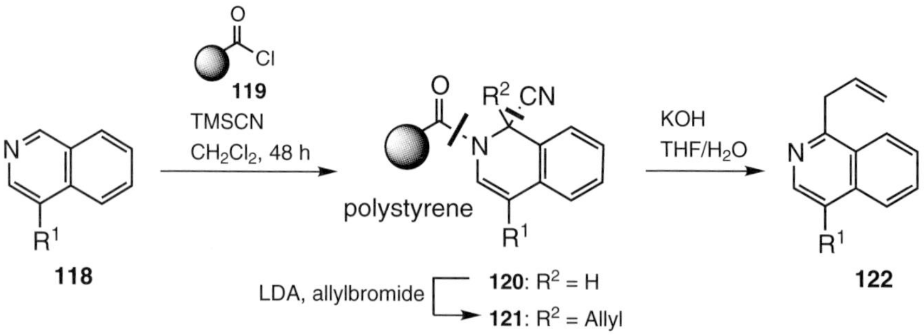

Scheme 6.1.30. The Reissert complex strategy.

One synthesis of benzofurans is based on cyclofragmentation. An appropriately substituted sulfone is used as a nucleophile in intramolecular ring opening of an epoxide. The resulting molecule loses a sulfinate and formaldehyde. By immobilization of the sulfinate on a resin this sequence can be used for the cleavage of benzofurans from solid supports [112] (Scheme 6.1.31).

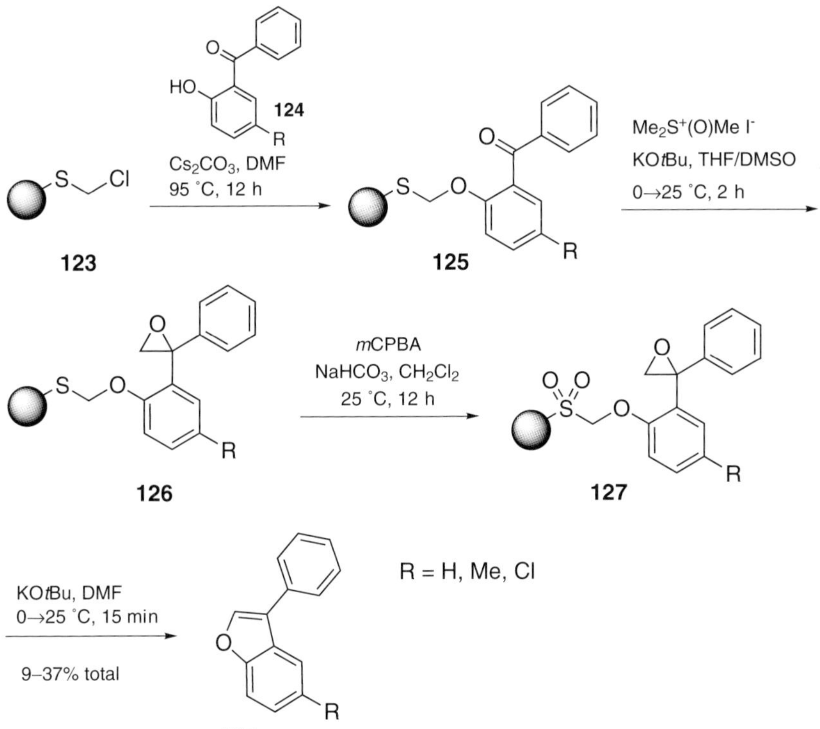

Scheme 6.1.31. Benzofuran synthesis.

6.1.6.4
Traceless Linkers

A traceless linker is a type of linker which leaves no residue on the compound after cleavage, i.e. it is replaced by hydrogen [7].

"Traceless linking" is regarded as "leaving no functionality" which, for arenes and alkanes, means that only a C–H bond remains at the original position of attachment (Scheme 6.1.32) (for reviews see Ref. [113]). This anchoring mode plays an important role in the design and synthesis of drug-like molecules, because the absence of any remaining part of the linker leads to an unbiased library.

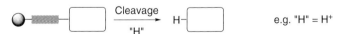

Scheme 6.1.32. Principle of the traceless linker.

To design a traceless linker one must start from a heteroatom–carbon bond which is labile towards protogenolytic, hydrogenolytic, or hydridolytic cleavage.

Because most heteroatom–carbon single bonds are less stable than carbon–carbon bonds, traceless linkers can be synthesized on the basis of nearly all heteroatoms. The enthalpies of C–X bonds are, however, only relevant for homolytic bond scission. Many linkers are cleaved heterolytically, and kinetic stability toward heterolytic bond cleavage is decisive in these linkers.

The most prominent anchors for traceless linkage of arenes are based on silyl linkers. Generation of a diverse benzodiazepine library [40] has clearly shown the advantages of this type of detachment because no additional functionality, which might bias the library, was preserved in the final molecules (Scheme 6.1.33).

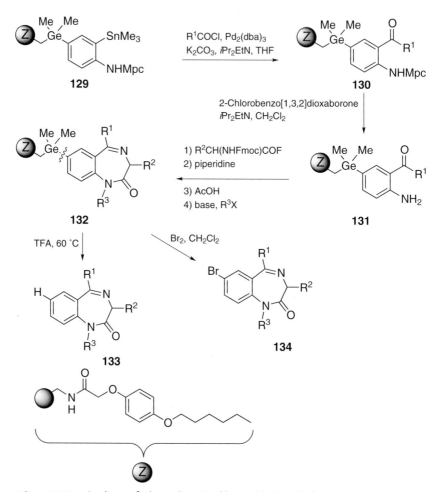

Scheme 6.1.33. Synthesis of a benzodiazepine library with the aid of a germanium linker.

Another possible means of conversion of functionalized arenes into the corresponding hydrocarbons is the reduction of diazonium compounds [49]. By application of this method cinnamic esters were synthesized in a sequence starting from the iodoarene resin **135**. Heck coupling with acrylates, using palladium catalysis, affords an immobilized cinnamate. This can be detached after a sequence of transformations, yielding to an allyl amine in a traceless fashion (Scheme 6.1.34).

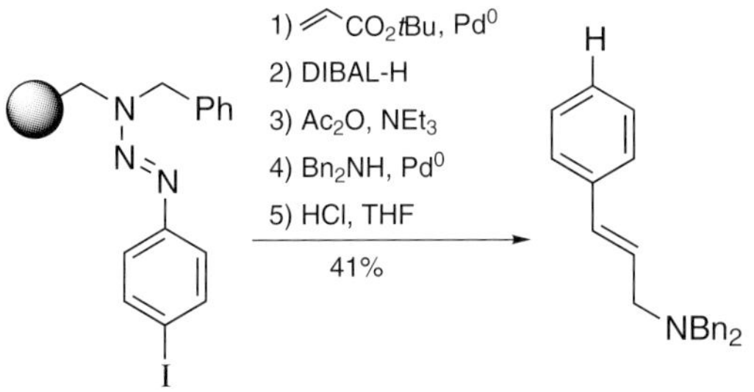

Scheme 6.1.34. The T1 linker for traceless cleavage.

6.1.6.5
Multifunctional Cleavage

Although traceless cleavage provides efficient access to hydrocarbon-like molecules, these monofunctional linkers provide only one type of compound in a library. So-called multifunctional cleavage [41] affords the important opportunity to incorporate additional diversity on cleavage (Scheme 6.1.35). Thus increasing the number of new functional groups increases the number of compounds produced. If the linker is amenable to different types of building block (for example nucleophile [A] and electrophiles [B]) incorporated during cleavage a substantial library of novel molecules can be prepared from one immobilized compound [114] (Scheme 6.1.36).

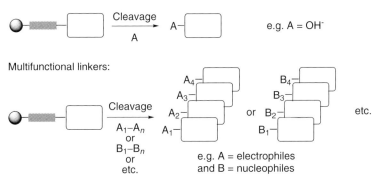

Scheme 6.1.35. Comparison of monofunctional and multifunctional cleavage.

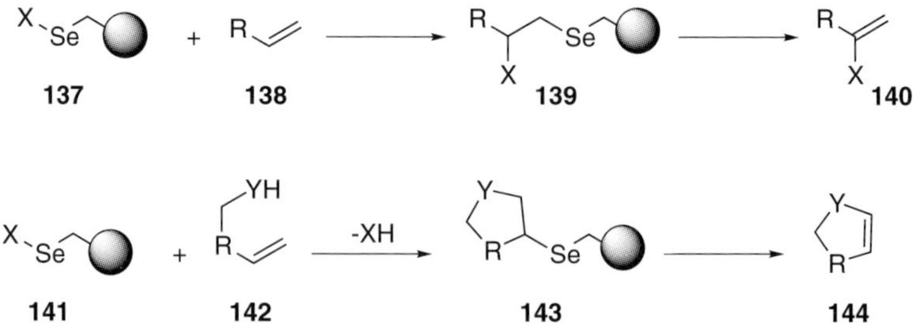

Scheme 6.1.36. Electrophilic selenium anchors as multifunctional linkers.

When considering the use of a multifunctional linker, however, one must take into account the nature of the cleavage reagent and the cleavage step. Cleavage consisting of, for example, addition of a Grignard reagent to an ester with a huge excess of the organometallic component requires aqueous work-up and, therefore, potential loss of valuable material. Excess reagents should, therefore, be easy to remove (volatile, low, or very high, solubility in certain solvents, easy to eliminate or to be removed by, for example, scavenger resins etc.) and should not interfere with the functionality of the library compounds.

6.1.7
Conclusion, Summary, and Outlook

In recent years different types of new linker have emerged (Table 6.1.7). For the synthesis of small molecules on solid supports, in particular, the design of a new anchoring group might be essential for success of the synthesis. Linker, cleavage conditions, and functional groups are selected for each other. The decision to use one specific linker type must, therefore, be balanced with the need of the library synthesized.

Although the "perfect" or "universal" linker is not in sight, and might prove unattainable, interesting new developments increase the flexibility of solid-phase synthesis by enabling traceless and multifunctional cleavage. Whereas traceless linkers provide access to unsubstituted compounds with "no memory" of solid-phase synthesis, multifunctional cleavage enables introduction of a variety of new functionality on cleavage from the resin. Backbone amide linkers afford new opportunities for solid-phase synthesis of small amidic structures. Cyclization–release strategies provide an opportunity to create novel carbo- and heterocyclic structures on cleavage.

An anchor for traceless linking might also have a safety-catch function or be suitable for multifunctional cleavage. Linker systems enable the introduction of certain atoms or molecule fragments and will play an important role in the development of diverse organic substance libraries. It is important to point out that the final diversity is achieved on cleavage, and not in an additional solution phase re-

Tab. 6.1.7. Short overview of the different linker families.

Functional group	Benzyl-type linkers	Ketal/ acetal linkers	Esters/ amide linkers	Silane linkers	Triazene linkers	Selenium/sulfur/ stannane linkers
R_3N	✓	✓	✓		✓	✓
ROH	✓	✓	✓	✓		
R_2NCOR	✓		✓		✓	
RH (traceless)	✓		✓	✓	✓	✓
RCO_2H	✓		✓			
Heterocycles	✓	✓			✓	
BAL	✓				✓	
RX					✓	
Safety catch option	✓		✓			✓
Multifunctional cleavage					✓	✓
Photo cleavage	✓				✓	

action step after cleavage. Only a few linker systems applicable to a wider range of substrates have yet been developed, however. Because these linker systems afford the widest possibilities of final diversity of the synthesized library they will experience increasing attention in the future.

References

1 N. K. Mathur, C. K. Narang, R. E. Williams, *Polymers as Aids in Organic Chemistry*, Academic Press, New York, 1980.
2 R. B. Merrifield, *J. Am. Chem. Soc.* 1963, 85, 2149–2154.
3 R. A. Fecik, K. E. Frank, E. J. Gentry, S. R. Menon, L. A. Mitscher, H. Telikepalli, *Med. Res. Rev.* 1998, 18, 149–185.
4 S. Bräse, C. Gil, K. Knepper, *Bioorg. Med. Chem.* 2002, 10, 2415–2437.
5 A. R. Brown, P. H. H. Hermkens, H. C. J. Ottenheijm, D. C. Rees, *Synlett* 1998, 817–827.
6 J. J. Scicinski, M. S. Congreve, C. Kay, S. V. Ley, *Curr. Med. Chem.* 2002, 9, 2103–2127.
7 D. Maclean, J. J. Baldwin, V. T. Ivanov, Y. Kato, A. Shaw, P. Schneider, E. M. Gordon, *Pure Appl. Chem.* 1999, 71, 2349–2365.
8 F. Guillier, D. Oran, M. Bradley, *Chem. Rev.* 2000, 100, 2091–2157.
9 B. J. Backes, J. Ellman, *Curr. Opin. Chem. Biol.* 1997, 1, 86–93.
10 S. Bräse, S. Dahmen, *Handbook of Combinatorial Chemistry*, K. C. Nicolaou, R. Hanko, W. Hartwig, eds., Wiley–VCH, Weinheim, 2002, pp. 59–169.
11 M. S. Bernatowicz, S. B. Daniels, H. Köster, *Tetrahedron Lett.* 1989, 30, 4645–4648.
12 C. T. Bui, F. A. Rasoul, F. Ercole, Y. Pham, N. J. Maeji, *Tetrahedron Lett.* 1998, 39, 9279–9282.
13 W. Rapp, *Combinatorial Peptide and Nonpeptide Libraries: A Handbook*, G. Jung, ed., VCH, Weinheim, 1996, pp. 425–464.
14 J. R. Hauske, P. Dorff, *Tetrahedron Lett.* 1995, 36, 1589–1592.
15 K. Akaji, Y. Kiso, L. A. Carpino, *J. Chem. Soc., Chem. Commun.* 1990, 584–586.
16 T. H. Chan, W. Q. Huang, *J. Chem. Soc., Chem. Commun.* 1985, 909–911.

17 J. T. Randolph, K. F. McClure, S. J. Danishefsky, *J. Am. Chem. Soc.* **1995**, *117*, 5712–5719.
18 W. J. Koot, *J. Comb. Chem.* **1999**, *1*, 467–473.
19 S.-S. Wang, *J. Am. Chem. Soc.* **1973**, *95*, 1328–1333.
20 B. Yan, N. Nguyen, L. Liu, G. Holland, B. Raju, *J. Comb. Chem.* **2000**, *2*, 66–74.
21 M. Mergler, R. Tanner, J. Gosteli, P. Grogg, *Tetrahedron Lett.* **1988**, *22*, 4005–4008.
22 M. Mergler, J. Gosteli, P. Grogg, P. Nyfeler, R. Tanner, *Chimia* **1999**, *53*, 29–34.
23 F. Albericio, G. Barany, *Tetrahedron Lett.* **1991**, *32*, 1015–1018.
24 E. Atherton, C. J. Logan, R. C. Sheppard, *J. Chem. Soc., Perkin Trans. 1* **1981**, 538–546.
25 S. Manabe, Y. Ito, T. Ogawa, *Synlett* **1998**, 628–630.
26 G.-S. Lu, S. Mojsov, J. P. Tam, R. B. Merrifield, *J. Org. Chem.* **1981**, *46*, 3433–3436.
27 T. L. Deegan, O. W. Gooding, S. Baudart, J. A. Porco, *Abstr., Pap. Am. Chem. Soc.* **1997**, *214*, 238-ORGN.
28 R. S. Garigipati, *Tetrahedron Lett.* **1997**, *38*, 6807–6810.
29 J. Alsina, T. S. Yokum, F. Albericio, G. Barany, *J. Org. Chem.* **1999**, *64*, 8761–8769.
30 N. Thieriet, F. Guibe, F. Albericio, *Org. Lett.* **2000**, *2*, 1815–1817.
31 H. Kunz, B. Dombo, *Angew. Chem. Int. Ed. Engl.* **1988**, *12*, 711–712; *Angew. Chem.* **1988**, *100*, 732–734.
32 H. Kunz, W. Kosch, J. März (Orpegen GmbH), US Pat. 5214195, **1990**.
33 C. Schumann, L. Seyfarth, G. Greiner, S. Reissmann, *J. Pept. Res.* **2000**, *55*, 428–435.
34 T. Johnson, R. C. Sheppard, *J. Chem. Soc., Chem. Commun.* **1991**, 1653–1655.
35 O. Seitz, H. Kunz, *J. Org. Chem.* **1997**, *62*, 813–826.
36 L. A. Thompson, J. A. Ellman, *Tetrahedron Lett.* **1994**, *35*, 9333–9336.
37 S. Q. Chen, K. D. Janda, *Tetrahedron Lett.* **1998**, *39*, 3943–3946.
38 Y. Kondo, T. Komine, M. Fujinami, M. Uchiyama, T. Sakamoto, *J. Comb. Chem.* **1999**, *1*, 123–126.
39 C. R. Millington, R. Quarrel, G. Lowe, *Tetrahedron Lett.* **1998**, *39*, 7201–7204.
40 M. J. Plunkett, J. A. Ellman, *J. Org. Chem.* **1997**, *62*, 2885–2893.
41 D. Obrecht, J. M. Villalgordo, *Solid-Supported Combinatorial and Parallel Synthesis of Small-Molecular-Weight Compound Libraries*, Elsevier, Oxford, **1998**, p. 98.
42 J. M. J. Frechet, L. J. Nuyens, E. Seymour, *J. Am. Chem. Soc.* **1979**, *101*, 432–436.
43 W. Li, K. Burgess, *Tetrahedron Lett.* **1999**, *40*, 6527–6530.
44 C. Garcia-Echeverria, *Tetrahedron Lett.* **1997**, *38*, 8933–8934.
45 H. Kuhn, W. P. Neumann, *Synlett* **1994**, 123–124.
46 K. C. Nicolaou, N. Winssinger, J. Pastor, F. Murphy, *Angew. Chem. Int. Ed.* **1998**, *37*, 2534–2537; *Angew. Chem.* **1998**, *110*, 2677–2680.
47 K. C. Nicolaou, J. Pastor, S. Barluenga, N. Winssinger, *Chem. Commun.* **1998**, 1947–1948.
48 T. Ruhland, K. Andersen, H. Pedersen, *J. Org. Chem.* **1998**, *63*, 9204–9211.
49 S. Bräse, D. Enders, J. Köbberling, F. Avemaria, *Angew. Chem. Int. Ed.* **1998**, *37*, 3413–3415; *Angew. Chem.* **1998**, *110*, 3614–3616.
50 S. Bräse, J. Köbberling, D. Enders, M. Wang, R. Lazny, S. Brandtner, *Tetrahedron Lett.* **1999**, *40*, 2105–2108.
51 S. Bräse, M. Schroen, *Angew. Chem. Int. Ed.* **1999**, *38*, 1071–1073; *Angew. Chem.* **1999**, *111*, 1139–1142.
52 M. E. P. Lormann, C. H. Walker, S. Bräse, *Chem. Commun.* **2002**, *12*, 1296–1297.
53 S. Bräse, S. Dahmen, J. Heuts, *Tetrahedron Lett.* **1999**, *40*, 6201–6203.
54 S. Bräse, S. Dahmen, M. Pfefferkorn, *J. Comb. Chem.* **2000**, *2*, 710–717.
55 S. Dahmen, S. Bräse, *Angew. Chem. Int. Ed.* **2000**, *39*, 3681–3683; *Angew. Chem.* **2000**, *112*, 3827–3830.

56 W. D. F. Meutermans, P. F. Alewood, *Tetrahedron Lett.* **1995**, *36*, 7709–7712.

57 R. B. Merrifield, L. D. Vizioli, H. G. Boman, *Biochemistry* **1982**, *21*, 5020–5031.

58 J. P. Tam, W. F. Heath, R. B. Merrifield, *J. Am. Chem. Soc.* **1983**, *105*, 6442–6455.

59 H. Yaijima, N. Fujii, H. Ogawa, H. Kawatani, *J. Chem. Soc., Chem. Commun.* **1974**, 107–108.

60 A. R. Mitchell, S. B. H. Kent, M. Engelhard, R. B. Merrifield, *J. Org. Chem.* **1978**, *43*, 2845–2852.

61 K. C. Nicolaou, N. Winssinger, J. Pastor, F. DeRoose, *J. Am. Chem. Soc.* **1997**, *119*, 449–450.

62 A. Mazurov, *Tetrahedron Lett.* **2000**, *41*, 7–10.

63 D. R. Englebretsen, C. T. Choma, G. T. Robillard, *Tetrahedron Lett.* **1998**, *39*, 4929–4932.

64 P. R. Hansen, C. E. Olsen, A. Holm, *Bioconj. Chem.* **1998**, *9*, 126–131.

65 Y. X. Han, G. Barany, *J. Org. Chem.* **1997**, *62*, 3841–3848.

66 M. C. Munson, G. Barany, *J. Am. Chem. Soc.* **1993**, *115*, 10203–10210.

67 G. Mezo, N. Mihala, G. Koczan, F. Hudecz, *Tetrahedron* **1998**, *54*, 6757–6766.

68 Z. H. Xu, C. R. McArthur, C. C. Leznoff, *Can. J. Chem.* **1983**, *61*, 1405–1409.

69 D. Limal, J. P. Briand, P. Dalbon, M. Jolivet, *J. Pept. Res.* **1998**, *52*, 121–129.

70 B. Yan, H. Gstach, *Tetrahedron Lett.* **1996**, *37*, 8325–8328.

71 J. Blake, C. H. Li, *J. Am. Chem. Soc.* **1968**, *90*, 5882–5884.

72 H. Rink, *Tetrahedron Lett.* **1987**, *28*, 3787–3790.

73 E. G. Mata, *Tetrahedron Lett.* **1997**, *38*, 6335–6338.

74 J. D. Winkler, W. McCoull, *Tetrahedron Lett.* **1998**, *39*, 4935–4936.

75 A. R. Brown, D. C. Rees, Z. Rankovic, J. R. Morphy, *J. Am. Chem. Soc.* **1997**, *119*, 3288–3295.

76 W. Bannwarth, J. Huebscher, R. Barner, *Bioorg. Med. Chem. Lett.* **1996**, *6*, 1525–1528.

77 B. K. Lorsbach, R. B. Miller, M. J. Kurth, *J. Org. Chem.* **1996**, *61*, 8716–8717.

78 J. D. Glass, A. Talansky, Z. Gronka, I. L. Schwartz, R. Walter, *J. Am. Chem. Soc.* **1974**, *96*, 6476–6480.

79 A. M. Bray, N. J. Maeji, A. G. Jhingran, R. M. Valerio, *Tetrahedron Lett.* **1991**, *32*, 6163–6166.

80 G. Faita, A. Paio, P. Quadrelli, F. Rancati, P. Seneci, *Tetrahedron Lett.* **2000**, *41*, 1265–1269.

81 A. J. Souers, A. A. Virgilio, S. S. Schürer, J. A. Ellman, T. P. Kogan, H. E. West, W. Ankener, P. Vanderslice, *Bioorg. Med. Chem. Lett.* **1998**, *8*, 2297–2302.

82 K. Kurokawa, H. Kumihara, H. Kondo, *Bioorg. Med. Chem. Lett.* **2000**, *10*, 1827–1830.

83 A. A. Virgilio, S. C. Schürer, J. A. Ellman, *Tetrahedron Lett.* **1996**, *37*, 6961–6964.

84 A. J. Souers, A. A. Virgilio, A. Rosenquist, W. Fenuik, J. A. Ellman, *J. Am. Chem. Soc.* **1999**, *121*, 1817–1825.

85 F. Berst, A. B. Holmes, M. Ladlow, P. J. Murray, *Tetrahedron Lett.* **2000**, *41*, 6649–6653.

86 F. Stieber, U. Grether, H. Waldmann, *Angew. Chem. Int. Ed.* **1999**, *38*, 1073–1077; *Angew. Chem.* **1999**, *111*, 1142–1145.

87 D. H. Rich, S. K. Gurwara, *J. Chem. Soc., Chem. Commun.* **1973**, 610–611.

88 A. Aiyaghosh, V. N. R. Pillai, *Tetrahedron* **1988**, *44*, 6661–6666.

89 U. Zehavi, A. Patchornik, *J. Am. Chem. Soc.* **1973**, *95*, 5673–5677.

90 D. L. Whitehouse, S. N. Savinov, D. J. Austin, *Tetrahedron Lett.* **1997**, *38*, 7851–7852.

91 D. J. Yoo, M. M. Greenberg, *J. Org. Chem.* **1995**, *60*, 3358–3364.

92 M. Rinnova, M. Novakova, V. Kasicka, J. Jiracek, *J. Pept. Sci.* **2000**, *6*, 355–365.

93 S. C. Schürer, S. Blechert, *Synlett* **1998**, 166–168.

94 C. Vanier, F. Lorgé, A. Wagner, C. Mioskowski, *Angew. Chem. Int. Ed.* **2000**, *39*, 1679–1683; *Angew. Chem.* **2000**, *112*, 1745–1749.

95 J. U. Peters, S. Blechert, *Synlett* **1997**, 348–350.
96 J. Pernerstorfer, M. Schuster, S. Blechert, *Chem. Commun.* **1997**, 1949–1950.
97 K. C. Nicolaou, N. Winssinger, J. Pastor, S. Ninkovic, F. Sarabia, Y. He, D. Vourloumis, Z. Yang, T. Li, P. Giannakakou, E. Hamel, *Nature* **1997**, *387*, 268–272.
98 A. D. Piscopio, J. F. Miller, K. Koch, *Tetrahedron Lett.* **1997**, *38*, 7143–7146.
99 A. D. Piscopio, J. F. Miller, K. Koch, *Tetrahedron Lett.* **1998**, *39*, 2667–2670.
100 A. D. Piscopio, J. F. Miller, K. Koch, *Tetrahedron* **1999**, *55*, 8189–8198.
101 J. H. Van Maarseveen, J. A. J. den Hartog, V. Engelen, E. Finner, G. Visser, C. G. Kruse, *Tetrahedron Lett.* **1996**, *37*, 8249–8252.
102 L. G. Melean, W.-C. Haase, P. H. Seeberger, *Tetrahedron Lett.* **2000**, *41*, 4329–4333.
103 S. Bräse, S. Dahmen, *Chem. Eur. J.* **2000**, *6*, 1899–1905.
104 B. J. Backes, J. A. Ellman, *J. Am. Chem. Soc.* **1994**, *116*, 11171–11172.
105 G. W. Kenner, J. R. McDermott, R. C. Sheppard, *J. Chem. Soc., Chem. Commun.* **1971**, 636–637.
106 H. S. Overkleeft, P. R. Bos, B. G. Hekking, E. J. Gordon, H. L. Ploegh, B. M. Kessler, *Tetrahedron Lett.* **2000**, *41*, 6005–6009.
107 K. C. Nicolaou, J. Pastor, N. Winssinger, F. Murphy, *J. Am. Chem. Soc.* **1998**, *120*, 5132–5133.
108 C. R. Johnson, B. R. Zhang, *Tetrahedron Lett.* **1995**, *36*, 9253–9256.
109 F. Camps, J. Castells, J. Pi, *Ann. Quim.* **1974**, *70*, 848–849.
110 H. P. Buchstaller, *Tetrahedron* **1998**, *54*, 3465–3470.
111 B. A. Lorsbach, J. T. Bagdanoff, R. B. Miller, M. J. Kurth, *J. Org. Chem.* **1998**, *63*, 2244–2250.
112 K. C. Nicolaou, S. A. Snyder, A. Bigot, J. A. Pfefferkorn, *Angew. Chem. Int. Ed.* **2000**, *39*, 1093–1096; *Angew. Chem.* **2000**, *112*, 1135–1138.
113 A. B. Reitz, *Curr. Opin. Drug. Disc. Develop.* **1999**, *2*, 358–364.
114 F. Zaragoza, *Angew. Chem. Int. Ed.* **2000**, *39*, 2077–2079; *Angew. Chem.* **2000**, *112*, 2158–2159.

6.2
Small Molecule Arrays

Rolf Breinbauer, Maja Köhn, and Carsten Peters

6.2.1
Introduction

The increasing pace of biological research is fed by the curiosity of researchers to answer the questions "What is life?" and "How does it work?". Parallelization and automation of experiments were key to the exponential gain in data and knowledge which ultimately might help us to find remedies against almost all diseases. One technique, which has just started to reveal its power, is arrays of biological probes such as DNA, RNA, small molecules, or whole cells. This review highlights the basic principles of these arrays, their fabrication and applications, with a special focus on the emerging field of small molecule microarrays. All arraying technologies involve three steps (Figure 6.2.1):

1. fabrication of a planar array of probes either by spotting or by spatially addressable synthesis;
2. incubation of the array with the analyte solution and washing away of unbound substrates; and
3. two-dimensional readout of the bound substrates and data processing.

6.2.2
Arrays

6.2.2.1
DNA Microarrays

The development of "DNA microarrays" or generally named "DNA chips" has been driven by modern approaches to analyze multiple gene mutations and expressed sequences [2]. The broad range of actual DNA chip applications includes the detection of pathogens, the measurement of differences in the expression of genes between different cell populations, and the analysis of genomic alterations

Highlights in Bioorganic Chemistry: Methods and Applications. Edited by Carsten Schmuck, Helma Wennemers.
Copyright © 2004 WILEY-VCH Verlag GmbH & Co. KGaA, Weinheim
ISBN: 3-527-30656-0

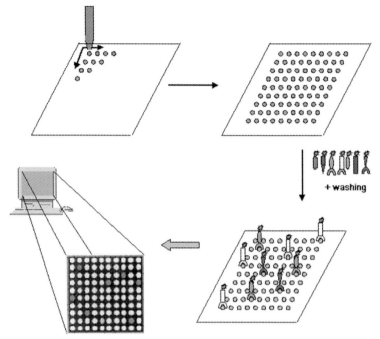

Fig. 6.2.1. Fabrication of an array by spotting or spatially addressable synthesis, probing with the analyte, and readout (modified after Ref. [1]).

(e.g. single nucleotide polymorphisms). In general the promise of DNA microarrays for cell biologists is to provide a more complete molecular view of cellular states and responses in complex tissues.

The analyte is most often fluorescence-labeled DNA or RNA molecules isolated from a cell extract. After hybridization with oligonucleotides immobilized on the microarray a fluorescence image can be generated which can be compared with one of an extract of cells in a different state (Figure 6.2.2). The underlying assumption is that genes at an upregulated level result in an increase in mRNA, which can be monitored by using this technology.

The generation of DNA microarrays can be achieved in two different ways: on the one hand one can immobilize either cDNA, generated by biochemical techniques from mRNA, or synthetic oligonucleotides on a glass slide using chemical functionalities introduced during conventional automated synthesis [3, 4]. These linking units – added at the 3′ end (starting point of the synthesis), internal positions, or the 5′ end (end of the synthesis) – can react with complementary functionalities on the glass surface (Table 6.2.1). Synthesis of the desired oligonucleotides directly on a glass slide is an alternative way in preparing arrays. One must distinguish between two different protecting group strategies:

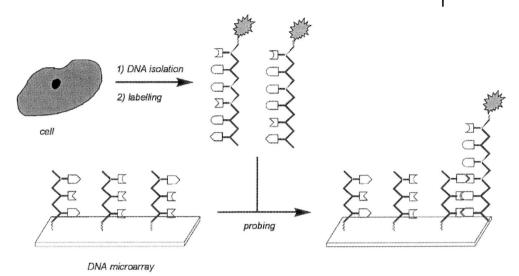

Fig. 6.2.2. Sample preparation and probing with a DNA microarray.

- the use of photolabile protecting groups like MeNPoc [5], DMBoc [6], NNEoc [7] that can be removed directly by irradiation with suitable wavelength (Figure 6.2.3) or, alternatively,
- use of conventional dimethoxytrityl (DMT) groups that can be removed by photochemically induced generation of acid at the site of irradiation [8].

In both of these, however, the patterning (and thus the "generation of spots") is done by means of photolithography. This kind of oligonucleotide synthesis has significantly higher spatial precision than spotting. Recently two new structuring methods similar to photolithography have been employed; these use digital assembly of micro-mirrors which are dynamically programmable [9, 10].

6.2.2.2
Protein Microarrays

As the focus has shifted from deciphering the human genome to understanding the more complex human proteome it has also become apparent that it would be desirable to adapt for this new field a technique which could become as powerful and important as gene-chip technology has become to genomics [11]. Not surprisingly, parallel to the increase in complexity of proteomics the design and fabrication of protein chips is also more challenging. In contrast with DNA, whose function can be deduced almost exclusively from its sequence, the biological activity of proteins is based on the three-dimensional arrangement of the sequence of 20 amino acid building blocks – and even this simplified picture neglects the importance of post-translational modifications such as lipidation, phosphorylation or

6.2 Small Molecule Arrays

Tab. 6.2.1. Immobilization reactions used for the preparation of DNA arrays, protein arrays, and small molecule arrays.

immobilization reaction	DNA	proteins	small molecules
Si–(CH₂)ₙ–epoxide + H₂N–R →(base) Si–(CH₂)ₙ–CH(OH)–CH₂–NH–R	[32], [33]	[51], [52]	
Si–(CH₂)ₙ–COOH + H₂N–R →(a) coupling reagent / b) adsorption) Si–(CH₂)ₙ–C(O)–NH–R	a) [34]	b) [53]	
Si–(CH₂)ₙ–CHO + H₂N–R → Si–(CH₂)ₙ–CH=N–R ; NaBH₄ → Si–(CH₂)ₙ–CH₂–NH–R	[34], [35]	[54], [15], [55] (on gel pads), [56]	
Si–ₙ(H₂C)–phenyl–NCS + H₂N–R →(pyridine/DCM) Si–(H₂C)ₙ–phenyl–NH–C(S)–NH–R	[36]	[57]	
Si–O–C(O)–NHS-ester (X) + H₂N–R → Si–O–C(O)–NH–R		X = H: [54], [57]; X = SO₃Na: [58]	
Si–(CH₂)ₙ–NH₂ + HO–C(O)–R →(a) coupling reagent / b) adsorption) Si–(CH₂)ₙ–NH–C(O)–R	a) [34], [37]	b) [14]	[64]
Si–(CH₂)ₙ–N(H)–C(O)–CHO + H₂NO–R →(pH 5.2) Si–(CH₂)ₙ–N(H)–C(O)–CH=N–O–R			[28]

6.2.2 Arrays

Tab. 6.2.1 (continued)

immobilization reaction	DNA	proteins	small molecules
Au|S-(CH₂)ₙ-NH-C(O)-CH₂-SH + maleimide-N-DNA → Au|S-(CH₂)ₙ-NH-C(O)-CH₂-S-succinimide-N-DNA	[41]		
|Si-(CH₂)ₙ-SH + maleimide-N-R → |Si-(CH₂)ₙ-S-succinimide-N-R	[42]		
|Si-(CH₂)ₙ-SH + R¹-S-S-R-DNA-oligomer pH 9 carbonate buffer → |Si-S-S-R-DNA-oligomer	[43]		
Au|S-(CH₂)ₙ-NH-C(O)-CH₂-S-S-pyridine + HS-R → Au|S-(CH₂)ₙ-NH-C(O)-CH₂-S-S-R	[41]		
Au|S-CH₂-(O(CH₂)₂)ₙ-cyclopentadiene + quinone-CH₂-C(O)-X-R H₂O → Au|S-CH₂-(O(CH₂)₂)ₙ-(Diels-Alder adduct)-C(O)-X-R			[66], [67], [68] X = NH; O
|Si-**oligonucleotide** + **PNA TAG**-R → hybridization			[29]
|Ni + protein **HisX6 tags** → coordination attachment		[15], [59]	
antibody/antigen pair/ PCR-product **adsorption on poly-l-lysine** coated microscope slides	[44]	[13]	

Method	Reference
a) polyacrylamide gel b) agarose gel c) |Si–X–C(=O)–N(H)–NH + **aldehyde** functionalized protein/oligonucleotide	a) [55] b) [60] c) [61]
|Si–X–C(=O)–N(H)–N=N–protein/oligonucleotide via a) + b) **hydrazine hydrate** → |gel/Si–X	a) [45]
|silicon-Au + pyrrole~oligonucleotide → electrochemical oxidation → |silicon-Au pyrrole-oligonucleotide	[46]
|silicon-Pt-agarose-**streptavidin** + 5'-**biotin**-DNA → streptavidin-biotin binding	[47]
|(**strept**)**avidin** + **biotinylated** protein → (strept)avidin-biotin binding	[62], [63]
|Si–(CH₂)₃–N(H)–C(=O)–CH₂–Br + ⁻S–P(=O)(OR)(OR') → |Si–(CH₂)₃–N(H)–C(=O)–CH₂–S–P(=O)(OR)(OR')	[48], [49]
electrospray deposition of charged products on an electrostatic substrate	[50]
|Si–X–O–N(succinimide)(C=O)₂ + **nucleophil**–R → |Si–X–C(=O)–nucleophil–R	X = several hetero-bifunctional cross-linkers: [61]
|Si-**nitrocellulose** + **dextran/inulin** → saline (0.9% NaCl) → glycoconjugate	[69]

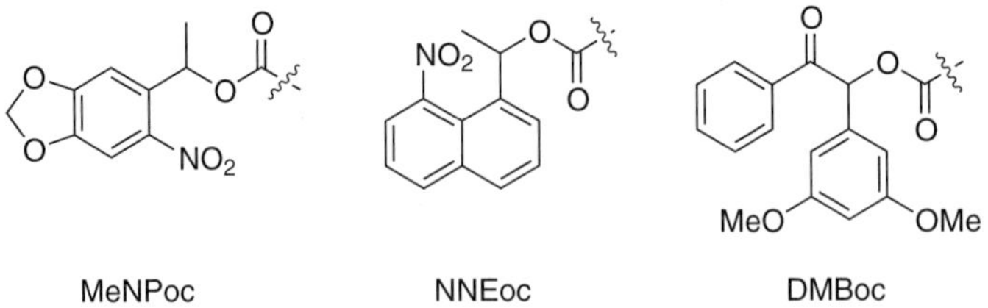

6.2 Small Molecule Arrays

Fig. 6.2.3. Photo-cleavable protecting groups used in light-directed synthesis of DNA chips. (MeNPoc, NNEoc, DMBoc)

glycosylation. It is, therefore, necessary that instead of a short sequence the whole folded protein has to be presented on the array to study its interaction with different partners (Table 6.2.1) [12].

Protein assays using ordered arrays emerged in the late 90s [13]. Recently Lahiri et al. introduced a technique which enables the fabrication of microarrays for membrane proteins [14]. Printed lipid microspots on γ-aminopropylsilane slides have high mechanical stability, independent of whether the lipid is in the gel- or fluid-phase. They immobilized G-protein coupled receptors and studied their interaction with a set of inhibitors.

An impressive example of how powerful this technique can be for the analysis of protein–protein and protein–small molecule interaction was given by Snyder et al. [15]. They attached 6566 protein preparations, representing 5800 different yeast proteins, through their HisX6 tags to Ni-coated glass slides using a commercial printer. The yeast proteome chip was probed with biotinylated calmodulin in the presence of calcium. The bound biotinylated protein was detected using fluorescence dye conjugate Cy3-labeled streptavidin. In addition to known partners, the calmodulin probe identified additional potential partners. Several of these binding partners contain a motif whose consensus is (I/L)QXXK(K/X)GB, where X is any residue and B is a basic residue. The yeast proteome chip was also screened against five types of phosphoinositide liposome, containing 1% of a biotinylated phospholipid. After visualization by use of Cy3-streptavidin a total of 150 different protein targets were identified; 52 corresponded to uncharacterized proteins.

So far the bottleneck in producing protein chips seems to be the preparation of the individual proteins, but for this heavily researched area solutions are on the horizon. The advantage of the protein chip approach is that a comprehensive set of individual proteins can be directly screened in vitro for a wide variety of activity, including protein–drug interactions, protein–lipid interactions, and enzymatic assays using a wide range of in vitro conditions – and faster and cheaper than with conventional methods.

6.2.2.3
Cell Arrays

Although DNA arrays and protein arrays provide valuable information, their model character remains. The cell as the smallest functional unit of life would provide a more realistic entity for biological investigations [16]. Ziauddin and Sabatini have recently introduced an elegant method of producing microarrays expressing defined cDNAs [17]. Nanoliter volumes of plasmid DNA in an aqueous gelatin solution were "printed" on a glass slide by use of a robotic arrayer. After exposure to a lipid transfection reagent the slide was then placed in a culture dish and covered with adherent mammalian cells in medium. The cells growing on the DNA and gelatin spots expressed the DNA and divided 2–3 times in the process of creating a microarray with features consisting of clusters of transfected cells. By screening transfected cell microarrays expressing 192 different cDNAs they identified proteins involved in tyrosine kinase signaling, apoptosis and cell adhesion.

6.2.3
Small Molecule Arrays

Identification of a plethora of new and interesting biological targets revealed that the traditional way of synthesizing compounds one after the other was not efficient enough to meet the challenge of identifying small-molecule modulators for these proteins. The technique of solid-phase synthesis (Boxes 14 and 25) in combination with combinatorial methods (Box 11, see also Chapter 3.1) provided the required gain in efficiency and productivity. The one-bead–one-compound approach (OBOC) developed by Lam et al. enables particularly rapid production of many compounds in a short time [18]. Because each compound-bead is spatially separable, the OBOC combinatorial library can be viewed as a huge microarray that is not addressable [19]. Over the last few years the first prototypes of spatially addressable small molecule microarrays have been reported in the literature [20].

6.2.3.1
Synthesis on Planar Supports

Direct synthesis on solid supports has proven to be a viable approach for production of DNA microarrays, because of the linear oligomeric structure based on a limited set of four building blocks. So far synthesis on planar supports has been limited to small molecules which can be built up iteratively by means of building blocks. It has, therefore, been used almost exclusively for synthesis of peptides [21, 22]. Before the development of light-directed, spatially addressable parallel chemical synthesis, Fodor and his group at Affymax implemented this approach for construction of a peptide library on a glass surface [23]. Photocleavable protecting groups on the modified surface of the attached amino-acid building blocks are cleaved unless they are covered by a mask. Spot synthesis, developed by Frank in 1992 [24], involves the positionally addressed delivery of small volumes of activated amino acid solutions directly to a coherent membrane sheet. The functional groups fixed on the membrane surface react with the pipetted reagents and conventional solid-phase syntheses occur. The purity of short peptides of up to 15 amino acids prepared by the Spot method are similar to those synthesized by solid-phase methods in reactors. A huge variety of biological tests has been conducted using cellulose-bound peptide arrays including the mapping of antibody epitopes and the study of protein–protein interactions [25].

6.2.3.2
Spotting of Small Molecules

As combinatorial chemistry has become an increasingly established method for the production of different sets of small molecules, a strategy was born to combine classic bead-based combinatorial synthesis with the preparation of arrays by spotting (Table 6.2.1). In 1999 Schreiber et al. were the first group to "print" small molecules on an inorganic planar support to study protein–ligand interactions [26]. They modified a standard microscope slide by attaching maleimide groups on its surface. They then spotted the dye rhodamine and three molecules with known protein-binding properties and containing free thiol groups on to the glass slide. The thiol groups reacted by 1,4-addition with the maleimide resulting in the desired covalent linkage to the surface. As expected, all three ligands participated in specific binding with their fluorescence-labeled protein partners – biotin (streptavidin), digoxigenin (monoclonal antibody DI-22), and a synthetic pipecolyl α-ketoamide (FKBP-12). By "printing" lower affinity ligands of FKBP12 they could show that the intensity of fluorescent spots correlated well with the affinity of the protein for the immobilized compounds. Being aware of the limitations imposed by the thiol group Schreiber et al. expanded their methodology to include alcohol-containing small molecules "printed" on glass slides (Scheme 6.2.1). With this strategy they could build upon well established solid-phase synthesis on polymeric supports using silyl linkers, cleave the small molecules by addition of HF-pyridine, and spot these stock solutions on activated glass slides [27].

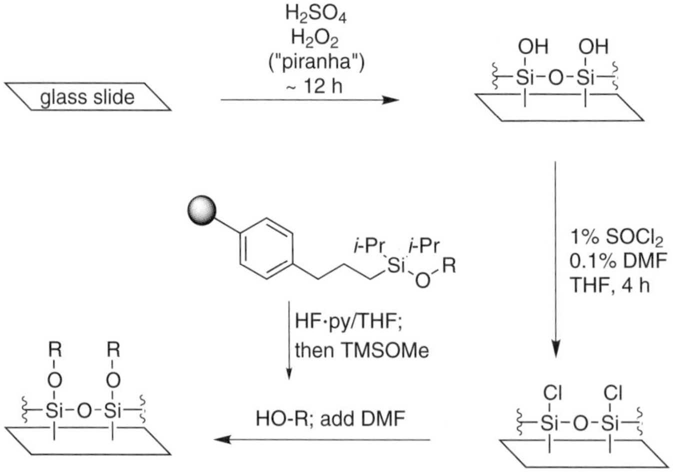

Scheme 6.2.1. Activation of glass slides and the covalent attachment of alcohols.

Lam et al. have introduced two new ligation methods for glyoxylic acid-modified glass slides with which they could immobilize biotin or peptide ligands derivatized either via an amino-oxy group by oxime formation or a 1,2-aminothiol group by thiazolidine ring formation [19, 28]. The microarray of immobilized ligands was analyzed in three different biological assays.

1. In a protein-binding assay with fluorescence detection a microarray of biotin, HPYPP-peptide and WSHHPQFEK-peptide was screened against streptavidin-Cy3 and avidin-Cy5. By following the same principle an anti-human insulin monoclonal antibody was also screened against a set of different peptides.
2. In a functional phosphorylation assay $[\gamma^{33}P]$-ATP and a specific protein kinase were used to label peptide substrate spots.
3. In an adhesion assay different areas of a slide with a certain peptide sequence were covered separately with different cell lines. After washing and staining only the WEHI 2312 cells were found to bind to the spotted area of the glass slide.

Schultz et al. devised an ingenious approach to convert a spatially separable, but not addressable split-pool (OBOC) library into a spatially addressable microarray (Scheme 6.2.2) [29]. Conventional solid-phase techniques were used for the synthesis of small molecules tethered to peptidonucleic acid (PNA) tags. The chemically robust and, by iterative amide formation, easily constructable PNA tags serve two purposes – first to encode the synthetic history of the small molecule and second to positionally encode the identity of the small molecule, by its location, on hybridization to a commercial oligonucleotide microarray. Overcoming the limitation of other screening technologies, which require fluorescence-labeled proteins or samples, they described an approach in which the fluorophore is conjugated to the PNA–small molecule entity. The library is then incubated with the target protein. Size-exclusion filtration enables one to separate unbound ligands from the protein–ligand mixture. After hybridization with the oligonucleotide array fluorescent spots encode molecules which have bound to the target protein. The feasibility of this approach has been shown by the screening of six mechanism-based cysteine protease inhibitors against cathepsin L.

Surpassing the mere proof-of-concept character of the work highlighted above, Schreiber et al. have prepared a high-density microarray of 3780 structurally complex 1,3-dioxane small molecules which have been synthesized by one-bead–one-stock-solution technology with the use of macrobeads [30]. By probing the array with fluorescence-labeled Ure2p, a protein which represses the transcription factors Gln3p and Nil1p, several compounds which bind Ure2p were identified. One of these compounds, named uretupamine, specifically activates a glucose-sensitive transcriptional pathway downstream of Ure2p.

So far undisclosed in the peer-reviewed literature are contributions by the company Graffinty (*www.graffinity.de*) [31]. It has built up a technology platform in which combinatorial libraries are generated by solid-phase methodology using an acid-labile S-trityl linker. After cleavage the free thiols of the small molecules react

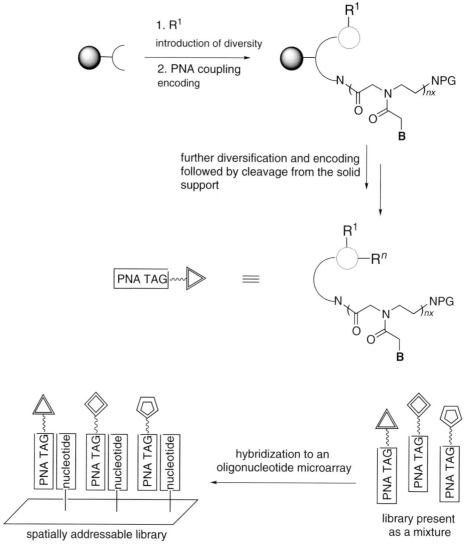

Scheme 6.2.2. Split-pool synthesis of PNA-encoded combinatorial libraries. R^n = element of diversity present in library, B = base of the peptidonucleic acid, x = number of bases encoding a single element of diversity, n = number of chemical diversity-introducing steps, PG = protecting group.

with maleimide-derivatized gold surfaces. Proprietary screening techniques enable Graffinity to perform spatially resolved surface plasmon resonance (SPR) experiments, enabling them to use any protein target without the need to label it first.

With these recent contributions the fundamental experimental challenges have been overcome. For this technology to mature for routine applications the fundamental aspects of activating small molecules for spotting (via orthogonal ligation reactions compatible with many functional groups) and biological screening (notably the problem of non-specific binding) require further optimization.

6.2.4
Outlook and Conclusions

DNA-chip technology has achieved a status which will be followed by establishment of the technique as a routine method in laboratories working in the field of microbiology, biochemistry, or chemical biology, although the quality of the gene chips has still to be improved and the cost of gene chips and arraying equipment must be reduced significantly. Protein-chip technology has the potential of becoming even more useful than the DNA chips, but up to that point many technological advances must still be made, in particular in the field of high-throughput protein expression.

The recently introduced technique of small molecule arrays must still prove its real advantage compared with conventional solution-based methods. For future applications it can, nevertheless, be speculated that biological laboratories using their, then routine, DNA or protein chip arrayer will be able to screen their newly discovered biological targets against arrays of diverse sets of small molecules. For this first-round screening known drugs, natural products, or other privileged structures would represent a perfect choice of compounds to be spotted on a glass slide. Screening against a chip containing an ensemble of widely used drugs could even give indications of side-effects during conventional treatment. On the other hand, a protein-detecting array composed of known protein and enzyme inhibitors and highly specific ligands to each protein could be capable of recognizing their target protein in complex biological solutions such as a cell extract [12].

References

1 C. M. NIEMEYER, D. BLOHM, *Angew. Chemie* 1999, *111*, 3039–3043; *Angew. Chem. Int. Ed.* 1999, *38*, 2865–2869.
2 D. J. LOCKHART, E. A. WINZELER, *Nature* 2000, *405*, 827–836 (Review).
3 M. C. PIRRUNG, *Angew. Chem.* 2002, *114*, 1326–1341; *Angew. Chem. Int. Ed.* 2002, *41*, 1276–1289.
4 C. M. NIEMEYER, *Angew. Chem.* 2001, *113*, 4254–4287; *Angew. Chem. Int. Ed.* 2001, *40*, 4128–4158.
5 G. H. MCGALL, A. D. BARONE, M. DIGGELMANN, S. P. A. FODOR, E. GENTALEN, N. NGO, *J. Am. Chem. Soc.* 1997, *119*, 5081–5090.
6 M. C. PIRRUNG, L. FALLON, G. H. MCGALL, *J. Org. Chem.* 1998, *63*, 241–246.
7 A. D. BARONE, J. E. BEECHER, P. BURY, C. CHEN, T. DOEDE, J. A. FIDANZA, G. H. MCGALL, *Nucleos. Nucleot. Nucl. Acids* 2001, *20*, 525–531.
8 G. WALLRAFF, J. LABADIE, P. BROCK, T. NGUYEN, T. HUYNH, W. HINSBERG, G. H. MCGALL, *Chemtech* 1997, 22–32.
9 S. SINGH-GASSON, R. D. GREEN, Y.

Yue, C. Nelson, F. Blattner, M. R. Sussman, F. Cerrina, *Nat. Biotechnol.* **1999**, *17*, 974–978.

10 E. LeProust, J. P. Pellois, P. Yu, H. Zhang, X. Gao, O. Srinnavit, E. Gulari, Z. Zhou, *J. Comb. Chem.* **2000**, *2*, 349–354.

11 R. E. Jenkins, S. R. Pennington, *Proteomics* **2001**, *1*, 13–29 (Review).

12 T. Kodadek, *Chem. Biol.* **2001**, *8*, 105–115.

13 B. B. Habb, M. J. Dunham, P. B. Brown, *Genome Biol.* **2000**, *1*, 1–22.

14 Y. Fang, A. G. Frutos, J. Lahiri, *J. Am. Chem. Soc.* **2002**, *124*, 2394–2395.

15 H. Zhu, M. Bilgin, R. Bangham, D. Hall, A. Casamyor, P. Bertone, N. Lan, R. Jansen, S. Bidlingmaier, T. Houfek, T. Mitchell, P. Miller, R. A. Dean, M. Gerstein, M. Snyder, *Science* **2001**, *293*, 2101–2105.

16 S. A. Sundberg, *Curr. Opin. Biotechnol.* **2000**, *11*, 47–53.

17 J. Ziauddin, D. M. Sabatini, *Nature* **2001**, *411*, 107–110.

18 K. S. Lam, S. E. Salmon, E. M. Hersh, V. J. Hruby, W. M. Kazmierski, R. J. Kanpp, *Nature* **1991**, *354*, 82–84.

19 K. S. Lam, M. Renil, *Curr. Opin. Chem. Biol.* **2002**, *6*, 353–358 (Review).

20 J. Khandurina, A. Guttman, *Curr. Opin. Chem. Biol.* **2002**, *6*, 359–366 (Review).

21 M. Lebl, *Biopolymers (Pept. Sci.)* **1998**, *47*, 397–404.

22 M. C. Pirrung, *Chem. Rev.* **1997**, *97*, 473–488.

23 S. P. A. Fodor, J. L. Read, M. C. Pirrung, L. Stryer, A. T. Lu, D. Solas, *Science* **1991**, *251*, 767–773.

24 R. Frank, *Tetrahedron* **1992**, *48*, 9217–9232.

25 H. Wenschuh, R. Volkmer-Engert, M. Schmidt, M. Schulz, J. Schneider-Mergener, U. Reineke, *Biopolymers (Pept. Sci.)* **2000**, *55*, 188–206.

26 G. MacBeath, A. N. Koehler, S. L. Schreiber, *J. Am. Chem. Soc.* **1999**, *121*, 7967–7968.

27 P. J. Hergenrother, K. M. Depew, S. L. Schreiber, *J. Am. Chem. Soc.* **2000**, *122*, 7849–7850.

28 J. R. Falsey, M. Renil, S. Park, S. Li, K. S. Lam, *Bioconjugate Chem.* **2001**, *12*, 346–353.

29 N. Winssinger, J. L. Harris, B. J. Backes, P. G. Schultz, *Angew. Chem. Int. Ed.* **2001**, *40*, 3152–3155.

30 F. G. Kuruvilla, A. F. Shamji, S. M. Sternson, P. J. Hergenrother, S. L. Schreiber, *Nature* **2002**, *416*, 653–657.

31 www.graffinity.de

32 (a) J. B. Lamture, K. L. Beattie, B. E. Burke, M. D. Eggers, D. J. Ehrlich, R. Fowler, M. A. Hollis, B. B. Kosicki, R. K. Reich, S. R. Smith, R. S. Varma, M. E. Hogan, *Nucleic Acids Res.* **1994**, *22*, 2121–2125; (b) M. D. Eggers, M. E. Hogan, R. K. Reich, J. B. Lamture, D. J. Ehrlich, M. Hollis, B. B. Kosicki, T. Powdrill, K. L. Beattie, S. Smith, *BioTechniques* **1994**, *17*, 516–525; (c) W. G. Beattie, L. Meng, S. L. Turner, R. S. Varma, D. D. Dao, K. L. Beattie, *Mol. Biotechnol.* **1995**, *4*, 213–225.

33 Y. Belosludtsev, B. Iverson, S. Lemeshko, R. Eggers, R. Wiese, S. Lee, T. Powdrill, M. Hogan, *Anal. Biochem.* **2001**, *292*, 250–256.

34 N. Zammatteo, L. Jeanmart, S. Hamels, S. Courtois, P. Louette, L. Hevesi, J. Remacle, *Anal. Biochem.* **2000**, *280*, 143–150.

35 www.arrayit.com

36 Z. Guo, R. A. Guifoyle, A. J. Thiel, R. Wang, L. M. Smith, *Nucleic Acids Res.* **1994**, *22*, 5456–5465.

37 B. Joos, H. Kuster, R. Cone, *Anal. Biochem.* **1997**, *247*, 96–101.

38 (a) T. Strother, W. Cai, X. Zao, R. J. Hamers, L. M. Smith, *J. Am. Chem. Soc.* **2000**, *122*, 1205–1209; (b) T. Strother, R. J. Hamers, L. M. Smith, *Nucleic Acids Res.* **2000**, *28*, 3535–3541.

39 T. Okamoto, T. Suzuki, N. Yamamoto, *Nat. Biotechnol.* **2000**, *18*, 438–441.

40 (a) A. G. Frutos, L. M. Smith, R. M. Corn, *J. Am. Chem. Soc.* **1998**, *120*, 10277–10282; (b) J. M. Brockmann, A. G. Frutos, R. M. Corn, *J. Am. Chem. Soc.* **1999**, *121*, 8044–8051.

41 E. A. Smith, M. J. Wanat, Y. Cheng,

S. V. P. Barreira, A. G. Frutos, R. M. Corn, *Langmuir* **2001**, *17*, 2502–2507.

42 (a) V. Benoit, A. Steel, M. Torres, Y.-Y. Yu, H. Yang, J. Cooper, *Anal. Chem.* **2001**, *73*, 2412–2420; (b) A. Steel, M. Torres, J. Hartwell, Y.-Y. Yu, N. Ting, G. Hoke, H. Yang, *Microarray Biochip Technology*, M. Schena, ed., Eaton, Natick, **2000**, pp. 87–118.

43 Y. H. Rogers, P. Jiang-Baucom, Z.-J. Huang, V. Bogdanov, S. Anderson, M. T. Boyce-Jacino, *Anal. Biochem.* **1999**, *266*, 23–30.

44 (a) M. Schena, D. Schalon, R. W. Davis, P. O. Brown, *Science* **1995**, *270*, 467–470; (b) D. A. Lashkari, J. L. DeRisi, J. H. McCusker, A. F. Namath, C. Gentile, S. Y. Hwang, P. O. Brown, R. W. Davis, *Proc. Natl. Acad. Sci. USA* **1997**, *94*, 13057–13062.

45 (a) J. Zlatanova, A. Mirzabekov, *Methods in Molecular Biology, Vol. 170: DNA arrays: methods and protocols*, J. B. Rampal, ed., Humana Press, Totowa, NJ, **2001**, 17–38; (b) G. Yershov, V. Barsky, A. Belgovskiy, E. Kirillov, E. Kreindlin, I. Ivanov, S. Parinov, D. Guschin, A. Drobishev, S. Dubiley, A. Mirzabekov, *Proc. Natl. Acad. Sci. USA* **1996**, *93*, 4913–4918.

46 (a) T. Livache, A. Roget, E. Dejean, C. Barthet, G. Bidan, R. Téoule, *Nucleic Acids Res.* **1994**, *22*, 2915–2921; (b) T. Livache, B. Fouque, A. Roget, J. Marchand, G. Bidan, R. Téoule, G. Mathis, *Anal. Biochem.* **1998**, *255*, 188–194.

47 (a) M. J. Heller, E. Tu, A. Holmsen, R. G. Sosnowski, J. O'Connel, *DNA Microarrays. A Practical Approach*, M. Schena, ed., Oxford Press, New York, **1999**, 167–185; P. N. Gilles, D. J. Wu, C. B. Foster, P. J. Dillon, S. J. Chanock, *Nat. Biotechnol.* **1999**, *17*, 365–370; (c) M. J. Heller, A. H. Forster, E. Tu, *Electrophoresis* **2000**, *21*, 157–164.

48 M. C. Pirrung, J. D. Davis, A. L. Odenbaugh, *Langmuir* **2000**, *16*, 2185–2191.

49 X. Zhao, S. Nampalli, A. J. Serino, S. Kumar, *Nucleic Acids Res.* **2001**, *29*, 955–959.

50 V. N. Morozov, T. Y. Morozova, *Anal. Chem.* **1999**, *71*, 3110–3117.

51 H. Zhu, M. Snyder, *Curr. Opin. Chem. Biol.* **2001**, *5*, 40–45.

52 H. Zhu, J. F. Klemic, S. Chang, P. Bertone, A. Casamyor, K. G. Klemic, D. Smith, M. Gerstein, M. A. Reed, M. Snyder, *Nat. Genet.* **2000**, *26*, 283–289.

53 K.-B. Lee, S.-J. Park, C. A. Mirkin, J. C. Smith, M. Mrksich, *Science* **2002**, *295*, 1702–1705.

54 G. MacBeath, S. L. Schreiber, *Science* **2000**, *289*, 1760–1763.

55 (a) D. Guschin, G. Yershov, A. Zaslavsky, A. Gemmel, A. Shick, D. Proudnikov, P. Arenkov, A. Mirzabekov, *Anal. Biochem.* **1997**, *250*, 203–211; (b) P. Arenkov, A. Kukthin, A. Gemmel, S. Voloshchuk, V. Chupeeva, A. Mirzabekov, *Anal. Biochem.* **2000**, *278*, 123–131.

56 K. Nakanishi, H. Muguruma, I. Karube, *Anal. Chem.* **1996**, *68*, 1695–1700.

57 R. Benters, C. M. Niemeyer, D. Wöhrle, *ChemBioChem* **2001**, *2*, 686–694.

58 S. C. Lin, F. G. Tseng, H. M. Huang, C. Y. Huang, C. C. Chieng, *Fresenius J. Anal. Chem.* **2001**, *371*, 202–208.

59 C. F. W. Becker, C. L. Hunter, R. P. Seidel, S. B. H. Kent, R. S. Goody, M. Engelhard, *Chem. Biol.* **2001**, *8*, 243–252.

60 R. Vankova, A. Gaudinova, H. Sussenbekova, P. Dobrev, M. Strnad, J. Holik, J. Lenfeld, *J. Chromatogr. A* **1998**, *811*, 77–84.

61 L. C. ShriverLake, B. Donner, R. Edelstein, K. Breslin, S. K. Bathia, F. S. Ligler, *Biosens. Bioelectron.* **1997**, *12*, 1101–1106 (comparison of heterobifunctional cross-linkers).

62 M.-L. Lesaicherre, R. Y. P. Lue, G. Y. J. Chen, Q. Zhu, S. Q. Yao, *J. Am. Chem. Soc.* **2002**, *124*, 8768–8769.

63 J. F. Mooney, A. J. Hunt, J. R. McIntosh, C. A. Liberko, D. M. Walba, C. T. Rogers, *Proc. Natl. Acad. Sci. USA* **1996**, *93*, 12287–12291.

64 G. Korbel, G. Lalic, M. D. Shair, *J. Am. Chem. Soc.* **2001**, *123*, 361–362.
65 O. Melnyk, X. Duburg, C. Olivier, F. Urbés, C. Auriault, H. Gras-Masse, *Bioconjugate Chem.* **2002**, *13*, 713–720.
66 B. T. Houseman, J. H. Huh, S. J. Kron, M. Mrksich, *Nat. Biotechnol.* **2002**, *20*, 270–274.
67 B. T. Houseman, M. Mrksich, *Trends Biotechnol.* **2002**, *20*, 279–281.
68 B. T. Houseman, M. Mrksich, *Chem. Biol.* **2002**, *9*, 443–454.
69 D. Wang, S. Liu, B. J. Trummer, C. Deng, A. Wang, *Nat. Biotechnol.* **2002**, *20*, 275–281.
70 S. Park, I. Shin, *Angew. Chem. Int. Ed.* **2002**, *41*, 3180–3182.

Selected Web Sites

Affibody AB: *http://www.affibody.com/*
Affymetrix Inc.: *http://www.affymetrix.com/*
Biacore International: *http://www.biacore.com/*
Cambridge Antibody Technology's: *http://www.cambridgeantibody.com/*
Ciphergen Biosystems: *http://www.ciphergen.com/*
Graffinity Pharmaceuticals AG: *http://www.graffinity.de/*
Milagen: *http://www.milagen.com/*
Pat Brown Laboratory: *http://cmgm.stanford.edu/pbrown/mguide/index.html*
Phylos: *http://www.phylos.com/*
Pierce Boston Technology: *http://www.piercenet.com/*
SomaLogic: *http://www.somalogic.com/*
Telechem International, Inc.: *http://www.arrayit.com/*
Zyomyx: *http://www.zyomyx.com/*

6.3
Biotechnological Production of D-Pantothenic Acid and its Precursor D-Pantolactone

Maria Kesseler

6.3.1
Introduction

D-Pantolactone (Figure 6.3.1) is an important intermediate in the production of D-pantothenic acid, also called vitamin B5. Deficiency of pantothenic acid can result in symptoms such as pathological changes of the skin and mucosa, disorders in the gastrointestinal tract and nervous system, organ changes, and hormonal disorders. Pantothenic acid is used mainly in feed for chicken and pigs and also as a vitamin supply in human nutrition. Its commercial form, the calcium salt, is produced worldwide on a multi-thousand ton scale.

In principle, the variety of synthetic methods used for production of optically active hydroxy acids can be divided into three types. Some acids are derived from the chiral pool of naturally occurring optically active substances. For example about 30,000 tons of L-tartaric acid are prepared from tartar; D-gluconic acid is produced by biocatalytic oxidation of D-glucose via gluconolactone in high yield, e.g. in an *Aspergillus* submers process [1]. Considering the diversity of optically active intermediates and products in the chemical and pharmaceutical industries, the chiral pool obviously gives access to only a limited number of compounds, so chemical processes were developed to overcome this limitation. These include the well known methods of asymmetric synthesis and the fractionated crystallization of diastereomeric salts of the acid. Examples are the hydrogenation of α-keto esters, for example acyl 2-oxo-4-phenylbutyrate (Ciba Geigy, see Ref. [2]) and the resolution of racemic mandelic acid by crystallization of R-mandelic acid with R-1-phenylethylamine by Yamakawa, Japan.

D-Pantothenic acid is also traditionally produced by chemical processes which involve efficient but troublesome and costly crystallization of diastereomeric salts of pantoate and chiral amines. After lactonization of the isolated D-pantoate, D-pantolactone is reacted with β-alanine to give D-pantothenate. Because the monovalent salts of pantothenic acid are highly hygroscopic, conversion into the calcium salt is essential for convenient formulation. The third class of synthetic processes for optically active compounds makes use of biotechnology. For natural com-

Highlights in Bioorganic Chemistry: Methods and Applications. Edited by Carsten Schmuck, Helma Wennemers.
Copyright © 2004 WILEY-VCH Verlag GmbH & Co. KGaA, Weinheim
ISBN: 3-527-30656-0

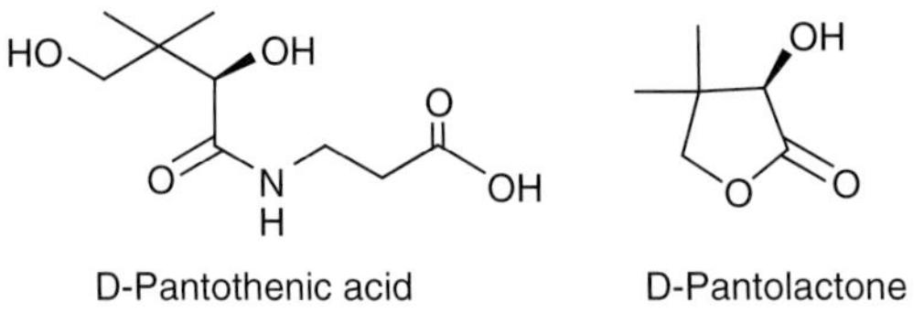

D-Pantothenic acid   D-Pantolactone

Fig. 6.3.1. D-Pantothenic acid and its chemical precursor D-pantolactone.

pounds, strains are generated that overproduce the desired metabolite (e.g. amino acids such as L-glutamate) during growth. For unnatural compounds, one or few biocatalytic steps are used to transform a precursor by isolated enzymes or whole cell catalysts into the desired product. In recent years, such new biotechnological preparation methods have arisen for the production of D-pantothenic acid. Here we review these reactions with a special focus on characteristics of the development of a typical biocatalyst.

6.3.2
Fermentative Production of D-Pantothenic Acid

Progress in the techniques of classical strain development and metabolic engineering (Box 24) have made a growing number of fermentation processes feasible and economically attractive. Beside the bulk amino acids, lactic acid, penicillins for the pharmaceutical market, and some vitamins, for example vitamin C (ascorbic acid [3]) and vitamin B5 (pantothenic acid), belong to the class of compound produced by fermentation. In the nineties Takeda, Japan, developed a fermentative process for the biosynthesis of D-pantothenic acid in *E. coli*. By adding a supply of β-alanine high titers (product concentrations achieved) and space–time yields (product concentrations achieved per hour) of >60 g L^{-1} pantothenic acid in 72 h were achieved [4]. Germany's Degussa AG is working on this process in cooperation with the University of Bielefeld and the Research Center of Jülich, Germany, using *E. coli* [5] and *Corynebacterium glutamicum* [6] as production organisms, respectively. The genes responsible for the biosynthesis of pantothenic acid (*panBCD*, *panE*; Figure 6.3.2) were overexpressed and production of the precursor α-ketoisovalerate was enhanced by overexpression of the genes *ilvBNCD* encoding the acetohydroxy acid synthase, isomerase reductase, and dihydroxy acid dehydratase and by deletion of *ilvA* encoding threonine dehydratase. Downstream processing included (optionally) separation of the cells and transformation of the mixed salt pantothenate solution into calcium D-pantothenate by addition of calcium hydroxide, evaporation of water, and spray drying [7].

In recent years, BASF AG, Ludwigshafen, and Omnigene Bioproducts, Cambridge, MA, USA, have developed a novel fermentation process for preparation of D-pantothenate with *Bacillus* [8]. *Bacillus subtilis* is inherently capable of producing

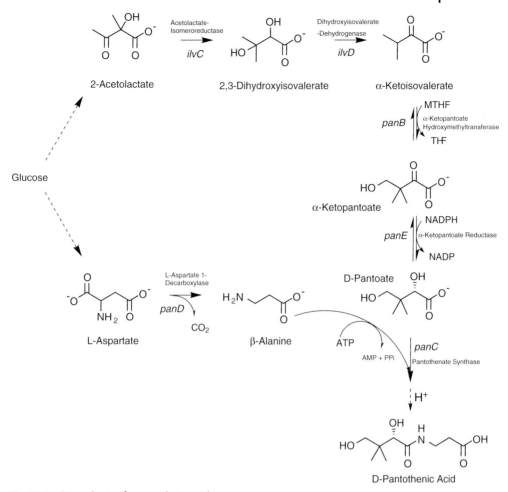

Fig. 6.3.2. Biosynthesis of D-pantothenic acid.

and excreting D-pantothenate – wild-type cells produce approximately 30 mg L^{-1} D-pantothenate in 24 h. For strains overexpressing the *panBCD* genes, the *panE* gene and the *ilvD* gene, and the *ilvBNCD* genes D-pantothenate titers increased by roughly three orders of magnitude in bench scale fermentation. By improving the availability of the second PanB substrate methylene tetrahydrofolate (MTHF) by overexpression of the genes *serA* and *glyA* encoding the 3-phosphoglycerate dehydrogenase and serine hydroxymethyltransferase, respectively, the titers could be further increased. Calcium D-pantothenate can be obtained by either adding calcium salts during or after the fermentation, or by ion exchange. Other downstream processing steps follow as usual.

6.3.3
Biocatalytic Production of D-Pantolactone

6.3.3.1
Biocatalytic Asymmetric Synthesis

Apart from microbial synthesis of D-pantothenic acid, the biocatalytic preparation of enantiomerically pure D-pantolactone is an attractive approach, because it not only fits into the old chemistry of D-pantothenic acid synthesis by reaction of β-alanine but also gives access to a second, related, product, D-pantothenol (D-N-pantoyl-3-propanolamine, (dex)panthenol), which is used in cosmetic industry. Two enzymatic methods for asymmetric synthesis are reported in the literature. One involves reduction of prochiral keto pantoic acid or keto pantolactone by bacteria and yeasts [9]. Although some strains afforded good yields and enantiomeric excesses, the biocatalytic reduction remained an elegant synthesis only on the laboratory scale, mainly because the keto forms are not as abundantly available as their reduced counterparts. Yamada and coworkers tried to overcome this drawback by using *rac*-pantolactone as starting compound for a mixed process, with enantioselective nocardial oxidation of L-pantolactone and subsequent D-specific agrobacterial reduction of keto pantoic acid which is spontaneously formed by chemical hydrolysis of keto pantolactone at pH 7 [10]. Although D-pantolactone was produced with 83% ee and 89% yield, product concentrations and space time yields were low (71 g L^{-1} in 8 days).

In a second approach hydrocyanic acid was added to hydroxypivaldehyde by use of (R)-selective hydroxynitrile lyase from almonds (PaHNL) [11]. (R)-Cyanohydrin was obtained in 84% yield and 89% ee, and was directly cyclized to give crude D-pantolactone by acid-catalyzed hydrolysis. Unfortunately, in contrast with O-protected hydroxy and halogenated pivalaldehydes, the technically available starting compound hydroxypivaldehyde requires use of purified enzyme (and high enzyme loading).

Finally, biocatalytic resolution was developed for more efficient production of D-pantolactone. Whereas the resolution of O-acyl pantolactone with lipases or esterases [12] did not lead to an industrially attractive process, the hydrolysis of *rac*-pantolactone by pantolactone hydrolases enabled development of a technically feasible and economic process.

6.3.3.2
Resolution of *rac*-Pantolactone by Fungal Hydrolysis of D-Pantolactone

In the nineties a process for the kinetic resolution of pantolactone by enantioselective hydrolysis was developed by Yamada and coworkers with Fuji Chemical Industries, Japan (now Daiichi Fine Chemical) using the fungus *Fusarium oxysporum* AKU3702 [13]. Like most of the *Fusarium* strains and strains of the related species *Gibberella*, *Cylindrocarpon*, *Penicillium*, and *Aspergillus*, *Fusarium oxysporum*

AKU3702 produces an aldonolactonase which not only hydrolyses a series of sugar lactones and some aromatic compounds, for example dihydrocoumarin, highly enantioselectively, but also d-pantolactone. The lactonohydrolase is a highly glycosylated protein dimer, with a subunit weight of 60 kDa, containing 1 mol calcium per subunit. The K_m and V_{max} for d-galactono-γ-lactone are 3.6 mM and 1440 U mg^{-1}, respectively (1 Unit = 1 µmol product formed per min; the Unit is an indicator of enzymes' catalytic activity); those for d-pantolactone are 123 mM and 653 U mg^{-1}, respectively. The enzyme was found to be very stable after immobilization of the fungal mycelium by embedding it in calcium alginate gel beads. It retained >90% of its activity after 180 batches at one batch per day. In the process d-pantolactone (350 L aqueous rac-pantolactone, 350 g L^{-1}) is hydrolyzed by 280 L immobilized mycelia (containing 15.2 kg wet cells) to d-pantoic acid, which is continuously titrated by addition of aqueous ammonia to keep the pH at 7.0 (Figure 6.3.3) [14]. d-Pantoate is obtained in high yields and optical purity (45–47.5% yield, ee 90–97% after 21 h) and can be separated from the remaining l-pantolactone by solvent extraction. Whereas l-pantolactone is racemized and re-introduced to the enzymatic resolution step, d-pantoate is converted to d-pantolactone by heating under acidic conditions. Extraction leads to the required synthon for production of d-pantothenic acid (and d-pantothenol, respectively). The aldonolactonase of the fungus *Fusarium oxysporum* is currently used by Fuji/Daiichi Chemicals for production of d-pantothenic acid on a multi-thousand-ton scale [15]. A similar process using *Fusarium moniliforme* SW-902 entrapped in potassium-carrageenan gels was recently developed in China [16].

6.3.3.3
Resolution of rac-Pantolactone by Bacterial Hydrolysis of l-Pantolactone: The Development of a Novel Biocatalyst

In recent years new pantolactone-hydrolyzing enzymes with enantioselectivity opposite to that of the fungal biocatalysts have been purified from bacteria (for a review of the diversity of lactonases see Ref. [17]).

For a kinetic resolution process use of both d- or an l-specific pantolactone hydrolase is possible in principle (Figure 6.3.3). If the unwanted l-form is hydrolyzed it might take longer for the remaining d-pantolactone to reach a sufficient enantiomeric excess; the process is, however, much more robust, e.g. towards competing spontaneous chemical hydrolysis.

From two independent screening approaches by Shimizu and coworkers and by BASF AG [18], agrobacterial pantolactone hydrolases were found to be highly active and enantioselective towards l-pantolactone. Because of low wild-type expression levels, the l-pantolactone hydrolase (Lph) from *Agrobacterium tumefaciens* Lu681 was cloned and expressed in *E. coli*. The corresponding *lph* gene was identified as an 807 bp open reading frame. Overexpression of the *lph* gene under control of the strong promoters P$_{tac}$ and P$_{rha}$ yields l-pantolactone hydrolase activity up to 600 U g^{-1} cell dry weight compared with 3 U g^{-1} in the wild type strain.

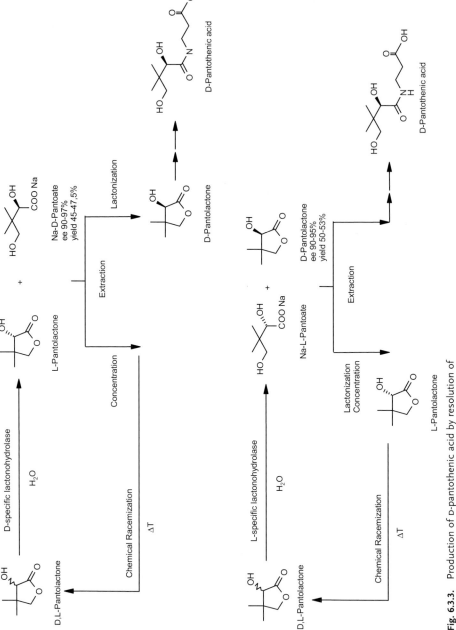

Fig. 6.3.3. Production of D-pantothenic acid by resolution of D,L-pantolactone by either a D- or an L-specific lactonohydrolase.

The enzyme has a monomer weight of 30 kDa and a K_m and V_{max} for L-pantolactone of 7 mM and 30 U mg^{-1}, respectively. X-ray fluorescence spectroscopy of crystals, and renaturation of urea/EDTA-denatured Lph in the presence of Zn^{2+}, Mn^{2+}, Co^{2+}, or Ni^{2+} indicated Lph to be a Zn^{2+}-hydrolase. Kinetic resolution of *rac*-pantolactone proceeds similarly to the fungal process mentioned above except that L-pantolactone is hydrolyzed and D-pantolactone is left behind. Repeated batches with isolated Lph and enzyme recovery by membrane filtration give D-pantolactone with 50% yield and 90–95% ee over 6 days.

When the kinetics of the hydrolysis of *rac*-pantolactone by Lph were investigated a decrease in the reaction velocity was observed; this was found to be because of competitive inhibition by D-pantolactone (Eq. 1) [19] and slight product inhibition of Lph. Under the same conditions of pH (7.5) and temperature (30 °C), L-pantolactone was completely converted to L-pantoic acid. This is certainly a disadvantage of Lph-catalyzed kinetic resolution, because space–time yields come to levels as low as 6 g L^{-1} h^{-1}.

$$\frac{1}{V} = \frac{1}{V_{max}} + \frac{K_m}{V_{max}} \left(1 + \frac{[D-PL]}{K_{D-PL}}\right) \frac{1}{[L-PL]} \tag{1}$$

For economic resolution of *rac*-pantolactone repeated use of the corresponding catalyst is required. Repeated batch conversions of D,L-pantolactone (30% *w/v*) with crude extracts of *E. coli* gave D-pantolactone in 50–53% yield with 90–95% ee over 6 days. The membrane filtration used for enzyme recycling from the product is, however, quite a costly separation step. While cell embedding into different types of (coated) alginate beads led to either diffusional limitations or dissolution of the beads, because of the high concentrations of monovalent cations introduced by titration, covalent immobilization of crude Lph to EupergitC (Röhm, Darmstadt, Germany) led to a stable biocatalyst easy to handle in a repeated batch process. The half-life of immobilized Lph was extended to 13 days. In the hydrolytic step the immobilisate was slowly ground to fine particles that tended to clog the filters during product recovery. Use of a lower solids content (<8% *w/v*) and optimization of the stirring geometry and velocity (while maintaining efficient titration of the pantoic acid synthesized) are expected to further increase the half-life.

After recombinant overexpression and immobilization and classical process engineering, the genetic engineering of an enzyme surely has great potential for improvement of biocatalysts (for a short review of genetic engineering issues see above). Because no structural information is yet available on Lph from *Agrobacterium*, it has not been possible to use a rational approach involving molecular modeling and site-directed mutagenesis. Future X-ray crystallization studies of Lph are expected to reveal molecular details of enzymatic L-pantolactone hydrolysis and to enable site-directed manipulations aiming at higher activity and elimination of competitive inhibition by excess D-pantolactone. For directed evolution (Box 20) of Lph a library of 11680 *lph* mutants has so far been generated by error-prone PCR. Improved variants of Lph were selected by high-throughput screening in microplates with nitrazinyellow as an indicator of pantoic acid synthesis. The activity of mutants F62S, K197D, and F100L was increased 2.3-, 1.7-, and 1.5-fold, respectively.

Although mutant F62S has excellent activity in the standard enzyme assay, its performance under process conditions (high concentrations of D,L-pantolactone) was poor. Because close adaptation of the high throughput screening assay to these conditions is very difficult for reasons of viscosity, this example shows the challenge of assay development for improvement of *industrially valuable* catalysts. It also shows that the factors of improvement achievable by only one round of random mutagenesis and screening strongly depend on the starting point, because obviously, as in this example, the specific activity of an already good biocatalyst is not as easily improved by factors in the range 10–100 as when the starting point is substantially lower [20]. Resolution of D,L-pantolactone by mutants K197D and F100L gave D-pantolactone with 90.4% ee at 51.8% conversion and 90.2% ee at 50.1% conversion, respectively, after 12 h using only 80% of the biomass compared with the unmutated control (90.5% ee at 52.3% conversion after 15 h). This corresponds to an increase of the productivity of the recombinant biocatalyst from 136 to 170 g g^{-1} (g D-pantolactone per g biomass) and slight shortening of the reaction time down to 12 h. Despite these great achievements the process has not reached commercial application, yet.

6.3.4
Conclusions

Several biotechnological synthetic methods for D-pantothenic acid and its precursor D-pantolactone have been developed over the past 15 years. Although all have reached preparative scale and might result in cost-effective production processes, they differ considerably in their process characteristics – for example educts and space–time yields, especially when a fermentation and biotransformation are compared. Compared with the chemical process, the biotechnological processes reduce waste production and provide the possibility of a more environmentally friendly yet still competitive means of production of D-pantothenic acid.

Acknowledgments

I am especially grateful to Professor Dr Bernhard Hauer for support and to him and Dr Edzard Scholten for critical reading of the manuscript. I would also like to thank my colleagues Dr Thomas Friedrich, Dr Hans Wolfgang Höffken, and Dr Markus Lohscheidt and Birgit Bender, August Blechschmidt, Elke Eichler, Hans-Jürgen Lang, Agnieszka Peters, and Andreas Schädler for committing themselves to the Lph project and for their technical assistance, respectively.

References

1 A. J. Vroemen, M. Beverini, WO 96/35800 (Gist Brocades B.V.) (13.5.96).
2 H. U. Blaser, M. Studer, *Chirality* **1999**, *11*, 459.
3 (a) T. Reichstein, A. Grüssner, *Helv. Chim. Acta* **1934**, *17*, 311–328; (b) R. D. Hancock, R. Viola, *Appl. Microbial. Biotechnol.* **2001**, *56*, 567–576.
4 (a) Y. Hikichi, T. Moriya, I. Nogami, H. Miki, T. Yamaguchi (Takeda Chemical Ind.), *EP 590857*, **1994**; (b) T. Moriya, Y. Hikichi, Y. Moriya, T. Yamaguchi (Takeda Chemical Ind.), *WO 97/10340*, **1997**.
5 (a) N. Dusch, A. Pühler, J. Kalinowski, *Appl. Environ. Microbiol.* **1999**, *65*, 1530–1539; (b) F. Elischewski, A. Pühler, J. Kalinowski, *J. Biotechnol.* **1999**, *75*, 135–146; (c) F. Elischewski, J. Kalinowski, A. Pühler, N. Dusch, J. Dohmen, M. Farwick, G. Thierbach (Degussa–Hüls AG), *EP 1001027*, **2000**; (d) N. Dusch, J. Kalinowski, A. Pühler (Degussa–Hüls AG), *EP 1006192*, **2000**; (e) M. Rieping, G. Thierbach, W. Pfefferle, N. Dusch, J. Kalinowski, A. Pühler (Degussa–Hüls AG), *EP 1006193*, **2000**.
6 (a) H. Sahm, L. Eggeling, *Appl. Environ. Microbiol.* **1999**, *65*, 1973–1979; (b) L. Eggeling, G. Thierbach, H. Sahm (Degussa–Hüls AG), *EP 1006189*, **2000**.
7 M. Binder, K.-E. Uffmann, I. Walger, U. Becker, W. Pfefferle, H. Friedrich, *EP 1050219 A1*, **2000**.
8 (a) R. R. Yocum, T. A. Patterson, T. Hermann, J. G. Pero (Omnigene Bioproducts Inc.), *WO 01/21772*, **2001**; (b) C. Beck, H.-P. Harz, D. Klein, M. Lohscheidt, E. Scholten, T. E. Hermann, T. A. Patterson, J. G. Pero, R. R. Yocum, *Metabolic Engineering IV*, Il Ciocco, Castelvecchio Pascoli, Italy, Oct 6–11, **2002**, United Engineering Foundation, New York; (c) T. Hermann, T. A. Patterson, J. G. Pero, R. R. Yocum, K.-U. Baldenius, C. Beck (Omnigene Bioproducts Inc.), *WO 02/57476*, **2002**.
9 (a) D. R. Wilken, R. E. Dyar, *Arch. Biochem. Biophys.* **1978**, *189*, 251–255; (b) S. Shimizu, H. Hata, H. Yamada, *Agric. Biol. Chem.* **1984**, *48*, 2285–2291; (c) H. Yamada, S. Shimizu, *Ann. N. Y. Acad. Sci. (Enzyme Engineering XI)* **1992**, *672*, 374–386.
10 (a) S. Shimizu, S. Hattori, H. Hata, H. Yamada, *Appl. Environ. Microbiol.* **1987**, *53*, 519–522; (b) M. Kataoka, S. Shimizu, H. Yamada, *Rec. Trav. Chim. Pays Bas* **1991**, *110*, 155–157.
11 (a) F. Effenberger, J. Eichhorn, J. Roos, *Tetrahedron Asymmetry* **1995**, *6*, 271–282; (b) F. Effenberger, J. Eichhorn, J. Roos (Degussa AG) *DE 19506728*, **1996**.
12 B. I. Glänzer, K. Faber, H. Griengl, *Enz. Microbiol. Technol.* **1988**, *10*, 689–690.
13 (a) S. Shimizu, M. Kataoka, K. Shimizu, M. Hirakata, K. Sakamoto, H. Yamada, *Eur. J. Biochem.* **1992**, *209*, 383–390; (b) M. Kataoka, K. Shimizu, K. Sakamoto, H. Yamada, S. Shimizu, *Appl. Microbiol. Biotechnol.* **1995**, *43*, 974–977.
14 S. Shimizu, M. Kataoka, *Ann. NY Acad. Sci. 799 (Enzyme Engineering XIII)* **1996**, 650–658.
15 (a) K. Sakamoto, H. Yamada, S. Shimizu (Fuji Yakuhin Kogyo Kabushiki Kaisha), *EP 436730 B1*, **1995**; (b) K. Sakamoto, H. Yamada, S. Shimizu (Fuji Yakuhin Kogyo Kabushiki Kaisha), *EP 504421 B1*, **1995**.
16 (a) L. Hua, Y. Tang, Z. Sun, X. Guo, J. Wang, *Gongye Weishengwu* **2001**, *31(4)*, 5–8; (b) Y. Tang, Z. Sun, L. Hua, X. Guo, J. Wang, *Gongye Weishengwu* **2002**, *31(3)*, 1–5; (c) Z. Sun (Xinfu Bio-Chemical Co.) *CN 1313402*, **2001**.
17 S. Shimizu, M. Kataoka, K. Honda, K. Sakamoto, *J. Biotechnol.* **2001**, *92*, 187–194.
18 (a) M. Kesseler, B. Hauer, T. Friedrich, R. Mattes (BASF AG) *DE 19952501*, **2000**; (b) M. Kataoka, J.-I. Nomura, M. Shinohara, K. Nose, K. Sakamoto, S. Shimizu, *Biosci.*

Biotechnol. Biochem. **2000**, *64*, 1255–1262; (c) M. Kesseler, T. Friedrich, H. W. Höffken, B. Hauer, *Adv. Synth. Catal.* **2002**, *344*, 1103–1110.

19 L. Stryer, *Biochemie*, Spektrum der Wissenschaft Verlagsgesellschaft mbH, Heidelberg, **1990**, p. 203.

20 (a) H. Uchiyama, T. Inaoka, T. Ohkuma-Soyejima, H. Togame, Y. Shibanaka, T. Yoshimoto, T. Kokubo, *J. Biochem.* **2000**, *128*, 441–447; for reviews see, e.g., (b) F. H. Arnold, P. L. Wintrode, K. Miyazaki, A. Gershenson, *Trends Biochem. Sci.* **2001**, *26*, 100–106; (c) M. T. Reetz, K.-E. Jaeger, *Top. Curr. Chem.: Biocatalysis – from Discovery to Application*, Vol. 200, W. D. Fessner, ed., Springer, Berlin, **1999**, pp. 31–57.

6.4
Microbially Produced Functionalized Cyclohexadiene-*trans*-diols as a New Class of Chiral Building Blocks in Organic Synthesis: On the Way to Green and Combinatorial Chemistry

Volker Lorbach,* Dirk Franke,* Simon Eßer, Christian Dose, Georg A. Sprenger, and Michael Müller

6.4.1
Introduction

The aim of a synthetically oriented chemist is usually not so much to build new molecules by connecting one atom to one other, but to use and, if necessary, generate methods and materials that lead him in an easy way to the desired goal. Therefore chemists are usually grateful when they can use enantiopure material provided by nature either as the starting substance or as a catalyst in asymmetric syntheses. A fact that limits the applicability of material provided by nature is often the large amount of sometimes rigid functionalization (starting material) or its high specificity (catalysts). Whereas in recent years new methods have emerged for overcoming problems arising from specificity of catalysts, as a result of new techniques of molecular biology, e.g. directed evolution of enzymes, rational design of novel enzyme properties, or the generation of catalytic antibodies, the generation of appropriate substances is much more sophisticated and there is no generally applicable approach to the generation of tailor-made products. To look for access to substances with high synthetic potential as starting material for new classes of substances is therefore an important task for chemists and biologists working at the interface between these two disciplines.

6.4.2
The Shikimate Pathway

The shikimate biosynthetic pathway occurs in bacteria, plants, and fungi (including yeasts) and is a major entry into the biosynthesis of primary and secondary metabolites, for example aromatic amino acids, menaquinones, vitamins, and antibiotics [1]. Starting from erythrose-4-phosphate (E4P) and phosphoenol-pyruvate

---

*These authors contributed equally to this manuscript.

Highlights in Bioorganic Chemistry: Methods and Applications. Edited by Carsten Schmuck, Helma Wennemers.
Copyright © 2004 WILEY-VCH Verlag GmbH & Co. KGaA, Weinheim
ISBN: 3-527-30656-0

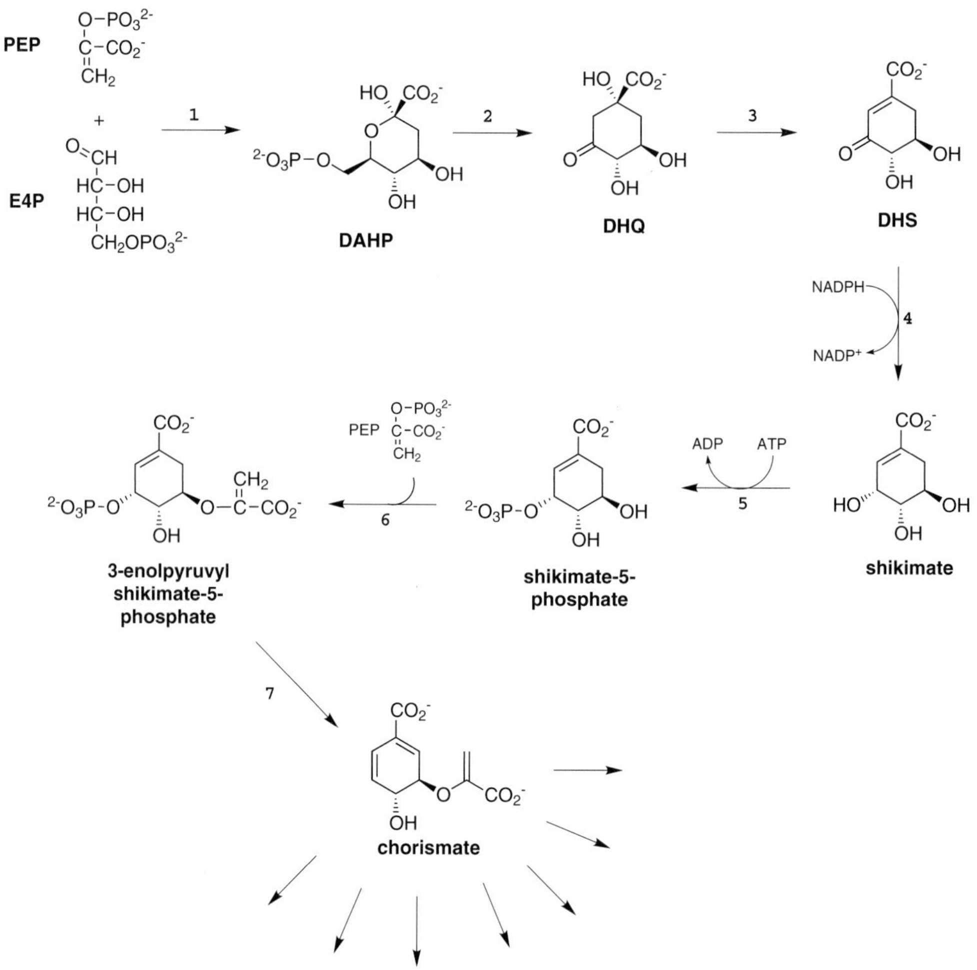

Scheme 6.4.1. The shikimate biosynthetic pathway. The enzymes involved are (1) 3-deoxy-D-arabino-heptulosonate 7-phosphate synthase, (2) dehydroquinate synthase, (3) 5-dehydroquinate dehydratase, (4) shikimate dehydrogenase, (5) shikimate kinase, (6) 3-enolpyruvylshikimate-5-phosphate synthase, and (7) chorismate synthase.

(PEP) shikimate is biosynthesized via 3-deoxy-D-arabino-heptulosonate 7-phosphate (DAHP), dehydroquinate (DHQ), and dehydroshikimate (DHS) (Scheme 6.4.1).

Chorismate is formally the last metabolite of the shikimate pathway and serves as a branch point towards different biosynthetic byways (Scheme 6.4.2) [2]. From an evolutionary standpoint chorismate was evolved not as a metabolite with a distinct cell function, but rather as a highly flexible building block. Because of the special character of 1,3-cyclohexadiene systems, with only a small energy barrier to aromatization, chorismate and its constitutional isomer isochorismate, which

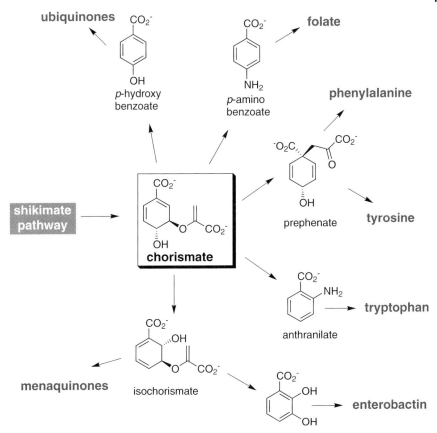

Scheme 6.4.2. Chorismate and isochorismate as branch points toward the synthesis of many different primary and secondary metabolites.

is formed from chorismate in one catalytic step, are ideal starting materials for different functionalized aromatic cell products [3]. Thus, biosynthetically both chorismate and isochorismate are characterized by their diversity-oriented bias. Because of their high functionalization these molecules are also very interesting substances chemically and biochemically, not only in terms of possible aromatization reactions and their products, but also use of the carboxylic and diol functionality and, especially, the cyclohexadiene system as reactive groups for diversity-oriented modifications.

In this contribution we would like to show the production and chemical use of chorismate-derived compounds as an example of the development of microbial production pathways resulting in highly flexible building blocks, which themselves should be applicable in diversity-oriented syntheses. Not only can naturally occurring metabolites be chosen as target molecules, non-natural compounds can also be produced.

6.4.3
Microbial Production of 2,3-trans-CHD

Functionalized cyclohexadiene-trans-diols (trans-CHD; dihydroxydihydrobenzenes) closely related to chorismate or isochorismate have been chosen as primary target molecules for the metabolic pathway engineering approach [4] (Box 24). The corresponding cyclohexadiene-cis-diols (cis-CHD) are known and established as valuable precursors for the production of natural products and bioactive materials. They are readily accessible via whole-cell bioconversion starting from aromatic compounds [5]. Mainly because of the limited availability of trans-CHD [6], their use as chiral synthetic building blocks is rare. All chemical preparations are characterized by tedious multistep syntheses [7]. Attempts by several working groups to prepare trans-CHD from cis-CHD have been characterized by low overall yield [8]. As an alternative, homochiral diols with the trans configuration from bacterial sources have been described as metabolites involved in metabolic pathways derived from the shikimate pathway, therefore, trans-CHD should, in principle, be accessible by metabolic engineering. Using techniques of metabolic pathway engineering in E. coli, J. W. Frost et al. were able to produce shikimate and quinate in concentrations up to 71 g L^{-1} and 13 g L^{-1}, respectively, using a whole-cell process in a stirred-tank reactor [9].

Pioneering work in the field of bacterial production of trans-CHD has been conducted by Leistner and co-workers, who used strains of Klebsiella pneumoniae as biological hosts [10]. Mutants with expressed plasmid-encoded genes (entC and/or entB) and defects in the postchorismate pathways have been shown to produce the two trans-CHD **1** and **2** in concentrations of up to 200 mg L^{-1}. (S,S)-2,3-Dihydroxy-2,3-dihydrobenzoic acid (2,3-trans-CHD, **2**) is the immediate hydrolytic product of isochorismate and occurs as an intermediate in the biosynthesis of the iron chelator enterobactin. In two steps chorismate is converted through isochorismate synthase (encoded by the gene entC) and isochorismatase (encoded by entB) to give 2,3-trans-CHD **2**. This compound is intracellularly transformed in an aromatization step catalyzed by 2,3-dihydroxybenzoate synthase (encoded by entA; Scheme 6.4.3) to give 2,3-dihydroxybenzoate **3**.

For our purpose we decided to use Escherichia coli K-12 as a safe and well-characterized host strain for microbial production of trans-CHD. For production of 2,3-trans-CHD **2** the metabolic flux toward catechol **3** has to be barred and the flux toward the desired compound should be increased. E. coli strains with a defective gene entA (entA$^-$) were transformed with plasmids containing entB/entC. This led to production of 2,3-trans-CHD **2** and excretion of the product into the culture medium [11, 12].

In the production of **2**, strains with entA$^-$-mutation have long-term stability and production rates are high. Typical cultivation of 2,3-trans-CHD **2** is performed at pH 6.8 and 37 °C in a stirred-tank reactor with glucose feeding in a fed-batch mode. Microbial production over a process time of 40 h affords 92 g **2** from 690 g glucose monohydrate in a 20-L cultivation experiment. This corresponds to a molar yield of 17% [13]. Advantageously, 2,3-trans-CHD **2** is the major product with no

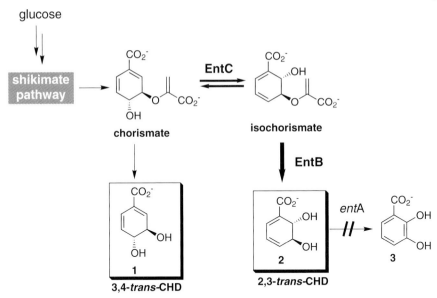

Scheme 6.4.3. Metabolic engineering for microbial production of 2,3-*trans*-CHD (bold arrows show enzymes expressed from the plasmid).

detectable impurities of other metabolites of the shikimate or enterobactin biosynthetic pathways [12]. It was found that 2,3-*trans*-CHD **2** can be produced with other renewable carbon sources, for example galactose or glycerol, that are not taken up by the phosphotransferase system.

Compound **2** was efficiently isolated by anion-exchange chromatography on Dowex 1 × 8 resin in a purification process similar to that devised for small amounts of chorismate [14]. Elution of the product occurs under moderately acidic conditions, enabling isolation without aromatization. This procedure enables an isolation of 2,3-*trans*-CHD on the high decagram scale with a purity of 95% after lyophilization [12].

6.4.4
Application of 2,3-*trans*-CHD in Natural-product Syntheses

Because of their suitable skeletal structure *trans*-CHD are applicable in the synthesis of compound classes of chemical, biological, and pharmaceutical interest. Regio- and stereoselective epoxidation of either one or two double bonds of **2**, for example, opens up an approach to a group of naturally occurring cyclohexane epoxides which have attracted considerable attention because of their unusual structures, biogenesis, and biological activity [15].

Subsequent nucleophilic opening of the resulting epoxides, or regio- and stereoselective dihydroxylation, affords entry to the class of carbasugars (Scheme 6.4.4); these are of great pharmaceutical interest because of their structural analogy to common sugars, but different biological decomposition. This is also true of amino carbasugars such as valienamine [7b], which, attached to 6-deoxyglucosamine, is the primary active moiety (acarviosine) of the α-glucosidase inhibitor acarbose, and which is also the structural component of other secondary metabolites, for example the adiposins, amylostatins, salbostatin, trestatins, and validamycins. The significance of 2,3-*trans*-CHD **2** as versatile building block in diversity-oriented synthesis has, therefore, been examined by these types of reaction.

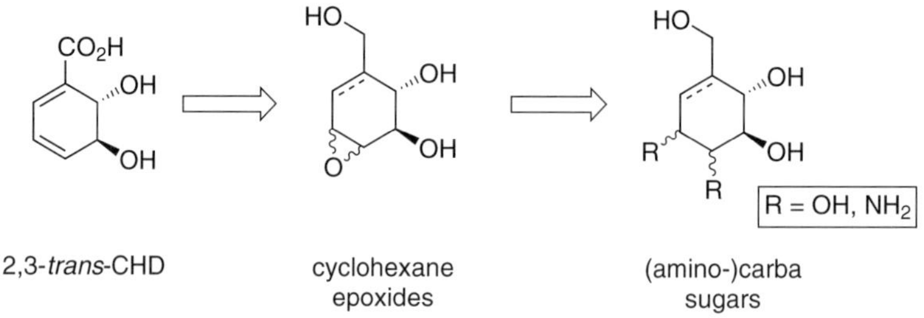

2,3-*trans*-CHD   cyclohexane epoxides   (amino-)carba sugars

Scheme 6.4.4. 2,3-*trans*-CHD **2** as a precursor of cyclohexane epoxides and (amino-) carbasugars.

6.4.5
Regio- and Stereoselective Epoxidation

Compound **2** was found to be stable under a wide range of thermal conditions and pH. Transformation to the ester **4** could be achieved easily under acidic conditions by use of trimethylsilylchloride or hydrochloride in methanol. In general, addition reactions to the conjugated diene system are controlled by the steric demand of the ring substituents and, furthermore, by the electron deficiency at the C5–C6 double bond, because of the electron-withdrawing effect of the carboxylic group [7b, 16, 17]. Thus, addition to **4** of *meta*-chloroperoxybenzoic acid (*m*-CPBA), used as the epoxidizing agent, occurs exclusively at double bond C3–C4. The stereochemistry of peroxide attack, and hence the configuration of the resulting epoxide, is directed by the functionality at C2. If this is an allylic hydroxyl group, *m*-CPBA coordinates by hydrogen bonding and forces the C2–C3 cis configuration of **5**, as is proven by X-ray structural analysis (Scheme 6.4.5, Figure 6.4.1) [18]. When the same conditions are applied to the protected diol **6**, which is easily obtained from **4** by silylation with TBDMS-triflate, the bulky 2-siloxy group shields the α-face of the cyclohexadiene plane hence attack of the peracid occurs regioselectively from the β-face resulting, after deprotection, in the formation of diastereomer **8** [17].

The reduction of the ester functionality of **6** to the corresponding alcohol **9** can be achieved efficiently by use of diisobutylaluminum hydride (DIBAL-H). Subse-

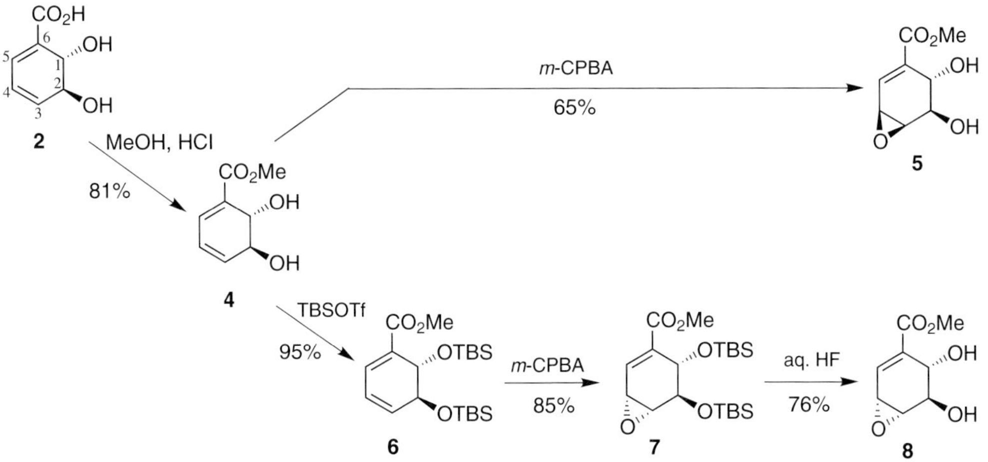

6.4.5 Regio- and Stereoselective Epoxidation | 517

Scheme 6.4.5. Regio- and stereoselective epoxidation of 2,3-*trans*-CHD.

quent epoxidation with equimolar amounts of *m*-CPBA at low temperature results exclusively in the monoepoxide **10** [19]. Under more rigorous conditions (raised temperature, prolonged reaction time, and excess of oxidizing agent) bisepoxide **11** is formed in high yield (Scheme 6.4.6) [17, 20].

Both epoxides **10** and **11** already have the core structure of *ent*-senepoxide and *iso*-crotepoxide. Benzoylation of the free hydroxyl group, cleavage of the siloxy groups, and subsequent acetylation of the diol unit gives the stereoisomers of the natural products (Scheme 6.4.7) [17]. This approach to stereoisomers of the naturally occurring senepoxide and crotepoxide is characterized by a small number of reaction steps and high overall yield. Investigation of the regio- and stereoselective

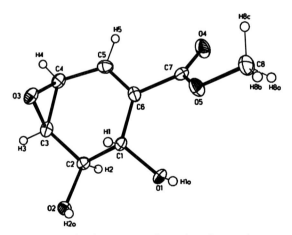

Fig. 6.4.1. Molecular structure of epoxide **5** showing the C2–C3 cis configuration.

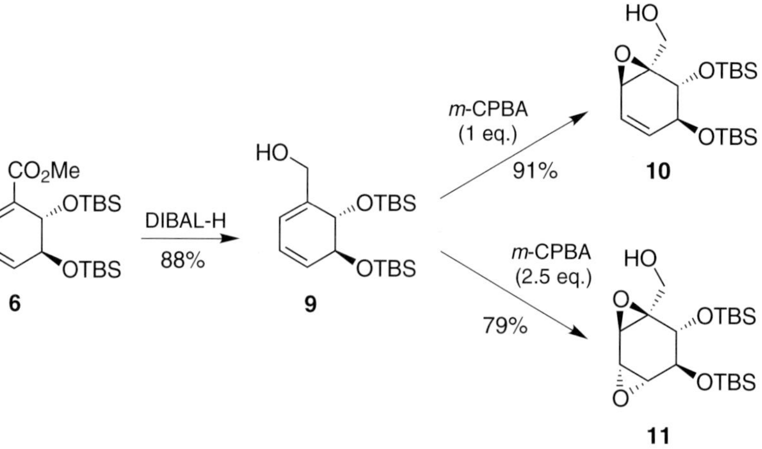

Scheme 6.4.6. Regio- and stereoselective mono- and bisepoxidation of alcohol **9**.

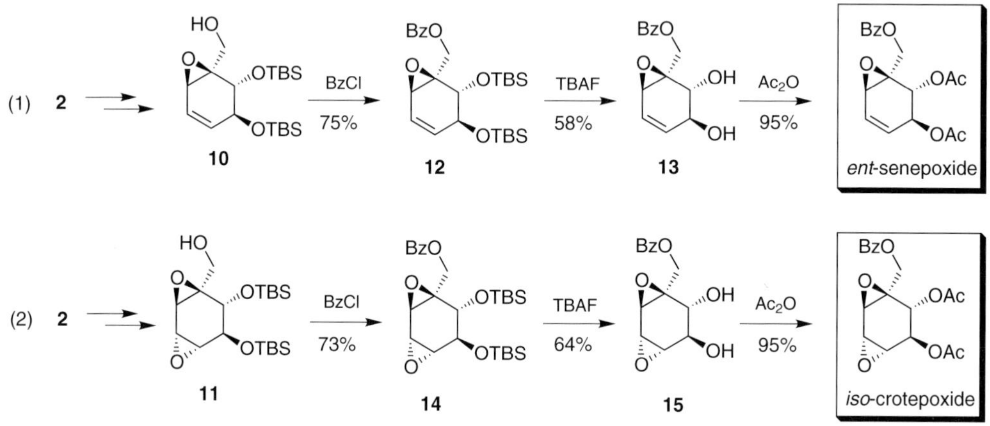

Scheme 6.4.7. Synthesis of *ent*-senepoxide and *iso*-crotepoxide.

epoxidation of double bonds of the cyclohexadiene ring holds further potential for syntheses of related carbasugar compounds.

6.4.6
Nucleophilic Opening of the Epoxides Obtained

Because of the large number of different nucleophiles that can be used, the opening of epoxides is a powerful means of gaining access to a vast variety of highly functionalized cyclitols. It also seems to be a valuable method for production of carbasugars and aminocarbasugars, by use of appropriate nucleophiles. For generating aminocarbasugars, methods for opening of the oxirane ring in **7** have been tested under different conditions, e.g. sodium azide [21], sodium azide with Lewis acid catalysts [22], or trimethylsilyl azide with titanium complexes [23], to give a C3/C4 azido alcohol functionality. The oxirane moiety proves to be remarkably stable under these conditions. Successful conversion of **7** to azido alcohol **16** is possible with trimethylsilyl azide and ytterbium tri-*iso*-propoxide, by following the procedure of Yamamoto et al. (Scheme 6.4.8; Eq. 1) [24, 25].

When the same conditions are applied to oxirane **5** azido alcohol **17** is formed; this can be selectively reduced with triphenylphosphine to give amino alcohol **18** in good yield (Scheme 6.4.8; Eq. 2) [11, 25, 26]. This compound and other, related, epoxides are a new approach to the chemistry of aminocarbasugars and their unsaturated analogs such as valienamine [27].

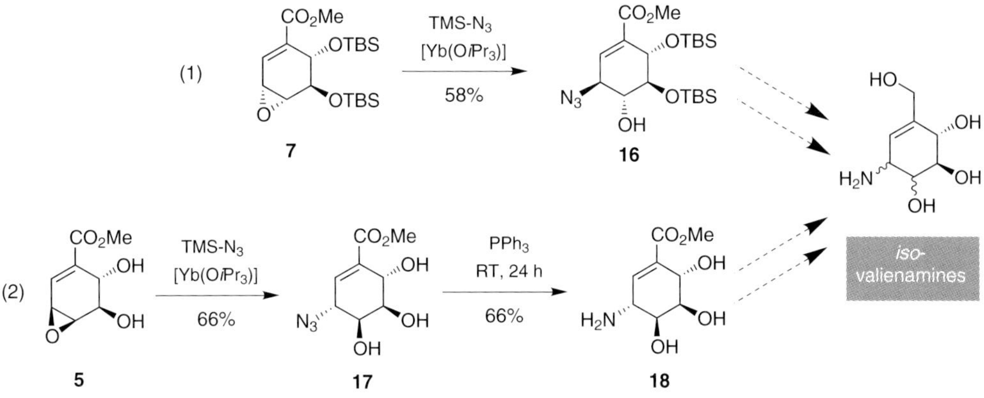

Scheme 6.4.8. Nucleophilic opening of epoxides **7** and **5**.

Because the carboxyl group of 2,3-*trans*-CHD **2** is ideally suited to covalent bonding to a polymeric matrix and, moreover, the transformations are performed under very mild conditions, a diversity-oriented approach to regio- and stereo-isomeric (amino)carbasugars, on a solid surface, is applicable in principle.

6.4.7
Regio- and Stereoselective Dihydroxylation

Besides the opening of the previously described epoxides with oxygen nucleophiles, which would result in the formation of a second diol unit with the trans configuration, an approach towards a diol unit with cis configuration is achievable by dihydroxylation reactions. The osmium tetroxide-catalyzed dihydroxylation of the unprotected diol **4** gives a diastereomeric mixture of **19** and **20** in a ratio of 5:1 (Scheme 6.4.9; Eq. 1). Compound **21** is obtained as a single diastereomer when using protected *trans*-CHD **7** as starting material [25]. Proof of the relative config-

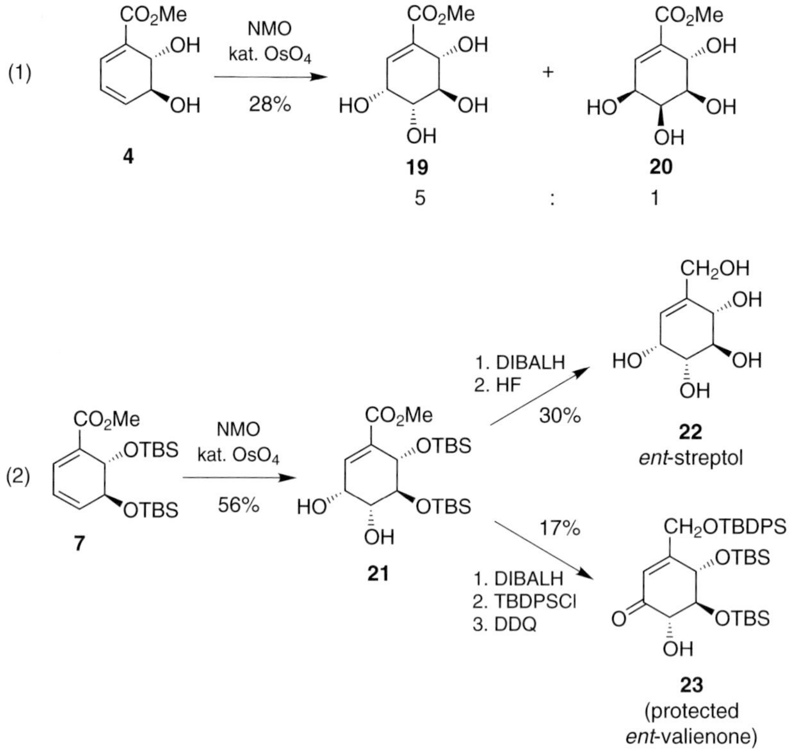

6.4 Microbially Produced Functionalized Cyclohexadiene-trans-diols

Scheme 6.4.9. Regio- and stereoselective dihydroxylation and subsequent modification.

uration was given by transforming **21**, via reduction and subsequent deprotection, to the enantiomer of the plant growth-inhibitor streptol **22** from *Streptomyces* sp. (Scheme 6.4.9; Eq. 2) [28]. Alternatively, reduction of the ester functionality of **21**, then protection of the primary hydroxy group and subsequent selective oxidation of the allylic hydroxy group, results in a protected derivative of *ent*-valienone **23** (Scheme 6.4.9; Eq. 2) [26].

It has been demonstrated that by alteration of regio- and stereoselective epoxidation, then hydrolytic opening, and selective dihydroxylation, a variety of different yet stereochemically defined carbasugars and aminocyclitols can be produced from enantiopure 2,3-*trans*-CHD.

6.4.8
Microbial Production of 3,4-*trans*-CHD

In contrast to 2,3-*trans*-CHD **2** the corresponding 3,4-*trans*-CHD **1** has not yet been found to have any function in the metabolism of *E. coli*. If, however, isochorismatase (encoded by *entB*), the enzyme that catalyzes the hydrolysis of isochorismate to 2,3-*trans*-CHD, is present at high concentration, 3,4-*trans*-CHD can be produced from chorismate in vitro and in vivo with strains of *Klebsiella pneumoniae* [10].

It has, nevertheless, been found that a sole overproduction of EntB up to 160-fold in *E. coli* did not result in excretion of 3,4-*trans*-CHD [11]. This can best be attributed to the meager k_{cat}/K_m value of EntB with chorismate as substrate, which cannot compete with k_{cat}/K_m values of chorismate lyase (UbiC), *p*-aminobenzoate synthase (PabA, PabB), anthranilate synthase (TrpE, TrpD), isochorismate synthase (EntC/MenF), and chorismate mutase together with prephenate dehydrogenase (TyrA) or prephenate dehydratase (PheA) which evolved naturally for chorismate as substrate [29]. Excretion of 3,4-*trans*-CHD was achieved when pathways to aromatic amino acid and folate production were blocked (Scheme 6.4.10). Blocking of the pathways to other metabolites starting from chorismate was not necessary because

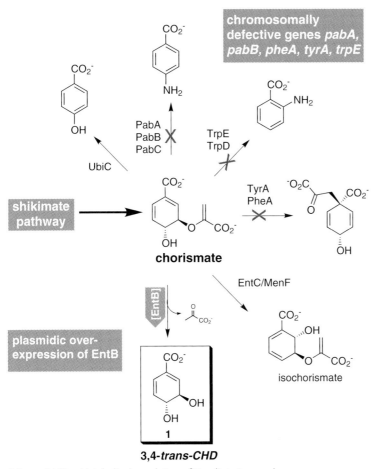

Scheme 6.4.10. Metabolic deregulation of *E. coli* strains used for the production of 3,4-*trans*-CHD.

of their low metabolic fluxes. Transformants of E. coli strain BN117, with deletions in the pathways to p-aminobenzoate, anthranilate, and prephenate (pabA$^-$, pabB$^-$, pheA$^-$, tyrA$^-$, trpE$^-$), produced 3,4-*trans*-CHD with a maximum of 19 mg h^{-1} g (dry cell mass)$^{-1}$, when EntB was overproduced [11, 30].

By use of synthetic medium the formation of 3,4-*trans*-CHD was maintained for a 36-h cultivation period, resulting in accumulation of up to 790 mg L^{-1} 3,4-*trans*-CHD **1** [11]. With this strain, however, it is necessary to separate the growth and production phases, because the substances for which E. coli strain BN117 is known to be auxotrophic, i.e. tryptophan, tyrosine, phenylalanine, proline, arginine, histidine, and p-aminobenzoate, also partially inhibit entry to the chorismate biosynthesis pathway (feedback inhibition of DAHP-synthase).

Here again the microbial product is free from metabolic by-products, and highly pure 3,4-*trans*-CHD **1** was isolated by anion-exchange chromatography. In contrast with its pyruvyl analog chorismate, **1** is reasonably stable at room temperature and neutral pH; this is an important prerequisite for its use as a starting material in organic syntheses. With the production of 3,4-*trans*-CHD **1**, both regioisomers of *trans*-CHD now complement the chiral pool substance spectrum of *cis*-CHD as starting material for organic chemistry.

6.4.9
Discussion

2,3-*trans*-CHD [(S,S)-2,3-dihydroxy-2,3-dihydrobenzoic acid] **2** and 3,4-*trans*-CHD [(R,R)-3,4-dihydroxy-3,4-dihydrobenzoic acid] **1** are examples of interesting chiral building blocks which can be readily prepared starting from glucose or other renewable carbon sources by metabolically deregulated, recombinant strains of *Escherichia coli*. Excretion of **1** and **2** into the culture medium occurs without metabolic by-products, enabling separation on a preparative scale by use of ion-exchange chromatography.

Because of their large amount of chemically different functionalization, *trans*-CHD are promising precursors especially for the synthesis of, e.g., carbasugars and their analogs with potential biological activity. The simple, but highly functionalized structure of *trans*-CHD also has a high potential for diversity-oriented syntheses that could lead to a vast variety of different structures in short chemical reaction sequences.

The combination of identifying metabolic compounds that are accessible by state-of-the-art techniques of metabolic pathway engineering, and examination of their synthetic application potential implies a number of advantages when compared with other current approaches to the production of new compound classes. On the one hand, this approach is based on sustainable access and the production of compounds by a safe and environmental friendly route (green chemistry). On the other hand, the amount of functionalization makes it possible to apply diversity-oriented chemical methods for modification (combinatorial chemistry).

Finally, since these compounds are of natural origin they and compound libraries derived thereof putatively posses an enhanced probability of biological activity [31].

References

1 (a) E. HASLAM, *Shikimic Acid: Metabolism and Metabolites*, Wiley, Chichester, **1993**; (b) A. R. KNAGGS, *Nat. Prod. Rep.* **2003**, *20*, 119–136.

2 (a) F. DOSSELAERE, J. VANDERLEYDEN, *Crit. Rev. Microbiol.* **2001**, *27*, 75–131; (b) C. T. WALSH, J. LIU, F. RUSNAK, M. SAKAITANI, *Chem. Rev.* **1990**, *90*, 1105–1129.

3 E. LEISTNER, *Comprehensive Natural Products Chemistry*, Vol. 1, K. NAKANISHI, D. H. R. BARTON, ed., Elsevier, Amsterdam, **1999**, pp. 609–622.

4 M. D. BURKART, *Org. Biomol. Chem.* **2003**, *1*, 1–4 and references cited therein.

5 T. HUDLICKY, D. GONZALEZ, D. T. GIBSON, *Aldrichim. Acta* **1999**, *32*, 35–62 and references cited therein.

6 D. R. BOYD, N. D. SHARMA, *J. Mol. Catal. B* **2002**, *19/20*, 31–42.

7 (a) T. K. M. SHING, *Tetrahedron Asymmetry* **1996**, *7*, 353–356; (b) B. M. TROST, L. S. CHUPAK, T. LÜBBERS, *J. Am. Chem. Soc.* **1998**, *120*, 1732–1740.

8 (a) T. HUDLICKY, G. SEOANE, T. PETTUS, *J. Org. Chem.* **1989**, *54*, 4239–4243; (b) D. R. BOYD, N. D. SHARMA, H. DALTON, D. A. CLARKE, *Chem. Commun.* **1996**, 45–46; (c) B. P. MCKIBBEN, G. S. BARNOSKY, T. HUDLICKY, *Synlett* **1995**, 806–807; (d) D. R. BOYD, N. D. SHARMA, C. R. O'DOWD, F. HEMPENSTALL, *Chem. Commun.* **2000**, 2151–2152.

9 (a) K. M. DRATHS, D. R. KNOP, J. W. FROST, *J. Am. Chem. Soc.* **1999**, *121*, 1603–1604; (b) J. M. GIBSON, P. S. THOMAS, J. D. THOMAS, J. L. BARKER, S. S. CHANDRAN, M. K. HARRUP, K. M. DRATHS, *Angew. Chem.* **2001**, *113*, 1999–2002; *Angew. Chem. Int. Ed.* **2001**, *40*, 1945–1948.

10 R. MÜLLER, M. BREUER, A. WAGENER, K. SCHMIDT, E. LEISTNER, *Microbiology* **1996**, *142*, 1005–1012.

11 D. FRANKE, *Dissertation*, University of Bonn, **2002**.

12 D. FRANKE, G. A. SPRENGER, M. MÜLLER, *Angew. Chem.* **2001**, *113*, 578–581; *Angew. Chem. Int. Ed.* **2001**, *40*, 555–557.

13 With regard to the theoretical maximum yield the relative molar yield was approx. 40%; cf. K. LI, J. W. FROST, *Biotechnol. Prog.* **1999**, *15*, 876–883.

14 Using methods of ion-exchange chromatography, *trans*-CHD have already been isolated from cultivation medium for analytical purposes: (a) G. I. YOUNG, F. GIBSON, C. G. MACDONALD, *Biochim. Biophys. Acta* **1969**, *61*, 62–72; (b) C. F. RIEGER, J. L. TURNBULL, *Prep. Biochem.* **1996**, *26*, 67–76; (c) Ref. [10].

15 Crotepoxide, for example, was isolated from fruits of *Croton macrostachys* and has been shown to have significant tumor-inhibitory activity for Lewis lung carcinoma in mice (LL) and Walker intramuscular carcinosarcoma in rats (WM).

16 S. OGAWA, T. TAKAGAKI, *Bull. Chem. Soc. Jpn.* **1988**, *61*, 1413–1415.

17 V. LORBACH, D. FRANKE, M. NIEGER, M. MÜLLER, *Chem. Commun.* **2002**, 494–495.

18 (a) H. B. HENBEST, R. A. L. WILSON, *J. Chem. Soc.* **1959**, 1958–1965; (b) K. B. SHARPLESS, T. R. VERHOEVEN, *Aldrichim. Acta* **1979**, *12*, 63; (c) M. FRECCERO, R. GANDOLFI, M. SARZI-AMADÈ, *J. Org. Chem.* **2000**, *65*, 8948–8959.

19 (a) R. H. SCHLESSINGER, A. LOPES, *J. Org. Chem.* **1981**, *46*, 5252–5253; (b) S. OGAWA, T. TOYOKUNI, M. ARA, M. SUETSUNG, T. SUAMI, *Bull. Chem. Soc. Jpn.* **1983**, *56*, 1710–1714; (c) H. A. J. CARLESS, O. Z. OAK, *Tetrahedron Lett.* **1991**, *32*, 1671–1674.

20 S. OGAWA, T. TAKAGAKI, *Bull. Chem. Soc. Jpn.* **1987**, *60*, 800–802.

21 M. J. MCMANUS, G. A. BERCHTOLD, D. M. JERINA, *J. Am. Chem. Soc.* **1985**, *107*, 2977–2978.

22 (a) B. ZIPPERER, D. HUNKLER, H. FRITZ, G. RIHS, H. PRINZBACH, *Angew. Chem.* **1984**, *96*, 296–297; *Angew. Chem. Int. Ed.* **1984**, *23*, 309–311; (b) M. G. BANWELL, N. HADDAD, T. HUDLICKY, T. C. NUGENT, M. F. MACKAY, S. L. RICHARDS, *J. Chem. Soc., Perkin Trans. 1* **1997**, 1779–1791.

23 (a) J. M. CHONG, K. B. SHARPLESS, *J. Am. Chem. Soc.* **1985**, *50*, 1560–1563;

(b) D. Sinou, M. Emziane, Tetrahedron Lett. **1986**, *27*, 4423–4426; (c) C. Blandy, R. Choukroun, D. Gervais, Tetrahedron Lett. **1983**, *24*, 4189–4192.
24 M. Meguro, N. Asao, Y. Yamamoto, J. Chem. Soc., Chem. Commun. **1995**, 1021–1022.
25 D. Franke, V. Lorbach, S. Esser, C. Dose, G. A. Sprenger, M. Halfar, J. Thömmes, R. Müller, R. Takors, M. Müller, Chem. Eur. J. **2003**, *9*, 4188–4196.
26 S. Esser, Diploma Thesis, University of Bonn, **2001**.
27 (a) S. Lee, I. Tornus, H. Dong, S. Gröger, J. Labelled Compd. Radiopharm. **1999**, *42*, 361–372; (b) H. Dong, T. Mahmud, I. Tornus, S. Lee, H. G. Floss, J. Am. Chem. Soc. **2001**, *123*, 2733–2742 and references cited therein.
28 A. Isogai, S. Sakuda, J. Nakayama, S. Watanabe, A. Suziki, Agric. Biol. Chem. **1987**, *51*, 2277–2279.
29 F. Rusnak, J. Liu, N. Quinn, G. A. Berchtold, C. T. Walsh, Biochemistry **1990**, *29*, 1425–1435.
30 D. Franke, G. A. Sprenger, M. Müller, Chembiochem. **2003**, *4*, 775–777.
31 R. Breinbauer, I. R. Vetter, H. Waldmann, Angew. Chem. **2002**, *114*, 3002–3015; Angew. Chem. Int. Ed. **2002**, *41*, 2878–2890.

B.24
Metabolic Pathway Engineering

Volker Lorbach, Dirk Franke, Georg A. Sprenger, Michael Müller

Since the first occurrence of living systems on earth environmental pressure has enforced adaptation to the given natural conditions. This evolutionary pressure has resulted in the development of modified or novel metabolic pathways providing compounds necessary for survival. As naturally produced compounds were investigated their potential and, therefore, the demand for these compounds for pharmaceutical, biological, and chemical purposes became obvious. Yet the fundamental basis of nature is not large-scale production that is adaptable for industrial production, but reproduction and survival, so metabolic compounds are usually produced on the minimum desired scale. Before knowledge of biosynthetic pathways was elucidated, improvement of production strains was achieved by random mutagenesis and screening (e.g. production of penicillin starting in the 1940s).

Recent exploration and development in modern microbiology has provided tools for a rational approach toward engineered microorganisms that can be used to produce potentially important and commercially valuable natural and non-natural metabolites. Thus in 1991 the late Professor James E. Bailey defined a new scientific field, metabolic engineering, as "A scientific field aiming at the directed modification of the enzymatic, regulatory, or transport activities of the cell to improve the cellular properties, with the use of recombinant DNA technology" [1]. The goal of metabolic engineering is to use the knowledge and methods of gene regulation and metabolic pathway regulation (allosteric control), from microbiology, to design improved microorganisms for production purposes.

Genetically, modifications might, for example, be concerned with:

- blockage of competing pathways;
- removal of bottlenecks in metabolic fluxes, including limiting transcriptional and allosteric controls;
- introduction or enhancement of biological transport systems (e.g. transport of substrates into the cell or the target compound out of the cell);
- improvement of energy-providing pathways; and
- introduction of new metabolic pathways (molecular breeding).

Application of metabolic engineering techniques can obviously result in higher yields and productivity, but also fewer side-products and improved stereospecificity. In contrast with competing chemical processes environmentally friendly production methods and renewable resources can often be used for production of high-value products.

Reference

1 J. E. BAILEY, Science **1991**, *252*, 1668–1675.

6.5
Artificial Molecular Rotary Motors Based on Rotaxanes

Thorsten Felder and Christoph A. Schalley

Abstract

This chapter is devoted to artificial mimics of natural molecular machines. It focuses on ATP synthase, a rotary motor which is a highly complex, although extremely elegant functional biochemical system. Chemists and biologists are striving to generate artificial minimal models serving as simpler functional analogs. Starting from a simplified woodcut description of the basic principles behind the natural system, this contribution discusses selected examples of man-made devices which can be regarded as the first steps towards rotary motors, mainly on the basis of rotaxanes and catenanes. The idea that unidirectional motion is closely related to (topological) chirality is pronounced. To conclude the chapter problems associated with these systems are summarized and future prospects are discussed.

6.5.1
"Molecular Machines" – Reality or Just a Fashionable Term?

During the last ten years or so, the terms "molecular machine" or "molecular device" have been used in so many research papers that one might be tempted to conclude this area of supramolecular and bioorganic chemistry to be an old field where there is not much left to do. As we will see, however, these phrases have most often been used to describe a vision rather than anything already achieved. Sometimes even an overstatement, they are nevertheless remarkable in the sense that they are indicative of a clear shift in interest of many chemists away from the world of beautiful structures and challenging syntheses towards the world of functional molecules and new materials. Nature again has served as the role model for many projects as it provides what we could coin "elegance in complexity". While the basic principles behind functional biomolecules often are of an astonishingly beautiful simplicity, the details of the patterns providing the structural basis are of frightening complexity and not predictable in detail. It is no surprise if many researchers aim at minimal systems realizing the basic principles in a much simpler way. Beauty is not, however, a scientific category in a strict sense and some analysis is indicated. In the following section an attempt is made to compress such

Highlights in Bioorganic Chemistry: Methods and Applications. Edited by Carsten Schmuck, Helma Wennemers.
Copyright © 2004 WILEY-VCH Verlag GmbH & Co. KGaA, Weinheim
ISBN: 3-527-30656-0

an analysis to a few pages' space – an almost futile operation. The reader will hopefully excuse an oversimplified view of the natural predecessors of the artificial systems that will be the major aim of this chapter.

6.5.2
Tracing Back ATP Synthesis in Living Cells

Adenosine triphosphate ATP (Scheme 6.5.1) is the general energy currency of living cells. Energy-demanding steps in metabolism are coupled to hydrolysis of the energy-rich phosphate bonds of the ATP molecule. Often a pyrophosphate is cleaved, yielding adenosine monophosphate. A further hydrolysis step with phosphate as the final product helps to shift the equilibrium completely to one side. The organism regenerates ATP via adenosine diphosphate ADP, the final step from ADP to ATP being mediated by a trans-membrane enzyme complex usually called F_0F_1 ATP synthase [1] (Figure 6.5.1).

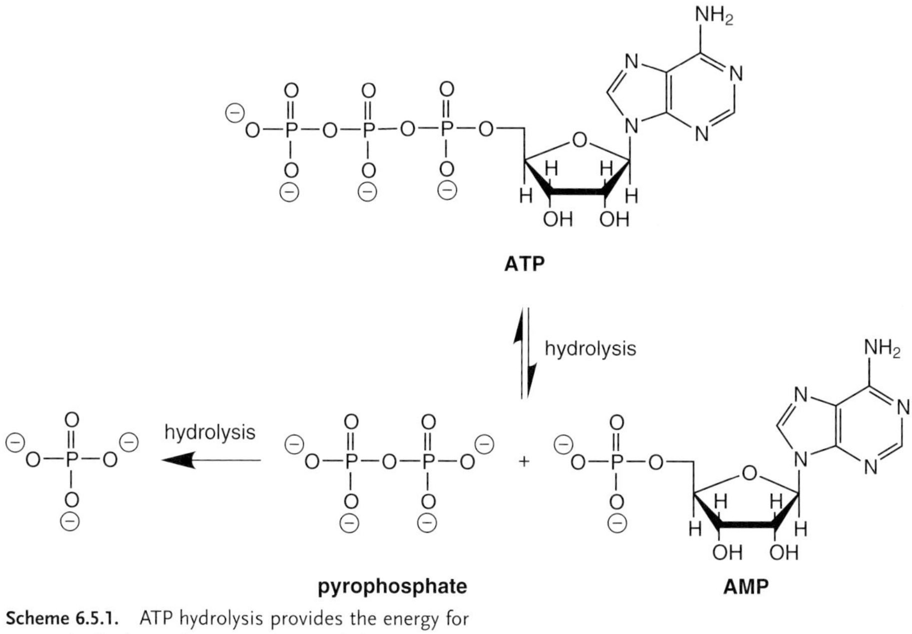

Scheme 6.5.1. ATP hydrolysis provides the energy for energetically demanding steps in metabolism.

In the F_1 part, a hexameric $\alpha_3\beta_3$ protein complex, the three β-units are catalytically active. In a three-step process, conformational changes of the active site induce the binding of ADP and phosphate inside a pocket of the enzyme, followed by formation of the triphosphate from these reactants and subsequent release of the new ATP molecule from the active site. The conformational changes necessary for passage through this three-step cycle are caused by a bent, rotating protein axle in

6.5 Artificial Molecular Rotary Motors Based on Rotaxanes

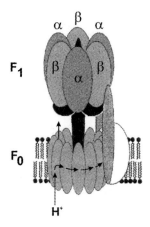

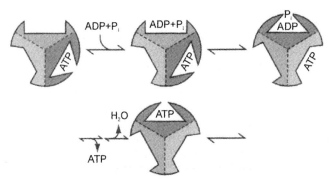

Fig. 6.5.1. Top: ATP synthase complex inserted into the membrane separating high from low proton concentration. The arrows are a schematic representation of the helical proton channel in the F_0 subunit of ATP synthase. Motion of the protons through the channel is coupled with rotation of the central rod. Bottom: Schematic drawing of the conformational changes induced by the rotating axle in the catalytically active β subunits of F_1 ATP synthase.

the center of the cyclic F_1 unit that moves in 120° steps and interacts with the three β-subunits in three different ways (Figure 6.5.1). Thus, each complete rotational cycle induces three different stages in the β-subunits producing a total of three ATP molecules.

But how does nature achieve rotation of this axle? What is its driving force? Figure 6.5.1 shows the F_0 part of ATP synthase as a multiprotein complex spanning the tylacoid membrane. The axle is again located at the center of this cyclic array. The F_0 ATP synthase bears a helical channel through which protons can flow. The protons enter on the side of the membrane distal from the F_1 part, then follow the channel for almost a complete circle around the axle, and finally leave the membrane at the other side, where the F_1 part is located. The proton flow is coupled with the rotation of the axle through temporary binding of protons to carboxylates appropriately positioned at the axle's lower section.

Several questions arise from these considerations. First, it is by no means clear how and why unidirectional rotation is achieved in the natural system. A hint has been given above. The proton channel is helical and is thus chiral. If one takes into account that a clockwise rotation is nothing other than a mirror image of a counterclockwise rotation, it becomes obvious why chirality is of such great importance. In the presence of an additional element of chirality the two enantiomeric senses of rotation become diastereomeric and, consequently, one might be preferred energetically over the other [2]. The second question is that of the driving force. Protons only flow through the proton channel if there is a concentration gradient between the plasma on one side of the membrane and the volume on the other side. Indeed, in natural ATP synthase this is realized and for most living organisms translocation of 12 protons is necessary to rotate the axle by 360°, producing three ATP molecules. These considerations now enable us to understand the importance of the membrane and the reason why ATP synthase is a trans-membrane protein complex. For correct function, it is necessary to build up the proton gradient by compartmentation and prevent it from leveling out. In this respect, a third aspect is pivotal – all ATP synthase multiprotein complexes must be inserted into the membrane in the same orientation, because otherwise each ATP synthase complex with the wrong orientation would consume ATP and accelerate proton migration through the membrane.

These few paragraphs on the natural ATP synthase rotary motor should suffice to give a rough impression of how it works. We have followed ATP synthesis back step by step from the final product through different energy conversion processes to a proton gradient as the final energy source. In principle ATP synthase can be regarded as a motor [3] (the F_0 subunit), converting chemical energy into, molecular motion coupled to a generator (the F_1 part), that uses the mechanical energy of the rotating axle to convert it back into chemical energy (ATP synthesis). Several key properties, which will lead to the discussion of artificial approaches to such machines in the following sections should be summarized here.

- A functional system for ATP generation, requires a membrane which separates areas of high proton concentration from those of lower acidity.
- The orientation of the trans-membrane ATP synthase protein complex in the membrane is of major importance.
- To provide a mechanical motor, an axle must be able to rotate inside a stator, and unidirectionality requires the presence of an element of chirality.
- Finally, a machine needs to be coupled to the motor to convert the mechanical energy back into chemical energy.

6.5.3
Rotaxanes as Artificial Analogs to Molecular Motors?

Judging from geometrical features, rotaxanes [4] (Scheme 6.5.2, top) seem to provide an elegant basis for construction of molecular motors [5]. In these compounds, an axle is threaded into a macrocyclic wheel and prevented from deslip-

6.5 Artificial Molecular Rotary Motors Based on Rotaxanes

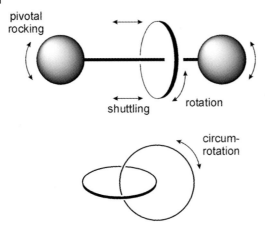

Scheme 6.5.2. Molecular motion feasible in rotaxanes (top) and catenanes (bottom).

ping by two bulky stopper groups that mechanically trap the axle. Consequently, the axle is free to move inside the wheel, because there is no covalent bond between the two components. Several different types of motion are feasible. Besides rotation of the axle inside the wheel, the wheel can move back and forth along the axle, if it is long enough. This process can be regarded as one-dimensional diffusion along the thread. Pivotal rocking of the axle is also possible, as indicated by the arrows in Scheme 6.5.2 (top).

For efficient rotary motion it is necessary to guide the axle properly [6]. Pivotal rocking and the shuttling along the axle should be reduced to a minimum, whereas rotation should be freely possible. It should, however, be mentioned here that the shuttling and the rotation in rotaxanes become identical in catenanes (Scheme 6.5.2, bottom), another kind of mechanically bound molecule in which two wheels are interlocked just like two members of a chain. The process of one wheel rotating through the other is called circumrotation. Because of this analogy of catenanes and rotaxanes, we will include both species in the following discussion.

6.5.4
Rotaxane Synthesis via Template Effects

The use of mechanically bound molecules as the basis for functional devices requires efficient strategies for their preparation. The statistical synthesis of rotaxanes, that is their accidental formation, is usually a low-yield process and thus does not permit their production on a larger scale. Template effects [7] must necessarily be used to thread the axle through the wheel of the rotaxane. Quite an arsenal of effects exist which enable synthesis of a large number of structurally different rotaxanes in high yield. Rotaxane synthesis is increasingly becoming routine. Among the different template effects are those involving metal coordination to a convex copper template [8] and those operating via π-donor–π-acceptor interactions between electron rich axles and electron-poor wheels (or vice versa) [9]. Some utilize hydrogen bonding between ammonium ions and crown ethers [9] or between amide groups in the axle and wheel [10]. Even the hydrophobic effect has been used to introduce an axle into a cyclodextrin [11].

Because earlier reviews deal explicitly with these template effects [4, 7, 12], the following discussion is restricted to a recently discovered anion template effect (Scheme 6.5.3). In non-competitive solvents, the tetralactam macrocycle strongly binds anions such as chloride, bromide, or phenolate. Phenolate stoppers bound to the macrocycle can act as "wheeled" nucleophiles and react through the wheel with a semi-axle generated in situ to yield ether rotaxanes in yields of 57% to 95% [13]. Consequently, the macrocycle not only represents an anion receptor, but as a concave template simultaneously provides the correct orientation of the guest for threading the axle into the wheel.

6.5.5
How to Achieve Unidirectional Rotation in Artificial Molecular Motors?

Any artificial rotaxane-based mimic of the natural ATP synthase motor must bear a chiral element that helps to define the direction of rotation. Chirality can, of course, be implemented in such molecules by adding chiral groups as the stoppers or the wheel. Rotaxanes with cyclodextrins as the wheels have been described [11], and rotaxanes with glucose-containing stoppers are known [14]. Rotaxanes with elements of planar chirality have also been realized [15].

More challenging, however, is the preparation of cycloenantiomeric, topologically chiral rotaxanes [16] and the resolution of their two enantiomers [17]. In these compounds, neither the axle nor the wheel is chiral but both together give a cycloenantiomeric species (Scheme 6.5.4). Such molecules can be prepared if both axle and wheel bear groups with a certain directionality. For example, the sulfonamide group in the wheel in Scheme 6.5.4 defines such a directionality and any rotaxane built from an axle with two different stoppers and the sulfonamide wheel exists in two enantiomeric forms, as shown in the diagram. Based on the same basic idea, a cyclodiastereomeric [3]rotaxane – a rotaxane with two wheels on one axle – has been synthesized that exists as a pair of enantiomers plus an additional meso form [18]. Separation into the stereoisomers by HPLC on a chiral stationary phase is often possible, and the CD spectra have been recorded for many of them. Topologically chiral rotaxanes offer an interesting future perspective to realize molecular rotors which not only have an axle free to rotate within the wheel, but which couple this rotation to chirality in order to provide unidirectionality. The problem of how to couple topological chirality to the rotational sense has not yet been solved satisfactorily, however. Thus, as an illustration of the principle, let us discuss briefly an artificial minimal system, which is not based on rotaxanes but provides continuous unidirectional motion. This system, which was developed by Feringa et al. [19], contains an overcrowded double bond with chiral substituents (Scheme

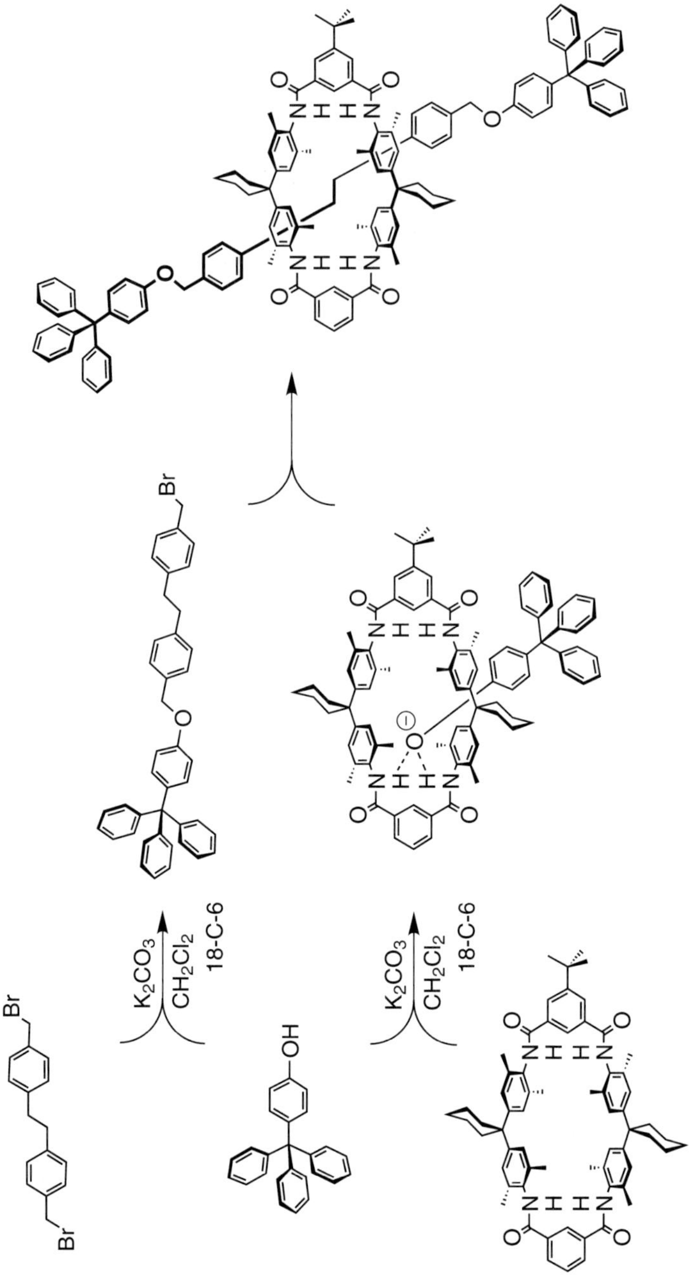

Scheme 6.5.3. Anion template effect for efficient rotaxane synthesis.

6.5.5 How to Achieve Unidirectional Rotation in Artificial Molecular Motors?

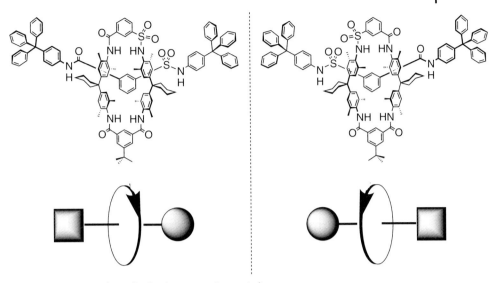

Scheme 6.5.4. A topologically chiral rotaxane. Arrows indicate the directionality imposed by the sulfonamide group in the wheel.

6.5.5). Photochemical $E \rightarrow Z$ isomerization produces a sterically strained state in which the methyl substituents are positioned in close proximity. The strain is released when the methyl groups and the naphthyl moieties switch their positions. This second step is followed by back-isomerization from the Z to the E form of the double bond and again steric hindrance provides an energetically unfavorable state which can reach an energetically more favorable state by a second flip of the sub-

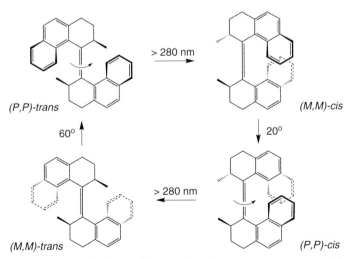

Scheme 6.5.5. Light-driven unidirectional rotation around an overcrowded double bond.

stituents. This system is driven by light inducing the isomerization reactions of the double bond. Each photoinduced step leads upwards in energy and is followed by a second comforting flip of the substituents.

6.5.6
The Fuel for Driving the Motor: Light, Electrons, and Chemical Energy

The Feringa system brings us to the question of how to drive molecular motion. What energy source should be employed [20]? How can such a compound be influenced by an external stimulus? In principle, at least three different sources of energy for driving rotaxane-based molecular motors can be imagined:

1. light [21], which is also the driving force for proton translocation in the natural photosynthetic system;
2. the reaction energy of any appropriate chemical transformation; and
3. electrical energy applied to the rotaxane or catenane by means of electrochemical methods.

Because these approaches can often be equally well applied to the shuttling motion of rotaxanes, the circumrotation of catenanes, and the threading/dethreading of pseudorotaxanes, we do not restrict this discussion to rotation only, but present a small selection of different examples.

Driven by light, the dethreading of a rod-like molecule from a pseudorotaxane can be realized, followed by reformation of the interlocked molecule after oxidation by air (Scheme 6.5.6) [22]. In the resting form, the (electron-rich) axle of this "piston–cylinder" machine is held inside the (electron-poor) bis-paraquat cyclophane wheel by π-donor–π-acceptor interactions. A photosensitizer built into the pseudorotaxane structure absorbs light of appropriate wavelength and transfers an electron to one of the paraquat units in the wheel. In its reduced form the non-covalent forces between axle and wheel are reduced and the axle dethreads if back-electron transfer is suppressed by reducing the photosensitizer with a sacrificial reductant such as oxalate. Addition of oxygen to the solution reoxidizes the cyclophane wheel and rethreading occurs, restoring the initial state.

Recently, Brower, Leigh and coworkers published a study [23] describing the light-controlled shuttling of a wheel along an axle (Scheme 6.5.7). In its resting state the rotaxane wheel is bound to the left diamide side by hydrogen bonding. Light-induced electron transfer, from a donor present in solution to the stopper on the right, generates an anion radical with better hydrogen bond acceptor qualities and consequently, the wheel moves surprisingly fast to this side. Back-conversion to the resting state by reduction of the anion radical causes the wheel to shift back to the left.

Utilizing the Cu(I)/Cu(II) redox pair, Sauvage and his group designed a molecular muscle (Scheme 6.5.8) [24]. Two filaments are each connected to a macro-

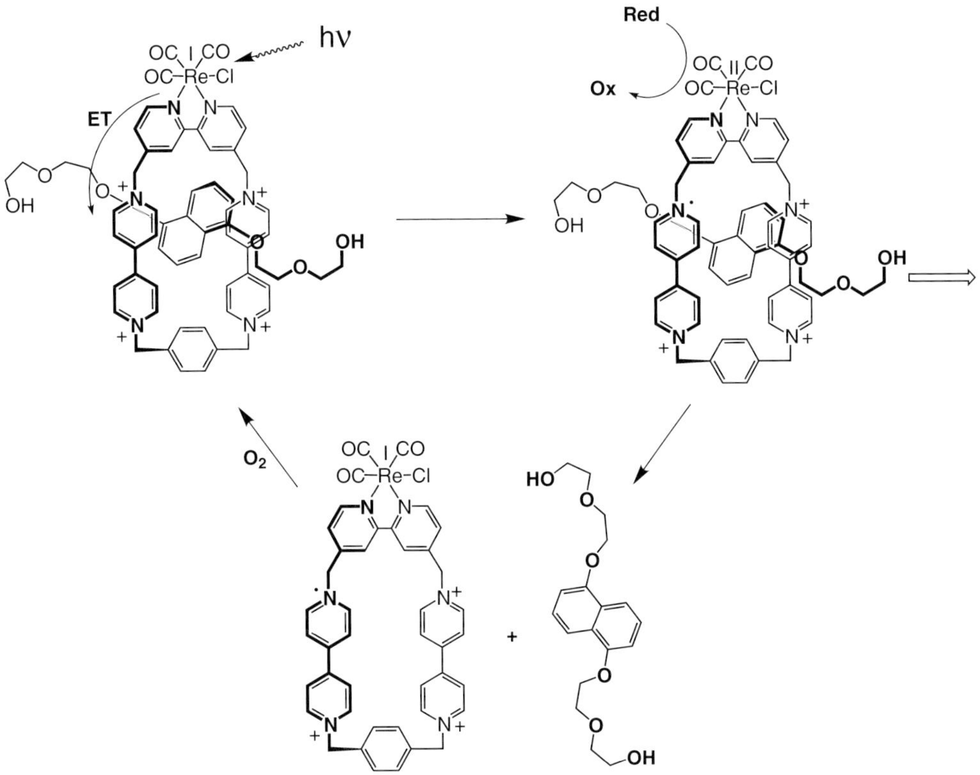

Scheme 6.5.6. A piston–cylinder machine based on light-induced electron-transfer processes within a pseudorotaxane.

cycle. By the Cu(I) ion template effect two of these molecules can be threaded into each other as depicted in Scheme 6.5.8. One of the particular features of copper is that it changes its preferred coordination geometry on oxidation from Cu(I) (tetrahedron) to Cu(II) (trigonal bipyramid). Incorporation of one phenanthroline unit in each of the wheels and both a phenanthroline and a terpyridine in each of the filaments enables switching between an extended and a contracted form. Cu(I) prefers to be bound to two phenanthroline ligands, but oxidation to Cu(II) provokes preferential binding to the terpyridine unit. This change necessarily involves an increase of the distance between the two macrocycles. Reduction of the Cu(II) ion to Cu(I) reverses the process.

The last example is a molecular shuttle with two different states between which one can switch electrochemically and chemically [25]. The more electron-rich part of the axle is the bisaniline moiety and, consequently, the electron-poor wheel tends to bind more strongly to this side (Scheme 6.5.9, center). On oxidation and/or protonation charge repulsion moves the wheel to the bisphenol part of the axle. Thus, the switching between the two states can be controlled by two different

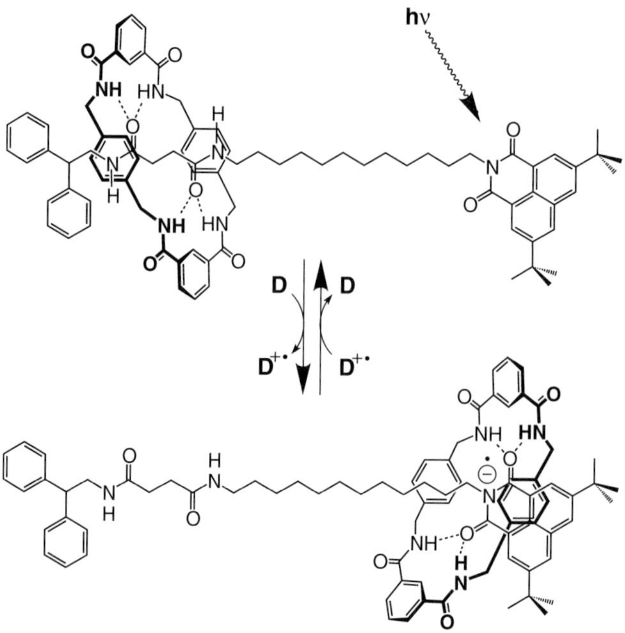

6.5 Artificial Molecular Rotary Motors Based on Rotaxanes

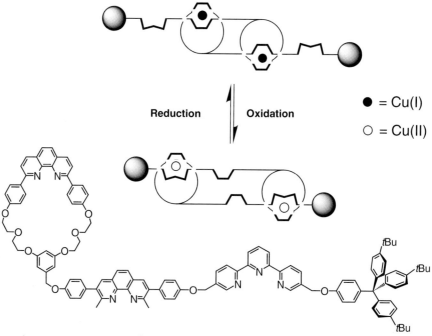

Scheme 6.5.7. Light-driven shuttling of a rotaxane with internal hydrogen bonding.

Scheme 6.5.8. Contraction of a molecular "muscle" mediated by electrochemical redox processes.

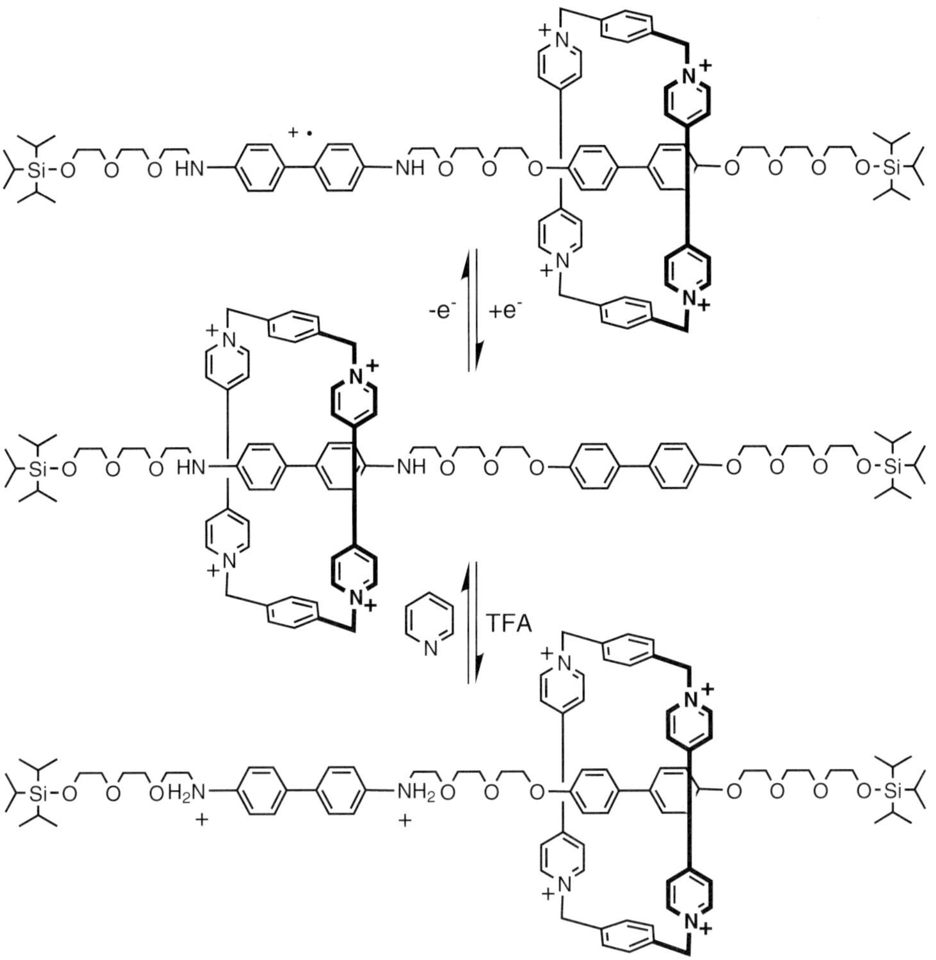

Scheme 6.5.9. An electrochemically and chemically driven molecular shuttle.

stimuli opening the field of nanoelectronics for such compounds beyond controlling their motion. Logic gates can be constructed that combine two different input signals (protonation/oxidation) with one output (shift of the wheel). From this point of view, this example would represent an OR gate.

6.5.7
Conclusions

These examples of artificial molecular motors illustrate that a real molecular motor does not yet exist except in nature. Despite all the progress made in this field and all the intriguing and wonderful chemistry discovered on the way to the final goal, it has not been possible to combine all the features necessary to generate one complete functional system. Nevertheless, research in this area is valuable and rewarding, because it gives detailed and fundamental insight into many aspects of supramolecular and bioorganic chemistry, nanotechnology, and materials sciences. The perfection of the natural motors demonstrate that they can, in principle, be made. The complexity of these systems may, however, indicate that an artificial model system should not be oversimplified. For successful implementation of a motor at the molecular level many variables must probably be optimized, making the necessary fine tuning of the system extremely complex. Finally, membranes and surfaces might help to order such molecules to achieve macroscopic effects.

References

1 (a) P. D. Boyer, Angew. Chem. **1998**, 110, 2424; Angew. Chem. Int. Ed. **1998**, 37, 2296; (b) J. E. Walker, Angew. Chem. **1998**, 110, 2438; Angew. Chem. Int. Ed. **1998**, 37, 2308; (c) C. Bustamante, D. Keller, G. Oster, Acc. Chem. Res. **2001**, 34, 412.

2 C. A. Schalley, K. Beizai, F. Vögtle, Acc. Chem. Res. **2001**, 34, 465.

3 G. Banting, S. J. Higgins, eds., Essays in Biochemistry: Molecular Motors, Portland Press, London, **2000**.

4 (a) G. Schill, Catenanes, Rotaxanes and Knots, Academic Press, New York, **1971**; (b) J.-P. Sauvage, C. Dietrich-Buchecker, eds., Molecular Catenanes, Rotaxanes, and Knots. Wiley–VCH, Weinheim, **1999**.

5 (a) V. Balzani, M. Gómez-López, J. F. Stoddart, Acc. Chem. Res. **1998**, 31, 405; (b) J.-P. Sauvage, Acc. Chem. Res. **1998**, 31, 611; (c) Z. Asfari, J. Vicens, J. Inclusion Phenom. Macrocyc. Chem. **2000**, 36, 103; (d) V. Balzani, A. Credi, F. M. Raymo, J. F. Stoddart, Angew. Chem. **2000**, 112, 3484; Angew. Chem. Int. Ed. **2000**, 39, 3348; (e) A. R. Pease, J. O. Jeppesen, J. F. Stoddart, Y. Luo, P. Collier, J. R. Heath, Acc. Chem. Res. **2001**, 34, 433.

6 A. Affeld, G. M. Hübner, C. Seel, C. A. Schalley, Eur. J. Org. Chem. **2001**, 2877.

7 (a) N. V. Gerbeleu, V. B. Arion, J. Burgess, Template Synthesis of Macrocyclic Compounds, Wiley–VCH, Weinheim, **1999**; (b) T. J. Hubin, A. G. Kolchinski, A. L. Vance, D. H. Busch, Template Control of Supramolecular Architectures, in: Advances in Supramolecular Chemistry, Vol. 5, JAI Press, **1999**, p. 237; (c) F. Diederich, P. J. Stang, Template Directed Synthesis, Wiley–VCH, Weinheim, **2000**.

8 J.-C. Chambron, C. O. Dietrich-Buchecker, J.-F. Nierengarten, J.-P. Sauvage, Pure Appl. Chem. **1994**, 66, 1543.

9 (a) D. B. Amabilino, J. F. Stoddart, Chem. Rev. **1995**, 95, 2725; (b) F. M. Raymo, J. F. Stoddart, Chem. Rev. **1999**, 99, 1643; (c) M. B. Nielsen, C. Lomholt, J. Becher, Chem. Soc. Rev. **2000**, 29, 153.

10 (a) F. Vögtle, T. Dünnwald, T. Schmidt, Acc. Chem. Res. **1996**, 29, 451; (b) R. Jäger, F. Vögtle, Angew. Chem. **1997**, 109, 966; Angew. Chem. Int. Ed. **1997**, 36, 930.

11 S. A. Nepogodiev, J. F. Stoddart, Chem. Rev. **1998**, 98, 1959–1976.

12 P. Linnartz, C. A. Schalley, Rotaxanes and Pseudorotaxanes. In: Encyclopedia of Supramolecular Chemistry, J. L. Atwood, J. W. Steed, eds., Dekker, New York, in press.

13 (a) G. M. Hübner, J. Gläser, C. Seel, F. Vögtle, Angew. Chem. **1999**, 111,

395; *Angew. Chem. Int. Ed.* **1999**, *38*, 383. (b) C. REUTER, W. WIENAND, G. M. HÜBNER, C. SEEL, F. VÖGTLE, *Chem. Eur. J.* **1999**, *5*, 2692; (c) C. SEEL, F. VÖGTLE, *Chem. Eur. J.* **2000**, *6*, 21; (d) C. A. SCHALLEY, G. SILVA, C. F. NISING, P. LINNARTZ, *Helv. Chim. Acta* **2002**, *85*, 1578.

14 T. SCHMIDT, R. SCHMIEDER, W. M. MÜLLER, B. KIUPEL, F. VÖGTLE, *Eur. J. Org. Chem.* **1998**, 2003.

15 P. R. ASHTON, J. A. BRAVO, F. M. RAYMO, J. F. STODDART, A. J. P. WHITE, D. J. WILLIAMS, *Eur. J. Org. Chem.* **1999**, 899.

16 For reviews on topological chirality, see: (a) J.-C. CHAMBRON, C. DIETRICH-BUCHECKER, J.-P. SAUVAGE, *Top. Curr. Chem.* **1993**, *165*, 131; (b) K. MISLOW, *Top. Stereochem.* **1999**, 1; (c) C. REUTER, R. SCHMIEDER, F. VÖGTLE, *Pure Appl. Chem.* **2000**, *72*, 2233.

17 (a) C. REUTER, C. SEEL, M. NIEGER, F. VÖGTLE, *Helv. Chim. Acta* **2000**, *83*, 630; (b) C. REUTER, A. MOHRY, A. SOBANSKI, F. VÖGTLE, *Chem. Eur. J.* **2000**, *6*, 1674; (c) C. REUTER, W. WIENAND, C. SCHMUCK, F. VÖGTLE, *Chem. Eur. J.* **2001**, *7*, 1728; (d) C. YAMAMOTO, Y. OKAMOTO, T. SCHMIDT, R. JÄGER, F. VÖGTLE, *J. Am. Chem. Soc.* **1997**, *119*, 10547.

18 R. SCHMIEDER, G. HÜBNER, C. SEEL, F. VÖGTLE, *Angew. Chem.* **1999**, *111*, 3741; *Angew. Chem. Int. Ed.* **1999**, *38*, 3528.

19 (a) A. M. SCHOEVAARS, W. KRUIZINGA, R. W. J. ZIJLSTRA, N. VELDMAN, A. L. SPEK, B. L. FERINGA, *J. Org. Chem.* **1997**, *62*, 4943; (b) N. KOUMURA, R. W. J. ZIJLSTRA, R. A. VAN DELDEN, N. HARADA, B. L. FERINGA, *Nature* **1999**, *401*, 152; (c) B. L. FERINGA, *Acc. Chem. Res.* **2001**, *34*, 504; (d) N. KOUMURA, E. M. GEERTSEMA, M. B. VAN GELDER, A. MEETSMA, B. L. FERINGA, *J. Am. Chem. Soc.* **2002**, *124*, 5037. See also T. R. KELLY, *Acc. Chem. Res.* **2001**, *34*, 514.

20 R. BALLARDINI, V. BALZANI, A. CREDI, M. T. GANDOLFI, M. VENTURI, *Acc. Chem. Res.* **2001**, *34*, 445.

21 A. C. BENNISTON, *Chem. Soc. Rev.* **1996**, *25*, 427.

22 P. R. ASHTON, V. BALZANI, O. KOCIAN, L. PRODI, N. SPENCER, J. F. STODDART, *J. Am. Chem. Soc.* **1998**, *120*, 11190.

23 A. M. BROUWER, C. FROCHOT, F. G. GATTI, D. A. LEIGH, L. MOTTIER, F. PAOLUCCI, S. ROFFIA, G. W. H. WURPEL, *Science* **2001**, *291*, 2124.

24 J.-P. COLLIN, C. DIETRICH-BUCHECKER, P. GAVIÑA, M. C. JIMENEZ-MOLERO, J.-P. SAUVAGE, *Acc. Chem. Res.* **2001**, *34*, 477.

25 R. A. BISSELL, E. CÓRDOVA, A. E. KAIFER, J. F. STODDART, *Nature* **1994**, *369*, 133.

6.6
Chemical Approaches for the Preparation of Biologically-inspired Supramolecular Architectures and Advanced Polymeric Materials

Harm-Anton Klok

6.6.1
Introduction

Since the foundation of the fundamental concepts of polymer science in the 1920s and 1930s by Staudinger and Carothers [1], the world consumption of this class of materials has grown exponentially and now far exceeds that of metals [2]. Polymers, or "plastics", are relatively cheap, lightweight, and durable, which makes them the material of choice for many applications, including, e.g., packaging materials, protective coatings, etc. Examples of polymers which are produced industrially on a large scale include, e.g., poly(ethylene), poly(propylene), and poly(styrene). These so-called "commodity" plastics are characterized by relatively low structural order and generally are amorphous or semi-crystalline materials. The lack of structural order is because of the heterogeneous nature of the chain length and composition of most synthetic polymers. For some applications lack of crystallinity is advantageous, because it can provide properties such as optical clarity and elasticity. The heterogeneous nature of synthetic polymers also prevents precise control of structure and properties at the molecular level, however; such control would be desirable for many advanced applications in optics, electronics, and biomedicine.

In contrast with synthetic polymers, proteins are characterized by very high levels of structural order. Unlike synthetic polymers, proteins are characterized by absolutely uniform chain lengths and well-defined monomer sequences (primary structure) [3]. These features are two of the requirements that enable folding of linear polypeptide chains into structurally well-defined and functional proteins. Proteins play an important role in numerous processes in biology, e.g. as carriers for small molecules and ions (examples are presented in Chapter 2.2), as catalysts, or as muscle fibers, and their exquisite properties are closely related to their well-defined three-dimensional structure [3].

Over the years the notion has grown that integration of biological design concepts in macromolecular chemistry might offer unprecedented opportunities for the development of novel polymeric materials [4]. The capacity of peptides to self-organize into well-defined hierarchically organized superstructures might enable the development of polypeptide (hybrid)materials with a level of structural control approaching that of natural proteins. The integration of biological design concepts is of interest not only structurally, but also functionally. First of all, there is a wealth of protein secondary structures, for example α-helices and β-sheets, whose conformation is sensitive to, e.g., temperature, pH, or ion-strength. Such secondary structural motifs (Box 8) can be regarded as switching elements that might enable reversible manipulation of the structure and properties of protein-inspired materials. Secondly, the combination of proteins/peptides with synthetic polymers might enable the development of bioactive materials that can communicate with their environment, e.g. with cells and proteins. It is obvious that such materials would be highly interesting for a variety of biomedical applications.

So far, peptide synthesis has mainly involved the preparation of biologically or pharmaceutically relevant substances, total chemical synthesis of natural proteins or protein engineering. The objective of this chapter is to show how the different methods that have been developed for peptide synthesis can be used to prepare biologically-inspired supramolecular architectures and polymeric materials, which might be of potential interest for a variety of advanced applications.

6.6.2
Ring-opening Polymerization of α-Amino Acid N-Carboxyanhydrides

The history of the synthesis and polymerization of α-amino acid N-carboxyanhydrides (NCA) goes back to the work of Herman Leuchs in the early 1900s [5]. NCA polymerization, which is outlined in Scheme 6.6.1, is a very straightforward process and enables the preparation of relatively high-molecular-weight polypeptides in multigram quantities. The first step involves the synthesis of the NCA monomers. In the original work by Leuchs, NCA were prepared by heating the corresponding N-alkoxycarbonyl α-amino acid chlorides [5]. Nowadays NCA synthesis is usually accomplished by treating the appropriate α-amino acid with phosgene [6], or more conveniently, diphosgene [7] or triphosgene [8]. NCA can be polymerized using a variety of nucleophiles and bases [9]; as for most polymerization reactions,

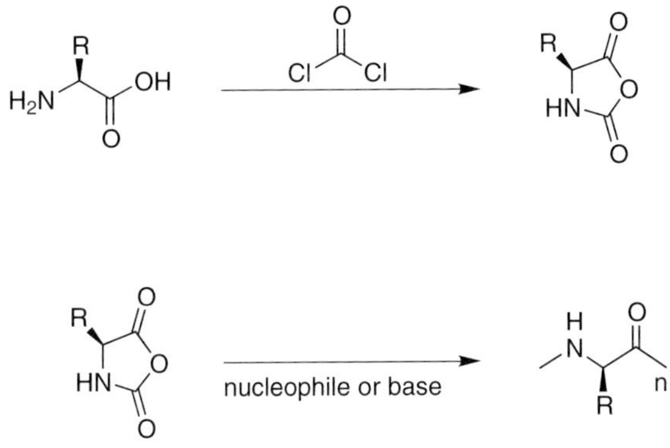

Scheme 6.6.1.

however, NCA polymerization neither affords perfectly monodisperse materials nor enables precise control of the composition and α-amino acid sequence. This is, of course, in great contrast with natural proteins, and automatically limits the utility of this approach for the generation of biologically-inspired or biomimetic architectures. Nevertheless, because of its simplicity, NCA polymerization has been widely used for preparation of a range of polypeptide homopolymers, random copolymers, (hybrid) block copolymers, and graft copolymers [9, 10].

Very recently, amphiphilic poly(butadiene)-b-poly(L-glutamic acid) copolymers have been reported which were obtained by ring-opening polymerization of γ-benzyl-L-glutamate N-carboxyanhydride (Bn-Glu NCA) by use of a poly(butadiene) macroinitiator, followed by a hydrogenation step to remove the benzyl ester protective groups (Scheme 6.6.2) [11]. Amphiphilic (ionic) block copolymers have been extensively studied and are well known to form a variety of supramolecular aggregates in dilute aqueous solution [12]. In contrast with most systems studied so far, the block copolymers shown in Scheme 6.6.2 have a hydrophilic block which can adopt a well-defined α-helical secondary structure. By adjusting the pH of the aqueous solution, the secondary structure of poly(L-glutamic acid) can be reversibly changed from a compact α-helix into a disordered coil, and vice versa [13]. Depending on the relative lengths of the blocks, the polymers shown in Scheme 6.6.2 were found to form spherical micelles or vesicular aggregates (Figure 6.6.1) [11]. The structure and size of these supramolecular aggregates was investigated by a combination of light- and neutron scattering experiments and freeze-fracture transmission electron microscopy. Because these micelles and vesicles bear some structural resemblance to compactly folded globular proteins, this work is a good illustration of the power of self-assembly to generate biologically inspired or biomimetic supramolecular architectures.

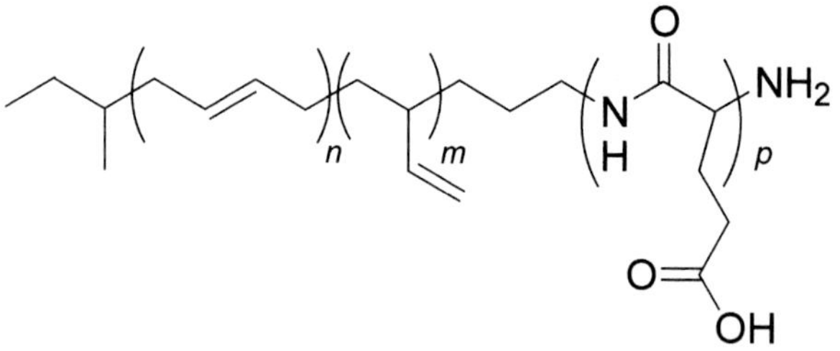

Scheme 6.6.2.

One factor that contributes to the heterogeneous nature of polypeptides produced by NCA polymerization is the multiple reactivity of the monomers [9]. NCA contain four reactive sites; two electrophilic carbonyl carbons and two nucleophilic centers after deprotonation of the α-CH and NH hydrogen atoms. By use of traditional nucleophilic or basic initiators NCA ring-opening polymerization can proceed simultaneously along different competing pathways, which broadens mo-

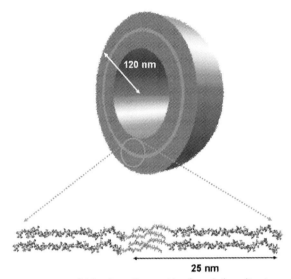

Fig. 6.6.1. Model for the self-assembly of a poly(butadiene)$_{40}$-b-poly(L-glutamic acid)$_{100}$ diblock copolymer into vesicular aggregates. (Subscripts indicate the number-average degree of polymerization of the blocks. Adapted from Ref. [11b]).

lecular weight distributions, restricts accurate control of the composition of the peptides, and hampers the formation of well-defined block copolymers. Several years ago, however, it was discovered that transition-metal complexes such as bpyNi(cod) (bpy = 2,2'-bipyridyl, cod = cycloocta-1,5-dienyl) and Co(PMe$_3$)$_4$ can overcome a number of these drawbacks and enable "living" polymerization of NCA with unprecedented control of chain length and narrow polydispersities [14]. This approach has been successfully used to prepare a variety of random and (hybrid) block copolypeptides [15].

Very recently, transition-metal-mediated NCA polymerization has been used to prepare a series of amphiphilic, ionic block copolypeptides [16]. While the total degree of polymerization was kept constant, the chemical composition and block length ratio of the block copolymers differed. The hydrophilic blocks of the copolymers comprised charged poly(L-lysine) or poly(L-glutamic acid) sequences with no regular secondary structure at neutral pH. The hydrophobic, water-insoluble blocks were based on L-leucine or L-valine, which are known to have a high propensity to form rod-like α-helices or crystalline β-strands, respectively. When attempts were made to prepare dilute aqueous solutions of the block copolymers, gelation was observed instead of the formation of vesicles or micelles. The gels were investigated by use of rheology, small-angle neutron scattering (SANS), laser-scanning confocal microscopy, and cryogenic transmission electron microscopy [16, 17]. Interestingly, it was found that the gelation behavior of the block copolypeptides was related to the secondary structure of the hydrophobic block, with α-

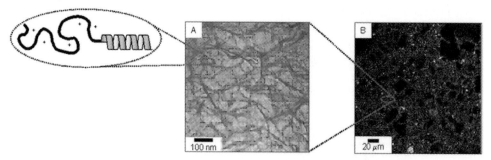

Fig. 6.6.2. Hierarchical self-assembly of an ionic poly(L-lysine)$_{160}$-b-poly(L-leucine)$_{40}$ block copolymer amphiphile into nanosized membraneous structures visualized by (a) cryogenic transmission electron microscopy and (b) laser-scanning confocal microscopy. (Subscripts indicate the number-average degree of polymerization of the blocks. Adapted from Ref. [16]).

helical blocks being slightly better than β-strands, which were better than random coils. SANS and microscopy experiments indicated that the gels were hierarchically structured and consisted of nanosized membraneous structures that were assembled into a microscale porous structure. This is illustrated in Figure 6.6.2, where cryogenic transmission electron microscopy (Figure 6.6.2a) and laser-scanning confocal microscopy images (Figure 6.6.2b) show the nanoscale and microscopic ordering, respectively, in the hydrogels [16]. These systems are another example of the use of NCA chemistry to generate block copolymers that can self-assemble into hierarchically organized structures by exploiting the capacity of peptide sequences both to adopt well-defined secondary structures and to direct the formation of higher-order assemblies.

6.6.3
Solid-phase Peptide Synthesis

Amphiphilic molecules, which consist of a hydrophilic water-soluble part and a hydrophobic water-insoluble part, can self-organize into a variety of supramolecular structures. This is not only true for polymeric amphiphiles as was discussed above (Section 6.6.2) [12], but applies also to low-molecular-weight amphiphiles. The size and shape of the supramolecular structures formed by these molecules in aqueous solution depend, among other factors, on the relative sizes of the hydrophilic and hydrophobic parts, the charge on the hydrophilic headgroup, and the geometry of the molecules [18]. Amphiphiles consisting of a hydrophilic peptide headgroup and a hydrophobic alkyl tail are a particularly interesting, although relatively unexplored, class of molecules [19]. The ability of solid phase peptide synthesis (SPPS, Box 25) to prepare peptides with well-defined α-amino acid sequences and chain lengths, however, offers the possibility of precise control over the size and charge of the hydrophilic peptide segment and also facilitates the incorporation of bioactive peptide sequences. Thus, SPPS affords unprecedented opportunities to tailor the structure and properties of supramolecular architectures and materials composed of peptide amphiphiles.

In two recent publications, Stupp and coworkers reported the synthesis and supramolecular organization of a series of novel peptide amphiphiles [20]. The molecules consisted of a hydrophilic peptide headgroup containing 8–12 α-amino acids and alkyl tails comprising 6–22 carbon atoms. The hydrophilic headgroup contained both ionizable α-amino acid residues to enable pH-control of the self-assembly process, and cysteine residues, which were explored to covalently capture supramolecular structures formed in aqueous solution. Because disulfide bonds can be reversibly formed and broken on oxidation and reduction, respectively, supramolecular aggregates can be transformed into covalently captured nano-objects, and vice versa. The self-assembly of the peptide amphiphiles was studied in detail with transmission electron microscopy (TEM) and is schematically illustrated in Figure 6.6.3. Acidification of aqueous solutions of the peptide amphiphiles resulted in the formation of precipitates with, according to TEM, a nanofibrillar structure. Oxidation of the thiol groups resulted in cross-linking of the supramolecular fibers and led to covalently captured nano-objects. Interestingly, these cross-linked nanofibers could direct the mineralization of hydroxyapatite to form a composite material structurally similar to the mineralized collagen fibrils found in bone [20]. At peptide amphiphile concentrations above 0.25 wt%, acidification resulted in the formation of pH-sensitive hydrogels, which could be reversibly formed and disassembled by adjusting the pH.

SPPS is not only useful for preparation of peptide amphiphiles based on unnatural α-amino acid sequences, as discussed above, but can also be used to generate

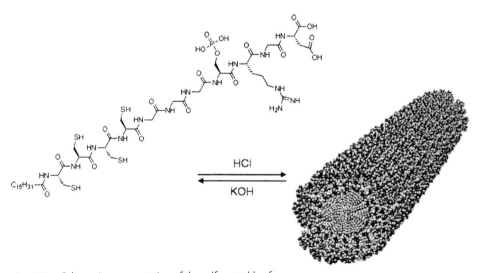

Fig. 6.6.3. Schematic representation of the self-assembly of a peptide amphiphile into cylindrical micellar nanofibers. (Adapted from Ref. [20a]).

hybrid block copolymers of synthetic macromolecules and α-amino acid sequences derived from protein-folding motifs. Protein-folding motifs, or supersecondary structures, consist of a small number of secondary structural elements such as α-helices or β-strands packed in close proximity in a well-defined geometric arrangement [3]. The specificity of the folding process, which essentially involves transformation of a linear unordered polypeptide chain into a protein with a well-defined three-dimensional structure and very specific properties, is in great contrast to the self-assembly of synthetic amphiphilic block copolymers. Polyalkyl-b-poly(ethylene oxide) and poly(propylene oxide)-b-poly(ethylene oxide) copolymers can, for example, form micellar type structures and hydrogels in aqueous media and are being intensively investigated for biomedical applications including drug delivery and tissue engineering [21]. The self-assembly of such block copolymers is exclusively driven by non-specific hydrophobic interactions. Synthetic block copolymers are usually polydisperse and do not contain well-defined monomer sequences. Consequently, the structure and properties of the resulting micelles and hydrogels can only be controlled and tailored to a limited extent.

In an attempt to overcome these limitations, we and others have recently started to explore peptide sequences derived from protein-folding motifs to enhance control of the supramolecular organization and association behavior of synthetic poly(ethylene oxide) (PEO)-based block copolymers [22]. If the capacity of peptide sequences to assemble into protein-folding motifs is retained on conjugation to PEO the supramolecular organization and association behavior of the block copolymers might be very precisely controlled. Improving control over the association behavior and supramolecular organization of PEO-based block copolymers might be of great interest in a variety of biomedical applications and could enable the development of micellar drug carriers and hydrogels whose formation and dissociation can be triggered and controlled with unprecedented precision. The coiled-coil motif, a supersecondary structure consisting of two or more α-helices wound around each other in a superhelical fashion, was explored in preliminary experiments [23]. Coiled-coils are among the most abundant folding motifs and are characterized by a primary structure consisting of a repetitive heptad repeat pattern (-[a-b-c-d-e-f-g]$_n$-). Within this heptad repeat pattern, positions *a* and *d* are usually occupied by hydrophobic α-amino acids.

PEO-b-peptide block copolymers were prepared as outlined Scheme 6.6.3. The primary structure of the peptide block was not derived from a natural protein-folding motif but was based on a de-novo designed coiled-coil [24]. Whereas the α-amino acid sequence of natural proteins can be rather irregular, de novo protein-folding motifs are based on a repetitive sequence of heptad repeat units containing a minimum number of different α-amino acids. The regular and repetitive nature of the α-amino acid sequence of such de-novo coiled-coils greatly facilitates systematic variation of block length. The association behavior and supramolecular organization of the PEO-b-peptide copolymers were investigated by means of circular dichroism (CD) and analytical ultracentrifugation (AUC). These experiments indicated that the capacity of the peptide blocks to assemble into well-defined higher-order structures is retained on conjugation to PEO and can be used to direct

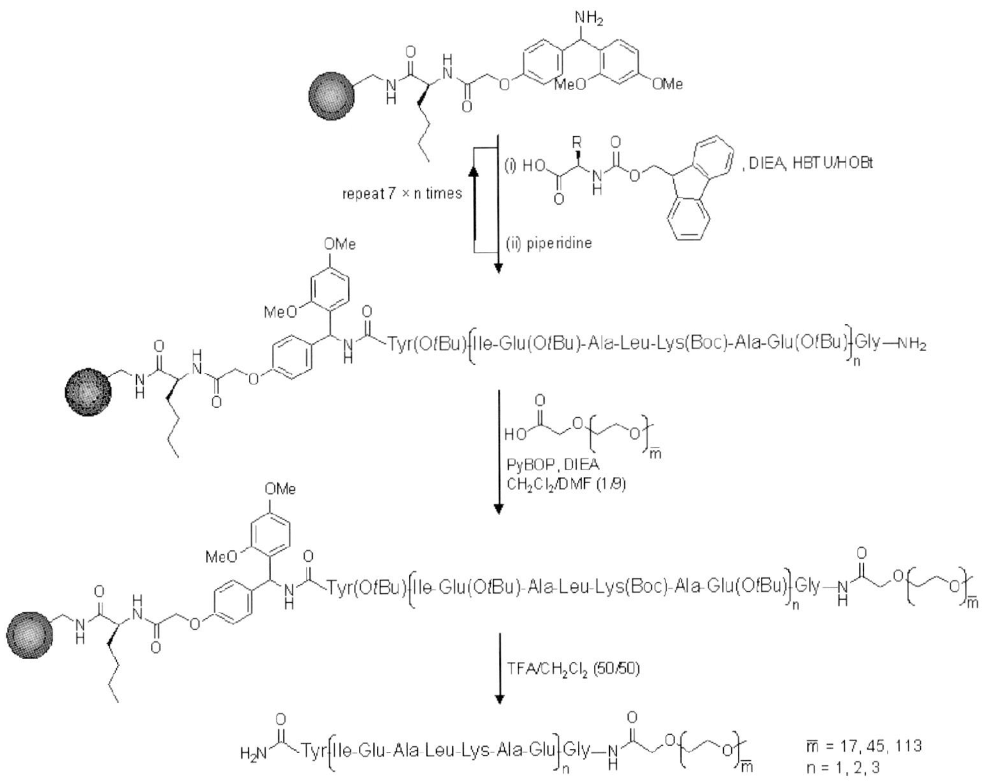

Scheme 6.6.3.

the self-organization of the block copolymers. The self-assembly of PEO-*b*-peptide block copolymers inspired by coiled-coil protein-folding motifs can be described as an equilibrium between unimeric block copolymer molecules with partially ordered peptide blocks and intermolecular coiled-coil dimers and tetramers (Figure 6.6.4). The folding and unfolding of the peptide blocks is sensitive to temperature and concentration; this affords prospects for the development of polypeptide hybrid materials whose structure and properties can be precisely controlled by external stimuli. The PEO-*b*-peptide block copolymers are not amphiphilic and their self-organization is exclusively driven by the propensity of the peptide segments to fold into higher-order structures. As a result, only discrete, in this instance mainly dimeric and tetrameric, aggregates are found, in contrast with the large and polydisperse structures which are formed on self-assembly of conventional amphiphilic block copolymers. Although the work performed so far is merely a proof of concept, the experiments clearly demonstrate the feasibility of using concepts from protein folding to develop novel block copolymer-type materials whose self-organization is mediated by the formation of well-defined hierarchically organized supramolecular structures.

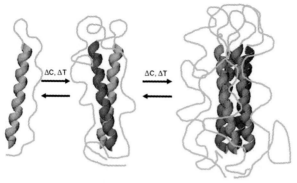

Fig. 6.6.4. Reversible self-assembly of PEO-*b*-peptide block copolymers based on protein-folding motifs.

6.6.4
Peptide Ligation

Although the yields in each of the coupling and deprotection steps during solid-phase peptide synthesis (SPPS) are usually very high (98–99%), they are not quantitative, which results in the formation of deletion sequences. These deletion sequences accumulate statistically throughout synthesis and limit the practical utility of SPPS to the preparation of peptides containing 50–60 α-amino acid residues. To illustrate this, consider a peptide composed of 60 α-amino acids. The synthesis of this peptide would involve 120 steps, one deprotection and one coupling reaction for each residue. Assuming that each of the reactions proceeds in 98% yield, SPPS would only afford $(0.98)^{120} = 8.9\%$ of the desired peptide. Because most proteins contain more than 60 α-amino acid residues, it is obvious that SPPS is of only limited utility for the chemical synthesis of naturally occurring proteins and for protein engineering.

One possible means of overcoming the limited chain lengths that can be achieved by use of SPPS is to use oligo- or polypeptides instead of simple α-amino acids as building blocks. As mentioned above, oligo- or polypeptides containing 50–60 α-amino acids are conveniently prepared by means of SPPS. It is obvious that the use of such peptides as building blocks dramatically increases the molecular weight of the products that can be prepared and facilitates the chemical synthesis of large naturally occurring proteins. Originally, this strategy involved the use of protected peptide segments which were coupled either in homogeneous solution or on solid supports and was named fragment or segment condensation [25]. However, because protected peptide segments are usually poorly soluble, and difficult to purify and characterize, fragment condensation was significantly hampered by several practical problems. This situation changed approximately 10 years ago when strategies became available that enable regiospecific coupling of unprotected peptide segments in aqueous solution [26]. Fragment condensation of unprotected peptide segments is commonly referred to as peptide ligation (Box 26). Because it is based on unprotected peptide building blocks, peptide ligation does not suffer the solubility, purification, and characterization problems encountered with the use of protected peptide segments. Among the different techniques that have been developed, the native chemical ligation (NCL) strategy, introduced by Kent et al. [27], has received particular widespread attention. The original NCL strategy involves coupling of a peptide-α-thioester with a second peptide segment containing an N-terminal cysteine (Cys) residue [27]. Because Cys residues amount to only ~1.7% of the α-amino acid residues in globular proteins, it is obvious that this requirement significantly restricts the practical utility of NCL for total chemical synthesis of natural proteins. Very recently, however, auxiliary groups have been developed that enable ligation of peptide segments at sites other than Xxx–Cys (Scheme 6.6.4) [28]. These auxiliaries are based on the 1-phenyl-2-mercaptoethyl moiety, which is linked to the N-terminus of one of the two peptide segments. The thiol group of the auxiliary acts as the nucleophile in the capture reaction. After

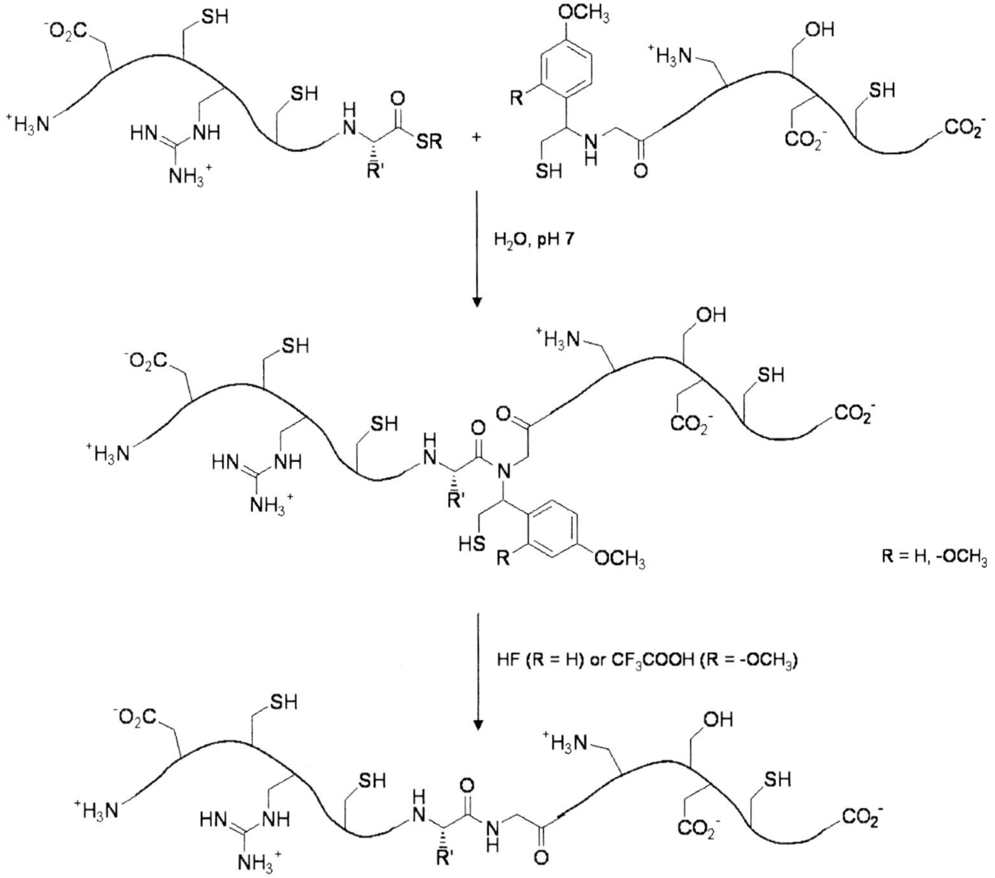

Scheme 6.6.4.

S,N-acyl transfer and removal of the auxiliary group an Xxx–Gly bond is generated to connect the two original peptide segments. NCL can be performed both in homogeneous solution and on solid supports. Solid-phase NCL is particularly attractive for sequential ligation of multiple peptide segments [29]. NCL has been used successfully both for the total chemical synthesis of proteins [30] and for protein-engineering purposes [31]. NCL is an extremely versatile strategy for protein engineering, because it enables site-specific incorporation of unnatural structures which are impossible or difficult to introduce by biosynthetic techniques.

Despite its many successful applications and the development of auxiliary moieties such as the 1-phenyl-2-mercaptoethyl group, NCL is essentially restricted to ligation of peptide segments via an Xxx–Cys or Xxx–Gly bond. It is obvious that the impact of techniques that enable chemoselective ligation via the formation of native peptide bonds would be even larger if they were completely independent of the α-amino acid sequence of the target polypeptide. A method that holds great promise as a potentially universal ligation method, irrespective of α-amino acid sequence, is based on the Staudinger reaction [32]. In the "Staudinger ligation", which is outlined in Scheme 6.6.5, a phosphinothiol is used to couple a peptide thioester and a peptide azide. After transthioesterification with the phosphinothiol, the putative mechanism involves covalent capture by reaction with the peptide azide, followed by an S- to N-acyl shift [33]. Hydrolysis of the intermediate amidophosphonium salt affords the final product in which the two original peptide segments are linked via a native peptide bond.

Peptide ligation strategies used so far have mainly been used for protein total synthesis and protein engineering purposes. The ability to prepare perfectly monodisperse and relatively high molar mass peptides with precise control over the α-amino acid sequence could also afford unprecedented opportunities for the development of novel biologically-inspired supramolecular architectures and materials. As a final, recent, example, the preparation of a protein[2]catenane using NCL is shown in Figure 6.6.5 [34].

6.6.5
Summary and Conclusions

The starting point of this chapter was the need for synthetic methods that enable integration of biological and, in particular, protein design concepts into organic and macromolecular chemistry as a means to enhance control of the structure and properties of organic and polymeric materials. Three chemical strategies for the preparation of (poly)peptides have been discussed. The objective was to show that the utility of these methods goes beyond the preparation of biologically and pharmaceutically relevant substances and that they are attractive tools for the development of novel biologically-inspired supramolecular architectures and polymeric materials.

The most straightforward method discussed for peptide synthesis was the NCA

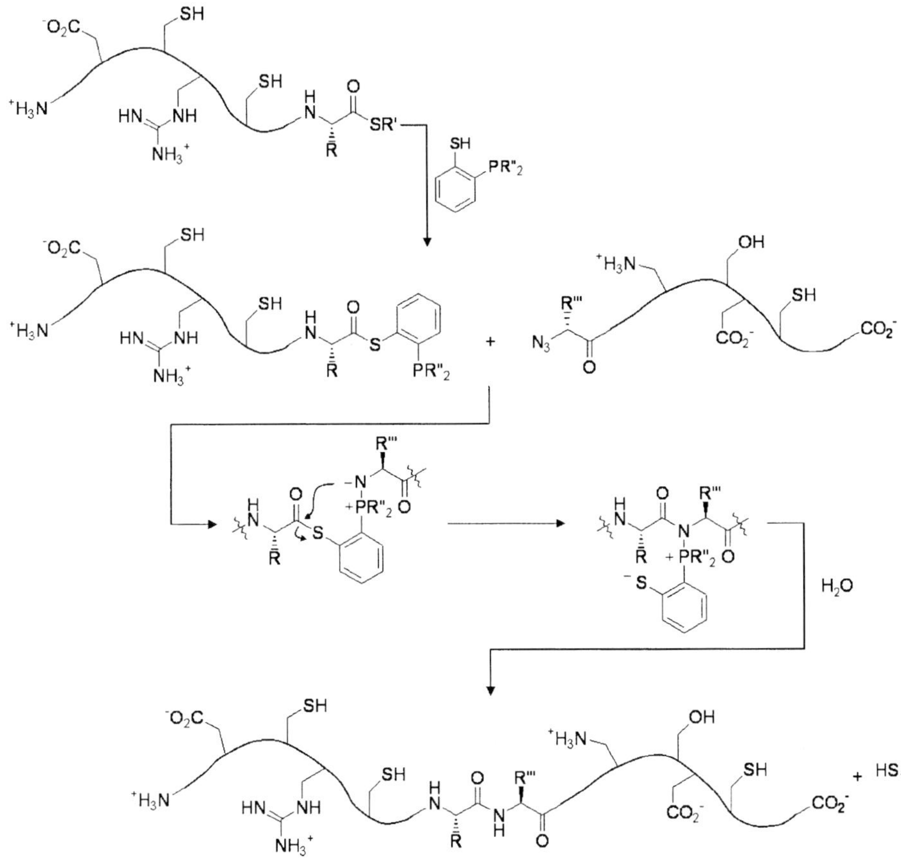

6.6.5 Summary and Conclusions

Scheme 6.6.5.

ring-opening polymerization. Because of its simplicity, this method has been extensively used for the preparation of hybrid block copolymers and other macromolecular architectures. However, because NCA polymerization is hampered by chain-breaking, -transfer and -termination reactions, it neither affords perfectly monodisperse materials nor enables preparation of peptides with predictable molecular weights. Also, NCA polymerization does not enable control of the α-amino acid sequence of the resulting polypeptides. The synthetic ease of NCA polymerization is, therefore, at the expense of structural perfection and this method is only of restricted use for the preparation of truly biomimetic supramolecular structures or polymeric materials.

SPPS is a second method that has found application outside the traditional area of peptide synthesis and which has been used successfully for the preparation of

CGGGEYFTLQIRGRERFEMFRELNEALELKDAQAGKEPGG

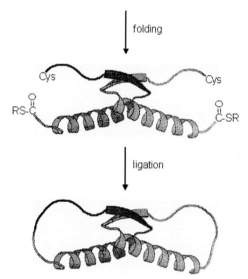

Fig. 6.6.5. Synthesis of a protein[2]catenane using native chemical ligation. (Adapted from Ref. [34]).

a variety of building blocks for biologically-inspired supramolecular architectures. In contrast with the NCA polymerization SPPS enables preparation of perfectly monodisperse peptides with precise control of α-amino acid sequence. For practical reasons, however, the utility of SPPS is restricted to peptides containing 50–60 α-amino acid residues. Because most proteins contain more than 60 α-amino acids, it is obvious that this limits the use of SPPS, e.g. for the total chemical synthesis of proteins.

A method that can bridge the gap between SPPS and NCA polymerization, i.e. which enables the preparation of high-molecular-weight and perfectly monodisperse peptides with well-defined α-amino acid sequences, is peptide ligation. Peptide ligation involves chemoselective coupling of unprotected peptide segments in aqueous media. Peptide ligation techniques have been used with great success in protein total synthesis and protein engineering. The advantage of peptide ligation in comparison with biological protein synthesis (which has not been considered in this chapter) is that it tolerates any unnatural α-amino acid and can also enable facile conjugation of peptides/proteins to synthetic non-biological oligomers/polymers. The potential of peptide ligation has so far been largely unnoticed outside the biochemical/chemical biology community. The possibility of preparing high-molecular-weight and perfectly monodisperse peptides with precisely defined α-amino acid sequences might, however, also enable access to unprecedented biologically-inspired supramolecular architectures and polymeric materials.

References

1. (a) H. Morawetz, *Polymers – The Origins and Growth of a Science*, John Wiley and Sons, New York, **1985**; (b) Y. Furukawa, *Inventing Polymer Science: Staudinger, Carothers and the Emergence of Macromolecular Chemistry*, University of Pennsylvania Press, **1998**.
2. J. A. Brydson, *Plastic Materials*, 5th edn., Butterworths, London, **1989**.
3. (a) B. Alberts, D. Bray, J. Lewis, M. Raff, K. Roberts, J. D. Watson, *Molecular Biology of the Cell*, 3rd edn., Garland Publishing, New York, **1994**; (b) C. Branden, J. Tooze, *Introduction to Protein Structure*, 2nd edn., Garland Publishing, New York, **1999**.
4. For recent reviews see, e.g., (a) J. C. M. van Hest, D. A. Tirrell, *Chem. Commun.* **2001**, 1897–1904; (b) H.-A. Klok, *Angew. Chem. Int. Ed.* **2002**, *41*, 1509–1513.
5. (a) H. Leuchs, *Ber. Dtsch. Chem. Ges.* **1906**, *39*, 857–861; (b) H. Leuchs, W. Manasse, *Ber. Dtsch. Chem. Ges.* **1907**, *40*, 3235–3249; (c) H. Leuchs, W. Geiger, *Ber. Dtsch. Chem. Ges.* **1908**, *41*, 1721–1726.
6. W. D. Fuller, M. S. Verlander, M. Goodman, *Biopolymers* **1976**, *15*, 1869–1871.
7. R. Katakai, Y. Iizuka, *J. Org. Chem.* **1985**, *50*, 715–716.
8. W. H. Daly, D. S. Poche, *Tetrahedron Lett.* **1988**, *29*, 5859–5862.
9. For comprehensive reviews see, e.g., (a) H. R. Kricheldorf, *α-Aminoacid N-Carboxyanhydrides and related Heterocycles*, Springer, Berlin, **1987**; (b) T. J. Deming, *J. Polym. Sci. Part A: Polym. Chem.* **2000**, *38*, 3011–3018.
10. (a) B. Gallot, *Prog. Polym. Sci.* **1996**, *21*, 1035–1088; (b) H.-A. Klok, S. Lecommandoux, *Adv. Mater.* **2001**, *13*, 1217–1229.
11. (a) H. Kukula, H. Schlaad, M. Antonietti, S. Förster, *J. Am. Chem. Soc.* **2002**, *124*, 1658–1663; (b) F. Chécot, S. Lecommandoux, Y. Gnanou, H.-A. Klok, *Angew. Chem. Int. Ed.* **2002**, *41*, 1339–1343.
12. For a recent review see S. Förster, T. Plantenberg, *Angew. Chem. Int. Ed.* **2002**, *41*, 688–714.
13. Y. P. Myer, *Macromolecules* **1969**, *2*, 624–628.
14. (a) T. J. Deming, *Nature* **1997**, *390*, 386–389; (b) T. J. Deming, *Macromolecules* **1999**, *32*, 4500–4502.
15. See, for example, (a) M. E. Yu, T. J. Deming, *Macromolecules* **1998**, *31*, 4739–4745; (b) K. R. Brzezinska, T. J. Deming, *Macromolecules* **2001**, *34*, 4348–4354; (c) K. R. Brzezinska, S. A. Curtin, T. J. Deming, *Macromolecules* **2002**, *35*, 2970–2976.
16. A. P. Nowak, V. Breedveld, L. Pakstis, B. Ozbas, D. J. Pine, D. Pochan, T. J. Deming, *Nature* **2002**, *417*, 424–428.
17. D. J. Pochan, L. Pakstis, B. Ozbas, A. P. Nowak, T. J. Deming, *Macromolecules* **2002**, *35*, 5358–5360.
18. J. N. Israelachvili, *Intermolecular and Surface Forces*, 2nd edn. Academic Press, New York, **1991**.
19. See, for example, (a) P. Berndt, G. B. Fields, M. Tirrell, *J. Am. Chem. Soc.* **1995**, *117*, 9515–9522; (b) Y.-C. Yu, P. Berndt, M. Tirrell, G. B. Fields, *J. Am. Chem. Soc.* **1996**, *118*, 12515–12520; (c) Y.-C. Yu, M. Tirrell, G. B. Fields, *J. Am. Chem. Soc.* **1998**, *120*, 9979–9987; (d) K. C. Lee, P. A. Carlson, A. S. Goldstein, P. Yager, M. H. Gelb, *Langmuir* **1999**, *15*, 5500–5508; (e) J. S. Martinez, G. P. Zhang, P. D. Holt, H.-T. Jung, C. J. Carrano, M. G. Haygood, A. Butler, *Science* **2000**, *287*, 1245–1247; (f) T. Gore, Y. Dori, Y. Talmon, M. Tirrell, H. Bianco-Peled, *Langmuir* **2001**, *17*, 5352–5360.
20. (a) J. D. Hartgerink, E. Beniash, S. I. Stupp, *Science* **2001**, *294*, 1684–1688; (b) J. D. Hartgerink, E. Beniash, S. I. Stupp, *Proc. Natl. Acad. Sci. USA* **2002**, *99*, 5133–5138.
21. See, for example, (a) M. J. Lawrence, *Chem. Soc. Rev.* **1994**, *23*, 417–424; (b) L. E. Bromberg, E. S. Ron, *Adv. Drug Deliver. Rev.* **1998**, *31*, 197–221; (c) A.

V. Kabanov, V. Y. Alakhov, Crit. Rev. Ther. Drug Carr. Syst. 2002, 19, 1–72.

22 (a) G. W. M. Vandermeulen, H.-A. Klok, Polym. Prepr. (Am. Chem. Soc. Div. Polym. Chem.) 2001, 42(2), 84–85; (b) M. Pechar, P. Kopecková, L. Joss, J. Kopecek, Macromol. Biosci. 2002, 2, 199–206; (c) H.-A. Klok, G. W. M. Vandermeulen, A. Rösler, Polym. Prepr. (Am. Chem. Soc. Div. Polym. Chem.) 2002, 43(2), 715–716.

23 For reviews see, e.g., (a) A. Lupas, Trends Biochem. Sci. 1996, 21, 375–382; (b) W. D. Kohn, R. S. Hodges, Trends Biotechnol. 1998, 16, 379–389; (c) P. Burkhard, J. Stetefeld, S. V. Strelkov, Trends Cell Biol. 2001, 11, 82–88.

24 J. Y. Su, R. S. Hodges, C. M. Kay, Biochemistry 1994, 33, 15501–15510.

25 See, e.g., (a) E. T. Kaiser, Acc. Chem. Res. 1989, 22, 47–54; (b) S. Aimoto, Curr. Org. Chem. 2001, 5, 45–87.

26 See, e.g., (a) P. E. Dawson, S. B. H. Kent, Annu. Rev. Biochem. 2000, 69, 923–960; (b) J. A. Borgia, G. B. Fields, Trends Biotechnol. 2000, 18, 243–251; (c) J. P. Tam, J. Xu, K. D. Eom, Biopolymers (Pept. Sci.) 2001, 60, 194–205.

27 P. E. Dawson, T. W. Muir, I. Clark-Lewis, S. B. H. Kent, Science 1994, 266, 776–779.

28 P. Botti, M. R. Carrasco, S. B. H. Kent, Tetrahedron Lett. 2001, 42, 1831–1833.

29 L. E. Canne, P. Botti, R. J. Simon, Y. Chen, A. E. Dennis, S. B. H. Kent, J. Am. Chem. Soc. 1999, 121, 8720–8727.

30 See, e.g., (a) G. G. Kochendoerfer, S. B. H. Kent, Curr. Opin. Chem. Biol. 1999, 3, 665–671; (b) D. W. Low, M. G. Hill, M. R. Carrasco, S. B. H. Kent, P. Botti, Proc. Natl. Acad. Sci. USA 2001, 98, 6554–6559.

31 For a review, see G. J. Cotton, T. W. Muir, Chem. Biol. 1999, 6, R247–R256.

32 H. Staudinger, J. Meyer, Helv. Chim. Acta 1919, 2, 635–646.

33 (a) B. L. Nilsson, L. L. Kiessling, R. T. Raines, Org. Lett. 2000, 2, 1939–1941; (b) E. Saxon, J. I. Armstrong, C. R. Bertozzi, Org. Lett. 2000, 2141–2143; (c) M. B. Soellner, B. L. Nilsson, R. T. Raines, J. Org. Chem. 2002, 67, 4993–4996.

34 L. Z. Yan, P. E. Dawson, Angew. Chem. Int. Ed. 2001, 40, 3625–3627.

B.25
Solid-phase Peptide Synthesis

Harm-Anton Klok

Solid-phase peptide synthesis (SPPS), first introduced by R. B. Merrifield [1], involves a repeated sequence of α-amino acid coupling and deprotection reactions performed on an insoluble solid support [2]. The support enables the use of excess α-amino acids and reagents, which helps to drive each of the reactions to completion. By-products and excess reagents are easily removed by filtration. After attachment of the last α-amino acid the linker connecting the support and the peptide, and protective groups that mask the side-chains of trifunctional α-amino acids, are cleaved, affording the crude product. Because of incomplete reactions, side-reactions, or impure reagents, the crude product usually contains defect sequences and other contaminants; these are usually removed by high-performance liquid chromatography (HPLC).

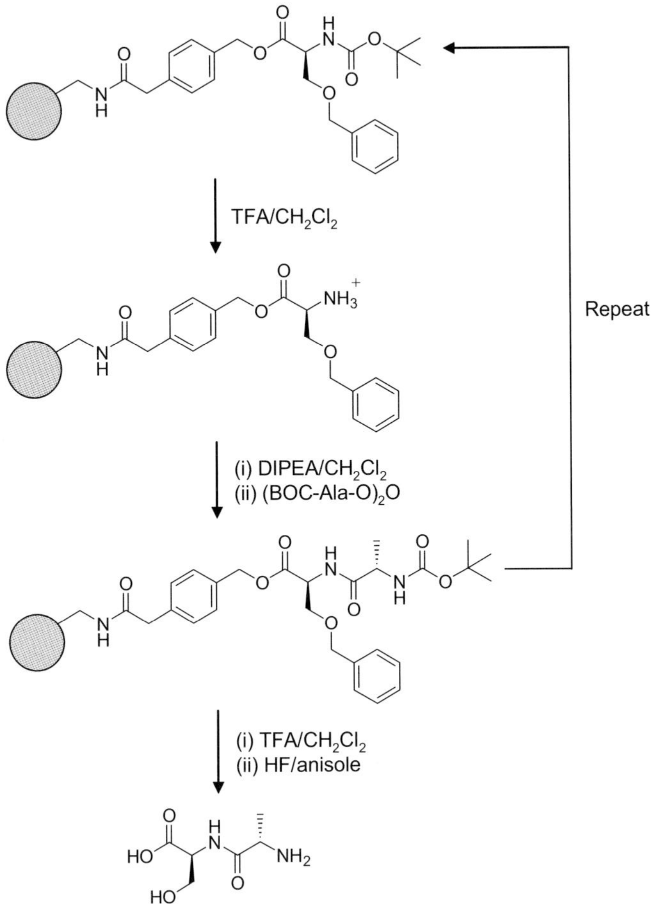

Scheme B.25.1.

Merrifield SPPS uses the *tert*-butoxycarbonyl (BOC) group to protect the α-amino group (Scheme B.25.1). After coupling to the N-terminus of the growing peptide chain the BOC groups are removed with trifluoroacetic acid (TFA). Before a subsequent coupling reaction the resulting trifluoroacetates are neutralized with diisopropylethylamine (DIPEA) in CH_2Cl_2. To protect the side-chains of trifunctional α-amino acids, a range of benzyl-based protective groups has been developed; these are cleaved, with the linker, by action of HF. Although the Merrifield method has proven very efficient, the use of highly toxic HF and the need for Teflon-lined equipment has prevented many from applying the technique.

An alternative method, which uses the base-labile 9-fluorenylmethoxycarbonyl (Fmoc) group to protect the α-amino group in combination with

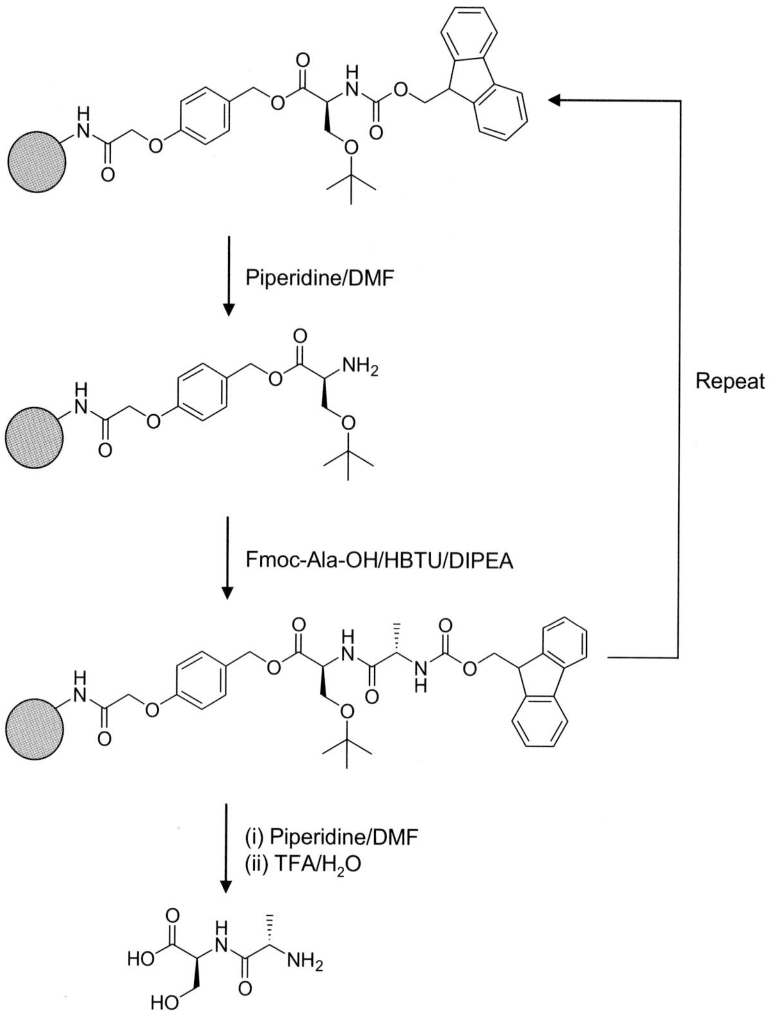

Scheme B.25.2.

an acid labile linker, and acid-labile protective groups to mask the side-chains of trifunctional α-amino acids was introduced in 1978 (Scheme B.25.2) [3, 4]. The advantage of Fmoc SPPS is that removal of the Fmoc group and cleavage of the linker and side-chain protective groups are performed under mild conditions and can be conducted in standard glassware. Along with Merrifield SPPS, Fmoc SPPS has become increasingly popular and has matured into a routine method.

References

1 R. B. MERRIFIELD, *J. Am. Chem. Soc.* **1963**, *85*, 2149–2154.
2 For general overviews, see e.g., (a) W. C. CHAN, P. D. WHITE, eds., *Fmoc Solid Phase Peptide Synthesis*, Oxford University Press, **2000**; (b) G. B. FIELDS, R. L. NOBLE, *Int. J. Peptide Protein Res.* **1990**, *35*, 161–214.
3 E. ATHERTON, H. FOX, D. HARKISS, C. J. LOGAN, R. C. SHEPPARD, B. J. WILLIAMS, *J. Chem. Soc. Chem. Commun.* **1978**, 537–539.
4 C. D. CHANG, J. MEIENHOFER, *Int. J. Peptide Prot. Res.* **1978**, *11*, 246–249.

B.26
Peptide Ligation

Harm-Anton Klok

Peptide ligation involves regiospecific coupling of unprotected peptide segments in aqueous solution. The concept of peptide ligation is outlined in Scheme B.26.1. It consists of two steps, a chemoselective capture reaction and an intramolecular acyl rearrangement [1]. Chemoselective ligation requires a suitable combination of electrophilic and nucleophilic groups

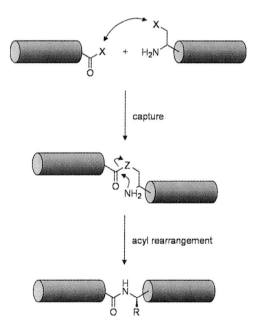

Scheme B.26.1.

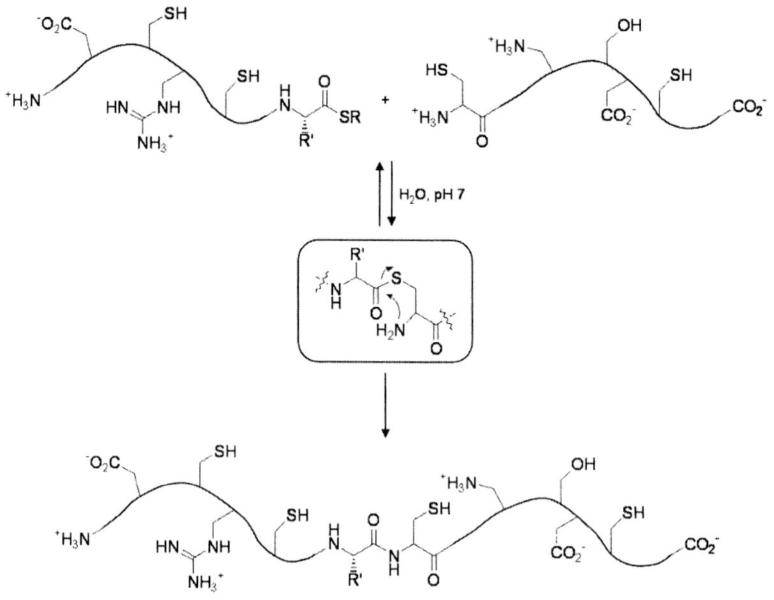

Scheme B.26.2.

located at the C- and N-termini of the peptide segments that need to be coupled. In Scheme B.26.1, X and Y represent an electrophile–nucleophile pair and Z the intermediate bond between the two peptide segments that originates after the capture reaction. Natural α-amino acids containing heteroatoms in their side-chains, e.g. Cys, Ser, Thr, His and Trp, can be used as N-terminal nucleophiles [1]. Typical C-terminal electrophiles include, e.g., O-glycoaldehyde esters and thioesters. In recent years a variety of peptide ligation methods has been developed; these differ in the nature of the electrophilic and nucleophilic groups involved in the chemoselective capture reaction [1].

A peptide ligation method that has attracted widespread attention is the so-called native chemical ligation (NCL) strategy introduced by Kent et al. (Scheme B.26.2) [2]. In NCL the thiol group of an N-terminal Cys residue acts as a nucleophile to attack the carbonyl carbon atom of a thioester moiety at the C-terminus of a second peptide segment. This results in the formation of a thioester intermediate which undergoes rapid intramolecular S,N-acyl migration to generate a peptide bond. This method has been successfully applied to the total chemical synthesis of the 72-amino acid protein interleukin 8 (IL-8) [2]. After purification, the correctly folded protein with properties identical with those of an authentic IL-8 sample was obtained.

References

1 (a) P. E. Dawson, S. B. H. Kent, *Annu. Rev. Biochem.* **2000**, *69*, 923–960; (b) J. A. Borgia, G. B. Fields, *Trends Biotechnol.* **2000**, *18*, 243–251; (c) J. P. Tam, J. Xu, K. D. Eom, *Biopolymers (Peptide Sci.)* **2001**, *60*, 194–205.

2 P. E. Dawson, T. W. Muir, I. Clark-Lewis, S. B. H. Kent, *Science* **1994**, *266*, 776–779.

Index

a

absorption and emission spectroscopy 172, 176
acceptor-donor-acceptor pattern 156
acetylcholine 125, 137
acridizinium 184
acridizinium derivatives 176 ff
acyl acceptor and donor 390 f
acyl-4-amidino and -4-guanidinophenyl esters (OGp) 391
acetylation 440
active transport 63, 79, 137 f
adenosine 253
adenosine triphosphate ATP 527
S-adenosyl methionine (SAM) 94 f
affinity 130, 221
affinity chromatography 249, 435
aggregation 160
aging 364
albothricin 70
alkaline-phosphatase 211
alkaloids 63, 70, 79
allosteric effect 204
allylic 1,2-interaction 22
allylic alkylations 438
Alzheimer 148, 155 f, 192, 262
amino acids *see also* peptides 63, 135, 140, 209, 389, 392, 502
– α-amino acid N-carboxyanhydrides 541
– Leuch's anhydrides 541
– non-proteinogenic amino acid 63 ff, 389
α-amino acids 68, 541 f, 545 f, 548
– aspartic acid 266, 276
– diamino acids 209
– glutamic acid 126
– hydroxyproline 134 f
– proline 18 ff, 29, 128 f, 133
– racemization 163
β-amino acids 63 ff

– β-alanine 64, 90 ff, 502
– β-amino-L-alanine 74
– β-aminoisobutyric acid 64, 93
– aminomutases 93
– β-arginine 69
– aromatic 66, 70 f
– biosynthesis 90 ff
– cispentacin 79
– β-Dopa 67, 70 ff, 99 f
– β-glutamate 69
– δ-hydroxylations 75
– isoserine 67
– β-leucine 69
– β-lysine 69
– nomenclature 65, 66 f, 76 f, 83
– β-phenylalanine 67, 70 ff, 98 f
– β-tyrosine 67, 70 ff, 99
α-amino acid coupling 554
9-aminoacridizinium bromide 176
3-aminobenzoic acid 126 ff
aminocarbasugars 516 f
aminoglycosides 69 f
aminomethylanthracene 441 f
aminomutases 93 ff
– reaction mechanism 94
6-aminopicolinic acid 132
aminopyrazole 156 ff
amphiphilic molecules 544
amyloid-β-peptide (Aβ) 147, 155
amyloid precursor protein (APP) 147, 167, 262 f, 264
anion complexation 129 f
anion recognition 131 f
anisotropy differences 313
anthracene 423 f
antibacterial 80
antifungal 80
antioxidant 197
– NO· 197

aptamers 249 ff
α-L arabinose 120
arrays 485 ff
– cell arrays 492
– DNA arrays 485 f
– protein arrays 487 f
– small molecule arrays 493 f
artificial lectin ligands 203 ff
artificial molecular motors 537
artificial molecular rotary motors 526
artificial nucleobase 379
artificial receptors 109 ff, 124 ff, 140 ff, 155 ff, 218 f
aryl diynes 237
aspartic protease 262 ff, 276 f
association constant = K_a value see also dissociation constant, binding constant, IC_{50} 112, 117, 143, 146, 151, 158 f, 162, 177
asialoglycoprotein receptor (ASGPR) 207
asthma 227 ff
ATP 527 ff
ATP synthesis 527
automated synthesizer 243
automation 437, 485
aza-Diels-Alder reaction 23

b
bacteria 63 ff
bacterial and viral infections 109
bacteriophage T7 333 f
β-barrels 137, 139
bathochromic shift 176
bestatin 285
binding constant 144, 146, 150, 178, 212
binding energy 154
binding motifs 109
binding pocket 302
binding selectivity 124 f, 140 f, 155 f, 203 f, 217 f, 235 f, 252 f, 262 f, 293 f, 317, 396
binding site 120, 178
BINOL 112 f, 117
biotechnological production 501 ff
biotin 211, 345 f, 425 ff, 492
boronic acid 109
bryostatin 53
building block 278

c
CaaX-box 218, 220
calixarenes 126 f
cancer 311, 352, 423
cancer metastasis 203
cancer therapeutics 219

carbasugars 518
carbohydrate 49, 109 ff, 203 ff
carbohydrate binding proteins 120, 203 ff
carbohydrate recognition 109 ff, 119 ff, 203 ff
– enthalpy 204
– entropy 204
multivalency 203 f, 212 f
carbohydrate-lectin interactions 119 ff, 203 ff
carboxylate binding 143 ff
carboxymethyl thioesters 392
carboxypeptidase A 131 f, 141 f
carcinogenesis 364, 369, 379
Caro's acid 283
catalysis see also enzymes 423
– biocatalysts 501 ff
– discovery of catalysts 436 ff
– fermentation 501 ff
– proteases 389 ff
– ribozymes 404 ff, 422 ff
catalyst discovery 436 ff, 508
catalytic triad 239, 397
catalytic turnover 429
catechol units 40
catenanes 526, 530
– protein[2]catenane 550, 552
cation-π interaction 126, 146
cation recognition 124 f
CD spectroscopy 26 f, 176 f, 253, 255, 531
cell-cell communication 203
cell-surface engineering 57
ceramide 48
chaotrophic anions 26
chaperones 169
charge transfer in DNA 172, 369 ff
chemical enzyme modifications 396
chip technology see microarrays 379, 485 ff
cholera toxin 50
chorismate 512 ff
chymotrypsin 240, 392, 394, 396
CIDNP 354
circular dichorism 26 f, 176 f, 253, 255, 531
cis amide conformations 133
cleavage of a phosphodiester bond 409
cleavage of the O-O bond 194
coiled-coils 546
co-immobilization 443 f
colorimetric and fluorescent screening 437
combinatorial biosynthesis 48
combinatorial chemistry 147, 208 f, 217 ff, 242 ff, 290, 292, 295, 330, 422 ff, 433 f, 436ff, 493, 511, 522
– catalyst discovery 436 ff
– combinatorial binding assay 149, 209, 225, 248

– dansyl group 148
– diversity-oriented synthesis 290, 513 f
– DNA shuffling 341 f, 401
– encoded combinatorial library 220, 225
– error prone PCR 341 f
– in-vitro evolution 217, 422
– mRNA display 341
– one-bead-one-compound approach 208, 225, 493
– parallel libraries 437 ff
– parallel synthesis 436
– phage display 217, 332, 341, 401
– RAS 218
– ribosome display 341
– RNA libraries 422
– saturation mutagenesis 342
– SELEX (systematic evolution of ligands by exponential enrichment) 217, 248 ff, 330, 422 ff, 433 ff
– split-and-mix libraries 148, 208 f, 225, 436 ff, 495
– spot synthesis 493
– yeast-two-hybrid 217
complex stability 128 f
complexation induced chemical shifts 129, 133, 143, 158
confocal fluorescence microscopy 348, 350
conformation 5, 18, 32, 45, 52, 109, 126, 128, 301, 541
– amyloid-β-peptide 147
– cone conformation 126
– DNA polymerases 299 ff, 329 ff
– peptides
– – β-sheets 18 ff, 155 ff
– – β-turn 18 ff
– – α-helix 18 ff
– – stabilization by metal complexes 31 ff
– lipids 48 ff
– RNA 3 ff, 404 ff, 422 ff
– self-aggregated β-sheets 148
conformational restriction 48, 51, 54
– lipids 52
– peptides 18 ff, 155 ff, 209
complementary binding sites 141
cooperative binding 135, 202
copolymers 542
corticosteroids 228
CpG islands 378
cross-linking 204
cross-linking unit 32
crown ether 125
cryogenic transmission electron microscopy 544
cyanobacteria 75

[2 + 2] cycloaddition 352
[2 + 4] cycloaddition 422
cyclobutane dimers see also thymine cyclobutane dimers 352
cyclodextrins 531
cyclohexadiene-*trans*-diols 511
cyclopeptides 41, 124, 126, 130 ff, 209
cyclophiline 30
cytosine methylase *HhaI* 376
cytotoxic compounds 80

d
dansyl group 148 f
degenerate oligonucleotide cassettes 331
de-novo design 18
desolvation 154
Dess-Martin reagent 282
1,2-diacylglycerol (DAG) 54
dielectric constant 154
Diels-Alder reactions 423
Diels-Alderase ribozyme 422 ff
diene 432
difluoroketones 271, 276
dihedral angles Φ and Ψ 25, 44
dihydroxylation 519 f
N^6,N^6-dimethyladenosines 11
dipeptide mimics 24
dipole-dipole interactions 175
directed evolution 329, 331, 341 ff, 402, 507
– degenerate oligonucleotide cassettes 331
– DNA shuffling 331, 341 f
– enzymes 511
– error prone DNA synthesis 331
– error prone PCR 341 f
– mRNA display 341
– phage display 341
– ribosome display 341
– saturation mutagenesis 342
directed molecular evolution 341 f
dissociation constants 121 f, 203, 253, 256
diversity see combinatorial chemistry
DNA 172 ff, 299 ff, 311 ff, 329 ff, 344 ff, 352 ff, 369 ff, 485 ff
– abasic sites 304
– absorption and emission spectroscopy 176
– base pairing 327
– binding 183
– – acridizinium derivatives 178 f
– – CD spectroscopy 180
– – groove binders 178, 188
– – intercalators 172 ff, 189, 318
– – LD spectroscopy 183
– – protein 375
– charge transfer 172, 369 ff

DNA (cont.)
– chip technology   379, 485, 492
– cleavage   73, 369
– damage   364, 378 f
– – G radical cation   373
– – photooxidants   373
– diagnostics   311 ff
– electrostatic interactions   173, 175
– enantiomeric nucleic acids   248
– groove binding   174, 312
– hairpin   318, 326, 379
– hybridization   318, 327
– intercalation   172 ff, 318
– major groove   375
– microarrays   485
– mutations   302, 329
– photolyase   352
– polymerase   250, 299 ff, 329 ff
– pyrenyl-modified DNA   383
– reductive electron transfer   358, 379
– repair   346, 354, 358, 360, 362
– shuffling   331, 341, 401
– π-stacking   373
– thymine cyclobutane dimers   352 ff
dNTP   300 ff
π-donor-π-acceptor interactions   534
drug development   51
drug discovery   218
duplex formation   315
dyes   172, 311 f

e

Edman degradation   211
eicosanoids   61
elastase   240
electron transfer   179, 312 ff, 352 ff, 381 ff, 441, 534
electrophoretic gel mobility shift   427
electrostatic interactions   129, 131 f, 142, 146, 159, 173, 175
ELISA   348
embryogenesis   50
emission and time-resolved transient absorption spectroscopy   376
emission spectroscopy   179
enantiomeric nucleic acids   248, 430
energy transfer   318, 353
engineered microorganisms   524
(en)Pd-fragment   33
enthalpy   51, 204 f, 354
– free binding   159
entropy   51, 153 f, 204 f
enzymes see also proteins   94, 352
– acetylcholinesterase   125

– active site   391
– aminomutases   93
– aspartic proteases   262
– carboxypeptidase A   141
– chymotrypsin   240
– DNA polymerase   250, 299, 329 ff, 335
– elastase   240
– engineering   396
– mutant enzymes   330
– pantolactone hydrolases   505
– photolyases   352
– proteases   389
– RNA polymerase   249, 329 ff
– rotamases   21, 30
– β-secretase   267
– specificity   390
– trypsin   240
enzyme-linked lectin assay (ELLA)   209
epoxidation   515 ff
– nucleophilic opening   516, 518
epoxides   519
equilibria of RNA conformations   7 ff
error prone PCR   341 f
error prone DNA synthesis   331
Escherichia coli   520 ff, 329 ff
ESI-MS (electron spray ionization mass spectrometry)   35
esterase ativity   397
estrogen-like activity   41
Eubacteria   64
evolutionary techniques   217, 248, 330, 341, 401, *see also* combinatorial chemistry
excited singlet state   353

f

farnesylation of RAS   217
fatty acid   49, 61
fatty alcohols   61
femtosecond transient absorption spectroscopy   380
fermentative production   502
FK506-binding proteins   30
flavin intercalator   379
fluorescein   414 f
fluorescence   179 f, 311 ff, 317 ff, 338, 442
– anisotropy   314
– assays   325, 423
– correlation spectroscopy   167, 335 f
– detection   311, 335
– enhancement   312 f, 326
– hybridization-induced fluorescence   314, 316
– intercalation   312
– label   148, 258, 335 f

– resonance energy transfer (FRET)   319 f, 344
fluorescent beads   147, 438
fluorescent pH indicators   441
fluorophore   315 f, 335 f
Fmoc group   556
Fmoc strategy   42, 148
folding of proteins   25, 44, 547
footprinting   3, 373
fragment condensation   392
FRET   318, 326
fungi   63 ff
fusion proteins   344

g
GalNAc-transferase   56
gangliosides   48, 56
gel electrophoresis   373
gelation behavior   543
gels   443
gel-shift analysis   7, 14
gene expression   172
gene therapy   404
gene-shuffling   331, 404
genesis   379
gene-targeting applications   328
genetic engineering   507
genetic enzyme modifications   398
globin fold   202
glucosyltransferase   54
glycerolipids   61
glycolipids   49
glycosylation   209
glycosylphosphatidylinositol   53
glycosyltransferase   48, 57 f
Golgi apparatus   58
gonadotropin-releasing hormone   256
G-protein linked receptors   219
gramicidin A   137
green chemistry   522
groove binding   174, 178, 188, 312
group I introns   405
guanidiniocarbonyl pyrroles   141 ff
guanidinium group   69 f, 131, 141
Gulliver effect   142

h
hAGT   350
– fluorescence labeling   348 f
– in-vivo labeling   345, 347
hAGT fusion proteins   345, 347 f
hairpin structure   3 ff, 318
H-bonds   18, 147, 153
– solvation   153
HBTU   42, 210

heat of reaction $\Delta H_r°$ 439
Heck reaction   55, 437, 469, 479
α-helix   25, 45, 169
helix bundles   33
helix capping   26 f
helix-coil transition   25
helix propensities   25 ff
hemoglobin   191, 201
high-throughput analysis technologies   80, 295, 444, 507
His tag   345
HIV-1 reverse transcriptase   303 f
hopping mechanism   369
host molecules see also receptor   124
human mast cell tryptase   227
hybridization   313, 318
hydration energies   131 f
hydrogels   545 f
hydrogen bond   20, 25, 33, 51, 110 f, 120 f, 128 ff, 131 ff, 143, 146, 153, 155, 159, 161, 202, 299, 303 f, 534, 536
hydrolysis of peptide bonds   389
hydrolysis or transesterification of phosphodiester bonds   422
hydrolytic activity   441
hydrophobic cavity   126
hydrophobic interactions   121, 142
hydrosilation of alkenes and imines   437
α-hydroxy-β-amino carbonyl moieties   289
hydroxyproline   134
hyperchromicity   325
hypervalent iodine compounds   282
hypochromicity   176

i
IC_{50} 207 f, 222, 269, 271
imide bonds   20
imino proton NMR spectroscopy   9, 14 f
indigo dye   441
inflammation   203
inhibition activity   236
inhibitors   229, 265, 267
– BACE   265
– Diels-Alderase   432
– HIV-protease   285
– hydroxamates   295
– peptide aldehydes   294 f
– peptide isostere   294 f
– protease   277
– secretase inhibition   265
– β-secretase   269
– γ-secretase   269
– substrate analogs   294
– suicide-inhibitors   294 f

interaction 111, 173, 175, 314
intercalation 178, 189, 312 f, 371
– single-base-mutation analysis 317
intermolecular interactions see molecular recognition
intramolecular hydrogen bonds 127, 157, 169
introns 378, 405, 407
in-vitro selection 4, 249 f, 342, 404, 422 ff, 433
in-vivo labeling 348 ff
ion channels 64, 138 f
ion complexation 124
ion pair 129, 143 ff
ion recognition 124 ff
ion transport 137
ion-exchange resins 292
ionophores 124 f, 137 ff
iron(III)peroxynitrito complex 193, 195
IR-thermography 439 f
isochorismate 513 f, 521
isoenergetic 30
isosbestic points 176 f, 180
isosteres
– nucleotide isosteres 303
– peptide isosteres 278, 303, 394
isosteric motif 285

j
Job plot 158
Juglon 30

k
K_a values 158
Karplus analyses 159
k_{cat}/K_m 521
Kemp's triacid 157, 159 f
K_i values 235 f
kinetic reaction control 36
kinetic resolution 504, 507
kinetics
– binding of nitrogen monoxide to hemoglobin and myoglobin 193 f
– Diels-Alder ribozyme 429
– hydrolysis of *rac*-pantolactone 507
Klenow fragment 303 ff, 335, 337

l
labeling 90, 344
β-lactam antibiotics 74
lactonization 501
laser-scanning confocal microscopy 543
LD spectroscopy 183
lectins 109, 121 f, 203 ff

Leuch's anhydrid 541
library 208 f, 277, 330, 342 see also combinatorial chemistry
ligase ribozymes 404 ff, 419 ff
ligation of peptides 557
light- and neutron scattering experiments 542
linker 277
– π-allyl-based linkers 469
– boron linkers 459
– cyclative cleavage 474 f
– Dde linkers 467
– definition 449
– ester linker 457
– fragmentation strategies 475
– HAL linker 454
– HYCRAM linker 456
– HYCRON 320, 456
– hydrazide resins 458
– Kenner safety-catch linker 474
– Ketal linker 456
– Knorr linker 452
– linker (PAL) 454
– multifunctional cleavage 479
– orthogonality between linkers 465
– PAL linker 455
– PEG linker 424
– photocleavable linkers 468
– REM linkers 467
– safety-catch linkers 474
– SASRIN linker 454
– SASRIN resin 453 f
– selenium linkers 461
– silyl linkers 458
– stannane linkers 460
– sulfur linkers 459
– traceless linkers 458, 477
– triazine linkers 461
– Trityl resins 455
– urethane type 209
– Wang resin 3 453 f
lipases 439
lipids 48, 52, 60
– lipidomimetics 59
– triacylglycerols 60
lipophilic contact 51
lock-and-key principle 31
lower oxidation potentials 376

m
macrocycles 124, 126
– cyclopeptides 126
macrocyclic host 126
major groove binding 173

maleimide (dienophile) 423, 425 ff, 431 f
α-D-mannose 122
MAP kinase cascade 219
marine organisms 63, 65 ff, 75 f
melting temperature 314, 323 ff
membranes 48, 125, 137, 195, 203, 527, 529
– ion transport 137
– proton gradient 529
– transport mechanism 137
metabolic pathway engineering 502, 522, 524
metabolism 50, 90 f, 520, 527
metabolites 65 ff, 513, 515
metal coordination 31 ff, 112 ff, 530
metathesis 439 f, 471
Met-enkephalin 53
methylated proteases 397
methylation of RNA 12 f
Mg^{2+} 5, 407
Michael addition 71 f
Michaelis complex 241
Michaelis-Menten 421, 426
microbial production 511 ff, 520
microbiological transformations 242
miniaturization 437, 485 ff
minor groove 365
minor or major groove binding 173
mismatched base pair 314
molecular beacons 318 ff, 322, 320, 325 f, 337
molecular forceps 220
molecular machines 526
molecular modelling 113, 146
– conformational searchers 157
– Monte Carlo conformational searchers 157 f
molecular recognition 31, 172
– carbohydrate recognition 109 ff, 119 ff, 203 ff
– cyclopeptides 124 ff
– peptide recognition 140 ff, 155 ff, 217 ff
– ion recognition 124 ff
– β-sheet recognition 155 ff
molybdenum 40
monodisperse materials 542 f
Monte Carlo conformational searches 157
morphine 53
mRNA display 341
multicomponent synthesis 21
multivalency 121 f, 203 ff
– parameter β 206
mutagenesis 333, 341, 364, 369, 508
mutagenesis strategies 331
mutations 10, 217, 302, 329, 333, 352, 368, 398, 400, 427

mutation analysis 311 ff, 432
myoglobin 191, 201
myristic acid 52

n

Na^+/K^+-ATP-ase 138
nanoelectronics 537
native chemical ligation 558
nitrate radicals 355 ff
nitrotyrosine 196
nitrogen monoxide (nitric oxide) 191
NMR spectroscopy 14 f, 22, 126 f, 143, 159
– binding studies 117, 144
– complexation induced chemical shifts 129, 133, 143, 158
– imino protons 9, 14 f
– Job plot 158, 161
– NOESY technique 171
– titrations 158, 162, 164
NO˙ 193 ff
non-covalent coordination bonds 46
non-covalent interaction see also molecular recognition 31, 124, 140, 142, 303
– amyloid-β-peptide (Aβ) 147
– attractive van der Waals 174
– electrostatic interaction 159
– hydrogen bond 140, 159, 174
– hydrophobic interactions 140 f, 174
– ion pairs 141
– salt bridges 140
– π-stacking 159, 174
– steric repulsion 159
non-polar interactions 121
nucleic acid polymerases 329 ff
nucleic acids 327
– DNA 172 ff, 299 ff, 311 ff, 329 ff, 344 ff, 352 ff, 369 ff, 485 ff
– RNA 3 ff, 248 ff, 329 ff, 404 ff, 422 ff
nucleobase methylations 11 ff
nucleophilic catalysts 440
nucleophilic hydroxy anion 276
nucleoside triphosphate (dNTP) 299, 302
nucleotide analogs 299, 303, 313 f, 329, 335

o

O^6-alkylguanine DNA 345
OGp esters 394
on-bead screening 147, 209 f, 218 f, 440 ff
one-bead-one-compound library 147, 208 f, 218 f, 440 ff, 493
one-electron oxidant 358
opioid alkaloids 52
optical rotation 35
orthogonality principle 451

overexpression of genes   502ff
oxidation   372, 376
oxidation of alcohols   279 ff
oxidation potential   180, 359 ff
– nucleic bases   180
oxidative cleavage   356, 363 f
oxidative splitting   354 ff, 360
ozonolysis   23

p

D-pantothenic acid (Vitamin B5)   501 ff
pantolactone hydrolases   505
parallel synthesis   436 ff, 485 ff
parvulines   30
PCR (polymerase chain reaction)   249, 312, 329, 341, 404, 424 f
pentacovalent transition state   309
peptidases see proteases
peptides   18 ff, 31 ff, 51, 63, 124 ff, 140 ff, 155 ff, 209 f, 217 f, 389 ff, 441 f, 540 ff
– aldehyde   278, 284 f
– amyloid-β-peptide (Aβ)   147, 155
– aminopyrazoles   160 f
– antiparallel β-sheet   33 f, 156
– binding of   140 ff, 155 ff, 217 ff
– cis/trans isomerisation of amide bonds   29 ff
– cleavage   396
– coiled-coils   546
– conformation   18, 32, 159, 541
– cyclopeptides   41, 63 ff, 124, 209 f
– de-novo design   18
– dihedral angles, ϕ and ψ 25, 34
– dipeptide mimics   24
– electrostatic interactions hydrophobic contacts   148
– epitopes   221
– foldamers   64
– gramicidin   137
– guanidiniocarbonyl pyrrole   147
– α-helix   18, 26, 31f, 541
– hydrogen bonds   20, 25, 33, 155, 161
– isosteres   278, 294
– ligation   548 f, 552, 557
– metal coordination   31 ff
– mimetics   28, 267
– neoglycopeptides   209 f
– nucleic acid, PNA   317, 327 f
– β-peptides   63
– pipecolic acid   18
– piperidine-2-carboxylic acid   21
– polymerisation   543 ff
– proteolytic stability of peptides   63
– pseudopeptide backbone   327
– Ramachandran   34
– Ramachandran's method   44
– recognition   140 ff, 155 ff, 217 ff
– secondary structures   18 f, 155
– β-sheets   19, 148, 150, 155 ff, 541 ff
– synthesis   397
– β-turns   156
peptide recognition   140 ff, 155 ff, 217 ff
peptidomimetics   18, 51
peptidylprolyl isomerases   30, 392
peptidyl-prolyl-cis/trans-isomerase   392
peroxynitrite   191
pH sensor   442
phage display   217, 332, 341, 349, 401
phorbol   53
phoshorimagery   373
phosphodiester bond   309, 404, 411
phosphonates   111
phosphorescence   179
photochemical or photophysical oxidation   373
photo-cleavable protecting group   492 f
[2 + 2] photocycloaddition   366
photoinduced cleavage of T-T dimers   379
photolithography   487
photooxidants   374
photophysical oxidation   373
photosensitizers   354 f, 357, 359, 362, 534
pipecolic acid   18, 22 f
pK_a values   143
plaque-related proteins   262
PNA (peptide nucleic acid)   315, 317, 320, 495
polyalanine   26
polyproline   30
poly(ethylene)   540
poly(propylene)   540
poly(styrene)   540
polyacrylamide gel electrophoresis (PAGE)   426
polymer supported reagents   277 ff, 290 f
polymerases   299 ff, 329 ff
polymerase chain reaction (PCR)   249, 312, 329, 341, 404, 424 f
polymer-assisted solution-phase (PASP)   277 ff, 290 ff
polymerization   540 ff
polystyrene resin   243, 291, 450
preorganization   129, 143
preorganized binding site   109
preorientation   157
primer template   302
prion protein scrapie   164
prodrugs   58

proline   18 ff, 29, 128 f, 133
– mimics of syn and anti proline peptide bonds   21
proteases   63, 293, 389
– acyl-4-amidino and -4-guanidinophenyl esters (OGp)   391
– aspartic protease   262, 266
– carboxymethyl thioesters   392
– enzyme's active site   391
– HIV protease   269
– inhibitor   277 f, 278, 293
– – libraries   277
– methylated proteases   397
– serine proteases   228, 239
– site-specific modifications   397
– specificities   389 ff
– substrate engineering   391 f
– subtilisin   398, 400 f
protecting group   209, 453, 487
protein-carbohydrate interactions   119 ff, 203 ff
protein engineering   63, 550
protein folding   21, 29, 156, 164, 203, 546 f
protein interactions   344
protein-protein interactions   41
protein secondary structures   541
– β-barrels   137, 139
– coiled-coils   546
– dihedral angles Φ and Ψ 45
– α-helices   45 f, 541
– helix bundles   33
– Ramachandran's method   18, 37 f, 169
– β-sheet   45 f, 541
proteins see also enzymes   31, 155, 217, 344 ff, 541, 550
– amyloid precursor protein   262 f, 270
– arrays   488
– CaaX-box   220
– carboxypeptidase A   141
– chymotrypsin   240
– cytosine methylase *HhaI*   376
– DNA polymerase   250, 299 f, 329 f
– elastase   240
– farnesyl transferase (FTase)   219
– folding   25, 44
– fusion   344
– HIV-1 reverse transcriptase   303
– human DNA repair protein O^6-alkylguanine DNA alkyltransferase (hAGT)   345
– human mast cell tryptase   227
– Lewis-acid activity   46
– microarrays   487
– Na^+/K^+-ATP-ase   138
– plaque-related proteins   262

– proteases *see there*
– RAS   217
– RNA polymerase   249, 299 f, 329 f
– β-secretase   263
– serine protease   228, 239
– signal transduction proteins   218
– transmembrane proteins   138
– trypsin   240
proton gradient   529 f
pyrene   304, 314, 379 ff
pyrophosphate   310, 527

r
racemization   163
radiation   365
radicals
– anion   353 f
– cation   353 ff, 360 ff
– DNA   190, 379 ff
– G radical cation   372
– hydroxyl radicals   358, 365
– nitrate ($NO_3^·$)   355 ff
– nitrogen monoxide ($NO^·$)   191 ff
– oxidative G damage   375
– peptide radicals   376 f
– pyrenyl radical cation   380
RAF protein kinase   219
Ramachandran method   18, 37 f, 63, 169
random mutagenesis   401, 436, 524
rapamycin   30
RAS   217 f
receptors   109 ff, 124 ff, 140 ff, 156 ff, 203 ff
– amino acids   140
– amino-pyrazole derivatives   156
– carbohydrates   109 ff, 203 ff
– carboxylate   143 ff
– molecular forceps   217
– peptides   140, 217, 157
red blod cells   201
redox potentials   373
reductive electron transfer   379
Rehm-Weller equation   179
replication
– error rates   333
resin   279, 283 f, 436, 449
– backbone   277
– ion-exchange resins   292
– polystyrene   291, 450
– Rink resin   454
– Tentagel resin   452
resin 1   279
reverse transcriptases   305, 339, 425
Rh(III) complex   373, 375
ribosome   420

ribosome display 341
ribozymes 3, 248, 404 ff, 419 ff, 422 ff
– Diels-Alderase ribozyme 422 ff
– engineering 425
– hairpin ribozyme 404 ff, 420
– hammerhead ribozyme 5, 248, 404 ff, 420
– hepatitis delta virus (HDV) ribozyme 4, 404
– in-vitro selection 249 ff, 422, 433
– SELEX 249, 422, 433
– self-splicing pre-ribosomal RNA 404 ff
– twin ribozyme 412 ff
ring-opening polymerization 541
RNA 3 ff, 248 ff, 329 ff, 404 ff, 422 ff
– bistable RNAs 15
– cleavage 407
– conformation 5 ff
– folding 3, 420
– footprinting methods 14
– imino proton NMR spectroscopy 14
– library 248, 422
– methylation 11 ff
– polymerase 249, 331, 339
– repair 406
– RNAi 5
– spiegelmers 248
– viruses 333
– world 248
RNase P 404
rotamases 21, 30
rotaxanes 526 ff, 530 f, 534, 536
– anion template effect 532
– topologically chiral 531, 533
Ru complexes 373 f

S

safety catch resin strategy 341
salt bridges *see* ion pairs
saturation mutagenesis 342
scaffolds *see also* templates 206 ff, 220, 230, 242
scavenger resin 288, 291
secondary interactions 146, 204
secondary metabolites 68 ff, 90 ff, 516
secondary structures 3, 18 f, 31, 150, 167 f, 414, 544
β-secretase 264
segetalins A and B 41, 43
selection 217, 331 f, 341 f, 424, 436 f
SELEX (systematic evolution of ligands by exponential enrichment) 217, 248ff, 330, 422 ff 433 ff
self-aggregate 111
self-assembly 113, 117, 545, 546, 548
self-assembled receptors 112

self-splicing group I and group II RNA 420
self-splicing RNA 404 ff
semi-crystalline materials 540
sensitizers 353
serine protease 228, 237, 239
shikimate pathway 511, 513 ff, 521
β-sheet 45, 150, 155, 169
sialic acid 50
sialyltransferase II 56
signal transduction 50, 54
single-bead analysis 209 f, 220, 225, 440
site-directed manipulation
site-directed mutagenesis 344, 400 f, 507
small molecule arrays 485 ff
solid supported reagents 277
solid-phase chemistry 147, 209, 218, 249, 277 ff, 290 ff, 320, 415, 436 ff, 449 ff, 485 ff, 554 f
– automated synthesizer 243
– linker 277, 449 ff
– peptide synthesis 147, 209, 218, 390, 441, 487, 554
– NMR spectroscopy 452
solvation 153 f
specificities *see* binding selectivities
spectrophotometric titrations 180
spermine 72, 90, 92
sphingolipids 48 ff
spiegelmers 248 ff, 430 f
split-and-mix library 148, 209 f, 220, 225 f, 440 f, 495
spotting 494 f
stability constant 127, 129 f, 134 f
stabilizing α-helical peptides 25
π-stacking 175, 189
steady-state fluorescence 381
stereoelectronic effects 362
stereoselectivity 430
steric constraints 353
steric repulsion 20, 146, 360
steroids 61, 242 ff
Stille couplings 469
streptavidin 348, 425, 492
structure-activity relationships 230
structure-binding relationship 151
substrates
– ions 124 ff
– carbohydrates 109 ff, 203 ff
– sugars 109 ff, 203 ff
– peptides 140 ff
– β-sheets 155 ff
– cationic dyes by DNA 172 ff
subtilisin 398, 400 f
superexchange mechanism 370

supramolecular assembly   31 f, 109 f, 141, 174, 540 ff
Suzuki coupling   470
symbionts   66 f, 75 ff, 79
syn-anti isomerization   29 ff
systematic evolution of ligands by exponential enrichment (SELEX)   248 ff, 330, 422 ff, 433 ff

t
taxol   71
telomeres   378
template effect   530, 535
template-directed   299
templates *see also* scaffolds   20, 156, 230 ff, 309
Tentagel resin   452
terpenes   61, 71
tert-butoxycarbonyl (BOC)   555
tether   205
tetracycline resistance   333 f
tetrahedral intermediate   240 f, 267, 276
tetramer disrupters   229
thermodynamically controlled system   37
thiazole orange   312 f
thioketal linker   243
thymine dimers   352 ff
titanium ions   35, 40
titration curve   117
topological chirality   531
transcription   424
transesterification of phosphodiester bonds   422
transition state   52
– Diels-Alder   432
transition-state analog (TSA)   434
transmission electron microscopy   542, 545
transport of dioxygen   201
triacylglycerols   60
trypanosomes   53

trypsin   240, 228
tryptase inhibitor   228
tunneling process   372
two-hybrid systems   342
two-metal-ion mechanism   309
tyrosine kinases   219
tyrosine nitration   192
β-turns   156

u
ultracentrifugation   166
UV absorbance   423
UV assay   149
UV irradiation   190, 352, 366
UV melting experiments   7 ff
UV-visible spectroscopy   176, 193

v
V8 protease   392
valienamine   519
valinomycin   137
β values   370 ff
vasopressin   254
viral infections   109
virus variability   333
Vitamin B5   501 ff

w
Watson-Crick base-paired   15, 301, 312, 323, 405
Western blots   348
wheat germ agglutinin (WGA)   209 f
Winterstein's acid   98

y
yeast-two-hybrid system   217, 344

z
Zimm-Bragg theory   25
zinc finger motiv   46